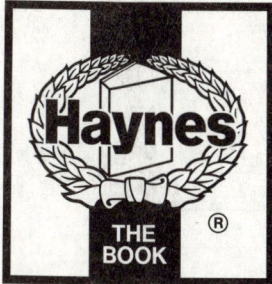

# Kawasaki ZX750 Fours
## Service and Repair Manual

## by Alan Ahlstrand
## and John H Haynes Member of the Guild of Motoring Writers

*(2054-336-3AE2)*

**UK models covered**
Kawasaki ZX750H (ZXR750). 748 cc. 1989 and 1990
Kawasaki ZX750J (ZXR750). 749 cc. 1991 and 1992
Kawasaki ZX750L (ZXR750). 749 cc. 1993 to 1996
Kawasaki ZX750K (ZXR750R). 749 cc. 1991 and 1992
Kawasaki ZX750M (ZXR750R). 749 cc. 1993 and 1994

**US models covered**
Kawasaki ZX750H (Ninja ZX-7). 748 cc. 1989 and 1990
Kawasaki ZX750J (Ninja ZX-7). 749 cc. 1991 and 1992
Kawasaki ZX750L (Ninja ZX-7). 749 cc. 1993 to 1995
Kawasaki ZX750K (Ninja ZX-7R). 749 cc. 1991 and 1992
Kawasaki ZX750M (Ninja ZX-7R). 749 cc. 1993 and 1994

**Note:** *This manual does not cover the ZX750P (Ninja ZX-7R) or ZX750N (Ninja ZX-7RR) introduced in 1996*

© Haynes Publishing 2005

ABCDE
FGHIJ
KLMN

A book in the **Haynes Service and Repair Manual Series**

ISBN 1 85960 285 1

**British Library Cataloguing in Publication Data**
A catalogue record for this book is available from the British Library

Library of Congress Catalog Card Number 98-72827

Printed in the USA

**Haynes Publishing**
Sparkford, Yeovil, Somerset BA22 7JJ, England

**Haynes North America, Inc**
861 Lawrence Drive, Newbury Park, California 91320, USA

**Editions Haynes**
4, Rue de l'Abreuvoir
92415 COURBEVOIE CEDEX, France

**Haynes Publishing Nordiska AB**
Box 1504, 751 45 UPPSALA, Sweden

# Contents

## LIVING WITH YOUR KAWASAKI ZX750

### Introduction

### Daily (pre-ride) checks

## MAINTENANCE

### Routine maintenance and servicing

# Contents

## REPAIRS AND OVERHAUL

## REFERENCE

# Kawasaki The Green Meanies

*by Julian Ryder*

## Kawasaki Heavy Industries

Kawasaki is a company of contradictions. It is the smallest of the big four Japanese manufacturers but the biggest company, it was the last of the four to make and market motorcycles yet it owns the oldest name in the Japanese industry, and it was the first to set up a factory in the USA. Kawasaki Heavy Industries, of which the motorcycle operation is but a small component, is a massive company with its heritage firmly in the old heavy industries like shipbuilding and railways; nowadays it is as much involved in aerospace as in motorcycles.

In fact it may be because of this that Kawasaki's motorcycles have always been quirky, you get the impression that they are designed by a small group of enthusiasts who are given an admirably free hand. More realistically, it may be that Kawasaki's designers have experience with techniques and materials from other engineering disciplines. Either way, Kawasaki have managed to be the factory who surprise us more than the rest. Quite often, they do this by totally ignoring a market segment the others are scrabbling over, but more often they hit us with pure, undiluted performance.

The origins of the company, and its name, go back to 1878 when Shozo Kawasaki set up a dockyard in Tokyo. By the late 1930s, the company was making its own steel in massive steelworks and manufacturing railway locos and rolling stock. In the run up to war, the Kawasaki Aircraft Company was set up in 1937 and it was this arm of the now giant operation that would look to motorcycle engine manufacture in post-war Japan.

They bought their high-technology experience to bear first on engines which were sold on to a number of manufacturers as original equipment. Both two- and four-stroke units were made, a 58 cc and 148 cc OHC unit. One of the customer companies was Meihatsu Heavy Industries, another company within the Kawasaki group, which in 1961 was shaken up and renamed Kawasaki Auto Sales. At the same time, the Akashi factory which was to be Kawasaki's main production facility until the Kobe earthquake of 1995, was opened. Shortly afterwards, Kawasaki took over the ailing Meguro company, Japan's oldest motorcycle maker, thus instantly obtaining a range of bigger bikes which were marketed as Kawasaki-Meguros. The following year, the first bike to be made and sold as a Kawasaki was produced, a 125 cc single called the B8 and in 1963 a motocross version, the B8M appeared.

## Model development

Kawasaki's first appearance on a road-race circuit came in 1965 with a batch of disc-valve 125 twins. They were no match for the opposition from Japan in the shape of Suzuki and Yamaha or for the fading force of the factory MZs from East Germany. Only after the other Japanese factories had pulled out of the class did Kawasaki win, with British rider Dave Simmonds becoming World 125 GP Champion in 1969 on a bike that looked astonishingly similar to the original racer. That same year Kawasaki reorganised once again, this time merging three

The H1 three cylinder two-stroke 500

companies to form Kawasaki Heavy Industries. One of the new organisation's objectives was to take motorcycle production forward and exploit markets outside Japan.

KHI achieved that target immediately and set out their stall for the future with the astonishing and frightening H1. This three-cylinder air-cooled 500 cc two-stroke was arguably the first modern pure performance bike to hit the market. It hypnotised a whole generation of motorcyclists who'd never before encountered such a ferocious, wheelie inducing power band or such shattering straight-line speed allied to questionable handling. And as for the 750 cc version ...

The triples perfectly suited the late '60s, fitting in well with the student demonstrations of 1968 and the anti-establishment ethos of the Summer of Love. Unfortunately, the oil crisis would put an end to the thirsty strokers but Kawasaki had another high-performance ace up their corporate sleeve. Or rather they thought they did.

The 1968 Tokyo Show saw probably the single most significant new motorcycle ever made unveiled: the Honda CB750. At Kawasaki it caused a major shock, for they also had a 750 cc four, code-named New York Steak, almost ready to roll and it was a double, rather than single, overhead cam motor. Bravely, they took the decision to go ahead - but with the motor taken out to 900 cc. The result was the Z1, unveiled at the 1972 Cologne Show. It was a bike straight out of the same mould as the H1, scare stories spread about unmanageable power, dubious straight-line stability and frightening handling, none of which stopped the sales graph rocketing upwards and led to the coining of the term 'superbike'. While rising fuel prices cut short development of the big two-strokes,

The first Superbike, Kawasaki's 900 cc Z1

the Z1 went on to found a dynasty, indeed its genes can still be detected in Kawasaki's latest products like the ZZ-R1100 (Ninja ZX-11).

This is another characteristic of the way Kawasaki operates. Models quite often have very long lives, or gradually evolve. There is no major difference between that first Z1 and the air-cooled GPz range. Add water-cooling and you have the GPZ900, which in turn metamorphosed into the GPZ1000RX and then the ZX-10 and the ZZ-R1100. Indeed, the last three models share the same 58 mm stroke. The bikes are obviously very different but it's difficult to put your finger on exactly why.

Other models have remained effectively untouched for over a decade: the KH and KE single-cylinder air-cooled two-stroke learner bikes, the GT550 and 750 shaft-drive hacks favoured by big city despatch riders and the GPz305 being prime examples. It's only when they step outside the performance field that Kawasakis seems less sure. Their first factory customs were dire, you simply got the impression that the team that designed them didn't have their heart in the job. Only when the Classic range appeared in 1995 did they get it right.

## Racing success

Kawasaki also have a more focused approach to racing than the other factories. The policy has always been to race the road bikes and with just a couple of exceptions that's what they've done. Even Simmonds' championship winner bore a strong resemblance to the twins they were selling in the late '60s and racing versions of the 500 and 750 cc triples were also sold as over-the-counter racers, the H1R and H2R. The 500 was in the forefront of the two-stroke assault on MV Agusta but wasn't a Grand Prix winner. It was the 750 that made the impact and carried the factory's image in F750 racing against the Suzuki triples and Yamaha fours.

The factory's decision to use green, usually regarded as an unlucky colour in sport, meant its bikes and personnel stood out and the phrase 'Green Meanies' fitted them perfectly. The Z1 motor soon became a full 1000 cc and powered Kawasaki's assault in F1 racing, notably in endurance which Kawasaki saw as

One of the two-stroke engined KH and KE range - the KH125EX

**The GT750 - a favourite hack for despatch riders**

being most closely related to its road bikes.

That didn't stop them dominating 250 and 350 cc GPs with a tandem twin two-stroke in the late '70s and early '80s, but their path-breaking monocoque 500 while a race winner never won a world title. When Superbike arrived, Kawasaki's road 750s weren't as track-friendly as the opposition's out-and-out race replicas. This makes Scott Russell's World title on the ZXR750 in 1993 even more praiseworthy, for the homologation bike, the ZXR750RR, was much heavier and much more of a road bike than the Italian and Japanese competition.

The company's Supersport 600 contenders have similarly been more sports-tourers than race-replicas, yet they too have been competitive on the track. Indeed, the flagship bike, the ZZ-R1100, is most definitely a sports tourer capable of carrying two people and their luggage at high speed in comfort all day and then doing it again the next day. Try that on one of the race replicas and you'll be in need of a course of treatment from a chiropractor.

Through doing it their way Kawasaki developed a brand loyalty for their performance bikes that kept the Z1's derivatives in production until the mid-'80s

and turned the bike into a classic in its model life. You could even argue that the Z1 lives on in the shape of the 1100 Zephyr's GPz1100-derived motor. And that's another Kawasaki invention, the retro bike. But when you look at what many commentators refer to as the retro boom, especially in

Japan, you find that it is no such thing. It is the Zephyr boom. Just another example of Japan's most surprising motorcycle manufacturer getting it right again.

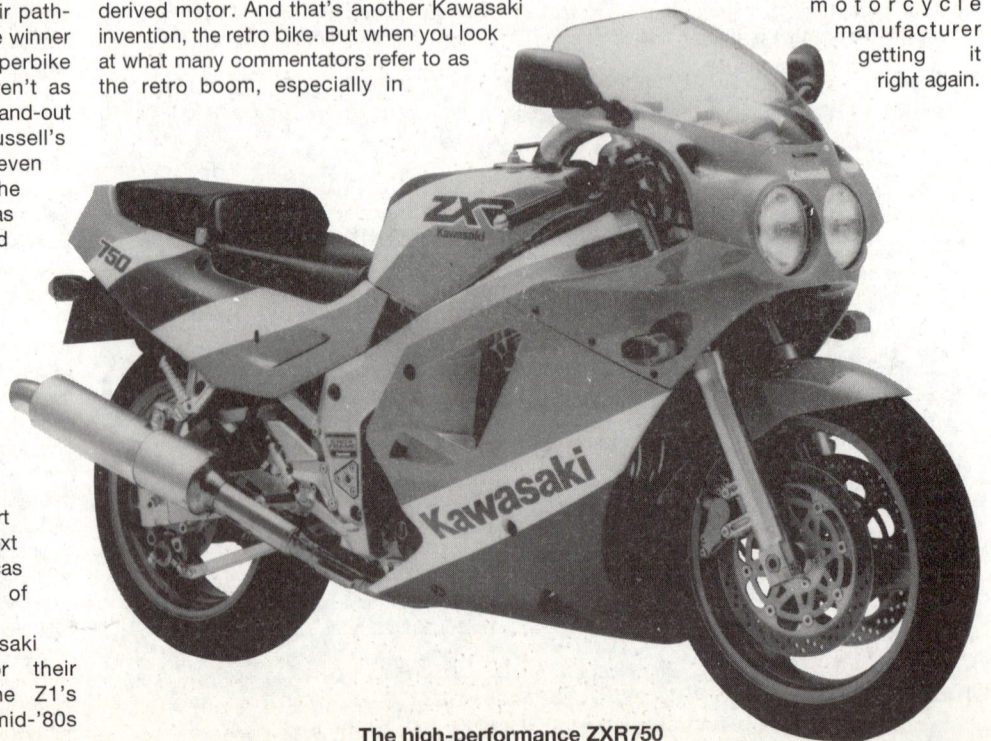

**The high-performance ZXR750**

## Return of the Green Meanies

My first ride on a ZXR750 (ZX-7) was at the launch at the Estoril race track in Portugal. Even at the speeds I rode at, one fact was obvious, Kawasaki's super-sports 750 was more road bike than racer, which mattered quite a lot as this was the model that would be the homologation bike for World Superbike racing. As a roadster, it had all you needed; as a racer it had too much weight despite the very sporty looks it inherited – among many other things – from the old F1 racers.

The ZX750H model was the first 750 Kawasaki with an aluminum frame but power was provided by what was essentially a tuned-up version of the old GPX's 68 x 51.5 mm motor. Even those trick looking inlet tubes from the scoops on the front of the fairing merely ducted cool air onto the cylinder head, not into a plenum chamber to feed the intake system slightly pressurized air. The H2 model got few tuning tricks, but looked a lot better for the addition of the big slab-sided swinging arm off the racers to replace the original box-section item.

Sports fans had to wait until 1991 for a really competitive Kawasaki when the J and K models were released. The former was a pure roadbike, the later was a specialized homologation model designed to let Superbike racers use more track-friendly parts – particularly flat-slide carburetors. These same parts tended to make the ZXR750R (ZX-7R) as it was known a complete pig to ride on the road but those carbs, the close-ratio gearbox, single seat, alloy fuel tank and higher state of tune were a great starting point for the racers.

The standard 750 roadbike got more than just the racer's looks. Visually, the first clues you see are the big upside-down front forks and those dummy intake hoses running into

The ZX750L2 model

the frame members. But what you can't see is much more important: under the fairing and housed in a lightened frame lurked a brand new 71 x 47.3 mm short-stroke 749 cc motor with drive to the camshafts moved from the center of the crank to the right-hand side which liberated enough space in the top end to allow the angle between intake and exhaust valves to be narrowed down to just 20°.

Further models followed in 1992, designated L for the roadster and M for the double-R homologation special, but the only mechanical change of any significance was in the official race kit which included inlet parts that made those cosmetic hoses the leading edge of a true ram-air system.

Immediately after the introduction of the 749 cc short-stroke engine, the ZXR750R (Ninja ZX-7R) was a contender for racing honors the world over, despite the basic bike still being much more of a roadster than the

opposition. In 1991 Kawasakis won the World Endurance Championship, the AMA SuperSport 750 Championship and the Australian Superbike Championship. In 1992 the Endurance crown went to Kawasaki again thanks to Carl Fogarty and Terry Rymer, Scott Russell won the Daytona 200 at record speed plus the AMA Superbike title, but most importantly the American along with New Zealander Aaron Slight won the Suzuka 8 Hours on a ZXR (Ninja), the first time a Green Meanie had won the most important single race in the Japanese factories' calendar. And in 1993 Scott Russell finally overcame his nemesis Carl Fogarty and the bigger, lighter, more powerful Ducati V-twins to win the World Superbike Championship, the first and so far only time a good old, simple in-line four has taken the top title in four-stroke racing. Not a bad record for an overweight road bike.

## Acknowledgments

Our thanks to Kawasaki Motors (UK) Ltd. for permission to reproduce certain illustrations used in this manual. We would also like to thank NGK Spark Plugs (UK) Ltd for supplying the spark plug condition photos and the Avon Rubber Company, who kindly supplied information and technical assistance on tire fitting.

Special thanks to Grand Prix Kawasaki/ Yamaha/Suzuki, Santa Clara, California, for supplying the bikes used in these photographs; to Mark Woodward, service manager, for arranging the facilities and fitting the teardown into his shop's busy schedule; and to Jim Holly, service technician, for his careful, precise mechanical work and thorough technical knowledge.

Thanks are also due to Paul Branson Motorcycles of Yeovil, Somerset for supplying the ZX750H model.

Thanks are also due to the Kawasaki Information Service and Kel Edge for supplying color transparencies. The introduction 'The Green Meanies' was written by Julian Ryder.

## About this manual

The aim of this manual is to help you get the best value from your motorcycle. It can do so in several ways. It can help you decide what work must be done, even if you choose to have it done by a dealer; it provides information and procedures for routine maintenance and servicing; and it offers diagnostic and repair procedures to follow when trouble occurs.

We hope you use the manual to tackle the work yourself. For many simpler jobs, doing it yourself may be quicker than arranging an appointment to get the motorcycle into a

dealer and making the trips to leave it and pick it up. More importantly, a lot of money can be saved by avoiding the expense the shop must pass on to you to cover its labor and overhead costs. An added benefit is the sense of satisfaction and accomplishment that you feel after doing the job yourself.

References to the left or right side of the motorcycle assume you are sitting on the seat, facing forward.

Professional mechanics are trained in safe working procedures. However enthusiastic you may be about getting on with the job at hand, take the time to ensure that your safety is not put at risk. A moment's lack of attention can result in an accident, as can failure to observe simple precautions.

There will always be new ways of having accidents, and the following is not a comprehensive list of all dangers; it is intended rather to make you aware of the risks and to encourage a safe approach to all work you carry out on your bike.

## Asbestos

● Certain friction, insulating, sealing and other products - such as brake pads, clutch linings, gaskets, etc. - contain asbestos. Extreme care must be taken to avoid inhalation of dust from such products since it is hazardous to health. If in doubt, assume that they do contain asbestos.

## Fire

● Remember at all times that petrol is highly flammable. Never smoke or have any kind of naked flame around, when working on the vehicle. But the risk does not end there - a spark caused by an electrical short-circuit, by two metal surfaces contacting each other, by careless use of tools, or even by static electricity built up in your body under certain conditions, can ignite petrol vapour, which in a confined space is highly explosive. Never use petrol as a cleaning solvent. Use an approved safety solvent.

● Always disconnect the battery earth terminal before working on any part of the fuel or electrical system, and never risk spilling fuel on to a hot engine or exhaust.
● It is recommended that a fire extinguisher of a type suitable for fuel and electrical fires is kept handy in the garage or workplace at all times. Never try to extinguish a fuel or electrical fire with water.

## Fumes

● Certain fumes are highly toxic and can quickly cause unconsciousness and even death if inhaled to any extent. Petrol vapour comes into this category, as do the vapours from certain solvents such as trichloro-ethylene. Any draining or pouring of such volatile fluids should be done in a well ventilated area.
● When using cleaning fluids and solvents, read the instructions carefully. Never use materials from unmarked containers - they may give off poisonous vapours.
● Never run the engine of a motor vehicle in an enclosed space such as a garage. Exhaust fumes contain carbon monoxide which is extremely poisonous; if you need to run the engine, always do so in the open air or at least have the rear of the vehicle outside the workplace.

## The battery

● Never cause a spark, or allow a naked light near the vehicle's battery. It will normally be giving off a certain amount of hydrogen gas, which is highly explosive.

● Always disconnect the battery ground (earth) terminal before working on the fuel or electrical systems (except where noted).
● If possible, loosen the filler plugs or cover when charging the battery from an external source. Do not charge at an excessive rate or the battery may burst.
● Take care when topping up, cleaning or carrying the battery. The acid electrolyte, evenwhen diluted, is very corrosive and should not be allowed to contact the eyes or skin. Always wear rubber gloves and goggles or a face shield. If you ever need to prepare electrolyte yourself, always add the acid slowly to the water; never add the water to the acid.

## Electricity

● When using an electric power tool, inspection light etc., always ensure that the appliance is correctly connected to its plug and that, where necessary, it is properly grounded (earthed). Do not use such appliances in damp conditions and, again, beware of creating a spark or applying excessive heat in the vicinity of fuel or fuel vapour. Also ensure that the appliances meet national safety standards.
● A severe electric shock can result from touching certain parts of the electrical system, such as the spark plug wires (HT leads), when the engine is running or being cranked, particularly if components are damp or the insulation is defective. Where an electronic ignition system is used, the secondary (HT) voltage is much higher and could prove fatal.

# Remember...

✗ **Don't** start the engine without first ascertaining that the transmission is in neutral.
✗ **Don't** suddenly remove the pressure cap from a hot cooling system - cover it with a cloth and release the pressure gradually first, or you may get scalded by escaping coolant.
✗ **Don't** attempt to drain oil until you are sure it has cooled sufficiently to avoid scalding you.
✗ **Don't** grasp any part of the engine or exhaust system without first ascertaining that it is cool enough not to burn you.
✗ **Don't** allow brake fluid or antifreeze to contact the machine's paintwork or plastic components.
✗ **Don't** siphon toxic liquids such as fuel, hydraulic fluid or antifreeze by mouth, or allow them to remain on your skin.
✗ **Don't** inhale dust - it may be injurious to health (see Asbestos heading).
✗ **Don't** allow any spilled oil or grease to remain on the floor - wipe it up right away, before someone slips on it.
✗ **Don't** use ill-fitting spanners or other tools which may slip and cause injury.
✗ **Don't** lift a heavy component which may be beyond your capability - get assistance.

✗ **Don't** rush to finish a job or take unverified short cuts.
✗ **Don't** allow children or animals in or around an unattended vehicle.
✗ **Don't** inflate a tyre above the recommended pressure. Apart from overstressing the carcass, in extreme cases the tyre may blow off forcibly.
✔ **Do** ensure that the machine is supported securely at all times. This is especially important when the machine is blocked up to aid wheel or fork removal.
✔ **Do** take care when attempting to loosen a stubborn nut or bolt. It is generally better to pull on a spanner, rather than push, so that if you slip, you fall away from the machine rather than onto it.
✔ **Do** wear eye protection when using power tools such as drill, sander, bench grinder etc.
✔ **Do** use a barrier cream on your hands prior to undertaking dirty jobs - it will protect your skin from infection as well as making the dirt easier to remove afterwards; but make sure your hands aren't left slippery. Note that long-term contact with used engine oil can be a health hazard.
✔ **Do** keep loose clothing (cuffs, ties etc. and long hair) well out of the way of moving mechanical parts.

✔ **Do** remove rings, wristwatch etc., before working on the vehicle - especially the electrical system.
✔ **Do** keep your work area tidy - it is only too easy to fall over articles left lying around.
✔ **Do** exercise caution when compressing springs for removal or installation. Ensure that the tension is applied and released in a controlled manner, using suitable tools which preclude the possibility of the spring escaping violently.
✔ **Do** ensure that any lifting tackle used has a safe working load rating adequate for the job.
✔ **Do** get someone to check periodically that all is well, when working alone on the vehicle.
✔ **Do** carry out work in a logical sequence and check that everything is correctly assembled and tightened afterwards.
✔ **Do** remember that your vehicle's safety affects that of yourself and others. If in doubt on any point, get professional advice.
● **If** in spite of following these precautions, you are unfortunate enough to injure yourself, seek medical attention as soon as possible.

## Frame and engine numbers

The frame serial number is stamped into the right side of the steering head and the engine serial number is stamped into the right engine case. Both of these numbers should be recorded and kept in a safe place so they can be furnished to law enforcement officials in the event of theft.

The frame serial number, engine serial number and carburetor identification number should also be kept in a handy place (such as with your driver's license) so they are always available when purchasing or ordering parts for your machine.

The following table is a breakdown of the initial frame numbers for each model and year of production:

### US and Canada models

| Prod yr | Model | Initial frame number |
|---------|-------|---------------------|
| 1989 | ZX750H1 (Ninja ZX-7) | JKAZXDH1*KA000001 |
| 1990 | ZX750H2 (Ninja ZX-7) | JKAZXDH1*LA015001 |
| 1991 | ZX750J1 (Ninja ZX-7) | JKAZXDJ1*MA000001 |
| 1992 | ZX750J2 (Ninja ZX-7) | JKAZXDJ1*NA013901 |
| 1993 | ZX750L1 (Ninja ZX-7) | JKAZXDL1*PA000001 |
| 1994 | ZX750L2 (Ninja ZX-7) | JKAZXDL1*RA020001 |
| 1995 | ZX750L3 (Ninja ZX-7) | JKAZXDL1*SA030001 |
| 1991 | ZX750K1 (Ninja ZX-7R) | JKAZXDK1*MA000001 |
| 1992 | ZX750K2 (Ninja ZX-7R) | JKAZXDK1*NA007001 |
| 1993 | ZX750M1 (Ninja ZX-7R) | JKAZXDM1*PA000001 |
| 1994 | ZX750M2 (Ninja ZX-7R) | JKAZXDM1*RA002001 |

*This digit changes with each motorcycle.

### UK models

| Prod yr | Model | Initial frame number |
|---------|-------|---------------------|
| 1989 | ZX750H1 (ZXR750) | ZX750H-000001 |
| 1990 | ZX750H2 (ZXR750) | ZX750H-015001 |
| 1991 | ZX750J1 (ZXR750) | ZX750J-000001 |
| 1992 | ZX750J2 (ZXR750) | ZX750J-013901 |
| 1993 | ZX750L1 (ZXR750) | ZX750L-000001 |
| 1994 | ZX750L2 (ZXR750) | ZX750L-020001 |
| 1995 | ZX750L3 (ZXR750) | ZX750L-030001 |
| 1991 | ZX750K1 (ZXR750R) | ZX750K-000001 |
| 1993 | ZX750M1 (ZXR750R) | ZX750M-000001 |
| 1994 | ZX750M2 (ZXR750R) | ZX750M-020001 |

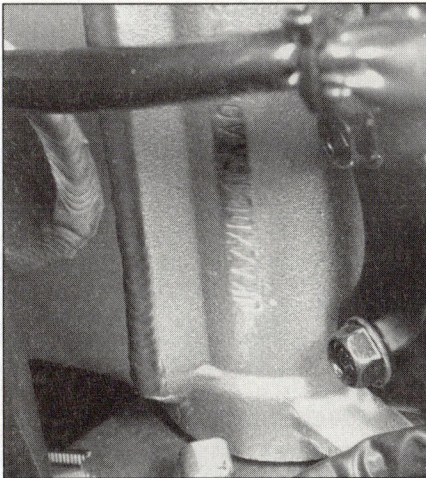

The frame number is stamped on the steering head . . .

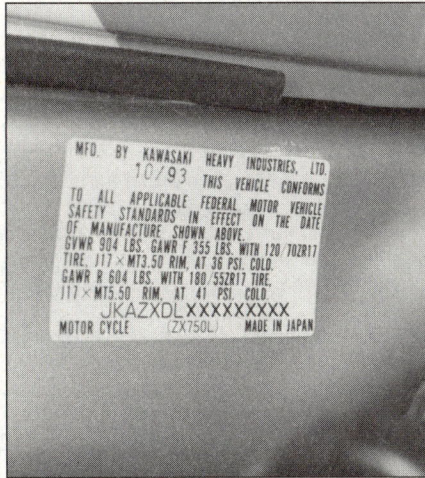

. . . and displayed on a decal

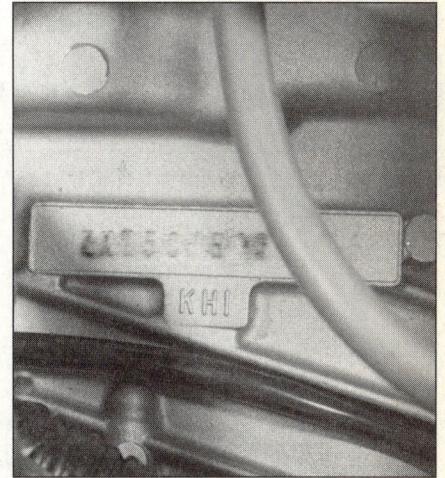

The engine number is located on the right side of the crankcase

## Buying spare parts

Once you have found all the identification numbers, record them for reference when buying parts. Since the manufacturers change specifications, parts and vendors (companies that manufacture various components on the machine), providing the ID numbers is the only way to be reasonably sure that you are buying the correct parts.

Whenever possible, take the worn part to the dealer so direct comparison with the new component can be made. Along the trail from the manufacturer to the parts shelf, there are numerous places that the part can end up with the wrong number or be listed incorrectly.

The two places to purchase new parts for your motorcycle – the accessory store and the franchised dealer – differ in the type of parts they carry. While dealers can obtain virtually every part for your cycle, the accessory dealer is usually limited to normal high wear items such as shock absorbers, tune-up parts, various engine gaskets, cables, chains, brake parts, etc. Rarely will an accessory outlet have major suspension components, cylinders, transmission gears, or cases.

Used parts can be obtained for roughly half the price of new ones, but you can't always be sure of what you're getting. Once again, take your worn part to the wrecking yard (breaker) for direct comparison.

Whether buying new, used or rebuilt parts, the best course is to deal directly with someone who specializes in parts for your particular make.

# 1 Engine/transmission oil level check

## Before you start:

✔ Take the motorcycle on a short run to allow it to reach normal operating temperature.

*Caution: Do not run the engine in an enclosed space such as a garage or shop.*

✔ Stop the engine and support the motorcycle in an upright position on level ground. Allow it to stand undisturbed for five minutes to allow the oil level to stabilise.

## Bike care:

● If you have to add oil frequently, you should check whether you have any oil leaks. If there is no sign of oil leakage from the joints and gaskets the engine could be burning oil (see *Fault Finding*).

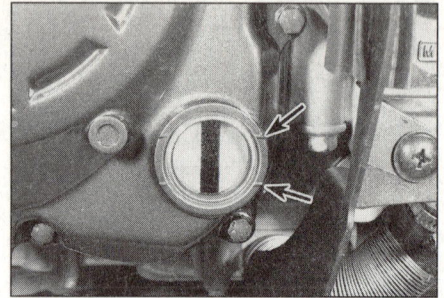

1 Check the oil level in the window located at the lower part of the right crankcase cover. The level should be between the Maximum and Minimum level marks arrowed.

## The correct oil

● Modern, high-revving engines place great demands on their oil. It is very important that the correct oil for your bike is used.
● Always top up with a good quality oil of the specified type and viscosity and do not overfill the engine.

| Oil type | API grade SE, SF or SG |
|---|---|
| **Oil viscosity** | |
| Cold climates | SAE 10W/40 or 10W/50 |
| Warm climates | SAE 20W/40 or 20W/50 |

2 If the level is below the Minimum mark, remove the oil filler cap from the right crankcase cover . . .

3 . . . and add enough oil of the recommended grade and type to bring the level up to the Maximum mark. o not overfill.

# 2 Clutch fluid level check

⚠ *Warning: Brake and clutch hydraulic fluid can harm your eyes and damage painted surfaces, so use extreme caution when handling and pouring it. Do not use fluid that has been standing open for some time, as it absorbs moisture from the air which can cause a loss of clutch effectiveness.*

## Before you start:

✔ Make sure you have the correct hydraulic fluid. DOT 4 is recommended.
✔ With the motorcycle supported in a level position, turn the handlebars until the top of the clutch master cylinder is as level as possible.
✔ Before removing the master cylinder cap, place rags beneath the reservoir (to protect the paint from brake fluid spills) and remove all dust and dirt from the area around the cap.

## Bike care:

● In order to ensure proper operation of the clutch hydraulic system, the fluid level in the master cylinder reservoir must be properly maintained. If the fluid level was low, inspect the clutch hydraulic system for leaks (see Chapter 1).
● Check the operation of the clutch. If the lever is spongy indicating the presence of air in the clutch line, bleed the system as described in Chapter 2.

1 The clutch fluid reservoir is located above the left front fork. The fluid level is visible through the reservoir. Make sure that the fluid level is above the Lower mark on the reservoir. If the level is low, the fluid must be replenished.

2 To top up, remove the cap retainer plate, the cap, the diaphragm retainer and rubber diaphragm to add fluid . . .

3 . . . and top up to the Upper mark with DOT 4 hydraulic fluid.

# 3 Brake fluid level checks

⚠️ **Warning: Brake and clutch hydraulic fluid can harm your eyes and damage painted surfaces, so use extreme caution when handling and pouring it and cover surrounding surfaces with rag. Do not use fluid that has been standing open for some time, as it absorbs moisture from the air which can cause a dangerous loss of braking effectiveness.**

## Before you start:
✔ Make sure you have the correct hydraulic fluid – DOT 4 is recommended.
✔ With the motorcycle supported in a level position, turn the handlebars until the top of the front brake master cylinder is as level as possible.
✔ On J, K, L and M models remove the right side cover to view the rear brake fluid reservoir. On H models the level marks can be seen through the slot in the right side cover.

## Bike care:
● In order to ensure proper operation of the hydraulic disc brakes, the fluid level in the master cylinder reservoir must be properly maintained. If the brake fluid level was low, inspect the brake system for leaks.

● The fluid in the master cylinder reservoir will drop slightly as the brake pads wear down.
● Check the operation of the brakes before taking the machine on the road; if there is evidence of air in the system (spongy feel to level or pedal), it must be bled as described in Chapter 7.

1 The fluid level is visible through the master cylinder reservoir. Make sure that the fluid level is above the Lower mark on the reservoir.

2 If the level is low, remove the cap retainer screw . . .

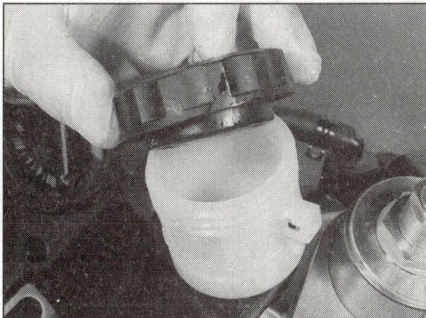

3 . . unscrew the cap and lift off the retainer and rubber diaphragm. **Note:** *Don't operate the brake lever with the cap removed.*

4 Add new, clean brake fluid of the recommended type until the level is above the inspection window. Don't mix different brands of brake fluid in the reservoir, as they may not be compatible.

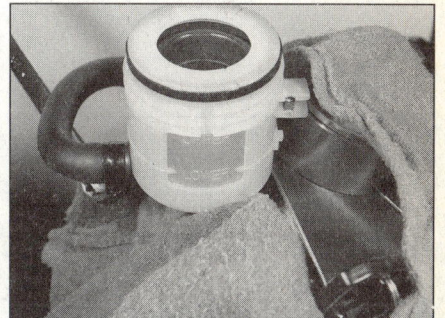

5 Reinstall the rubber diaphragm and retainer. Install the cap retainer and the cap and tighten its screw securely, but do not overtighten it. Wipe any spilled fluid off the reservoir body.

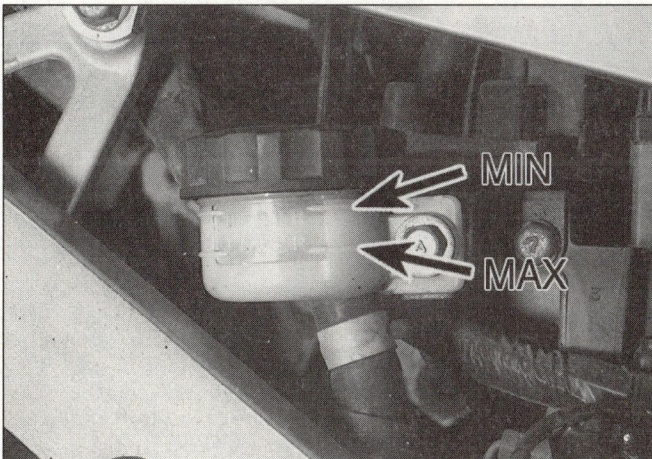

6 Rear brake fluid reservoir location – H models. Unscrew the cap and add fluid as described for the front brake.

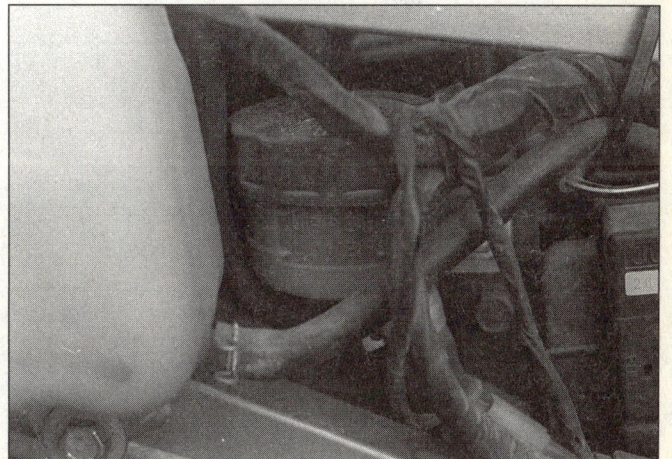

7 Rear brake fluid reservoir location – J, K, L and M models. Unscrew the cap and add fluid as described for the front brake.

## 4 Coolant level check

### Before you start:
✔ Make sure you have a supply of coolant available (a mixture of 50% distilled water and 50% corrosion inhibited ethylene glycol antifreeze).

✔ The engine must be cold for the results to be accurate, so always perform this check before starting the engine for the first time each day. Make sure the motorcycle is on level ground.

### Bike care:
● Use only the specified coolant mixture. It is important that antifreeze is used in the cooling system all year round, not just during the winter months. Don't top-up with water alone, as the antifreeze will become too diluted.
● Do not overfill the coolant reservoir. The coolant level is satisfactory if it is between the Low and Full marks on the reservoir.
● If the coolant level seems to be consistently low, check the entire cooling system for leaks.

**1** The coolant reservoir is located behind the right middle fairing panel on H models . . .

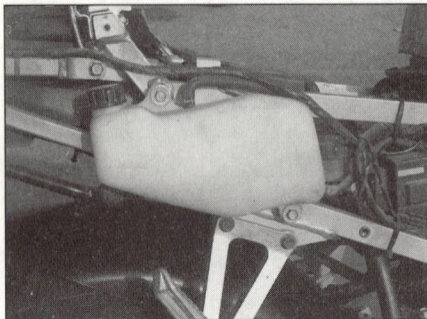

**2** . . . and under the right rear side cover on J, K, L and M models. The reservoir tank marks should be visible from below and to the rear of the cover.

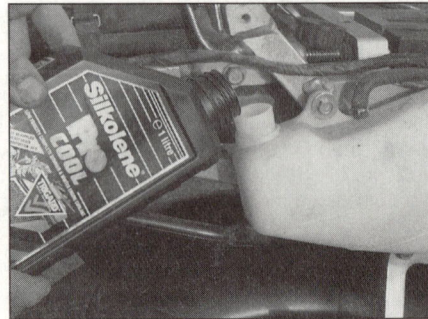

**3** If the level is at or below the Low mark, remove the filler cap and add the recommended coolant mixture until the Full level is reached.

## 5 Suspension, steering and drive chain checks

### Suspension and Steering:
● Make sure the steering operates smoothly, without looseness and without binding.
● Check front and rear suspension for smooth operation.
● Check that the suspension is adjusted as required.

### Drive chain:
● Make sure the drive chain isn't out of adjustment (Chapter 1).
● If the chain looks dry, lubricate it (Chapter 1).

## 6 Legal and safety checks

### Lighting and signalling:
● Take a minute to check that the headlight, tail light, brake light, instrument lights and turn signals all work correctly.
● Check that the horn sounds when the switch is operated.
● A working speedometer graduated in mph is a statutory requirement in the UK.

### Safety:
● Check that the throttle grip rotates smoothly and snaps shut when released, and that it does so in all steering positions. Also check for the correct amount of freeplay (see Chapter 1).
● Make sure the sidestand returns to its fully up position and stays there under spring pressure.
● Make sure the engine kill switch works properly.

### Fuel:
● This may seem obvious, but check that you have enough fuel to complete your journey. If you notice signs of fuel leakage - rectify the cause immediately.
● Ensure you use the correct grade unleaded or low lead fuel (subject to local regulations) – minimum 91 octane (RON).

## 7 Tire checks

### Tire care:

● Check the tires carefully for cuts, tears, embedded nails or other sharp objects and excessive wear. Operation of the motorcycle with excessively worn tires is extremely hazardous, as traction and handling are directly affected.
● Check the condition of the tire valve and ensure the dust cap is in place.
● Pick out any stones or nails which may have become embedded in the tire tread. If left, they will eventually penetrate through the casing and cause a puncture.
● If tire damage is apparent, or unexplained loss of pressure is experienced, seek the advice of a tire fitting specialist without delay.

### Tire tread depth:

● At the time of writing UK law requires that tread depth must be at least 1 mm over ¾ of the tread breadth all the way around the tire, with no bald patches. Many riders, however, consider 2 mm tread depth minimum to be a safer limit. Kawasaki recommend the following minimum tread depths.

| Regular speed | Front | Rear |
| --- | --- | --- |
| Up to 80 mph (130 kmh) | 1 mm | 2 mm |
| Above 80 mph (130 kmh) | 1 mm | 3 mm |

● Many tires now incorporate wear indicators in the tread. Identify the triangular pointer, or TWI marking, on the tire sidewall to locate the indicator bar and replace the tire if the tread has worn down to the bar.

**1** Check the tire pressures when the tires are cold and keep them properly inflated.

**2** Measure the tread depth at the center of the tire using a tread depth gauge.

**3** Tire tread wear indicators bars and location marking on the sidewall (arrows).

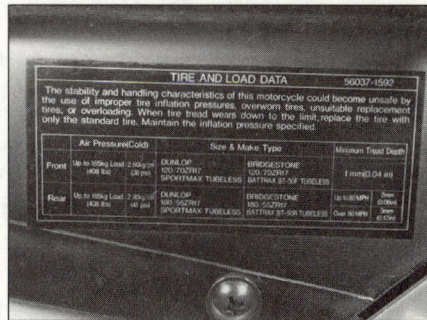

**4** The tire information lable provides tire size, pressure and tread depth details

### The correct pressures:

● The tires must be checked when **cold**, not immediately after riding. Note that low tire pressures may cause the tire to slip on the rim or come off. High tire pressures will cause abnormal tread wear and unsafe handling.
● Use an accurate pressure gauge.
● Proper air pressure will increase tire life and provide maximum stability and ride comfort.
● Note that tire pressures vary according to the type of tires fitted – refer to the tire information label on the bike, or if non-standard tires are fitted, refer to the tire manufacturer for the correct pressures.

| Tire pressures | H, J and K models | L and M models |
| --- | --- | --- |
| Front | 36 psi (2.48 bars) | 33 psi (2.27 bars) |
| Rear | 41 psi (2.82 bars) | 36 psi (2.48 bars) |

**Notes**

# Chapter 1
# Routine maintenance and servicing

## Contents

## Degrees of difficulty

| Easy, suitable for novice with little experience | Fairly easy, suitable for beginner with some experience | Fairly difficult, suitable for competent DIY mechanic | Difficult, suitable for experienced DIY mechanic | Very difficult, suitable for expert DIY or professional |
|---|---|---|---|---|

## Specifications

### Engine

Spark plugs
  Type
    US H, J, K, L1, M1 models ............................... NGK C9E or ND U27ES-N
    US L2, L3, M2 models ................................. NGK CR9E or ND U27ESR-N
    UK models ........................................... NGK CR9E or ND U27ESR-N
  Gap ..................................................... 0.7 to 0.8 mm (0.028 to 0.031 inch)
Engine idle speed
  H models
    All except California models ........................... 1000 ± 50 rpm
    California models ...................................... 1250 ± 50 rpm
  J, K, L, M models
    All except California models ........................... 1100 ± 50 rpm
    California models ...................................... 1300 ± 50 rpm
Valve clearances (COLD engine)
  H models
    Intake ................................................ 0.15 to 0.24 mm (0.006 to 0.009 inch)
    Exhaust ............................................... 0.20 to 0.29 mm (0.008 to 0.011 inch)
  J, K, L, M models
    Intake ................................................ 0.18 to 0.23 mm (0.007 to 0.009 inch)
    Exhaust ............................................... 0.25 to 0.30 mm (0.010 to 0.012 inch)
Cylinder compression pressure acceptable range
  H and L models ........................................... 139 to 213 psi (9.57 to 14.67 bars)
  J models ................................................. 119 to 185 psi (8.2 to 12.74 bars)
  K and M models ........................................... 145 to 220 psi (9.99 to 15.15 bars)
Carburetor synchronization (vacuum difference between cylinders) ... Less than 2 cm (0.39 inch) Hg
Cylinder numbering (from left side to right side of bike) ............. 1-2-3-4
Firing order .................................................. 1-2-4-3

## Miscellaneous

Brake pad lining thickness
New . . . . . . . . . . . . . . . . . . . . . . . . . . . . . . . . . 4 mm (0.16 inch)
Wear limit . . . . . . . . . . . . . . . . . . . . . . . . . . . . . 1 mm (0.04 inch)
Rear brake pedal height
H models . . . . . . . . . . . . . . . . . . . . . . . . . . . . . . . Approximately 55 mm (2.2 inches) below top of footpeg
J, K, L, M models . . . . . . . . . . . . . . . . . . . . . . . . . Approximately 70 mm (2.8 inches) below top of footpeg
Freeplay adjustments
Throttle grip . . . . . . . . . . . . . . . . . . . . . . . . . . . . . 2 to 3 mm (0.08 to 0.12 inch)
Choke cable . . . . . . . . . . . . . . . . . . . . . . . . . . . . . 2 to 3 mm (0.08 to 0.12 inch)
Drive chain
Slack
H models
Standard . . . . . . . . . . . . . . . . . . . . . . . . . . . . 20 to 25 mm (0.79 to 0.98 inch)
Limit . . . . . . . . . . . . . . . . . . . . . . . . . . . . . . . 20 to 30 mm (0.79 to 0.118 inch)
J, K, L, M models
Standard . . . . . . . . . . . . . . . . . . . . . . . . . . . . 10 to 15 mm (0.39 to 0.59 inch)
Limit . . . . . . . . . . . . . . . . . . . . . . . . . . . . . . . 10 to 20 mm (0.39 to 0.79 inch)
20-link length
Standard . . . . . . . . . . . . . . . . . . . . . . . . . . . . . . 317.5 to 318.2 mm (12.5 to 12.53 inch)
Limit . . . . . . . . . . . . . . . . . . . . . . . . . . . . . . . . . 323 mm (12.73 inch)
Battery electrolyte specific gravity . . . . . . . . . . . . . . . . . 1.260 minimum
Minimum tire tread depth . . . . . . . . . . . . . . . . . . . . . . See 'Daily (pre-ride) checks'
Tire pressures . . . . . . . . . . . . . . . . . . . . . . . . . . . . . . See 'Daily (pre-ride) checks'

## Torque settings

Oil drain plug . . . . . . . . . . . . . . . . . . . . . . . . . . . . . . 20 Nm (14.5 ft-lbs)
Oil filter (all except H2 models) . . . . . . . . . . . . . . . . . . . 9.8 Nm (87 inch-lbs)
Oil filter mounting bolt (H2 models) . . . . . . . . . . . . . . . . 20 Nm (14.5 ft-lbs)
Coolant bleed valve
H models . . . . . . . . . . . . . . . . . . . . . . . . . . . . . . . 9.8 Nm (87 inch-lbs)
J, K, L, M models . . . . . . . . . . . . . . . . . . . . . . . . . . 7.8 Nm (69 inch-lbs)
Coolant drain bolt
H models (in cylinder block) . . . . . . . . . . . . . . . . . . . . 7.8 Nm (69 inch-lbs)
J, K, L, M models (in water pump) . . . . . . . . . . . . . . . . 12 Nm (104 inch-lbs)
Spark plugs . . . . . . . . . . . . . . . . . . . . . . . . . . . . . . . . 14 Nm (10 ft-lbs)

## Recommended lubricants and fluids

Engine/transmission oil
Type . . . . . . . . . . . . . . . . . . . . . . . . . . . . . . . . . . . API grade SE, SF or SG multigrade oil
Viscosity
In cold climates . . . . . . . . . . . . . . . . . . . . . . . . . . . SAE 10W40 or 10W50
In warm climates . . . . . . . . . . . . . . . . . . . . . . . . . . SAE 20W40 or 20W50
Capacity
H models
With filter change . . . . . . . . . . . . . . . . . . . . . . . . 3.0 liters (3.2 US qt, 5.3 Imp pt)
Oil change only . . . . . . . . . . . . . . . . . . . . . . . . . 2.6 liters (2.7 US qt, 4.6 Imp pt)
J, K, L, M models
With filter change . . . . . . . . . . . . . . . . . . . . . . . . 3.5 liters (3.7 US qt, 6.1 Imp pt)
Oil change only . . . . . . . . . . . . . . . . . . . . . . . . . 3.4 liters (3.6 US qt, 6.0 Imp pt)
Coolant
Type . . . . . . . . . . . . . . . . . . . . . . . . . . . . . . . . . . . 50/50 mixture of water and ethylene glycol antifreeze containing corrosion inhibitors for aluminum engines
Capacity (including full reservoir tank)
H1 models . . . . . . . . . . . . . . . . . . . . . . . . . . . . . . 2.0 liters (2.1 US qt)
H2 models . . . . . . . . . . . . . . . . . . . . . . . . . . . . . . 2.3 liters (2.4 US qt)
J, K, L, M models . . . . . . . . . . . . . . . . . . . . . . . . . 2.8 liters (3.0 US qt)
Brake/clutch fluid . . . . . . . . . . . . . . . . . . . . . . . . . . . DOT 4
Fork oil . . . . . . . . . . . . . . . . . . . . . . . . . . . . . . . . . . See Chapter 6

## Miscellaneous

Wheel bearings . . . . . . . . . . . . . . . . . . . . . . . . . . . . . Medium weight, lithium-based multi-purpose grease
Swingarm pivot bearings . . . . . . . . . . . . . . . . . . . . . . . Medium weight, lithium-based multi-purpose grease
Cables and lever pivots . . . . . . . . . . . . . . . . . . . . . . . Chain and cable lubricant or 10W30 motor oil
Sidestand pivot . . . . . . . . . . . . . . . . . . . . . . . . . . . . . Chain and cable lubricant or 10W30 motor oil
Brake pedal/shift lever pivots . . . . . . . . . . . . . . . . . . . . Chain and cable lubricant or 10W30 motor oil
Throttle grip . . . . . . . . . . . . . . . . . . . . . . . . . . . . . . . Multi-purpose grease or dry film lubricant

**Note:** *The pre-ride inspection outlined at the front of this manual covers checks and maintenance that should be carried out on a daily basis. Always perform the pre-ride* *inspection at every maintenance interval (in addition to the procedures listed). The intervals listed below are the shortest intervals recommended by the manufacturer for each* *particular operation during the model years covered in this manual. Your owner's manual may have different intervals for your model.*

## Daily (pre-ride)
☐ See *'Daily (pre-ride) checks'* beginning on page 0•10

## Every 200 miles (300 km)
*Carry out all of the 'Daily (pre-ride) checks' plus the following:*
☐ Lubricate the drive chain (Section 1)

## After the initial 500 miles (800 km)
**Note:** *This check is usually performed by a Kawasaki dealer after the first 500 miles (800 km) from new. Thereafter, maintenance is carried out according to the following intervals of the schedule.*

## Every 500 miles (800 km)
*Carry out all of the 'Daily (pre-ride) checks' plus the following:*
☐ Check/adjust the drive chain slack (Section 2)

## Every 3000 miles (5000 km)
*Carry out all of the 200 and 500 miles (300 and 800 km) maintenance tasks plus the following:*
☐ Clean and gap the spark plugs and replace (if necessary) (Section 3)
☐ Check the operation of the air suction valves (if equipped) (Section 4)
☐ Check/adjust the carburetor synchronization and idle speed (Section 5)
☐ Check the evaporative emission control system (California models) (Section 6)
☐ Check the drive chain and sprockets for wear (Section 7)
☐ Check the brake fluid level (Section 8)
☐ Check the adjustment of the brake light switch (Section 8)
☐ Check/adjust the brake pedal position (Section 8)
☐ Check the brake discs and pads (Section 9)
☐ Check the clutch fluid level (Section 10)
☐ Lubricate all cables (Section 11)
☐ Lubricate the clutch and brake lever pivots (Section 11)
☐ Lubricate the shift/brake lever pivots and the sidestand pivot (Section 11)
☐ Check the steering (Section 12)
☐ Check the tires and wheels (Section 13)
☐ Check the battery electrolyte level (H models only) (Section 14) – see Note 1
☐ Check the exhaust system for leaks (Section 15)
☐ Check the tightness of all fasteners (Section 16)
☐ Check/adjust the throttle and choke operation/grip freeplay (Section 17)

## Every 6000 miles (10,000 km)
*Carry out all of the 3000 miles (5000 km) maintenance tasks plus the following:*
☐ Change the engine oil and oil filter (Section 18) – see Note 2
☐ Clean the air filter element (Section 19)
☐ Check the cleanliness of the fuel system and the condition of the fuel and vacuum hoses (Section 20)
☐ Lubricate the swingarm needle bearings and Uni-trak linkage (Section 21)
☐ Check the cylinder compression (Section 22)
☐ Check/adjust the valve clearances (Section 23)

## Every 12,000 miles (20,000 km)
*Carry out all of the 6000 miles (10,000 km) maintenance tasks plus the following:*
☐ Replace the air filter element (Section 24) – see Note 3
☐ Replace the fuel filter (Section 25)
☐ Change the clutch fluid (Section 26) – see Note 4
☐ Lubricate the steering head bearings (Section 27)
☐ Change the brake fluid (Section 28) – see Note 4

## Every 18,000 miles (30,000 km)
*Carry out all of the 6000 miles (10,000 km) maintenance tasks plus the following:*
☐ Check the cooling system and replace the coolant (Section 29) – see Note 3
☐ Change the fork oil (Section 30)

## Every two years
☐ Rebuild the clutch master and release cylinders (Section 31)
☐ Rebuild the brake calipers and master cylinders (Section 32)

## Every four years
☐ Replace the fuel hoses (Section 33)
☐ Replace the brake hoses (Section 34)

## Notes
☐ Note 1 – Or once a month, whichever comes first
☐ Note 2 – Or once a year, whichever comes first
☐ Note 3 – Or after every five cleanings
☐ Note 4 – Or every two years, whichever comes first

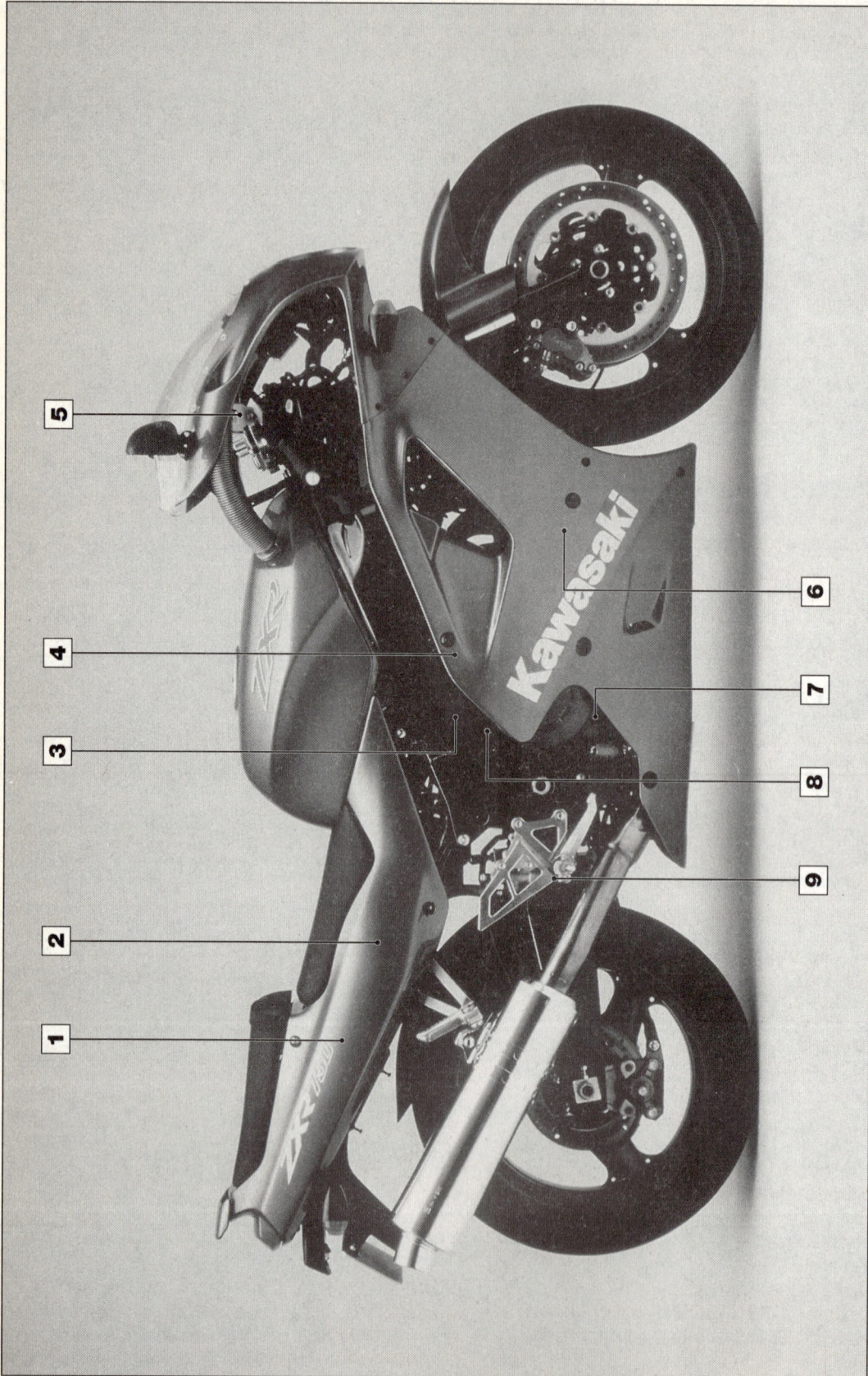

**Component locations on right-hand side**

1 Coolant reservoir (J, K, L and M models)
2 Rear brake fluid reservoir
3 Idle speed adjuster
4 Coolant reservoir (H models)
5 Front brake fluid reservoir
6 Timing rotor cover
7 Engine oil level window
8 Engine oil filler cap
9 Rear brake pedal height adjuster and brake light switch

**Component locations on left-hand side**

1  Clutch fluid reservoir
2  Steering head bearings
3  Air filter element (J, K, L and M models)
4  Air filter element (H models)
5  Battery
6  Drive chain adjuster
7  Engine oil filter (H2 models)
8  Coolant drain plug and bleed valve on water pump
9  Oil drain bolt
10 Engine oil filter (H1, J, K, L and M models)

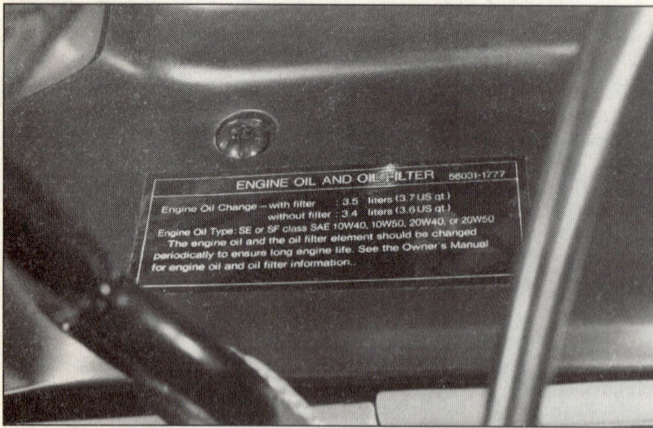
Decals on the motorcycle include oil type and capacity . . .

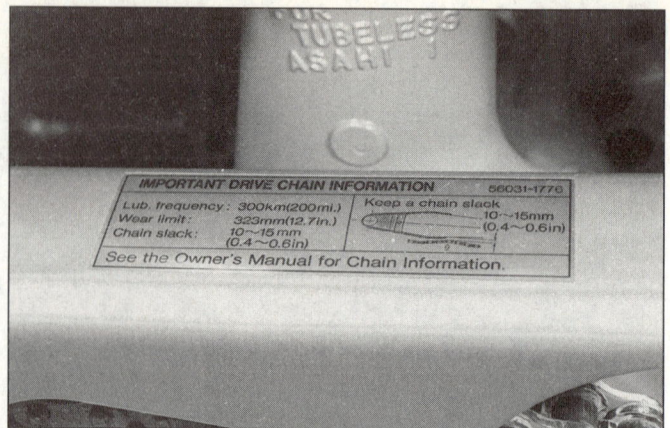
. . . and drive chain information

# Introduction

This Chapter covers in detail the checks and procedures necessary for the tune-up and routine maintenance of your motorcycle and includes the routine maintenance schedule, which is designed to keep the machine in proper running condition and prevent possible problems. The remaining Sections contain detailed procedures for carrying out the items listed on the maintenance schedule, as well as additional maintenance information designed to increase reliability. Maintenance information is also printed on decals, which are mounted in various locations on the motorcycle (see illustrations). Where information on the decals differs from that presented in this Chapter, use the decal information.

Since routine maintenance plays such an important role in the safe and efficient operation of your motorcycle, it is presented here as a comprehensive check list. For the rider who does all his own maintenance, these lists outline the procedures and checks that should be done on a routine basis.

Deciding where to start or plug into the routine maintenance schedule depends on several factors. If you have a motorcycle whose warranty has recently expired, and if it has been maintained according to the warranty standards, you may want to pick-up routine maintenance as it coincides with the next mileage or calendar interval. If you have owned the machine for some time but have never performed any maintenance on it, then you may want to start at the nearest interval and include some additional procedures to ensure that nothing important is overlooked. If you have just had a major engine overhaul, then you may want to start the maintenance routine from the beginning. If you have a used machine and have no knowledge of its history or maintenance record, you may desire to combine all the checks into one large service initially and then settle into the maintenance schedule prescribed.

The Sections which actually outline the inspection and maintenance procedures are written as step-by-step comprehensive guides to the actual performance of the work. They explain in detail each of the routine inspections and maintenance procedures on the check list. References to additional information in applicable Chapters is also included and should not be overlooked.

Before beginning any actual maintenance or repair, the machine should be cleaned thoroughly, especially around the oil filter housing, spark plugs, cylinder head cover, side covers, carburetors, etc. Cleaning will help ensure that dirt does not contaminate the engine and will allow you to detect wear and damage that could otherwise easily go unnoticed.

# Every 200 miles (300 km)

## 1 Drive chain – lubrication

**Note:** If the chain is extremely dirty, it should be removed and cleaned before it is lubricated (see Chapter 6).

**1** The best time to lubricate the chain is after the motorcycle has been ridden. When the chain is warm, the lubricant will penetrate the joints between the side plates, pins, bushings and rollers to provide lubrication of the internal load bearing areas.

**2** Use a good quality chain lubricant and apply it to the area where the side plates overlap – not the middle of the rollers. After applying the lubricant, let it soak in a few minutes before wiping off any excess.

**Caution: Make sure the chain lubricant is compatible with O-ring chains – some kinds may damage the O-rings.**

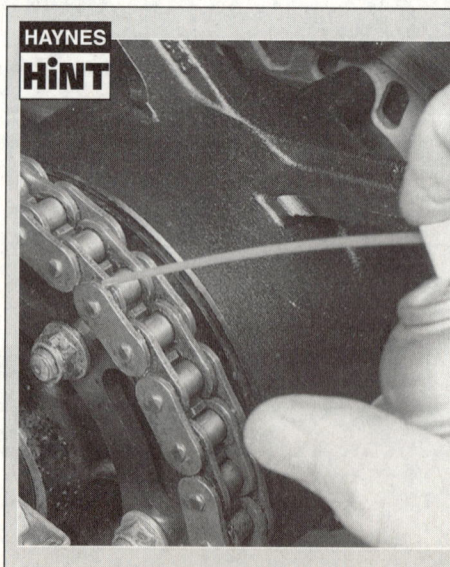

HAYNES HiNT

*Apply chain lubricant to the joints between the side plates and the rollers – not in the center of the rollers (with the bike supported securely upright and the rear wheel off the ground, hold the plastic nozzle near the edge of the chain and turn the wheel by hand as the lubricant sprays out – repeat the procedure on the inside edge of the chain)*

# Every 500 miles (800 km)

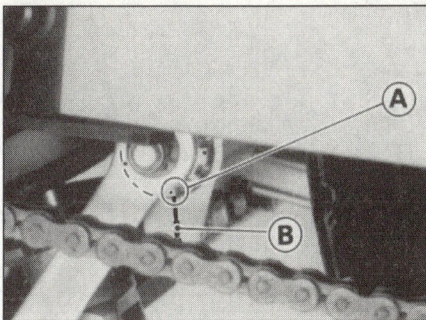

**2.2 On H1 models, align the middle mark with the split in the swingarm**

A  *Middle mark*  B  *Split*

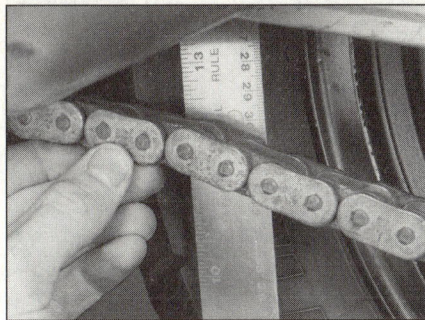

**2.3 Push up on the bottom run of the chain and measure how far it deflects – if it's not within the specified limits, adjust the slack in the chain**

**2.5 Remove the rubber cap (arrow) and the cotter pin (if equipped); loosen the nut on the torque link bolt**

## 2  Drive chain – check and adjustment

### Check

1 A neglected drive chain won't last long and can quickly damage the sprockets. Routine

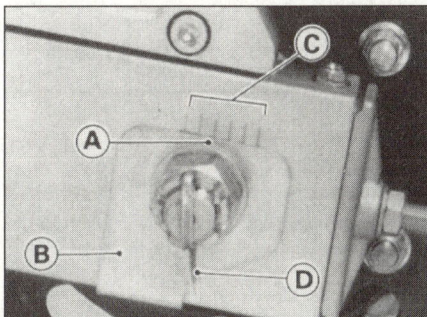

**2.6a  Left side chain adjuster details (H1/1989) models**

A  *Adjuster plate notch*
B  *Adjuster plate*
C  *Adjustment marks*
D  *Axle nut cotter pin*

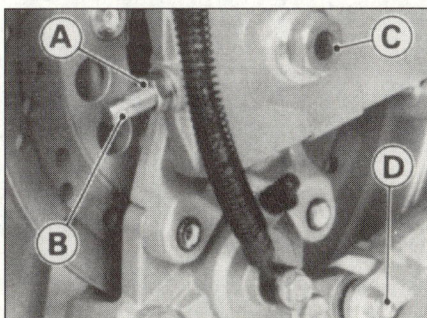

**2.6b  Right side chain adjuster details (H1/1989) models**

A  *Locknut*
B  *Adjuster*
C  *Axle*
D  *Torque link nut*

chain adjustment and lubrication isn't difficult and will ensure maximum chain and sprocket life.

2 To check the chain, support the bike on its sidestand and shift the transmission into neutral. Make sure the ignition is OFF. If you are working on an H1 (1989 model) check that the middle mark on the suspension linkage eccentric cam aligns with the split in the swingarm. **(see illustration)**.

3 Push up on the bottom run of the chain and measure the slack at a point midway between the sprockets **(see illustration)**, then compare your measurements to the value listed in this Chapter's specifications. As wear occurs the chain will stretch, which means adjustment is necessary to remove excess slack. Chains rarely wear evenlyl so rotate the rear wheel to allow several different sections the chain to be checked.

### Adjustment

4 Rotate the rear wheel until the chain is positioned with the least amount of slack present.

5 Remove the rubber cap and cotter pin (if equipped) from the torque link-to-rear caliper bracket bolt and nut. Loosen the nut **(see illustration)**.

6 Loosen and back-off the locknuts on the adjuster bolts **(see illustrations)**.

7 Remove the cotter pin and loosen the axle nut **(see illustration 2.6a or 2.6c)**.

8 Turn the axle adjusting bolts on both sides of the swingarm until the proper chain tension is obtained. Be sure to turn the adjusting bolts evenly to keep the rear wheel in alignment. If the adjusting bolts reach the end of their travel, the chain is excessively worn and should be replaced with a new one (see Chapter 6).

9 When the chain has the correct amount of slack, make sure the marks on the adjusters correspond to the same relative marks on each side of the swingarm **(see illustrations 2.6a through 2.6d)**. Tighten the axle nut to the torque listed in Chapter 7 Specifications, then install a new cotter pin, bending its ends around the nut to secure it. If necessary, turn the nut an additional amount to line up the cotter pin hole with the castellations in the nut – don't loosen the nut to do this.

10 Tighten the locknuts and the torque link nut securely. Install a new cotter pin in the torque link bolt (bending its ends over to secure the nut) and install the rubber cap on the end of the bolt (if equipped).

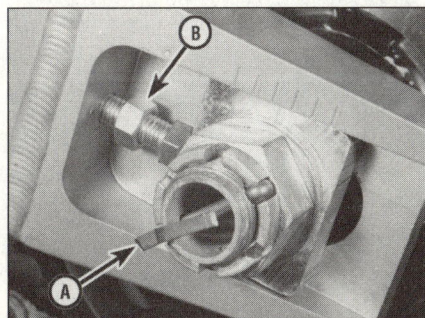

**2.6c  On H2 and J, K, L, M models, remove the cotter pin (A) and loosen the axle nut; the adjuster bolt locknut (B) should be turned back, away from the swingarm**

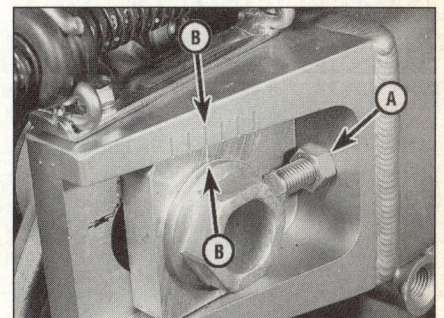

**2.6d  On the right side, loosen the adjuster locknut (A); use the adjustment marks (B) to make sure both sides are adjusted evenly**

# Every 3000 miles (5000 km)

**3.1 A 16 mm plug socket is required for spark plug removal**

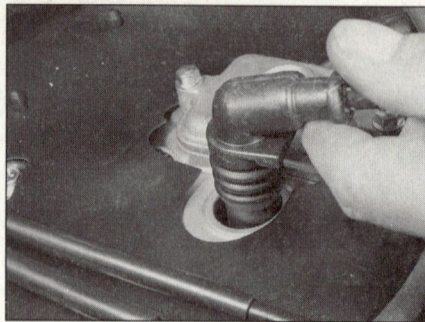
**3.2a Rotate the plug caps back and forth to break the seal . . .**

**3.2b . . . then pull them off the plugs and check for brittleness or deterioration**

## 3  Spark plugs – replacement

**1** This motorcycle is equipped with spark plugs that have a 16 mm wrench hex **(see illustration)**. Make sure your spark plug socket is the correct size before attempting to remove the plugs.

**2** Remove the fuel tank (see Chapter 4), then disconnect the spark plug caps from the spark plugs **(see illustrations)**. If available, use compressed air to blow any accumulated debris from around the spark plugs. Remove the plugs **(see illustrations)**.

**3** Inspect the electrodes for wear. Both the center and side electrodes should have square edges and the side electrode should be of uniform thickness. Look for excessive deposits and evidence of a cracked or chipped insulator around the center electrode. Compare your spark plugs to the spark plug reading chart on the inside rear cover of this manual. Check the threads, the washer and the ceramic insulator body for cracks and other damage.

**4** If the electrodes are not excessively worn, and if the deposits can be easily removed with a wire brush, the plugs can be regapped and reused (if no cracks or chips are visible in the insulator). If in doubt concerning the condition of the plugs, replace them with new ones, as the expense is minimal.

**5** Cleaning spark plugs by sandblasting is permitted, provided you clean the plugs with a high flash-point solvent afterwards.

**6** Before installing new plugs, make sure they are the correct type and heat range. Check the gap between the electrodes, as they are not preset. For best results, use a wire-type gauge rather than a flat gauge to check the gap. If the gap must be adjusted, bend the side electrode only and be very careful not to chip or crack the insulator nose **(see illustrations)**. Make sure the washer is in place before installing each plug.

**7** Since the cylinder head is made of aluminum, which is soft and easily damaged, thread the plugs into the head by hand **(see Haynes Hint)**.

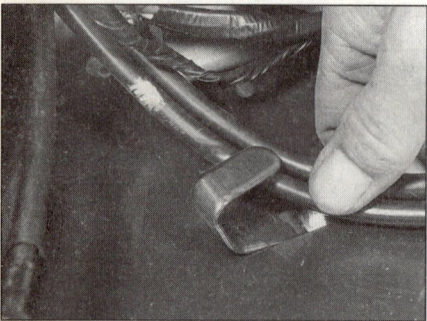
**3.2c Remove the plug wires (HT leads) from the clips; note the routing so they can be reinstalled in the clips**

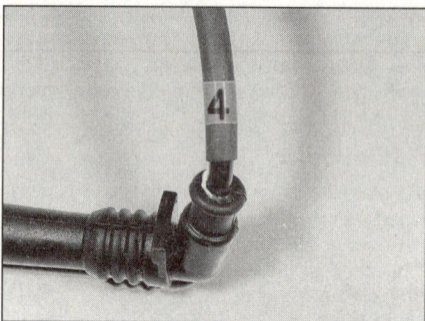
**3.2d The plug wires (HT leads) are labeled according to cylinder number (counting from the left side of the motorcycle)**

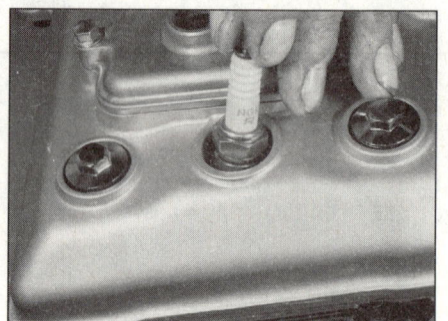
**3.2e If the spark plug didn't stay in the wrench rubber insert, lift it out once it's loose . . .**

**3.2f . . . a gripping tool like this can be used to pull the plug out of its bore**

**3.6a Spark plug manufacturers recommend using a wire type gauge when checking the gap**

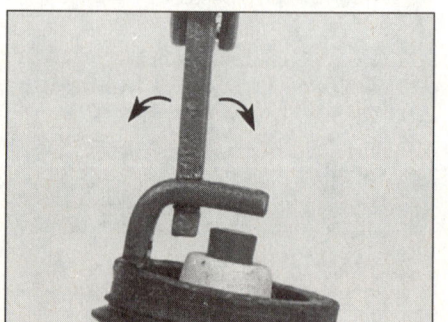
**3.6b To change the gap, bend the side electrode only, as indicated by the arrows**

**HAYNES HINT**

*Since the plugs are quite recessed, slip a short length of hose over the end of the plug to use as a tool to thread it into place. The hose will grip the plug well enough to turn it, but will start to slip if the plug begins to cross-thread in the hole – this will prevent damaged threads and the accompanying repair costs.*

8 Once the plugs are finger tight, the job can be finished with a socket. If a torque wrench is available, tighten the spark plugs to the torque listed in this Chapter's Specifications. If you do not have a torque wrench, tighten the plugs finger tight (until the washers bottom on the cylinder head) then use a wrench to tighten them an additional 1/4 turn. Regardless of the method used, do not over-tighten them.

9 Reconnect the spark plug caps and reinstall all components removed for access.

## 4  Air suction valves – check

1 The air suction valves, installed on US and some European models only, are one-way check valves that allow fresh air to flow into the exhaust ports. The suction developed by the exhaust pulses pulls the air from the air cleaner, through a hose to the vacuum switch valve, through a pair of hoses and two pairs of reed valves, and finally into the exhaust ports. The introduction of fresh air helps ignite any fuel that may not have been burned by the normal combustion process.

2 Remove the fuel tank (see Chapter 4).

3 Disconnect the hoses from the air suction valves **(see illustration)**. Remove the bolts and lift off the covers.

4 Check the valve for cracks, warping, burning or other damage. Check the area where the reeds contact the valve holder for scratches, separation and grooves **(see illustration)**. If any of these conditions are found, replace the valve.

5 Wash the valves with solvent if carbon has accumulated between the reed and the valve holder.

6 Installation of the valves is the reverse of removal. Be sure to use a new gasket.

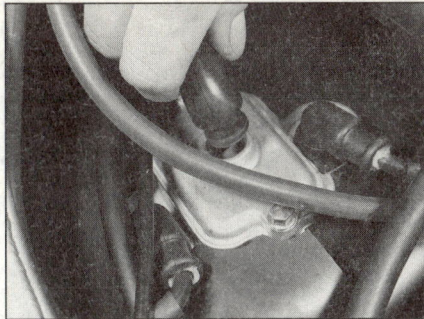

4.3 Disconnect the hoses from the air suction valves, then remove the two bolts and lift the valve off the engine

## 5  Carburetors – synchronization and idle speed

⚠️ *Warning: Gasoline (petrol) is extremely flammable, so take extra precautions when you work on any part of the fuel system. Don't smoke or allow open flames or bare light bulbs near the work area, and don't work in a garage where a natural gas-type appliance (such as a water heater or clothes dryer) is present. If you spill any fuel on your skin, rinse it off immediately with soap and water. When you perform any kind of work on the fuel system, wear safety glasses and have a fire extinguisher suitable for class B type fires (flammable liquids) on hand.*

### Synchronization

1 Carburetor synchronization is simply the process of adjusting the carburetors so they pass the same amount of fuel/air mixture to each cylinder. This is done by measuring the vacuum produced in each cylinder. Carburetors that are out of synchronization will result in decreased fuel mileage, increased engine temperature, less than ideal throttle response and higher vibration levels.

2 To properly synchronize the carburetors, you will need some sort of vacuum gauge

5.6a Disconnect the hoses (if equipped) from the vacuum fittings – L models (shown) are on the bottom of the intake manifold boots; H and J models are on the top . . .

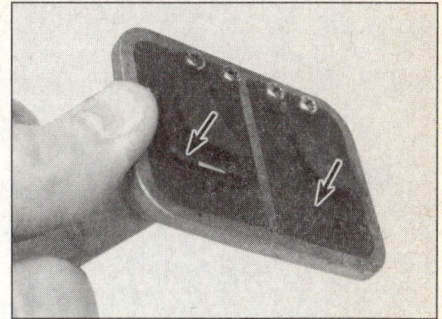

4.4 Check the reeds (arrows) on the air suction valve for damage and carbon build-up

setup, preferably with a gauge for each cylinder, or a manometer, which is a calibrated tube arrangement to indicate engine vacuum.

⚠️ *Warning: Some manometers use mercury as the liquid – since this is extremely toxic, extra precautions must be taken during use and storage of the instrument.*

3 Because of the nature of the synchronization procedure and the need for special instruments, most owners leave the task to a dealer service department or a reputable motorcycle repair shop. But if you want to tackle this job, here is how it is done:

4 Start the engine and let it run until it reaches normal operating temperature, then shut it off.

### H, J and L models

5 If you're working on an H model, detach the fuel tank and move it back to provide access, but leave the fuel line attached (see Chapter 4). If you're working on a J or L model, remove the fuel tank and air cleaner housing (see Chapter 4). Connect an auxiliary fuel supply.

6 Detach the vacuum hoses (if equipped) from the fittings on the carburetor intake manifolds **(see illustration)**, then hook up the vacuum gauge set or the manometer according to the manufacturer's instructions. On models with plugs in the vacuum ports instead of vacuum hoses, remove the plugs and install threaded vacuum hose fittings **(see illustration)**. Make sure there are no leaks in the setup, as false readings will result.

5.6b . . . if there are plugs in the vacuum fittings holes, unscrew them and thread in fittings like this (carburetors removed for clarity)

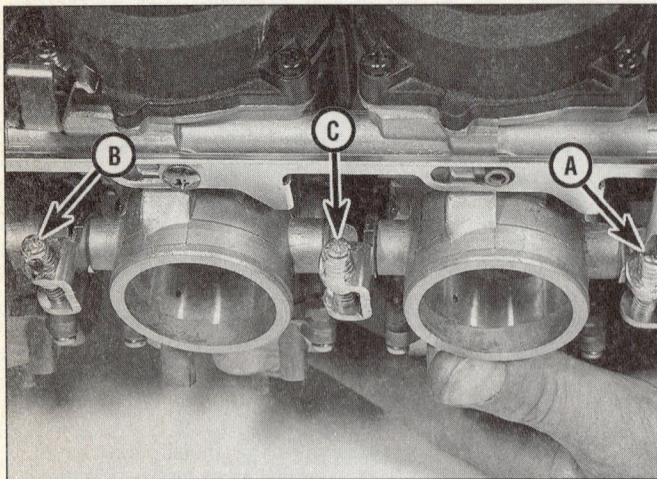

**5.9 Adjusting screws for nos. 1 and 2 carburetors (A), nos. 3 and 4 carburetors (B), both pairs (C)**

**5.21 Loosen the locknut (arrow) and turn the screw to adjust throttle plate position**

**7** Start the engine and make sure the idle speed is correct.

**8** The vacuum readings for all of the cylinders should be the same, or at least within the tolerance listed in this Chapter's Specifications. If the vacuum readings vary, adjust as necessary.

**9** To perform the adjustment, synchronize the carburetors for cylinders 1 and 2 by turning the adjusting screw between those two carburetors, as needed, until the vacuum is identical or nearly identical for those two cylinders **(see illustration)**.

**10** Next, synchronize the carburetors for cylinders 3 and 4, using the adjusting screw situated between those two carburetors.

**11** Finally, synchronize the carburetors for cylinders 1 and 2 to the carburetors for cylinders 3 and 4 by turning the center adjusting screw.

**12** When the adjustment is complete, recheck the vacuum readings and idle speed, then stop the engine. Remove the vacuum gauge or manometer and attach the hoses to the fittings on the carburetors. Reinstall the fuel tank and all components removed for access.

### K and M models

**13** Remove the fuel tank (see Chapter 4) and connect an auxiliary fuel supply.

**14** Turn the idle speed screw in about three threads, to the point where it's just about to move the throttle pulley.

**15** Start the engine and warm it up. Check idle speed, adjust as necessary and turn the engine off.

**16** Remove the air cleaner housing and the carburetors (see Chapter 4).

**17** Detach the vacuum hoses (if equipped) from the fittings on the carburetor intake manifolds, then hook up the vacuum gauge set or the manometer according to the manufacturer's instructions. On models with plugs in the vacuum ports instead of vacuum hoses, remove the plugs and install threaded vacuum hose fittings. Make sure there are no leaks in the setup, as false readings will result.

**18** Start the engine and make sure idle speed is correct.

**19** The vacuum readings for all of the cylinders should be the same, or at least within the tolerance listed in this Chapter's Specifications. If the vacuum readings vary, adjust as necessary.

**20** To make the adjustment, remove the top covers from no. 1, no. 2 and no. 3 carburetors, but don't remove the cover from carburetor no. 4 (the carburetors are numbered from the left side of the engine).

**21** Loosen the locknut on each balance adjusting screw and turn the screw to synchronize it with no. 4 **(see illustration)**. Turning the screw in closes the throttle.

**22** If the carburetors can't be synchronized with the balance adjusting screws, shut the engine off, then remove the carburetors (see Chapter 4).

**23** Check the carburetors for obvious contamination which could be blocking jets and passageways and overhaul them as necessary.

**24** Look into the throttle bores and check the throttle valve clearance **(see illustration)**. Set the throttle valve clearances of carburetors no. 1, 2 and 3, using the balance adjusting screws, so they're the same as carburetor no. 4.

**25** Reinstall the carburetors and check synchronization again. If the carburetors still won't synchronize, the pilot screw settings may be at fault. Refer to Chapter 4 for pilot screw removal and adjustment procedures.

### *Idle speed*

**26** The idle speed should be checked and adjusted after the carburetors are synchronized and when it is obviously too high or too low. Before adjusting the idle speed, make sure the valve clearances and spark plug gaps are correct. Also, turn the handlebars back-and-forth and see if the idle speed changes as this is done. If it does, the throttle cable may not be adjusted correctly, or it may be worn out. Be sure to correct this problem before proceeding.

**27** The engine should be at normal operating temperature, which is usually reached after 10 to 15 minutes of stop and go riding. Support the motorcycle securely upright and make sure the transmission is in Neutral.

**28** Turn the throttle stop screw **(see illustration)** until the idle speed listed in this Chapter's Specifications is obtained.

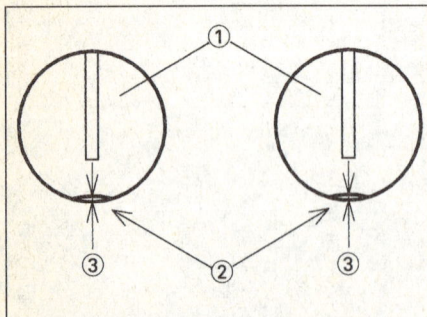

**5.24 All four throttle plates should have the same gap at the bottom**

*1 Throttle plates*    *3 Gaps*
*2 Throttle bores*

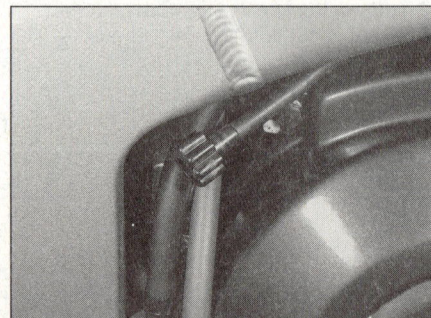

**5.28 Idle speed is adjusted with the throttle stop screw (L model shown; others similar)**

**29** Snap the throttle open and shut a few times, then recheck the idle speed. If necessary, repeat the adjustment procedure.
**30** If a smooth, steady idle can't be achieved, the fuel/air mixture may be incorrect. Refer to Chapter 4 for additional carburetor information.

## 6 Evaporative emission control system (California models only) – check

**1** This system, installed on California models to conform to stringent emission control standards, routes fuel vapors from the fuel system into the engine to be burned, instead of letting them evaporate into the atmosphere. When the engine isn't running, vapors are stored in a carbon canister.
**2** Periodic inspection consists of checking the hoses for looseness, kinking or deterioration, and the liquid/vapor separator for cracks or other obvious damage. Refer to Chapter 4 for system details. Drain the catch tank periodically as described in Section 19 of this Chapter.

## 7 Drive chain and sprockets – check for wear

**1** Unbolt the chain guard. Check the entire length of the chain for damaged rollers, loose links and pins. In some cases where lubrication has been neglected, corrosion and galling may cause the links to bind and kink, which effectively shortens the chain's length. If the chain is tight between the sprockets, rusty or kinked, it's time to replace it with a new one.
**2** Hang a 20-lb (10 kg) weight on the bottom run of the chain and measure the length of 20 links along the top run. Rotate the wheel and repeat this check at several places on the chain, since it may wear unevenly. Compare your measurements with the maximum 20-link length listed in this Chapter's Specifications. If any of your measurements exceed the maximum, replace the chain.
*Caution: Never install a new chain on old sprockets, and never use the old chain if you install new sprockets – replace the chain and sprockets as a set.*
**3** Remove the engine sprocket cover (see Chapter 6). Check the teeth on the engine sprocket and the rear sprocket for wear **(see illustration)**. Refer to Chapter 6 for the sprocket diameter measurement procedure if the sprockets appear to be worn excessively.

## 8 Brake system – general check

**1** A routine general check of the brakes will ensure that any problems are discovered and remedied before the rider's safety is jeopardized.
**2** Check the brake lever and pedal for loose connections, excessive play, bends, and other damage. Replace any damaged parts with new ones (see Chapter 7).
**3** Make sure all brake fasteners are tight. Check the brake pads for wear (see Section 9) and make sure the fluid level in the reservoir is correct (see *'Daily (pre-ride) checks'*). Look for leaks at the hose connections and check for cracks in the hoses. If the lever is spongy, bleed the brakes as described in Chapter 7.

### Brake light switch

**4** Make sure the brake light operates when the front brake lever is depressed.
**5** Make sure the brake light is activated when

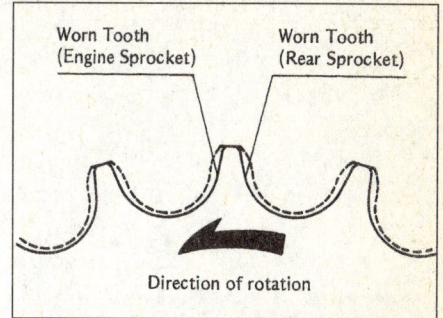

**7.3 Check the sprockets in the areas indicated to see if they're worn excessively**

the rear brake pedal is depressed approximately 10 mm (3/8 in).
**6** If adjustment is necessary, hold the switch and turn the adjusting nut on the switch body **(see illustration)** until the brake light is activated when required. Raising the switch will cause the brake light to come on sooner; lowering it will cause the brake light to come on later. If the switch doesn't operate the brake lights, check it as described in Chapter 9.
**7** The front brake light switch is not adjustable. If it fails to operate properly, replace it with a new one (see Chapter 9).

### Brake pedal position

**8** Rear brake pedal position should be set at the height listed in this Chapter's Specifications. Measure vertically up from the pedal tip to the top of the footpeg.
**9** If you're working on a J, K, L or M model, unbolt the footpeg bracket and rotate it out of the way (see Chapter 8).
**10** To adjust the position of the pedal, loosen the locknut on the master cylinder clevis, then rotate the hex on the master cylinder pushrod **(see illustration)**.
**11** If necessary, adjust the brake light switch (see Steps 5 and 6 above).

**8.6 Hold the switch body so it doesn't turn, and rotate the plastic nut (arrow) to adjust the rear brake light switch**

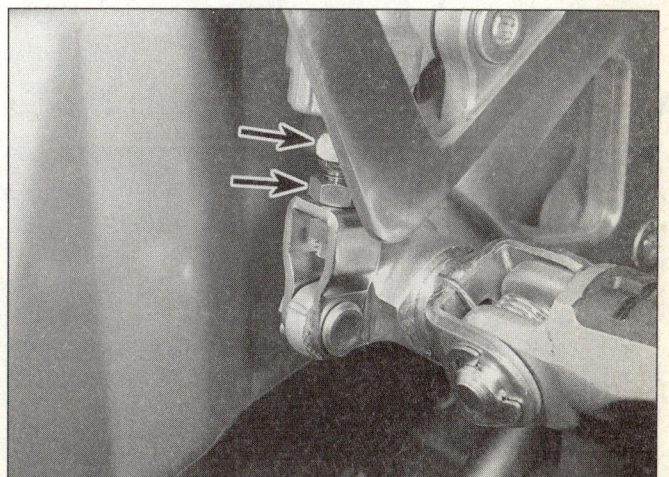

**8.10 Loosen the locknut (lower arrow) and turn the adjusting hex (upper arrow) to adjust brake pedal position**

**9.2a Remove the screws (arrows) and lift off the pad spring to inspect the front brake pads**

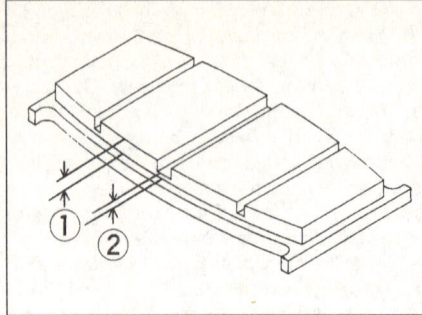

**9.2b On H models, replace the pads when the friction material is worn to the specified minimum thickness**

*1 Friction material thickness when new*
*2 Wear limit*

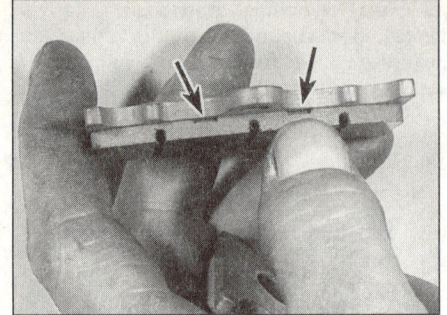

**9.2c J, K, L, M models have wear cutouts (arrows) in the friction material; when the material is worn so that the cutouts are exposed, it's time to replace the pads**

## 9 Brake pads – wear check

**1** The brake pads should be checked at the recommended intervals and replaced with new ones when worn beyond the limit listed in this Chapter's Specifications.

**2** To check the front brake pads, remove the pad spring so you can see clearly into the brake caliper. The brake pads should have at least the specified minimum amount of lining material remaining on the metal backing plate **(see illustrations)**. Be sure to check the pads in both calipers.

**3** Remove the rear brake caliper to check the pads (see Chapter 7). Remove the pad cover (if equipped) to expose the pads **(see illustration)**.

**4** If the pads are worn excessively, they must be replaced with new ones (see Chapter 7).

## 10 Clutch – check

**1** The hydraulic clutch mechanism on these models eliminates the need for freeplay adjustment. No means of manual adjustment is provided. However, the distance from the clutch lever to the handlebar can be adjusted with the span adjuster **(see illustration)**.

**2** Check the fluid level (see *'Daily (pre-ride) checks'*). Check for fluid leaks around the master cylinder, at the fluid line connections and at the release cylinder on the left side of the engine **(see illustration)**. If leaks are found, refer to Chapter 2 for repair procedures.

**3** Start the bike, release the clutch and ride off, noting the position of the clutch lever when the clutch begins to engage. If it's too close to the handlebar, there may be air in the clutch fluid (the air compresses, rather than transmitting lever force to the release mechanism). Refer to Chapter 2 and bleed the system.

## 11 Lubrication – general

**1** Since the controls, cables and various other components of a motorcycle are exposed to the elements, they should be lubricated periodically to ensure safe and trouble-free operation.

**2** The footpegs, clutch and brake levers, brake pedal, shift lever and sidestand pivots should be lubricated frequently. In order for the lubricant to be applied where it will do the most good, the component should be disassembled. However, if chain and cable lubricant is being used, it can be applied to the pivot joint gaps and will usually work its

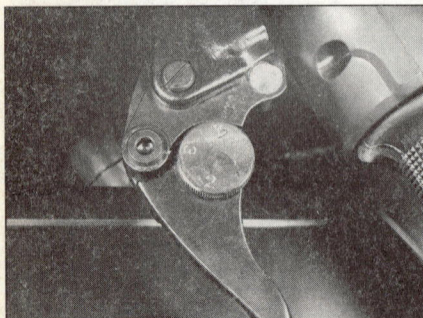

**9.3 Remove the pad cover (if equipped) from the rear caliper to inspect the pads**

way into the areas where friction occurs. If motor oil or light grease is being used, apply it sparingly as it may attract dirt (which could cause the controls to bind or wear at an accelerated rate). **Note:** *One of the best lubricants for the control lever pivots is a dry-film lubricant (available from many sources by different names).*

**3** The throttle and choke cables should be removed and treated with motor oil or a commercially available cable lubricant which is specially formulated for use on motorcycle control cables. Small adapters for pressure lubricating the cables with spray can lubricants are available and ensure that the cable is lubricated along its entire length **(see illustration)**. If motor oil is being used, tape a funnel-shaped piece of heavy paper or plastic

**10.1 Turn the span adjuster to change the clutch lever distance from the handlebar**

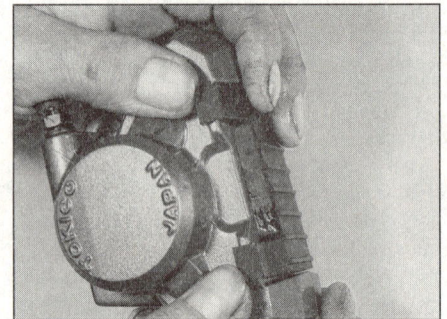

**10.2 Look for fluid leaks around the clutch release cylinder and make sure its mounting bolts are tight**

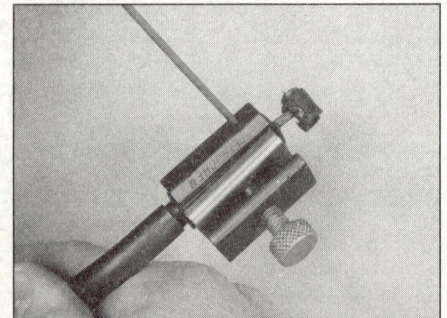

**11.3a Lubricating a cable with a pressure lube adapter (make sure the tool seats around the inner cable)**

to the end of the cable, then pour oil into the funnel and suspend the end of the cable upright **(see illustration)**. Leave it until the oil runs down into the cable and out the other end. When attaching the cable to the lever, be sure to lubricate the barrel-shaped fitting at the end with multi-purpose grease.

4 To lubricate the throttle and choke cables, disconnect the cables at the lower end, then lubricate the cable with a pressure lube adapter **(see illustration 11.3a)**. See Chapter 4 for the cable removal procedure.

5 The speedometer cable should be removed from its housing and the lower end of the inner cable lubricated with a thin coat of multi-purpose grease.

6 Refer to Chapter 6 for the swingarm needle bearing and Uni-trak linkage lubrication procedures.

## 12 Steering head bearings – check and adjustment

1 This vehicle is equipped with ball-and-cone type steering head bearings which can become dented, rough or loose during normal use of the machine. In extreme cases, worn or loose steering head bearings can cause steering wobble that is potentially dangerous.

2 To check the bearings, support the motorcycle securely upright and block the machine so the front wheel is in the air.

3 Point the wheel straight ahead and slowly move the handlebars from side-to-side. Dents or roughness in the bearing races will be felt and the bars will not move smoothly. **Note:** *Make sure any hesitation in movement is not being caused by the cables and wiring harnesses that run to the handlebars.*

4 Next, grasp the fork legs and try to move the wheel forward and backward **(see illustration)**. Any looseness in the steering head bearings will be felt. If play is felt in the bearings, refer to Chapter 6 and adjust them.

5 Lubrication of the steering head bearings should be carried out at the recommended intervals (see Maintenance schedule) and is detailed in Chapter 6.

## 13 Tires/wheels – general check

### Tires

1 Check the tires carefully for damage, tread depth and correct pressure as described in the *'Daily (pre-ride) checks'*.

### Wheels

2 The cast wheels used on this machine are virtually maintenance free, but they should be kept clean and checked periodically for cracks and other damage. Never attempt to repair damaged cast wheels; they must be replaced with new ones.

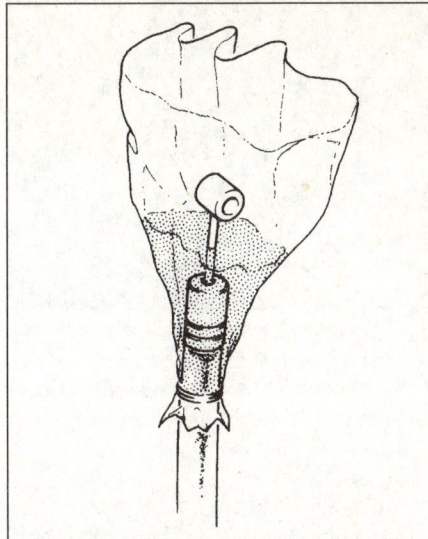

**11.3b Lubricating a control cable with a makeshift funnel and motor oil**

3 Check the valve stem locknuts to make sure they're tight. Also, make sure the valve stem cap is in place and tight. If it is missing, install a new one made of metal or hard plastic.

## 14 Battery electrolyte level/specific gravity (H models) – check

⚠ *Warning: Be extremely careful when handling or working around the battery. The electrolyte is very caustic and an explosive gas (hydrogen) is given off when the battery is charging.*

1 This procedure only applies to H models, which use fillable batteries. Later models use maintenance-free batteries, which don't require the addition of water.

2 Remove the seat and both side covers (see Chapter 8). Remove the battery (see Chapter 9).

3 Check the electrolyte level, which is visible through the translucent battery case – it should be between the Upper and Lower level marks.

4 If it is low, remove the cell caps and fill each cell to the upper level mark with distilled water. Do not use tap water (except in an emergency), and do not overfill. Never add electrolyte (diluted sulfuric acid) to a battery in service. If the level is within the marks on the case, additional water is not necessary.

**HAYNES HiNT** *The cell holes are quite small, so it may help to use a plastic squeeze bottle with a small spout to add the water.*

5 Next, check the specific gravity of the electrolyte in each cell with a small hydrometer made especially for motorcycle batteries (if the electrolyte level is known to be sufficient it won't be necessary to remove the battery).

**12.4 Checking for play in the steering head bearings**

These are available from most dealer parts departments or motorcycle accessory stores.

6 Remove the caps, draw some electrolyte from the first cell into the hydrometer **(see illustration)** and note the specific gravity. Compare the reading to the Specifications listed in this Chapter. **Note:** *Add 0.004 points to the reading for every 10° above 20°C (68°F) – subtract 0.004 points from the reading for every 10° below 20°C (68°F).* Return the electrolyte to the appropriate cell and repeat the check for the remaining cells. When the check is complete, rinse the hydrometer thoroughly with clean water.

7 If the specific gravity of the electrolyte in each cell is as specified, the battery is in good condition and is apparently being charged by the machine's charging system.

8 If the specific gravity is low, the battery is not fully charged. This may be due to corroded battery terminals, a dirty battery case, a malfunctioning charging system, or loose or corroded wiring connections. On the other hand, it may be that the battery is worn out, especially if the machine is old, or that infrequent use of the motorcycle prevents normal charging from taking place.

9 Be sure to correct any problems and charge the battery if necessary. Refer to Chapter 9 for additional battery maintenance and charging procedures.

10 Install the battery cell caps, tightening them securely.

11 Install the battery. Be sure to refer to the safety precautions regarding battery installation in Chapter 9.

**14.6 Check the specific gravity with a hydrometer**

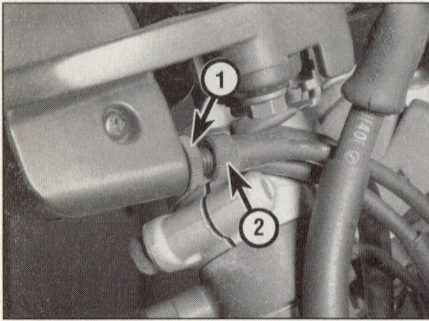

**17.3 Loosen the accelerator cable lockwheel (1) and turn the adjuster (2) in or out to obtain the correct throttle freeplay**

## 15 Exhaust system – check

1 Periodically check all of the exhaust system joints for leaks and loose fasteners. The lower fairing will have to be removed to do this properly (see Chapter 8). If tightening the clamp bolts fails to stop any leaks, replace the gaskets with new ones (a procedure which requires disassembly of the system).

2 The exhaust pipe flange nuts at the cylinder head are especially prone to loosening, which could cause damage to the head. Check them frequently and keep them tight.

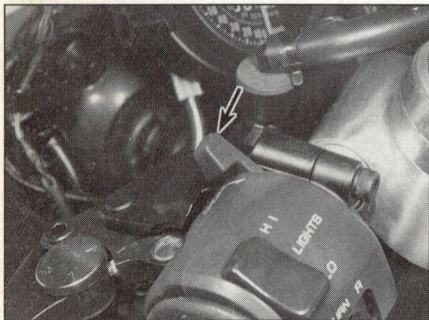

**17.8a Operate the choke lever (arrow) to take up cable slack . . .**

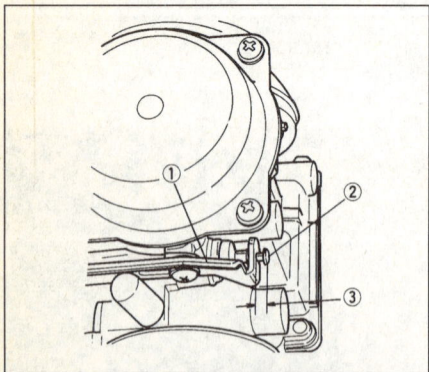

**17.8b . . . and check freeplay at the choke plunger lever**

*1 Choke plunger lever*  *2 Choke plunger*
*3 Freeplay*

**17.5a Throttle cable details (H models)**

*A Accelerator cable*   *B Decelerator cable*
*C Lockwheels*

## 16 Fasteners – check

1 Since vibration of the machine tends to loosen fasteners, all nuts, bolts, screws, etc. should be periodically checked for proper tightness.

2 Pay particular attention to the following:

*Spark plugs*
*Engine oil drain plug*
*Oil filter mounting bolt (H2 model)*
*Gearshift lever*
*Footpegs and sidestand*
*Engine mount bolts*
*Shock absorber mount bolts*
*Uni-trak linkage bolts*
*Front axle and clamp bolt*
*Rear axle nut*

3 If a torque wrench is available, use it along with the torque specifications at the beginning of this, or other Chapters.

## 17 Throttle and choke operation/grip freeplay – check and adjustment

### Throttle check

1 Make sure the throttle grip rotates easily from fully closed to fully open with the front

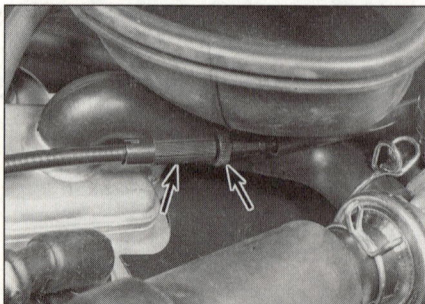

**17.9 Loosen the lockwheel (right arrow) and turn the adjuster (left arrow) to adjust choke cable freeplay (L model shown; others similar)**

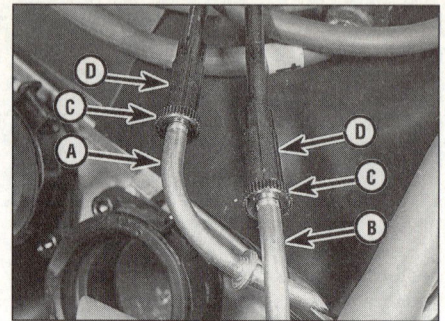

**17.5b Throttle cable details (J, K, L, M models)**

*A Accelerator cable*   *C Lockwheels*
*B Decelerator cable*   *D Adjusters*

wheel turned at various angles. The grip should return automatically from fully open to fully closed when released. If the throttle sticks, check the throttle cables for cracks or kinks in the housings. Also, make sure the inner cables are clean and well-lubricated.

2 Check for a small amount of freeplay at the grip (measured in terms of grip rotation) and compare the freeplay to the value listed in this Chapter's Specifications.

### Throttle adjustment

**Note:** *These motorcycles use two throttle cables – an accelerator cable and a decelerator cable.*

3 Freeplay adjustments can be made at the throttle grip end of the accelerator cable. Loosen the lockwheel on the cable **(see illustration)** and turn the adjuster until the desired freeplay is obtained, then retighten the lockwheel.

4 If the freeplay can't be adjusted at the grip end, adjust the cables at the lower ends. To do this, first remove the fuel tank (see Chapter 4).

5 Loosen the locknuts on both throttle cables **(see illustrations)**, then turn both adjusting nuts in completely.

6 Turn out the adjusting nut of the decelerator cable until the inner cable becomes tight, then tighten the locknut.

7 Turn the accelerator adjusting nut until the desired freeplay is obtained, then tighten the locknut. Make sure the throttle linkage lever contacts the idle adjusting screw when the throttle grip is at rest.

### Choke check

8 Remove the fuel tank (see Chapter 4). Operate the choke lever while you watch the choke plunger at the carburetor assembly **(see illustrations)**. The amount the plunger lever travels before it contacts the plunger is choke cable freeplay. Compare with the value listed in this Chapter's Specifications.

### Choke adjustment

9 If freeplay is incorrect, locate the midline cable adjuster **(see illustration)**. Loosen the lockwheel and turn the adjuster to obtain the correct freeplay, then tighten the lockwheel.

# Every 6000 miles (10,000 km)

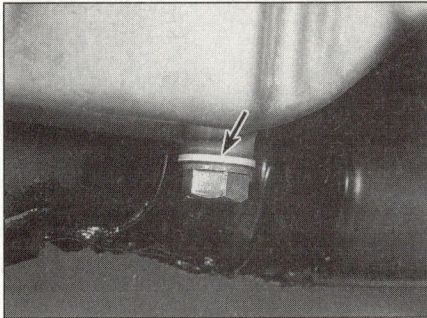
18.6 Remove the oil pan drain plug (arrow) (J through M models shown; H models similar)

18.7a Unscrew the oil filter with an oil filter wrench; the square opening allows a torque wrench to be used for installation . . .

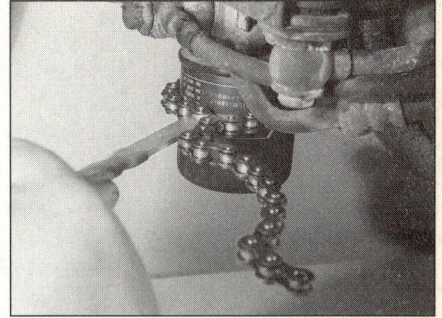
18.7b . . . a chain wrench or strap wrench can be used if the special tool isn't available (H1 model)

## 18 Engine oil/filter – change

1 Consistent routine oil and filter changes are the single most important maintenance procedure you can perform on a motorcycle. The oil not only lubricates the internal parts of the engine, transmission and clutch, but it also acts as a coolant, a cleaner, a sealant, and a protectant. Because of these demands, the oil takes a terrific amount of abuse and should be replaced often with new oil of the recommended grade and type.

**HAYNES HINT** *Saving a little money on the difference in cost between a good oil and a cheap oil won't pay off if the engine is damaged.*

2 Before changing the oil and filter, warm up the engine so the oil will drain easily. Be careful when draining the oil, as the exhaust pipes, the engine, and the oil itself can cause severe burns.
3 Support the motorcycle securely upright over a clean drain pan.

### H1, J, K, L and M models

4 If you're working on an H1 model, remove the front lower fairing. If you're working on a J, K, L or M model, remove the lower left fairing panel (see Chapter 8).
5 Remove the oil filler cap to vent the crankcase and act as a reminder that there is no oil in the engine.
6 Next, remove the drain plug from the engine (see illustration) and allow the oil to drain into the pan. Do not lose the sealing washer on the drain plug.
7 As the oil is draining, remove the oil filter with an oil filter wrench (see illustrations). If additional maintenance is planned for this time period, check or service another component while the oil is allowed to drain completely.

8 Wipe any remaining oil off the filter sealing area of the crankcase.
9 Check the condition of the drain plug threads and the sealing washer.
10 Coat the gasket on the new filter with clean engine oil. Install the filter and tighten it to the torque listed in this Chapter's Specifications (see illustrations).
11 Slip a new sealing washer over the drain plug, then install and tighten the plug to the torque listed in this Chapter's Specifications. Avoid overtightening, as damage to the engine case will result.

### H2 models

12 Remove the left and right lower fairing panels (see Chapter 8).
13 Remove the oil cooler (see Chapter 3).
14 Remove the radiator (see Chapter 3).
15 Remove the muffler/silencer (see Chapter 4).
16 Support the motorcycle securely upright over a clean drain pan. Remove the oil filler cap to vent the crankcase and to serve as a reminder that there's no oil in the engine.
17 Next, remove the drain plug from the bottom of the crankcase and allow the oil to drain into the pan. Don't lose the sealing washer on the drain plug.
18 As the oil is draining, remove the oil filter mounting bolt and lower the filter out of the crankcase (see illustrations). Separate the filter from the mounting bolt, then remove the washer, spring and oil fence (see illustration).

18.10a Check the installation instructions printed on the filter

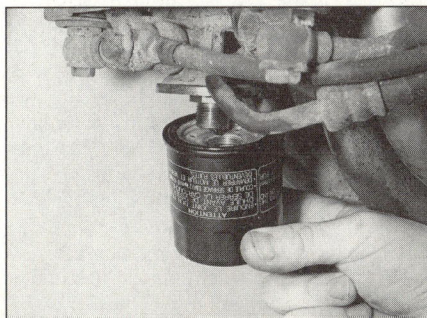
18.10b Coat the filter gasket with clean engine oil and thread the filter on (H1 model)

18.18a Unscrew the oil filter mounting bolt (arrow) . . .

18.18b . . . then lower the filter from the crankcase

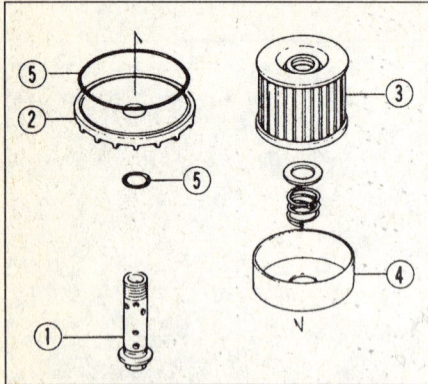

**18.18c Oil filter details – H2 models**

1  Mounting bolt/
   bypass valve
   assembly
2  Filter cover
3  Oil filter
4  Oil fence
5  O-ring

If additional maintenance is planned for this time period, check or service another component while the oil is allowed to drain completely.

**19** Clean the filter cover and housing with solvent or clean rags. Wipe any remaining oil off the oil filter cover sealing area of the crankcase.

**20** Remove the mounting bolt from the filter cover. The oil filter bypass valve is located inside the mounting bolt. Wash the mounting bolt in solvent and check the bypass valve for damage. If the valve is full of sludge, drive out the retaining pin and remove the spring and steel ball **(see illustrations)**.

**21** Clean the components and check them for damage – especially, be sure to check the spring for distortion. If any damage is found, replace the mounting bolt/bypass valve assembly. If the components are okay, reassemble the valve and install the retaining pin.

**22** Check the condition of the drain plug threads and the sealing washer. Use a new O-ring on the filter housing when it's installed.

**23** Install a new O-ring on the mounting bolt, lubricate the mounting bolt with clean engine oil and insert the bolt through the filter cover. Place the oil fence over the mounting bolt.

**24** Install a new O-ring on the filter cover,

**18.24 Lubricate the mounting bolt and twist the new filter onto it, making sure the grommets aren't dislodged from the filter**

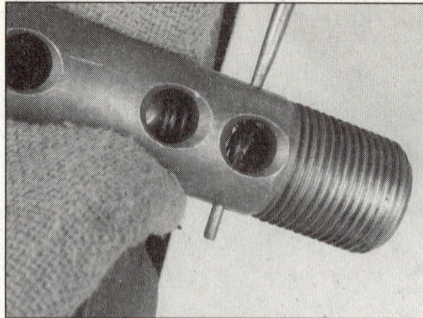

**18.20a To disassemble the bypass valve, drive out the retaining pin . . .**

then install the spring and washer. Twist the new filter down the mounting bolt, making sure the rubber grommets on the filter don't slip out of place **(see illustration)**.

**25** Guide the oil filter assembly up into the crankcase. Tighten the mounting bolt with your fingers until the filter cover contacts the crankcase, making sure the O-ring on the filter cover stays in its groove and seals properly. Tighten the mounting bolt to the torque listed in this Chapter's Specifications.

**26** Slip the sealing washer over the drain plug, then install and tighten the plug to the torque listed in this Chapter's Specifications. Avoid overtightening, as damage to the engine case will result.

### All models

**HAYNES HINT** *Before refilling the engine, check the old oil carefully. If the oil was drained into a clean pan, small pieces of metal or other material can be easily detected. If the oil is very metallic colored, then the engine is experiencing wear from break-in (new engine) or from insufficient lubrication. If there are flakes or chips of metal in the oil, then something is drastically wrong internally and the engine will have to be disassembled for inspection and repair. If there are pieces of fiber-like material in the oil, the clutch is experiencing excessive wear and should be checked.*

**27** If the inspection of the oil turns up nothing unusual, refill the crankcase to the proper level with the recommended oil and install the filler cap. Start the engine and let it run for two or three minutes. Shut it off, wait a few minutes, then check the oil level. If necessary, add more oil to bring the level up to the Maximum mark. Check around the drain plug and filter housing for leaks.

**28** The old oil drained from the engine cannot be reused in its present state and should be disposed of. Check with your local refuse disposal company, disposal facility or environmental agency to see whether they will accept the oil for recycling. Don't pour used oil into drains or onto the ground. After the oil

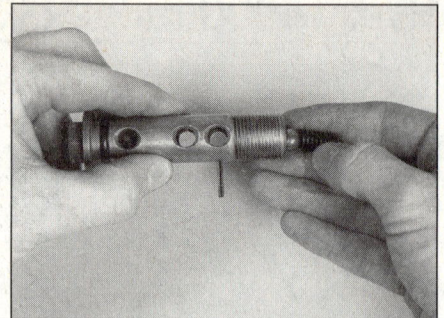

**18.20b . . . then remove the spring and ball from the mounting bolt**

has cooled, it can be drained into a suitable container (capped plastic jugs, topped bottles, milk cartons, etc.) for transport to one of these disposal sites.

**OIL CARE** — FOLLOW THE CODE
OIL BANK LINE
**0800 66 33 66**

*Note: It is antisocial and illegal to dump oil down the drain. To find the location of your local oil recycling bank, call this number free.*

*In the US note that any oil supplier must accept used oil for recycling.*

### 19 Air filter element – cleaning

#### H models

**1** Remove the front seat (see Chapter 8).

**2** Remove the bolts that secure the air filter retainer, then grasp the knob and pull out the filter element **(see illustration)**.

#### J and K models

**3** Remove the side covers (J models) or the tailpiece (K models) (see Chapter 8).

**4** Remove the fuel tank mounting bolts (see Chapter 4). Lift up the fuel tank and pull out the filter element **(see illustration)**.

#### L and M models

**5** Remove the side covers (see Chapter 8) and the fuel tank (see Chapter 4).

**6** Pull the retainer off the slot in the air filter housing and pull out the filter element **(see illustrations)**.

#### All models

**7** Wipe out the housing with a clean rag.

**8** Clean the filter element in high-flash point solvent. Soak a towel in SAE 30 SE or SF engine oil, then pad the towel on the foam side of the filter element to saturate it with the oil.

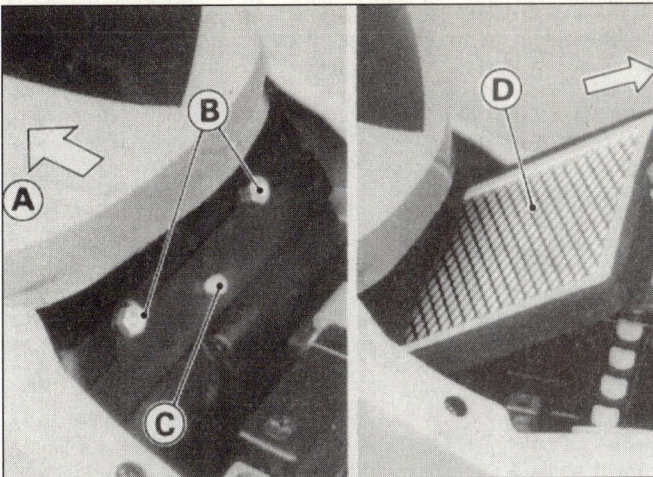

**19.2 Air filter element details – H models**

A  *Front of motorcycle*  C  *Pull knob*
B  *Retainer bolts*  D  *Filter element*

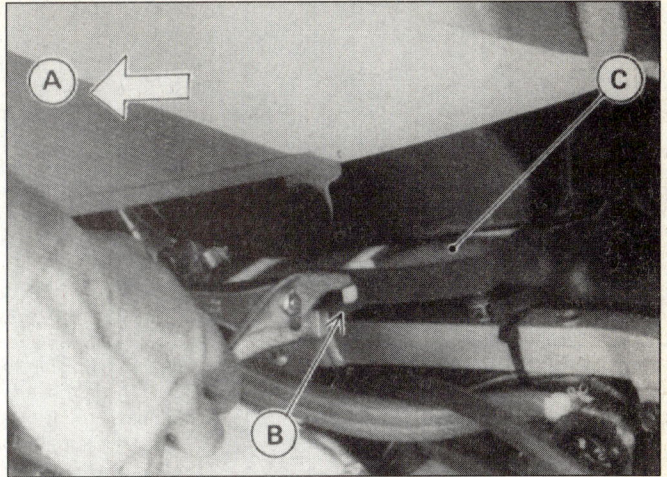

**19.4 Air filter element details – J and K models**

A  *Front of motorcycle*  C  *Filter element*
B  *Pull knob*

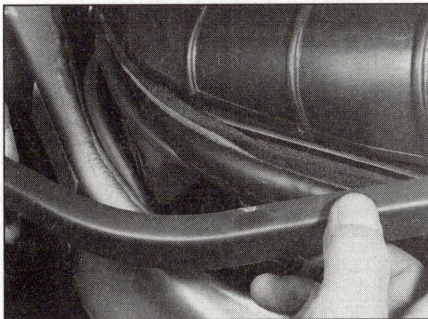

**19.6a On L and M models, pull out the filter element retainer . . .**

**19.6b . . . and slide the element out of the air filter housing**

## 20 Fuel system – check

**Warning: Gasoline (petrol) is extremely flammable, so take extra precautions when you work on any part of the fuel system. Don't smoke or allow open flames or bare light bulbs near the work area, and don't work in a garage where a natural gas-type appliance (such as a water heater or clothes dryer) is present. If you spill any fuel on your skin, rinse it off immediately with soap and water. When you perform any kind of work on the fuel system, wear safety glasses and have a fire extinguisher suitable for class B type fires (flammable liquids) on hand.**

9 Reinstall the filter by reversing the removal procedure. Reinstall all components removed for access.

10 Check the catch tank for accumulated water and oil **(see illustration)**. California models use two catch tanks, one for the air cleaner housing and one for the evaporative emission control system **(see illustration)**. If oil or water has built up in the catch tank, remove the plug from the bottom of the hose and let it drain. Be sure to reinstall the plug in the hose or air may be drawn in while the bike is running.

1 Check the fuel tank, the fuel tap, the lines and the carburetors for leaks and evidence of damage.

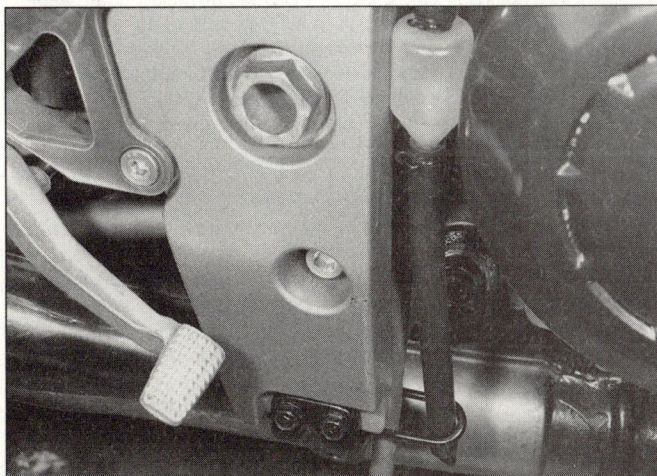

**19.10a Be sure the UP mark on the air filter housing catch tank is upright and the plug is installed in the bottom of the hose**

**19.10b On California models, there's a catch tank for the evaporative emission control system (L model shown)**

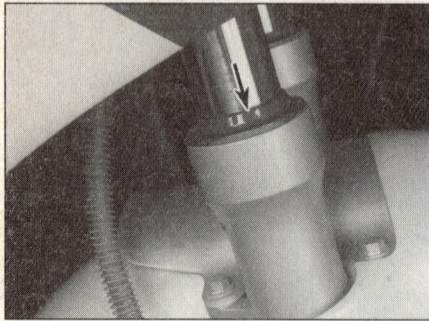

**21.3 If oil is leaking past the fork seals (arrow), replace them (upright type fork shown)**

2 If carburetor gaskets are leaking, the carburetors should be disassembled and rebuilt by referring to Chapter 4.

3 If the fuel tap is leaking, tightening the screws may help. If leakage persists, the tap should be disassembled and repaired or replaced with a new one.

4 If the fuel lines are cracked or otherwise deteriorated, replace them with new ones.

5 Check the vacuum hose connected to the fuel tap. If it is cracked or otherwise damaged, replace it with a new one.

6 The fuel filter may become clogged and should be inspected periodically. If there's any visible dirt inside the filter or if it looks dark, it should be replaced (see Section 25).

## 21 Suspension – check

1 The suspension components must be maintained in top operating condition to ensure rider safety. Loose, worn or damaged suspension parts decrease the vehicle's stability and control.

2 While standing alongside the motorcycle, lock the front brake and push on the handlebars to compress the forks several times. See if they move up-and-down smoothly without binding. If binding is felt, the forks should be disassembled and inspected as described in Chapter 6.

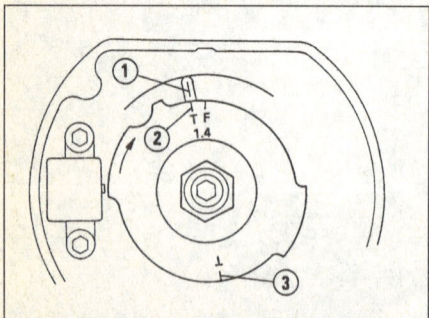

**23.6a Timing mark details (H models)**

1  Crankcase mark
2  Timing mark for cylinders 1 and 4
3  Timing mark for cylinders 2 and 3

3 Carefully inspect the area around the fork seals for any signs of fork oil leakage **(see illustration)**. If leakage is evident, the seals must be replaced as described in Chapter 6.

4 Check the tightness of all front suspension nuts and bolts to be sure none have worked loose.

5 Inspect the rear shock absorber for fluid leakage and tightness of the mounting nuts. If leakage is found, the shock should be replaced.

6 Support the bike securely upright with its rear wheel off the ground. Grab the swingarm on each side, just ahead of the axle. Rock the swingarm from side to side – there should be no discernible movement at the rear. If there's a little movement or a slight clicking can be heard, make sure the pivot shaft is tight. If the pivot shaft is tight but movement is still noticeable, the swingarm will have to be removed and the bearings replaced as described in Chapter 6.

7 Inspect the tightness of the rear suspension nuts and bolts.

## 22 Cylinder compression – check

1 Among other things, poor engine performance may be caused by leaking valves, incorrect valve clearances, a leaking head gasket, or worn pistons, rings and/or cylinder walls. A cylinder compression check will help pinpoint these conditions and can also indicate the presence of excessive carbon deposits in the cylinder heads.

2 The only tools required are a compression gauge and a spark plug wrench. Depending on the outcome of the initial test, a squirt-type oil can may also be needed.

3 Refer to the procedure under the *'Fault Finding Equipment'* heading in the Reference

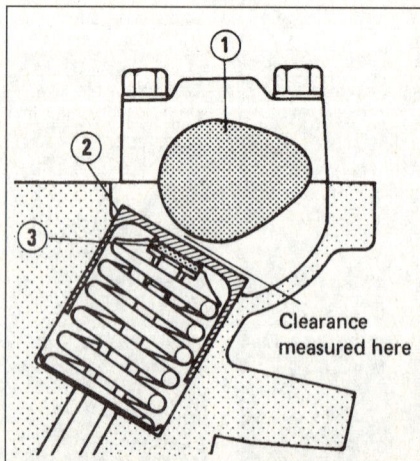

**23.6b On H models, measure clearance between the cam lobe and valve lifter**

1  Cam lobe
2  Valve lifter
3  Valve adjusting shim

Section. The cylinder compression figure is given in the Specifications at the beginning of this Chapter.

## 23 Valve clearances – check and adjustment

1 The engine must be completely cool for this maintenance procedure, so let the machine sit overnight before beginning.

2 Disconnect the cable from the negative terminal of the battery.

3 Refer to Chapter 4 and remove the fuel tank.

4 Remove the valve cover (see Chapter 2).

5 Remove the pick-up coil cover (see Chapter 5).

### H models

6 Position the no. 1 piston (on the left side of the engine) at Top Dead Center (TDC) on the compression stroke. Do this by turning the crankshaft, with a wrench placed on the timing rotor hex, until the TDC mark on the rotor (T 1.4) is aligned with the timing mark on the crankcase **(see illustration)**.
*Caution: Don't try to turn the engine with the timing rotor Allen bolt or the bolt may snap off.*
Now, check the position of the no. 1 cylinder camshaft lobes; they should not be depressing the valve lifters for either the intake valves or the exhaust valves **(see illustration)**. If they are, turn the crankshaft one complete revolution and realign the timing rotor mark. Piston number 1 is now at TDC compression.

7 With the engine in this position, all of the valves for cylinder no. 1 can be checked, as well as the exhaust valves for cylinder no. 2 and the intake valves for cylinder no. 3 **(see illustration)**.

8 Start with the no. 1 intake valve clearance. Insert a feeler gauge of the thickness listed in this Chapter's Specifications between the valve stem and cam lobe. Pull the feeler gauge out slowly – you should feel a slight drag. If there's no drag, the clearance is too loose. If there's a heavy drag, the clearance is too tight.

9 If the clearance is incorrect, write down the actual measured clearance. You'll need this information later to select a new valve adjusting shim.

10 Now measure the no. 1 exhaust valves, following the same procedure you used for the intake valves. Make sure to use a feeler gauge of the specified thickness and write down the actual clearances of any valves that aren't within the Specifications.

11 Proceed to measure the clearances of the no. 2 exhaust valves and the no. 3 intake valves. Again, write down the measured clearances of any valves that aren't within the Specifications.

12 Rotate the crankshaft one complete revolution and align the TDC mark on the rotor

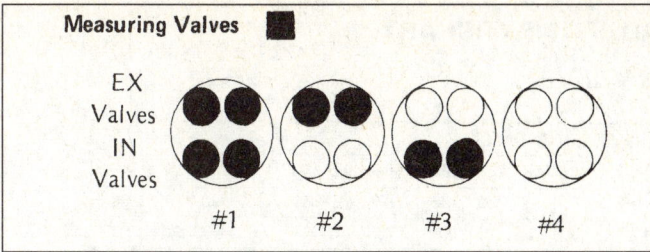

**Measuring Valves**

EX Valves
IN Valves

#1    #2    #3    #4

**23.7  With cylinder no. 1 at TDC compression, the shaded valves can be adjusted**

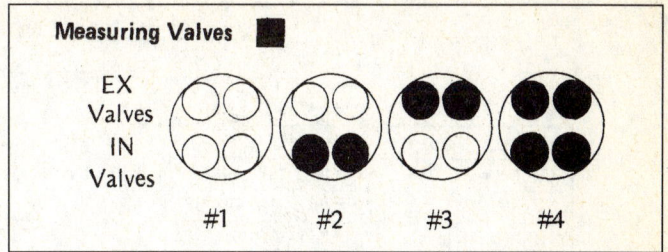

**Measuring Valves**

EX Valves
IN Valves

#1    #2    #3    #4

**23.12  With cylinder no. 4 at TDC compression, the shaded valves can be adjusted**

(T 1.4) with the timing mark on the crankcase, which will position piston no. 4 at TDC compression. Measure all four valves on cylinder no. 4, followed by the no. 3 exhaust valves and the no. 2 intake valves **(see illustration)**. Again, write down the measured clearances of any valves that aren't within the Specifications.

**13** If any of the clearances need to be adjusted, go to Step 14. If all of the clearances are within the Specifications, go to Step 20.

**14** Refer to Chapter 2 and remove the camshafts. Remove the valve lifters and adjusting shims from any valves that needed adjustment. Be sure to keep the lifters and shims in order so they can be returned to the locations they were removed from.

**15** Determine the thickness of the shim(s) you removed. It should be marked on the bottom of the shim, but the ideal way is to measure it with a micrometer.

**16** If the clearance was too large, you need a thicker shim. If the clearance was too small, you need a thinner shim. Calculate the thickness of the replacement shim by referring to the accompanying charts **(see illustrations)**.

**17** To use the charts, find the actual measured clearance of the valve in the left-hand column, and the thickness of the existing shim in the top row. Follow across

## VALVE CLEARANCE ADJUSTMENT CHART

### INLET VALVE

| | | PRESENT SHIM ↓ Example | | | | | | | | | | | | | | | | | | | |
|---|---|---|---|---|---|---|---|---|---|---|---|---|---|---|---|---|---|---|---|---|---|
| PART No. (92025-) | 1870 | 1871 | 1872 | 1873 | 1874 | 1875 | 1876 | 1877 | 1878 | 1879 | 1880 | 1881 | 1882 | 1883 | 1884 | 1885 | 1886 | 1887 | 1888 | 1889 | 1890 |
| MARK | 0 | 5 | 10 | 15 | 20 | 25 | 30 | 35 | 40 | 45 | 50 | 55 | 60 | 65 | 70 | 75 | 80 | 85 | 90 | 95 | 00 |
| THICKNESS (mm) | 2.00 | 2.05 | 2.10 | 2.15 | 2.20 | 2.25 | 2.30 | 2.35 | 2.40 | 2.45 | 2.50 | 2.55 | 2.60 | 2.65 | 2.70 | 2.75 | 2.80 | 2.85 | 2.90 | 2.95 | 3.00 |

| VALVE CLEARANCE (mm) | | | | | | | | | | | | | | | | | | | | | |
|---|---|---|---|---|---|---|---|---|---|---|---|---|---|---|---|---|---|---|---|---|---|
| 0.00 – 0.02 | | | | | 2.00 | 2.05 | 2.10 | 2.15 | 2.20 | 2.25 | 2.30 | 2.35 | 2.40 | 2.45 | 2.50 | 2.55 | 2.60 | 2.65 | 2.70 | 2.75 | 2.80 |
| 0.03 – 0.05 | | | | | | 2.00 | 2.05 | 2.10 | 2.15 | 2.20 | 2.25 | 2.30 | 2.35 | 2.40 | 2.45 | 2.50 | 2.55 | 2.60 | 2.65 | 2.70 | 2.75 | 2.80 | 2.85 |
| 0.06 – 0.10 | | | | 2.00 | 2.05 | 2.10 | 2.15 | 2.20 | 2.25 | 2.30 | 2.35 | 2.40 | 2.45 | 2.50 | 2.55 | 2.60 | 2.65 | 2.70 | 2.75 | 2.80 | 2.85 | 2.90 |
| 0.11 – 0.14 | | | 2.00 | 2.05 | 2.10 | 2.15 | 2.20 | 2.25 | 2.30 | 2.35 | 2.40 | 2.45 | 2.50 | 2.55 | 2.60 | 2.65 | 2.70 | 2.75 | 2.80 | 2.85 | 2.90 | 2.95 |
| 0.15 – 0.24 | SPECIFIED CLEARANCE / NO CHANGE REQUIRED | | | | | | | | | | | | | | | | | | | | |
| 0.25 – 0.29 | 2.05 | 2.10 | 2.15 | 2.20 | 2.25 | 2.30 | 2.35 | 2.40 | 2.45 | 2.50 | 2.55 | 2.60 | 2.65 | 2.70 | 2.75 | 2.80 | 2.85 | 2.90 | 2.95 | 3.00 |
| 0.30 – 0.34 | 2.10 | 2.15 | 2.20 | 2.25 | 2.30 | 2.35 | 2.40 | 2.45 | 2.50 | 2.55 | 2.60 | 2.65 | 2.70 | 2.75 | 2.80 | 2.85 | 2.90 | 2.95 | 3.00 |
| 0.35 – 0.39 | 2.15 | 2.20 | 2.25 | 2.30 | 2.35 | 2.40 | 2.45 | 2.50 | 2.55 | 2.60 | 2.65 | 2.70 | 2.75 | 2.80 | 2.85 | 2.90 | 2.95 | 3.00 |
| 0.40 – 0.44 | 2.20 | 2.25 | 2.30 | 2.35 | 2.40 | 2.45 | 2.50 | 2.55 | 2.60 | 2.65 | 2.70 | 2.75 | 2.80 | 2.85 | 2.90 | 2.95 | 3.00 |
| 0.45 – 0.49 | 2.25 | 2.30 | 2.35 | 2.40 | 2.45 | 2.50 | 2.55 | 2.60 | 2.65 | 2.70 | 2.75 | 2.80 | 2.85 | 2.90 | 2.95 | 3.00 |
| 0.50 – 0.54 | 2.30 | 2.35 | 2.40 | 2.45 | 2.50 | 2.55 | 2.60 | 2.65 | 2.70 | 2.75 | 2.80 | 2.85 | 2.90 | 2.95 | 3.00 |
| 0.55 – 0.59 | 2.35 | 2.40 | 2.45 | 2.50 | 2.55 | 2.60 | 2.65 | 2.70 | 2.75 | 2.80 | 2.85 | 2.90 | 2.95 | 3.00 |
| 0.60 – 0.64 | 2.40 | 2.45 | 2.50 | 2.55 | 2.60 | 2.65 | 2.70 | 2.75 | 2.80 | 2.85 | 2.90 | 2.95 | 3.00 |
| 0.65 – 0.69 | 2.45 | 2.50 | 2.55 | 2.60 | 2.65 | 2.70 | 2.75 | 2.80 | 2.85 | 2.90 | 2.95 | 3.00 |
| 0.70 – 0.74 | 2.50 | 2.55 | 2.60 | 2.65 | 2.70 | 2.75 | 2.80 | 2.85 | 2.90 | 2.95 | 3.00 |
| 0.75 – 0.79 | 2.55 | 2.60 | 2.65 | 2.70 | 2.75 | 2.80 | 2.85 | 2.90 | 2.95 | 3.00 |
| 0.80 – 0.84 | 2.60 | 2.65 | 2.70 | 2.75 | 2.80 | 2.85 | 2.90 | 2.95 | 3.00 |
| 0.85 – 0.89 | 2.65 | 2.70 | 2.75 | 2.80 | 2.85 | 2.90 | 2.95 | 3.00 |
| 0.90 – 0.94 | 2.70 | 2.75 | 2.80 | 2.85 | 2.90 | 2.95 | 3.00 |
| 0.95 – 0.99 | 2.75 | 2.80 | 2.85 | 2.90 | 2.95 | 3.00 |
| 1.00 – 1.04 | 2.80 | 2.85 | 2.90 | 2.95 | 3.00 |
| 1.05 – 1.09 | 2.85 | 2.90 | 2.95 | 3.00 |
| 1.10 – 1.14 | 2.90 | 2.95 | 3.00 |
| 1.15 – 1.19 | 2.95 | 3.00 |
| 1.20 – 1.24 | 3.00 |

INSTALL THE SHIM OF THIS THICKNESS (mm)

**23.16a  Intake valve shim selection chart - H models**

# VALVE CLEARANCE ADJUSTMENT CHART

## EXHAUST VALVE

Example → PRESENT SHIM

| PART No. (92025-) | 1870 | 1871 | 1872 | 1873 | 1874 | 1875 | 1876 | 1877 | 1878 | 1879 | 1880 | 1881 | 1882 | 1883 | 1884 | 1885 | 1886 | 1887 | 1888 | 1889 | 1890 |
|---|---|---|---|---|---|---|---|---|---|---|---|---|---|---|---|---|---|---|---|---|---|
| MARK | 0 | 5 | 10 | 15 | 20 | 25 | 30 | 35 | 40 | 45 | 50 | 55 | 60 | 65 | 70 | 75 | 80 | 85 | 90 | 95 | 00 |
| THICKNESS (mm) | 2.00 | 2.05 | 2.10 | 2.15 | 2.20 | 2.25 | 2.30 | 2.35 | 2.40 | 2.45 | 2.50 | 2.55 | 2.60 | 2.65 | 2.70 | 2.75 | 2.80 | 2.85 | 2.90 | 2.95 | 3.00 |

VALVE CLEARANCE (mm) — Example

INSTALL THE SHIM OF THIS THICKNESS (mm)

| VALVE CLEARANCE | 2.00 | 2.05 | 2.10 | 2.15 | 2.20 | 2.25 | 2.30 | 2.35 | 2.40 | 2.45 | 2.50 | 2.55 | 2.60 | 2.65 | 2.70 | 2.75 | 2.80 | 2.85 | 2.90 | 2.95 | 3.00 |
|---|---|---|---|---|---|---|---|---|---|---|---|---|---|---|---|---|---|---|---|---|---|
| 0.00 – 0.02 | /// | /// | /// | /// | /// | 2.00 | 2.05 | 2.10 | 2.15 | 2.20 | 2.25 | 2.30 | 2.35 | 2.40 | 2.45 | 2.50 | 2.55 | 2.60 | 2.65 | 2.70 | 2.75 |
| 0.03 – 0.05 | /// | /// | /// | /// | 2.00 | 2.05 | 2.10 | 2.15 | 2.20 | 2.25 | 2.30 | 2.35 | 2.40 | 2.45 | 2.50 | 2.55 | 2.60 | 2.65 | 2.70 | 2.75 | 2.80 |
| 0.06 – 0.10 | /// | /// | /// | 2.00 | 2.05 | 2.10 | 2.15 | 2.20 | 2.25 | 2.30 | 2.35 | 2.40 | 2.45 | 2.50 | 2.55 | 2.60 | 2.65 | 2.70 | 2.75 | 2.80 | 2.85 |
| 0.11 – 0.15 | /// | /// | 2.00 | 2.05 | 2.10 | 2.15 | 2.20 | 2.25 | 2.30 | 2.35 | 2.40 | 2.45 | 2.50 | 2.55 | 2.60 | 2.65 | 2.70 | 2.75 | 2.80 | 2.85 | 2.90 |
| 0.16 – 0.19 | /// | 2.00 | 2.05 | 2.10 | 2.15 | 2.20 | 2.25 | 2.30 | 2.35 | 2.40 | 2.45 | 2.50 | 2.55 | 2.60 | 2.65 | 2.70 | 2.75 | 2.80 | 2.85 | 2.90 | 2.95 |
| 0.20 – 0.29 | SPECIFIED CLEARANCE / NO CHANGE REQUIRED | | | | | | | | | | | | | | | | | | | | |
| 0.30 – 0.34 | 2.05 | 2.10 | 2.15 | 2.20 | 2.25 | 2.30 | 2.35 | 2.40 | 2.45 | 2.50 | 2.55 | 2.60 | 2.65 | 2.70 | 2.75 | 2.80 | 2.85 | 2.90 | 2.95 | 3.00 | |
| 0.35 – 0.39 | 2.10 | 2.15 | 2.20 | 2.25 | 2.30 | 2.35 | 2.40 | 2.45 | 2.50 | 2.55 | 2.60 | 2.65 | 2.70 | 2.75 | 2.80 | 2.85 | 2.90 | 2.95 | 3.00 | | |
| 0.40 – 0.44 | 2.15 | 2.20 | 2.25 | 2.30 | 2.35 | 2.40 | 2.45 | 2.50 | 2.55 | 2.60 | 2.65 | 2.70 | 2.75 | 2.80 | 2.85 | 2.90 | 2.95 | 3.00 | | | |
| 0.45 – 0.49 | 2.20 | 2.25 | 2.30 | 2.35 | 2.40 | 2.45 | 2.50 | 2.55 | 2.60 | 2.65 | 2.70 | 2.75 | 2.80 | 2.85 | 2.90 | 2.95 | 3.00 | | | | |
| 0.50 – 0.54 | 2.25 | 2.30 | 2.35 | 2.40 | 2.45 | 2.50 | 2.55 | 2.60 | 2.65 | 2.70 | 2.75 | 2.80 | 2.85 | 2.90 | 2.95 | 3.00 | | | | | |
| 0.55 – 0.59 | 2.30 | 2.35 | 2.40 | 2.45 | 2.50 | 2.55 | 2.60 | 2.65 | 2.70 | 2.75 | 2.80 | 2.85 | 2.90 | 2.95 | 3.00 | | | | | | |
| 0.60 – 0.64 | 2.35 | 2.40 | 2.45 | 2.50 | 2.55 | 2.60 | 2.65 | 2.70 | 2.75 | 2.80 | 2.85 | 2.90 | 2.95 | 3.00 | | | | | | | |
| 0.65 – 0.69 | 2.40 | 2.45 | 2.50 | 2.55 | 2.60 | 2.65 | 2.70 | 2.75 | 2.80 | 2.85 | 2.90 | 2.95 | 3.00 | | | | | | | | |
| 0.70 – 0.74 | 2.45 | 2.50 | 2.55 | 2.60 | 2.65 | 2.70 | 2.75 | 2.80 | 2.85 | 2.90 | 2.95 | 3.00 | | | | | | | | | |
| 0.75 – 0.79 | 2.50 | 2.55 | 2.60 | 2.65 | 2.70 | 2.75 | 2.80 | 2.85 | 2.90 | 2.95 | 3.00 | | | | | | | | | | |
| 0.80 – 0.84 | 2.55 | 2.60 | 2.65 | 2.70 | 2.75 | 2.80 | 2.85 | 2.90 | 2.95 | 3.00 | | | | | | | | | | | |
| 0.85 – 0.89 | 2.60 | 2.65 | 2.70 | 2.75 | 2.80 | 2.85 | 2.90 | 2.95 | 3.00 | | | | | | | | | | | | |
| 0.90 – 0.94 | 2.65 | 2.70 | 2.75 | 2.80 | 2.85 | 2.90 | 2.95 | 3.00 | | | | | | | | | | | | | |
| 0.95 – 0.99 | 2.70 | 2.75 | 2.80 | 2.85 | 2.90 | 2.95 | 3.00 | | | | | | | | | | | | | | |
| 1.00 – 1.04 | 2.75 | 2.80 | 2.85 | 2.90 | 2.95 | 3.00 | | | | | | | | | | | | | | | |
| 1.05 – 1.09 | 2.80 | 2.85 | 2.90 | 2.95 | 3.00 | | | | | | | | | | | | | | | | |
| 1.10 – 1.14 | 2.85 | 2.90 | 2.95 | 3.00 | | | | | | | | | | | | | | | | | |
| 1.15 – 1.19 | 2.90 | 2.95 | 3.00 | | | | | | | | | | | | | | | | | | |
| 1.20 – 1.24 | 2.95 | 3.00 | | | | | | | | | | | | | | | | | | | |
| 1.25 – 1.29 | 3.00 | | | | | | | | | | | | | | | | | | | | |

**23.16b Exhaust valve shim selection chart - H models**

23.22a Turn the engine with a wrench on the timing rotor hex (don't use the Allen bolt or it may snap off) . . .

23.22b . . . and align the no. 1 cylinder timing mark with the crankcase mark . . .

**23.22c** . . . the cam lobes should be positioned like this

**23.24** Insert a feeler gauge between the rocker arm and cam lobe to measure the clearance

and down to where the row and column meet; the shim listed at that point is the one you need. For example:

a) *If the actual measured clearance was 0.28 mm, find the 0.25 to 0.29 mm entry on the left-hand side of the chart.*

b) *If the existing shim thickness is 2.60 mm, find the 2.60 entry in the top row of the chart.*

c) *Follow the lines across and down from these two entries until they meet. The number listed in that space (2.65 mm) is the needed shim thickness. Note: In addition to the shims listed in the charts, shims are available in the following thicknesses: 2.43 mm, 2.48 mm, 2.53 mm and 2.58 mm.*

**18** Perform Steps 15 through 17 to select a new shim for each of the valves that needed adjustment.

**19** Install the shims, lifters and camshafts and recheck the clearances (see Chapter 2 and Steps 6 through 12 above).

**20** With all of the clearances within the Specifications, install the valve cover and all of the components that had to be removed to get it off.

**21** Install the fuel tank and reconnect the cable to the negative terminal of the battery.

## J, K, L and M models

**22** Position the number 1 piston (on the left side of the engine) at Top Dead Center (TDC)

on the compression stroke. Do this by turning the crankshaft, with a wrench placed on the timing rotor hex, until the TDC mark on the rotor is aligned with the timing mark on the crankcase **(see illustrations)**.

***Caution: Don't try to turn the engine with the timing rotor Allen bolt or the bolt may snap off.***

Now, check the position of the no. 1 cylinder camshaft lobes; they should not be depressing the rocker arms for either the intake valves or the exhaust valves **(see illustration)**. If they are, turn the crankshaft one complete revolution and realign the timing rotor mark. Piston number 1 is now at TDC compression.

**23** With the engine in this position, all of the valves for cylinder no. 1 can be checked, as well as the exhaust valves for cylinder no. 2 and the intake valves for cylinder no. 3 **(see illustration 23.7)**.

**24** Start with the no. 1 intake valve clearance. Insert a feeler gauge of the thickness listed in this Chapter's Specifications between the adjusting shim and rocker arm **(see illustration)**. Pull the feeler gauge out slowly – you should feel a slight drag. If there's no drag, the clearance is too loose. If there's a heavy drag, the clearance is too tight.

**25** If the clearance is incorrect, write down the actual measured clearance. You'll need this information later to select a new valve adjusting shim.

**26** Now measure the no. 1 exhaust valves, following the same procedure you used for the intake valves. Make sure to use a feeler gauge of the specified thickness and write down the actual clearances of any valves that aren't within the Specifications.

**27** Proceed to measure the clearances of the no. 2 exhaust valves and the no. 3 intake valves. Again, write down the measured clearances of any valves that aren't within the Specifications.

**28** Rotate the crankshaft one complete revolution and align the TDC mark on the rotor with the timing mark on the crankcase, which will position piston no. 4 at TDC compression. Measure all four valves on cylinder no. 4, followed by the no. 3 exhaust valves and the no. 2 intake valves **(see illustration 23.12)**. Again, write down the measured clearances of any valves that aren't within the Specifications.

**29** If any of the clearances need to be adjusted, go to Step 30. If all of the clearances are within the Specifications, go to Step 36.

**30** Slide the rocker arms aside and remove the adjusting shims from any valves that needed adjustment **(see illustrations)**. Be sure to keep the shims in order so they can be returned to the locations they were removed from.

**31** Determine the thickness of the shim(s) you removed. It should be marked on the bottom of the shim **(see illustration)**, but the ideal way is to measure it with a micrometer.

**23.30a** Push the rocker arm aside . . .

**23.30b** . . . and remove the valve adjusting shim (arrow) with a magnet

**23.31** The shim size is printed on the underside of the shim

## VALVE CLEARANCE ADJUSTMENT CHART – INLET VALVE

**PRESENT SHIM** — Example

| PART No. (92180 -) | 1014 | 1016 | 1018 | 1020 | 1022 | 1024 | 1026 | 1028 | 1030 | 1032 | 1034 | 1036 | 1038 | 1040 | 1042 | 1044 | 1046 | 1048 | 1050 | 1052 | 1054 |
|---|---|---|---|---|---|---|---|---|---|---|---|---|---|---|---|---|---|---|---|---|---|
| MARK | 50 | 55 | 60 | 65 | 70 | 75 | 80 | 85 | 90 | 95 | 00 | 05 | 10 | 15 | 20 | 25 | 30 | 35 | 40 | 45 | 50 |
| THICKNESS (mm) | 2.50 | 2.55 | 2.60 | 2.65 | 2.70 | 2.75 | 2.80 | 2.85 | 2.90 | 2.95 | 3.00 | 3.05 | 3.10 | 3.15 | 3.20 | 3.25 | 3.30 | 3.35 | 3.40 | 3.45 | 3.50 |

VALVE CLEARANCE MEASUREMENT (Example)

| Measurement | 2.50 | 2.55 | 2.60 | 2.65 | 2.70 | 2.75 | 2.80 | 2.85 | 2.90 | 2.95 | 3.00 | 3.05 | 3.10 | 3.15 | 3.20 | 3.25 | 3.30 | 3.35 | 3.40 | 3.45 | 3.50 |
|---|---|---|---|---|---|---|---|---|---|---|---|---|---|---|---|---|---|---|---|---|---|
| 0.00 ~ 0.03 |  |  |  |  | 2.50 | 2.55 | 2.60 | 2.65 | 2.70 | 2.75 | 2.80 | 2.85 | 2.90 | 2.95 | 3.00 | 3.05 | 3.10 | 3.15 | 3.20 | 3.25 | 3.30 |
| 0.04 ~ 0.08 |  |  |  | 2.50 | 2.55 | 2.60 | 2.65 | 2.70 | 2.75 | 2.80 | 2.85 | 2.90 | 2.95 | 3.00 | 3.05 | 3.10 | 3.15 | 3.20 | 3.25 | 3.30 | 3.35 |
| 0.09 ~ 0.13 |  |  | 2.50 | 2.55 | 2.60 | 2.65 | 2.70 | 2.75 | 2.80 | 2.85 | 2.90 | 2.95 | 3.00 | 3.05 | 3.10 | 3.15 | 3.20 | 3.25 | 3.30 | 3.35 | 3.40 |
| 0.14 ~ 0.17 |  | 2.50 | 2.55 | 2.60 | 2.65 | 2.70 | 2.75 | 2.80 | 2.85 | 2.90 | 2.95 | 3.00 | 3.05 | 3.10 | 3.15 | 3.20 | 3.25 | 3.30 | 3.35 | 3.40 | 3.45 |
| 0.18 ~ 0.23 | SPECIFIED CLEARANCE/NO CHANGE REQUIRED | | | | | | | | | | | | | | | | | | | | |
| 0.24 ~ 0.28 | 2.55 | 2.60 | 2.65 | 2.70 | 2.75 | 2.80 | 2.85 | 2.90 | 2.95 | 3.00 | 3.05 | 3.10 | 3.15 | 3.20 | 3.25 | 3.30 | 3.35 | 3.40 | 3.45 | 3.50 |  |
| 0.29 ~ 0.33 | 2.60 | 2.65 | 2.70 | 2.75 | 2.80 | 2.85 | 2.90 | 2.95 | 3.00 | 3.05 | 3.10 | 3.15 | 3.20 | 3.25 | 3.30 | 3.35 | 3.40 | 3.45 | 3.50 |  |  |
| 0.34 ~ 0.38 | 2.65 | 2.70 | 2.75 | 2.80 | 2.85 | 2.90 | 2.95 | 3.00 | 3.05 | 3.10 | 3.15 | 3.20 | 3.25 | 3.30 | 3.35 | 3.40 | 3.45 | 3.50 |  |  |  |
| 0.39 ~ 0.43 | 2.70 | 2.75 | 2.80 | 2.85 | 2.90 | 2.95 | 3.00 | 3.05 | 3.10 | 3.15 | 3.20 | 3.25 | 3.30 | 3.35 | 3.40 | 3.45 | 3.50 |  |  |  |  |
| 0.44 ~ 0.48 | 2.75 | 2.80 | 2.85 | 2.90 | 2.95 | 3.00 | 3.05 | 3.10 | 3.15 | 3.20 | 3.25 | 3.30 | 3.35 | 3.40 | 3.45 | 3.50 |  |  |  |  |  |
| 0.49 ~ 0.53 | 2.80 | 2.85 | 2.90 | 2.95 | 3.00 | 3.05 | 3.10 | 3.15 | 3.20 | 3.25 | 3.30 | 3.35 | 3.40 | 3.45 | 3.50 |  |  |  |  |  |  |
| 0.54 ~ 0.58 | 2.85 | 2.90 | 2.95 | 3.00 | 3.05 | 3.10 | 3.15 | 3.20 | 3.25 | 3.30 | 3.35 | 3.40 | 3.45 | 3.50 |  |  |  |  |  |  |  |
| 0.59 ~ 0.63 | 2.90 | 2.95 | 3.00 | 3.05 | 3.10 | 3.15 | 3.20 | 3.25 | 3.30 | 3.35 | 3.40 | 3.45 | 3.50 |  |  |  |  |  |  |  |  |
| 0.64 ~ 0.68 | 2.95 | 3.00 | 3.05 | 3.10 | 3.15 | 3.20 | 3.25 | 3.30 | 3.35 | 3.40 | 3.45 | 3.50 |  |  |  |  |  |  |  |  |  |
| 0.69 ~ 0.73 | 3.00 | 3.05 | 3.10 | 3.15 | 3.20 | 3.25 | 3.30 | 3.35 | 3.40 | 3.45 | 3.50 |  |  |  |  |  |  |  |  |  |  |
| 0.74 ~ 0.78 | 3.05 | 3.10 | 3.15 | 3.20 | 3.25 | 3.30 | 3.35 | 3.40 | 3.45 | 3.50 |  |  |  |  |  |  |  |  |  |  |  |
| 0.79 ~ 0.83 | 3.10 | 3.15 | 3.20 | 3.25 | 3.30 | 3.35 | 3.40 | 3.45 | 3.50 |  |  |  |  |  |  |  |  |  |  |  |  |
| 0.84 ~ 0.88 | 3.15 | 3.20 | 3.25 | 3.30 | 3.35 | 3.40 | 3.45 | 3.50 |  |  |  |  |  |  |  |  |  |  |  |  |  |
| 0.89 ~ 0.93 | 3.20 | 3.25 | 3.30 | 3.35 | 3.40 | 3.45 | 3.50 |  |  |  |  |  |  |  |  |  |  |  |  |  |  |
| 0.94 ~ 0.98 | 3.25 | 3.30 | 3.35 | 3.40 | 3.45 | 3.50 |  |  |  |  |  |  |  |  |  |  |  |  |  |  |  |
| 0.99 ~ 1.03 | 3.30 | 3.35 | 3.40 | 3.45 | 3.50 |  |  |  |  |  |  |  |  |  |  |  |  |  |  |  |  |
| 1.04 ~ 1.08 | 3.35 | 3.40 | 3.45 | 3.50 |  |  |  |  |  |  |  |  |  |  |  |  |  |  |  |  |  |
| 1.09 ~ 1.13 | 3.40 | 3.45 | 3.50 |  |  |  |  |  |  |  |  |  |  |  |  |  |  |  |  |  |  |
| 1.14 ~ 1.18 | 3.45 | 3.50 |  |  |  |  |  |  |  |  |  |  |  |  |  |  |  |  |  |  |  |
| 1.19 ~ 1.23 | 3.50 |  |  |  |  |  |  |  |  |  |  |  |  |  |  |  |  |  |  |  |  |

INSTALL THE SHIM OF THIS THICKNESS (mm)

**23.32a  Intake valve shim selection chart - J, K, L and M models**

**32** If the clearance was too large, you need a thicker shim. If the clearance was too small, you need a thinner shim. Calculate the thickness of the replacement shim by referring to the accompanying charts (see illustrations).
**33** To use the charts, find the actual measured clearance of the valve in the left-hand column, and the thickness of the existing shim in the top row. Follow across and down to where the row and column meet; the shim listed at that point is the one you

need. For example:
a) If the actual measured clearance was 0.43 mm, find the 0.41 to 0.45 mm entry on the left-hand side of the chart.
b) If the existing shim thickness is 3.10 mm, find the 3.10 entry in the top row of the chart.
c) Follow the lines across and down from these two entries until they meet. The number listed in that space (3.25 mm) is the needed shim thickness.

**34** Perform Steps 31 through 33 to select a new shim for each of the valves that needed adjustment.
**35** Install the shims, reposition the rocker arms and recheck the clearances (see Steps 22 through 28 above).
**36** Install the valve cover and all of the components that had to be removed to get it off.
**37** Install the fuel tank and reconnect the cable to the negative terminal of the battery.

## VALVE CLEARANCE ADJUSTMENT CHART – EXHAUST VALVE

| PART No. (92180 -) | 1014 | 1016 | 1018 | 1020 | 1022 | 1024 | 1026 | 1028 | 1030 | 1032 | 1034 | 1036 | 1038 | 1040 | 1042 | 1044 | 1046 | 1048 | 1050 | 1052 | 1054 |
|---|---|---|---|---|---|---|---|---|---|---|---|---|---|---|---|---|---|---|---|---|---|
| MARK | 50 | 55 | 60 | 65 | 70 | 75 | 80 | 85 | 90 | 95 | 00 | 05 | 10 | 15 | 20 | 25 | 30 | 35 | 40 | 45 | 50 |
| THICKNESS (mm) | 2.50 | 2.55 | 2.60 | 2.65 | 2.70 | 2.75 | 2.80 | 2.85 | 2.90 | 2.95 | 3.00 | 3.05 | 3.10 | 3.15 | 3.20 | 3.25 | 3.30 | 3.35 | 3.40 | 3.45 | 3.50 |

VALVE CLEARANCE MEASUREMENT (PRESENT SHIM above; install the shim of this thickness (mm)):

| Measurement | 2.50 | 2.55 | 2.60 | 2.65 | 2.70 | 2.75 | 2.80 | 2.85 | 2.90 | 2.95 | 3.00 | 3.05 | 3.10 | 3.15 | 3.20 | 3.25 | 3.30 | 3.35 | 3.40 | 3.45 | 3.50 |
|---|---|---|---|---|---|---|---|---|---|---|---|---|---|---|---|---|---|---|---|---|---|
| 0.00 ~ 0.05 | | | | | | 2.50 | 2.55 | 2.60 | 2.65 | 2.70 | 2.75 | 2.80 | 2.85 | 2.90 | 2.95 | 3.00 | 3.05 | 3.10 | 3.15 | 3.20 | 3.25 |
| 0.06 ~ 0.10 | | | | | 2.50 | 2.55 | 2.60 | 2.65 | 2.70 | 2.75 | 2.80 | 2.85 | 2.90 | 2.95 | 3.00 | 3.05 | 3.10 | 3.15 | 3.20 | 3.25 | 3.30 |
| 0.11 ~ 0.15 | | | | 2.50 | 2.55 | 2.60 | 2.65 | 2.70 | 2.75 | 2.80 | 2.85 | 2.90 | 2.95 | 3.00 | 3.05 | 3.10 | 3.15 | 3.20 | 3.25 | 3.30 | 3.35 |
| 0.16 ~ 0.20 | | | 2.50 | 2.55 | 2.60 | 2.65 | 2.70 | 2.75 | 2.80 | 2.85 | 2.90 | 2.95 | 3.00 | 3.05 | 3.10 | 3.15 | 3.20 | 3.25 | 3.30 | 3.35 | 3.40 |
| 0.21 ~ 0.24 | | 2.50 | 2.55 | 2.60 | 2.65 | 2.70 | 2.75 | 2.80 | 2.85 | 2.90 | 2.95 | 3.00 | 3.05 | 3.10 | 3.15 | 3.20 | 3.25 | 3.30 | 3.35 | 3.40 | 3.45 |
| 0.25 ~ 0.30 | SPECIFIED CLEARANCE/NO CHANGE REQUIRED | | | | | | | | | | | | | | | | | | | | |
| 0.31 ~ 0.35 | 2.55 | 2.60 | 2.65 | 2.70 | 2.75 | 2.80 | 2.85 | 2.90 | 2.95 | 3.00 | 3.05 | 3.10 | 3.15 | 3.20 | 3.25 | 3.30 | 3.35 | 3.40 | 3.45 | 3.50 | |
| 0.36 ~ 0.40 | 2.60 | 2.65 | 2.70 | 2.75 | 2.80 | 2.85 | 2.90 | 2.95 | 3.00 | 3.05 | 3.10 | 3.15 | 3.20 | 3.25 | 3.30 | 3.35 | 3.40 | 3.45 | 3.50 | | |
| 0.41 ~ 0.45 | 2.65 | 2.70 | 2.75 | 2.80 | 2.85 | 2.90 | 2.95 | 3.00 | 3.05 | 3.10 | 3.15 | 3.20 | 3.25 | 3.30 | 3.35 | 3.40 | 3.45 | 3.50 | | | |
| 0.46 ~ 0.50 | 2.70 | 2.75 | 2.80 | 2.85 | 2.90 | 2.95 | 3.00 | 3.05 | 3.10 | 3.15 | 3.20 | 3.25 | 3.30 | 3.35 | 3.40 | 3.45 | 3.50 | | | | |
| 0.51 ~ 0.55 | 2.75 | 2.80 | 2.85 | 2.90 | 2.95 | 3.00 | 3.05 | 3.10 | 3.15 | 3.20 | 3.25 | 3.30 | 3.35 | 3.40 | 3.45 | 3.50 | | | | | |
| 0.56 ~ 0.60 | 2.80 | 2.85 | 2.90 | 2.95 | 3.00 | 3.05 | 3.10 | 3.15 | 3.20 | 3.25 | 3.30 | 3.35 | 3.40 | 3.45 | 3.50 | | | | | | |
| 0.61 ~ 0.65 | 2.85 | 2.90 | 2.95 | 3.00 | 3.05 | 3.10 | 3.15 | 3.20 | 3.25 | 3.30 | 3.35 | 3.40 | 3.45 | 3.50 | | | | | | | |
| 0.66 ~ 0.70 | 2.90 | 2.95 | 3.00 | 3.05 | 3.10 | 3.15 | 3.20 | 3.25 | 3.30 | 3.35 | 3.40 | 3.45 | 3.50 | | | | | | | | |
| 0.71 ~ 0.75 | 2.95 | 3.00 | 3.05 | 3.10 | 3.15 | 3.20 | 3.25 | 3.30 | 3.35 | 3.40 | 3.45 | 3.50 | | | | | | | | | |
| 0.76 ~ 0.80 | 3.00 | 3.05 | 3.10 | 3.15 | 3.20 | 3.25 | 3.30 | 3.35 | 3.40 | 3.45 | 3.50 | | | | | | | | | | |
| 0.81 ~ 0.85 | 3.05 | 3.10 | 3.15 | 3.20 | 3.25 | 3.30 | 3.35 | 3.40 | 3.45 | 3.50 | | | | | | | | | | | |
| 0.86 ~ 0.90 | 3.10 | 3.15 | 3.20 | 3.25 | 3.30 | 3.35 | 3.40 | 3.45 | 3.50 | | | | | | | | | | | | |
| 0.91 ~ 0.95 | 3.15 | 3.20 | 3.25 | 3.30 | 3.35 | 3.40 | 3.45 | 3.50 | | | | | | | | | | | | | |
| 0.96 ~ 1.00 | 3.20 | 3.25 | 3.30 | 3.35 | 3.40 | 3.45 | 3.50 | | | | | | | | | | | | | | |
| 1.01 ~ 1.05 | 3.25 | 3.30 | 3.35 | 3.40 | 3.45 | 3.50 | | | | | | | | | | | | | | | |
| 1.06 ~ 1.10 | 3.30 | 3.35 | 3.40 | 3.45 | 3.50 | | | | | | | | | | | | | | | | |
| 1.11 ~ 1.15 | 3.35 | 3.40 | 3.45 | 3.50 | | | | | | | | | | | | | | | | | |
| 1.16 ~ 1.20 | 3.40 | 3.45 | 3.50 | | | | | | | | | | | | | | | | | | |
| 1.21 ~ 1.25 | 3.45 | 3.50 | | | | | | | | | | | | | | | | | | | |
| 1.26 ~ 1.30 | 3.50 | | | | | | | | | | | | | | | | | | | | |

INSTALL THE SHIM OF THIS THICKNESS (mm)

**23.32b Exhaust valve shim selection chart - J, K, L and M models**

# Every 12,000 miles (20,000 km)

## 24 Air filter element – replacement

**Note:** *Replace the air filter element every five cleanings (or more frequently, if the bike is operated in dusty conditions).*
Refer to Section 19 for details.

## 25 Fuel filter – replacement

**1** Remove the fuel tank and air filter housing (see Chapter 4).
**2** Free the filter from its mounting bracket **(see illustration)**. Have shop towels handy to catch any spilled fuel, then squeeze the hose clamps and slide them down the hoses, away from the filter. Twist and pull the hoses off the filter fittings. If they won't come easily, carefully pry them off with a screwdriver.
**3** Connect the hoses to the new filter and install it in its mounting bracket. **Note:** *The filter must be installed with the arrow on its body pointing toward the fuel pump, ie in the direction of fuel flow.*
**4** Install all components removed for access.

## 26 Clutch fluid – renewal

**1** Refer to the clutch bleeding procedure in Chapter 2A, Section 20 (H models), or Chapter 2B, Section 20 (J, K, L and M models). Bleed the fluid out of the system via the bleed valve until it is almost drained from the fluid reservoir. At this point fill the reservoir with new fluid up to the MAX line and continue the bleeding operation until the new fluid is seen

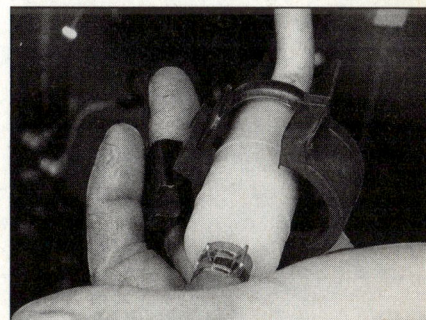

**25.2 Separate the filter from its mounting bracket and disconnect the hoses**

emerging from the bleed valve. When all traces of air bubbles have been expelled from the system, tighten the bleed valve and top the fluid reservoir up to the correct level (see *Daily (pre-ride) checks*).

> **HAYNES HINT** *Old hydraulic fluid is invariably much darker in color than new fluid, making it easy to see when all old fluid has been expelled from the system.*

### 27 Steering head bearings – lubrication

Refer to the relevant Section in Chapter 6.

### 28 Brake fluid – change

**1** Refer to the brake bleeding procedure in

Chapter 7. Bleed the fluid out of the system via the bleed valve until it is almost drained from the fluid reservoir. At this point fill the reservoir with new fluid up to the MAX line and continue the bleeding operation until the new fluid is seen emerging from the bleed valve. When all traces of air bubbles have been expelled from the system and the brake lever (front) or pedal (rear) is firm, not spongy, when applied, tighten the bleed valve and top the fluid reservoir up to the correct level (see *Daily (pre-ride) checks*). See *Haynes Hint*.

# Every 18,000 miles (30,000 km)

### 29 Cooling system – servicing

## Check

> ⚠ *Warning: The engine must be cool before beginning this procedure.*

**Note:** *Refer to 'Daily (pre-ride) checks' at the beginning of this manual and check the coolant level before performing this check.*

**1** The entire cooling system should be checked carefully at the recommended intervals. Look for evidence of leaks, check the condition of the coolant, check the radiator for clogged fins and damage and make sure the fan operates when required.

**2** Examine each of the rubber coolant hoses along its entire length. Look for cracks, abrasions and other damage. Squeeze each hose at various points. They should feel firm, yet pliable, and return to their original shape when released. If they are dried out or hard, replace them with new ones. On J, K, L and M models, also check the oil cooler hoses.

**3** Check for evidence of leaks at each cooling system joint. Tighten the hose clamps carefully to prevent future leaks.

**4** Check the radiator for evidence of leaks and other damage (remove the fairings if necessary – see Chapter 8). Leaks in the radiator leave tell-tale scale deposits or coolant stains on the outside of the core

below the leak. If leaks are noted, remove the radiator (refer to Chapter 3) and have it repaired professionally or replace it with a new one.
*Caution: Do not use a liquid leak stopping compound to try to repair leaks.*
**5** Check the radiator fins for mud, dirt and insects, which may impede the flow of air through the radiator. If the fins are dirty, force water or low pressure compressed air through the fins from the backside. If the fins are bent or distorted, straighten them carefully with a screwdriver.

**6** Remove the radiator cap by turning it counterclockwise until it reaches a stop. If you hear a hissing sound (indicating there is still pressure in the system), wait until it stops. Now, press down on the cap with the palm of your hand and continue turning the cap anti-clockwise until it can be removed **(see illustration)**. Check the condition of the coolant in the radiator. If it is rust colored or if accumulations of scale are visible in the radiator, drain, flush and refill the system with new coolant. Check the cap gaskets for cracks and other damage. If its condition is in doubt, have the cap tested by a dealer service department or replace it with a new one. Install the cap by turning it clockwise until it reaches the first stop, then push down on the cap and continue turning until it can turn no further.

**7** Check the antifreeze content of the coolant with an antifreeze hydrometer **(see illustration)**. Sometimes coolant may look like

it's in good condition, but might be too weak to offer adequate protection. If the hydrometer indicates a weak mixture, drain, flush and refill the cooling system.
**8** Start the engine and let it reach normal operating temperature, then check for leaks again. As the coolant temperature increases, the fan should come on automatically and the temperature should begin to drop. If it doesn't, refer to Chapter 3 and check the fan and fan circuit carefully.

**9** If the coolant level is consistently low, and no evidence of leaks can be found, have the entire system pressure checked by a Kawasaki dealer service department, motorcycle repair shop or service station.

## Draining

> ⚠ *Warning: Allow the engine to cool completely before performing this maintenance operation. Also, don't allow antifreeze to come into contact with your skin or painted surfaces of the vehicle. Rinse off spills immediately with plenty of water. Antifreeze is highly toxic if ingested. Never leave antifreeze lying around in an open container or in puddles on the floor; children and pets are attracted by its sweet smell and may drink it. Check with local authorities about disposing of used antifreeze. Many communities have collection centers which will see that antifreeze is disposed of safely. Antifreeze is also combustible, so don't store or use it near open flames.*

**10** Loosen the radiator cap **(see illustration 29.6)**. Place a large, clean drain pan under the left side of the engine.

**11** Remove the lower fairing (see Chapter 8).

## H models

**12** Remove the rubber cap from the bleed valve/drain fitting **(see illustration)**. Slip a length of rubber tubing over the fitting and place the other end in a drain pan. Turn the valve anti-clockwise to open it and let the coolant drain.

**13** Move the drain pan beneath the drain plug at each front corner of the cylinder block. Remove the plugs and let the coolant drain **(see illustration)**.

**29.6 Loosen the radiator cap slowly and let all pressure escape before removing it**

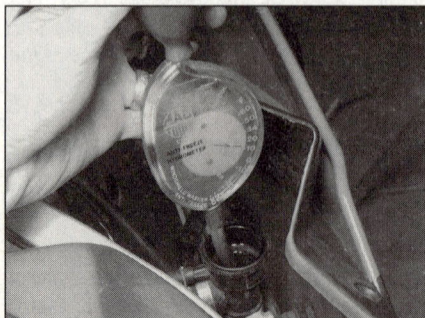

**29.7 An antifreeze hydrometer is helpful in determining the condition of the coolant**

29.12 On H models, coolant is drained through the bleed valve (A) on the water pump cover (B)

29.13 A coolant drain plug is located on each front corner of the cylinder block

### J, K, L and M models

14 Remove the drain bolt from the side of the water pump cover **(see illustration)** and allow the coolant to drain into the pan. **Note:** *The coolant will rush out with considerable force, so position the drain pan accordingly. Remove the radiator cap completely to ensure that all of the coolant can drain.*

### All models

15 Drain the coolant reservoir. Refer to Chapter 3 for the reservoir removal procedure. Wash the reservoir out with water.

### *Flushing*

16 Flush the system with clean tap water by inserting a garden hose in the radiator filler neck. Allow the water to run through the system until it is clear when it exits the drain(s). If the radiator is extremely corroded, remove it by referring to Chapter 3 and have it cleaned at a radiator shop.

17 Check the drain bolt gasket (s). Replace as necessary.

18 Clean the hole(s), then install the drain bolt(s) and tighten to the torque listed in this Chapter's Specifications. If you're working on an H model, close the bleed valve.

19 Fill the cooling system with clean water mixed with a flushing compound. Make sure the flushing compound is compatible with aluminum components, and follow the manufacturer's instructions carefully.

20 Start the engine and allow it to reach normal operating temperature. Let it run for about ten minutes, then stop the engine. Let the machine cool for awhile, then cover the radiator cap with a heavy rag and turn it counterclockwise (anti-clockwise) to the first stop, releasing any pressure that may be present in the system. Once the hissing stops, push down on the cap and remove it completely.

21 Drain the system once again.

22 Fill the system with clean water, then repeat Steps 19, 20 and 21.

### *Refilling*

23 Fill the system with the proper coolant mixture (see this Chapter's Specifications). Add coolant slowly through the radiator filler cap opening, allowing time for air to escape. Add coolant until the system is full (all the way up to the top of the radiator cap filler neck).

24 Place a pan beneath the bleed valve. Remove the rubber cap (if you haven't already done so), then open the valve **(see illustration 29.12 and the accompanying illustrations)**, allowing mixed air and coolant to drain. Let the coolant run out until air bubbles are no longer visible, then close the bleed valve.

25 Start the engine and let it idle with the radiator cap removed. While the engine runs, tap the radiator hoses with a screwdriver handle or similar tool to shake loose any air bubbles inside the hoses. When bubbles are no longer visible in the radiator filler neck, shut the engine off.

26 Slowly add coolant to the radiator filler opening until it reaches the top, then install the radiator cap.

27 Fill the coolant reservoir to the upper level line (but no higher) with coolant.

28 Run the engine and check the system for leaks. **Note:** *Do not dispose of the old coolant by pouring it down a drain. Instead, pour it*

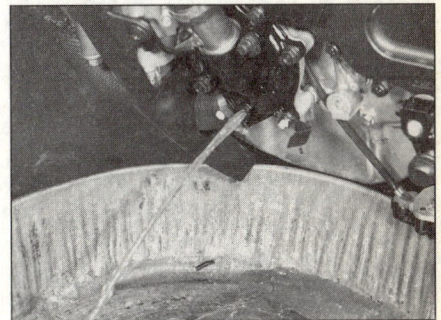

29.14 On J, K, L and M models, remove the drain bolt from the water pump to drain the coolant and be prepared for the coolant to spurt out

*into a heavy plastic container, cap it tightly and take it to an authorized disposal site or a service station.*

### 30 Fork oil – replacement

Changing the fork oil requires removal and partial disassembly of the forks. Refer to Chapter 6 for details.

29.24a To bleed the cooling system, remove the rubber cap from the bleed valve (L model shown) . . .

29.24b . . . and open the bleed valve with a box wrench (ring spanner)

# Every two years

## 31 Clutch master and release cylinders – seal replacement

1 Hydraulic seals will deteriorate over a period of time and lose their effectiveness, leading to sticking operation or fluid loss, or allowing the ingress of air and dirt.
2 Refer to the relevant Section in Chapter 2 and dismantle the components for seal replacement.

## 32 Brake calipers and master cylinders – seal replacement

1 Hydraulic seals will deteriorate over a period of time and lose their effectiveness, leading to sticking operation or fluid loss, or allowing the ingress of air and dirt.
2 Refer to the relevant Section in Chapter 7 and dismantle the components for seal replacement.

# Every four years

## 33 Fuel hoses – replacement

1 Fuel system hoses will deteriorate with age and must be replaced with new ones.
2 Refer to Chapter 4.

## 34 Brakes hoses – replacement

1 Brake system hoses will deteriorate with age and must be replaced with new ones.
2 Refer to Chapter 7 for details.

# Chapter 2  Part A:
# Engine, clutch and transmission (748 cc H models)

## Contents

## Degrees of difficulty

| Easy, suitable for novice with little experience | Fairly easy, suitable for beginner with some experience | Fairly difficult, suitable for competent DIY mechanic | Difficult, suitable for experienced DIY mechanic | Very difficult, suitable for expert DIY or professional |
|---|---|---|---|---|

## Specifications

### General

Bore ...................................... 68.0 mm (2.677 inch)
Stroke .................................... 51.5 mm (2.028 inch)
Displacement ............................. 748 cc
Compression ratio ......................... 11.3 : 1
Cylinder numbering ........................ 1-2-3-4 (from left end of engine)
Firing order .............................. 1-2-4-3

## Camshafts

Lobe height – H1 model
  Intake
    Standard . . . . . . . . . . . . . . . . . . . . . . . . . . . . . . . . . . . . . . . . . . . 36.95 to 37.05 mm (1.455 to 1.459 inch)
    Minimum . . . . . . . . . . . . . . . . . . . . . . . . . . . . . . . . . . . . . . . . . . . 36.85 mm (1.451 inch)
  Exhaust
    Standard . . . . . . . . . . . . . . . . . . . . . . . . . . . . . . . . . . . . . . . . . . . 36.55 to 36.65 mm (1.439 to 1.443 inch)
    Minimum . . . . . . . . . . . . . . . . . . . . . . . . . . . . . . . . . . . . . . . . . . . 36.45 mm (1.435 inch)
Lobe height – H2 model
  Intake
    Standard . . . . . . . . . . . . . . . . . . . . . . . . . . . . . . . . . . . . . . . . . . . 37.25 to 37.35 mm (1.467 to 1.470 inch)
    Minimum . . . . . . . . . . . . . . . . . . . . . . . . . . . . . . . . . . . . . . . . . . . 37.15 mm (1.463 inch)
  Exhaust
    Standard . . . . . . . . . . . . . . . . . . . . . . . . . . . . . . . . . . . . . . . . . . . 36.95 to 37.05 mm (1.455 to 1.459 inch)
    Minimum . . . . . . . . . . . . . . . . . . . . . . . . . . . . . . . . . . . . . . . . . . . 36.85 mm (1.451 inch)
Bearing oil clearance*
  Journals 1 and 4
    Standard . . . . . . . . . . . . . . . . . . . . . . . . . . . . . . . . . . . . . . . . . . . 0.048 to 0.091 mm (0.0019 to 0.0036 inch)
    Maximum . . . . . . . . . . . . . . . . . . . . . . . . . . . . . . . . . . . . . . . . . . . 0.18 mm (0.0071 inch)
  Journals 2 and 3
    Standard . . . . . . . . . . . . . . . . . . . . . . . . . . . . . . . . . . . . . . . . . . . 0.078 to 0.121 mm (0.0031 to 0.0048 inch)
    Maximum . . . . . . . . . . . . . . . . . . . . . . . . . . . . . . . . . . . . . . . . . . . 0.21 mm (0.0083 inch)
Journal diameter*
  Journals 1 and 4
    Standard . . . . . . . . . . . . . . . . . . . . . . . . . . . . . . . . . . . . . . . . . . . 23.930 to 23.952 mm (0.9421 to 0.9430 inch)
    Minimum . . . . . . . . . . . . . . . . . . . . . . . . . . . . . . . . . . . . . . . . . . . 23.900 mm (0.9409 inch)
  Journals 2 and 3
    Standard . . . . . . . . . . . . . . . . . . . . . . . . . . . . . . . . . . . . . . . . . . . 23.900 to 23.922 mm (0.9409 to 0.9446 inch)
    Minimum . . . . . . . . . . . . . . . . . . . . . . . . . . . . . . . . . . . . . . . . . . . 23.87 mm (0.9398 inch)
Bearing journal inside diameter
  Standard . . . . . . . . . . . . . . . . . . . . . . . . . . . . . . . . . . . . . . . . . . . 24.000 to 24.021 mm (0.9449 to 0.9457 inch)
  Maximum . . . . . . . . . . . . . . . . . . . . . . . . . . . . . . . . . . . . . . . . . . . 24.08 mm (0.9480 inch)
Camshaft runout . . . . . . . . . . . . . . . . . . . . . . . . . . . . . . . . . . . . . . . 0.02 mm (0.0008 inch) or less
Cam chain 20-link length
  Standard . . . . . . . . . . . . . . . . . . . . . . . . . . . . . . . . . . . . . . . . . . . 127.0 to 127.4 mm (5.0 to 5.0157 inch)
  Maximum . . . . . . . . . . . . . . . . . . . . . . . . . . . . . . . . . . . . . . . . . . . 128.9 mm (5.0748 inch)
*Camshaft bearing journals are numbered 1 through 4 from left side of the engine*

## Cylinder head, valves and valve springs

Cylinder head warpage limit . . . . . . . . . . . . . . . . . . . . . . . . . . . . . . 0.05 mm (0.0020 inch)
Valve clearances . . . . . . . . . . . . . . . . . . . . . . . . . . . . . . . . . . . . . . . See Chapter 1
Valve stem runout limit . . . . . . . . . . . . . . . . . . . . . . . . . . . . . . . . . . 0.01 mm (0.0004 inch) or less
Valve stem diameter
  Intake valve
    Standard . . . . . . . . . . . . . . . . . . . . . . . . . . . . . . . . . . . . . . . . . . . 4.975 to 4.990 mm (0.1959 to 0.1965 inch)
    Minimum . . . . . . . . . . . . . . . . . . . . . . . . . . . . . . . . . . . . . . . . . . . 4.95 mm (0.1949 inch)
  Exhaust valve
    Standard . . . . . . . . . . . . . . . . . . . . . . . . . . . . . . . . . . . . . . . . . . . 4.955 to 4.970 mm (0.1951 to 0.1957 inch)
    Minimum . . . . . . . . . . . . . . . . . . . . . . . . . . . . . . . . . . . . . . . . . . . 4.93 mm (0.1941 inch)
Valve guide inside diameter (intake and exhaust)
  Standard . . . . . . . . . . . . . . . . . . . . . . . . . . . . . . . . . . . . . . . . . . . 5.000 to 5.012 mm (0.1969 to 0.1973 inch)
  Maximum . . . . . . . . . . . . . . . . . . . . . . . . . . . . . . . . . . . . . . . . . . . 5.08 mm (0.2 inch)
Valve stem-to-guide clearance
  Intake valve
    Standard . . . . . . . . . . . . . . . . . . . . . . . . . . . . . . . . . . . . . . . . . . . 0.033 to 0.144 mm (0.0013 to 0.0057 inch)
    Maximum . . . . . . . . . . . . . . . . . . . . . . . . . . . . . . . . . . . . . . . . . . . 0.34 mm (0.0134 inch)
  Exhaust valve
    Standard . . . . . . . . . . . . . . . . . . . . . . . . . . . . . . . . . . . . . . . . . . . 0.095 to 0.181 mm (0.0037 to 0.0071 inch)
    Maximum . . . . . . . . . . . . . . . . . . . . . . . . . . . . . . . . . . . . . . . . . . . 0.40 mm (0.0157 inch)
Valve seat width (intake and exhaust) . . . . . . . . . . . . . . . . . . . . . . . 0.5 to 1.0 mm (0.0197 to 0.0394 inch)
Valve spring free length
  Inner spring
    Standard . . . . . . . . . . . . . . . . . . . . . . . . . . . . . . . . . . . . . . . . . . . 36.5 mm (1.4370 inch)
    Maximum . . . . . . . . . . . . . . . . . . . . . . . . . . . . . . . . . . . . . . . . . . . 34.8 mm (1.3700 inch)
  Outer spring
    Standard . . . . . . . . . . . . . . . . . . . . . . . . . . . . . . . . . . . . . . . . . . . 41.4 mm (1.6299 inch)
    Maximum . . . . . . . . . . . . . . . . . . . . . . . . . . . . . . . . . . . . . . . . . . . 39.7 mm (1.5630 inch)

## Cylinder block

Bore diameter
   Standard . . . . . . . . . . . . . . . . . . . . . . . . . . . . . . . . . . . . . . . . . . .   68.000 to 68.012 mm (2.6772 to 2.6776 inch)
   Maximum . . . . . . . . . . . . . . . . . . . . . . . . . . . . . . . . . . . . . . . . . . .   68.10 mm (2.6812 inch)

## Pistons

Piston diameter
   Standard . . . . . . . . . . . . . . . . . . . . . . . . . . . . . . . . . . . . . . . . . . .   67.942 to 67.958 mm (2.6749 to 2.6755 inch)
   Minimum . . . . . . . . . . . . . . . . . . . . . . . . . . . . . . . . . . . . . . . . . . .   67.80 mm (2.6693 inch)
Piston-to-cylinder clearance . . . . . . . . . . . . . . . . . . . . . . . . . . . . . .   0.042 to 0.070 mm (0.0017 to 0.0028 inch)
Top ring side clearance
   Standard . . . . . . . . . . . . . . . . . . . . . . . . . . . . . . . . . . . . . . . . . . .   0.03 to 0.07 mm (0.0012 to 0.0028 inch)
   Maximum . . . . . . . . . . . . . . . . . . . . . . . . . . . . . . . . . . . . . . . . . . .   0.17 mm (0.0067 inch)
Second (middle) ring side clearance
   Standard . . . . . . . . . . . . . . . . . . . . . . . . . . . . . . . . . . . . . . . . . . .   0.02 to 0.06 mm (0.0008 to 0.0024 inch)
   Maximum . . . . . . . . . . . . . . . . . . . . . . . . . . . . . . . . . . . . . . . . . . .   0.16 mm (0.0063 inch)
Top ring groove width – H1 model
   Standard . . . . . . . . . . . . . . . . . . . . . . . . . . . . . . . . . . . . . . . . . . .   1.02 to 1.04 mm (0.0402 to 0.0410 inch)
   Maximum . . . . . . . . . . . . . . . . . . . . . . . . . . . . . . . . . . . . . . . . . . .   1.12 mm (0.0441 inch)
Second (middle) ring groove width – H1 model
   Standard . . . . . . . . . . . . . . . . . . . . . . . . . . . . . . . . . . . . . . . . . . .   1.01 to 1.03 mm (0.0398 to 0.0406 inch)
   Maximum . . . . . . . . . . . . . . . . . . . . . . . . . . . . . . . . . . . . . . . . . . .   1.11 mm (0.0437 inch)
Top ring groove width – H2 model
   Standard . . . . . . . . . . . . . . . . . . . . . . . . . . . . . . . . . . . . . . . . . . .   0.82 to 0.84 mm (0.0323 to 0.0331 inch)
   Maximum . . . . . . . . . . . . . . . . . . . . . . . . . . . . . . . . . . . . . . . . . . .   0.92 mm (0.0362 inch)
Second (middle) ring groove width – H2 model
   Standard . . . . . . . . . . . . . . . . . . . . . . . . . . . . . . . . . . . . . . . . . . .   0.81 to 0.83 mm (0.0319 to 0.0327 inch)
   Maximum . . . . . . . . . . . . . . . . . . . . . . . . . . . . . . . . . . . . . . . . . . .   0.91 mm (0.0358 inch)
Oil ring groove width – H2 model
   Standard . . . . . . . . . . . . . . . . . . . . . . . . . . . . . . . . . . . . . . . . . . .   2.01 to 2.03 mm (0.0791 to 0.0799 inch)
   Maximum . . . . . . . . . . . . . . . . . . . . . . . . . . . . . . . . . . . . . . . . . . .   2.11 mm (0.0831 inch)
Top and second (middle) ring thickness – H1 model
   Standard . . . . . . . . . . . . . . . . . . . . . . . . . . . . . . . . . . . . . . . . . . .   0.97 to 0.99 mm (0.0382 to 0.0390 inch)
   Minimum . . . . . . . . . . . . . . . . . . . . . . . . . . . . . . . . . . . . . . . . . . .   0.90 mm (0.0354 inch)
Top and second (middle) ring thickness – H2 model
   Standard . . . . . . . . . . . . . . . . . . . . . . . . . . . . . . . . . . . . . . . . . . .   0.77 to 0.79 mm (0.0303 to 0.0311 inch)
   Minimum . . . . . . . . . . . . . . . . . . . . . . . . . . . . . . . . . . . . . . . . . . .   0.70 mm (0.0276 inch)
Top and second (middle) ring end gap
   Standard . . . . . . . . . . . . . . . . . . . . . . . . . . . . . . . . . . . . . . . . . . .   0.20 to 0.35 mm (0.0079 to 0.0138 inch)
   Maximum . . . . . . . . . . . . . . . . . . . . . . . . . . . . . . . . . . . . . . . . . . .   0.70 mm (0.0276 inch)
Oil ring end gap – H2 model
   Standard . . . . . . . . . . . . . . . . . . . . . . . . . . . . . . . . . . . . . . . . . . .   0.20 to 0.70 mm (0.0079 to 0.0276 inch)
   Maximum . . . . . . . . . . . . . . . . . . . . . . . . . . . . . . . . . . . . . . . . . . .   1.0 mm (0.0394 inch)

## Crankshaft and bearings

Crankshaft endplay
   Standard . . . . . . . . . . . . . . . . . . . . . . . . . . . . . . . . . . . . . . . . . . .   0.05 to 0.2 mm (0.002 to 0.0079 inch)
   Maximum . . . . . . . . . . . . . . . . . . . . . . . . . . . . . . . . . . . . . . . . . . .   0.4 mm (0.0157 inch)
Crankshaft runout . . . . . . . . . . . . . . . . . . . . . . . . . . . . . . . . . . . . . .   0.03 mm (0.0019 inch) or less
Main bearing oil clearance
   Standard . . . . . . . . . . . . . . . . . . . . . . . . . . . . . . . . . . . . . . . . . . .   0.020 to 0.044 mm (0.0008 to 0.0017 inch)
   Maximum . . . . . . . . . . . . . . . . . . . . . . . . . . . . . . . . . . . . . . . . . . .   0.08 mm (0.0031 inch)
Crankcase main bearing bore diameter
   '0' mark on crankcase . . . . . . . . . . . . . . . . . . . . . . . . . . . . . . . . . .   37.000 to 37.008 mm (1.4567 to 1.4570 inch)
   No mark on crankcase . . . . . . . . . . . . . . . . . . . . . . . . . . . . . . . . . .   37.009 to 37.016 mm (1.4570 to 1.4573 inch)
Main bearing journal diameter
   No mark on crank throw . . . . . . . . . . . . . . . . . . . . . . . . . . . . . . . .   33.984 to 33.992 mm (1.338 to 1.3383 inch)
   '1' mark on crank throw . . . . . . . . . . . . . . . . . . . . . . . . . . . . . . . .   33.993 to 34.000 mm (1.3383 to 1.3386 inch)
Connecting rod side clearance
   Standard . . . . . . . . . . . . . . . . . . . . . . . . . . . . . . . . . . . . . . . . . . .   0.13 to 0.38 mm (0.0051 to 0.015 inch)
   Maximum . . . . . . . . . . . . . . . . . . . . . . . . . . . . . . . . . . . . . . . . . . .   0.60 mm (0.0236 inch)
Connecting rod bearing oil clearance
   Standard . . . . . . . . . . . . . . . . . . . . . . . . . . . . . . . . . . . . . . . . . . .   0.046 to 0.076 mm (0.0018 to 0.003 inch)
   Maximum . . . . . . . . . . . . . . . . . . . . . . . . . . . . . . . . . . . . . . . . . . .   0.11 mm (0.0043 inch)
Connecting rod inside diameter
   No mark on side of rod . . . . . . . . . . . . . . . . . . . . . . . . . . . . . . . . .   38.000 to 38.008 mm (1.4961 to 1.4964 inch)
   '0' mark on side of rod . . . . . . . . . . . . . . . . . . . . . . . . . . . . . . . . .   38.009 to 38.016 mm (1.4964 to 1.4967 inch)
Connecting rod journal (crankpin) diameter
   No mark on crank throw . . . . . . . . . . . . . . . . . . . . . . . . . . . . . . . .   34.984 to 34.992 mm (1.3773 to 1.3776 inch)
   '0' mark on crank throw . . . . . . . . . . . . . . . . . . . . . . . . . . . . . . . .   34.993 to 35.000 mm (1.3777 to 1.378 inch)

## Oil pump and relief valve

Oil pressure @ 4000 rpm . . . . . . . . . . . . . . . . . . . . . . . . . . . . . . . . . . .  38 to 53 psi (2.6 to 3.7 Bars)
Relief valve opening pressure . . . . . . . . . . . . . . . . . . . . . . . . . . . . . . . .  46 to 78 psi (3.2 to 5.4 Bars)

## Clutch

Spring free length
    Standard . . . . . . . . . . . . . . . . . . . . . . . . . . . . . . . . . . . . . . . . . . . . .  33.2 mm (1.3071 inch)
    Minimum . . . . . . . . . . . . . . . . . . . . . . . . . . . . . . . . . . . . . . . . . . . . .  32.1 mm (1.2638 inch)
Friction and steel plate warpage
    Standard . . . . . . . . . . . . . . . . . . . . . . . . . . . . . . . . . . . . . . . . . . . . .  0.2 mm (0.0079 inch) or less
    Maximum . . . . . . . . . . . . . . . . . . . . . . . . . . . . . . . . . . . . . . . . . . . . .  0.3 mm (0.0118 inch)
Pressure plate freeplay
    With new plates . . . . . . . . . . . . . . . . . . . . . . . . . . . . . . . . . . . . . . . .  0.2 to 0.5 mm (0.0079 to 0.0197 inch)
    With used plates . . . . . . . . . . . . . . . . . . . . . . . . . . . . . . . . . . . . . . . .  0.3 to 0.9 mm (0.0118 to 0.0354 inch)

## Transmission

Gear ratios
    1st gear . . . . . . . . . . . . . . . . . . . . . . . . . . . . . . . . . . . . . . . . . . . . . .  2.923 : 1 (38/13T)
    2nd gear . . . . . . . . . . . . . . . . . . . . . . . . . . . . . . . . . . . . . . . . . . . . .  2.125 : 1 (34/16T)
    3rd gear . . . . . . . . . . . . . . . . . . . . . . . . . . . . . . . . . . . . . . . . . . . . .  1.666 : 1 (35/21T)
    4th gear . . . . . . . . . . . . . . . . . . . . . . . . . . . . . . . . . . . . . . . . . . . . .  1.380 : 1 (29/21T)
    5th gear . . . . . . . . . . . . . . . . . . . . . . . . . . . . . . . . . . . . . . . . . . . . .  1.217 : 1 (28/23T)
    6th gear . . . . . . . . . . . . . . . . . . . . . . . . . . . . . . . . . . . . . . . . . . . . .  1.083 : 1 (26/24T)
Shift fork groove width in gears
    Standard . . . . . . . . . . . . . . . . . . . . . . . . . . . . . . . . . . . . . . . . . . . . .  5.05 to 5.15 mm (0.1988 to 0.2028 inch)
    Maximum . . . . . . . . . . . . . . . . . . . . . . . . . . . . . . . . . . . . . . . . . . . . .  5.3 mm (0.2087 inch)
Shift fork ear thickness
    Standard . . . . . . . . . . . . . . . . . . . . . . . . . . . . . . . . . . . . . . . . . . . . .  4.9 to 5.0 mm (0.1929 to 0.1969 inch)
    Minimum . . . . . . . . . . . . . . . . . . . . . . . . . . . . . . . . . . . . . . . . . . . . .  4.8 mm (0.189 inch)
Shift fork guide pin diameter
    Standard . . . . . . . . . . . . . . . . . . . . . . . . . . . . . . . . . . . . . . . . . . . . .  7.9 to 8.0 mm (0.3110 to 0.315 inch)
    Minimum . . . . . . . . . . . . . . . . . . . . . . . . . . . . . . . . . . . . . . . . . . . . .  7.8 mm (0.3071 inch)
Shift drum groove width
    Standard . . . . . . . . . . . . . . . . . . . . . . . . . . . . . . . . . . . . . . . . . . . . .  8.05 to 8.20 mm (0.3169 to 0.3228 inch)
    Maximum . . . . . . . . . . . . . . . . . . . . . . . . . . . . . . . . . . . . . . . . . . . . .  8.3 mm (0.3268 inch)

## Torque specifications

Valve cover bolts . . . . . . . . . . . . . . . . . . . . . . . . . . . . . . . . . . . . . . . . .  9.8 Nm (87 in-lbs)
Camshaft bearing cap bolts . . . . . . . . . . . . . . . . . . . . . . . . . . . . . . . . .  12 Nm (104 in-lbs)
Camshaft sprocket bolts . . . . . . . . . . . . . . . . . . . . . . . . . . . . . . . . . . .  15 Nm (11 ft-lbs)
Cam chain tensioner cap . . . . . . . . . . . . . . . . . . . . . . . . . . . . . . . . . . .  4.9 Nm (43 in-lbs)
Cam chain tensioner mounting bolts . . . . . . . . . . . . . . . . . . . . . . . . . .  12 Nm (104 in-lbs)
Cylinder head bolts
    Initial torque . . . . . . . . . . . . . . . . . . . . . . . . . . . . . . . . . . . . . . . . . .  20 Nm (14.5 ft-lbs)
    Final torque
        New bolts, washers and cylinder head . . . . . . . . . . . . . . . . . . . . .  54 Nm (40 ft-lbs)
        Used bolts, washers and cylinder head . . . . . . . . . . . . . . . . . . . . .  49 Nm (36 ft-lbs)
Clutch cover bolts . . . . . . . . . . . . . . . . . . . . . . . . . . . . . . . . . . . . . . . . .  9.8 Nm (87 in-lbs)
Clutch cover noise damper bolts . . . . . . . . . . . . . . . . . . . . . . . . . . . . . .  9.8 Nm (87 in-lbs)
Clutch spring bolts . . . . . . . . . . . . . . . . . . . . . . . . . . . . . . . . . . . . . . . .  8.8 Nm (78 in-lbs)
Clutch hub nut . . . . . . . . . . . . . . . . . . . . . . . . . . . . . . . . . . . . . . . . . . .  135 Nm (100 ft-lbs)
Clutch master cylinder clamp bolts . . . . . . . . . . . . . . . . . . . . . . . . . . . .  8.8 Nm (78 in-lbs)
Clutch slave cylinder bleed valve . . . . . . . . . . . . . . . . . . . . . . . . . . . . .  7.8 Nm (69 in-lbs)
Clutch hose banjo bolts . . . . . . . . . . . . . . . . . . . . . . . . . . . . . . . . . . . .  25 Nm (18 ft-lbs)
Clutch hose in-line connector . . . . . . . . . . . . . . . . . . . . . . . . . . . . . . . .  18 Nm (13 ft-lbs)
Oil pan bolts . . . . . . . . . . . . . . . . . . . . . . . . . . . . . . . . . . . . . . . . . . . . .  12 Nm (104 in-lbs)
Oil pump mounting bolts . . . . . . . . . . . . . . . . . . . . . . . . . . . . . . . . . . . .  12 Nm (104 in-lbs)
Oil pump cover screws . . . . . . . . . . . . . . . . . . . . . . . . . . . . . . . . . . . . .  9.8 Nm (87 in-lbs)
Oil hose-to-cylinder head banjo bolts . . . . . . . . . . . . . . . . . . . . . . . . . .  12 Nm (104 in-lbs)
Oil hose-to-oil cooler banjo bolts . . . . . . . . . . . . . . . . . . . . . . . . . . . . .  25 Nm (18 ft-lbs)
Pressure relief valve . . . . . . . . . . . . . . . . . . . . . . . . . . . . . . . . . . . . . . .  15 Nm (11 ft-lbs)
Breather cover bolts . . . . . . . . . . . . . . . . . . . . . . . . . . . . . . . . . . . . . . .  12 Nm (104 in-lbs)
Crankcase bolts (see illustrations)
    T1 . . . . . . . . . . . . . . . . . . . . . . . . . . . . . . . . . . . . . . . . . . . . . . . . . . .  12 Nm (104 in-lbs)
    T5 . . . . . . . . . . . . . . . . . . . . . . . . . . . . . . . . . . . . . . . . . . . . . . . . . . .  22 Nm (16 ft-lbs)
    T6 . . . . . . . . . . . . . . . . . . . . . . . . . . . . . . . . . . . . . . . . . . . . . . . . . . .  27 Nm (20 ft-lbs)

UPPER CRANKCASE

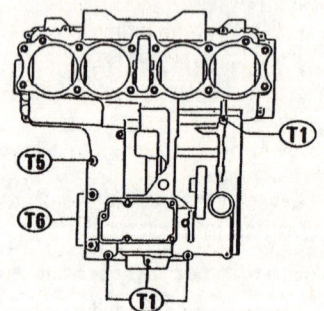

LOWER CRANKCASE

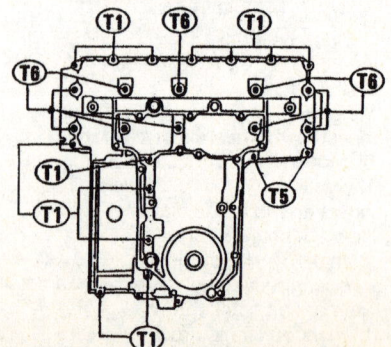

## Torque specifications (continued)

Connecting rod nuts
  H1 model . . . . . . . . . . . . . . . . . . . . . . . . . . . . . . . . . . . . . . 36 Nm (27 ft-lbs)
  H2 model (new connecting rod) . . . . . . . . . . . . . . . . . . . . . . . 18 Nm (13 ft-lbs), then 120°
  H2 model (used connecting rod) . . . . . . . . . . . . . . . . . . . . . . 25 Nm (18 ft-lbs), then 120°
Alternator pulley bolt . . . . . . . . . . . . . . . . . . . . . . . . . . . . . . . 39 Nm (29 ft-lbs)
Alternator/starter chain guide bolts . . . . . . . . . . . . . . . . . . . . 12 Nm (104 in-lbs)
Engine mountings
  Downtube upper and lower bolts . . . . . . . . . . . . . . . . . . . . . . 44 Nm (33 ft-lbs)
  Downtube-to-engine front mount bolts . . . . . . . . . . . . . . . . . . 20 Nm (14.5 ft-lbs)
  Engine front mount throughbolt nuts . . . . . . . . . . . . . . . . . . . 44 Nm (33 ft-lbs)
  Rear pivot bolt nuts . . . . . . . . . . . . . . . . . . . . . . . . . . . . . . . . 44 Nm (33 ft-lbs)
  Rear pivot bolt threaded sleeves . . . . . . . . . . . . . . . . . . . . . . 9.8 Nm (87 in-lbs)

## 1  General information

The ZX750H engine/transmission unit is of the water-cooled, inline, four-cylinder design, installed transversely across the frame. The sixteen valves are operated by double overhead camshafts which are chain driven off the crankshaft.

The engine/transmission assembly is constructed from aluminum alloy. The crankcase is divided horizontally.

The crankcase incorporates a wet sump, pressure-fed lubrication system which uses a chain-driven, dual-rotor oil pump, an oil filter and by-pass valve assembly, a relief valve and an oil pressure switch. Also contained in the crankcase is the idler shaft and the starter motor clutch. The idler shaft turns the alternator pulley, which in turn drives the alternator through a ribbed belt.

Power from the crankshaft is routed to the transmission via the clutch, which is of the wet, multi-plate type and is gear-driven off the crankshaft. The transmission is a six-speed, constant-mesh unit.

## 2  Operations possible with the engine in the frame

The components and assemblies listed below can be removed without having to remove the engine from the frame. If, however, a number of areas require attention at the same time, removal of the engine is recommended.

*Gear selector mechanism external
components
Water pump
Starter motor
Alternator and its pulley
Clutch assembly and slave cylinder
Oil pan, oil pump and relief valve
Valve cover, camshafts and lifters
Cam chain tensioner*

## 3  Operations requiring engine removal

It is necessary to remove the engine/transmission assembly from the frame and separate the crankcase halves to gain access to the following components:

*Cylinder head
Cylinder block and pistons
Crankshaft, connecting rods and bearings
Transmission shafts
Shift drum and forks
Idler shaft and starter motor clutch
Camshaft chain
Alternator/starter chain*

## 4  Major engine repair – general note

1 It is not always easy to determine when or if an engine should be completely overhauled, as a number of factors must be considered.
2 High mileage is not necessarily an indication that an overhaul is needed, while low mileage, on the other hand, does not preclude the need for an overhaul. Frequency of servicing is probably the single most important consideration. An engine that has regular and frequent oil and filter changes, as well as other required maintenance, will most likely give many miles of reliable service. Conversely, a neglected engine, or one which has not been broken in properly, may require an overhaul very early in its life.
3 Exhaust smoke and excessive oil consumption are both indications that piston rings and/or valve guides are in need of attention. Make sure oil leaks are not responsible before deciding that the rings and guides are bad. Refer to Chapter 1 and perform a cylinder compression check to determine for certain the nature and extent of the work required.
4 If the engine is making obvious knocking or rumbling noises, the connecting rod and/or main bearings are probably at fault.
5 Loss of power, rough running, excessive valve train noise and high fuel consumption rates may also point to the need for an overhaul, especially if they are all present at the same time. If a complete tune-up does not remedy the situation, major mechanical work is the only solution.
6 An engine overhaul generally involves restoring the internal parts to the specifications of a new engine. During an overhaul the piston rings are replaced and the cylinder walls are bored and/or honed. If a rebore is done, then new pistons are also required. The main and connecting rod bearings are generally replaced with new ones and, if necessary, the crankshaft is also replaced. Generally the valves are serviced as well, since they are usually in less than perfect condition at this point. While the engine is being overhauled, other components such as the carburetors and the starter motor can be rebuilt also. The end result should be a like-new engine that will give as many trouble free miles as the original.
7 Before beginning the engine overhaul, read through all of the related procedures to familiarize yourself with the scope and requirements of the job. Overhauling an engine is not all that difficult, but it is time consuming. Plan on the motorcycle being tied up for a minimum of two weeks. Check on the availability of parts and make sure that any necessary special tools, equipment and supplies are obtained in advance.
8 Most work can be done with typical shop hand tools, although a number of precision measuring tools are required for inspecting parts to determine if they must be replaced. Often a dealer service department or motorcycle repair shop will handle the inspection of parts and offer advice concerning reconditioning and replacement.

**HAYNES HiNT** *As a general rule, time is the primary cost of an overhaul so it doesn't pay to install worn or substandard parts.*

9 As a final note, to ensure maximum life and minimum trouble from a rebuilt engine, everything must be assembled with care in a spotlessly clean environment.

5.1a Auxiliary stand used to support motorcycle securely upright; the stand mounts on the swingarm left . . .

5.1b . . . and right side pivots

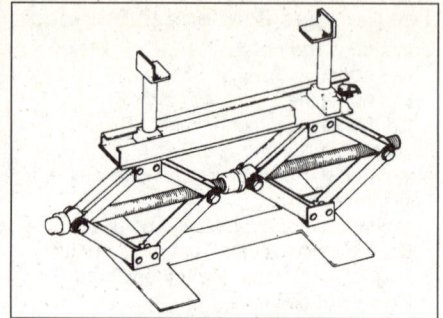

5.1c Kawasaki motorcycle jack (part no. 57001-1238)

## 5 Engine – removal and installation

**Note:** *Engine removal and installation should be done with the aid of an assistant to avoid damage or injury that could occur if the engine is dropped. A hydraulic floor jack is advised to support the engine under the oil pan, and some means must be devised of supporting the motorcycle securely upright (see below).*

### Removal

**1** The motorcycle must be supported securely in an upright position for engine removal. The accompanying photos show the motorcycle supported on an auxiliary stand which uses the swingarm pivot ends for

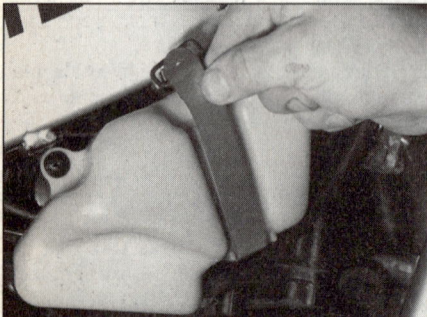

5.5a The coolant reservoir is retained by a band . . .

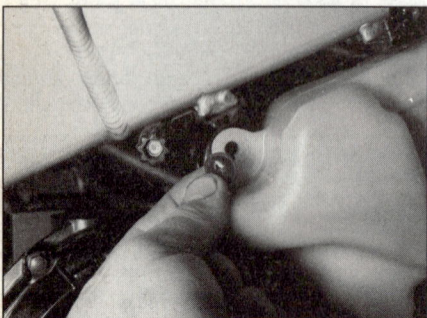

5.5b . . . and a single screw

support. Alternatively the Kawasaki jack can be used (part no. 57001-1238); this locates on the frame downtube lower mounting points and requires that the sidestand be removed **(see illustrations)**.
**2** Remove the fairing lower panels, seat and side covers (see Chapter 8).
**3** Disconnect the battery leads.

⚠️ *Warning: Always disconnect the battery negative lead first and reconnect it last.*

If the motorcycle is going to be out of use for a while, remove the battery and give it regular refresher charges (see Chapter 9). Unbolt the engine ground (earth) wire from the top of the crankcase, near the oil filler cap.
**4** Drain the engine oil (see Chapter 1).
**5** Drain the coolant (see Chapter 1). Release the band retaining the reservoir tank to the frame, followed by the mounting screw **(see illustrations)**. Move the reservoir tank clear of the engine and support it with wire to prevent coolant loss.
**6** Remove the fuel tank, air filter housing and carburetors (see Chapter 4). Plug the cylinder head openings with clean rags to prevent the entry of dirt.
**7** Remove the vacuum switching valve and hoses.
**8** Remove the baffle plate.
**9** Pull the spark plug caps off the plugs and tie the plug wires to the top of the frame.
**Note:** *If the plug wires are not numbered to indicate their plug location, label them accordingly.*

**10** Disconnect and remove the oil cooler (see Chapter 3).
**11** Remove the radiator (see Chapter 3). Slacken its clamp and pull the coolant hose off the water pump.
**12** Remove the exhaust system (see Chapter 4).
**13** Leaving its hydraulic hose attached, remove the clutch slave cylinder bolts and withdraw the slave cylinder from the casing **(see illustration)**. Push the piston as far into the slave cylinder as possible with hand pressure and hold it there while squeezing the clutch lever towards the handlebar; hold the lever against the handlebar with several strong elastic bands to prevent the slave cylinder piston being expelled.
**14** Punch mark the end of the gearshift shaft level with the split in the clamp as a reference to installation. Disconnect the gearshift pedal link rod from the gearshift shaft end **(see illustration)**.
**15** Remove the drive chain cover and disconnect the drive chain from the engine sprocket. Remove the sprocket, but note that the drive chain must be freed from the gearbox shaft as the engine is removed from the frame (see Chapter 6).
**16** Remove the alternator (see Chapter 9).
**17** Disconnect the lead from the starter motor **(see illustration)**. Screw the nut and washers back on for safekeeping.
**18** Locate their wiring connectors behind the left side cover (see *Wiring diagrams* at the end of this manual for wire identification), and

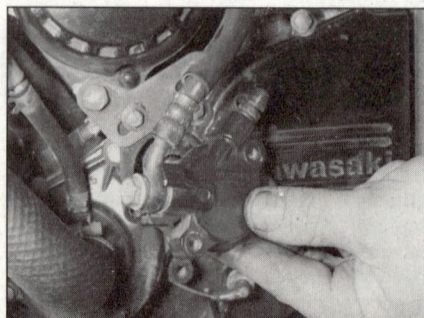

5.13 Remove its two bolts and withdraw the clutch slave cylinder from the engine case

5.14 Remove the clamp bolt fully and pull the shift pedal off its shaft

**5.17 Unscrew the terminal nut and detach the starter motor lead**

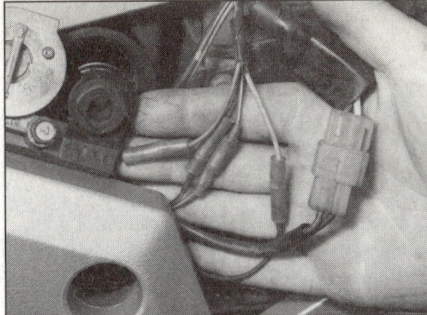

**5.18a Location of wire connectors for the alternator, neutral switch, sidestand switch and oil pressure switch**

**5.18b Release the igniter unit from its mounting and disconnect the pickup coil wires**

disconnect the wires for the alternator, neutral switch, sidestand switch and oil pressure switch **(see illustration)**. Disconnect the pickup coil wiring from the igniter unit under the left side cover **(see illustration)**.

**19** At this point, make sure the engine is supported securely underneath the oil pan and that the motorcycle is held securely upright (see Step 1). Make sure no wires or hoses are still attached to the engine assembly.

**20** Remove the sidestand switch **(see illustrations)**.

**21** Remove the two bolts at each joint of the downtube-to-main frame and downtube brace-to-engine front mounts **(see illustrations)**.

**5.20a Release its two retaining screws . . .**

**5.20b . . . and remove the sidestand switch from the stand bracket**

**5.21a Downtube-to-frame upper left bolts . . .**

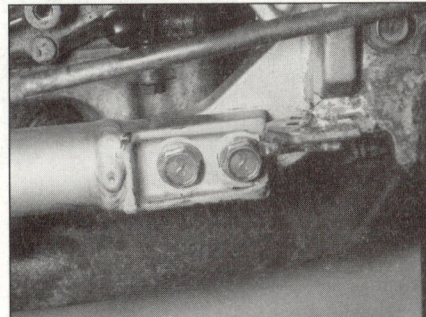

**5.21b . . . lower left bolts . . .**

**5.21c . . . upper right bolts . . .**

**5.21d . . . and lower right bolts**

**5.21e Downtube-to-front engine mounting bolts (right . . .**

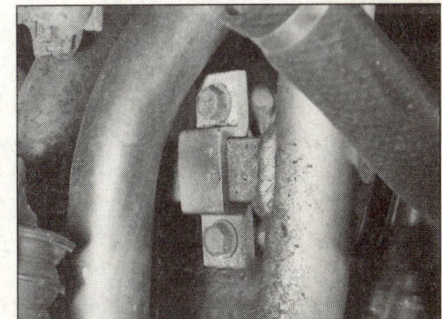

**5.21f . . . and left side)**

**5.22a Engine rear mounting upper nut on right side . . .**

**5.22b . . . and lower nut on right side**

**5.22c Engine rear mounting upper pivot bolt on left side . . .**

22 Remove the nuts from the two rear engine mounting bolts on the right side and withdraw the mounting bolts **(see illustrations)**. Back off the engine mounting adjuster sleeves located on both sides of the frame at the top mounting, and on the right side at the bottom mounting.

23 With the aid of an assistant, ease the engine forward and disengage the drive chain from the transmission output shaft. Lift the engine unit to the work area.

⚠ *Warning: Beware! as the engine unit is heavy and may cause injury if it falls.*

## Installation

24 Before installing the engine, inspect the rubber mounts on the crankcase front **(see illustration)**. If the rubber dampers have deteriorated, unscrew the through-bolt and nut which retain them to the crankcase. Remove the end caps and press the dampers out of their brackets. When installing new dampers apply a soap and water solution to them to aid installation; do not use oil or the damper will deteriorate. Install the caps and through-bolt and tighten the nut to the specified torque.

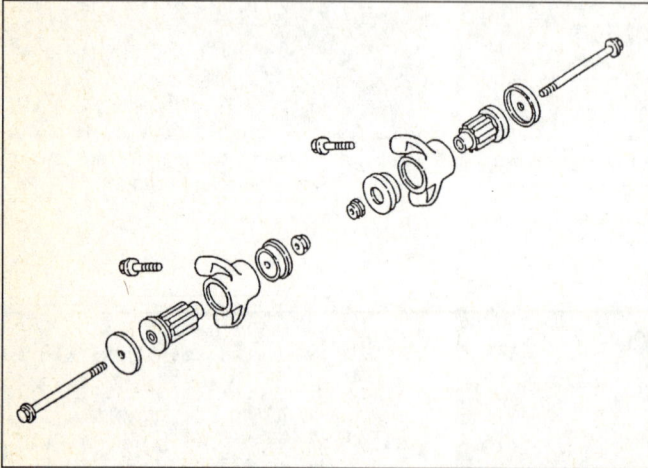

**5.24 Engine front mount details**

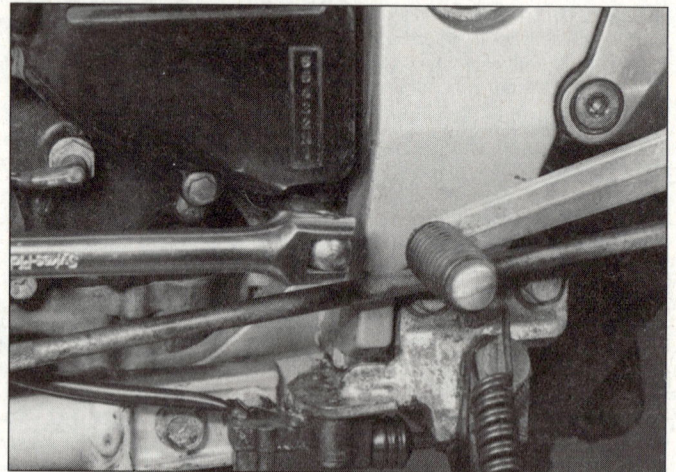

**5.22d . . . and lower pivot bolt on left side**

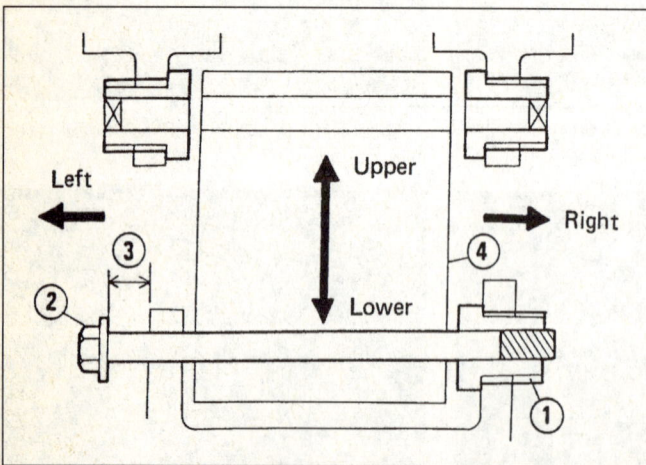

**5.27 Engine rear lower bolt installation**

| | |
|---|---|
| 1 *Right side threaded sleeve* | 3 *10 mm gap* |
| 2 *Lower bolt* | 4 *Engine* |

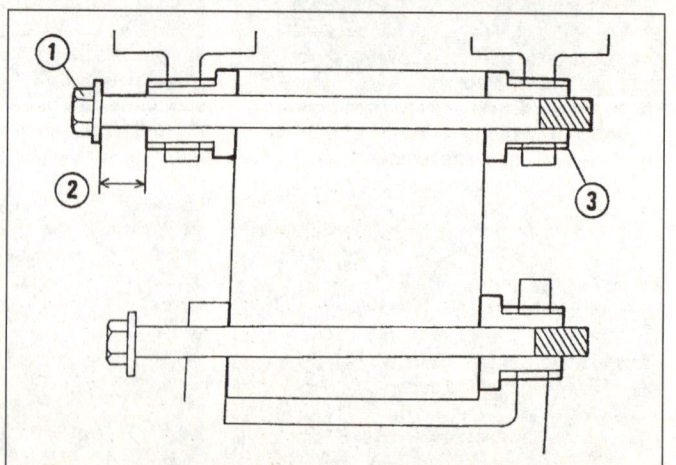

**5.28 Engine rear upper bolt installation**

| | |
|---|---|
| 1 *Upper bolt* | 3 *Right side threaded* |
| 2 *10 mm gap* | *sleeve* |

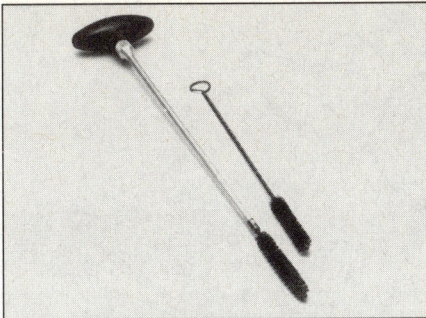

6.2a A selection of brushes is required for cleaning holes and passages in the engine components

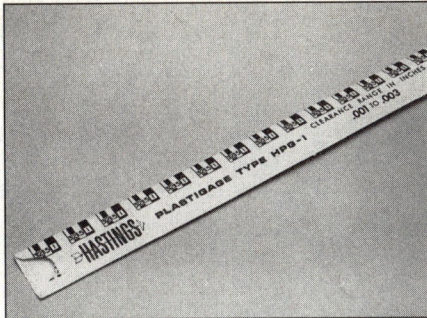

6.2b Type HPG-1 Plastigauge needed to check the crankshaft, connecting rod and camshaft oil clearances

6.3 An engine stand made from short lengths of 2 x 4 lumber and lag bolts or nails

25 Thread the engine mounting adjuster sleeves back into the frame to allow plenty of clearance for engine installation; there are two adjuster sleeves on the top rear bolt location and one on the lower bolt right side.

26 Position the engine in the frame and loop the drive chain over the output shaft. Place a support under the oil pan.

27 Start by installing the lower rear bolt from the left side, yet leave a 10 mm (0.4 inch) gap between the bolt head flange and the frame **(see illustration)**. Screw the adjuster sleeve in against the engine, tightening it to the specified torque.

28 Screw the left top rear mounting adjuster sleeve in towards the engine, tightening it to the specified torque. Install the top mounting bolt from the left side of the frame and leave a gap of 10 mm (0.4 inch) between its flanged head and the frame **(see illustration)**. Screw in the top mounting right adjuster sleeve and tighten to the specified torque.

29 Push the top and lower rear mounting bolts fully into the frame and install their nuts. Do not tighten them fully at this stage.

30 Offer up the downtube to the frame and install all upper, lower and front mounting bolts.

31 Tighten the two rear mounting nuts and all downtube bolts to the specified torques.

32 The rest of the engine installation procedure is basically the reverse of removal. Note the following points:

a) Use new gaskets at all exhaust pipe connections.

b) If the sidestand was removed, apply thread locking compound to the bracket bolts.

c) Tighten all fasteners to the specified torque settings.

d) Adjust the drive chain, throttle cables and choke cable following the procedures in Chapter 1.

e) Make sure all electrical connections are correctly made and secured by the ties provided.

f) Fill the cooling system with fresh coolant and fill the engine with the specified type and quantity of oil (see Chapter 1 for quantities and Daily (pre-ride) checks for topping up).

## 6 Engine disassembly and reassembly – general information

**Note:** Refer to the 'Tools and Workshop Tips' Section at the end of this manual for further information.

1 Before disassembling the engine, clean the exterior with a degreaser and rinse it with water. A clean engine will make the job easier and prevent the possibility of getting dirt into the internal areas of the engine.

2 In addition to the precision measuring tools mentioned earlier, you will need a torque wrench, a valve spring compressor, oil gallery brushes, a piston ring removal and installation tool, a piston ring compressor, a pin-type spanner wrench and a clutch holder tool (which is described in Section 19). Some new, clean engine oil of the correct grade and type, some engine assembly lube (or moly-based grease), a tube of Kawasaki Bond liquid gasket (part no. 56019-120) or equivalent, and a tube of RTV (silicone) sealant will also be required. Although it may not be considered a tool, some Plastigauge (type HPG-1) should also be obtained to use for checking bearing oil clearances **(see illustrations)**.

3 An engine support stand made from short lengths of 2 x 4's bolted together will facilitate the disassembly and reassembly procedures **(see illustration)**. The perimeter of the mount should be just big enough to accommodate the engine oil pan. If you have an automotive-type engine stand, an adapter plate can be made from a piece of steel plate, some angle iron and some nuts and bolts.

4 When disassembling the engine, keep 'mated' parts together (including gears, cylinders, pistons, etc. that have been in contact with each other during engine operation). These 'mated' parts must be reused or replaced as an assembly.

5 Engine/transmission disassembly should be done in the following general order with reference to the appropriate Sections.

Remove the valve cover
Remove the cylinder head
Remove the cylinder block
Remove the pistons
Remove the water pump (see Chapter 3)
Remove the alternator (see Chapter 9)
Remove the starter motor (see Chapter 9)
Remove the clutch
Remove the oil pan
Remove the external shift mechanism
Remove the ignition pickup/rotor (see Chapter 5)
Separate the crankcase halves
Remove the idler shaft and starter motor clutch
Remove the crankshaft and connecting rods
Remove the transmission shafts/gears
Remove the shift drum/forks

6 Reassembly is accomplished by reversing the general disassembly sequence.

## 7 Valve cover – removal and installation

**Note:** The valve cover can be removed with the engine in the frame. If the engine has been removed, ignore the steps which don't apply.

### Removal

1 Support the motorcycle securely on its stand.

2 Remove the fuel tank (see Chapter 4).

3 Unbolt the ignition HT coils from the frame (see Chapter 5).

4 Remove the air suction valve and the vacuum switching valve (see Chapter 1). Remove the single bolt from the air baffle plate at the front of the valve cover and pull off the spark plug caps.

5 Remove the valve cover bolts **(see illustration)**.

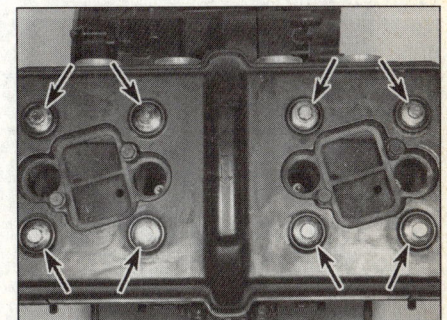

7.5 Remove the bolts (arrows) and lift the valve cover off

7.7 The cam chain upper guide is mounted inside the valve cover

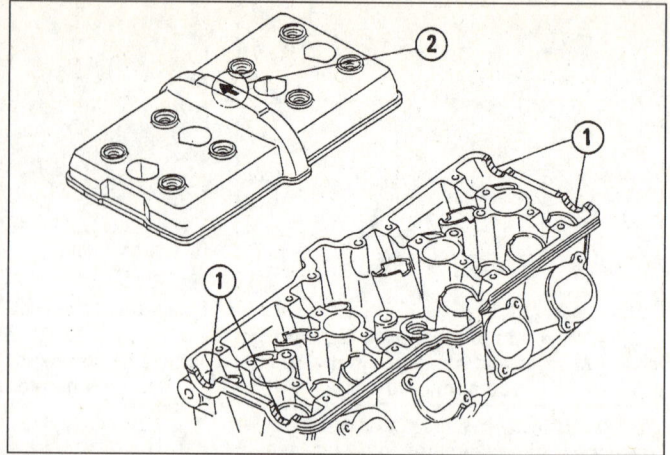

7.8 Apply a thin film of sealant to the half-circle cutouts (1); the arrow mark (2) must face forward

6 Lift the cover off the cylinder head. If it's stuck, don't attempt to pry it off – tap around the sides of it with a plastic hammer to dislodge it. Maneuver the cover out from under the throttle cables, if necessary disconnecting the cables from the carburetor pulley (see Chapter 4).

### Installation

7 Check the chain guide in the center of the cover – if it's excessively worn, pry it out and install a new one (see illustration). Peel the rubber gasket from the cover. If it's cracked, hardened, has soft spots or shows signs of general deterioration, replace it wit new one.
8 Clean the mating surfaces of the cylinder head and the valve cover with lacquer thinner, acetone or brake system cleaner. Apply a thin film of RTV sealant to the half-circle cutouts on each side of the head (see illustration).
9 Install the gasket to the cover. Position the cover on the cylinder head, making sure the gasket doesn't slip out of place. **Note:** *The arrow mark cast in the top of the valve cover must face forward* (see illustration 7.8).

10 Check the rubber seals on the valve cover bolts, replacing them if necessary. Install the bolts, tightening them evenly to the torque list in this Chapter's Specifications.
11 The remainder of installation is the reverse of removal.

### 8 Camshaft chain tensioner – removal and installation

### Removal

*Caution: Once you start to remove the tensioner bolts, you must remove the tensioner all the way and reset it before tightening the bolts. The tensioner extends and locks in place, so if you loosen the bolts partway and then retighten them, the tensioner or cam chain will be damaged.*
1 Loosen the tensioner cap bolt while the tensioner is still installed (see illustration).
2 Remove the tensioner mounting bolts and take it off the engine (see illustration).

3 Remove the tensioner cap bolt and O-ring (see illustration).

### Installation

4 Check the O-ring on the tensioner body for cracks or hardening. It's a good idea to replace this O-ring whenever the tensioner is removed.

### Original tensioner

5 Place the tensioner mounting bolts where you can reach them with one hand while the other hand holds the tensioner in position in Step 7.
6 Press the end of the rod that contacts the chain into the tensioner body. At the same time, turn the other end of the rod clockwise with a screwdriver (or a fabricated holder tool – see Step 9) until the rod stops.
*Caution: Don't turn the rod counter-clockwise (anti-clockwise) or it may separate from the tensioner. If this happens it can't be reassembled.*
7 Place the tensioner in position on the engine. Push it firmly against the engine, remove the screwdriver, and install the mounting bolts finger-tight.

8.1 Loosen the tensioner cap bolt (center arrow), then remove the tensioner mounting bolts (outer arrows) . . .

8.2 . . . and take the tensioner off

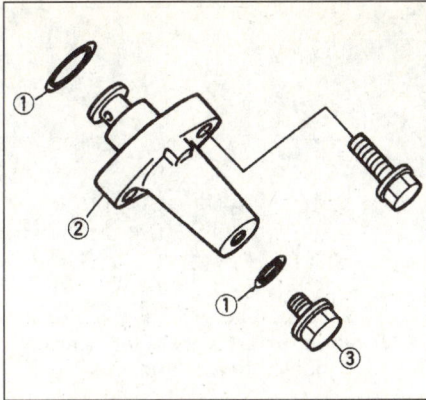

**8.3 Cam chain tensioner details**

1 O-rings
2 Tensioner body
3 Tensioner cap bolt

*Caution: If the tensioner moves away from the engine before you tighten the bolts, the rod will extend too far. If this happens (or you think it might have happened), remove the tensioner and repeat Step 6, then continue with Step 7.*

**8** Tighten the mounting bolts to the torque listed in this Chapter's Specifications.

### New tensioner

**9** New tensioners come with a keeper that fits in the tensioner rod slot and holds the rod in the correct position for installation **(see illustrations)**.

**10** Place the tensioner on the engine. Install the mounting bolts and tighten them to the torque listed in this Chapter's Specifications.

**11** Pull the keeper out with needle nosed pliers. **Note:** *Save the keeper and place it in your toolbox for future use. You can use it to hold the tensioner rod in position next time you install the tensioner, leaving both hands free.*

### Original or new tensioner

**12** Install the tensioner cap and O-ring. Tighten the cap to the torque listed in this Chapter's Specifications.

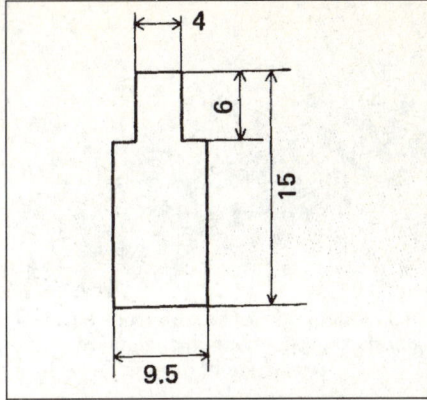

**8.9a  Tensioner keeper dimensions (millimeters)**

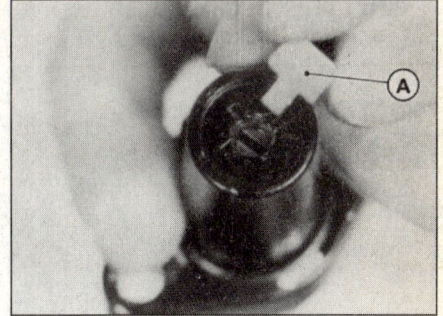

## 9  Camshafts and lifters – removal, inspection and installation

**Note:** *This procedure can be performed with the engine in the frame.*

### *Camshafts*

#### Removal

**1** Remove the fairing lower covers (see Chapter 8).
**2** Remove the valve cover (see Section 7).
**3** Remove the pickup coil cover from the right side of the engine (see Chapter 5). Using a wrench on the large engine turning hex on the end of the crankshaft, rotate the engine clockwise until the T 1.4 mark on the timing rotor aligns with the static index mark on the casing **(see illustration)**.
**4** Remove the cam chain tensioner (see Section 8).
**5** Unscrew the bearing cap bolts evenly, a little at a time, until they are all loose, then lift off the bearing caps complete with the connecting oil pipes.
*Caution: If the bearing cap bolts aren't loosened evenly, the camshaft may bind.*

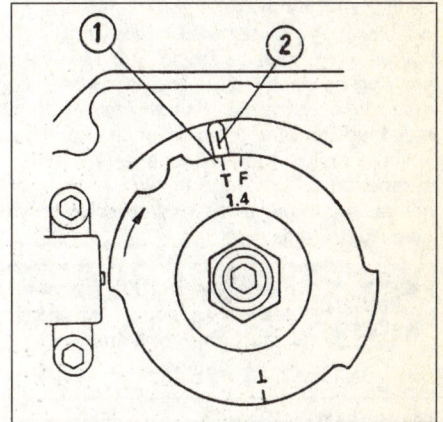

**8.9b  Install the keeper (A) in the tensioner rod slot to hold it in the correct position**

**9.3  Rotate engine until T1.4 mark (1) aligns exactly with static index mark (2)**

Note the numbers on the bearing caps which indicate location on the cylinder head **(see illustration 9.21)**. When you reinstall the caps, be sure to install them in the correct positions.
**6** Pull up on the camshaft chain and carefully guide the camshafts out **(see illustration)**. Look for marks on the camshafts. The intake camshaft should have an IN mark and the exhaust camshaft should have an EX mark **(see illustration)**. If you can't find these marks, label the camshafts to ensure they are installed in their original locations.

**9.6a  Pull up on the cam chain, disengage it from the sprocket and guide the camshaft out**

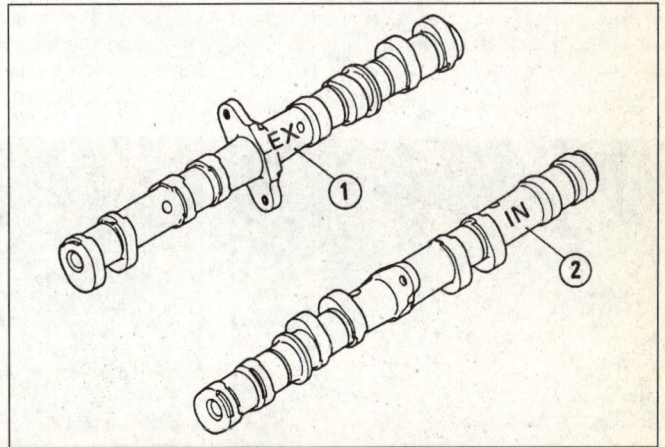

**9.6b  EX mark denotes exhaust camshaft (1) and IN denotes intake camshaft (2)**

**9.7 Wire the cam chain up so it doesn't slip down off the crankshaft sprocket**

**9.9a Check the lobes for wear - here's a good example of damage which requires replacement or repair**

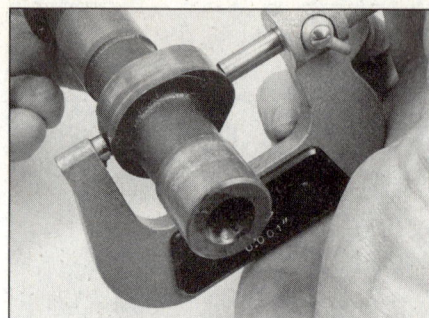

**9.9b Measure the height of the camshaft lobes with a micrometer**

**Note:** *Don't remove the sprockets from the camshafts unless absolutely necessary.*

7 While the camshafts are out, don't allow the chain to go slack – if you do, it will become detached from the gear on the crankshaft and may bind between the crankshaft and case, which could cause damage to these components. Wire the chain to another component to prevent it from dropping down **(see illustration)**.

> **HAYNES HINT** *Cover the top of the cylinder head with a rag to prevent foreign objects from falling into the engine.*

### Inspection

**Note:** *Before replacing camshafts or the cylinder head and bearing caps because of damage, check with local machine shops specializing in motorcycle engine work. In the case of the camshafts, it may be possible for cam !obes to be welded, reground and hardened, at a cost far lower than that of a new camshaft. If the bearing surfaces in the cylinder head are damaged, it may be possible for them to be bored out to accept bearing inserts. Due to the cost of a new cylinder head it is recommended that all options be explored before condemning it as trash!*

8 Inspect the cam bearing surfaces of the head and the bearing caps. Look for score marks, deep scratches and evidence of spalling (a pitted appearance).
9 Check the camshaft lobes for heat discoloration (blue appearance), score marks,

chipped areas, flat spots and spalling **(see illustration)**. Measure the height of each lobe with a micrometer **(see illustration)** and compare the results to the minimum lobe height listed in this Chapter's Specifications. If damage is noted or wear is excessive, the camshaft must be replaced. Also, be sure to check the condition of the followers as described later in this Section.
10 Next, check the camshaft bearing oil clearances. Clean the camshafts, the bearing surfaces in the cylinder head and bearing caps with a clean lint-free cloth, then lay the cams in place in the cylinder head. Engage the cam chain with the sprockets, so the camshafts don't turn as the bearing caps are tightened.
11 Cut eight strips of Plastigauge (type HPG-1) and lay one piece on each bearing journal, parallel with the camshaft centerline **(see illustration)**. Install the bearing caps in their proper positions **(see illustration 9.21)** and install the bolts. Tighten the bolts evenly in a criss-cross pattern until the specified torque is reached. While doing this, don't let the camshafts rotate.
12 Now unscrew the bolts a little at a time, and carefully lift off the bearing caps.
13 To determine the oil clearance, compare the crushed Plastigauge (at its widest point) on each journal to the scale printed on the Plastigauge container **(see illustration)**. Compare the results to this Chapter's Specifications, noting that the figures differ depending on the cylinder number. If the oil clearance is greater than specified, measure

the diameter of the cam bearing journal with a micrometer **(see illustration)**. If the journal diameter is less than the specified limit, replace the camshaft with a new one, and recheck the clearance. If the clearance is still too great, replace the cylinder head and bearing caps with new parts.
14 Except in cases of oil starvation, the cam chain wears very little. If the chain has stretched excessively, which makes it difficult to maintain proper tension, replace it with a new one (see Section 28 for chain stretch measurement and replacement).
15 Check the sprockets for wear, cracks and other damage, replacing them if necessary. If the sprockets are worn, the cam chain is also worn, and also the sprocket on the crankshaft. If wear this severe is apparent, the entire engine should be disassembled for inspection.
16 Unbolt the sprockets from the camshafts if renewal is required. Install the new sprocket so that its marked side faces outwards, to the right of the engine **(see illustration)**. Apply thread locking compound to the sprocket bolts and tighten them to the specified torque.
17 The front (exhaust side) cam chain guide can be removed once the cylinder head has been removed, and the rear (intake side) guide can be removed once the cylinder block has been removed. Refer to Section 28 for inspection details.

### Installation

18 Make sure the bearing surfaces in the cylinder head and the bearing caps are clean, then apply engine oil to each of them.

**9.11 Lay a strip of Plastigauge on each bearing journal, parallel with the camshaft centerline**

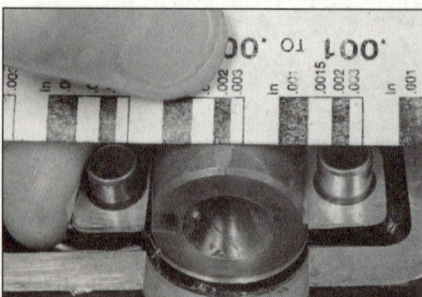

**9.13a Compare the width of the crushed Plastigauge to the scale printed on the Plastigauge container**

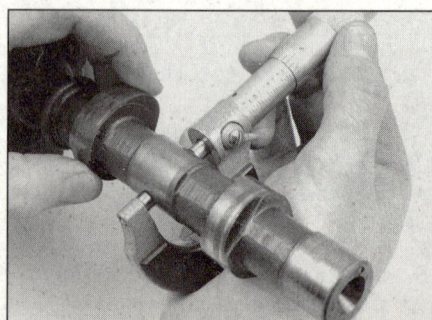

**9.13b Measure the cam bearing journal with a micrometer**

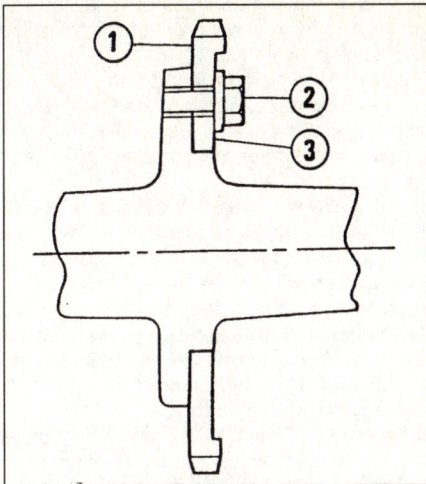

**9.16  Correct installation of camshaft sprockets**

1  Sprocket        3  Marked face
2  Bolt

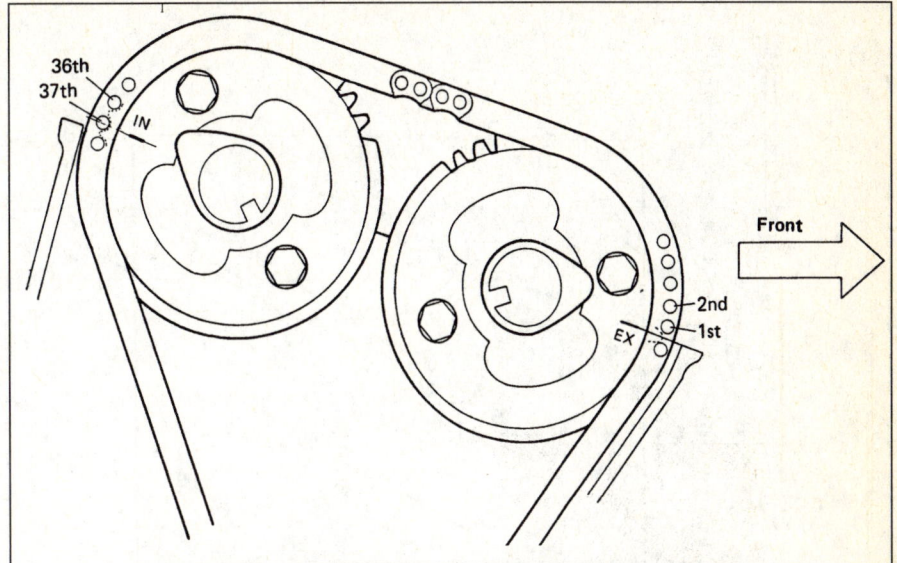

**9.19  Camshaft sprocket (valve timing) alignment marks**

**19**  Apply engine oil (or a coat of moly-based grease if new camshafts are being fitted) to the lobes. Make sure the camshaft bearing journals are clean, then lay the exhaust camshaft, followed by the intake camshaft, in the cylinder head (do not mix them up – **see illustration 9.6b**). Check that the T 1.4 mark is still aligned (see Step 3) and align the marks on the cam sprockets exactly with the cylinder head surface **(see illustration)**.

**20**  Make sure that the timing marks are aligned as described in Step 19, then mesh the chain with the camshaft sprockets. Count the number of chain link pins between the EX mark and the IN mark **(see illustration 9.19)**. There should be no slack in the chain between the two sprockets.

**21**  Before installing the bearing caps, check that the O-rings are in position where the rear caps join the cylinder head oilways. Also install the dowels in their locations. Carefully set the bearing caps in place in their proper positions **(see illustration)** and install the bolts. Snug all of the bolts evenly, then tighten them in a criss-cross pattern to the torque listed in this Chapter's Specifications.

**22**  Insert your finger or a wood dowel into the cam chain tensioner hole and apply pressure to the cam chain. Check the timing marks to make sure they are aligned (see Step 19) and there are still the correct number of link pins between the EX and IN marks on the cam sprockets. If necessary, change the position of the sprocket(s) on the chain to bring all of the marks into alignment.
*Caution: If the marks are not aligned exactly as described, the valve timing will be incorrect and the valves may contact the pistons, causing extensive damage to the engine.*

**23**  Install the cam chain tensioner (see Section 8).

**24**  Check the valve clearances (see Chapter 1) and install the pickup coil cover (see Chapter 5).

**25**  Install the valve cover (see Section 7).

**26**  Install the fairing lower panels (see Chapter 8).

## Lifters

### Removal

**27**  Remove the camshafts.

**28**  Obtain a container which is divided into eight compartments and label each compartment with the number of its corresponding valve in the cylinder head.

**29**  Using a magnet if necessary, lift each follower out of the cylinder head and store it in its corresponding compartment in the container. Note that the shim is likely to stick to the inside of the follower so take great care not to lose it as the follower is removed. Remove the shims and store each one with its respective follower.

### Inspection

**30**  No figures are available to determine wear of the lifters or bore. Inspect the lifters for signs of excessive wear or scoring, and if necessary replace them. Wear of their bores will necessitate cylinder head replacement.

### Installation

**31**  Fit each shim to the top of its correct valve making sure it is correctly seated in the valve spring retainer. **Note:** *It is essential that the shims are returned to their original valves, otherwise the valve clearances will be inaccurate.*

**32**  Lubricate the surface of the followers with engine oil, then install them in their respective positions in the cylinder head, making sure each one squarely enters its bore.

**33**  Install the camshafts and check the valve clearances.

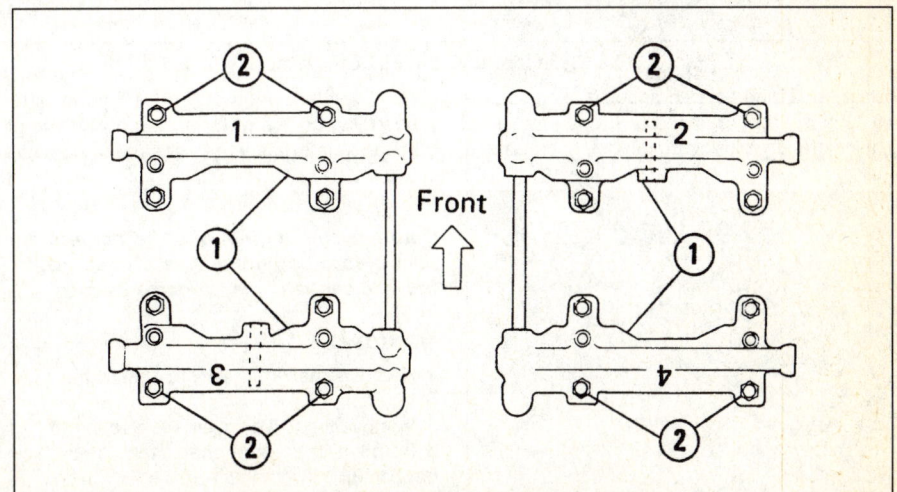

**9.21  Camshaft bearing cap identification**

1  Bearing caps                2  Position of longer bolts

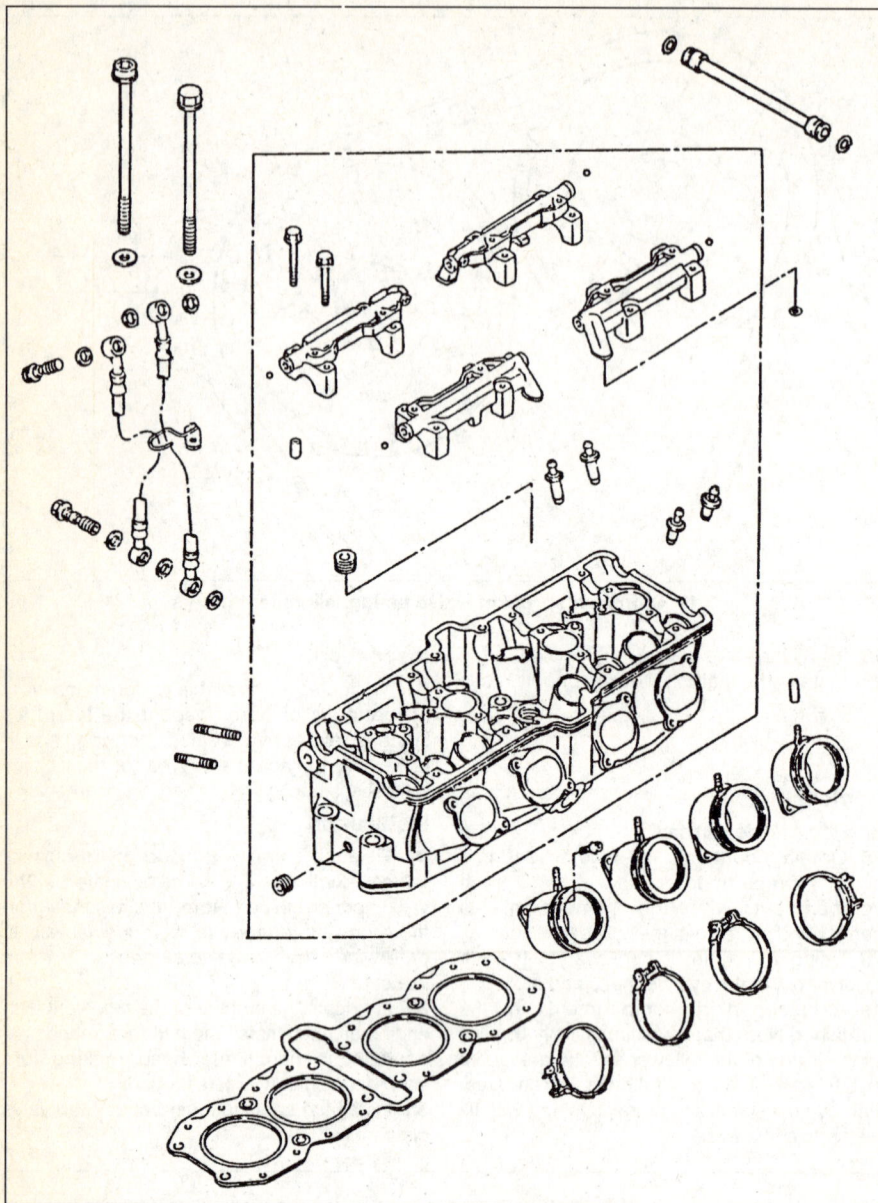

**10.5  Cylinder head and oil hose detail**

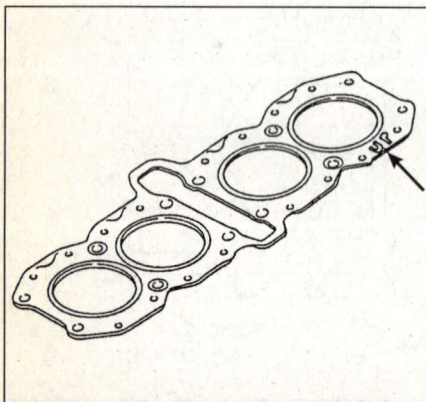

**10.13  Position cylinder head gasket UP mark on the right side of the engine**

6 Slacken the cylinder head bolts, a little at a time, using the **reverse** order of the tightening sequence in illustration 10.16.

7 Lift out the cylinder head bolts and their washers. **Note:** *The four hex head bolts are 10 mm longer than the Allen head bolts; these are installed in the outer holes (holes 9, 10, 11 and 12 in illustration 10.16).*

8 Pull the cylinder head off the cylinder block. If the head is stuck, tap around the side of the head with a rubber mallet to jar it loose, or use two wooden dowels inserted into the intake or exhaust ports to lever the head off. *Caution: Don't attempt to pry the head off by inserting a screwdriver between the head and the cylinder block – you'll damage the sealing surfaces.*

9 Stuff a clean rag into the cam chain tunnel to prevent the entry of debris. Remove all of the head bolt washers from their seats, using a pair of needle-nose pliers.

10 Remove the two dowel pins from the cylinder block. Peel off the old cylinder head gasket and discard it; a new one must be fitted on installation.

11 Clean all traces of old gasket material from the cylinder head and block. Be careful not to let any of the gasket material fall into the crankcase, the cylinder bores or the water passages.

12 Check the cylinder head gasket and the mating surfaces on the cylinder head and block for leakage, which could indicate warpage. Refer to Section 12 and check the flatness of the cylinder head.

## Installation

13 Install the two dowel pins into their locations in the cylinder block **(see illustration 13.5b)**, then lay the new gasket in place on the cylinder block. Make sure the UP mark on the gasket is uppermost and positioned on the right-hand side of the engine **(see illustration)**. Never reuse the old gasket and don't use any type of gasket sealant.

14 Carefully lower the cylinder head onto the cylinder block. It's helpful to have an assistant support the camshaft chain with a piece of wire so it doesn't fall and become kinked or detached from the crankshaft. When the head is resting against the cylinder block, wire the cam chain to another component to keep tension on it.

15 Lubricate both sides of the head bolt washers with engine oil and place them on the bolts.

16 Install the head bolts, making sure the four longer (hex head) bolts are installed in the outer holes. Using the proper sequence **(see illustration)**, tighten the bolts to the initial the torque listed in this Chapter's Specifications.

17 Using the same sequence, tighten the bolts to the torque listed this Chapter's Specifications. **Note:** *The final torque is dependent on whether you're working on new or used bolts, washers and cylinder head.*

## 10 Cylinder head – removal and installation

*Caution: The engine must be completely cool before beginning this procedure, or the cylinder head may become warped.*

### Removal

1 Remove the engine from the frame (see Section 5).
2 Remove the valve cover (see Section 7).
3 Remove the cam chain tensioner (see Section 8).
4 Remove the camshafts (see Section 9).
5 Detach the oil hoses from the front of the cylinder head **(see illustration)**.

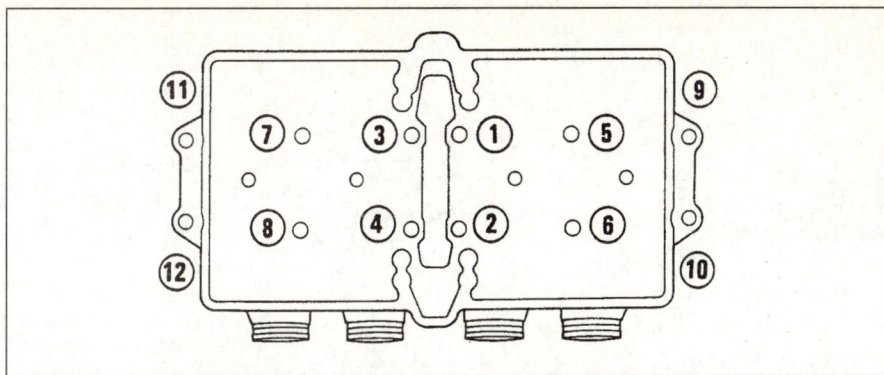

**10.16 Cylinder head bolt TIGHTENING sequence**

**18** The remainder of installation is the reverse of the removal steps. When installing the oil hoses to the cylinder head, use new sealing washers on each side of the banjo hose union and tighten the union bolts to the specified torque (see this Chapter's Specifications).

---

## 11 Valves/valve seats/ valve guides – servicing

**1** Because of the complex nature of this job and the special tools and equipment required, servicing of the valves, the valve seats or the valve guides (commonly known as a valve job) is best left to a professional.

**2** The home mechanic can, however, remove and disassemble the head, do the initial cleaning and inspection, then reassemble and deliver the head to a dealer service department or properly equipped motorcycle repair shop for the actual valve servicing.

**3** The dealer service department will remove the valves and springs, recondition or replace the valves and valve seats, replace the valve guides, check and replace the valve springs, spring retainers and keepers/collets (as necessary), replace the valve seals with new ones and assemble the valve components.

**4** After the valve job has been performed, the head will be in like new condition. When the head is returned, be sure to clean it again very thoroughly before installation on the engine to remove any metal particles or abrasive grit that may still be present from the valve service operations. Use compressed air, if available, to blow out all the holes and passages.

---

## 12 Cylinder head and valves – disassembly, inspection and reassembly

**1** As mentioned in the previous Section, valve servicing and valve guide replacement should be left to a dealer service department or motorcycle repair shop. However, disassembly, cleaning and inspection of the valves and related components can be done (if the necessary special tools are available) by the home mechanic. This way no expense is incurred if the inspection reveals that service work is not required.

**2** To properly disassemble the valve components without the risk of damaging them, a valve spring compressor is absolutely necessary. This special tool can usually be rented, but if it's not available, have a dealer service department or motorcycle repair shop handle the entire process of disassembly, inspection, service or repair (if required) and reassembly of the valves.

### Disassembly

**3** Remove the lifters and their shims if you haven't already done so (see Section 9). Store the components in such a way that they can be returned to their original locations without getting mixed up.

**4** Before the valves are removed, scrape away any traces of gasket material from the head gasket sealing surface. Work slowly and do not nick or gouge the soft aluminum of the head. Gasket removing solvents, which work very well, are available at most motorcycle shops and auto parts stores.

**5** Carefully scrape all carbon deposits out of the combustion chamber area. A hand held wire brush or a piece of fine emery cloth can be used once the majority of deposits have been scraped away.

*Caution: Do not use a wire brush mounted in a drill motor, or one with extremely stiff bristles, as the head material is soft and may be eroded away or scratched by the wire brush.*

**6** Before proceeding, arrange to label and store the valves along with their related components so they can be kept separate and reinstalled in the same valve guides they are removed from (labeled plastic bags work well for this).

**7** Compress the valve spring on the first valve with a spring compressor, then remove the keepers (collets) and the retainer from the valve assembly **(see illustrations)**. *Note: Take great care not to mark the cylinder head lifter bore with the spring compressor. Do not compress the springs any more than is absolutely necessary. Carefully release the valve spring compressor and remove the springs and the valve from the head. If the valve binds in the guide (won't pull through), push it back into the head and deburr the area around the keeper (collet) groove with a very fine file or whetstone* **(see illustration).**

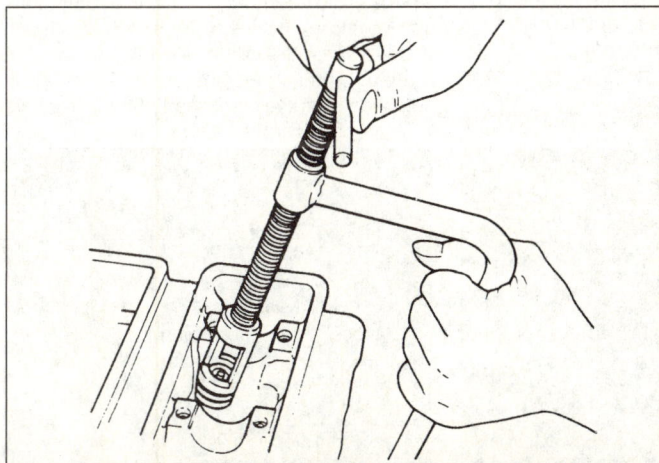

**12.7a Compressing the valve springs with a valve spring compressor**

**12.7b Remove the keepers (collets) with needle-nose pliers**

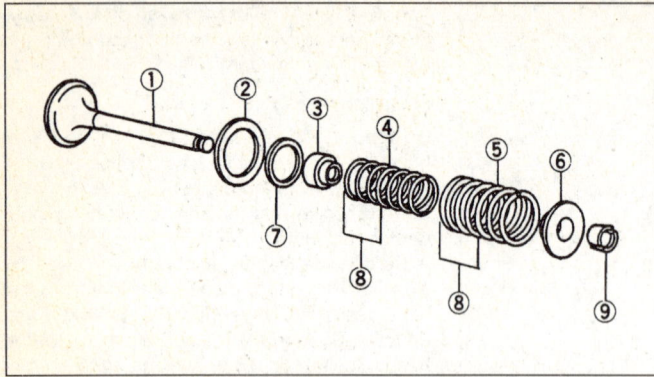

**12.7c Valve components - exploded view**

| | | | | | |
|---|---|---|---|---|---|
| 1 | Valve | 4 | Inner spring | 7 | Inner spring seat |
| 2 | Outer spring seat | 5 | Outer spring | 8 | Tightly wound coils |
| 3 | Oil seal | 6 | Valve spring retainer | 9 | Keepers (collets) |

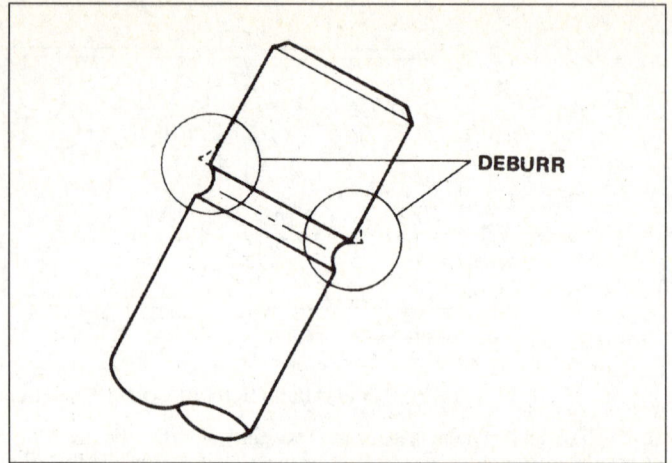

**12.7d If the valve binds in the guide, deburr the area above the keeper (collet) groove**

8 Repeat the procedure for the remaining valves. Remember to keep the parts for each valve together so they can be reinstalled in the same location.

9 Once the valves have been removed and labeled, pull off the valve stem seals with pliers and discard them (the old seals should never be reused), then remove the spring seats.

10 Next, clean the cylinder head with solvent and dry it thoroughly. Compressed air will speed the drying process and ensure that all holes and recessed areas are clean.

11 Clean all of the valve springs, keepers (collets), retainers and spring seats with solvent and dry them thoroughly. Do the parts from one valve at a time so that no mixing of parts between valves occurs.

12 Scrape off any deposits that may have formed on the valve, then use a motorized wire brush to remove deposits from the valve heads and stems. Again, make sure the valves do not get mixed up.

### Inspection and reassembly

13 These procedures are the same as for 749 cc engines. Refer to Part B of this Chapter for procedures, and Part A for Specifications.

14 After assembly, install the shims and the lifters (see Step 9).

### 13 Cylinder block – removal, inspection and installation

#### Removal

1 Remove the cylinder head (see Section 10). Make sure the crankshaft is positioned at Top Dead Center (TDC) for cylinders 1 and 4.

2 Remove the water pipe from the rear of the cylinder block (see illustrations).

3 Lift out the cam chain front guide (see illustration).

4 Lift the cylinder block straight up off the pistons to remove it. If it's stuck, tap around its perimeter with a soft-faced hammer. Don't attempt to pry between the block and the crankcase, as you will ruin the sealing surfaces.

5 Remove the dowel pins from the mating surface of the crankcase (see illustrations). Be careful not to let these drop into the engine. Stuff rags around the pistons and remove the gasket and all traces of old gasket material from the surfaces of the cylinder block and the cylinder head.

6 If required the cam chain rear (intake side) guide can be removed from the crankcase surface.

### Inspection

7 This is the same as for 749 cc engines. Refer to Part B of this Chapter for procedures, and Part A for specifications.

### Installation

8 Lubricate the cylinder bores with plenty of clean engine oil. Apply a thin film of moly-based grease to the piston skirts. If the cam chain rear (intake side) guide was removed, install it at this stage (see Section 28).

9 Install the dowel pins, then lower a new cylinder base gasket over the studs, with the

**13.2a Remove the Allen bolt (arrow) from the upper side of the water pipe . . .**

**13.2b . . . and from the lower side (arrow), then pull the pipe fittings out of the block**

**13.3 Lift out the cam chain front guide**

**13.5a Dowel pin for the cylinder block at each front corner of the crankcase (arrow)**

**13.5b  Cylinder block details** ▶

| | |
|---|---|
| 1  Block | 6  Base gasket |
| 2  Liner | 7  Dowel pin |
| 3  O-rings | 8  Piston rings |
| 4  Coolant drain | 9  Piston |
|    plug | 10 Piston pin |
| 5  Dowel pin | 11 Circlip |

UP mark on the right-hand side of the engine. Some gaskets also have an arrow, which must point to the front of the engine

**10** Slowly rotate the crankshaft until all of the pistons are at the same level. Slide lengths of welding rod or pieces of a straightened-out coat hanger under the pistons, on both sides of the connecting rods **(see illustration)**; this will help keep the pistons level as the cylinder block is lowered onto them.

**11** Attach four piston ring compressors to the pistons and compress the piston rings **(see illustration)**. Large hose clamps can be used instead – just make sure they don't scratch the pistons, and don't tighten them too much.

**12** Make sure the intake side cam chain guide is installed **(see illustration 13.10)**.

**13** Position the cylinder block over the engine and carefully lower it down until the piston crowns fit into the cylinder liners. While doing this, pull the camshaft chain up, using a hooked tool or a piece of coat hanger. Push down on the cylinder block, making sure the pistons don't get cocked sideways, until the bottom of the cylinder liners slide down past the piston rings. A wood or plastic hammer handle can be used to gently tap the block down, but don't use too much force or the pistons will be damaged.

**14** Remove the piston ring compressors or hose clamps, being careful not to scratch the pistons. Remove the rods from under the pistons.

**15** Install the cam chain front guide, noting that it must engage its cup in the crankcase **(see illustration 28.12a)**.

**16** The remainder of installation is the reverse of removal.

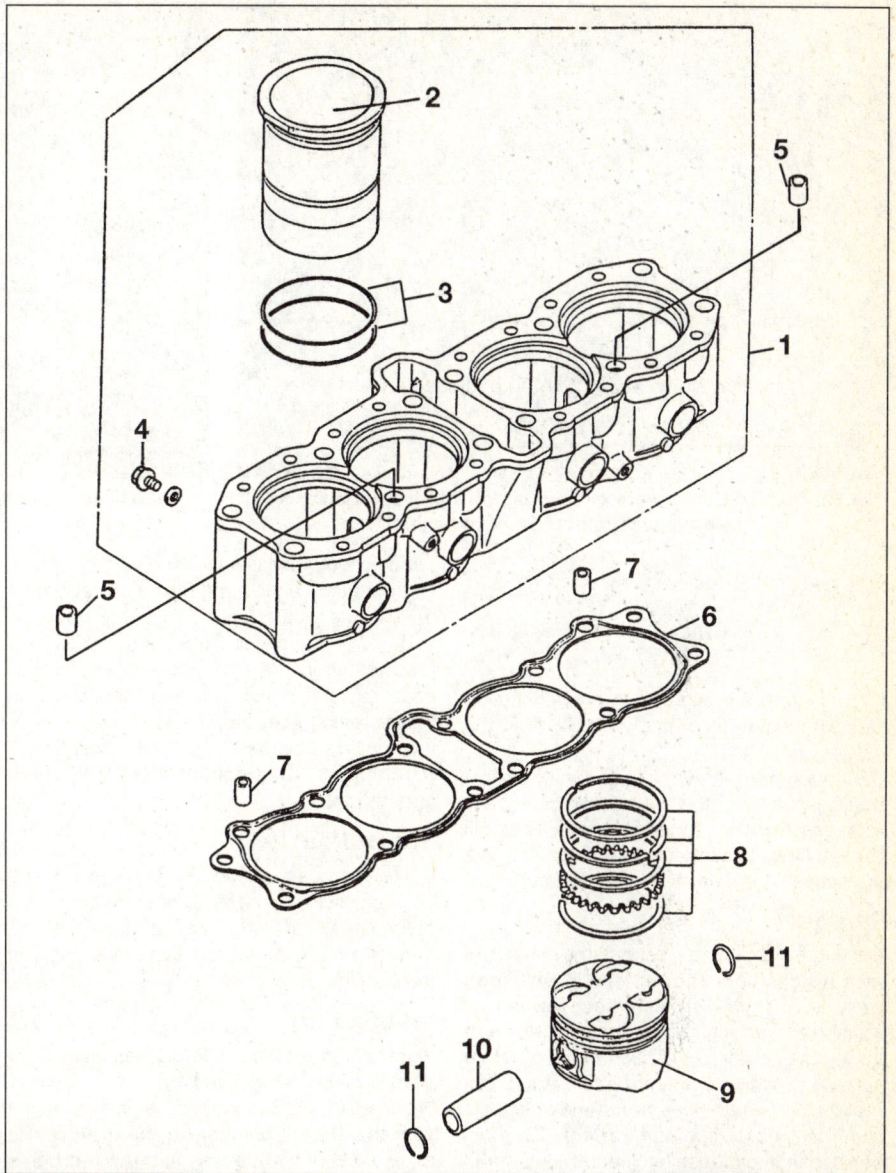

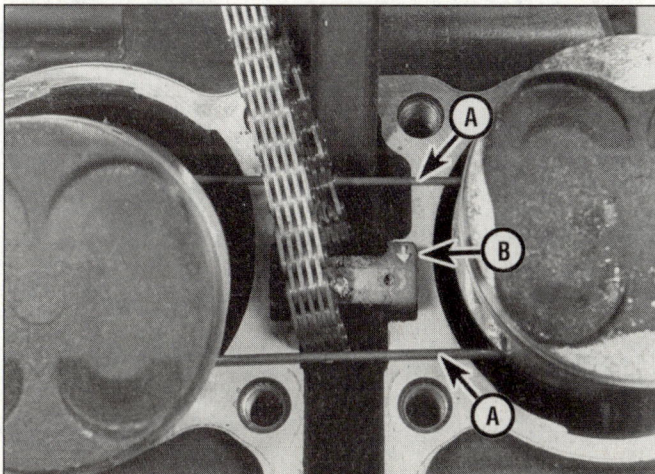

**13.10  Slip a rod (A) beneath the front and rear of the pistons; the rear (intake side) cam chain guide must be in place (B)**

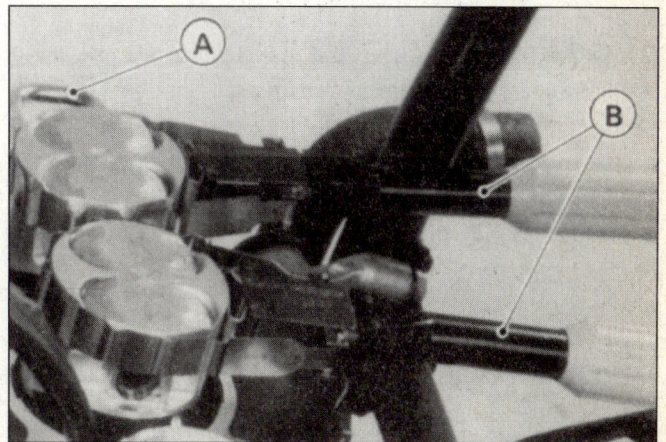

**13.11  You can guide the cylinder block over the pistons with a screwdriver, but it's a good idea to use ring compressors**

*A  Piston support rods*          *B  Ring compressors*

14.3a Scratch the number of each piston into the piston crown (arrow must point to front of engine)

14.3b Using needle-nose pliers remove the piston pin circlips

## 14 Pistons –
removal, inspection and installation

1 The pistons are attached to the connecting rods with piston pins that are a slip fit in the pistons and rods.
2 Before removing the pistons from the rods, stuff a clean shop towel into each crankcase hole, around the connecting rods. This will prevent the circlips from falling into the crankcase if they are inadvertently dropped.

### Removal

3 Using a sharp scribe, scratch the number of each piston into its crown (see illustration). Each piston should also have an arrow pointing toward the front of the engine. If not, scribe an arrow into the piston crown before removal. Support the first piston, grasp the circlip with needle-nose pliers and remove it from the groove (see illustration). If the pin won't come out, use a special piston pin

removal tool (Kawasaki tool no. 57001-910 or a home-made equivalent) (see illustration).
4 Push the piston pin out from the opposite end to free the piston from the rod. You may have to deburr the area around the groove to enable the pin to slide out (use a triangular file for this procedure). Repeat the procedure for the remaining pistons. Use large rubber bands to support the connecting rods (see illustration).

### Inspection

5 These procedures are the same as for 749 cc engines. Refer to Part B of this Chapter for procedures, and Part A for Specifications.
6 Refer to Section 15 and install the rings on the pistons.

### Installation

Note: *When installing the pistons, install the pistons for cylinders 2 and 3 first.*
7 Install the pistons in their original locations with the arrows pointing to the front of the engine. Lubricate the pins and the rod bores

with clean engine oil. Install new circlips in the grooves in the inner sides of the piston (don't reuse the old circlips). Push the pins into position from the opposite side and install new circlips. Compress the circlips only enough for them to fit in the piston. Make sure the clips are properly seated in the groove.

## 15 Piston rings –
installation

1 Before installing the new piston rings, the ring end gaps must checked.
2 Lay out the pistons and the new ring sets so the rings will be matched with the same piston and cylinder during the end gap measurement procedure and engine assembly.
3 Insert the top (No. 1) ring into the bottom of the first cylinder and square it up with the cylinder walls by pushing it in with the top of the piston. The ring should be about one inch above the bottom edge of the cylinder. To measure the end gap, slip a feeler gauge between the ends of the ring (see illustration) and compare the measurement to the Specification.

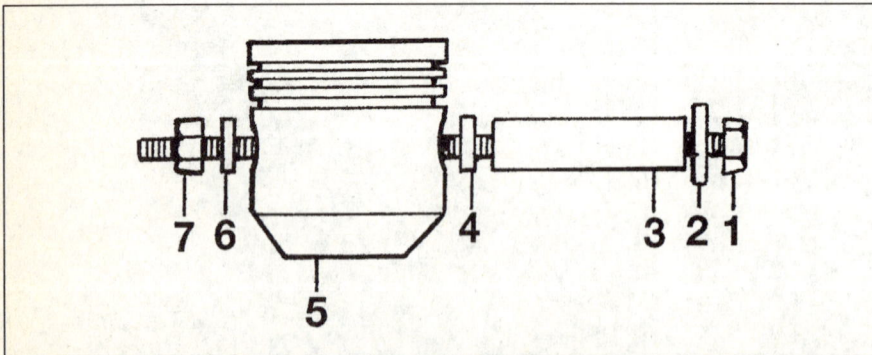

14.3c If the piston pins won't come out, this removal tool can be fabricated from readily available parts

| | | | |
|---|---|---|---|
| 1 Bolt | 3 Pipe (A) | 5 Piston | 7 Nut (B) |
| 2 Washer | 4 Padding (A) | 6 Washer (B) | |

*A Large enough for piston pin to fit inside*
*B Small enough to fit through piston pin bore*

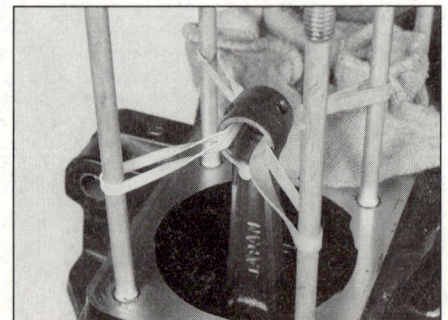

14.4 Use rubber bands to keep the connecting rods from flopping around after the pistons are removed

**15.3 Check the piston ring end gap with a feeler gauge**

**15.5 If the end gap is too small, clamp a file in a vise and file the ring ends to enlarge the gap slightly**

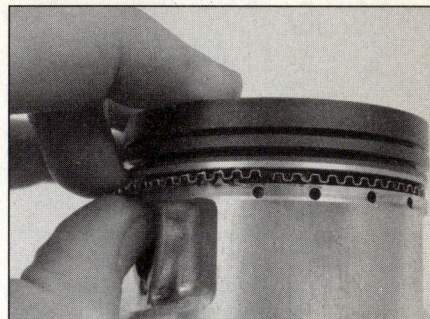

**15.9a Installing the oil ring expander - make sure the ends don't overlap**

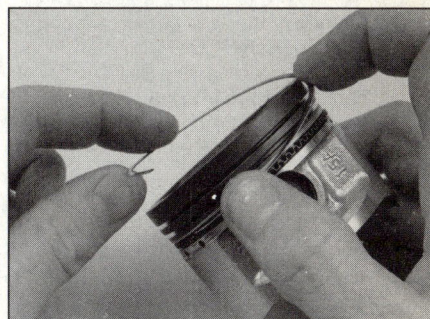

**15.9b Installing an oil ring side rail - don't use a ring installation tool to do this**

**4** If the gap is larger or smaller than specified, double check to make sure that you have the correct rings before proceeding.

**5** If the gap is too small, it must be enlarged or the ring ends may come in contact with each other during engine operation, which can cause serious damage. The end gap can be increased by filing the ring ends very carefully with a fine file **(see illustration)**. When performing this operation, file only from the outside in.

**6** Excess end gap is not critical unless it is greater than 1 mm (0.04 inch). Again, double check to make sure you have the correct rings for your engine.

**7** Repeat the procedure for each ring that will be installed in the first cylinder and for each ring in the remaining cylinder. Remember to keep rings, pistons and cylinders matched up.

**8** Once the ring end gaps have been checked/corrected, the rings be installed on the pistons.

**9** The oil control ring (lowest on the piston) is installed first. It is composed of three separate components. Slip the expander into the groove then install the upper side rail **(see illustrations)**. Do not use a piston ring installation tool on the oil ring side rails as they may be damaged. Instead, place one end of the side rail into the groove between the spacer expander and the ring land. Hold it firmly in place and slide a finger around the piston while pushing the rail into the groove. Next, install the lower side rail in the same manner.

**10** After the three oil ring components have been installed, check to make sure that both the upper and lower side rails can be turned smoothly in the ring groove.

**11** Install the No. 2 (middle) ring next. It can be readily distinguished from the top ring by its cross-section shape **(see illustration)**. Do not mix the top and middle rings.

**12** To avoid breaking the ring, use a piston ring installation tool and make sure that the identification mark (N on H1 models and R on H2 models) is facing up **(see illustration 15.11)**. Fit the ring into the middle groove on the piston. Do not expand the ring any more than is necessary to slide it into place.

**13** Finally, install the No. 1 (top) ring in the same manner. Again make sure the identifying mark is facing up.

**14** Repeat the procedure for the remaining piston and rings. Be very careful not to confuse the No. 1 and No. 2 rings.

**15** Once the rings have been properly installed, stagger the end gaps, including those of the oil ring side rails **(see illustration)**.

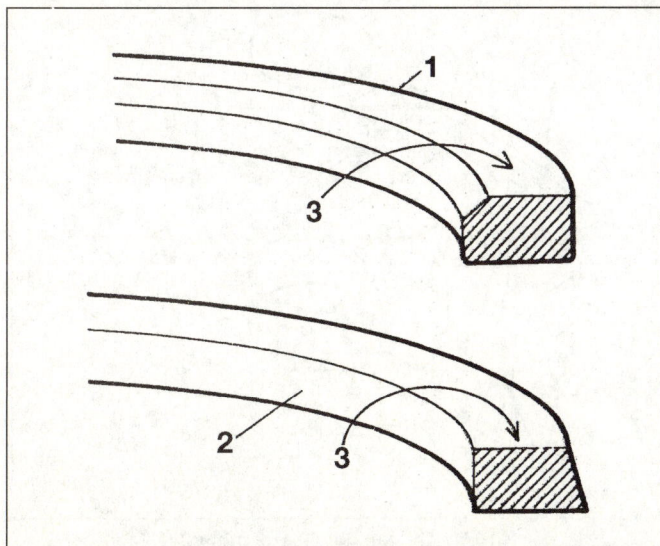

**15.11 Don't confuse the top ring with the middle (second) ring**

*1 Top ring*  *3 Identification marking*
*2 Middle (second) ring*

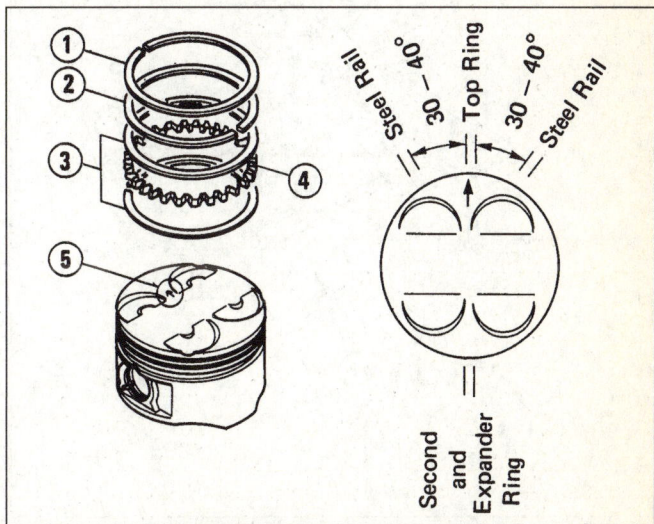

**15.15 Piston ring end gap positions**

*1 Top ring*  *4 Expander*
*2 Middle (second) ring*  *5 Piston*
*3 Oil ring side rails*

## 16 Oil pan –
### removal and installation

**Note:** *The oil pan can be removed with the engine in the frame.*

### Removal

**1** Ensure that the motorcycle is well supported, using an auxiliary stand if necessary.
**2** Remove both fairing lower panels (see Chapter 8).
**3** Drain the engine oil (see Chapter 1). On H2 models, also remove the oil filter.
**4** Remove the oil cooler, disconnecting its hoses from the oil pan (see Chapter 3). On H1 models, unbolt the oil cooler mounting bracket from the frame tubes.
**5** Remove the exhaust system (see Chapter 4).
**6** Pull back the rubber cover from the oil pressure switch and disconnect the wire from the switch **(see illustrations)**.

**7** Ensuring that the motorcycle and the engine are well supported, remove the frame's front downtube assembly (see Section 5).
**8** Remove the oil pan bolts (noting their correct position – they are different lengths on H2 models) and detach the pan from the crankcase. Recover the gasket and seals.

### Installation

**Note:** *Before installing the oil pan it is advised to check the oil strainer for blockage and clean it if necessary* (see Section 17).
**9** Remove all traces of old gasket material from the mating surfaces of the oil pan and crankcase.
**10** Check the small O-rings in the oil passages in the crankcase and on H2 models also check the large O-ring around the oil filter hole (in the pan) for cracking and general deterioration. Replace them if necessary. The O-ring between the oil pan and oil pump holder must be installed with its flat side facing the crankcase.

**11** Position a new gasket on the oil pan. A thin film of RTV sealant can be used to hold the gasket in place. Install the oil pan and bolts, making sure the bolts are in the correct holes. Tighten the bolts to the torque listed in this Chapter's Specifications, using a criss-cross pattern.
**12** The remainder of installation is the reverse of removal, noting the following:
  a) *Tighten the frame downtube bolts to the specified torque.*
  b) *Use new O-rings at the flanged oil hose connections and tighten the bolts to the specified torque. On H1 models, use new sealing washers on each side of the oil line banjo fittings and tighten the banjo bolts to the specified torque.*
  c) *Install a new filter on H2 models. On all models, fill the crankcase with oil (see Chapter 1) and check the level (see Daily (pre-ride) checks), then run the engine and check that there are no leaks.*

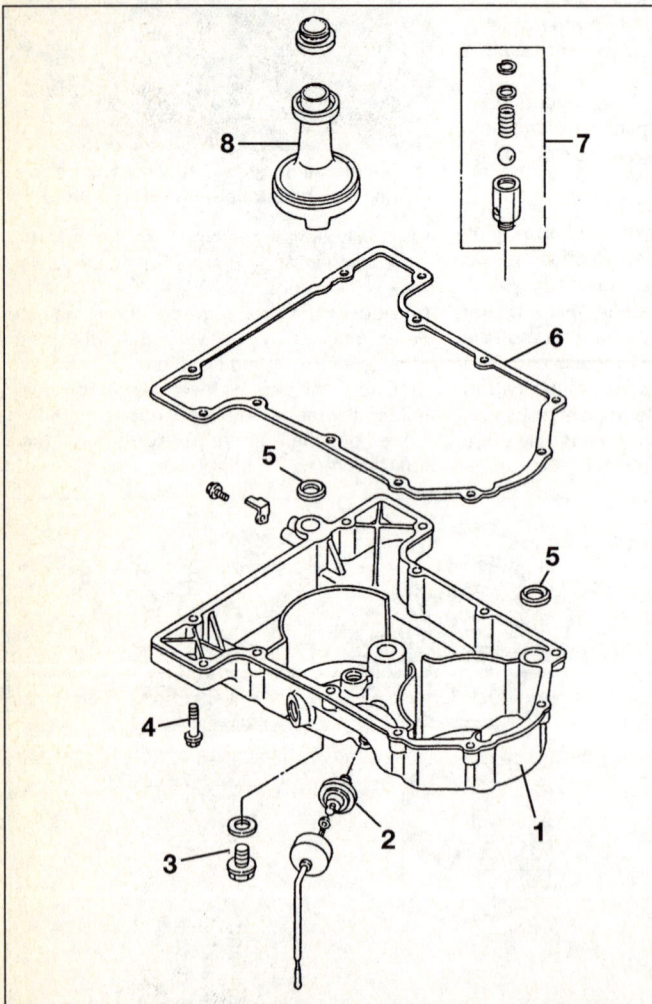

**16.6a Oil pan details (H1 model)**

| 1 | Oil pan | 4 | Bolt | 7 | Pressure relief |
| 2 | Oil pressure switch | 5 | O-ring | | valve |
| 3 | Oil drain bolt | 6 | Gasket | 8 | Oil strainer/pickup |

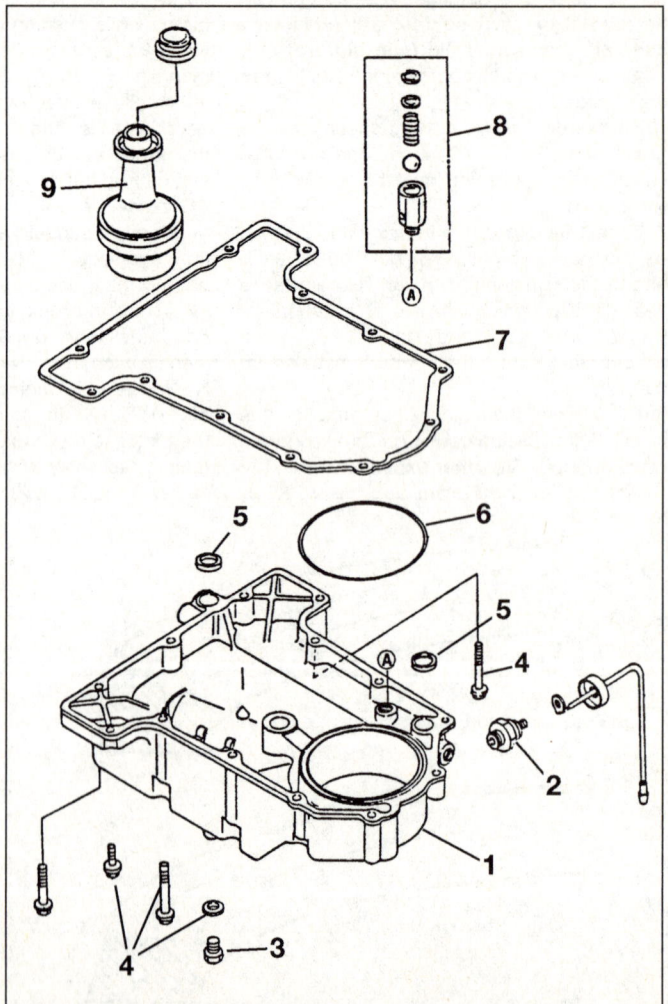

**16.6b Oil pan details (H2 model)**

| 1 | Oil pan | 4 | Bolt | 7 | Gasket |
| 2 | Oil pressure switch | 5 | O-ring | 8 | Pressure relief valve |
| 3 | Oil drain bolt | 6 | Large O-ring | 9 | Oil strainer/pickup |

17.2 Remove the oil gallery plug (arrow) and install a pressure gauge

17.9 Work the pickup O-ring free of the crankcase and take the pickup/strainer out

17.10 Place the oil pump drive tab (arrow) and the water pump drive slot in a vertical position

17.11a Remove the oil pump sprocket bolt (arrow) located behind the clutch housing

17.11b Use a socket (arrow) to remove the chain guide bolt . . .

17.11c . . . lift out the chain guide . . .

### 17 Oil pump –
pressure check, removal, inspection and installation

**Note:** *The oil pump can be removed with the engine in the frame.*

## Pressure check

⚠️ **Warning: If the oil gallery plug is removed when the engine is hot, hot oil will drain out – wait until the engine is cold before beginning this check (it must be cold to perform the relief valve opening pressure check anyway).**

1 Remove the right lower fairing panel (see Chapter 8).
2 Remove the plug at the bottom of the crankcase and install an oil pressure gauge **(see illustration)**.
3 Start the engine and watch the gauge while varying the engine rpm. The pressure should stay within the relief valve opening pressure listed in this Chapter's Specifications. If the pressure is too high, the relief valve is stuck closed. To check it, see Section 18.
4 If the pressure is lower than the standard, either the relief valve is stuck open, the oil pump is faulty, or there is other engine damage. Begin diagnosis by checking the relief valve (see Section 18), then the oil pump. If those items check out okay, chances are the bearing oil clearances are excessive and the engine needs to be overhauled.
5 If the pressure reading is in the desired range, allow the engine to warm up to normal operating temperature and check the pressure again, at the specified engine rpm. Compare your findings with this Chapter's Specifications.
6 If the pressure is significantly lower than specified, check the relief valve, oil pump and main bearings.

## Removal

7 Remove the oil pan (see Section 16).
8 Remove the clutch assembly (see Section 19).
9 Remove the oil pickup screen **(see illustration)**.
10 Turn the crankshaft so the drive slot and tab in the oil pump shaft and water pump shaft are vertical **(see illustration)**.
11 Remove the oil pump sprocket bolt and chain guide bolt, then take out the sprocket **(see illustrations)**.
12 Remove the oil pump mounting bolts, then take the pump and holder out **(see illustration)**.

## Inspection

13 Remove the oil pump cover screws and lift off the cover **(see illustration)**. Thoroughly clean the mating surfaces.

17.11d . . . then disengage the oil pump sprocket from the chain and lift it out

17.12 Loosen the holder-to-oil pump bolt (A) if you plan to separate the pump and holder, then remove the mounting bolts (B)

17.13 Remove the cover screws (arrows); use an impact driver if necessary

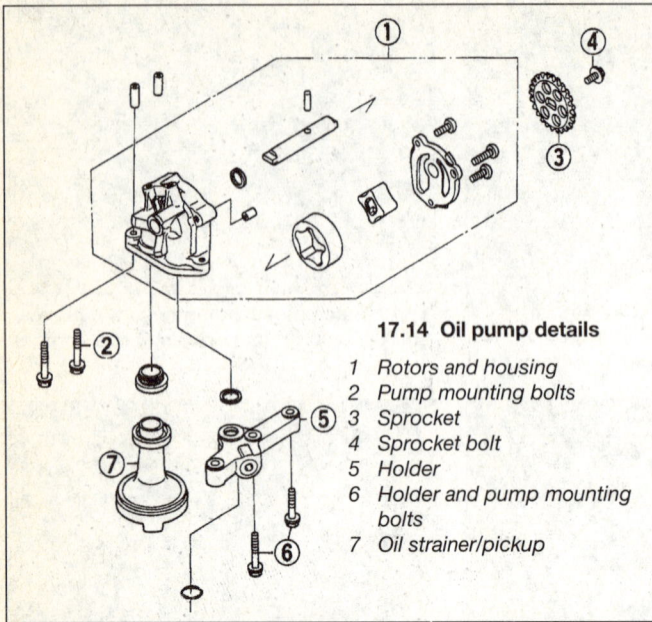

**17.14 Oil pump details**

1 *Rotors and housing*
2 *Pump mounting bolts*
3 *Sprocket*
4 *Sprocket bolt*
5 *Holder*
6 *Holder and pump mounting bolts*
7 *Oil strainer/pickup*

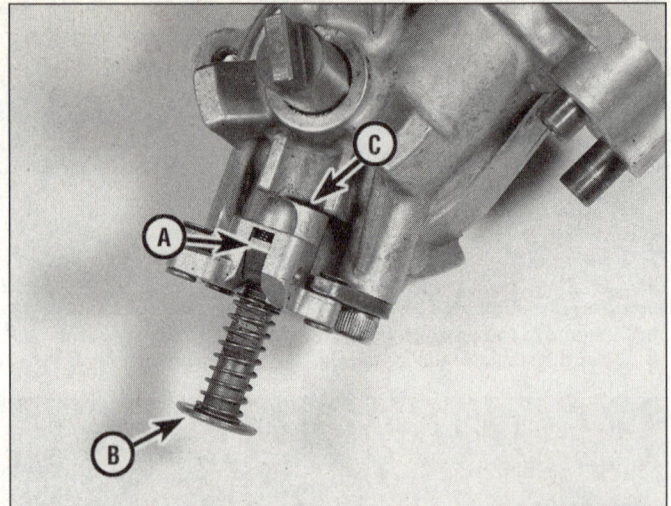

**17.18 Lift the latch (A) clear of the grooves in the tensioner pushrod, then compress the pushrod (B) against the spring - when the small hole appears at point C, slip a thin piece of wire into it to hold the pushrod compressed**

14 Remove the oil pump shaft, pin, inner rotor and outer rotor from the pump **(see illustration)**. Mark the rotors so they can be installed in the same relative positions.

15 Wash all the components in solvent, then dry them off. Check the pump body, the rotors and the cover for scoring and wear. Kawasaki doesn't publish clearance specifications, so if any damage or uneven or excessive wear is evident, replace the pump. If you are rebuilding the engine, it's a good idea to install a new oil pump. Clean the pickup screen with solvent.

16 Reassemble the pump by reversing the removal steps, but before installing the cover, pack the cavities between the rotors with petroleum jelly – this will ensure the pump develops suction quickly and begins oil circulation as soon as the engine is started. Be sure to replace all seals with new ones.

### Installation

17 If you removed the oil pump sprocket, engage it with the chain before installing the oil pump.

18 Hold the oil pump and unlatch the stopper

**17.20 Install the strainer/pickup with its flat side (arrow) toward the oil pump**

from the alternator chain tensioner pushrod **(see illustration)**. Press the alternator chain tensioner pushrod into the oil pump, then slip a piece of thin wire in the pushrod hole to hold the pushrod in the compressed position (the wire will be removed after the oil pump is installed).

19 Position the oil pump on the engine, meshing its shaft with the water pump shaft. Coat the threads of the oil pump bolts with non-permanent thread locking agent, then install the bolts and tighten them to the torque listed in this Chapter's Specifications. Remove the wire from the tensioner to allow it to take up chain slack.

20 Install the oil pickup so the flat side is toward the oil pump holder **(see illustration)**.

21 The remainder of installation is the reverse of the removal steps.

### 18 Oil pressure relief valve – removal, inspection and installation

### Removal

1 Remove the oil pan (see Section 16).

2 Unscrew the relief valve from the oil pan **(see illustration 16.6a or 16.6b)**.

### Inspection

3 Clean the valve with solvent and dry it, using compressed air if available.

4 Using a wood or plastic tool, depress the steel ball inside the valve and see if it moves smoothly. Make sure it returns to its seat completely. If it doesn't, replace it with a new one (don't attempt to disassemble and repair it).

### Installation

5 Apply a non-hardening thread locking

compound to the threads of the valve and install it into the oil pan, tightening it to the torque listed in this Chapter's Specifications.

6 The remainder of installation is the reverse of removal.

### 19 Clutch – removal, inspection and installation

**Note:** *The clutch can be removed with the engine in the frame.*

### Removal

1 Support the motorcycle securely and remove the right lower fairing (see Chapter 8).

2 Drain the engine oil (see Chapter 1).

3 Remove the coolant reservoir (see Chapter 3).

4 Remove the ten clutch cover bolts (noting the location of the pickup coil wiring clamp) and take the cover off **(see illustration)**. If the cover is stuck, tap around its perimeter with a soft hammer. Recover the clutch cover gasket.

**19.4 The clutch cover is retained by ten bolts; note the pickup wire clamp on top right bolt**

**19.5a Remove the six bolts . . .**

**19.5b . . . spring retainers . . .**

**19.5c . . . and springs . . .**

**19.5d . . . to free the clutch pressure plate**

**TOOL TiP**

*Another way to hold the clutch is to drill holes in a friction plate and a steel plate and bolt them together as they would be when installed. Slip the bolted plates into their installed positions; the clutch hub will be locked to the clutch housing. Shift the transmission into a low gear and have an assistant hold the rear brake on. Unscrew the nut and remove it.*

**5** Remove the six clutch spring bolts, slackening them evenly in a diagonal sequence (see illustration). Remove the spring retainers and withdraw the springs (see illustrations). Lift out the clutch pressure plate (see illustration).
**6** Remove the clutch friction and steel plates as a set.
**7** Remove the clutch hub nut, using a special holding tool (Kawasaki tool no. 57001-1243) to prevent the clutch housing from turning (see illustration). An alternative to this tool can be fabricated from some steel strap, bent at the ends and bolted together in the middle (see illustration and Tool Tip).
**8** After the nut has been removed, withdraw the torque limiter spring, collar, splined hub and hub center. Remove the spacer ring and thick thrust washer from the mainshaft. Thread two 8 mm bolts into the needle bearing sleeve and pull it out (see illustration overleaf).
**9** Remove the clutch housing and needle bearing, followed by the thick thrust washer and spacer ring from the mainshaft end.

## Inspection

**Note:** *Kawasaki supplies clutch steel (plain) plates in three different thicknesses (2.0 mm, 2.3 mm and 2.6 mm) to allow for adjustment of pressure plate freeplay. Since measurement of freeplay requires that the clutch be set up off the motorcycle on a spare transmission driveshaft, it is advised that it is carried out by a Kawasaki dealer. If there is insufficient freeplay, excessive engine braking and rear wheel hop will occur, and conversely, too much freeplay will result in a spongy or pulsating feel to the clutch lever.*

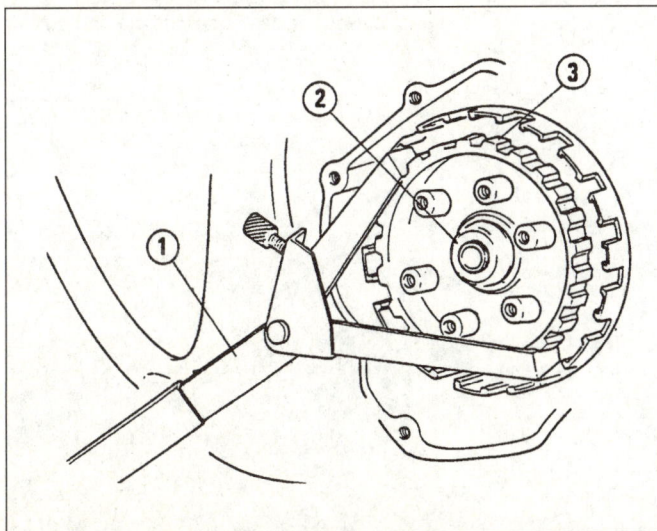

**19.7a Kawasaki tool (part no. 57001-1243) used to hold clutch while the hub nut is removed**

*1 Tool*     *2 Hub nut*     *3 Hub*

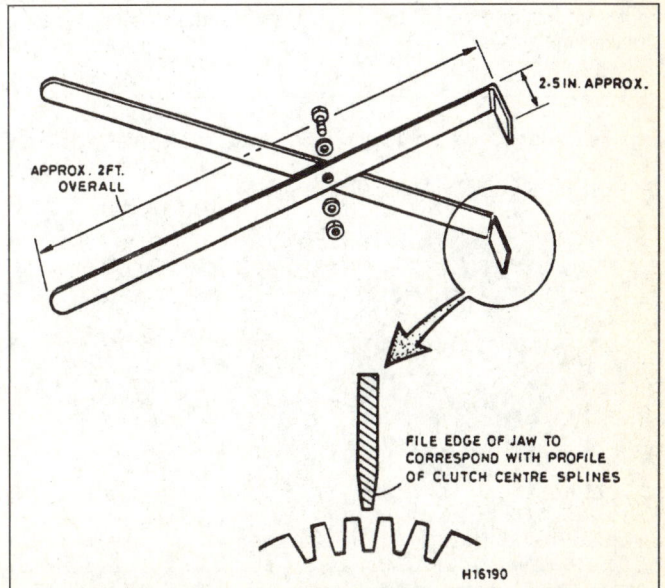

2·5 IN. APPROX.

APPROX. 2 FT. OVERALL

FILE EDGE OF JAW TO CORRESPOND WITH PROFILE OF CLUTCH CENTRE SPLINES

H16190

**19.7b Home-made version of Kawasaki clutch holding tool**

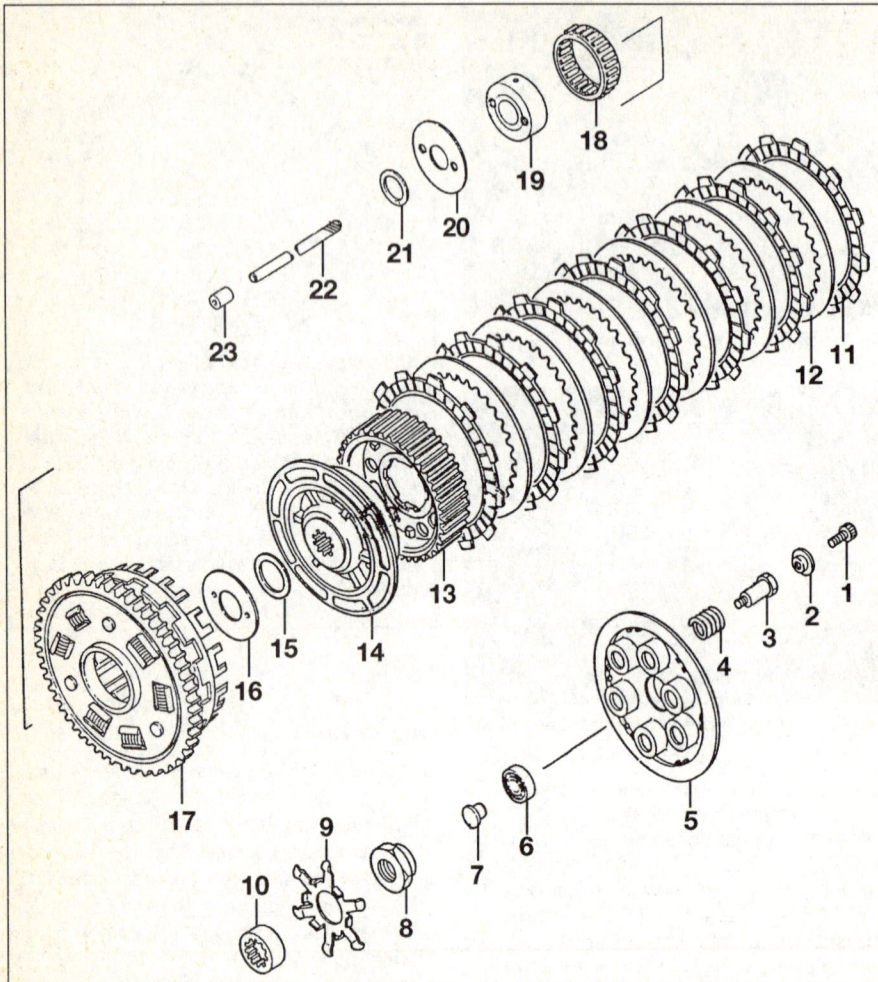

**19.8 Clutch detail**

| | | | |
|---|---|---|---|
| 1 | Bolt | 12 | Steel plate |
| 2 | Spring retainer | 13 | Splined hub |
| 3 | Spring post | 14 | Hub center |
| 4 | Spring | 15 | Spacer ring |
| 5 | Pressure plate | 16 | Thrust washer |
| 6 | Thrust bearing | 17 | Housing |
| 7 | Pushrod end | 18 | Needle roller |
| | piece | | bearing |
| 8 | Hub nut | 19 | Sleeve |
| 9 | Torque limiter | 20 | Thrust washer |
| | spring | 21 | Spacer ring |
| 10 | Collar | 22 | Pushrod |
| 11 | Friction plate | 23 | Pushrod end cap |

smells burnt or if it is glazed, new parts are required. If the steel plates are scored or discolored, they must be replaced with new ones.

**13** Lay all plates, one at a time, on a perfectly flat surface (such as a piece of plate glass) and check for warpage by tying to slip a 0.3 mm (0.0118 inch) feeler blade between the flat surface and the plate **(see illustration)**. Do this at several intervals around the plate's circumference. If the feeler blade can be slipped under the plate, it is warped and should be replaced with a new one.

**14** Check the tabs on the friction plates for excessive wear and mushroomed edges. They can be cleaned up with a file if the deformation is not severe.

**15** Check the edges of the slots in the clutch housing for indentations made by the friction plate tabs. If the indentations are deep they can prevent clutch release, so the housing should be replaced with a new one. If however, the indentations can be easily removed with a file, the life of the housing can be prolonged to an extent. Also, check the teeth of the primary driven gear on the back of the clutch housing for cracks, chips and excessive wear. If the gear is worn or damaged, the clutch housing must be replaced with a new one. Check the

**10** Examine the splines on the outside of the splined hub and the inside of the hub center. If any wear is evident, replace the hub components. Visually inspect the torque limiter spring and damper cam and follower **(see illustration)**.

**11** Measure the free length of the clutch springs and compare the results to the Specifications **(see illustration)**. If the springs have sagged, or if cracks are noted, replace them with new ones as a set.

**12** If the lining material of the friction plates

**19.10 Inspect the splined hub damper cams (A) and hub center followers (B)**

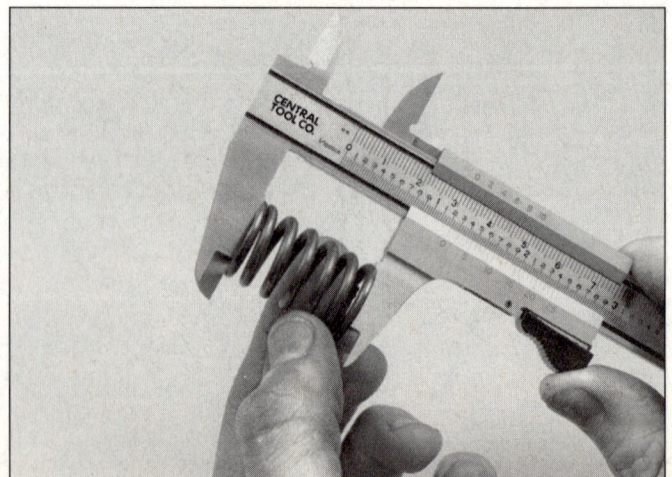

**19.11 Measure the clutch spring free length**

**19.13 Check all plates for warpage**

**19.17 Clutch cover noise damper is retained by four bolts**

**19.21 Use a new nut to secure the clutch hub; torque limiter spring tangs (arrows) must engage as shown**

needle bearing for score marks, scratches and excessive wear.

**16** Inspect the pressure plate for signs of cracks, replacing it if necessary. If the thrust bearing set in the pressure plate is noisy when rotated, or shows signs of excessive wear, support the pressure plate around the bearing housing and drift the bearing out. Tap the new bearing into position, using a drift which bears only on the outer race of the bearing.

**17** Clean all traces of gasket from the clutch cover and crankcase. Clean inside and out the oil level sight glass set in the cover. A noise damper is housed inside the clutch cover; if removal is required, remove its four retaining screws, followed by the plate and damper **(see illustration)**.

### Installation

**18** Install the spacer ring over the mainshaft, followed by the thick thrust washer.

**19** Lubricate the sleeve and the needle bearing with engine oil and slide them over the mainshaft.

**20** Install the clutch housing over the needle bearing and install the second thrust washer and spacer ring.

**21** Slide the hub center and splined hub onto the mainshaft, engaging the damper cams on

**19.22a Install a friction plate first . . .**

**19.22b . . . followed by a steel plate**

the rear face of the splined hub with the followers set in the hub center. Install the collar, followed by the torque limiter spring; arrange the spring so that its tangs engage the cutouts of the splined hub. Install a new clutch nut and using the method of retention employed on removal, tighten to the specified torque **(see illustration)**.

**22** If new clutch plates are being installed, coat them with engine oil. Install the clutch plates, starting with a friction plate and alternating them **(see illustrations)**. There are seven friction plates and six steel plates.

**Note:** *The last (outermost) friction plate tangs should slot into the shallow cutouts in the clutch housing* **(see illustration)**.

**23** If the pushrod was removed, smear it with moly-based grease and install it into the mainshaft **(see illustration)**. Take off the slave cylinder (see Section 20) and check that the pushrod end has correctly located in the end cap.

**24** Smear moly-based grease on the pushrod end piece and insert it into the pressure plate thrust bearing. Install the pressure plate, springs, retainers and spring bolts; tighten the

**19.22c Install the final friction plate with its outer tabs in alternative slots in hub center (A), and not the slots used for other plates (B)**

**19.23 Smear the pushrod with moly-based grease and insert it through the transmission mainshaft**

**19.24 Tighten the pressure plate bolts evenly in a criss-cross pattern to the specified torque**

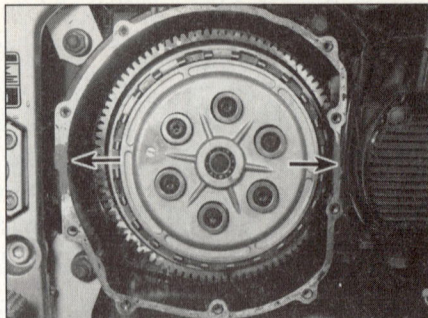

**19.25a Apply sealant to the crankcase joint on both sides (arrows)**

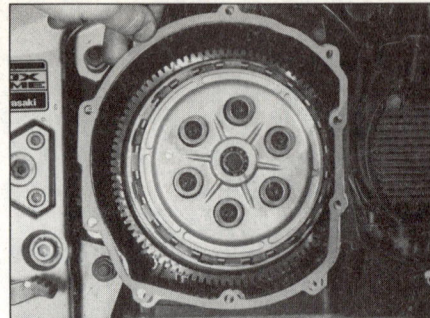

**19.25b Install a new clutch cover gasket**

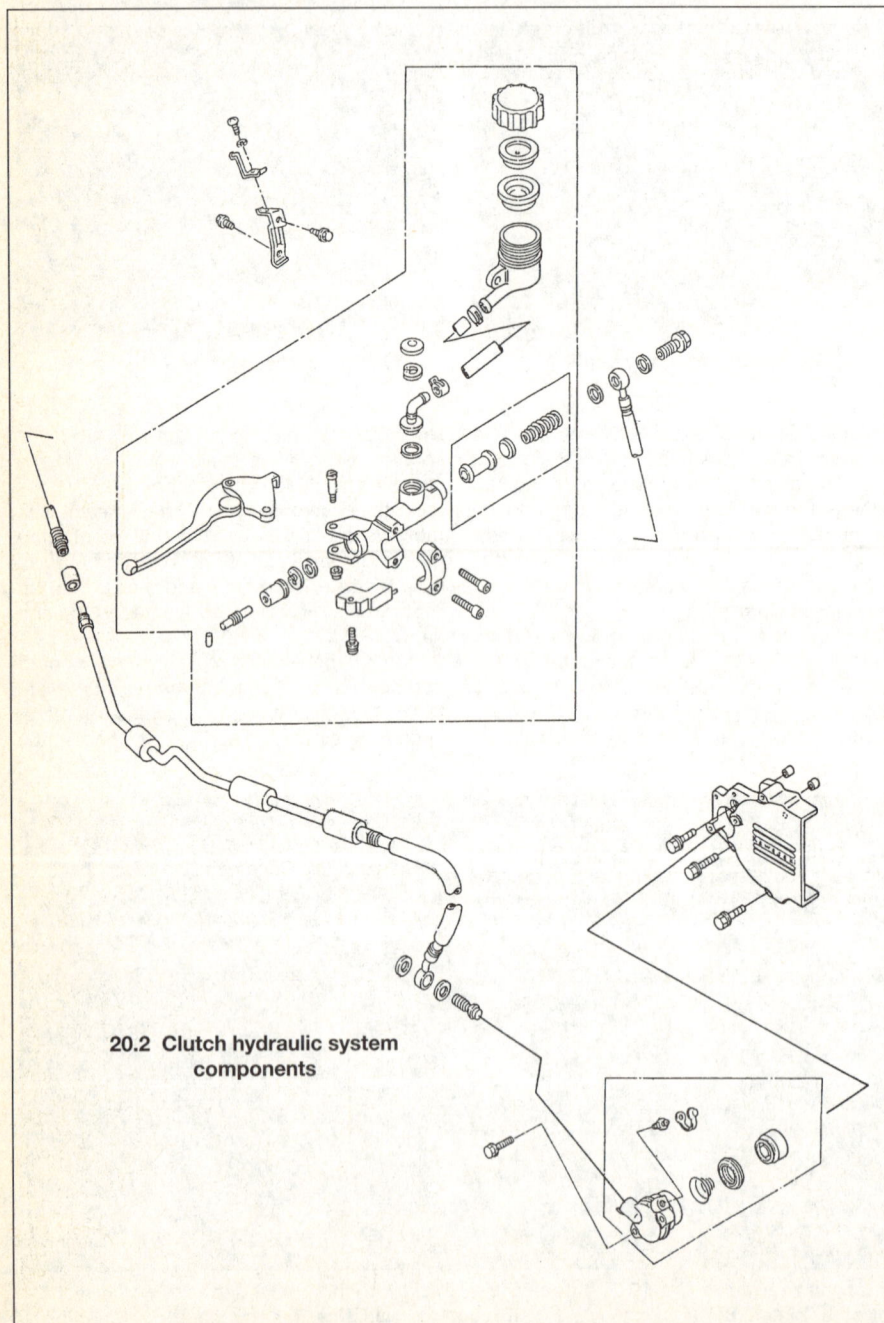

**20.2 Clutch hydraulic system components**

bolts evenly in a diagonal sequence until the specified torque is reached **(see illustration)**.
**25** Apply silicone sealant approximately 25 mm (1 inch) each side of the crankcase halves joint **(see illustration)**. Before installing the clutch cover, pull the clutch lever fully in against the handlebar grip and hold it there with a band or have an assistant hold it in this position while you push the end of the pushrod end piece inward. Install a new clutch cover gasket **(see illustration)**, the cover and bolts (not forgetting the pickup coil wire clamp – **see illustration 19.4**), then release the clutch lever. Tighten the bolts, in a criss-cross pattern, to the torque listed in this Chapter's Specifications.
**26** Install the engine coolant reservoir and top up the coolant (see *'Daily (pre-ride) checks'*).
**27** Fill the crankcase with the recommended type and amount of engine oil (see Chapter 1 for quantity and *Daily (pre-ride) checks* for topping up).
**28** Install the lower fairing panel (see Chapter 8).

## 20 Clutch hydraulic system – removal, overhaul, installation and system bleeding

### *Master cylinder and reservoir*

#### Removal

**1** Disconnect the electrical connector for the clutch switch beneath the master cylinder.
**2** Place a towel under the master cylinder to catch any spilled fluid, then remove the union bolt from the master cylinder fluid line **(see illustration)**.
*Caution: Brake fluid will damage paint. Wipe up any spills immediately and wash the area with soap and water.*
**3** Remove the master cylinder clamp bolts and take the cylinder body off the handlebar.
**4** Remove the cap securing bracket to allow the reservoir cap to be unscrewed. The reservoir is held to the upper triple clamp by a bolt **(see illustration)**.

#### Overhaul

**5** Remove the lever pivot bolt and nut and take off the lever.

20.4 Master cylinder fluid reservoir cap retainer and mounting bolt

20.16 Master cylinder clamp markings (arrow)

6 Remove the rubber boot and pushrod from the master cylinder.

7 Remove the snap-ring, then dump out the piston assembly and spring. If they won't come out, blow compressed air into the fluid line hole.

⚠️ **Warning: The piston may shoot out forcefully enough to cause injury. Point the piston at a block of wood or a pile of rags inside a box and apply air pressure gradually. Never point the end of the cylinder at yourself, including your fingers.**

8 Thoroughly clean all of the components in clean brake fluid (don't use any type of petroleum-based solvent).

9 Check the piston and cylinder bore for wear, scratches and rust. If the piston assembly shows these conditions, replace it. If the cylinder bore has any defects, replace the entire master cylinder.

10 Install the spring in the cylinder bore, wide end first (see illustration 20.2).

11 Coat the new primary cup with brake fluid and install it in the cylinder, wide end first.

12 Coat the piston with brake fluid and install it in the cylinder.

13 Install the washer. Press the piston into the bore and install the snap-ring to hold it in place.

14 Install the rubber boot and pushrod.

15 When you install the clutch lever, align the hole in the lever bushing with the pushrod.

### Installation

16 Installation is the reverse of the removal steps, with the following additions:
  a) Make sure the UP mark on the master cylinder clamp is upright and the arrow points upward (see illustration).
  b) Tighten the clamp bolts to the torque listed in this Chapter's Specifications. Tighten the upper bolt first, then the lower bolt. There will be a small gap between the master cylinder and the clamp at the bottom.
  c) Use a new sealing washer on each side of the banjo bolt fitting and tighten the banjo bolt to the torque listed in this Chapter's Specifications.
  d) Fill and bleed the master cylinder as described below. Operate the clutch and check for fluid leaks.

### *Slave cylinder*

#### Removal

17 If you're removing the slave cylinder for overhaul, loosen the banjo fitting bolt while the slave cylinder is still mounted on the engine. If you're just removing it for access to other components, leave the hydraulic line connected.

18 Remove the mounting bolts and take the slave cylinder off (see illustration 5.13).

19 If you're removing the slave cylinder for overhaul, remove the banjo fitting bolt and detach the hydraulic line. Place the end of the line in a container to catch dripping fluid.

*Caution: Brake fluid will damage paint. Wipe up any spills immediately and wash the area with soap and water.*

20 If the hydraulic line is still connected, push the piston as far into the bore as it will go. Hold the piston in, slowly squeeze the clutch lever to the handlebar and tie the clutch lever in that position. Otherwise, the slave cylinder piston will fall out.

### Overhaul

21 Let the pressure of the slave cylinder spring push the piston out of the cylinder, then remove the spring.

22 Separate the spring from the piston.

23 Thoroughly clean all parts in clean brake fluid (don't use any type of petroleum-based solvent).

24 Check the cylinder bore and piston for wear, scratches and rust. If the piston shows these conditions, replace it and the seal as a set. If the cylinder bore has any defects, replace the entire slave cylinder. If the piston and bore are good, carefully remove the seal from the piston and install a new one with its lip facing into the bore (see illustration).

25 Clean out the slave cylinder cavity in the casing. Remove the pushrod end cap and apply a smear of moly-based grease to the pushrod end (see illustration).

### Installation

26 Installation is the reverse of the removal procedure with the following additions:
  a) Use new sealing washers on the clutch fluid hose.
  b) Tighten the mounting bolts securely and the fluid line banjo bolt to the torque listed in this Chapter's Specifications.
  c) Bleed the clutch (see below).
  d) Operate the clutch and check for fluid leaks.

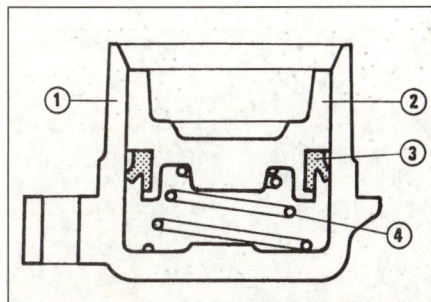

20.24 Clutch slave cylinder cross-section

  1  Cylinder body    3  Seal
  2  Piston           4  Spring

20.25 Pushrod end cap fits over pushrod

20.29 Clutch bleeding equipment

21.9 Detach neutral switch wire from the switch end

21.11 Remove the cover screws (arrows), if necessary using an impact driver

## System bleeding

**27** Ensure the motorcycle is securely supported on its stand. Ideally, have an assistant hold the motorcycle upright so that the fluid level in the reservoir is more easily maintained during the procedure.

**28** Remove the fluid reservoir cap (see Step 4), and lift out the plate and diaphragm. Top up the reservoir with fluid to the upper level line.

**29** Remove the cap from the bleed valve on the slave cylinder. Place a box wrench (ring spanner) over the bleed valve. Attach a rubber tube to the valve fitting and put the other end of the tube in a container **(see illustration)**. Pour enough clean brake fluid into the container to cover the end of the tube.

**30** Slowly squeeze the clutch lever several times. At the same time, tap on the clutch fluid hose, starting at the bottom and working your way to the top. Stop when no more air bubbles can be seen rising from the bottom of the reservoir.

**31** Squeeze the clutch lever several times until you feel an increase in the effort required to pull the lever, then hold it in.

**32** With the lever held in, quickly open the bleed valve to let air and fluid escape, then close it.

**33** Repeat Steps 31 and 32 until there aren't any more bubbles in the fluid flowing into the container. **Note:** *Keep an eye on the fluid level in the reservoir. If it drops too low, air will be sucked into the hose and the procedure will have to be repeated.*

**34** Replenish the reservoir with fluid, then reinstall the diaphragm, plate and cap.

## 21 External shift mechanism – removal, inspection and installation

### Removal

#### Shift pedal

**1** Support the motorcycle securely on its stand.

**2** Remove the left lower fairing panel (see Chapter 8).

**3** Punch mark the end of the shift shaft level with the split of the shift pedal clamp so that it can be used as a reference on installation.

**4** Remove the clamp pinch bolt and ease the clamp off the shift shaft splines.

**5** The rear end of the shift pedal pivots on the rider's left footpeg. Unbolt the footpeg bracket and remove the Allen bolt from inside the bracket to free the shift pedal and footpeg.

#### Shift mechanism

**6** Remove the shift pedal (see Steps 1 through 4 above).

**7** Drain the engine oil and coolant (see Chapter 1).

**8** Remove the clutch slave cylinder; there's no need to disconnect its hydraulic hose (see Section 20).

**9** Remove the engine sprocket cover and the engine sprocket (see Chapter 6). Disconnect the wire from the neutral switch **(see illustration)**. Remove the water pump (see Chapter 3).

**10** Remove the alternator (see Chapter 9).

**11** Position a drain pan under the shift mechanism cover. Remove the screws **(see illustration)** and detach the cover from the crankcase.

> **HAYNES HiNT**
> *The two flathead screws at the rear of the cover can be very difficult to remove, even with an impact driver. If you need to get the bike back in service quickly, it would be a good idea to start early on a day when your local Kawasaki dealership is open so you can buy new screws if the old ones are ruined during removal.*

**12** Pull the shift mechanism arm (against the pressure of the pawl spring) toward the shift shaft until the pointed tips clear the shift drum, then pull the mechanism and shaft off **(see illustrations)**.
*Caution: Don't pull the shift rod out of the crankcase – the shift forks will fall into the oil pan, and the crankcase will have to be separated to reinstall them.*

**13** Note the positions of the gear and neutral positioning levers and their springs **(see illustration)**. Remove the nuts that secure the levers and lift them off **(see illustrations)**. The levers are interchangeable, but it's a good idea to label them for return to their original locations, since they will take a wear pattern during use.

### Inspection

**14** Check the shift shaft for bends and

21.12a Compress the shift mechanism arm until the tips are clear . . .

21.12b . . . then take the shift mechanism out

21.13a Remove the nuts (arrows) that secure the gear and neutral positioning levers . . .

21.13b . . . remove the collar . . .

21.13c . . . the lever . . .

21.13d . . . the spring . . .

21.13e . . . and the washer

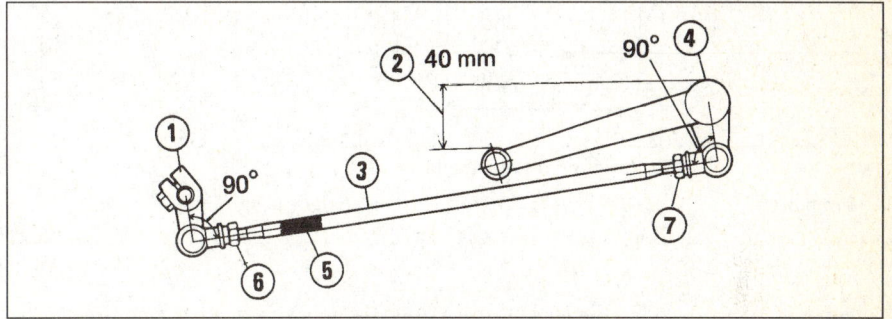
21.19 Shift pedal linkage adjustment

| 1 | Clamp | 4 | Pedal pivot | 6 | Locknut (left-hand threads) |
|---|---|---|---|---|---|
| 2 | Height | 5 | Knurled area | 7 | Locknut (right-hand threads) |
| 3 | Link rod | | | | |

damage to the splines. If the shaft is bent, you can attempt to straighten it, but if the splines are damaged it will have to be replaced.

**15** Check the condition of the return spring and the pawl spring. Replace them if they are cracked or distorted.

**16** Check the shift mechanism arm and the overshift limiter for cracks, distortion and wear. If any of these conditions are found, replace the shift mechanism.

**17** Make sure the return spring pin isn't loose in the crankcase. If it is, unscrew it, apply a non-hardening locking compound to the threads, then reinstall it and tighten it securely.

**18** Check the condition of the seals in the cover. If they have been leaking, drive them out with a hammer and punch. New seals can be installed by driving them in with a socket.

### Installation

#### Shift pedal

**19** Installation is a reverse of the removal procedure, noting the following:
a) *Align the shift shaft punch mark, made on removal, with the split in the clamp.*
b) *Check the height setting of the shift pedal. The top of the pedal should be approximately 40 mm (1.6 inch) lower than the top of the rider's footpeg* (**see illustration**). *To adjust, back off the link rod locknuts and rotate the rod to alter the height – note that the front locknut, near the knurling on the link rod is a left-hand thread.*

### Shift mechanism

**20** Slide the external shift mechanism into place, lifting the shift arm and the overshift limiter to clear the shift drum (**see illustration**).

**21** Install the gear and neutral positioning levers and make sure the springs are positioned correctly (**see illustration**).

**22** Apply high-temperature grease to the lips of the seals. Wrap the splines of the shift shaft with electrical tape, so the splines won't damage the seal as the cover is installed.

**23** Apply a thin coat of silicone sealant to the cover mating areas on the crankcase, where the halves of the crankcase join (**see illustration**).

**24** Carefully guide the cover into place. Apply non-permanent thread locking agent to the threads of the two flathead screws along the rear of the cover, then install all of the screws,

21.20 The shift mechanism should look like this when it's installed

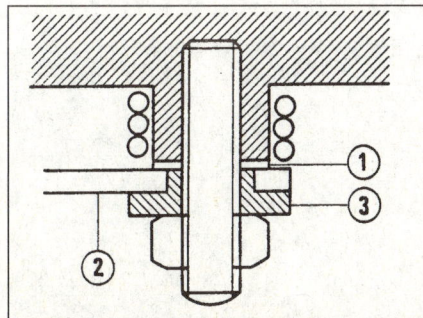
21.21 Ensure the gear and neutral positioning levers are assembled correctly
1 Washer   2 Lever   3 Collar

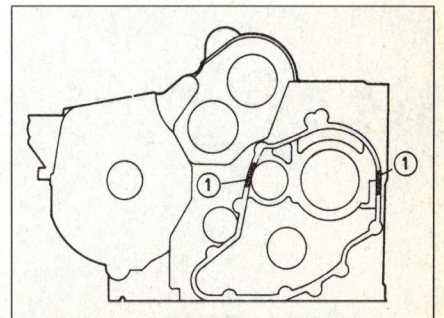
21.23 Apply sealant to the crankcase joint areas (1)

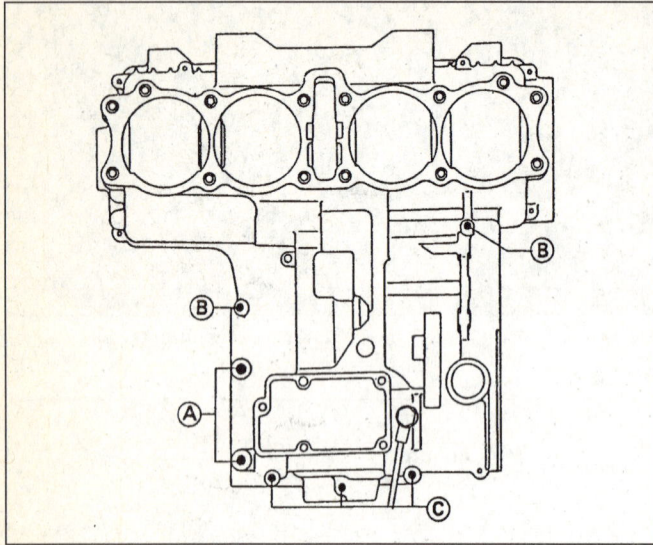

**22.7 Upper crankcase bolts**

A  8 mm bolts          B  6 mm bolts (white)          C  6 mm bolts (black)

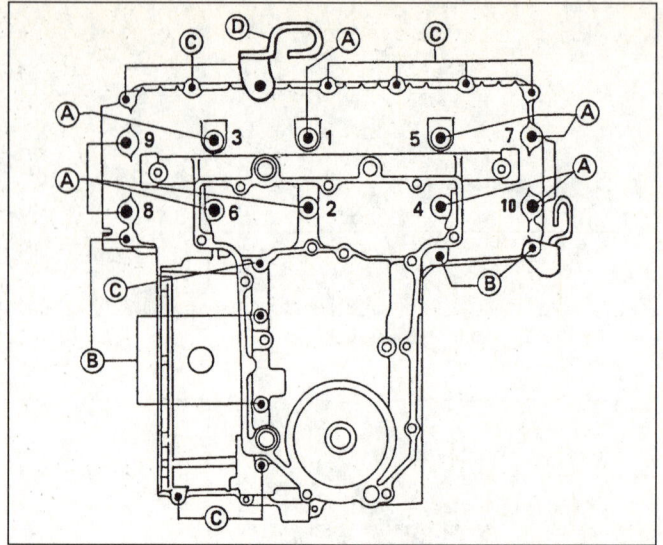

**22.9 Lower crankcase bolts**

A  8 mm bolts                    C  6 mm bolts (black)
B  6 mm bolts (white)            D  Clamp

*Numbers indicate tightening sequence*

tightening them securely. Reconnect the neutral switch wire.

**25** Install the water pump (see Chapter 3).

**26** Install the engine sprocket and chain (see Chapter 6) and engine sprocket cover.

**27** Install the shift pedal (see Step 19).

**28** Check the engine oil level and add some, if necessary (see 'Daily (pre-ride) checks').

## 22 Crankcase – disassembly and reassembly

### Disassembly

**1** To examine and repair or replace the crankshaft, connecting rods, bearings, transmission components and idler shaft/starter motor clutch, the engine must be removed (see Section 5) and the crankcase must be split into two parts.

**2** Remove the alternator and starter motor (see Chapter 9).

**3** On the right side of the engine, remove the clutch (see Section 19), pickup coil cover and pickup coils (see Chapter 5).

**4** On the left side of the engine, remove the water pump (see Chapter 3) and external shift mechanism (see Section 21).

**5** On the bottom of the engine, remove the oil filter, oil pan, pump and pickup strainer (see Sections 16 and 17).

**6** If you're planning to remove the crankshaft, remove the cylinder head, cylinder block and pistons (see Sections 10, 13 and 14).

**7** Remove the upper crankcase bolts **(see illustration)**.

**8** Unbolt the oil pump sprocket, then slide the sprocket and chain off together. Remove the chain guide and separate the sprocket from the chain (see Section 17).

**9** Turn the crankcase over and remove the lower crankcase bolts **(see illustration)**. Loosen the 8 mm bolts that secure the crankshaft in the reverse order of the tightening sequence.

**10** Pry the crankcase halves apart at the pry

points only **(see illustrations)**.
*Caution: Don't pry between the crankcase halves or the mating surfaces will be gouged, resulting in an oil leak.*

**11** Lift the lower crankcase half off the upper half **(see illustration)**.

**12** Refer to Sections 23 through 31 for information on the internal components of the crankcase.

### Reassembly

**13** Remove all traces of sealant from the crankcase mating surfaces. Be careful not to let any fall into the case as this is done.

**14** Check to make sure the two dowel pins are in place in their holes in the mating surface of the upper crankcase half.

**15** Pour some engine oil over the transmission gears, the crankshaft main bearings and the shift drum. Don't get any oil on the crankcase mating surface.

**16** Apply a thin, even bead of silicone sealant to the indicated areas of the crankcase mating surfaces **(see illustration)**.

**22.10a  There's a pry point at each end of the crankcase . . .**

**22.10b  . . . pry only at these points to prevent damaging the mating surfaces**

**22.11  Lift the lower crankcase half off the upper half**

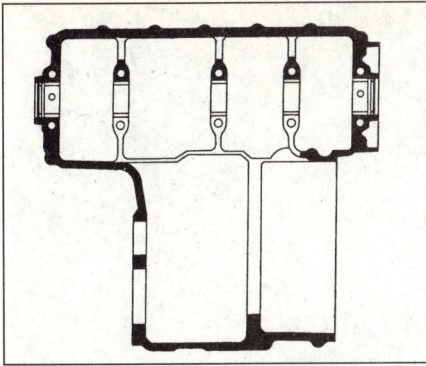

**22.16  Apply sealant to the black areas shown**

**22.17  Place the shift drum and forks in the neutral position**

**22.18  Make sure the shift forks engage the gear grooves (arrows)**

17  Check the position of the shift drum – make sure it's in the neutral position **(see illustration)**.

18  Carefully place the lower crankcase half onto the upper crankcase half. While doing this, make sure the shift forks fit into their gear grooves **(see illustration)**.

19  Install the 8 mm lower crankcase half bolts and tighten them to approximately half of their final torque value, following the numbered sequence **(see illustration 22.9)**. Secure them to the final torque value by repeating the sequence. Turn the crankcase over and install the two 8 mm bolts, tightening them to the specified torque **(see illustration 22.7)**.

20  Install the two 6 mm white-coded bolts into the upper crankcase half and the three in the lower half, tightening them to the specified torque. Finally install the 6 mm black-coded bolts in the lower and upper halves and tighten to the specified torque **(see illustrations 22.7 and 22.9)**.

21  Turn the mainshaft and the driveshaft to make sure they turn freely. Install the shift pedal on the shift shaft and, while turning the output shaft, shift the transmission through the gears, from first to sixth, then back to first. The positive neutral finder prevents the transmission from being shifted past neutral into second gear unless the output shaft is turning at a fairly high rate of speed, which can be difficult. If the transmission doesn't shift properly, the case will have to be separated again to correct the problem. Also make sure the crankshaft turns freely.

22  The remainder of installation is the reverse of removal.

## 23  Crankcase – inspection and servicing

1  After the crankcases have been separated and the crankshaft, shift drum and forks and transmission components removed, the crankcases should be cleaned thoroughly with new solvent and dried with compressed air. All oil passages should be blown out with compressed air and all traces of old gasket sealant should be removed from the mating surfaces.

*Caution: Be very careful not to nick or gouge the crankcase mating surfaces or leaks will result. Check both crankcase sections very carefully for cracks and other damage.*

2  Check the cam chain and alternator/starter chain guides for wear – one is in the upper case half and the other is in the lower case half. If they appear to be worn excessively, replace them (see Section 28).

3  Check the ball bearings in the case. If they don't turn smoothly, drive them out with a bearing driver or a socket having an outside diameter slightly smaller than that of the bearing. Before installing them, allow them to sit in the freezer overnight, and about fifteen minutes before installation, place the case half in an oven, set to about 94°C (200° F), and allow it to heat up. The bearings are an interference fit, and this will ease installation.

⚠ *Warning: Before heating the case, wash it thoroughly with soap and water so no explosive fumes are present and remove all detachable components, engine mountings, etc. which were not removed as part of the engine disassembly process. Also, don't use a flame to heat the case.*

4  Remove the breather cover and lift out the separator screen **(see illustrations)**. Clean the screen with solvent. Replace it if it's clogged or damaged.

5  If any damage is found that can't be repaired, replace the crankcase halves as a set.

6  Refer to Section 25 for details of crankshaft endplay measurement. If it exceeds the specified figure, the crankcases must be replaced.

## 24  Main and connecting rod bearings – general note

1  Even though main and connecting rod bearings are generally replaced with new ones during the engine overhaul, the old bearings should be retained for close examination as they may reveal valuable information about the condition of the engine.

2  Bearing failure occurs mainly because of lack of lubrication, the presence of dirt or other foreign particles, overloading the engine and/or corrosion. Regardless of the cause of bearing failure, it must be corrected before the engine is reassembled to prevent it from happening again.

**23.4a  Remove the breather cover bolts (arrows) . . .**

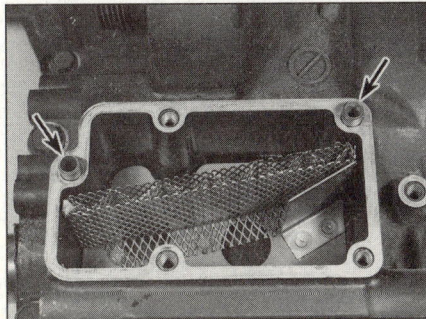

**23.4b  . . . and lift off the cover, noting the locations of the dowels (arrows) . . .**

**23.4c  . . . then lift out the screen**

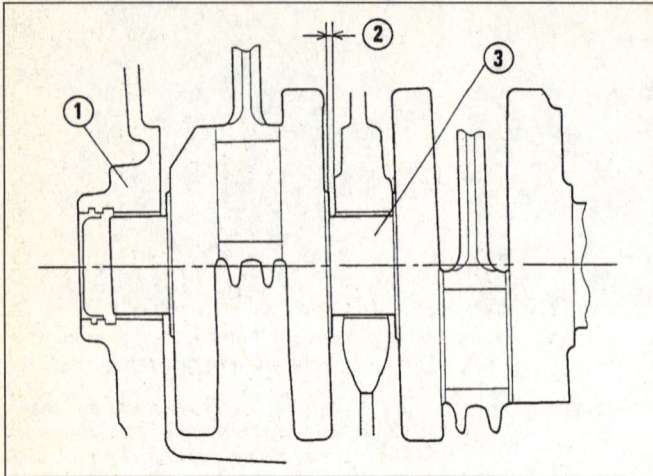

**25.1 Crankshaft end play measurement**

1 Crankcase
2 Measurement point
3 No. 2 crankshaft journal

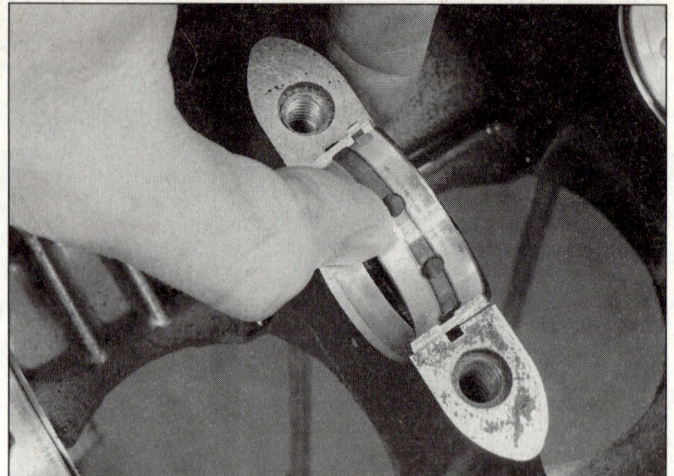

**25.2 To remove a main bearing insert, push it sideways and lift it out**

3 When examining the bearings, remove the main bearings from the case halves and the rod bearings from the connecting rods and caps and lay them out on a clean surface in the same general position as their location on the crankshaft journals. This will enable you to match any noted bearing problems with the corresponding crankshaft journal.

4 Dirt and other foreign particles get into the engine in a variety of ways. It may be left in the engine during assembly or it may pass through filters or breathers. It may get into the oil and from there into the bearings. Metal chips from machining operations and normal engine wear are often present. Abrasives are sometimes left in engine components after reconditioning operations such as cylinder honing, especially when parts are not thoroughly cleaned using the proper cleaning methods. Whatever the source, these foreign objects often end up imbedded in the soft bearing material and are easily recognized. Large particles will not imbed in the bearing and will score or gouge the bearing and journal. The best prevention for this cause of bearing failure is to clean all parts thoroughly and keep everything spotlessly clean during engine reassembly. Frequent and regular oil and filter changes are also recommended.

5 Lack of lubrication or lubrication breakdown has a number of interrelated causes. Excessive heat (which thins the oil), overloading (which squeezes the oil from the bearing face) and oil leakage or throw off from excessive bearing clearances, worn oil pump or high engine speeds all contribute to lubrication breakdown. Blocked oil passages will also starve a bearing and destroy it. When lack of lubrication is the cause of bearing failure, the bearing material is wiped or extruded from the steel backing of the bearing. Temperatures may increase to the point where the steel backing and the journal turn blue from overheating.

6 Riding habits can have a definite effect on bearing life. Full throttle low speed operation, or lugging (labouring) the engine, puts very high loads on bearings, which tend to squeeze out the oil film. These loads cause the bearings to flex, which produces fine cracks in the bearing face (fatigue failure). Eventually the bearing material will loosen in pieces and tear away from the steel backing. Short trip riding leads to corrosion of bearings, as insufficient engine heat is produced to drive off the condensed water and corrosive gases produced. These products collect in the engine oil, forming acid and sludge. As the oil is carried to the engine bearings, the acid attacks and corrodes the bearing material.

7 Incorrect bearing installation during engine assembly will lead to bearing failure as well. Tight fitting bearings which leave insufficient bearing oil clearances result in oil starvation. Dirt or foreign particles trapped behind a bearing insert result in high spots on the bearing which lead to failure.

8 To avoid bearing problems, clean all parts thoroughly before reassembly, double check all bearing clearance measurements and lubricate the new bearings with clean engine oil during installation.

## 25 Crankshaft and main bearings – removal, inspection and installation

### Removal

1 Crankshaft removal is a simple matter of lifting it out of place once the crankcase has been separated and the starter motor clutch/secondary sprocket assembly has been removed. Before removing the crankshaft check the endplay. This can be done with a dial indicator mounted in-line with the crankshaft, or feeler gauges inserted between the no. 2 crankcase main journal (see illustration). Compare your findings with this Chapter's Specifications. If the endplay is excessive, the case halves must be replaced.

2 The main bearing inserts can be removed from their saddles by pushing their centers to the side, then lifting them out (see illustration). Keep the bearing inserts in order. The main bearing oil clearance should be checked, however, before removing the inserts (see Step 8).

### Inspection

3 Mark and remove the connecting rods from the crankshaft (see Section 26).

4 Clean the crankshaft with solvent, using a rifle-cleaning brush to scrub out the oil passages. If available, blow the crank dry with compressed air. Check the main and connecting rod journals for uneven wear, scoring and pits. Rub a penny across the journal several times – if a journal picks up copper from the penny, it's too rough; replace the crankshaft.

5 Check the camshaft chain gear and the primary chain gear on the crankshaft for chipped teeth and other wear. If any undesirable conditions are found, replace the crankshaft. Check the chains as described in Section 28.

6 Check the rest of the crankshaft for cracks and other damage. It should be magnafluxed to reveal hidden cracks – a dealer service department or motorcycle machine shop will handle the procedure.

7 Set the crankshaft on V-blocks and check the runout with a dial indicator touching the center main journal, comparing your findings with Chapter's Specifications. If the runout exceeds the limit, replace crankshaft.

### Main bearing selection

8 To check the main bearing oil clearance, clean off the bearing inserts (and reinstall

25.8 Lay the Plastigauge strips (arrow) on the journals, parallel to the crankshaft centerline

25.10 Measuring the width of the crushed Plastigauge

them, if they've been removed from the case) and lower the crankshaft into the upper half of the case. Cut five pieces of Plastigauge (type HPG-1) and lay them on the crankshaft main journals, parallel with the journal axis **(see illustration)**.

**9** Very carefully, guide the lower case half down onto the upper half. Install the large (8 mm) bolts and tighten them, using the recommended sequence, to the torque listed in this Chapter's Specifications (see Section 22). Don't rotate the crankshaft!

**10** Now, remove the bolts and carefully lift the lower case half off. Compare the width of the crushed Plastigauge on each journal to the scale printed on the Plastigauge envelope to obtain the main bearing oil clearance **(see illustration)**. Write down your findings, then remove all traces of Plastigauge from the journals using your fingernail or the edge of a credit card.

**11** If the oil clearance falls into the specified range, no bearing replacement is required (provided they are in good shape). If the clearance is between 0.044 mm (0.0017 inch) and the service limit (0.08 mm/0.0031 inch), replace the bearing inserts with inserts that have blue paint marks, then check the oil clearance once again **(see illustration)**. Always replace all of the inserts at the same time.

**12** The clearance might be slightly greater than the standard clearance, but that doesn't matter, as long as it isn't greater than the maximum clearance or less than the minimum clearance.

**13** If the clearance is greater than the service limit listed in this Chapter's Specifications, measure the diameter of the crankshaft journals with a micrometer **(see illustration)** and compare your findings with this Chapter's Specifications. Also, by measuring the diameter at a number of points around each journal's circumference, you'll be able to determine whether or not the journal is out-of-round. Take the measurement at each end of the journal, near the crank throws, to determine if the journal is tapered.

**14** If any crank journal has worn down past the service limit, replace the crankshaft.

**15** If the diameters of the journals aren't less than the service limit but differ from the original markings on the crankshaft **(see illustration)**, apply new marks with a hammer and punch.

*If the journal measures between 33.984 and 33.992 mm (1.3380 to 1.3383 inch) don't make any marks on the crank (there shouldn't be any marks there, anyway).*

*If the journal measures between 33.993 and 34.000 mm (1.3383 to 1.3386 inch),*

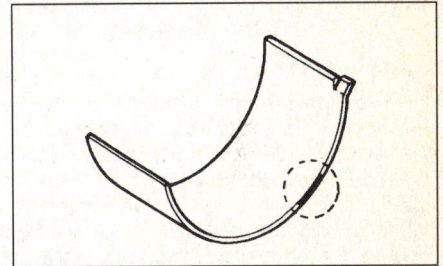

25.11 Location of bearing insert color code

*make a '1' mark on the crank in the area indicated (if it's not already there).*

**16** Remove the main bearing inserts and assemble the case halves (see Section 22). Using a telescoping gauge and a micrometer, measure the diameters of the main bearing bores, then compare the measurements with the marks on the upper case half **(see illustration)**.

*If the bores measure between 37.000 and 37.008 mm (1.4567 to 1.4570 inch), there should be a '0' mark in the indicated areas.*

*If the bores measure between 37.009 and 37.016 mm (1.4570 to 1.4573 inch), there shouldn't be any marks in the indicated areas.*

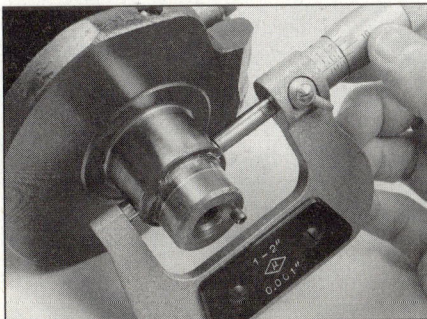

25.13 Measure the diameter of each crankshaft journal at several points

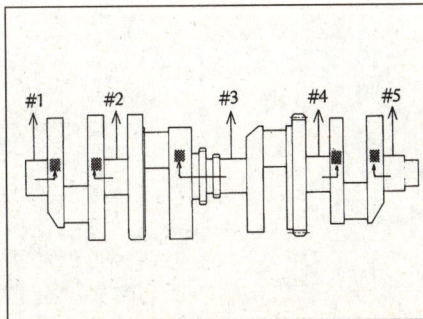

25.15 Crankshaft journal size marking locations; use in conjunction with . . .

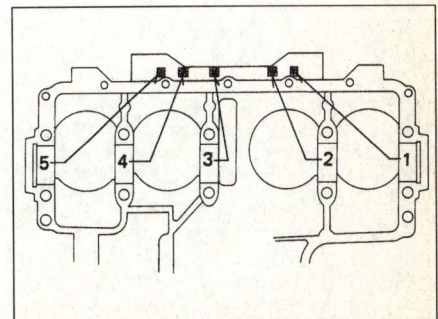

25.16 . . . crankcase markings to determine insert size

| Crankcase Main Bearing Bore Diameter Mark | Crankshaft Main Journal Diameter Mark | Crankshaft Bearing Insert* | | |
|---|---|---|---|---|
| | | Size Color | Part Number | Journal Nos. |
| O | 1 | Brown | 92028-1447 | 1, 3 ,5 |
| | | | 92028-1444 | 2, 4 |
| None | 1 | Black | 92028-1446 | 1, 3, 5 |
| O | None | | 92028-1443 | 2, 4 |
| None | None | Blue | 92028-1445 | 1, 3, 5 |
| | | | 92028-1442 | 2, 4 |

*The bearing inserts for Nos. 2 and 4 journals have an oil groove, respectively.

**25.17 Main bearing insert size selection table**

**25.18 Make sure the tabs on the bearing inserts fit into the notches in the web**

17 Using the marks on the crank and the marks on the case, determine the bearing sizes required by referring to the accompanying bearing selection chart (see illustration).

### Installation

18 Separate the case halves once again. Clean the bearing saddles in the case halves, then install the bearing inserts in their webs in the case (see illustration). The bearing inserts for journals 2 and 4 have oil grooves. When installing the bearings, use your hands only – don't tap them into place with a hammer.
19 Lubricate the bearing inserts with engine assembly lube or moly based grease
20 Install the connecting rods, if they were removed (see Section 26).
21 Install new oil seals to the ends of the crankshaft (the lips of the seals must face the crankshaft). Be sure to lubricate the lips of the seals with high-temperature grease before sliding them into place.
22 Loop the camshaft chain and the alternator/starter chain over the crankshaft and lay them onto their gears.
23 Carefully lower the crankshaft into place, making sure the ribs on the seal outer diameters seat in the grooves in the case (see illustration).
24 Assemble the case halves (see Section 22) and check to make sure the crankshaft and the transmission shafts turn freely.

### 26 Connecting rods and bearings – removal, inspection and installation

### Removal

1 Before removing the connecting rods from the crankshaft, measure the side clearance of each rod with a feeler gauge (see illustration). If clearance on any rod is greater than that listed in this Chapter's Specifications, that rod will have to be replaced with a new one.
2 Using a center punch, mark the position of each rod and cap, relative to its position on the crankshaft (see illustration).
3 Unscrew the bearing cap nuts, separate the cap from the rod, then detach the rod from the crankshaft. If the cap is stuck, tap on the end of the rod bolts with a soft faced hammer to free them.
4 Separate the bearing inserts from the rods and caps, keeping them in order so they can be reinstalled in their original locations. Wash the parts solvent and dry them with compressed air, if available.

### Inspection

5 Check the connecting rods for cracks and other obvious damage. Lubricate the piston pin for each rod, install it in the proper rod and check play (see illustration). If it is loose,

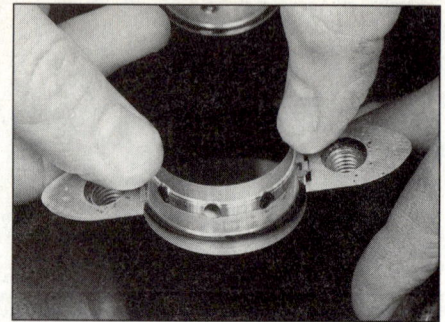

**25.23 Make sure the ribs on each seal seat in the grooves in the case (arrows)**

replace the connecting rod and/or the pin. Note: If connecting rod replacement is necessary, refer to the Note in Step 24 below).
6 Refer to Section 24 and examine the connecting rod bearing inserts. If they are scored, badly scuffed or appear to have been seized, new bearings must be installed. Always replace the bearings in the connecting rods as a set. If they are badly damaged, check the corresponding crankshaft journal.

> **HAYNES HINT** *Evidence of extreme heat, such as discoloration, indicates that lubrication failure has occurred. Be sure to thoroughly check the oil pump and pressure relief valve as well as all oil holes and passages before reassembling the engine.*

**26.1 Check the connecting rod side clearance with a feeler gauge**

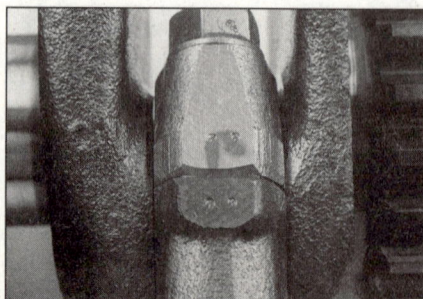

**26.2 Using a hammer and punch, make matching marks in the connecting rod and its cap**

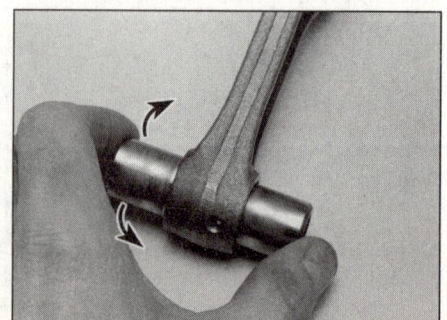

**26.5 Checking the piston pin and connecting rod bore for wear**

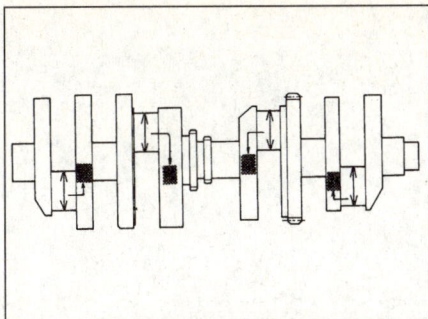

**26.18 Connecting rod journal size marking locations; use in conjunction with . . .**

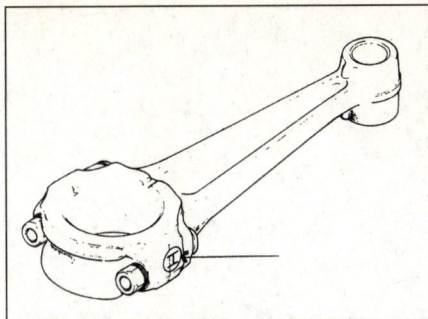

**26.20a . . . the mark (or no mark) on the connecting rod (arrow) to determine insert size**

*The letter is a weight grade mark*

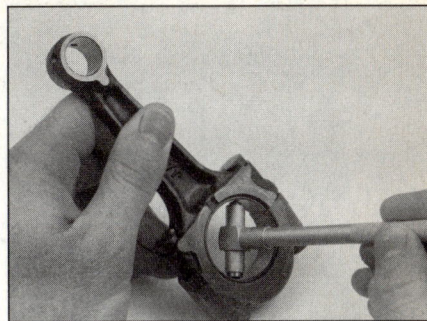

**26.20b Measure the diameter of the connecting rod with a telescoping gauge, then measure the gauge**

7 Have the rods checked for twist and bending at a dealer service department or other motorcycle repair shop.

### Bearing selection

8 If the bearings and journals appear to be in good condition, check the oil clearances as follows:

9 Start with the rod for the No. 1 cylinder. Wipe the bearing inserts and the connecting rod and cap clean, using a lint-free cloth.

10 Install the bearing inserts in the connecting rod and cap. Make sure the tab on the bearing engages with the notch in the rod or cap.

11 Wipe off the connecting rod journal with a lint-free cloth. Lay a strip of Plastigauge (type HPG-1) across the top of the journal, parallel with the journal axis **(see illustration 25.8)**.

12 Position the connecting rod on the bottom of the journal, then install the rod cap and nuts. Tighten the nuts to the torque listed in this Chapter's Specifications (noting the comments in Step 25 for H2 models), but don't allow the connecting rod to rotate at all.

13 Unscrew the nuts and remove the connecting rod and cap from the journal, being very careful not to disturb the Plastigauge. Compare the width of the crushed Plastigauge to the scale printed in the Plastigauge envelope **(see illustration 25.10)** to determine the bearing oil clearance.

14 If the clearance is within the range listed in this Chapter's Specifications and the bearings are in perfect condition, they can be reused. If the clearance is between 0.076 mm (0.0030 inch) and the service limit 0.11 mm (0.0043 inch), replace the bearing inserts with inserts that have blue paint marks **(see illustration 25.11)**, then check the oil clearance once again. Always replace all of the inserts at the same time.

15 The clearance might be slightly greater than the standard clearance, but that doesn't matter, as long as it isn't greater than the maximum clearance or less than the minimum clearance.

16 If the clearance is greater than the service limit listed in this Chapter's Specifications, measure the diameter of the connecting rod

journal with a micrometer and compare your findings with this Chapter's Specifications. Also, by measuring the diameter at a number of points around the journal's circumference, you'll be able to determine whether or not the journal is out-of-round. Take the measurement at each end of the journal to determine if the journal is tapered.

17 If any journal has worn down past the service limit, replace the crankshaft.

18 If the diameter of the journal isn't less than the service limit but differs from the original markings on the crankshaft **(see illustration)**, apply new marks with a hammer and punch.

*If the journal measures between 34.984 and 34.992 mm (1.3773 to 1.3776 inch) don't make any marks on the crankshaft (there shouldn't be one there anyway).*

*If the journal measures between 34.993 and 35.000 mm (1.3777 to 1.3780 inch), make a '0' mark on the crankshaft in the area indicated (if not already there).*

19 Remove the bearing inserts from the connecting rod and cap, then assemble the cap to the rod. Tighten the nuts to the torque listed in this Chapter's Specifications.

20 Using a telescoping gauge and a micrometer, measure the inside diameter of the connecting rod **(see illustration)**. The mark on the connecting rod (if any) should coincide with the measurement, but if it doesn't, make a new mark **(see illustration)**.

*If the inside diameter measures between 38.000 and 38.008 mm (1.4961 to 1.4964 inch), don't make any mark on the rod (there shouldn't be any there anyway).*

*If the inside diameter measures between 38.009 and 38.016 mm (1.4964 to 1.4967 inch), make a '0' mark on the rod (it should already be there).*

21 By referring to the accompanying chart **(see illustration)**, select the correct connecting rod bearing inserts.

22 Repeat the bearing selection procedure for the remaining connecting rods.

### Installation

**Caution: The connecting rod bolts on H2 models are designed to stretch in use and it is essential that both the bolts and nuts are replaced with new ones whenever they are disturbed.**

23 Wipe off the bearing inserts and connecting rods and caps. Apply moly-based grease to the back of the upper insert and install it in the connecting rod; use your hands only and make sure the tab on the insert engages the notch in the rod. Install the lower insert in the rod cap in the same way, but don't use any lubricant on the back of the insert. When the inserts are installed, lubricate their working surfaces with engine oil, taking care not to get any lubricant on the mating surfaces of the rod or cap.

24 Assemble each connecting rod to its proper journal, making sure the previously applied matchmarks correspond to each other **(see illustration 26.2)**. **Note:** *The letter present at the rod/cap seam on one side of the connecting rod is a weight mark. If new rods are being installed ensure that the weight mark for cylinders 1 and 2 is identical, and*

| Con-Rod Big End Bore Diameter Marking | Crankpin Diameter Mark | Bearing Insert | |
|---|---|---|---|
| | | Size Color | Part Number |
| O | None | Blue | 92028-1407 |
| None | None | Black | 92028-1408 |
| O | O | | |
| None | O | Brown | 92028-1409 |

**26.21 Connecting rod bearing insert selection table**

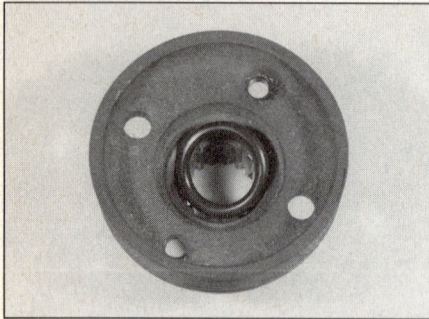

27.3 There's an O-ring ring behind the pulley bolt washer

27.4 Thread a pair of 6 mm bolts into the holes in the pulley then tighten them against the case to push the pulley off

27.9a Remove the oil nozzle bolts (arrows) . . .

27.9b . . . and work the O-ring free of the case . . .

27.9c . . . then unbolt the bearing retainer (arrow)

27.11 Support the starter clutch and remove the idler shaft

that the weight mark for cylinders 3 and 4 is identical. This will minimize vibration.

**25** When you're sure the rods are positioned correctly, install the bolts and tighten the nuts to the torque listed in this Chapter's Specifications. **Note:** *On H2 models, apply a small amount of engine oil to the threads of the bolts and the seating surface of the nuts before installation. After securing to the specified torque (see Specifications), tighten them an additional 120°.*

**26** Turn the rods on the crankshaft. If any of them feel tight, tap on the bottom of the connecting rod caps with a hammer – this should relieve stress and free them up. If it doesn't, recheck the bearing clearance.

**27** As a final step, recheck the connecting rod side clearances (see Step 1). If the clearances aren't correct, find out why before proceeding with engine assembly.

27.12a Lift the starter clutch out

## 27 Idler pulley, shaft and starter motor clutch – removal, inspection and installation

### Pulley removal and installation

**1** Remove the alternator (see Chapter 9). Where fitted, remove the alternator belt guard.
**2** Keep the pulley from turning. If the engine is in the bike, shift the transmission into first gear and have an assistant hold the rear brake on. If the engine has been removed, hold the pulley from turning with a strap wrench.
**3** Unscrew the pulley bolt, then remove the washer and O-ring **(see illustration)**.
**4** Thread a pair of 6 mm bolts into the threaded holes in the pulley and tighten them against the engine to push the pulley off **(see illustration)**.

27.12b Disengage the idler shaft sprocket from the chain and remove the sprocket

**5** On installation, apply thread locking compound to the pulley bolt and belt guard bolts. Tighten the pulley bolt to the specified torque.
**6** Install the alternator and set the belt tension (see Chapter 9).

### Idler shaft, gear and starter clutch

#### Removal

**7** Remove the engine, separate the crankcase halves and remove the transmission shafts (see Sections 5, 22 and 29).
**8** Remove the oil pump drive chain guide (see Section 17).
**9** Remove the idler shaft oil nozzle and bearing retainer **(see illustrations)**.
**10** Remove the idler shaft pulley (see Steps 2 through 4 above).
**11** Support the starter clutch chain with one hand and pull the idler shaft out of the crankcase **(see illustration)**.
**12** Lift the starter clutch and idler shaft sprocket out of the crankcase **(see illustrations)**.
**13** Remove the snap-ring from the idle gear shaft **(see illustration)**. Pull the shaft out of the crankcase and lift out the starter idle gear.

#### Inspection

**14** Hold the starter motor clutch and attempt to turn the starter motor clutch gear back and forth **(see illustration)**. It should turn in one direction only.
**15** If the starter motor clutch gear turns freely in both directions, or if it's locked up, replace it.

**27.13 Starter idle gear details**

A  Starter idle gear       C  Short collar
B  Long collar             D  Snap-ring

**27.14 Hold the starter clutch and try to rotate the gear**

**16**  Remove the snap-ring and washer and slide off the gear **(see illustration)**. Check the rollers for scoring and pitting and the retainers for damage. Check the gear teeth for cracks and chips. Check the needle roller bearing in the gear for wear or damage. Replace parts as necessary.

**17**  Inspect the teeth on the starter motor idle gear for cracks and chips. Turn the idle gear to make sure it spins freely. If the idle gear exhibits any undesirable conditions, replace it.

Remove the idle gear as described in Step 13. Coat the shaft with engine assembly lube or moly-based grease before installing it.

**18**  Check the splines, sprocket, and threads on the idler shaft for wear or damage **(see illustration)**.

**19**  Turn the bearing and feel for tight spots and roughness. If necessary, remove the snap-ring and washer from the end of the shaft and pull the bearing off the shaft, using a bearing puller. The new bearing can be

tapped onto the shaft, using a piece of pipe with an inside diameter large enough to fit over the shaft and contact the inner race of the bearing.

**20**  Check the idler shaft bearing in the crankcase **(see illustration)**. Replace it if it's worn or damaged.

**Installation**

**21**  Installation is the reverse of the removal steps, with the following additions:

a) Coat all parts with clean engine oil.
b) Be sure to place the long and short idle gear collars on the correct sides of the gear **(see illustration 27.13)**.

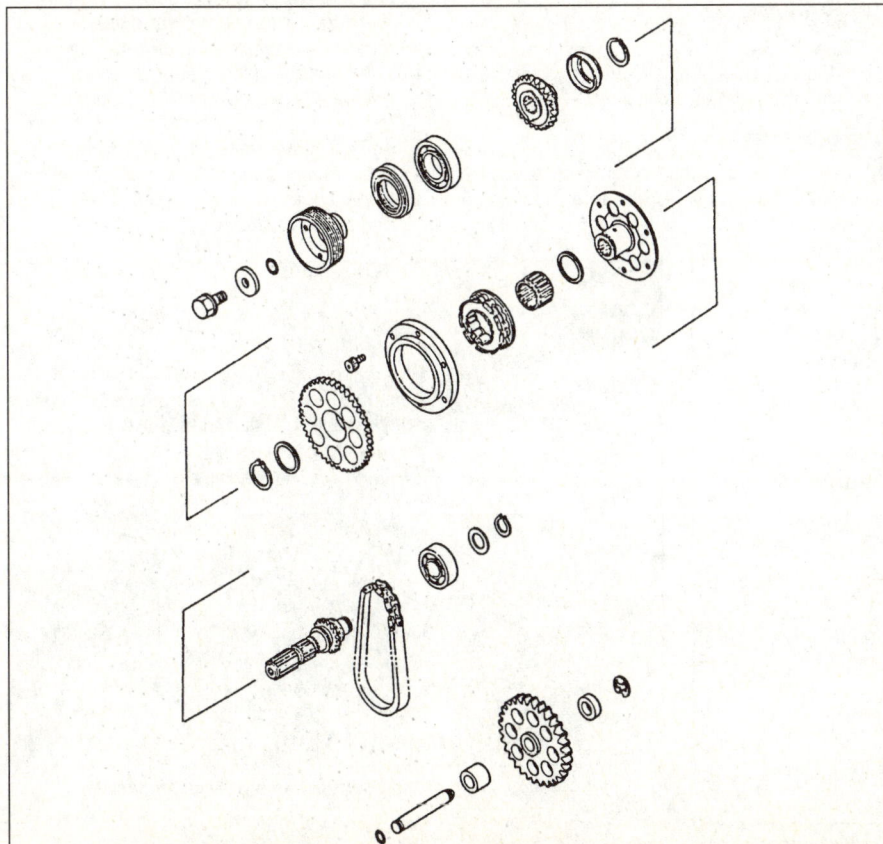

**27.18 Inspect the starter idler shaft splines and sprocket; if the bearing is worn or damaged, have it pressed off**

**27.16 Starter clutch, idler shaft and idle gear detail**

**27.20 Replace the idler shaft bearing if it's worn or damaged**

**28.4 Remove the chains from the crankshaft**

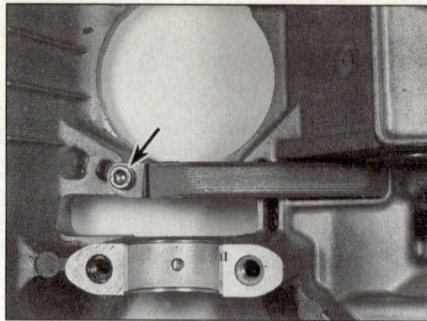

**28.7a The upper crankcase alternator/starter chain guide is retained by a single bolt (arrow) . . .**

**28.7b . . . and the guide in the lower crankcase is retained by two Allen bolts (arrows)**

## 28 Alternator/starter chain, camshaft chain and guides – removal, inspection and installation

### Removal

#### Alternator/starter chain and camshaft chain

**1** Remove the engine (see Section 5).
**2** Separate the crankcase halves (see Section 22).
**3** Lift the crankshaft out of the upper case (see Chapter 2).
**4** Remove the chains from the crankshaft **(see illustration)**.

#### Chain guides

**5** The cam chain front guide can be lifted from the cylinder block after the head has been removed (see Section 10).
**6** The cam chain rear guide is retained to the crankcase upper surface by a holder and pin. Remove the cylinder block (see Section 13) for access.

**7** The alternator/starter chain guide in the upper case half is secured by a single Allen bolt **(see illustration)**. Its guide in the lower case is fastened by a bracket and two bolts **(see illustration)**.

### Inspection

#### Camshaft chain

**8** Pull the chain tight to eliminate all slack and measure the length of twenty links, pin-to-pin **(see illustration)**. Compare your findings to this Chapter's Specifications.
**9** Also check the chain for binding and obvious damage.
**10** If the twenty-link length is not as specified, or there is visible damage, replace the chain.

#### Chain guides

**11** Check the guides for deep grooves, cracking and other obvious damage, replacing them if necessary.

### Installation

**12** Installation of these components is the reverse of the removal procedure, noting the following:

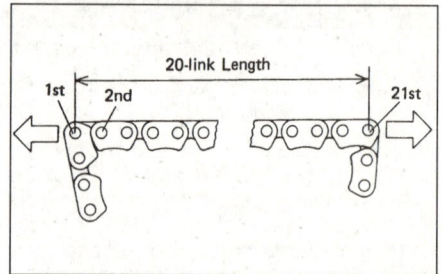

**28.8 Cam chain stretch measurement**

a) When installing the cam chain front guide, ensure that its end engages the cup in the crankcase **(see illustration)**. The arrow markings on the chain rear guide holder must face forward **(see illustration)**.
b) When installing the starter/alternator chain guides, apply a non-hardening thread locking compound to the threads of the bolts. Tighten the bolts to the torque listed in this Chapter's Specifications.
c) Apply engine oil to the faces of the guides and to the chains.

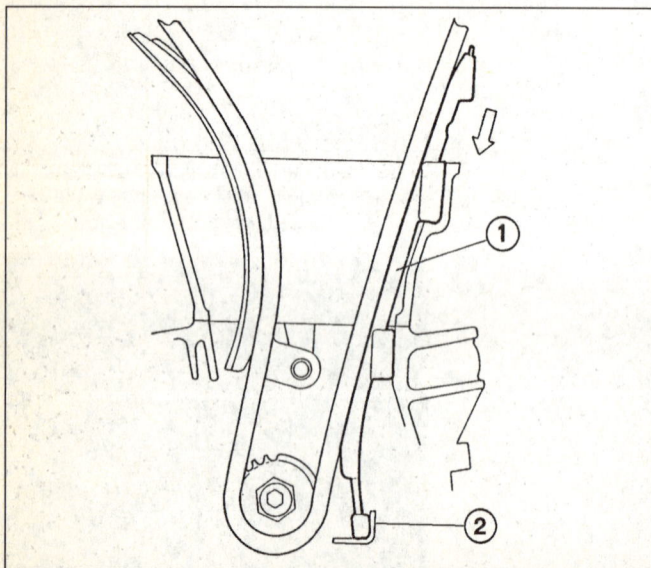

**28.12a Cam chain front (exhaust side) guide (1) must engage cup (2) in crankcase**

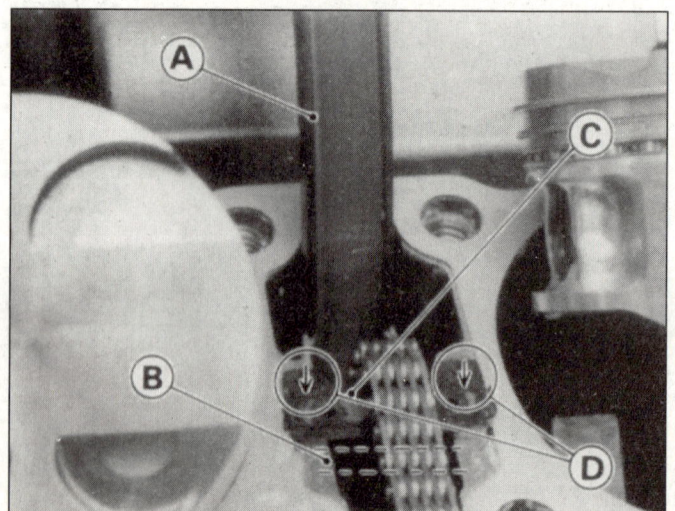

**28.12b Cam chain rear (intake side) guide detail**

A  Guide
B  Pin
C  Holder
D  Arrow markings

29.2a  Lift the transmission mainshaft . . .

29.2b  . . . and driveshaft out of the casing

29.4  Be sure the set pins and rings (arrows) are in position

## 29  Transmission shafts – removal and installation

### Removal

1  Remove the engine and clutch, then separate the case halves (see Sections 5, 22 and 29).

2  The transmission shafts can simply be lifted out of the upper half of the case **(see illustrations)**. If they are stuck, use a soft-face hammer and gently tap on the bearings on the ends of the shafts to free them. The shaft nearest the rear of the case is the driveshaft (output shaft) – the other shaft is the mainshaft (input shaft).

3  Refer to Section 30 for information pertaining to transmission shaft service and Section 31 for information pertaining to the shift drum and forks.

### Installation

4  Check to make sure the set pins and rings are present in the upper case half, where the shaft bearings seat **(see illustration)**.

5  Carefully lower each shaft into place. The holes in the needle bearing outer races must engage with the set pins, and the grooves in the ball bearing outer races must engage with the set rings.

6  The remainder of installation is the reverse of removal.

## 30  Transmission shafts – disassembly, inspection and reassembly

1  Remove the shafts from the case (see Section 29).

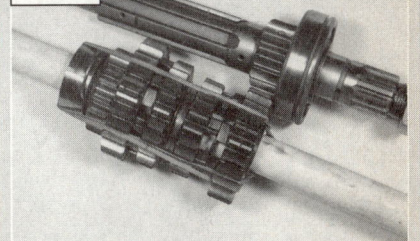

*When disassembling the transmission shafts, place the parts on a long rod or thread a wire through them to keep them in order and facing the proper direction.*

### Mainshaft

#### Disassembly

2  Remove the needle bearing outer race, then remove the snap-ring from the end of the shaft and slide the needle bearing off **(see illustrations)**.

3  Remove the thrust washer and slide second gear off the shaft **(see illustration)**.

4  Remove sixth gear and bushing **(see illustrations)**.

5  Slide the toothed washer off and remove the snap-ring **(see illustration)**. To keep the snap-ring from bending as it's expanded, hold the back of it with pliers **(see illustration)**.

30.2a  Slide off the needle bearing outer race . . .

30.2b  . . . then remove the snap-ring and bearing

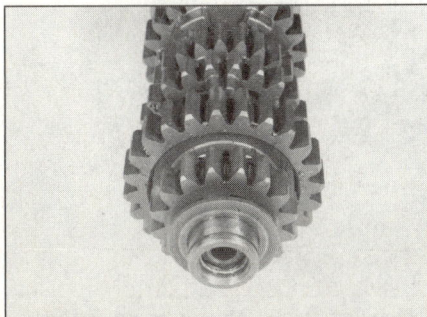
30.3  Remove the thrust washer and second gear

30.4a  Slide off sixth gear . . .

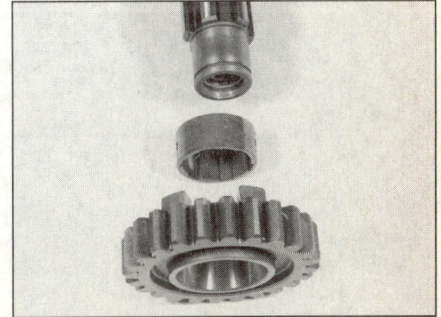
30.4b  . . . and its bushing

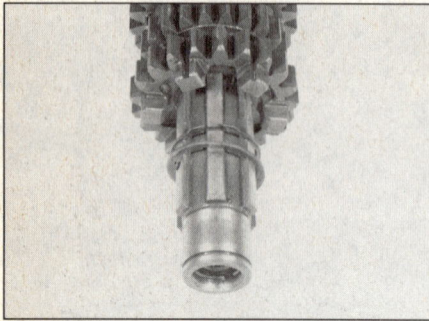

30.5a Remove the toothed washer and snap-ring . . .

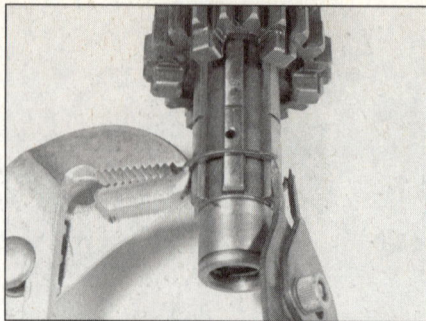

30.5b . . . holding the back of the snap-ring with pliers to prevent it from twisting

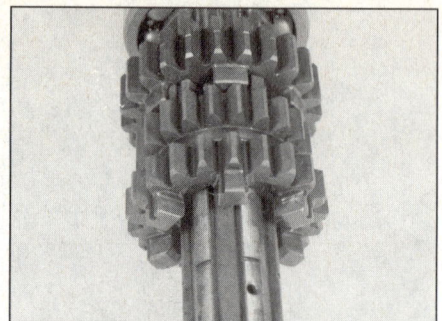

30.6 Slide the third/fourth gear off the shaft

30.7a Remove the snap-ring . . .

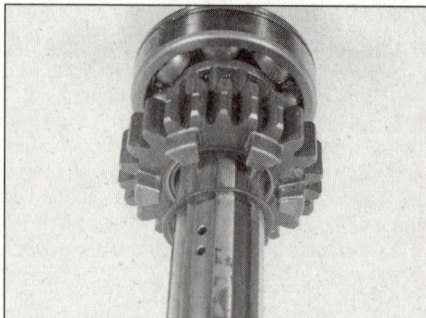

30.7b . . . the toothed washer and fifth gear . . .

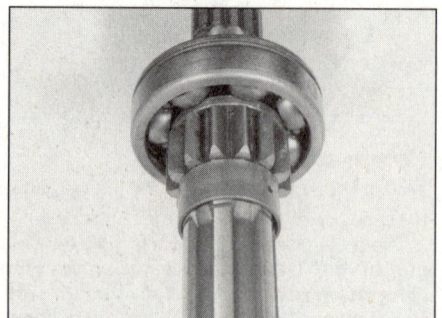

30.7c . . . and its bushing

6 Remove the third/fourth gear cluster from the shaft (see illustration).

7 Remove the next snap-ring, then slide the washer, fifth gear and its bushing off the shaft (see illustrations).

### Inspection

8 Wash all of the components in clean solvent and dry them off. Rotate the ball bearing on the shaft, feeling for tightness, rough spots, excessive looseness and listening for noises. If any of these conditions are found, replace the bearing. This will require the use of a hydraulic press or a bearing puller setup. If you don't have access to these tools, take the shaft and bearing to a Kawasaki dealer or other motorcycle repair shop and have them press the old bearing off the shaft and install the new one.

9 Measure the shift fork groove between third and fourth gears (see illustration). If the groove width exceeds the figure listed in this Chapter's Specifications, replace the third/fourth gear assembly, and also check the third/fourth gear shift fork (see Section 31).

10 Check the gear teeth for cracking and other obvious damage. Check the bushing and surface in the inner diameter of the fifth and sixth gears for scoring or heat discoloration. If either one is damaged, replace it.

11 Inspect the dogs (see illustration 30.9) and the dog holes in the gears for excessive wear (see illustration). Replace the paired gears as a set if necessary.

12 Check the needle bearing and race for wear or heat discoloration and replace them if necessary.

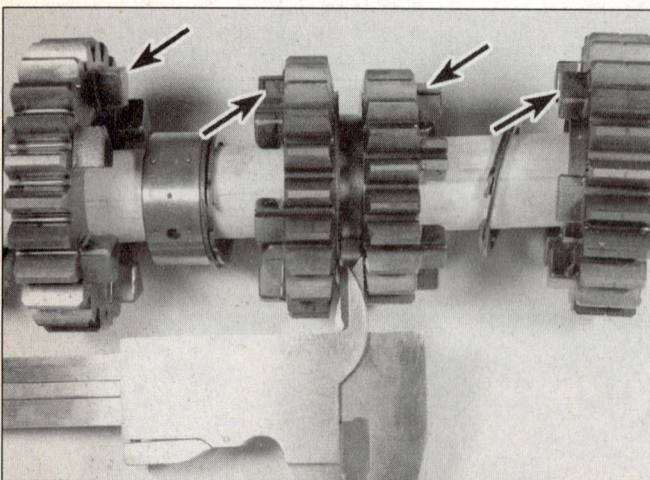

30.9 Measure the gear grooves; if they're too wide or if the dogs (arrows) are worn, replace the gear

30.11 Inspect the bushing (left arrow) in gears so equipped; if worn or if the edges of the slots (right arrow) are rounded, replace the gear

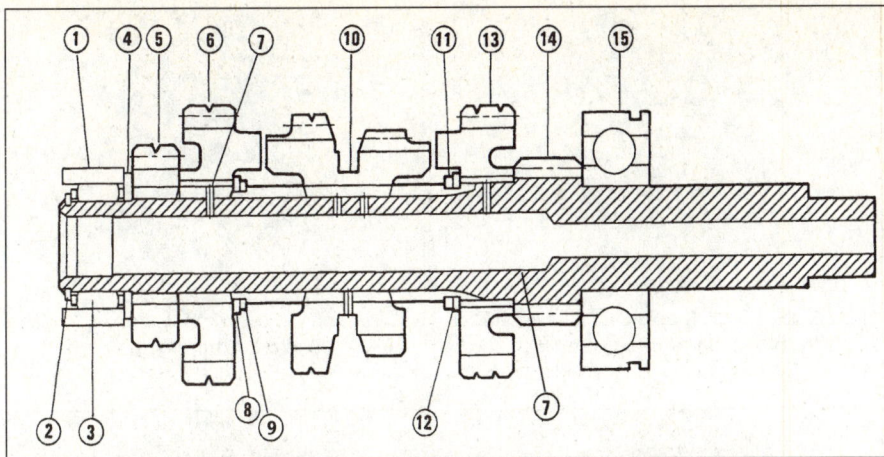

30.13a  Mainshaft gear details

1  Needle bearing outer race
2  Snap-ring
3  Needle bearing
4  Thrust washer
5  2nd gear
6  6th gear
7  Bushing
8  Toothed washer
9  Snap-ring
10 3rd/4th gear
11 Snap-ring
12 Toothed washer
13 5th gear
14 1st gear
15 Ball bearing

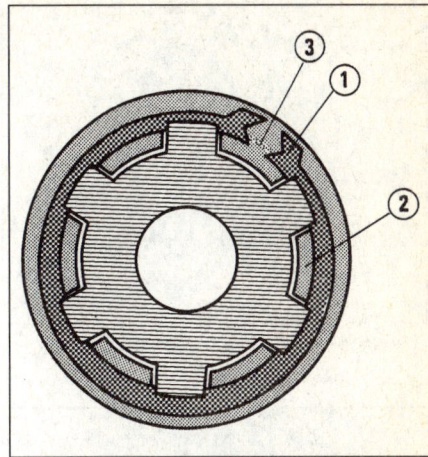

30.13b  When installing snap-rings (1), align the opening (3) with a spline groove (2)

### Reassembly

**13** Reassembly is the basically the reverse of the disassembly procedure, but take note of the following points **(see illustration)**:

a) *Always use new snap-rings and align the opening of the ring with a spline groove* **(see illustration)**. *Face the sharp side of the snap-ring toward the gear being secured; the rounded side faces away from the gear.*

b) *When installing the gear bushings on the shaft, align the oil hole in the shaft with the oil hole in the bushing (see illustration).*

c) *Lubricate the components with engine oil before assembling them.*

### Driveshaft

#### Disassembly

**14** Remove the needle bearing outer race and slide the needle bearing off **(see illustrations)**.

**15** Remove the thrust washer and first gear from the shaft **(see illustration)**.

**16** Remove fifth gear from the shaft. Fifth gear has three steel balls in it for the positive neutral finder mechanism. These lock fifth gear to the shaft unless it is spun rapidly enough to fling the balls outward. To remove fifth gear, grasp third gear and hold the shaft in a vertical position with one hand, and with the other hand, spin the shaft back and forth, holding onto fifth gear and pulling up **(see illustration)**. After fifth gear is removed, collect the three steel balls **(see illustration)**.

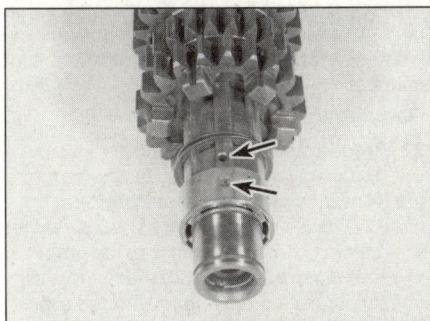

30.13c  Be sure to align the bushing oil hole with the shaft oil hole (arrows)

30.14a  Slide off the bearing outer race . . .

30.14b  . . . and the bearing

30.15  Remove the thrust washer (arrow) and first gear

30.16a  Hold third gear (A) with one hand and spin the transmission shaft while lifting up on fifth gear (B)

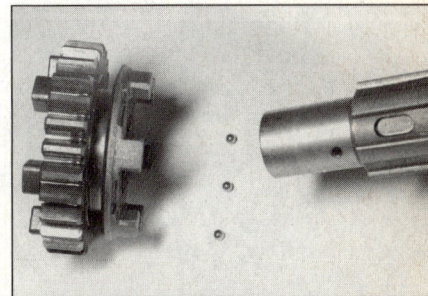

30.16b  These balls ride in slots in the shaft; they must be flung outward by centrifugal force to remove fifth gear

30.17a  Remove the snap-ring . . .

30.17b  . . . the toothed washer . . .

30.17c  . . . third gear . . .

30.17d  . . . its bushing and fourth gear

30.18a  Remove the toothed washer . . .

30.18b  . . . the snap-ring . . .

**Caution: Don't pull the gear up too hard or fast – the balls will fly out of the gear.**

**17**  Remove the snap-ring, toothed washer, third gear, bushing and fourth gear from the shaft **(see illustrations)**.

**18**  Remove the toothed washer, snap-ring and sixth gear **(see illustrations)**.

**19**  Remove the next snap-ring, toothed washer, second gear and its bushing **(see illustrations)**.

**20**  The ball bearing and collar can remain on the shaft unless they need to be replaced **(see illustration)**.

### Inspection

**21**  Refer to Steps 8 through 12 for the inspection procedures. They are the same, except when checking the shift fork groove width you'll be checking it on fifth and sixth gears.

### Reassembly

**22**  Reassembly is the basically the reverse of the disassembly procedure, but take note of the following points **(see illustration)**:

a) *Always use new snap-rings and align the opening of the ring with a spline groove* **(see illustration 30.13b)**. *Face the sharp side of each snap-ring toward the gear being secured; face the rounded side of snap-ring away from the gear.*

30.18c  . . . and sixth gear

30.19a  Remove the snap-ring . . .

30.19b  . . . the toothed washer and second gear . . .

30.19c  . . . and its bushing

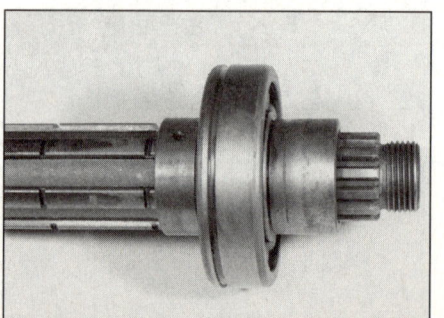

30.20  The bearing and collar can be left on the shaft unless worn or damaged

### 30.22 Driveshaft gear details ▶

| 1 Collar | 12 3rd gear |
|---|---|
| 2 Ball bearing | 13 Toothed washer |
| 3 Driveshaft | 14 Snap-ring |
| 4 2nd gear | 15 Steel ball |
| 5 Toothed washer | 16 5th gear |
| 6 Snap-ring | 17 1st gear |
| 7 6th gear | 18 Thrust washer |
| 8 Snap-ring | 19 Needle bearing |
| 9 Toothed washer | 20 Needle bearing |
| 10 4th gear | outer race. |
| 11 Bushing | |

b) When installing the bushing for third and fourth gear and second gear, align the oil hole in the bushing with the hole in the shaft.

c) When installing fifth gear, don't use grease to hold the balls in place – to do so would impair the positive neutral finder mechanism. Just set the balls in their holes (the holes that they can't pass through), keep the gear in a vertical position and carefully set it on the shaft (engine oil will help keep them in place). The spline grooves that contain the holes with the balls must be aligned with the slots in the shaft spline grooves. Lubricate the components with engine oil before assembling them.

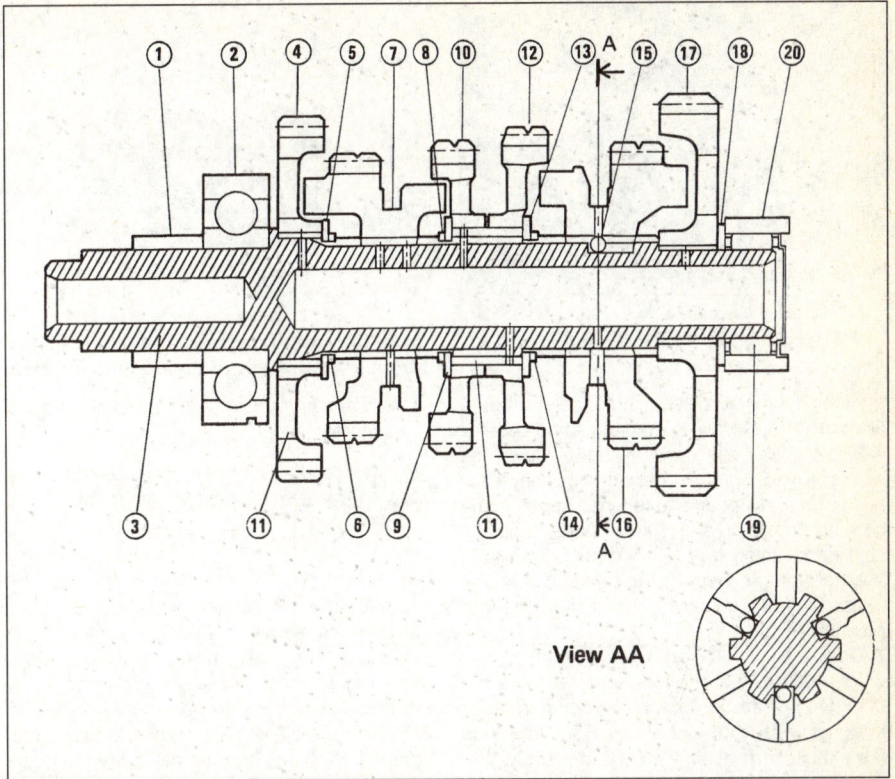

## 31 Shift drum and forks – removal, inspection and installation

### Removal

1 Remove the engine, separate the crankcase halves and remove the external shift mechanism (see Sections 5, 22 and 29).
2 Remove the retaining plates for the shift drum and shift rod (see illustrations).
3 Support the shift forks and pull the shift rods out (see illustrations). The driveshaft shift forks and the shift rods are interchangeable, but it's a good idea to assemble them as they were in the engine so they can be returned to their original positions (see illustration).

4 Slide the shift drum out of the crankcase (see illustration).

### Inspection

5 Check the edges of the grooves in the drum for signs of excessive wear (see illustration). Measure the widths of the grooves and compare your findings to this Chapter's Specifications.
6 Remove the Phillips screw from the end of

**View AA**

31.2a Remove the shift drum retainer bolts (arrows) . . .

31.2b . . . and the shift rod retainer bolt on the other side of the case (arrow)

31.3a Slide out the shift rod with the single fork . . .

31.3b . . . and the shift rod with two forks . . .

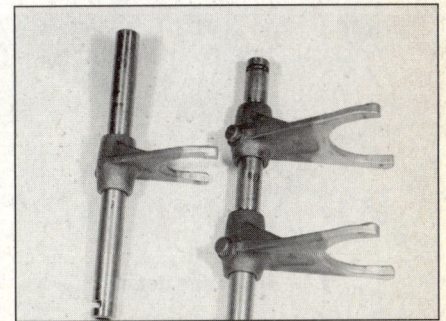

31.3c . . . and reassemble them so they can be returned to their original positions

**31.4 Slide the shift drum out of the crankcase**

the shift drum and disassemble the drum (see illustrations). Check the pin plate and pins for wear or damage and replace them as necessary. Spin the bearing and check for roughness, noise or looseness. Replace the bearing if defects are found. Reassemble the shift drum, making sure the dowel aligns with the hole in the bearing holder and the one longer pin fits in the recess in the pin plate (see illustration).

7 Check the shift forks for distortion and wear, especially at the fork ears. Measure the thickness of the fork ears and compare your findings with this Chapter's Specifications (see illustration). If they are discolored or severely worn they are probably bent. If damage or wear is evident, check the shift fork groove in the corresponding gear as well. Inspect the guide pins and the shaft bore for excessive wear and distortion and replace any defective parts with new ones.

8 Check the shift fork shafts for evidence of wear, galling and other damage. Make sure the shift forks move smoothly on the shafts. If the shafts are worn or bent, replace them with new ones.

### Installation

9 Installation is the reverse of removal, noting the following points:
a) Lubricate all parts with engine oil before installing them.
b) Use non-permanent thread locking agent on the threads of the shift drum and shift

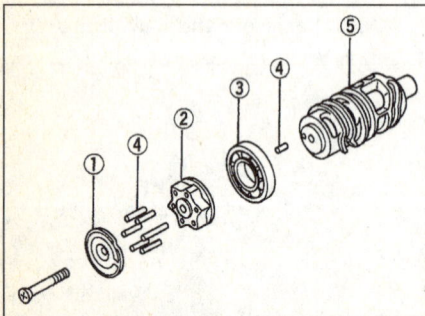

**31.5 Check the edges of the fork grooves for wear, especially at their points (arrows)**

rod retaining plate bolts. Tighten the bolts securely.

## 32 Initial start-up after overhaul

1 Make sure the cooling system is checked carefully (especially the coolant level) before starting and running the engine (see *Daily (pre-ride) checks*). Make sure the engine oil level is correct (see *Daily (pre-ride) checks*).

2 Pull off the spark plug caps and insert four spare spark plugs into the caps; lay the spark plug bodies against the cylinder head – make sure they make good ground (earth) contact with the head metal otherwise the ignition system may be damaged.

⚠ **Warning: Do not remove the spark plugs from the engine to do this because atomized fuel from the spark plug holes could ignite.**

3 Turn the kill switch to the RUN position and turn the ignition ON. Crank the engine over on the starter motor until the oil pressure indicator light goes off (which indicates that oil pressure exists). Turn the ignition OFF and remove the spare spark plugs. Install the spark plug caps.

4 Make sure there is fuel in the tank, then push the button on the fuel tap several times to prime the carburetors and operate the choke.

5 Start the engine and allow it to run at a moderately fast idle until it reaches operating temperature.

⚠ **Warning: If the oil pressure indicator light doesn't go off, or it comes on while the engine is running, stop the engine immediately.**

6 Check carefully for oil leaks and make sure the transmission and controls, especially the brakes, function properly before road testing the machine. Refer to Section 33 for the recommended break-in procedure.

7 Upon completion of the road test, and after the engine has cooled down completely, recheck the valve clearances (see Chapter 1).

## 33 Recommended break-in procedure

1 Any rebuilt engine needs time to break-in, even if parts have been installed in their original locations. For this reason, treat the machine gently for the first few miles to make sure oil has circulated throughout the engine and any new parts installed have started to seat.

2 Even greater care is necessary if the engine has been rebored or a new crankshaft has been installed. In the case of a rebore, the engine will have to be broken in as if the machine were new. This means greater use of the transmission and a restraining hand on the throttle until at least 500 miles (800 km) have been covered. There's no point in keeping to any set speed limit – the main idea is to keep from lugging (labouring) the engine and to gradually increase performance until the 500 mile (800 km) mark is reached. These recommendations can be lessened to an extent when only a new crankshaft is installed. Experience is the best guide, since it's easy to tell when an engine is running freely.

3 If a lubrication failure is suspected, stop the engine immediately and try to find the cause. If an engine is run without oil, even for a short period of time, irreparable damage will occur.

**31.6a Shift drum details**

1 *Pin plate*          4 *Pins*
2 *Bearing holder*     5 *Shift drum*
3 *Bearing*

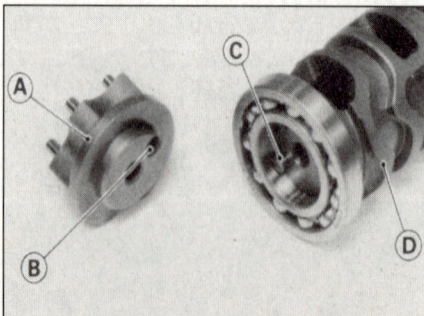

**31.6b Align the hole in the bearing holder with the shift drum pin**

A *Bearing holder*    C *Dowel*
B *Hole*              D *Shift drum*

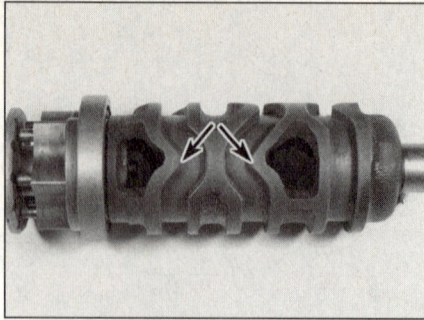

**31.7 Measure the thickness of the fork ears and replace the forks if they're worn**

# Chapter 2 Part B:
# Engine, clutch and transmission
# (749 cc J, K, L and M models)

## Contents

## Degrees of difficulty

**Easy,** suitable for novice with little experience | **Fairly easy,** suitable for beginner with some experience | **Fairly difficult,** suitable for competent DIY mechanic | **Difficult,** suitable for experienced DIY mechanic | **Very difficult,** suitable for expert DIY or professional

## Specifications

### General
Bore . . . . . . . . . . . . . . . . . . . . . . . . . . . . . . . . . . . . . . . . . . . . . . . . . 71 mm (2.795 inches)
Stroke . . . . . . . . . . . . . . . . . . . . . . . . . . . . . . . . . . . . . . . . . . . . . . . . 47.3 mm (1.862 inches)
Displacement . . . . . . . . . . . . . . . . . . . . . . . . . . . . . . . . . . . . . . . . . . 749 cc
Cylinder numbering . . . . . . . . . . . . . . . . . . . . . . . . . . . . . . . . . . . . . 1-2-3-4 (from left end of engine)
Firing order . . . . . . . . . . . . . . . . . . . . . . . . . . . . . . . . . . . . . . . . . . . . 1-2-4-3
Engine mounts
    Right front adjuster protrusion . . . . . . . . . . . . . . . . . . . . . . . . . . . 11.5 mm (0.453 inch)
    Left front adjuster protrusion . . . . . . . . . . . . . . . . . . . . . . . . . . . . 16.5 mm (0.649 inch)

## Camshaft and rocker arms

Lobe height
  J and L models ..................................................... 35.39 to 35.53 mm (1.3933 to 1.3988 inch)
  K and M models
    Intake ............................................................ 36.67 to 36.81 mm (1.4437 to 1.4492 inch)
    Exhaust .......................................................... 36.48 to 36.62 mm (1.4362 to 1.4417 inch)
Bearing oil clearance
  No. 1 and no. 4 bearing caps
    Standard ......................................................... 0.048 to 0.091 mm (0.0019 to 0.0035 inch)
    Maximum ......................................................... 0.18 mm (0.007 inch)
  No. 2 and no. 3 bearing caps
    Standard ......................................................... 0.078 to 0.121 mm (0.0030 to 0.0047 inch)
    Maximum ......................................................... 0.21 mm (0.0082 inch)
Journal diameter
  No. 1 and no. 4 caps
    Standard ......................................................... 23.930 to 23.952 mm (0.9421 to 0.9429 inch)
    Minimum .......................................................... 23.90 mm (0.9409 inch)
  No. 2 and no. 3 caps
    Standard ......................................................... 23.900 to 23.922 mm (0.9409 to 0.9418 inch)
    Minimum .......................................................... 23.87 mm (0.9398 inch)
Bearing journal inside diameter
  Standard ........................................................... 24.000 to 24.021 mm (0.9448 to 0.9457 inch)
  Maximum ........................................................... 24.08 mm (0.9480 inch)
Camshaft runout
  Standard ........................................................... 0.02 mm or less (0.0008 inch or less)
  Maximum ........................................................... 0.1 mm (0.0039 inch)
Cam chain 20-link length
  Standard ........................................................... 155.5 to 155.9 mm (6.122 to 6.1377 inch)
  Maximum ........................................................... 158.0 mm 6.2204 inch)
Rocker arm inside diameter
  Standard ........................................................... 12.000 to 12.018 mm (0.4724 to 0.4731 inch)
  Maximum ........................................................... 12.34 mm (0.4858 inch)
Rocker shaft diameter
  Standard ........................................................... 11.966 to 1.984 mm (0.4711 to 0.4718 inch)
  Minimum ........................................................... 11.94 mm (0.4700 inch)

## Cylinder head, valves and valve springs

Cylinder head warpage limit ............................................ 0.05 mm 0.002 inch)
Valve stem runout
  Standard ........................................................... 0.01 mm (0.0039 inch)
  Maximum ........................................................... 0.05 mm (0.002 inch)
Valve stem diameter
  Intake
    Standard ......................................................... 4.475 to 4.490 mm (0.1762 to 0.1767 inch)
    Minimum .......................................................... 4.46 mm (0.1755 inch)
  Exhaust
    Standard ......................................................... 4.455 to 4.470 mm (0.1754 to 0.1759 inch)
    Minimum .......................................................... 4.44 mm (0.1748 inch)
Valve head margin thickness
  Intake
    Standard ......................................................... 0.4 to 0.6 mm 0.0157 inch)
    Minimum .......................................................... 0.25 mm (0.0098 inch)
  Exhaust
    Standard ......................................................... 0.7 to 0.9 mm (0.0275 to 0.0354 inch)
    Maximum ......................................................... 0.5 mm 0.0196 inch)
Valve guide inside diameter (intake and exhaust)
  Standard ........................................................... 4.500 to 4.512 mm 0.1771 to 0.1776 inch)
  Maximum ........................................................... 4.58 mm (0.1803 inch)
Valve seat width (intake and exhaust) ................................ 0.5 to 1.0 mm (0.0196 to 0.0394 inch)
Valve spring free length (with inner and outer springs)
  Inner spring
    Standard ......................................................... 36.5 mm (1.4370 inch)
    Minimum .......................................................... 34.7 mm (1.3661 inch)
  Outer spring
    Standard ......................................................... 42.7 mm (1.6811 inch)
    Minimum .......................................................... 40.9 mm (1.6102 inch)
Valve spring free length (with single spring)
  Standard ........................................................... 42.7 mm (1.6811 inch)
  Minimum ........................................................... 41.3 mm (1.6259 inch)

## Cylinder block
Bore diameter
  Standard . . . . . . . . . . . . . . . . . . . . . . . . . . . . . . . . . . . . . . . . . . . . . . . . . 71.000 to 71.012 mm (2.7953 to 2.7957 inch)
  Maximum . . . . . . . . . . . . . . . . . . . . . . . . . . . . . . . . . . . . . . . . . . . . . . . . 71.1 mm (2.7992 inch)
Taper and out-of-round limits . . . . . . . . . . . . . . . . . . . . . . . . . . . . . . . . . not specified

## Pistons
Piston diameter
  Standard . . . . . . . . . . . . . . . . . . . . . . . . . . . . . . . . . . . . . . . . . . . . . . . . . 70.942 to 70.958 mm (2.7927 to 2.7936 inch)
  Minimum . . . . . . . . . . . . . . . . . . . . . . . . . . . . . . . . . . . . . . . . . . . . . . . . . 70.80 mm (2.7874 inch)
Piston-to-cylinder clearance . . . . . . . . . . . . . . . . . . . . . . . . . . . . . . . . . . 0.043 to 0.070 mm (0.0015 to 0.0027 inch)
Oversize pistons and rings . . . . . . . . . . . . . . . . . . . . . . . . . . . . . . . . . . . + 0.5 mm (0.020 inch) (one oversize only)
Ring side clearance
  Top
    Standard . . . . . . . . . . . . . . . . . . . . . . . . . . . . . . . . . . . . . . . . . . . . . 0.05 to 0.09 mm (0.0019 to 0.0035 inch)
    Maximum . . . . . . . . . . . . . . . . . . . . . . . . . . . . . . . . . . . . . . . . . . . . 0.19 mm (0.0074 inch)
  Second
    Standard . . . . . . . . . . . . . . . . . . . . . . . . . . . . . . . . . . . . . . . . . . . . . 0.03 to 0.07 mm (0.0012 to 0.0027 inch)
    Maximum . . . . . . . . . . . . . . . . . . . . . . . . . . . . . . . . . . . . . . . . . . . . 0.17 mm (0.0066 inch)
Ring groove width
  Top
    Standard . . . . . . . . . . . . . . . . . . . . . . . . . . . . . . . . . . . . . . . . . . . . . 0.84 to 0.86 mm (0.0330 to 0.0338 inch)
    Maximum . . . . . . . . . . . . . . . . . . . . . . . . . . . . . . . . . . . . . . . . . . . . 0.94 mm (0.0370 inch)
  Second
    Standard . . . . . . . . . . . . . . . . . . . . . . . . . . . . . . . . . . . . . . . . . . . . . 0.82 to 0.84 mm (0.0323 to 0.0338 inch)
    Maximum . . . . . . . . . . . . . . . . . . . . . . . . . . . . . . . . . . . . . . . . . . . . 0.92 mm (0.0362 inch)
  Oil . . . . . . . . . . . . . . . . . . . . . . . . . . . . . . . . . . . . . . . . . . . . . . . . . . . . . not specified
Ring thickness (top and second)
  Standard . . . . . . . . . . . . . . . . . . . . . . . . . . . . . . . . . . . . . . . . . . . . . . . . . 0.77 to 0.79 mm (0.0303 to 0.0311 inch)
  Minimum . . . . . . . . . . . . . . . . . . . . . . . . . . . . . . . . . . . . . . . . . . . . . . . . . 0.72 mm (0.0283 inch)
Ring end gap
  Top and second
    Standard . . . . . . . . . . . . . . . . . . . . . . . . . . . . . . . . . . . . . . . . . . . . . 0.20 to 0.35 mm (0.0078 to 0.0137 inch)
    Maximum . . . . . . . . . . . . . . . . . . . . . . . . . . . . . . . . . . . . . . . . . . . . 0.7 mm (0.0275 inch)
  Oil
    Standard . . . . . . . . . . . . . . . . . . . . . . . . . . . . . . . . . . . . . . . . . . . . . 0.20 to 0.70 mm (0.0078 to 0.0275 inch)
    Maximum . . . . . . . . . . . . . . . . . . . . . . . . . . . . . . . . . . . . . . . . . . . . 1.3 mm (0.0511 inch)

## Crankshaft and bearings
Main bearing oil clearance
  Standard . . . . . . . . . . . . . . . . . . . . . . . . . . . . . . . . . . . . . . . . . . . . . . . . . 0.012 to 0.036 mm (0.0005 to 0.0014 inch)
  Maximum . . . . . . . . . . . . . . . . . . . . . . . . . . . . . . . . . . . . . . . . . . . . . . . . 0.066 mm (0.0025 inch)
Main bearing journal diameter
  No mark on crank throw . . . . . . . . . . . . . . . . . . . . . . . . . . . . . . . . . . . . . 31.984 to 31.992 mm (1.2592 to 1.2595 inch)
  '1' mark on crank throw . . . . . . . . . . . . . . . . . . . . . . . . . . . . . . . . . . . . 31.993 to 32.000 mm (1.2596 to 1.2598 inch)
Crankshaft side clearance (endplay)
  J and K models
    Standard . . . . . . . . . . . . . . . . . . . . . . . . . . . . . . . . . . . . . . . . . . . . . 0.13 to 0.38 mm (0.0051 to 0.0149 inch)
    Maximum . . . . . . . . . . . . . . . . . . . . . . . . . . . . . . . . . . . . . . . . . . . . 0.58 mm (0.0228 inch)
  L and M models
    Standard . . . . . . . . . . . . . . . . . . . . . . . . . . . . . . . . . . . . . . . . . . . . . 0.05 to 0.20 mm (0.0019 to 0.0078 inch)
    Maximum . . . . . . . . . . . . . . . . . . . . . . . . . . . . . . . . . . . . . . . . . . . . 0.4 mm (0.0157 inch)
Crankshaft bend
  Standard . . . . . . . . . . . . . . . . . . . . . . . . . . . . . . . . . . . . . . . . . . . . . . . . . 0.02 mm (0.0008 inch) or less
  Maximum . . . . . . . . . . . . . . . . . . . . . . . . . . . . . . . . . . . . . . . . . . . . . . . . 0.05 mm (0.0019 inch)
Connecting rod side clearance
  Standard . . . . . . . . . . . . . . . . . . . . . . . . . . . . . . . . . . . . . . . . . . . . . . . . . 0.13 to 0.38 mm (0.0051 to 0.0149 inch)
  Maximum . . . . . . . . . . . . . . . . . . . . . . . . . . . . . . . . . . . . . . . . . . . . . . . . 0.6 mm (0.0236 inch)
Connecting rod bearing oil clearance
  Standard . . . . . . . . . . . . . . . . . . . . . . . . . . . . . . . . . . . . . . . . . . . . . . . . . 0.036 to 0.066 mm (0.0014 to 0.0025 inch)
  Maximum . . . . . . . . . . . . . . . . . . . . . . . . . . . . . . . . . . . . . . . . . . . . . . . . 0.10 mm (0.0039 inch)
Connecting rod big-end bore diameter
  No mark on side of rod . . . . . . . . . . . . . . . . . . . . . . . . . . . . . . . . . . . . . 37.000 to 37.008 mm (1.4567 to 1.4570 inch)
  '0' mark on side of rod . . . . . . . . . . . . . . . . . . . . . . . . . . . . . . . . . . . . 37.009 to 37.016 mm (1.4570 to 1.4573 inch)
Connecting rod journal diameter
  No mark on crank throw . . . . . . . . . . . . . . . . . . . . . . . . . . . . . . . . . . . . 33.984 to 33.992 mm (1.3379 to 1.3382 inch)
  '0' mark on crank throw . . . . . . . . . . . . . . . . . . . . . . . . . . . . . . . . . . . 33.993 to 34.000 mm (1.3383 to 1.3385 inch)
Connecting rod bolt stretch
  New connecting rod . . . . . . . . . . . . . . . . . . . . . . . . . . . . . . . . . . . . . . . 0.20 to 0.31 mm 0.0078 to 0.0122 inch)
  Use connecting rod . . . . . . . . . . . . . . . . . . . . . . . . . . . . . . . . . . . . . . . 0.26 to 0.28 mm (0.0102 to 0.0110 inch)

## Oil pump and relief valve
Oil pressure (warm) . . . . . . . . . . . . . . . . . . . . . . . . . . . . . . . . . . . . . . . . .   1.65 to 2.27 Bars (24 to 33 psi) @ 4000 rpm
Relief valve opening pressure . . . . . . . . . . . . . . . . . . . . . . . . . . . . . . . . .   4.34 to 5.85 Bars (63 to 85 psi)

## Clutch
Spring free length
  Standard . . . . . . . . . . . . . . . . . . . . . . . . . . . . . . . . . . . . . . . . . . . . . . .   34.7 mm (1.366 inch)
  Minimum . . . . . . . . . . . . . . . . . . . . . . . . . . . . . . . . . . . . . . . . . . . . . . .   33.5 mm (1.318 inch)
Friction plate thickness . . . . . . . . . . . . . . . . . . . . . . . . . . . . . . . . . . . . . .   not specified
Steel plate thicknesses . . . . . . . . . . . . . . . . . . . . . . . . . . . . . . . . . . . . . .   2.0, 2.3, 2.6 mm (0.0787, 0.0905, 0.1023 inch)
Friction and steel plate warpage limit . . . . . . . . . . . . . . . . . . . . . . . . . .   0.2 mm (0.0078 inch)

## Transmission
Gear ratios (no. of teeth)
  First . . . . . . . . . . . . . . . . . . . . . . . . . . . . . . . . . . . . . . . . . . . . . . . . . . .   2.375 : 1 (38/16)
  Second . . . . . . . . . . . . . . . . . . . . . . . . . . . . . . . . . . . . . . . . . . . . . . . . .   1.894 : 1 (36/19)
  Third . . . . . . . . . . . . . . . . . . . . . . . . . . . . . . . . . . . . . . . . . . . . . . . . . . .   1.691 : 1 (34/21)
  Fourth . . . . . . . . . . . . . . . . . . . . . . . . . . . . . . . . . . . . . . . . . . . . . . . . . .   1.409 : 1 (31/22)
  Fifth . . . . . . . . . . . . . . . . . . . . . . . . . . . . . . . . . . . . . . . . . . . . . . . . . . .   1.291 : 1 (31/24)
  Sixth . . . . . . . . . . . . . . . . . . . . . . . . . . . . . . . . . . . . . . . . . . . . . . . . . . .   1.200 : 1 (30/25)
Shift fork groove width
  Standard . . . . . . . . . . . . . . . . . . . . . . . . . . . . . . . . . . . . . . . . . . . . . . .   6.05 to 6.15 mm (0.2381 to 0.2421 inch)
  Maximum . . . . . . . . . . . . . . . . . . . . . . . . . . . . . . . . . . . . . . . . . . . . . . .   6.25 mm (0.2460 inch)
Shift fork ear thickness
  Standard . . . . . . . . . . . . . . . . . . . . . . . . . . . . . . . . . . . . . . . . . . . . . . .   5.9 to 6.0 mm (0.0323 to 0.2362 inch)
  Minimum . . . . . . . . . . . . . . . . . . . . . . . . . . . . . . . . . . . . . . . . . . . . . . .   5.8 mm (0.2283 inch)
Shift fork guide pin diameter
  Standard . . . . . . . . . . . . . . . . . . . . . . . . . . . . . . . . . . . . . . . . . . . . . . .   7.9 to 8.0 mm (0.3110 to 0.3149 inch)
  Minimum . . . . . . . . . . . . . . . . . . . . . . . . . . . . . . . . . . . . . . . . . . . . . . .   not specified
Shift drum groove width
  Standard . . . . . . . . . . . . . . . . . . . . . . . . . . . . . . . . . . . . . . . . . . . . . . .   8.05 to 8.20 mm (0.3169 to 0.3228 inch)
  Maximum . . . . . . . . . . . . . . . . . . . . . . . . . . . . . . . . . . . . . . . . . . . . . . .   8.3 mm (0.3267 inch)

## Torque specifications
Valve cover bolts . . . . . . . . . . . . . . . . . . . . . . . . . . . . . . . . . . . . . . . . . . .   9.8 Nm (87 inch-lbs)
Camshaft bearing cap bolts . . . . . . . . . . . . . . . . . . . . . . . . . . . . . . . . . .   12 Nm (104 inch-lbs)
Oil hose banjo bolt (to oil passage cover) . . . . . . . . . . . . . . . . . . . . .   34 Nm (25 ft-lbs)
Oil hose lower fitting bolts . . . . . . . . . . . . . . . . . . . . . . . . . . . . . . . . . . .   9.8 Nm (87 inch-lbs)
Camshaft chain tensioner cap (L and M models) . . . . . . . . . . . . . . .   8.3 Nm (74 inch-lbs)
10 mm cylinder head bolts
  Initial torque . . . . . . . . . . . . . . . . . . . . . . . . . . . . . . . . . . . . . . . . . . . .   20 Nm (14.5 ft-lbs)
  Final torque
    New bolts . . . . . . . . . . . . . . . . . . . . . . . . . . . . . . . . . . . . . . . . . . . . .   54 Nm (40 ft-lbs)
    Reused bolts . . . . . . . . . . . . . . . . . . . . . . . . . . . . . . . . . . . . . . . . . . .   49 Nm (36 ft-lbs)
6 mm cylinder head bolts . . . . . . . . . . . . . . . . . . . . . . . . . . . . . . . . . . . .   12 Nm (104 inch-lbs)
Coolant passage to cylinder block bolts . . . . . . . . . . . . . . . . . . . . . . .   9.8 Nm (87 inch-lbs)
Clutch cover bolts . . . . . . . . . . . . . . . . . . . . . . . . . . . . . . . . . . . . . . . . .   9.8 Nm (87 inch-lbs) (see note 1)
Pressure plate bolts . . . . . . . . . . . . . . . . . . . . . . . . . . . . . . . . . . . . . . . .   8.8 Nm (78 inch-lbs)
Pressure plate posts . . . . . . . . . . . . . . . . . . . . . . . . . . . . . . . . . . . . . . .   25 Nm (18 ft-lbs) (see note 2)
Clutch hub nut . . . . . . . . . . . . . . . . . . . . . . . . . . . . . . . . . . . . . . . . . . . .   135 Nm (100 ft-lbs) (see note 2)
Clutch fluid reservoir cap retainer screw . . . . . . . . . . . . . . . . . . . . . .   1.2 Nm (10 inch-lbs)
Master cylinder clamp bolts . . . . . . . . . . . . . . . . . . . . . . . . . . . . . . . . .   8.8 Nm (78 inch-lbs)
Clutch lever pivot screw . . . . . . . . . . . . . . . . . . . . . . . . . . . . . . . . . . . .   9.8 Nm (87 inch-lbs)
Clutch lever pivot screw locknut . . . . . . . . . . . . . . . . . . . . . . . . . . . . .   5.9 Nm (52 inch-lbs)
Clutch hose banjo bolts . . . . . . . . . . . . . . . . . . . . . . . . . . . . . . . . . . . .   25 Nm (18 ft-lbs)
Slave cylinder bleed valve . . . . . . . . . . . . . . . . . . . . . . . . . . . . . . . . . . .   7.8 Nm (69 inch-lbs)
Oil pan bolts . . . . . . . . . . . . . . . . . . . . . . . . . . . . . . . . . . . . . . . . . . . . . .   12 Nm (104 inch-lbs
Oil pump bolts . . . . . . . . . . . . . . . . . . . . . . . . . . . . . . . . . . . . . . . . . . . .   12 Nm (104 inch-lbs) (see note 3)
Oil return hose fitting bolts . . . . . . . . . . . . . . . . . . . . . . . . . . . . . . . . . .   9.8 Nm (87 inch-lbs) (see note 3)
Relief valves-to-crankcase . . . . . . . . . . . . . . . . . . . . . . . . . . . . . . . . . .   15 Nm (11 ft-lbs) (see note 3)
Engine mounting adjusters . . . . . . . . . . . . . . . . . . . . . . . . . . . . . . . . . .   9.8 Nm (87 inch-lbs)
Engine mounting adjuster locknuts . . . . . . . . . . . . . . . . . . . . . . . . . . .   49 Nm (36 ft-lbs)
Engine mounting bolts/nuts . . . . . . . . . . . . . . . . . . . . . . . . . . . . . . . . .   59 Nm (43 ft-lbs)
Crankcase bolts
  6 mm bolts . . . . . . . . . . . . . . . . . . . . . . . . . . . . . . . . . . . . . . . . . . . . .   20 Nm (14.5 ft-lbs)
  8 mm bolts . . . . . . . . . . . . . . . . . . . . . . . . . . . . . . . . . . . . . . . . . . . . .   27 Nm (20 ft-lbs)
Connecting rod nuts (see note 4)
  New rods . . . . . . . . . . . . . . . . . . . . . . . . . . . . . . . . . . . . . . . . . . . . . . .   25 Nm (19 ft-lbs) + 120°
  Used rods . . . . . . . . . . . . . . . . . . . . . . . . . . . . . . . . . . . . . . . . . . . . . .   18 Nm (13 ft-lbs) + 120°

## Torque specifications (continued)

| | |
|---|---|
| Crankcase oil passage plugs | 20 Nm (14.5 ft-lbs) (see note 5) |
| Alternator chain sprocket oil nozzle bolt | 12 Nm (104 inch-lbs) (see note 3) |
| Alternator chain tensioner bolts | 12 Nm (104 inch-lbs) (see note 3) |
| Alternator chain tensioner locknut | 20 Nm (14.5 ft-lbs) |
| Transmission oil pipe bolt | 12 Nm (104 inch-lbs) (see note 3) |
| Alternator shaft bearing retainer bolts | 12 Nm (104 inch-lbs) (see note 3) |
| Alternator coupling bolt | 25 Nm (18 ft-lbs) |
| External shift linkage return spring post | 25 Nm (18 ft-lbs) (see note 3) |
| Gear positioning lever bolt | 9.8 Nm (87 inch-lbs) (see note 3) |
| External shift linkage cover bolts/screws | |
|     Upper front bolt | 9.8 Nm (87 inch-lbs) (see note 3) |
|     Remaining bolts | 9.8 Nm (87 inch-lbs) |
|     Screws | 4.9 Nm (43 inch-lbs) (see note 3) |
| Shift drum cam bolt | 20 Nm (14.5 ft-lbs) |
| Shift rod retainer bolt | 12 Nm (104 inch-lbs) (see note 3) |
| Shift drum bearing holder bolts | 12 Nm (104 inch-lbs) (see note 3) |
| Starter clutch Allen bolts | 12 Nm (104 inch-lbs) (see note 3) |

**Notes:**

*(1) Apply non-permanent thread locking agent to the threads of the two forward bolts.*
*(2) Use a new nut.*
*(3) Apply non-permanent thread locking agent to the threads.*
*(4) Use new bolts. Apply engine oil to the threads, the undersides of the bolt heads and the nut surfaces that contact the connecting rods.*
*(5) Apply silicone sealant to the threads.*

## 1  General information

The engine/transmission unit is of the water-cooled, in-line, four-cylinder design, installed transversely across the frame. The sixteen valves are operated by double overhead camshafts which are chain driven off the crankshaft. The engine/transmission assembly is constructed from aluminum alloy. The crankcase is divided horizontally.

The crankcase incorporates a wet sump, pressure-fed lubrication system which uses a gear-driven, dual-rotor oil pump, an oil filter and by-pass valve assembly, a relief valve and an oil pressure switch. Also contained in the crankcase is the alternator shaft and the starter motor clutch.

Power from the crankshaft is routed to the transmission via the clutch, which is of the wet, multi-plate type and is gear-driven off the crankshaft. The transmission is a six-speed, constant-mesh unit.

## 2  Operations possible with the engine in the frame

The components and assemblies listed below can be removed without having to remove the engine from the frame. If, however, a number of areas require attention at the same time, removal of the engine is recommended.

*Gear selector mechanism external*
*components*
*Water pump*
*Starter motor*
*Alternator and alternator chain*
*Clutch assembly*
*Cam chain tensioner*
*Camshafts*
*Oil pan and pump*

## 3  Operations requiring engine removal

It is necessary to remove the engine/transmission assembly from the frame and separate the crankcase halves to gain access to the following components:

*Rocker arms and shafts*
*Cylinder head and block*
*Crankshaft, connecting rods and bearings*
*Transmission shafts*
*Shift drum and forks*
*Alternator shaft and starter motor clutch*
*Camshaft chain*

## 4  Major engine repair – general note

1  It is not always easy to determine when or if an engine should be completely overhauled, as a number of factors must be considered.

2  High mileage is not necessarily an indication that an overhaul is needed, while low mileage, on the other hand, does not preclude the need for an overhaul. Frequency of servicing is probably the single most important consideration. An engine that has regular and frequent oil and filter changes, as well as other required maintenance, will most likely give many miles of reliable service. Conversely, a neglected engine, or one which has not been broken in properly, may require an overhaul very early in its life.

3  Exhaust smoke and excessive oil consumption are both indications that piston rings and/or valve guides are in need of attention. Make sure oil leaks are not responsible before deciding that the rings and guides are bad. Refer to Chapter 1 and perform a cylinder compression check to determine for certain the nature and extent of the work required.

4  If the engine is making obvious knocking or rumbling noises, the connecting rod and/or main bearings are probably at fault.

5  Loss of power, rough running, excessive valve train noise and high fuel consumption rates may also point to the need for an overhaul, especially if they are all present at the same time. If a complete tune-up does not remedy the situation, major mechanical work is the only solution.

6  An engine overhaul generally involves restoring the internal parts to the specifications of a new engine. During an overhaul the piston rings are replaced and the cylinder walls are bored and/or honed. If a rebore is done, then new pistons are also required. The main and connecting rod bearings are generally replaced with new ones and, if necessary, the crankshaft is also replaced. Generally the valves are serviced as well, since they are usually in less than perfect condition at this point. While the engine is being overhauled, other components such as the carburetors and the starter motor can be rebuilt also. The end result should be a like-new engine that will give as many trouble-free miles as the original.

7  Before beginning the engine overhaul, read through all of the related procedures to familiarize yourself with the scope and requirements of the job. Overhauling an engine is not all that difficult, but it is time consuming. Plan on the motorcycle being tied up for a minimum of two weeks. Check on the availability of parts and make sure that any necessary special tools, equipment and supplies are obtained in advance.

**5.8 Leave the radiator support brackets bolted to the engine**

**5.10 Remove the engine sprocket, but leave the drive chain draped over the output shaft**

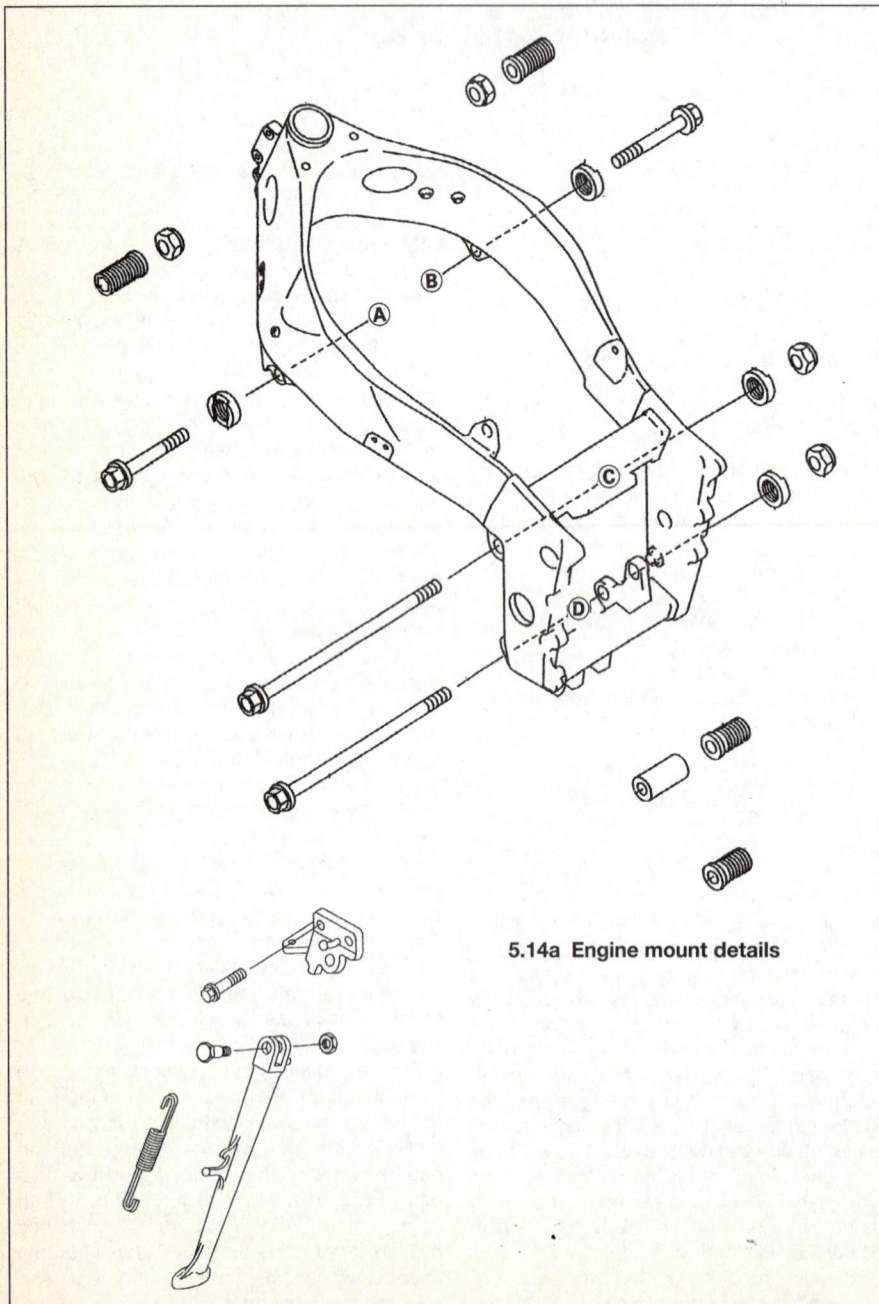

**5.14a Engine mount details**

**8** Most work can be done with typical shop hand tools, although a number of precision measuring tools are required for inspecting parts to determine if they must be replaced. Often a dealer service department or motorcycle repair shop will handle the inspection of parts and offer advice concerning reconditioning and replacement.

> **HAYNES HiNT** *As a general rule, time is the primary cost of an overhaul so it doesn't pay to install worn or substandard parts.*

**9** As a final note, to ensure maximum life and minimum trouble from a rebuilt engine, everything must be assembled with care in a spotlessly clean environment.

## 5 Engine – removal and installation

**Note:** *Engine removal and installation should be done with the aid of an assistant to avoid damage or injury that could occur if the engine is dropped. A hydraulic floor jack should be used to support and lower the engine if possible (they can be rented at low cost).*

### Removal

**1** Support the bike securely upright so it can't be knocked over during this procedure.
**2** Remove the seat(s) and the upper and lower fairing panels (see Chapter 8).
**3** Remove the clutch slave cylinder (see Section 20).
**4** Drain the coolant and the engine oil (see Chapter 1).
**5** Remove the fuel tank, air cleaner housing, carburetors and exhaust system (see Chapter 4). Plug the carburetor intake openings with rags.
**6** Disconnect the spark plug wires (see Chapter 1).
**7** Remove the baffle plate from the valve cover (see Section 7). Disconnect the air suction valve hoses (if equipped) from the valve cover.
**8** Disconnect the oil cooler hoses and remove the radiator (see Chapter 3). Remove the radiator mounting bracket, but leave its support brackets bolted to the engine **(see illustration)**.
**9** Remove the shift pedal (see Section 21).
**10** Remove the engine sprocket (see Chapter 6). Leave the chain draped over the output shaft **(see illustration)**.
**11** Remove the air suction valve and the vacuum switching valve (if equipped) (see Chapter 4).
**12** Disconnect the negative cable from the battery (see Chapter 9).

> ⚠ **Warning: Always disconnect the battery negative lead first and reconnect it last.**

**13** Label and disconnect the following wires (refer to Chapters 5 or 9 for component location if necessary):

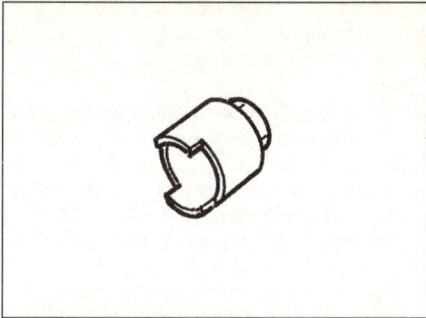

5.14b  This special tool is required to loosen and tighten the locknuts on the engine mounting adjusters

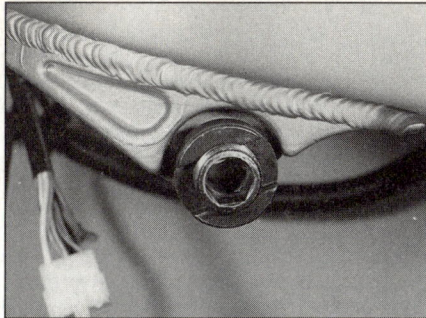

5.14c  Loosen the locknuts on the mounting adjusters accessible from inside the frame . . .

5.14d  . . . and the nuts on the upper rear mounting bolt . . .

5.14e  . . . on the lower rear mounting bolt . . .

5.14f  . . . on the right front mounting bolt (arrow) . . .

5.14g  . . . and on the left front mounting bolt (arrow)

5.18  Support the engine with a jack; place a block of wood between the jack and oil pan to protect the oil pan

*Pickup coil*
*Alternator*
*Oil pressure, sidestand and neutral switches*
*Starter cable*

14  Loosen the locknuts on the engine mounting adjusters and the nuts on the engine mounting bolts **(see illustrations)**.
15  Remove the sidestand (see Chapter 8).
16  Securely support the rear portion of the frame on jackstands or equivalent. Make sure whatever support you use is strong and stable so the bike can't be knocked over and the engine can't fall as it's removed.
17  Slowly squeeze the brake lever to lock the front wheel, then secure the brake lever in the locked position with a bungee cord or similar tie.

18  Support the engine with a floor jack **(see illustration)**. It's a good idea to use a block of soft wood between the jack and the oil pan to protect the pan from damage.
19  Remove the engine mounting bolts and the spacer on the upper rear bolt **(see illustration)**.
20  Turn all of the engine mounting adjusters anti-clockwise so they're clear of the engine **(see illustration)**.
21  Raise the engine slightly with the jack. Move it to the right and unhook the chain from the output shaft **(see illustration 5.10)**.
22  Slowly lower the engine to the floor. Have an assistant help you guide it so it doesn't fall, then move the engine to a suitable work surface **(see illustration)**.

5.19  Remove the spacer on the upper rear mounting bolt (arrow)

5.20  Turn the adjusters anti-clockwise to back them away from the engine

5.22  Have an assistant help support the engine; lower it slowly and carefully with the jack

**5.23 Turn the adjusters all the way anti-clockwise so they look like this when viewed from inside the frame**

## Installation

23 Before installing the engine, install the mounting adjusters and their locknuts. Leave the locknuts loose and turn the adjusters all the way anti-clockwise **(see illustration)**.

24 Lift the engine up into its installed position in the frame. As you go, hang the drive chain over the output shaft **(see illustration 5.10)**.

25 Install the two rear mounting bolts from the left side **(see illustration)**. Don't push them all the way in yet; leave them protruding about 55 mm (2¼ inches).

26 Use the jack to raise or lower the engine and align the front mounting bolt holes.

27 Screw the right front mounting adjuster in with an Allen bolt bit until it protrudes from the inside of the frame the distance listed in this Chapter's Specifications **(see illustration)**.

⚠️ **Warning: It's important to set the mounting adjusters to the correct length. Otherwise the engine may be incorrectly positioned in the frame, which will cause rapid drive chain wear and possible unstable handling.**

28 Screw the left front mounting adjuster in with an Allen bolt bit until it protrudes from the inside of the frame the distance listed in this Chapter's Specifications **(see illustration)**.

29 Tighten both front adjusters evenly to the torque listed in this Chapter's Specifications.

30 Install both front mounting bolts and their locknuts and tighten them slightly **(see illustration)**.

31 Turn the rear mounting adjusters in until they touch the engine **(see illustration)**, then tighten them to the torque listed in this Chapter's Specifications.

32 Push both of the rear mounting bolts the rest of the way in.

33 Tighten the front engine mounting bolts and nuts to the torque listed in this Chapter's Specifications.

34 Tighten the rear engine mounting bolts and nuts to the torque listed in this Chapter's Specifications.

35 Tighten the front mounting adjuster locknuts, then the rear mounting adjuster locknuts, to the torque listed in this Chapter's Specifications.

36 The remainder of installation is the reverse of the removal steps, with the following additions:
a) Tighten the sidestand bracket bolts to the torque listed in Chapter 8.
b) Use new gaskets at all exhaust pipe connections.
c) Adjust the drive chain, throttle cables and choke cable following the procedures in Chapter 1.
d) Fill the engine with oil and the cooling system with a 50/50 mixture of antifreeze and water, also following the procedures in Chapter 1. Run the engine and check for leaks.

## 6 Engine disassembly and reassembly – general information

**Note:** Refer to the 'Tools and Workshop Tips' Section at the end of this manual for further information.

1 Before disassembling the engine, clean the exterior with a degreaser and rinse it with

**5.25 Slip the rear mounting bolts partway into position**

**5.27 Set the adjusters to the specified length; this is the right front adjuster . . .**

**5.28 . . . and this is the left front adjuster**

**5.30 Install both front mounting bolts and their locknuts and tighten them slightly**

**5.31 Turn the rear mounting adjusters in until they touch the engine**

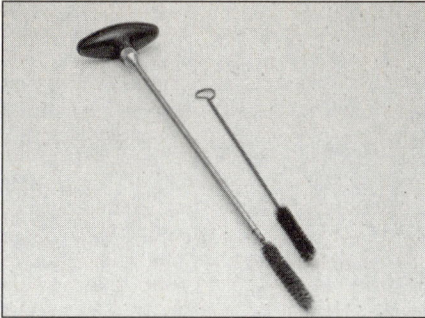

**6.2a  A selection of brushes is required for cleaning holes and passages in the engine components**

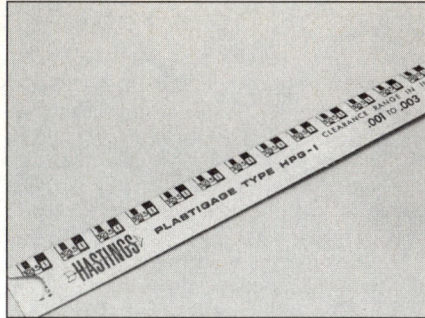

**6.2b  Type HPG-1 Plastigauge needed to check main bearing, connecting rod bearing and camshaft oil clearances**

**6.3  An engine stand can be made from short lengths of lumber and lag bolts or nails**

water. A clean engine will make the job easier and prevent the possibility of getting dirt into the internal areas of the engine.

2  In addition to the precision measuring tools mentioned earlier, you will need a torque wrench, a valve spring compressor, oil gallery brushes, a piston ring removal and installation tool, a piston ring compressor, a pin-type spanner wrench and a clutch holder tool (which is described in Section 19). Some new, clean engine oil of the correct grade and type, some engine assembly lube (or moly-based grease), a tube of Kawasaki Bond liquid gasket (part no. 92104-1003) or equivalent, and a tube of RTV (silicone) sealant will also be required. Although it may not be considered a tool, some Plastigauge (type HPG-1) should also be obtained to use for checking bearing oil clearances **(see illustrations)**.

**7.7a  Lift the baffle plate off the engine . . .**

3  An engine support stand made from short lengths of 2 x 4's bolted together will facilitate the disassembly and reassembly procedures **(see illustration)**. The perimeter of the mount should be just big enough to accommodate the engine oil pan. If you have an automotive-type engine stand, an adapter plate can be made from a piece of plate, some angle iron and some nuts and bolts.

4  When disassembling the engine, keep 'mated' parts together (including gears, cylinders, pistons, etc. that have been in contact with each other during engine operation). These "mated" parts must be reused or replaced as an assembly.

5  Engine/transmission disassembly should be done in the following general order with reference to the appropriate Sections.

*Remove the camshaft and rocker arms*
*Remove the cylinder head*
*Remove the cylinder block*
*Remove the pistons*
*Remove the clutch*
*Remove the oil pan*
*Remove the external shift mechanism*
*Remove the alternator chain*
*Remove the alternator coupling*
*Separate the crankcase halves*
*Remove the alternator shaft and starter motor clutch*
*Remove the crankshaft and connecting rods*
*Remove the transmission shafts/gears*
*Remove the shift drum/forks*

6  Reassembly is accomplished by reversing the general disassembly sequence.

## 7  Valve cover – removal and installation

**Note:** *The valve cover can be removed with the engine in the frame. If the engine has been removed, ignore the steps which don't apply.*

### Removal

1  Support the bike securely upright.
2  Remove the fuel tank (see Chapter 4).
3  If you're working on an L or M model, remove the air cleaner housing, ram air duct and air intake duct (see Chapters 4 and 8).
4  Remove the throttle and choke cables (see Chapter 4).
5  Remove the air suction valve (if equipped) (see Chapter 4).
6  Remove the ignition coils, along with the spark plug wires (see Chapter 5).
7  Remove the baffle plate **(see illustrations)**.
8  Remove the valve cover bolts and rubber grommets **(see illustrations)**.
9  Lift the cover off the cylinder head. If it's stuck, don't attempt to pry it off – tap around the sides of it with a plastic hammer to dislodge it. Check the chain guide in the end of the cover – if it's excessively worn, pry it out and install a new one **(see illustration)**.

### Installation

10  Peel the rubber gasket from the head **(see illustration)**. If it is cracked, hardened, has soft spots or shows signs of general deterioration, replace it with a new one.

**7.7b  . . . removing wire harness retainers and trim as necessary**

**7.8a  Remove the valve cover bolts . . .**

**7.8b  . . . and their rubber grommets**

7.9 Replace the chain guide inside the valve cover if worn or damaged

7.10 Lift the valve cover gasket off the cylinder head

7.11a Replace the seals on top of the camshaft brackets (arrows) if deteriorated or damaged . . .

7.11b . . . and replace the air suction tube O-rings if necessary (L3 models use two O-rings at the bottom end of each tube)

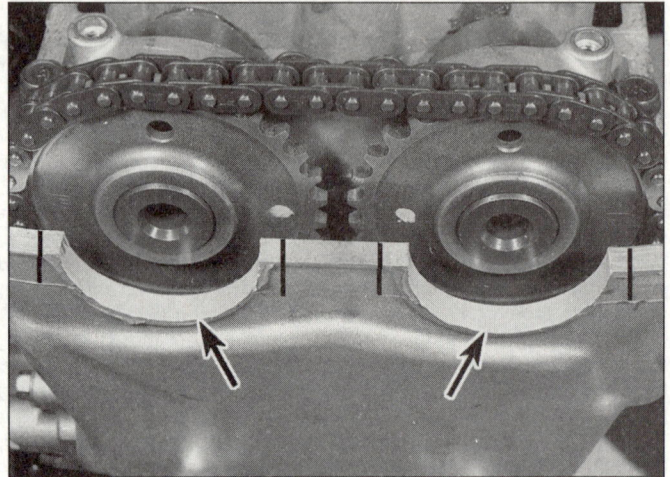

7.12 Apply sealant to the semicircular cutouts (arrows), extending outward to the lines

11 Inspect the seals on the camshaft brackets and, on models equipped with an air suction system, the O-rings on the air suction tubes **(see illustration)**. If the seals are cracked, hardened or deteriorated, pull them off and install new ones. If the O-rings show the same conditions, lift out the air suction tubes, install new O-rings and reinstall the tubes **(see illustration)**.

12 Clean the mating surfaces of the cylinder head and the valve cover with lacquer thinner, acetone or brake system cleaner. Apply a thin film of RTV sealant to the half-circle cutouts on each side of the head **(see illustration)**. If

you're working on an L or M model, also apply sealant to the top and bottom of the gasket along its straight runs (at front and rear of the gasket).

13 Position the gasket on the cylinder head. Position the valve cover on top of the gasket **(see illustrations)**.

14 Make sure the rubber seals are on the valve cover bolts. Install the bolts and tighten them evenly to the torque listed in this Chapter's Specifications.

15 The remainder of installation is the reverse of removal.

## 8 Camshaft chain tensioner – removal and installation

**Note:** *The camshaft chain tensioner can be removed with the engine in the frame. If the engine has been removed, ignore the steps which don't apply.*

### Removal

1 Support the bike securely upright.

2 Remove the fuel tank and carburetors (see Chapter 4).

3 If you're working on an L or M model, loosen the tensioner cap bolt **(see illustrations)**.

4 Remove the tensioner mounting bolts and detach it from the cylinder block **(see illustration)**.

*Caution: The tensioner piston locks in place as it extends. Once the tensioner bolts have been loosened, the tensioner must be removed from the engine and the piston reset before the bolts are tightened. If the bolts are loosened partway and then retightened without resetting the tensioner piston, the piston will be forced against the cam chain, damaging the tensioner or the chain.*

7.13a Install the gasket on the cylinder head . . .

7.13b . . . then install the valve cover on the gasket

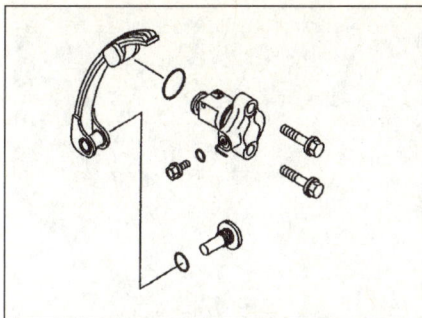

8.3a  Camshaft chain tensioner details

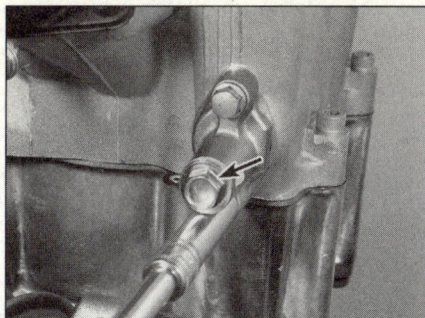

8.3b  Loosen the tensioner cap bolt (arrow) and the mounting bolts . . .

8.4  . . . and take the tensioner off the engine

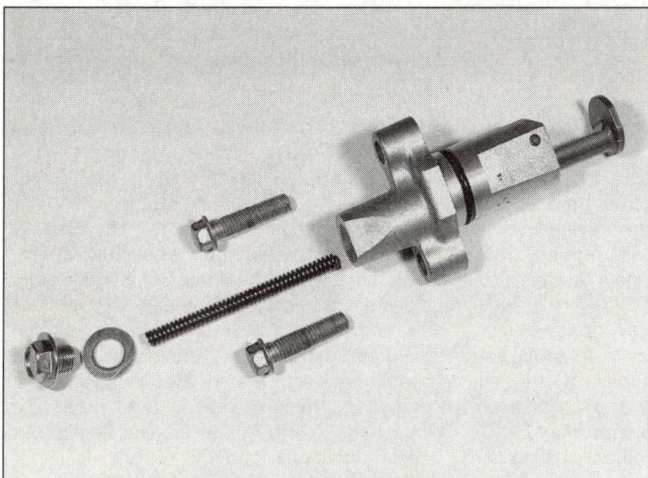

8.5  Camshaft chain tensioner components (L/M model shown; J/K similar)

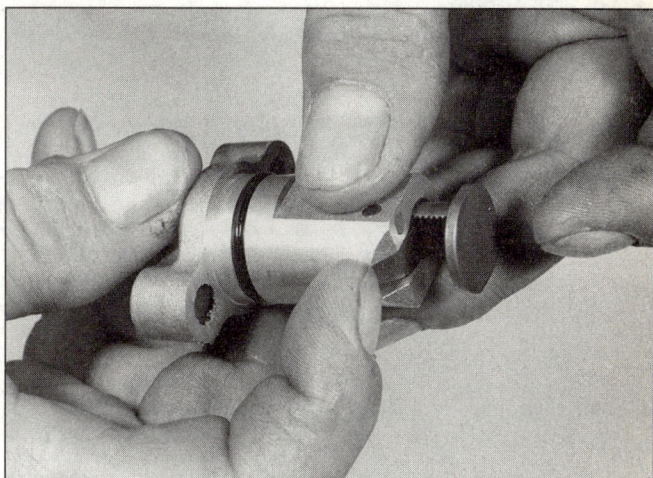

8.8  Lift the latch and push the tensioner piston into the body

**5** Remove the cap bolt (L and M models), sealing washer and spring from the tensioner body **(see illustration 8.3a and the accompanying illustration)** and wash them with solvent.

### Installation

**6** Lubricate the friction surfaces of the components with moly-based grease.
**7** Check the O-ring on the tensioner body for cracks or hardening. It's a good idea to replace this O-ring as a matter of course.

**8** Lift the tensioner latch, compress the tensioner piston into the body and release the latch **(see illustration)**. If you're working on a J or K model, insert a thin piece of wire into the cap bolt hole to lock the tensioner piston in position.
**9** Apply non-permanent thread locking agent to the threads of the tensioner mounting bolts. Position the tensioner body on the cylinder block and install the bolts, tightening them securely **(see illustration)**. On L and M

models, the flat side of the tensioner body must face the frame.
**10** If you're working on a J or K model, pull the wire out of the cap bolt hole. Install the cap bolt with a new O-ring.
**11** Install the spring (L and M models), sealing washer and cap bolt, tightening it to the torque listed in this Chapter's Specifications **(see illustrations)**.
**12** The remainder of installation is the reverse of removal.

8.9  Install the tensioner with the piston retracted

8.11a  Install the spring . . .

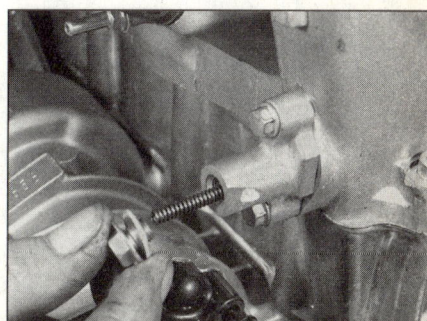

8.11b  . . . and the cap bolt with a new sealing washer (L/M model shown)

9.6a Loosen the bearing cap bolts evenly; note the cap number marks (arrow)

9.6b Lift off the bearing caps . . .

| 9 | Camshafts, rocker arm shafts and rocker arms – removal, inspection and installation |

**Note:** *The camshafts can be removed with the engine in the frame, but engine removal is necessary to remove the rocker arms and their shafts.*

## Camshafts

### Removal

**1** Remove the valve cover following the procedure given in Section 7.

**2** Remove the lower right fairing panel (see Chapter 8).
**3** Remove the pickup coil cover (see Chapter 5).
**4** Refer to the valve adjustment procedure in Chapter 1 and position no. 1 cylinder at top dead center (TDC) on its compression stroke.
**5** Remove the camshaft chain tensioner (see Section 8).
**6** Unscrew the bearing cap bolts evenly, a little at a time, until they are all loose **(see illustration)**.
*Caution: If the bearing cap bolts aren't loosened evenly, the camshaft may bind.*
Remove the bolts and lift off the bearing caps **(see illustrations)**. Note the numbers on the bearing caps which correspond to the

numbers on the cylinder head **(see illustration 9.6a)**. When you reinstall the caps, be sure to install them in the correct positions.
**7** Pull up on the camshaft chain and carefully guide the camshaft out **(see illustration)**. With the chain still held taut, remove the other camshaft **(see illustration)**. Look for marks on the camshafts. The intake camshaft should have an IN mark and the exhaust camshaft should have an EX mark **(see illustration)**. If you can't find these marks, label the camshafts to ensure they are installed in their original locations.
**8** Pull the exhaust side (front) cam chain guide out of its slot **(see illustrations)**.

9.6c . . . the grooves in the right-hand cap fit over the bearing retainer rings

9.7a Lift out one camshaft . . .

9.7b . . . then support the chain and lift out the other camshaft

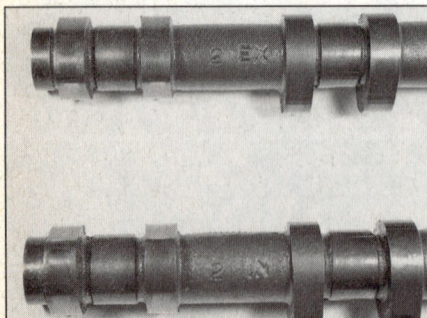

9.7c The camshafts are marked with IN (intake) and EX (exhaust)

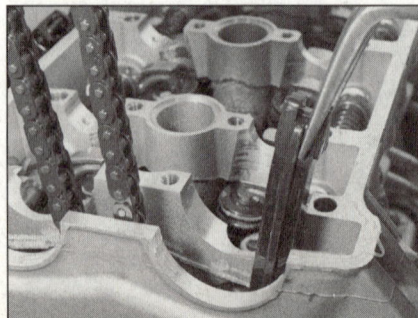

9.8a Lift out the front chain guide . . .

9.8b . . . the bottom end of the chain guide fits into a notch in the crankcase

9.9a Support the cam chain so it won't drop into the engine . . .

9.9b . . . if the chain is allowed to drop and bunch up like this, turning the crankshaft could cause damage

9.10 Install new bearing cap O-rings

**9** While the camshafts are out, wire the chain to another component to prevent it from dropping down **(see illustration)**. If the chain is allowed to drop, it will become wedged between the crankshaft and case, which could cause damage to these components if the crankshaft is turned with the camshafts removed **(see illustration)**.

**10** Inspect the bearing cap O-rings **(see illustration)**. Replace them if they're flattened, brittle or deteriorated. It's a good idea to replace them as a matter of course whenever the cam bearing caps are removed.

**11** Cover the top of the cylinder head with a rag to prevent foreign objects from falling into the engine.

## Inspection

**Note:** *Before replacing camshafts or the cylinder head and bearing caps because of damage, check with local machine shops specializing in motorcycle engine work. In the case of the camshafts, it may be possible for cam lobes to be welded, reground and hardened, at a cost far lower than that of a new camshaft. If the bearing surfaces in the cylinder head are damaged, it may be possible for them to be bored out to accept bearing inserts. Due to the cost of a new cylinder head it is recommended that all options be explored before condemning it as trash!*

9.13a Check camshaft lobes for wear – here's a good example of damage which will require replacement or repair

9.13b Measuring cam lobe height

**12** Inspect the cam bearing surfaces of the head and the bearing caps. Look for score marks, deep scratches and evidence of spalling (a pitted appearance).

**13** Check the camshaft lobes for heat discoloration (blue appearance), score marks, chipped areas, flat spots and spalling **(see illustration)**. Measure the height of each lobe **(see illustration)** and compare the results to the minimum lobe height listed in this Chapter's Specifications. If damage is noted or wear is excessive, the camshaft must be replaced. Also, be sure to check the condition of the rocker arms, as described later in this Section.

**Note:** *No. 1 cylinder must still be at TDC compression for this step.*

**14** Next, check the camshaft bearing oil clearances. Clean the camshafts, the bearing surfaces in the cylinder head and the bearing caps with a clean, lint-free cloth, then lay the cams in place in the cylinder head, with the IN and EX marks on the sprockets facing away from each other and level with the valve cover gasket surface of the cylinder head **(see illustrations)**. Engage the cam chain with the cam sprockets, so the camshafts don't turn as the bearing caps are tightened.

9.14a Align the 1.4 mark on the timing rotor with the crankcase pointer . . .

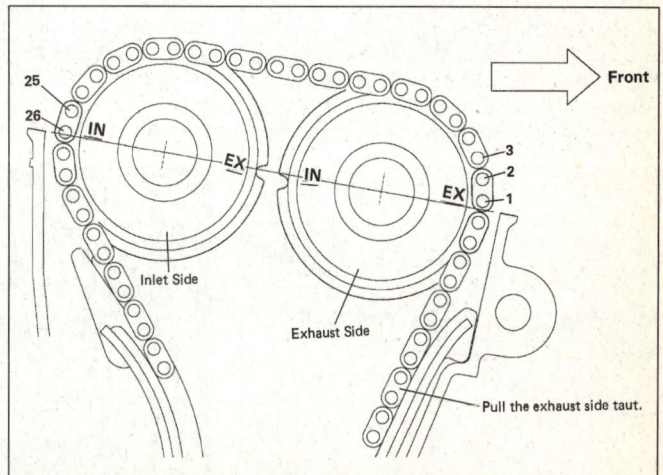

9.14b . . . and position the camshafts like this

**9.15 Lay a strip of Plastigauge on the camshaft journal, parallel to the camshaft centerline**

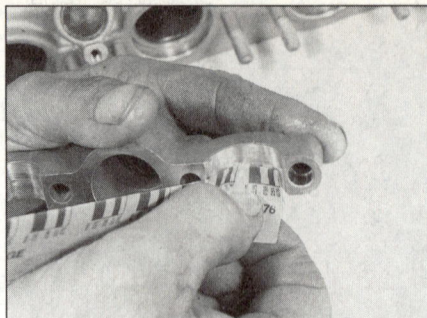

**9.19a Measure the width of the crushed Plastigauge with the scale on the container**

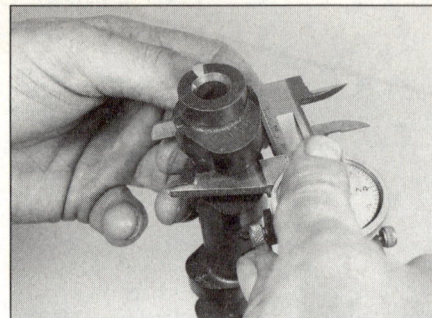

**9.19b Measuring the cam bearing journal diameter**

**15** Cut ten strips of Plastigauge (type HPG-1) and lay one piece on each bearing journal, parallel with the camshaft centerline **(see illustration)**.

**16** Install the bearing caps in their proper positions. The numbers on the caps must be toward the front of the engine. The first four caps are numbered one through 4, counting from the left side of the engine. The no. 5 cap can be identified by the size of its bearing journals (they fit the ball bearings on the ends of the camshafts).

**17** Install the cap bolts. Tighten the bolts on no. 5 cap (at the cam chain end) first, then tighten the remaining cap bolts evenly, in several stages, to the torque listed in this Chapter's Specifications. While doing this, do not let the camshafts rotate.

**18** Now unscrew the bolts, a little at a time, and carefully lift off the bearing caps.

**9.22 Check the ball bearing at the end of each camshaft for looseness, roughness or noise**

**19** To determine the oil clearance, compare the crushed Plastigauge (at its widest point) on each journal to the scale printed on the Plastigauge container **(see illustration)**. Compare the results to this Chapter's Specifications. If the oil clearance is greater than specified, measure the diameter of the cam bearing journal with a micrometer **(see illustration)**. If the journal diameter is less than the specified limit, replace the camshaft with a new one and recheck the clearance. If the clearance is still too great, replace the cylinder head and bearing caps with new parts (see the Note that precedes Step 12).

**20** Except in cases of oil starvation, the camshaft chain wears very little. If the chain has stretched excessively, which makes it difficult to maintain proper tension, replace it with a new one (see Section 29).

**21** Check the sprockets for wear, cracks and other damage, replacing them if necessary. If the sprockets are worn, the chain is also worn, and possibly the sprocket on the crankshaft (which can only be remedied by replacing the crankshaft). If wear this severe is apparent, the entire engine should be disassembled for inspection.

**22** Spin the ball bearing at the end of each camshaft and check for looseness or rough movement **(see illustration)**. The camshafts must be replaced if the bearings are defective.

**23** Check the chain guides for wear or damage. If they are worn or damaged, replace them.

### Installation

**24** Make sure the bearing surfaces in the cylinder head and the bearing caps are clean **(see illustration)**. Apply a coat of clean engine oil to the bearing surfaces and camshaft journals. If you're reinstalling the old camshafts, lubricate their lobes with clean engine oil; if you're installing new camshafts, lubricate the lobes with a thin coat of molybdenum disulfide grease.

**25** Lay the camshafts in the cylinder head (do not mix them up), ensuring the marks on the cam sprockets are aligned properly **(see illustration 9.14b)**.

**26** Make sure the timing marks are aligned as described in Step 11, then mesh the chain with the camshaft sprockets. Count the number of chain link pins between the EX mark and the IN mark **(see illustration 9.14b)**. There should be no slack in the chain between the two sprockets.

**27** Make sure the bearing cap dowels (and the locating half-rings on no. 5 bearing cap) are in position **(see illustrations)**. Install the bearing caps **(see illustrations 9.6a, 9.6b and 9.6c)**. *Caution: Be sure to install the caps in the correct locations; they're machined as a set with the cylinder head, so installing a cap in the wrong location may cause the camshaft to seize.*

**28** Install the cap bolts. Tighten the bolts on no. 5 cap (at the cam chain end) first, then tighten the remaining cap bolts evenly, in several stages, to the torque listed in this Chapter's Specifications.

**9.24 Check the cam bearing surfaces in the cap and head to make sure they're clean**

**9.27a Make sure the bearing cap dowels are in place . . .**

**9.27b . . . and install the bearing retainer rings**

**9.35 Thread a bolt (see text for details) into the end of the rocker shaft and pull on the bolt to remove the shaft**

**9.36 Remove the rocker arms and springs**

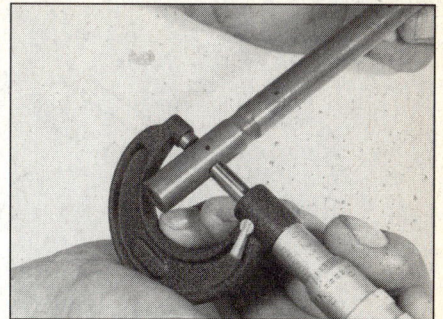

**9.39a Measuring the rocker shaft diameter**

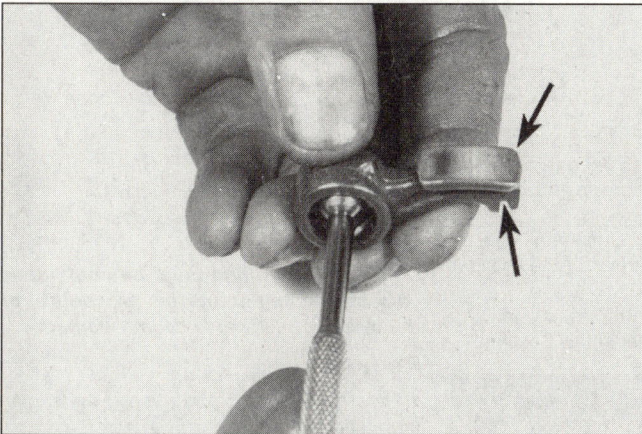

**9.39b Check the cam contact and valve contact surfaces (arrows) for wear and damage; measure the inside diameter of the rocker arm with a small hole gauge . . .**

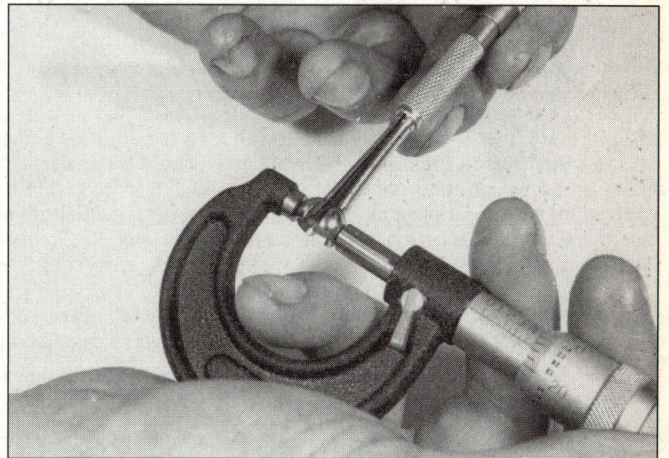

**9.39c . . . then measure the diameter of the gauge with a micrometer**

29 Insert your finger or a wood dowel into the cam chain tensioner hole and apply pressure to the cam chain. Check the timing marks to make sure they are aligned (see Step 14) and there are still the correct number of link pins between the EX and IN marks on the cam sprockets **(see illustration 9.14b)**. If necessary, change the position of the sprocket(s) on the chain to bring all of the marks into alignment.
*Caution: If the marks are not aligned exactly as described, the valve timing will be incorrect and the valves may contact the pistons, causing extensive damage to the engine.*
30 Install the cam chain tensioner as described in Section 8.
31 Adjust the valve clearances (see Chapter 1).
32 The remainder of installation is the reverse of removal.

### Rocker arm shafts and rocker arms

#### Removal

**Note:** *Removal of the engine is necessary to allow the rocker arm shafts to be withdrawn.*
33 Remove the engine (see Section 5). Remove the camshafts following the procedure earlier in this Section. Be sure to keep tension on the camshaft chain.

34 Remove the oil passage cover from the left end of the cylinder head (see Section 16).
35 Thread a bolt (M8 x 1.25 pitch x 20 mm) into the end of one of the rocker arm shafts and use it as a handle to pull the shaft out **(see illustration)**.
36 Remove the rocker arms and springs **(see illustration)**. Either slip them onto the rocker shaft in the order of removal, or label them so they can be returned to their original positions. Make sure the valve adjustment shims remain in the valve spring retainers.
37 Repeat the above Steps to remove the other rocker arm shaft and rocker arms. Keep all of the parts in order so they can be reinstalled in their original locations.

#### Inspection

38 Clean all of the components with solvent and dry them off. Blow through the oil passages in the rocker arms with compressed air, if available. Inspect the rocker arm faces for pits, spalling, score marks and rough spots **(see illustration 9.36)**. Check the rocker arm-to-shaft contact areas as well. Look for cracks in each rocker arm. If the faces of the rocker arms are damaged, the rocker arms and the camshafts should be replaced as a set.
39 Measure the diameter of the rocker arm shafts, in the area where the rocker arms ride,

and compare the results with this Chapter's Specifications **(see illustration)**. Also measure the inside diameter of the rocker arms and compare the results with this Chapter's Specifications **(see illustrations)**. If either the shaft or the rocker arms are worn beyond the specified limits, replace them as a set.

#### Installation

40 Position the rocker arm and springs in the cylinder head **(see illustration)**. If you're reinstalling the original rocker arms and springs, return them to their original locations.

**9.40 Position the rocker arms and springs in the cylinder head, then slide the shafts in**

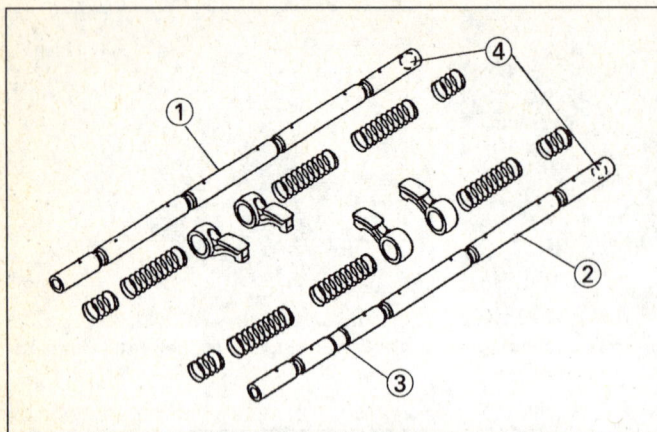

**9.41a Rocker assembly details**

1  *Exhaust rocker shaft*
2  *Intake rocker shaft*
3  *Blue paint mark or groove*
4  *Shaft end plugs*

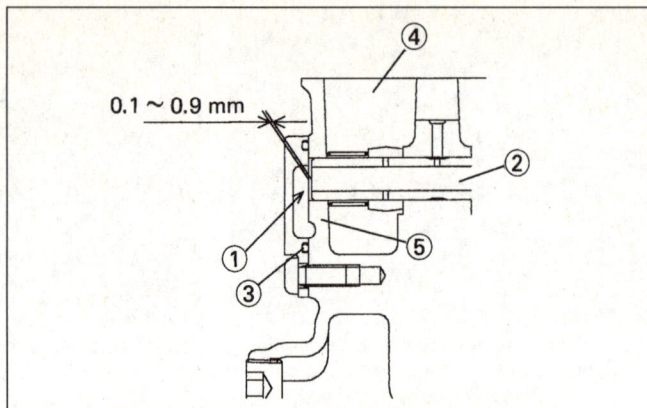

**9.41b Rocker shaft installation details**

1  *Left cylinder head cover*
2  *Rocker shaft*
3  *O-ring*
4  *Cylinder head*
5  *Rocker shaft support*

41 Lubricate the rocker arm shafts with engine oil and slide them into the cylinder head and through the rocker arms and springs. Install the rocker arm shaft with the blue mark and groove on the intake side **(see illustration)**. Insert both rocker arm shafts from the left side of the head (the same side they were removed from), with the plugged end of each shaft going in first. When the rocker shafts are pushed all the way in, they should be recessed 0.1 to 0.9 mm (0.004 to 0.036 inch) inside the bore in the cylinder head **(see illustration)**.

42 Refer to Section 16 and install the oil passage cover on the left end of the cylinder head.

43 Install the camshafts following the procedure described earlier in this Section.

## 10 Cylinder head – removal and installation

*Caution: The engine must be completely cool before beginning this procedure, or the cylinder head may become warped.*

### Removal

1 Refer to Section 5 and remove the engine from the motorcycle.

2 Remove the valve cover (see Section 7).

3 Remove the cam chain tensioner (Section 8).

4 Remove the camshafts and the front side cam chain guide (see Section 9).

5 Remove the oil passage cover from the left end of the cylinder head (see Section 16).

6 Remove the two 6 mm bolts from the cam chain end of the cylinder head **(see illustration)**. Loosen the remaining ten 10 mm bolts evenly, in two or three stages, starting with the outer bolts and working inward to the center of the cylinder head **(see illustration)**.

**10.6a Remove the two 6 mm bolts at the end of the cylinder head (arrows)**

**10.6b Then remove the 10 mm bolts evenly, starting with the outer bolts and working inward**

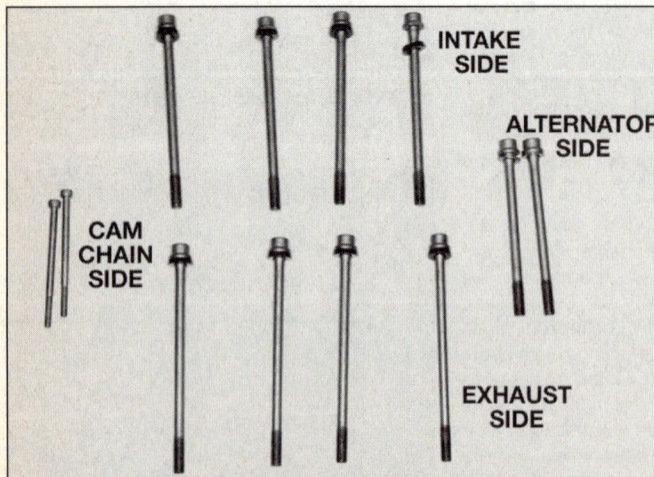

**10.6c Label the bolts; there are four different combinations of bolt size, head shape and washer thickness**

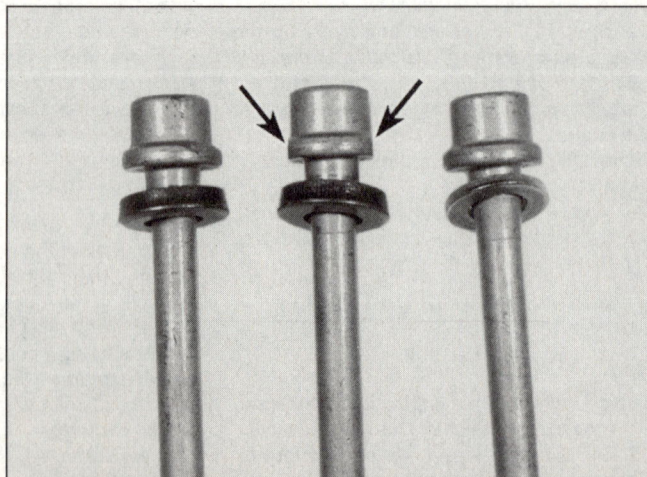

**10.6d The bolts on the rear (intake side) of the head have cutouts (arrows); the alternator side bolts have thinner washers**

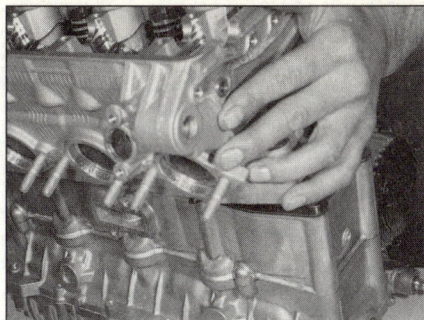

10.7  Lift the cylinder head off

10.8  Stuff a clean rag into the chain cavity (arrow) so nothing is dropped into it

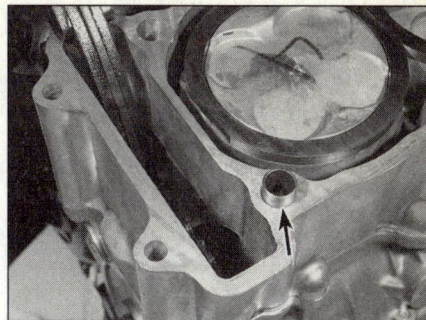

10.9a  There's a dowel at the right front corner (arrow) . . .

Label the bolts so they can be returned to their correct locations (see illustrations). The cam chain side bolts are smaller than the other ten; the alternator side bolts have thinner washers; the intake side bolts have cutaway areas on either side of the bolt head.

7  Lift the cylinder head off the cylinder block (see illustration). If the head is stuck, tap around the side of the head with a rubber mallet to jar it loose, or use two wooden dowels inserted into the intake or exhaust ports to lever the head off. Once the head is off, remove the gasket.

Caution: Don't attempt to pry the head off by inserting a screwdriver between the head and the cylinder block – you'll damage the sealing surfaces.

8  Support the cam chain so it won't drop into the cam chain tunnel, and stuff a clean rag into the cam chain tunnel to prevent the entry of debris (see illustration). Remove any head bolt washers that didn't come off with the bolts, using a pair of needle-nose pliers.

9  Remove the two dowel pins from the cylinder block (see illustrations).

10  Check the cylinder head gasket and the mating surfaces on the cylinder head and block for leakage, which could indicate warpage. Refer to Section 12 and check the flatness of the cylinder head.

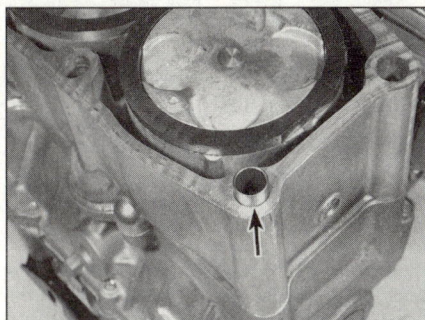

10.9b  . . . and at the left front corner (arrow)

11  Clean all traces of old gasket material from the cylinder head and block. Be careful not to let any of the gasket material fall into the crankcase, the cylinder bores or the water passages.

## Installation

12  Install the two dowel pins, then lay the new gasket in place on the cylinder block. Make sure the UP mark on the gasket is positioned on the intake (rear) side of the engine and is right side up (see illustration). If the word UP reads backwards, the gasket is upside down. Never reuse the old gasket and don't use any type of gasket sealant.

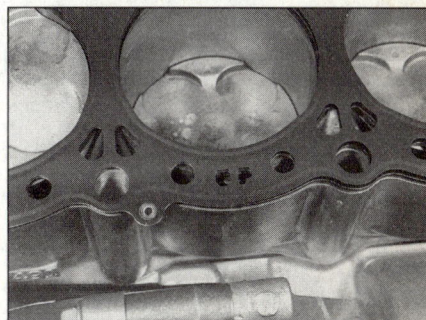

10.12  The UP mark should be upright when the head gasket is installed

13  Carefully lower the cylinder head over the dowels. It is helpful to have an assistant support the camshaft chain with a piece of wire so it doesn't fall and become kinked or detached from the crankshaft. When the head is resting against the cylinder block, wire the cam chain to another component to keep tension on it.

14  Lubricate both sides of the head bolt washers with engine oil and place them over the 10 mm head bolts. The two thin washers go on the bolts at the alternator end of the engine.

15  Install the head bolts (see illustration). Using the proper sequence (see illustration),

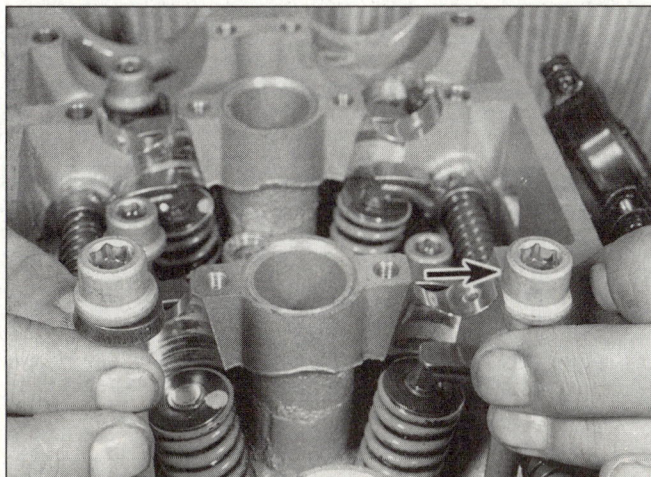

10.15a  Install the head bolts; the bolts with the cutouts (arrow) go on the rear (intake side)

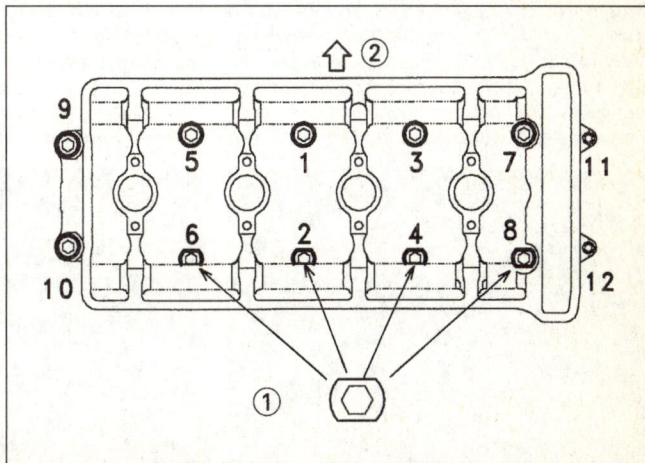

10.15b  Cylinder head bolt TIGHTENING sequence

1  Intake side bolt washers          2  Front

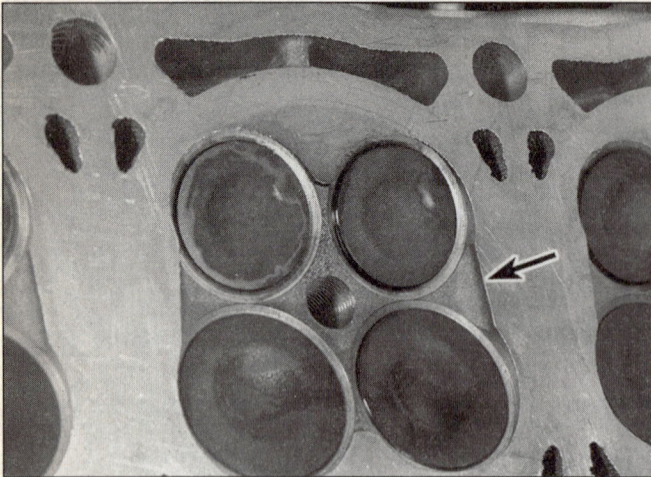

12.5 Thoroughly clean the combustion chamber (arrow), but don't scratch it

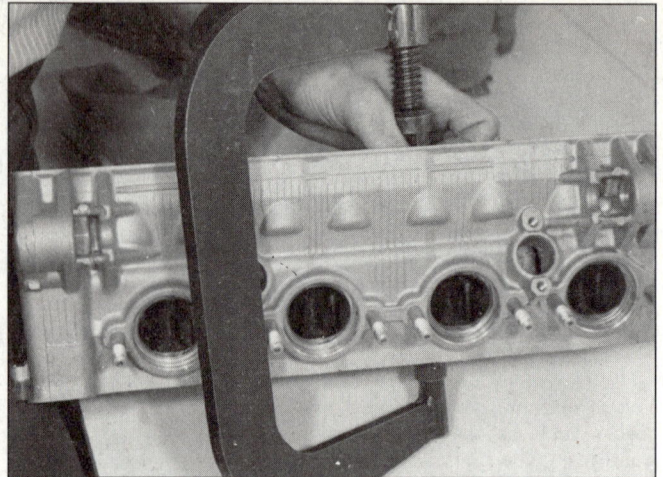

12.7a Install a valve spring compressor . . .

tighten the 10 mm bolts to the initial torque listed in this Chapter's Specifications.

**16** Using the same sequence, tighten all of the bolts to the final torque listed in this Chapter's Specifications.

**17** Install all parts removed for access, making sure that the cam chain front guide engages the notch in the crankcase **(see illustration 9.8b)**.

**18** Change the engine oil (see Chapter 1).

## 11 Valves/valve seats/ valve guides – servicing

**1** Because of the complex nature of this job and the special tools and equipment required, servicing of the valves, the valve seats and the valve guides (commonly known as a valve job) is best left to a professional.

**2** The home mechanic can, however, remove and disassemble the head, do the initial cleaning and inspection, then reassemble and deliver the head to a dealer service department or properly equipped motorcycle repair shop for the actual valve servicing. Refer to Section 12 for those procedures.

**3** The dealer service department will remove the valves and springs, recondition or replace the valves and valve seats, replace the valve

guides, check and replace the valve springs, spring retainers and keepers (as necessary), replace the valve seals with new ones and reassemble the valve components.

**4** After the valve job has been performed, the head will be in like-new condition. When the head is returned, be sure to clean it again very thoroughly before installation on the engine to remove any metal particles or abrasive grit that may still be present from the valve service operations. Use compressed air, if available, to blow out all the holes and passages.

## 12 Cylinder head and valves – disassembly, inspection and reassembly

**1** As mentioned in the previous Section, valve servicing and valve guide replacement should be left to a dealer service department or motorcycle repair shop. However, disassembly, cleaning and inspection of the valves and related components can be done (if the necessary special tools are available) by the home mechanic. This way no expense is incurred if the inspection reveals that service work is not required at this time.

**2** To properly disassemble the valve components without the risk of damaging them, a valve spring compressor is absolutely

necessary. If the special tool is not available, have a dealer service department or motorcycle repair shop handle the entire process of disassembly, inspection, service or repair (if required) and reassembly of the valves.

### Disassembly

**3** Remove the rocker arm shafts and rocker arms (see Section 9). Store the components in such a way that they can be returned to their original locations without getting mixed up (labeled plastic bags work well). Pick the valve clearance adjustment shims out of the valve spring retainers, using a magnet if necessary, and store them with their relative rocker arm.

**4** Before the valves are removed, scrape away any traces of gasket material from the head gasket sealing surface. Work slowly and do not nick or gouge the soft aluminum of the head. Gasket removing solvents, which work very well, are available at most motorcycle shops and auto parts stores.

**5** Carefully scrape all carbon deposits out of the combustion chamber area **(see illustration)**. A hand held wire brush or a piece of fine emery cloth can be used once the majority of deposits have been scraped away. *Caution: Do not use a wire brush mounted in a drill motor, or one with extremely stiff bristles, as the head material is soft and may be eroded away or scratched by the wire brush.*

**6** Before proceeding, arrange to label and store the valves along with their related components so they can be kept separate and reinstalled in the same valve guides they are removed from (again, plastic bags work well for this).

**7** Compress the valve spring(s) on the first valve with a spring compressor, then remove the keepers/collets and the retainer from the valve assembly **(see illustrations)**. Do not compress the spring(s) any more than is absolutely necessary. Carefully release the valve spring compressor and remove the spring(s), spring seat and valve from the head **(see illustrations)**. If the valve binds in the

12.7b . . . compress the spring and remove the keepers (collets)

12.7c . . . the valve spring retainer . . .

12.7d . . . the valve spring(s) . . .

12.7e . . . the spring seat . . .

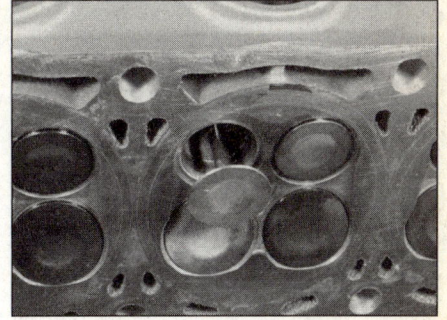
12.7f . . . and the valve

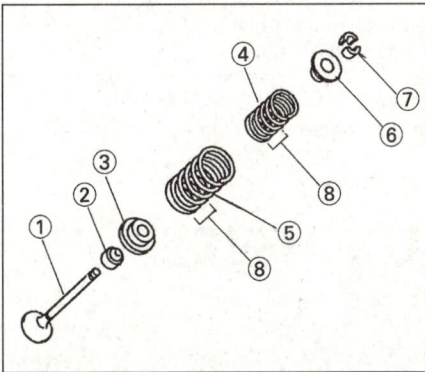
12.7g Valve details

1  Valve stem
2  Oil seal
3  Spring seat
4  Inner valve spring (except J models)
5  Outer valve spring
6  Valve spring retainer
7  Keepers (collets)
8  Tightly wound coils

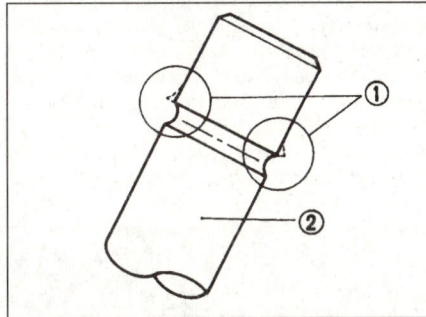
12.7h Check the area around the keeper/collet groove for burrs

1  Burrs (removed)        2  Valve stem

guide (won't pull through), push it back into the head and deburr the area around the keeper/collet groove with a very fine file or whetstone (see illustration).

8 Repeat the procedure for the remaining valves. Remember to keep the parts for each valve together so they can be reinstalled in the same location.

9 Once the valves have been removed and labeled, pull off the valve stem seals with pliers and discard them (the old seals should never be reused).

10 Next, clean the cylinder head with solvent and dry it thoroughly. Compressed air will speed the drying process and ensure that all holes and recessed areas are clean.

11 Clean all of the valve springs, keepers/collets, retainers and spring seats with solvent and dry them thoroughly. Do the parts from one valve at a time so that no mixing of parts between valves occurs.

12 Scrape off any deposits that may have formed on the valve, then use a motorized wire brush to remove deposits from the valve heads and stems. Again, make sure the valves do not get mixed up.

### Inspection

13 Inspect the head very carefully for cracks and other damage. If cracks are found, a new head will be required. Check the cam bearing surfaces for wear and evidence of seizure. Check the camshafts and rocker arms for wear as well (see Section 9).

14 Using a precision straight-edge and a feeler gauge, check the head gasket mating surface for warpage. Lay the straight-edge lengthwise, across the head and diagonally (corner-to-corner), intersecting the head bolt holes, and try to slip a 0.002 in (0.05 mm) feeler gauge under it, on either side of each combustion chamber (see illustration). If the feeler gauge can be inserted between the head and the straightedge, the head is warped and must either be machined or, if warpage is excessive, replaced with a new one.

15 Examine the valve seats in each of the combustion chambers. If they are pitted, cracked or burned, the head will require valve service that is beyond the scope of the home mechanic. Measure the valve seat width (see illustration) and compare it to this Chapter's

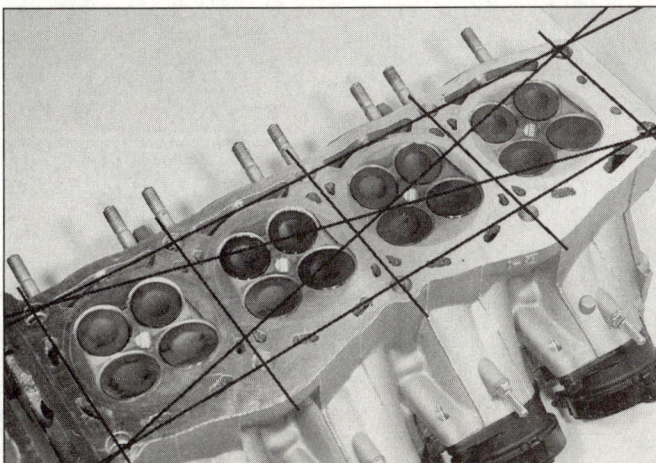
12.14 Check the gasket surface for flatness with a straight-edge and feeler gauge; measure in the directions shown

12.15 Measuring valve seat width

**12.16a Measure the valve guide inside diameter with a hole gauge . . .**

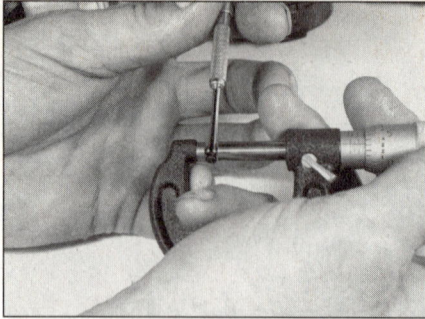

**12.16b . . . then measure the gauge with a micrometer**

Specifications. If it is not within the specified range, or if it varies around its circumference, valve service work is required.

16  Clean the valve guides to remove any carbon buildup, then measure the inside diameters of the guides (at both ends and the center of the guide) with a small hole gauge and a 0-to-1-inch micrometer **(see illustrations)**. Record the measurements for future reference. The guides are measured at the ends and at the center to determine if they are worn in a bell-mouth pattern (more wear at the ends). If they are, guide replacement is an absolute must.

17  Carefully inspect each valve face for cracks, pits and burned spots. Check the valve stem and the keeper groove area for cracks **(see illustration)**. Rotate the valve and check for any obvious indication that it is bent. Check the end of the stem for pitting and excessive wear and make sure the bevel is the specified width. The presence of any of the above conditions indicates the need for valve servicing.

18  Measure the valve stem diameter **(see illustration)**. If the diameter is less than listed in this Chapter's Specifications, the valves will have to be replaced with new ones. Also check the valve stem for bending. Set the valve in a V-block with a dial indicator touching the middle of the stem **(see illustration)**. Rotate the valve and note the reading on the gauge. If the stem runout exceeds the value listed in this Chapter's Specifications, replace the valve. Measure the valve head margin thickness **(see illustration)**. Replace the valve if it's less than the thickness listed in this Chapter's Specifications.

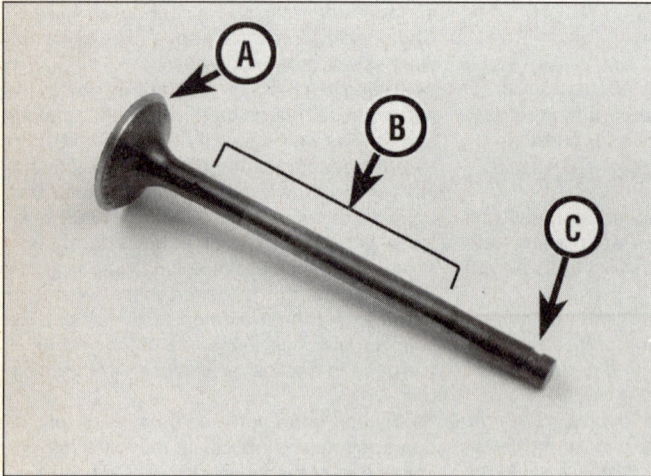

**12.17 Check the valve face (A), stem (B) and keeper/collet groove (C) for wear and damage**

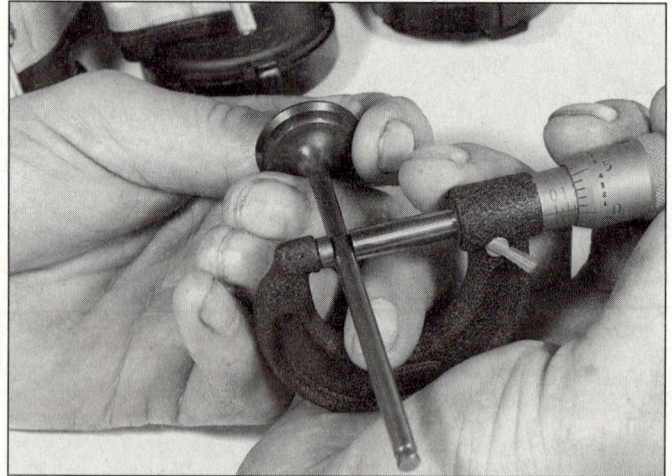

**12.18a Measuring valve stem diameter**

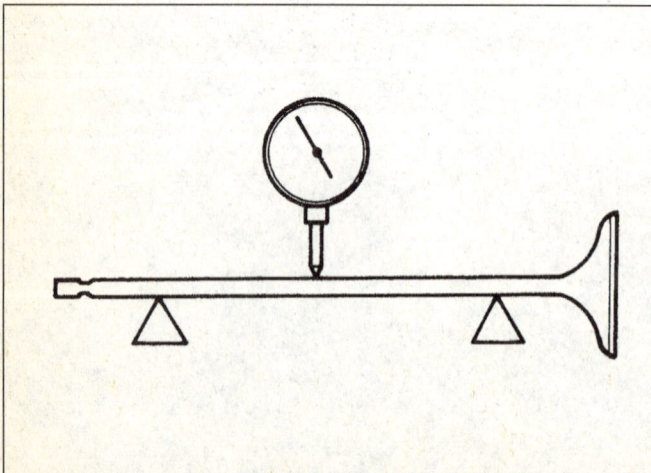

**12.18b Check the valve stem for bends with a V-block (or V-blocks, as shown here) and a dial indicator**

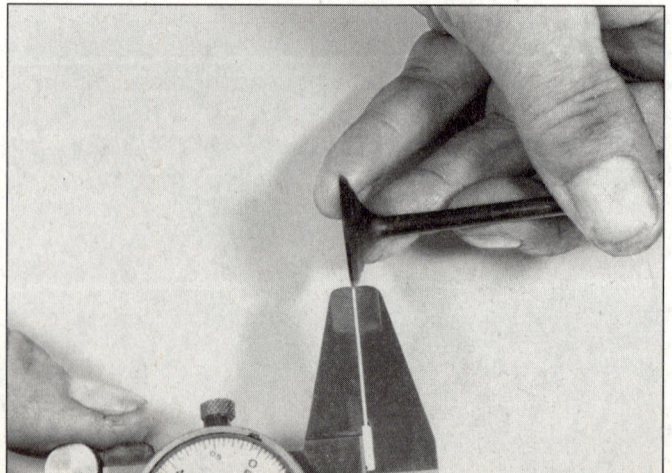

**12.18c Measuring valve head margin thickness**

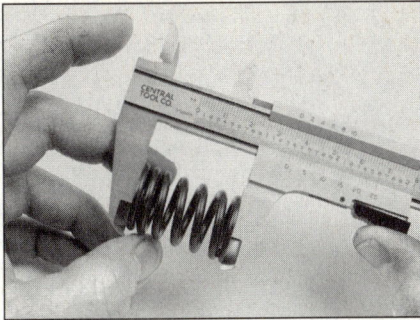

12.19a Measuring the free length of the valve springs

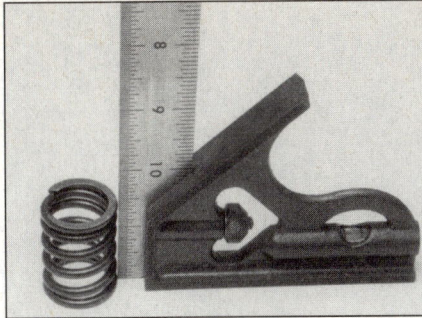

12.19b Checking the valve springs for squareness

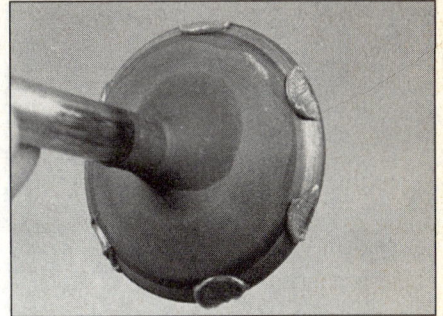

12.23 Apply the lapping compound very sparingly, in small dabs, to the valve face

19 Check the end of each valve spring for wear and pitting. Measure the free length **(see illustration)** and compare it to this Chapter's Specifications. Any springs that are shorter than specified have sagged and should not be reused. Stand the spring on a flat surface and check it for squareness **(see illustration)**.
20 Check the spring retainers and keepers/collets for obvious wear and cracks. Any questionable parts should not be reused, as extensive damage will occur in the event of failure during engine operation.
21 If the inspection indicates that no service work is required, the valve components can be reinstalled in the head.

### Reassembly

22 Before installing the valves in the head, they should be lapped to ensure a positive seal between the valves and seats. This procedure requires coarse and fine valve lapping compound (available at auto parts stores) and a valve lapping tool.

**TOOL TiP** *If a lapping tool is not available, a piece of rubber or plastic hose can be slipped over the valve stem (after the valve has been installed in the guide) and used to turn the valve.*

23 Apply a small amount of coarse lapping compound to the valve face **(see illustration)**, then slip the valve into the guide.

**Caution: Make sure the valve is installed in the correct guide and be careful not to get any lapping compound on the valve stem.**
24 Attach the lapping tool (or hose) to the valve and rotate the tool between the palms of your hands. Use a back-and-forth motion rather than a circular motion **(see illustration)**. Lift the valve off the seat and turn it at regular intervals to distribute the lapping compound properly. Continue the lapping procedure until the valve face and seat contact area is of uniform width and unbroken around the entire circumference of the valve face and seat **(see illustration)**. Once this is accomplished, lap the valves again with fine lapping compound.
25 Carefully remove the valve from the guide and wipe off all traces of lapping compound. Use solvent to clean the valve and wipe the seat area thoroughly with a solvent soaked cloth. Repeat the procedure for the remaining valves.
26 Lay the spring seats in place in the cylinder head, then install new valve stem seals on each of the guides. Use an appropriate size deep socket to push the seals into place until they are properly seated. Don't twist or cock them, or they will not seal properly against the valve stems. Also, don't remove them again or they will be damaged.
27 Coat the valve stems with assembly lube or moly-based grease, then install one of

them into its guide. Next, install the spring seats, springs and retainers, compress the springs and install the keepers/collets. **Note:** *Install the springs with the tightly wound coils at the bottom (next to the spring seat). When compressing the springs with the valve spring compressor, depress them only as far as is absolutely necessary to slip the keepers/collets into place. Apply a small amount of grease to the keepers/collets* **(see illustration)** *to help hold them in place as the pressure is released from the springs. Make certain that the keepers/collets are securely locked in their retaining grooves.*
28 Support the cylinder head on blocks so the valves can't contact the workbench top, then very gently tap each of the valve stems with a soft-faced hammer. This will help seat the keepers (collets) in their grooves.

**HAYNES HiNT** *Once all of the valves have been installed in the head, check for proper valve sealing by pouring a small amount of solvent into each of the valve ports. If the solvent leaks past the valve(s) into the combustion chamber area, disassemble the valve(s) and repeat the lapping procedure, then reinstall the valve(s) and repeat the check. Repeat the procedure until a satisfactory seal is obtained.*

12.24a After lapping, the valve face should exhibit a uniform, unbroken contact pattern (arrow) . . .

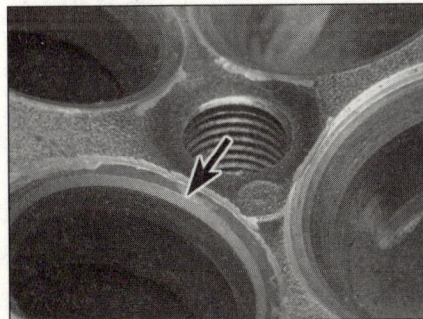

12.24b . . . and the seat should be the specified width (arrow) with a smooth, unbroken appearance

12.27 A dab of grease will hold the keepers/collets in place while the spring compressor is released

13.3a Unbolt the coolant passage from the cylinder block . . .

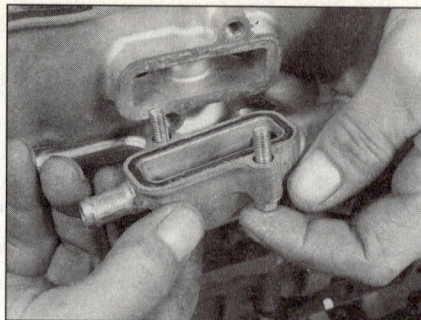

13.3b . . . and replace its O-ring

13.5 Lift the cylinder block off the crankcase

## 13 Cylinder block –
removal, inspection and installation

### Removal

1 Remove the engine from the bike (see Section 5).

2 Following the procedure given in Section 10, remove the cylinder head. Make sure the crankshaft is positioned at Top Dead Center (TDC) for cylinders 1 and 4.

3 Detach the oil cooler coolant hoses from the coolant passage on the front of the cylinder block, then remove the coolant passage from the cylinder block (see illustrations). Check its O-ring for brittleness or deterioration. It's a good idea to replace the O-ring whenever the passage is removed.

4 Lift out the cam chain front guide (see illustration 9.8a). Refer to Section 29 and remove the cam chain rear guide.

5 Lift the cylinder block straight up to remove it (see illustration). If it's stuck, tap around its perimeter with a soft-faced hammer. Don't attempt to pry between the block and the crankcase, as you'll ruin the sealing surfaces.

6 Remove the dowel pins from the mating surface of the crankcase (see illustrations). Be careful not to let these drop into the

13.6a There's a dowel at the left front corner . . .

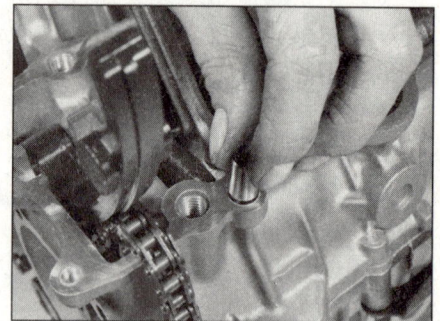

13.6b . . . and at the right front corner

engine. Stuff rags around the pistons and remove the gasket and all traces of old gasket material from the surfaces of the cylinder block and the cylinder head.

### Inspection

Caution: Don't attempt to separate the liners from the cylinder block.

7 Check the cylinder walls carefully for scratches and score marks.

8 Using the appropriate precision measuring tools, check each cylinder's diameter at the two specified distances from the top of the cylinder bore, parallel to the crankshaft axis (see illustration). Next, measure each cylinder's diameter at the same two locations across the crankshaft axis. Compare the results to this Chapter's Specifications. If the cylinder walls are tapered, out-of-round, worn beyond the specified limits, or badly scuffed or scored, have them rebored and honed by a dealer service department or a motorcycle repair shop. If a rebore is done, oversize pistons and rings will be required as well. Note: Kawasaki supplies pistons in one oversize only – +0.5 mm (+0.020 inch).

9 As an alternative, if the precision measuring tools are not available, a dealer service department or motorcycle repair shop will make the measurements and offer advice concerning servicing of the cylinders.

10 If they are in reasonably good condition and not worn to the outside of the limits, and if the piston-to-cylinder clearances can be maintained properly (see Section 14), then the cylinders do not have to be rebored; honing is all that is necessary.

11 To perform the honing operation you will need the proper size flexible hone with fine stones, or a "bottle brush" type hone, plenty of light oil or honing oil, some shop towels and an electric drill motor. Hold the cylinder block in a vise (cushioned with soft jaws or wood blocks) when performing the honing operation. Mount the hone in the drill motor, compress the stones and slip the hone into the cylinder. Lubricate the cylinder thoroughly, turn on the drill and move the hone up and down in the cylinder at a pace which will produce a fine crosshatch pattern on the cylinder wall with the crosshatch lines intersecting at approximately a 60° angle (see illustration). Be sure to use plenty of lubricant and do not take off any more material than is absolutely necessary to produce the desired effect. Do not withdraw the hone from the cylinder while it is running. Instead, shut off the drill and continue moving the hone up and down in the cylinder until it

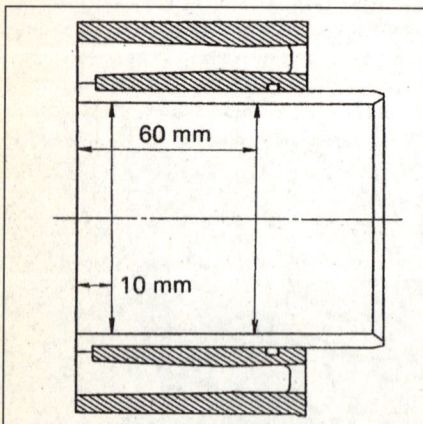

13.8 Measure the diameter 10 mm and 60 mm from the top, in front-to-rear and side-to-side directions

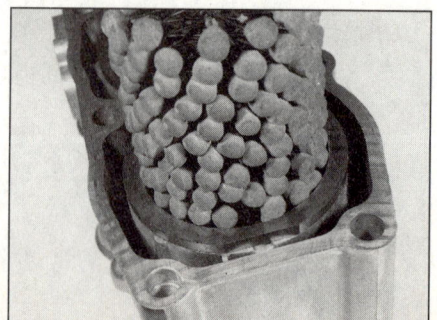

13.11 Honing the cylinders with a bottle-brush type hone

**13.14 Install the cylinder block base gasket and both dowels on the crankcase**

**13.17 The cylinder block can be installed over the rings without ring compressors, but compressors are recommended**

comes to a complete stop, then compress the stones and withdraw the hone. Wipe the oil out of the cylinder and repeat the procedure on the remaining cylinder. Remember, do not remove too much material from the cylinder wall. If you do not have the tools, or do not desire to perform the honing operation, a dealer service department or motorcycle repair shop will generally do it for a reasonable fee.

**12** Next, the cylinders must be thoroughly washed with warm soapy water to remove all traces of the abrasive grit produced during the honing operation. Be sure to run a brush through the bolt holes and flush them with running water. After rinsing, dry the cylinders thoroughly and apply a coat of light, rust-preventative oil to all machined surfaces.

### Installation

**13** Lubricate the cylinder bores with plenty of clean engine oil. Apply a thin film of moly-based grease to the piston skirts.
**14** Install the dowel pins, then lower a new cylinder base gasket over them **(see illustration)**.
**15** Slowly rotate the crankshaft until all of the pistons are at the same level. Slide lengths of welding rod or pieces of a straightened-out coat hanger under the pistons, on both sides of the connecting rods. This will help keep the

pistons level as the cylinder block is lowered onto them.
**16** Attach four piston ring compressors to the pistons and compress the piston rings. Large hose clamps can be used instead – just make sure they don't scratch the pistons, and don't tighten them too much.
**17** Install the cylinder block over the studs and carefully lower it down until the piston crowns fit into the cylinder liners **(see illustration)**. While doing this, pull the camshaft chain up, using a hooked tool or a piece of coat hanger. Push down on the cylinder block, making sure the pistons don't get cocked sideways, until the bottom of the cylinder liners slide down past the piston rings. A wood or plastic hammer handle can be used to gently tap the block down, but don't use too much force or the pistons will be damaged.
**18** Remove the piston ring compressors or hose clamps, being careful not to scratch the pistons. Remove the rods from under the pistons.
**19** Install the cam chain front and rear guides (see Section 29).
**20** Install the coolant passage on the front of the cylinder block, using a new O-ring. Tighten its bolts to the torque listed in this Chapter's Specifications.

**21** Install the cylinder head and tighten the bolts (see Section 10).
**22** Connect the coolant hose to the passage on the front of the cylinder block. Refer to 'Daily (pre-ride) checks' and top up the cooling system.

### 14 Pistons –
removal, inspection and installation

**1** The pistons are attached to the connecting rods with piston pins that are a slip fit in the pistons and rods.
**2** Before removing the pistons from the rods, stuff a clean shop towel into each crankcase hole, around the connecting rods. This will prevent the circlips from falling into the crankcase if they are inadvertently dropped.

### Removal

**3** Using a sharp scribe (or a felt pen if the piston is clean enough), mark the number of each piston on its crown **(see illustration)**. Each piston should also have an arrowhead pointing toward the front of the engine. If not, scribe an arrow into the piston crown before removal. Support the piston and pry the circlip out with a pointed tool, then push the piston pin out from the opposite side **(see illustrations)**.

**14.3a Mark the top of each piston with its cylinder number**

**14.3b Wear eye protection and pry the circlip out of its groove with a pointed tool**

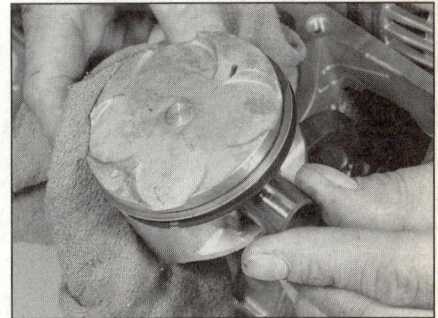

**14.3c Push the piston pin partway out, then pull it the rest of the way**

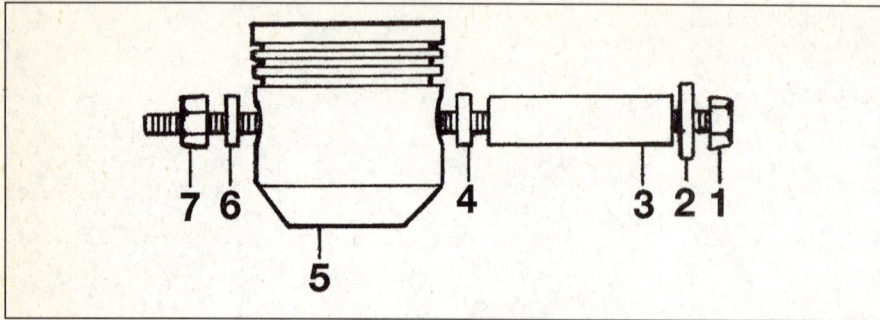

**14.3d  If the pins won't come out with hand pressure this removal tool can be fabricated from readily available parts**

| | | | |
|---|---|---|---|
| 1  Bolt | 3  Pipe (A) | 5  Piston | 7  Nut (B) |
| 2  Washer | 4  Padding (A) | 6  Washer (B) | |

A  Large enough for piston pin to fit inside
B  Small enough to fit through piston pin bore

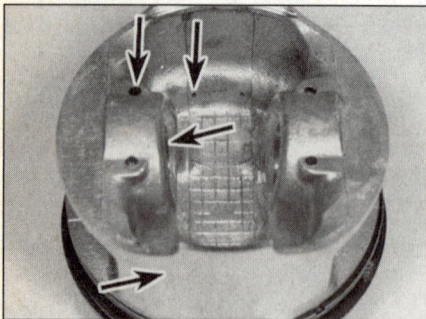

**14.6  Remove the piston rings with a ring removal and installation tool**

If the pin won't come out, you can fabricate a piston pin removal tool from a long bolt, a nut, a piece of tubing and washers **(see illustration)**.

**4** Push the piston pin out from the opposite end to free the piston from the rod. You may have to deburr the area around the groove to enable the pin to slide out (use a triangular file for this procedure). Repeat the procedure for the remaining pistons. Use large rubber bands to support the connecting rods.

## Inspection

**5** Before the inspection process can be carried out, the pistons must be cleaned and the old piston rings removed.

**6** Using a piston ring removal and installation tool, carefully remove the rings from the pistons **(see illustration)**. Do not nick or gouge the pistons in the process.

**7** Scrape all traces of carbon from the tops of the pistons. A hand-held wire brush or a piece of fine emery cloth can be used once the majority of the deposits have been scraped away.

*Caution: Do not, under any circumstances, use a wire brush mounted in a drill motor to remove deposits from the pistons; the piston material is soft and will be eroded away by the wire brush.*

**8** Use a piston ring groove cleaning tool to remove any carbon deposits from the ring grooves. If a tool is not available, a piece broken off the old ring will do the job. Be very careful to remove only the carbon deposits. Do not remove any metal and do not nick or gouge the sides of the ring grooves.

**9** Once the deposits have been removed, clean the pistons with solvent and dry them thoroughly. Make sure the oil return holes below the oil ring grooves are clear.

**10** If the pistons are not damaged or worn excessively and if the cylinders are not rebored, new pistons will not be necessary. Normal piston wear appears as even, vertical wear on the thrust surfaces of the piston and slight looseness of the top ring in its groove. New piston rings, on the other hand, should always be used when an engine is rebuilt.

**11** Carefully inspect each piston for cracks around the skirt, at the pin bosses and at the ring lands **(see illustration)**.

**12** Look for scoring and scuffing on the thrust faces of the skirt, holes in the piston crown and burned areas at the edge of the crown. If the skirt is scored or scuffed, the engine may have been suffering from overheating and/or abnormal combustion, which caused excessively high operating temperatures. The oil pump and cooling system should be checked thoroughly. A hole in the piston crown, an extreme to be sure, is an indication that abnormal combustion (pre-ignition) was occurring. Burned areas at the edge of the piston crown are usually evidence of spark knock (detonation). If any of the above problems exist, the causes must be corrected or the damage will occur again.

**13** Measure the piston ring-to-groove clearance by laying a new piston ring in the ring groove and slipping a feeler gauge in beside it **(see illustration)**. Check the clearance at three or four locations around the groove. Be sure to use the correct ring for each groove; they are different. If the clearance is greater than specified, new pistons will have to be used when the engine is reassembled.

**14** Check the piston-to-bore clearance by measuring the bore (see Section 13) and the piston diameter **(see illustration)**. Make sure that the pistons and cylinders are correctly matched. Measure the piston across the skirt on the thrust faces at a 90° angle to the piston pin, at the specified distance up from the bottom of the skirt. Subtract the piston diameter from the bore diameter to obtain the clearance. If it is greater than specified, the cylinders will have to be rebored and new oversized pistons and rings installed. If the appropriate precision measuring tools are not available, the piston-to-cylinder clearances can be obtained, though not quite as accurately, using feeler gauge stock. Feeler gauge stock comes in 12-inch lengths and various thicknesses and is generally available at auto parts stores. To check the clearance, slip a piece of feeler gauge stock of the same thickness as the specified piston clearance into the cylinder along with the appropriate piston.

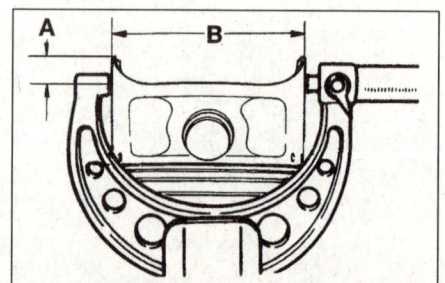

**14.11  Check the piston pin bore and the skirt for wear, and make sure the oil holes are clear (arrows)**

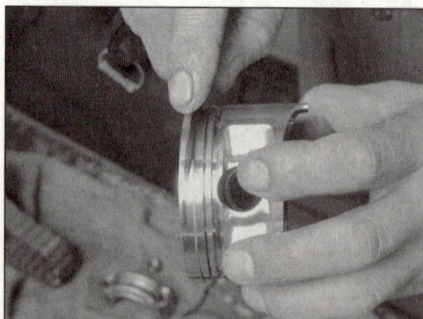

**14.13  Measure the piston ring-to-groove clearance with a feeler gauge**

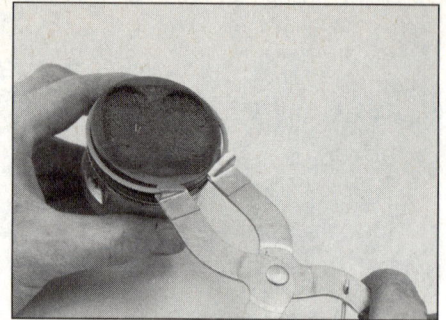

**14.14  Measure the piston diameter with a micrometer**

A  5 mm (0.2 inch) from bottom of piston
B  Piston diameter

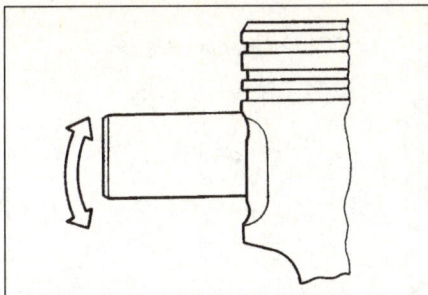

**14.15 Slip the pin into the piston and try to wiggle it back-and-forth to check for freeplay**

**15.3 Check the piston ring end gap with a feeler gauge**

**15.5 Clamp a file in a vise and file the ring ends (from the outside in only) to enlarge the gap slightly**

The cylinder should be upside down and the piston must be positioned exactly as it normally would be. Place the feeler gauge between the piston and cylinder on one of the thrust faces (90° to the piston pin bore). The piston should slip through the cylinder (with the feeler gauge in place) with moderate pressure. If it falls through, or slides through easily, the clearance is excessive and a new piston will be required. If the piston binds at the lower end of the cylinder and is loose toward the top, the cylinder is tapered, and if tight spots are encountered as the piston/feeler gauge is rotated in the cylinder, the cylinder is out-of-round. Repeat the procedure for the remaining pistons and cylinders. Be sure to have the cylinders and pistons checked by a dealer service department or a motorcycle repair shop to confirm your findings before purchasing new parts.

15  Apply clean engine oil to the pin, insert it into the piston and check for freeplay by rocking the pin back-and-forth **(see illustration)**. If the pin is loose, new pistons and possibly new pins must be installed.

16  Refer to Section 15 and install the rings on the pistons.

### Installation

**Note:** *When installing the pistons, install the pistons for cylinders 2 and 3 first.*

17  Install the pistons in their original locations with the arrows pointing to the front of the engine. Lubricate the pins and the rod bores with clean engine oil. Install new circlips in the grooves in the inner sides of the pistons

(don't reuse the old circlips). Push the pins into position from the opposite side and install new circlips. Compress the circlips only enough for them to fit in the piston. Make sure the clips are properly seated in the grooves.

### 15  Piston rings – installation

1  Before installing the new piston rings, the ring end gaps must be checked.

2  Lay out the pistons and the new ring sets so the rings will be matched with the same piston and cylinder during the end gap measurement procedure and engine assembly.

3  Insert the top (No. 1) ring into the bottom of the first cylinder and square it up with the cylinder walls by pushing it in with the top of the piston. The ring should be about one inch above the bottom edge of the cylinder. To measure the end gap, slip a feeler gauge between the ends of the ring **(see illustration)** and compare the measurement to the Specifications.

4  If the gap is larger or smaller than specified, double check to make sure that you have the correct rings before proceeding.

5  If the gap is too small, it must be enlarged or the ring ends may come in contact with each other during engine operation, which can cause serious damage. The end gap can be increased by filing the ring ends very carefully with a fine file **(see illustration)**. When performing this operation, file only from the outside in.

6  Excess end gap is not critical unless it is greater than 1 mm (0.040 inch). Again, double check to make sure you have the correct rings for your engine.

7  Repeat the procedure for each ring that will be installed in the first cylinder and for each ring in the remaining cylinders. Remember to keep the rings, pistons and cylinders matched up.

8  Once the ring end gaps have been checked/corrected, the rings can be installed on the pistons.

9  The oil control ring (lowest on the piston) is installed first. It is composed of three separate components. Slip the expander into the groove, then install the upper side rail **(see illustrations)**. Do not use a piston ring installation tool on the oil ring side rails as they may be damaged. Instead, place one end of the side rail into the groove between the spacer expander and the ring land. Hold it firmly in place and slide a finger around the piston while pushing the rail into the groove. Next, install the lower side rail in the same manner.

10  After the three oil ring components have been installed, check to make sure that both the upper and lower side rails can be turned smoothly in the ring groove.

11  Install the no. 2 (middle) ring next. It can be readily distinguished from the top ring by its cross-section shape **(see illustration)**. Do not mix the top and middle rings.

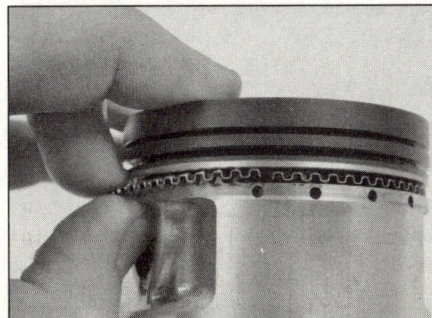

**15.9a Installing the oil ring expander – make sure the ends don't overlap**

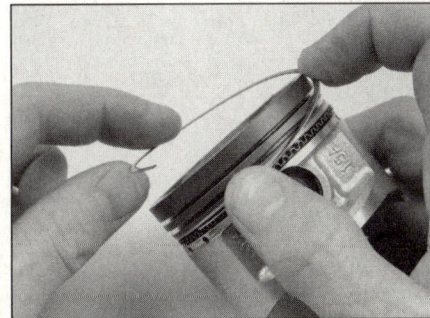

**15.9b Installing an oil ring side rail – don't use a ring installation tool to do this**

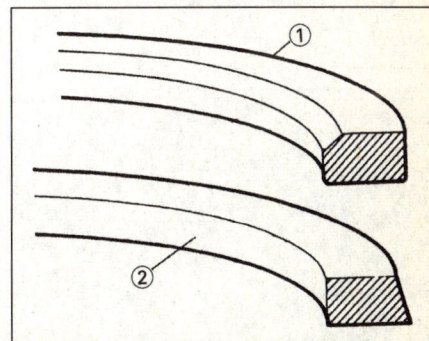

**15.11 The top and second (middle) rings can be told apart by their profiles**

*1  Top ring*
*2  Second (middle) ring*

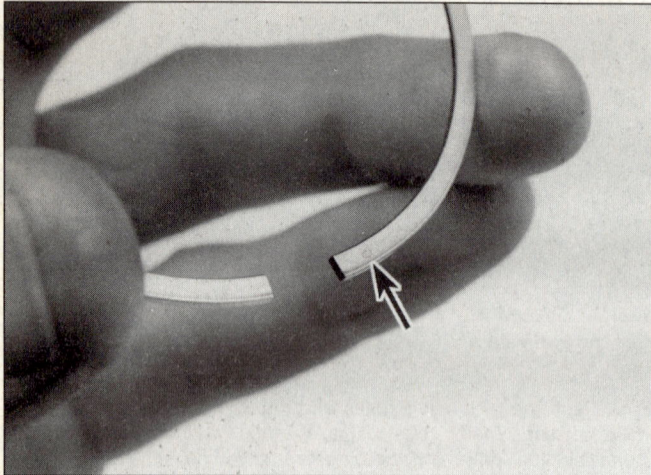

15.12 The marks on the rings (arrow) must face up when the rings are installed on the pistons

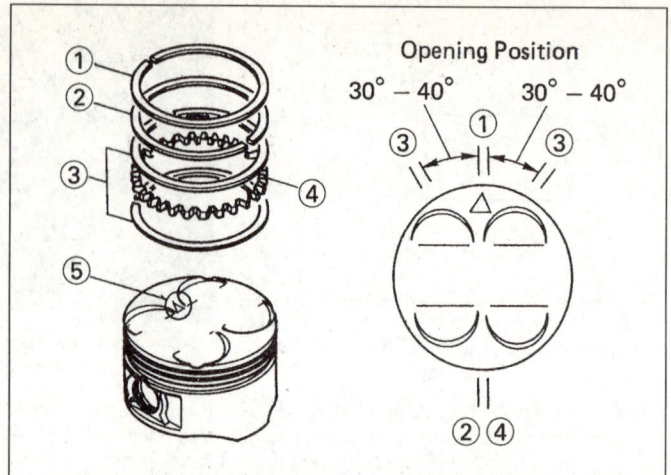

15.15 Ring gap positioning details

1 Top ring
2 Second (middle) ring
3 Oil ring side rails
4 Oil ring expander
5 Arrowhead mark
  (points to front of engine)

12 To avoid breaking the ring, use a piston ring installation tool and make sure that the identification mark is facing up (see illustration). Fit the ring into the middle groove on the piston. Do not expand the ring any more than is necessary to slide it into place.
13 Finally, install the no. 1 (top) ring in the same manner. Make sure the identifying mark is facing up.
14 Repeat the procedure for the remaining pistons and rings. Be very careful not to confuse the no. 1 and no. 2 rings.

15 Once the rings have been properly installed, stagger the end gaps, including those of the oil ring side rails (see illustration).

## 16 Oil pan and hoses – removal and installation

Note: Some of the following procedures are shown with the engine removed for clarity.

The oil pan and hoses can be removed and installed with the engine in the frame.

### Oil hoses

#### Removal

1 Support the bike securely upright.
2 Remove the lower fairing panels (see Chapter 8).
3 Drain the engine oil (see Chapter 1). This is a good time to replace the filter if you're close to a maintenance interval.
4 To remove the hose at the right front lower corner of the engine, loosen its hose clamps (see illustration). Slide the clamps back along the hose, then detach it from its fittings and take it out. If necessary, unbolt the hose fitting and take it off the engine (see illustration).
5 To remove the return hose on the left side of the engine, remove the union bolt at the top and unbolt the fitting at the bottom (see illustrations). Replace the O-ring on the bottom fitting and the O-ring inside the union bolt if they have been leaking or show signs of deterioration. Replace the sealing washers at the upper fitting whenever the fitting is loosened.

16.4a Loosen the hose clamps (arrows) and take the hose off

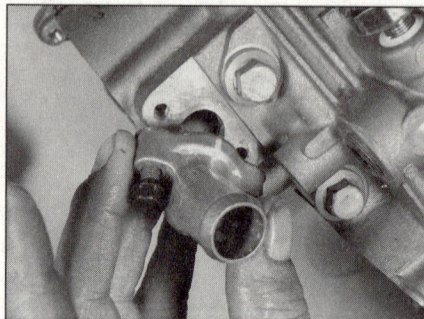

16.4b If necessary, unbolt the hose fitting from the base of the engine

16.5a Remove the bolt (arrow) . . .

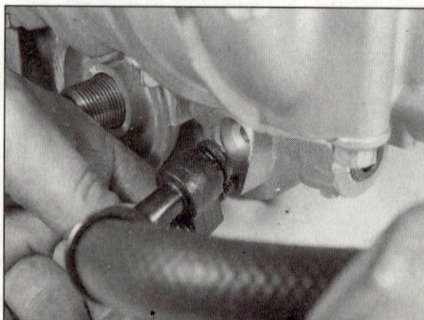

16.5b . . . and pull the hose out of the engine; its O-ring should be replaced

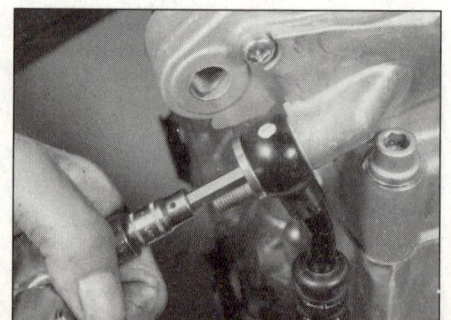

16.5c Remove the Allen-head banjo bolt . . .

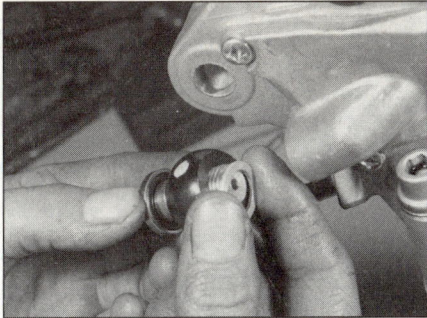

16.5d . . . and detach the upper fitting; use a new O-ring and sealing washer

16.6a Remove the oil passage cover screws with an impact driver . . .

16.6b . . . then take the cover off; use a new O-ring on installation

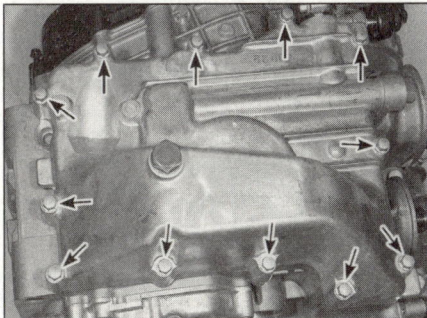

16.10a Remove the oil pan bolts (arrows) . . .

16.10b . . . and lift the oil pan off the engine

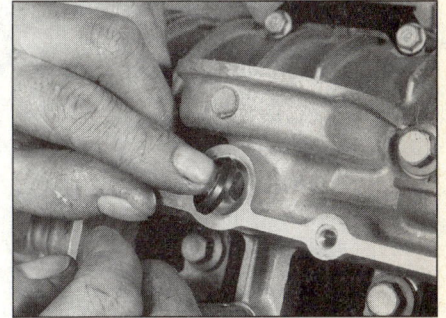

16.11 Remove the O-ring; its flat side goes toward the crankcase

**6** If you need to remove the oil passage cover from the left side of the cylinder head (to remove the rocker arm shafts, for example), undo its screws with an impact driver and take the cover off **(see illustrations)**.

### Installation

**7** Installation is the reverse of the removal steps, with the following additions:

a) *Tighten the union bolt and the bottom fitting bolt to the torques listed in this Chapter's Specifications.*

b) *If the lower right hose fitting was removed from the engine, apply non-permanent*

*thread locking agent to the threads of the bolts and tighten them to the torque listed in this Chapter's Specifications.*

c) *If the oil passage cover was removed, use a new O-ring on installation and tighten the screws evenly.*

### Oil pan, strainer and pipe

#### Removal

**Note:** *The oil pan can be removed with the engine in the frame.*

**8** Perform Steps 1 through 4 at the beginning of this Section.

**9** Remove the exhaust system (see Chapter 4).

**10** Remove the oil pan bolts and detach the pan from the crankcase **(see illustrations)**.

**11** Remove the oil pan O-ring **(see illustration)**. It's a good idea to replace it whenever the oil pan is removed. It should definitely be replaced if it shows signs of cracking or deterioration.

**12** Work the oil strainer seal free of its passage and lift the strainer out **(see illustration)**.

**13** Work the oil pipe O-rings free of their passages and lift the oil pipe out **(see illustration)**. Apply even pressure to all three fittings so the pipe isn't bent as it's pulled out.

16.12 Work the strainer loose from the crankcase; on installation, align its notch with the cast tab on the crankcase

16.13 Work the three O-rings clear of the crankcase passages and lift the oil pipe out

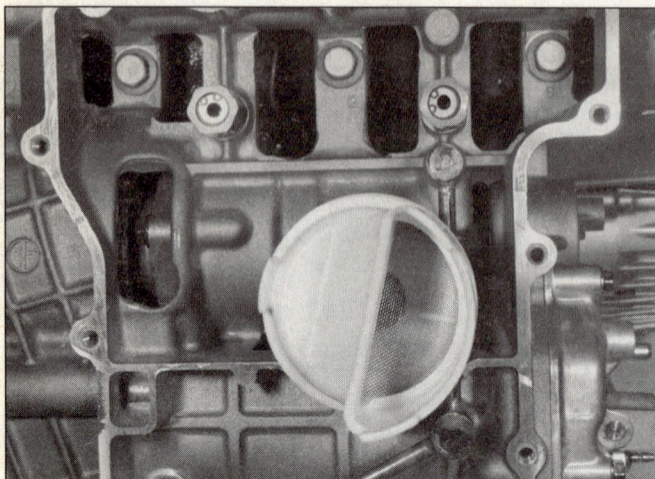

16.17 The oil strainer is installed like this

16.19 Install a new gasket on the crankcase

14  Remove all traces of old gasket material from the mating surfaces of the oil pan and crankcase.

### Installation

15  Install new O-rings on the oil pipe and install a new seal on the oil strainer.
16  Install the oil pipe, pressing its O-rings firmly into their passages.
17  Align the notch in the oil strainer with the tab on the crankcase and press the oil strainer into position **(see illustration 16.12 and the accompanying illustration)**.
18  Install the oil pan O-ring with its flat side facing the crankcase **(see illustration 16.11)**.
19  Position a new gasket on the oil pan **(see illustration)**. A thin film of RTV sealant can be used to hold the gasket in place. Install the oil pan and bolts, tightening the bolts to the torque listed in this Chapter's Specifications, using a criss-cross pattern.
20  The remainder of installation is the reverse of removal. Install a new filter and fill the crankcase with oil (see Chapter 1 and *Daily (pre-ride) checks*), then run the engine and check for leaks.

## 17  Oil pump –
pressure check, removal, inspection and installation

**Note:** *The oil pump can be removed with the engine in the frame.*

### Pressure check

⚠ **Warning: If the oil pressure switch is removed when the engine is hot, hot oil will drain out – wait until the engine is cold before beginning this check (it must be cold to perform the relief valve opening pressure check, anyway).**

1  Refer to Chapter 9 and remove the oil pressure switch.
2  Install an oil pressure gauge in the oil pressure switch hole.
3  Start the engine and watch the gauge while varying the engine rpm. The pressure should stay within the relief valve opening pressure listed in this Chapter's Specifications. If the pressure is too high, the left (main) relief valve is stuck closed. To check it, see Section 18.

4  If the pressure is lower than the standard, either the relief valve is stuck open, the oil pump is faulty, or there is other engine damage. Begin diagnosis by checking the relief valve (see Section 18), then the oil pump. If those items check out okay, chances are the bearing oil clearances are excessive and the engine needs to be overhauled.
5  If the pressure reading is in the desired range, allow the engine to warm up to normal operating temperature and check the pressure again, at the specified engine rpm. Compare your findings with this Chapter's Specifications.
6  If the pressure is significantly lower than specified, check the left (main) relief valve and the oil pump.

### Removal

7  Remove the clutch assembly (Section 19).
8  Remove the alternator drive chain and oil pump sprocket (see Section 27).
9  Remove the mounting bolts, then grasp the pump shaft and pull the pump out of the engine **(see illustrations)**.

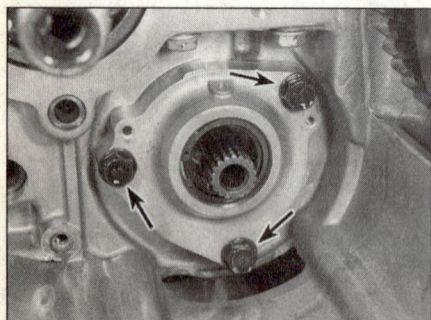

17.9a Remove the oil pump mounting bolts (arrows)

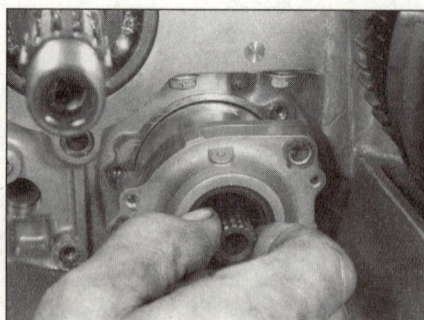

17.9b Pull on the shaft to detach the oil pump from the crankcase . . .

17.9c . . . the outer main rotor (arrow) may remain in the engine when the pump is removed

17.10a  The locating pin fits through the pump cover and body and into the crankcase

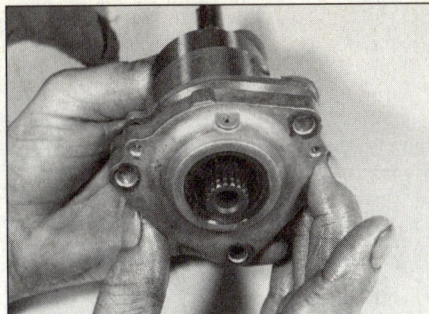

17.10b  Remove the pump cover

17.11  Remove the drive pin from the pump shaft

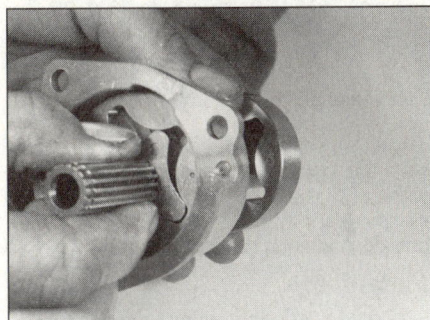

17.12a  Pull the shaft and inner sub-rotor from the pump body, then remove the inner sub-rotor from the shaft

17.12b  Oil pump details

## Inspection

**10**  Remove the locating pin and remove the pump cover **(see illustrations)**.

**11**  Slide the shaft down to expose the drive pin, then remove the drive pin from the shaft **(see illustration)**.

**12**  Take off the inner main rotor, pull the shaft and inner sub-rotor out of the pump body, then remove the inner sub-rotor from the shaft **(see illustrations)**.

**13**  Take the outer sub-rotor out of the pump body **(see illustration)**.

**14**  Wash all the components in solvent, then dry them off. Check the pump body, the rotors and the cover for scoring and wear. Make sure the pick-up screen isn't clogged. Kawasaki doesn't publish clearance specifications, so if any damage or uneven or

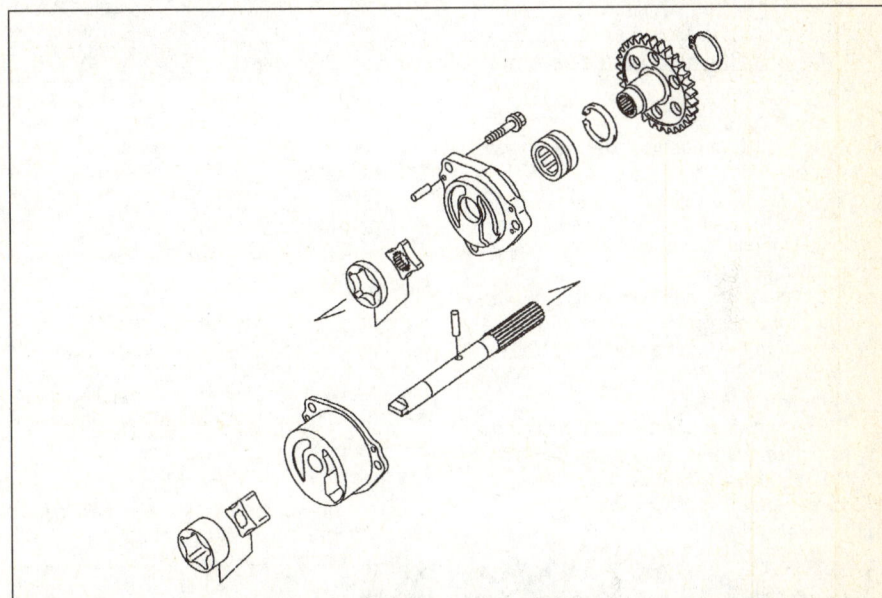

excessive wear is evident, replace the pump. If you are rebuilding the engine, it's a good idea to install a new oil pump.

**15**  Check the needle roller bearing inside the pump cover for wear or damage **(see illustration)**. If there's any doubt about its condition (and you don't plan to replace the entire pump), remove the snap-ring, replace the bearing and install a new snap-ring.

**16**  Reassemble the pump by reversing the disassembly steps. Be sure the inner main rotor fits over the drive pin **(see illustration)**. Align the mark on the sub-rotor with the drive pin **(see illustration)**. Before installing the cover, pack the cavities between the rotors with petroleum jelly – this will ensure the pump develops suction quickly and begins oil circulation as soon as the engine is started.

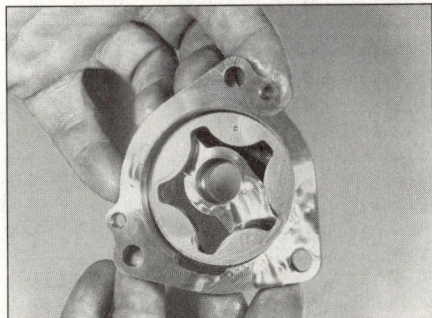

17.13  Remove the outer sub-rotor from the pump body

17.15  Replace the needle roller bearing in the pump cover if you plan to reuse the pump

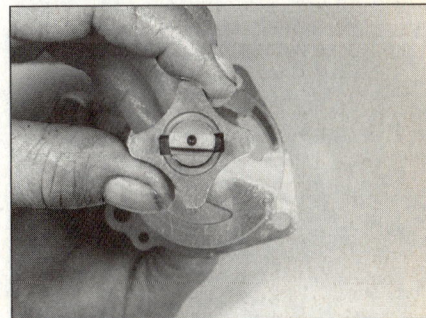

17.16a  Make sure the inner main rotor is pulled down over the drive pin

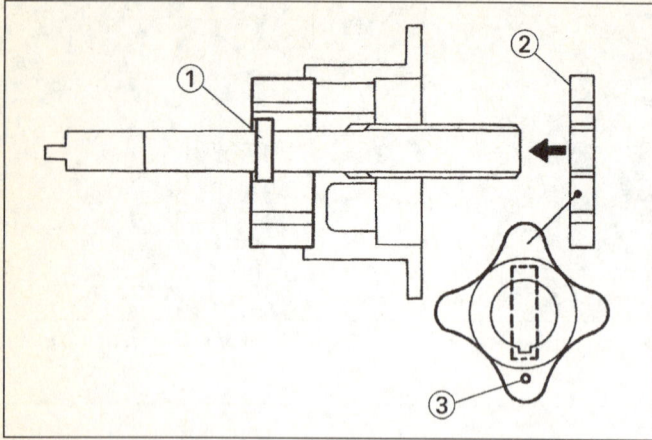

**17.16b  Align the mark on the inner sub-rotor with the drive pin**

| 1 Pin | 2 Sub-rotor | 3 Mark |
|---|---|---|

**18.2  Unscrew the relief valves (arrows) from the crankcase**

Also, lubricate the needle roller bearing with clean engine oil.

### Installation

17 Installation is the reverse of removal, with the following additions:

a) *Align the tooth on the end of the oil pump shaft with the slot in the water pump.*

b) *Make sure the locating pin is in place before installing the oil pump bolts (see illustration 17.10a).*

c) *Clean all oil from the oil pump bolts, apply non-permanent thread locking agent to their threads and tighten them to the torque listed in this Chapter's Specifications.*

## 18 Oil pressure relief valves –
removal, inspection and installation

### Removal

1 Remove the oil pan (see Section 16).
2 Unscrew the relief valve(s) from the crankcase **(see illustration)**. The left relief valve (viewed with the engine installed) is for the main oil passage; the right relief valve is for the oil cooler.

### Inspection

3 Clean the valve with solvent and dry it, using compressed air if available.

4 Using a wood or plastic tool, depress the steel ball inside the valve and see if it moves smoothly. Make sure it returns to its seat completely. If it doesn't, replace it with a new one (don't attempt to disassemble and repair it).

### Installation

5 Apply a non-hardening thread locking compound to the threads of the valve and install it into the case, tightening it to the torque listed in this Chapter's Specifications.
6 The remainder of installation is the reverse of removal.

## 19 Clutch –
removal, inspection and installation

**Note:** *The clutch can be removed with the engine in the frame.*

### Removal

1 Support the bike securely upright. Remove the right lower fairing panel (see Chapter 8).
2 Drain the engine oil (see Chapter 1).
3 Remove the clutch cover bolts and take the cover off **(see illustrations)**. If the cover is stuck, tap around its perimeter with a soft-face hammer.
4 Remove the pressure plate bolts **(see illustration)**. To prevent the assembly from

**19.3a  Remove the clutch cover bolts (arrows) . . .**

**19.3b  . . . taking note of any hose brackets or wiring harness retainers . . .**

**19.3c  . . . and take the cover off**

**19.4a  Remove the pressure plate bolts . . .**

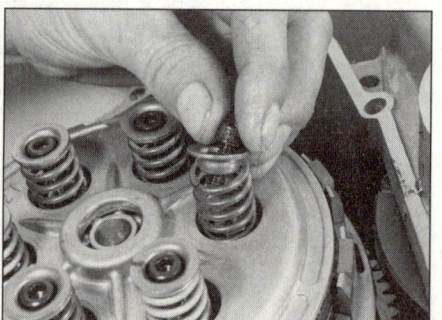

**19.4b  . . . the spring seats and the springs . . .**

**19.4c** . . . take the pressure plate off; the tabs of the last friction plate (arrow) fit in the shorter notch in the clutch housing

**19.4d** The clutch pushrod can be removed from the opposite side of the engine if the slave cylinder has been removed

turning, thread one of the cover mounting bolts into the case and wedge a screwdriver between the bolt and the clutch housing. Remove the clutch pressure plate with its bearing and push piece, then pull out the pushrod **(see illustrations)**.

**5** Remove the clutch friction and metal plates from the clutch housing **(see illustration)**.

**6** Loosen the clutch hub nut, using a special holding tool (Kawasaki tool no. 57001-1243) to prevent the clutch housing from turning **(see illustrations)**. An alternative to this tool can be fabricated from some steel strap, bent at the ends and bolted together in the middle (see illustration 19.7b in Chapter 2A). If the clutch plates will be replaced, you can drill holes in a friction plate and steel plate and bolt them together as they would be when installed (refer to *Tool Tip* in Chapter 2A).

### 19.4e Clutch – exploded view

1 Clutch cover
2 Gasket
3 Dampers
4 Damper cover
5 Collar
6 Clutch pushrod
7 Pushrod end piece
8 Slave cylinder
9 Friction plates
10 Metal plates
11 Sub-clutch hub
12 Clutch hub
13 Collar
14 Clutch housing
15 Pressure plate bolt
16 Spring seat
17 Post
18 Spring
19 Pressure plate
20 Ball bearing
21 Push piece
22 Clutch hub nut
23 Torque limiter spring
24 Splined spacer

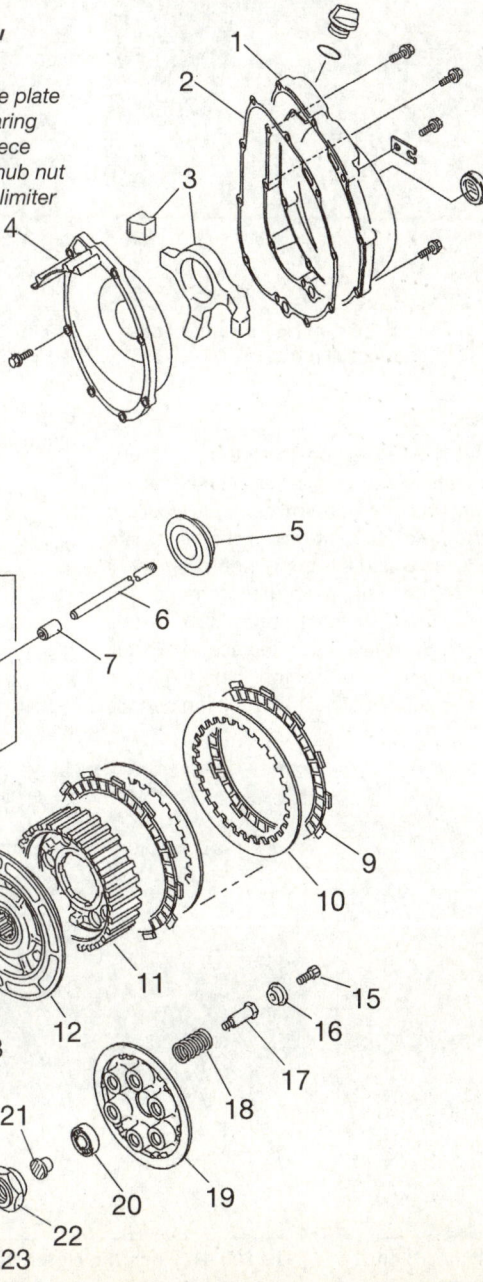

**19.5** Take the friction plates and metal plates off

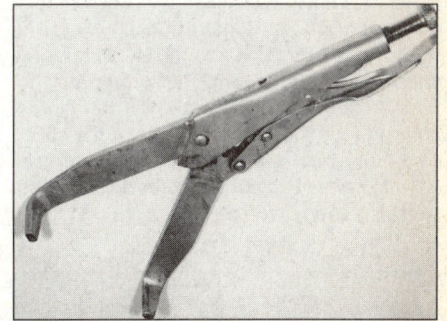

**19.6a** Special tool used to keep the clutch hub from turning

**19.6b** The ends of the tool fit the slots in the sub-clutch hub

**19.7 Remove the nut (noting which side faces the engine) and the torque limiter spring**

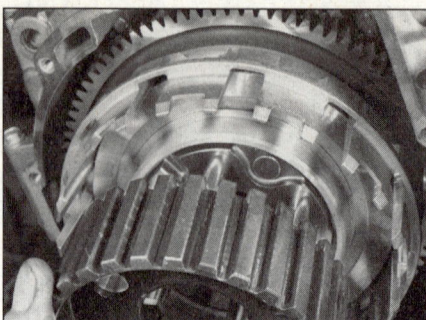

**19.8a Take off the sub-clutch hub . . .**

**19.8b . . . the splined spacer . . .**

**19.8c . . . and the clutch hub**

**19.9a Thread a 6 mm bolt into the collar and pull it out . . .**

**19.9b . . . then remove the clutch housing**

**7** Once the nut is loose, unscrew it and remove the torque limiter spring **(see illustration)**.

**8** Remove the sub-clutch hub, splined spacer and clutch hub **(see illustrations)**.

**9** Thread a 6 mm bolt into the clutch housing collar **(see illustration)**. Pull on the bolt to remove the collar, then take off the clutch housing **(see illustration)**. **Note:** *If the collar cocks sideways when you try to pull it off, use a 6 mm bolt in each of the collar holes so you can apply even pressure.*

### Inspection

**10** Check the screw posts on the clutch hub for looseness and check the internal splines and damper cam followers for wear or damage **(see illustration)**. If the posts are loose, unscrew them, apply non-permanent thread locking agent to their threads and retighten them to the torque listed in this Chapter's Specifications. Replace the clutch hub if the splines or damper cam followers show any defects. Check the friction surface

of the clutch hub for wear or scoring **(see illustration)** and replace it if any defects are found.

**11** Check the damper cams and splines on the sub-clutch hub for wear or damage **(see illustrations)**. Replace the sub-clutch hub if any defects are found.

**12** Measure the free length of the clutch springs **(see illustration)** and compare the results to this Chapter's Specifications. If the springs have sagged, or if cracks are noted, replace them with new ones as a set.

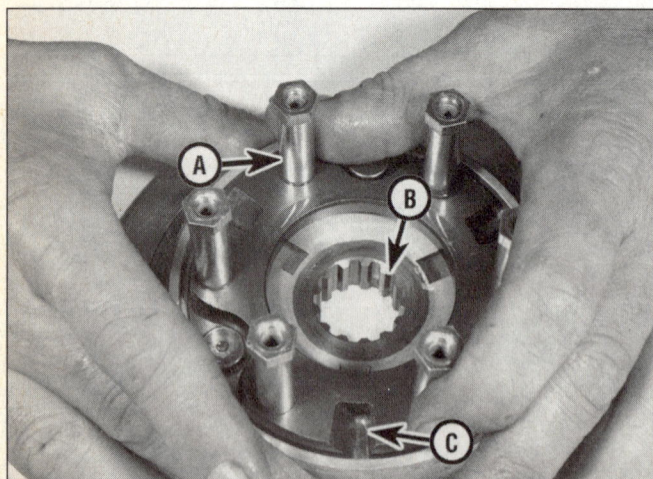

**19.10a Check the posts (A) for looseness, the splines (B) and damper cam followers (C) for wear or damage**

**19.10b Replace the clutch hub if its friction surface is worn or damaged**

19.11a  Inspect the sub-clutch hub damper cams . . .

19.11b  . . . and also check the splines; replace the clutch hub if any are worn or damaged

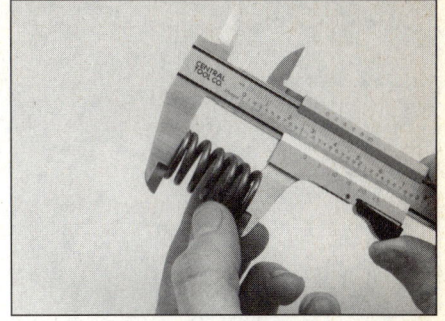

19.12  Measure the clutch spring free length

19.14  Check the metal plates for warpage

19.16a  Check the slots in the clutch housing for notches worn by the friction plate tabs

19.16b  Replace the clutch housing if the springs are broken or the gear teeth are worn or damaged

13  If the lining material of the friction plates smells burnt or if it is glazed, new parts are required. If the metal clutch plates are scored or discolored, they must be replaced with new ones. Replace with new parts any friction plates that are worn.

14  Lay the metal plates, one at a time, on a perfectly flat surface (such as a piece of plate glass) and check for warpage by trying to slip a feeler gauge between the flat surface and the plate (see illustration). The feeler gauge should be the same thickness as the maximum warp listed in this Chapter's Specifications. Do this at several places around the plate's circumference. If the feeler gauge can be slipped under the plate, it is warped and should be replaced with a new one.

15  Check the tabs on the friction plates for excessive wear and mushroomed edges. They can be cleaned up with a file if the deformation is not severe. Check the friction plates for warpage as described in Step 14.

16  Check the edges of the slots in the clutch housing for indentations made by the friction plate tabs (see illustration). If the indentations are deep they can prevent clutch release, so the housing should be replaced with a new one. If the indentations can be removed easily with a file, the life of the housing can be prolonged to an extent. Also, check the primary gear teeth for cracks, chips and excessive wear and the springs for

breakage (see illustration). If the gear is worn or damaged or the springs are broken, the clutch housing must be replaced with a new one. Check the bearing for score marks, scratches and excessive wear (see illustration).

17  Check the bearing journal on the transmission mainshaft for score marks, heat discoloration and evidence of excessive wear.

18  Check the clutch pressure plate for wear and damage. Rotate the push piece inside the bearing and check for roughness, looseness or excessive noise (see illustration).

19  Make sure the pushrod is not bent (roll it on a perfectly flat surface or use V-blocks and a dial indicator).

20  Insufficient pressure plate freeplay will cause abrupt engine braking, resulting in wheel hop on deceleration. Excessive freeplay

will cause the clutch lever to feel spongy or pulsate when it's pulled. Checking involves assembling the clutch assembly (without pressure plate springs or bolts) on a spare transmission shaft, then mounting the assembly in a vise, rotating it back-and-forth and measuring pressure plate freeplay with a dial indicator. If the bike had these symptoms and inspection didn't reveal a likely cause (such as worn metal or friction plates), have pressure plate freeplay checked by a Kawasaki dealer.

21  Unbolt the damper cover from the inside of the clutch cover. Check the damper for signs of deterioration or brittleness and replace as necessary. Apply gasket sealant to the indicated areas and assemble the damper cover to the clutch cover, using a new gasket (see illustrations).

19.16c  Inspect the bearing in the center of the clutch housing

19.18  Check the pressure plate bearing and push piece for wear or damage

19.21a Apply gasket sealant to the rearmost surface of the cover, between the arrows . . .

19.21b . . . and install a new gasket

19.23a Align the sprocket pins with the holes in the back of the clutch housing . . .

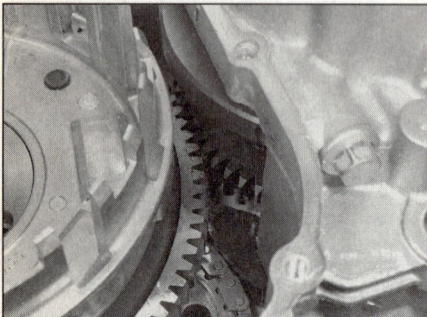

19.23b . . . engage the teeth of the clutch housing gear with the crankshaft primary gear

19.24 Install the collar

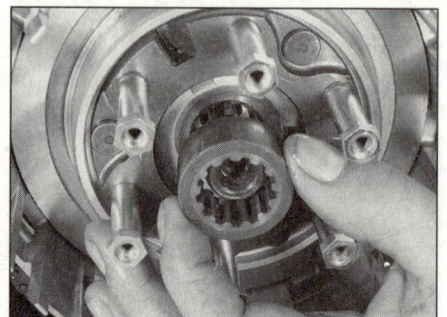

19.25a Install the clutch hub and splined spacer

22 Clean all traces of old gasket material from the clutch cover, damper cover and engine.

## Installation

23 Coat the bearing in the center of the clutch housing with engine oil. Install the clutch housing on the transmission mainshaft, making sure the sprocket pins engage the holes in the clutch housing and the gear teeth engage the primary gear on the crankshaft (see illustrations).

24 Coat the clutch housing collar with engine oil and install it in the center of the clutch housing (see illustration).

25 Install the clutch hub in the clutch housing and slip the splined spacer onto the transmission mainshaft (see illustration). Install the sub-clutch hub, making sure its damper cams engage the damper cam followers on the clutch hub (see illustration).

26 Assemble the two pieces of the torque limiter spring. Install the spring and thread a new clutch hub nut onto the transmission mainshaft (see illustrations).

27 Tighten the new hub nut to the torque listed in this Chapter's Specifications. Use the technique described in Step 6 to prevent the hub from turning. Stake the nut against the transmission mainshaft after it's tightened (see illustration).

28 Coat the clutch friction plates with engine oil. Install the clutch plates, starting with a friction plate, then a metal plate, and alternating them. Position the tabs of the last friction plate in the shorter slots in the clutch housing (see illustration 19.4c).

19.25b When installing the sub-clutch hub, engage the damper cams (A) with the cam followers (B)

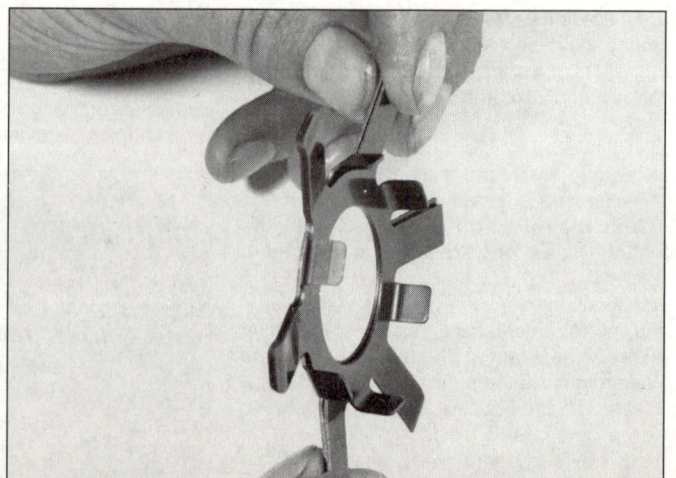

19.26a Assemble the halves of the torque limiter spring

**19.26b Install the spring and the clutch hub nut; tighten the nut to the specified torque . . .**

**19.27 . . . and stake it with a hammer and punch**

**19.30a Install the lower dowel . . .**

**19.30b . . . and the upper dowel**

**19.30c Apply non-permanent thread locking agent to the two forward clutch cover bolts**

29 Lubricate the pushrod and install it in the clutch assembly, then install the slave cylinder. Mount the pressure plate to the clutch assembly and install the springs and bolts, tightening them to the torque listed in this Chapter's Specifications in a criss-cross pattern.

30 Make sure the clutch cover dowels are in position **(see illustrations)**. Wipe a thin smear of gasket sealant across the cover mating areas, where the halves of the crankcase join. Install the clutch cover and bolts, using a new gasket. Apply non-permanent thread locking

agent to the two bolts at the forward edge of the cover **(see illustration)**, then install and tighten all of the bolts, in a criss-cross pattern, to the torque listed in this Chapter's Specifications. Don't forget to install the hose guide under the upper rear bolt **(see illustration 19.3b)**.

31 Fill the crankcase with the recommended type and amount of engine oil (see Chapter 1 and *Daily (pre-ride) checks*).

32 Install the right lower fairing panel (see Chapter 8).

## 20 Clutch hydraulic system – removal, overhaul, installation and system bleeding

### *Master cylinder removal*

1 Loosen the mounting bolt for the clutch fluid reservoir **(see illustration)**, so it can be unscrewed easily with fingers later.

2 Place a towel under the master cylinder to catch any spilled fluid, then remove the union bolt from the master cylinder fluid line **(see illustration)**.

*Caution: Brake fluid will damage paint. Wipe up any spills immediately and wash the area with soap and water.*

3 Remove the master cylinder mounting bolts and separate the halves of the master cylinder from the handlebar **(see illustrations)**. Disconnect the clutch switch electrical connector **(see illustration)**, remove the reservoir mounting bolt and take the master cylinder and reservoir off the bike.

**20.1 Remove the reservoir mounting bolt (arrow)**

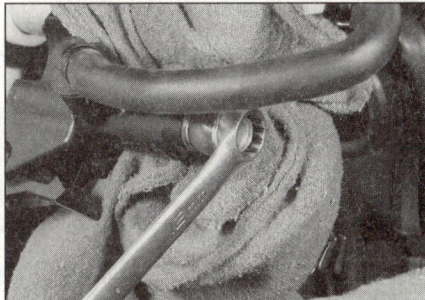

**20.2 Undo the clutch hose banjo bolt**

**20.3a Remove the master cylinder mounting bolts . . .**

**20.3b . . . and separate the master cylinder from the clamp . . .**

**20.3c . . . then unplug the clutch switch connector**

**20.4 Remove the clutch switch**

**20.5a Remove the pivot screw locknut (arrow) . . .**

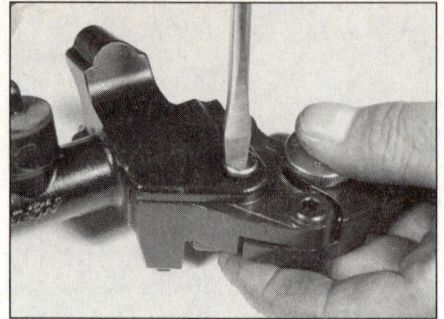

**20.5b . . . undo the pivot screw . . .**

**20.5c . . . and remove the clutch lever and pivot bushing**

## Master cylinder overhaul

**4** Remove the clutch switch from the master cylinder (**see illustration**).

**5** Remove the locknut from the lever pivot screw, then remove the screw (**see illustrations**). Remove the lever and pivot bushing (**see illustration**).

**6** Squeeze the hose clamp and slide it down the fluid feed hose, then disconnect the hose from the master cylinder fitting (**see illustration**).

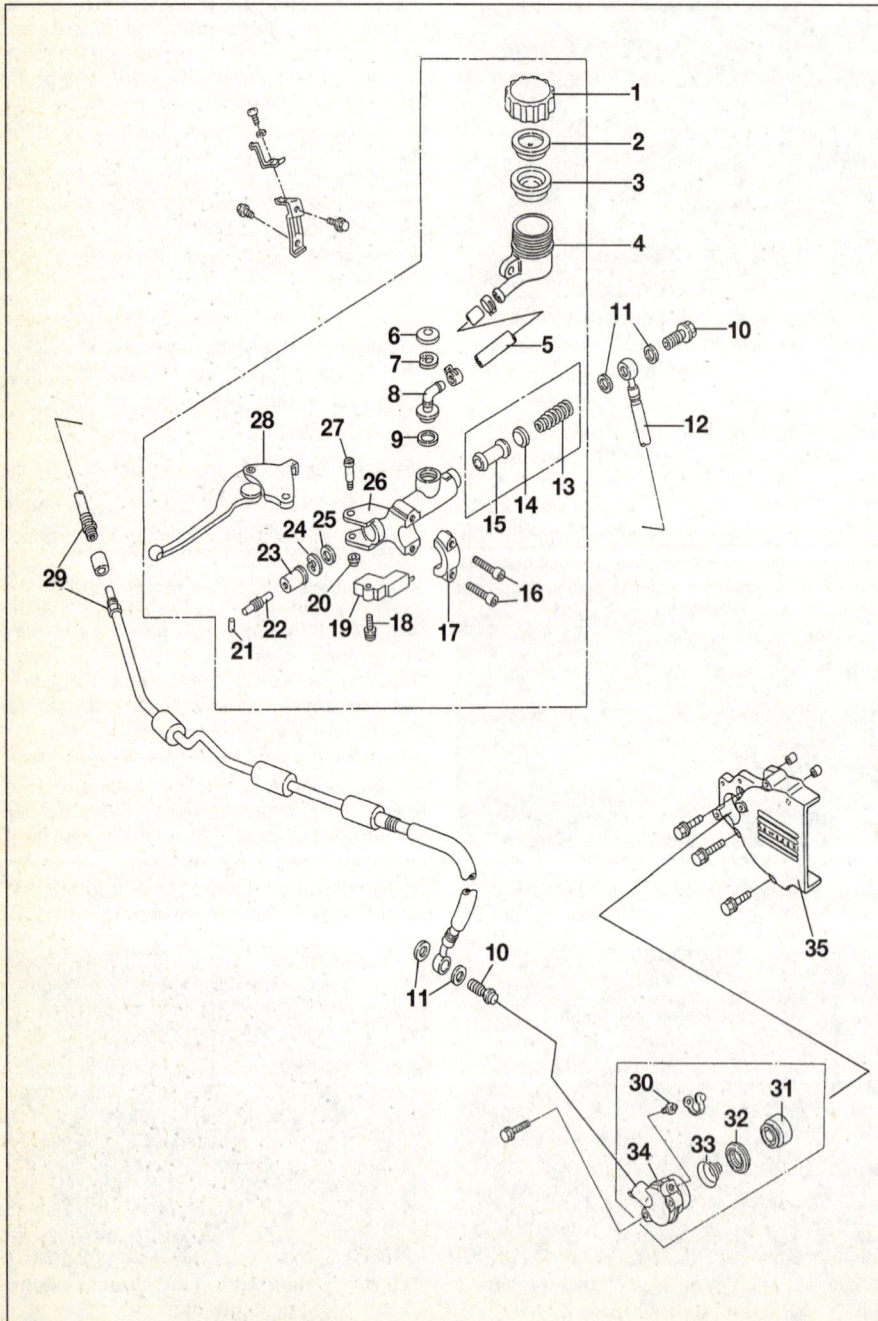

◄ **20.5d Clutch hydraulic system details**

| | |
|---|---|
| 1  Reservoir cap | 20 Pivot screw |
| 2  Retainer |     locknut |
| 3  Diaphragm | 21 Pivot bushing |
| 4  Fluid reservoir | 22 Pushrod |
| 5  Fluid feed hose | 23 Rubber boot |
| 6  Rubber cap | 24 Snap-ring |
| 7  Snap-ring | 25 Washer |
| 8  Fluid feed fitting | 26 Master cylinder |
| 9  O-ring |     body |
| 10 Banjo bolt | 27 Pivot screw |
| 11 Sealing washers | 28 Clutch lever |
| 12 Upper fluid | 29 Lower fluid hose |
|     hose | 30 Bleed valve |
| 13 Spring | 31 Slave cylinder |
| 14 Primary cup |     piston |
| 15 Piston | 32 Piston seal |
| 16 Clamp bolts | 33 Spring |
| 17 Clamp | 34 Slave cylinder |
| 18 Clutch switch |     body |
|     screw | 35 Engine sprocket |
| 19 Clutch switch |     cover |

**20.6  Remove the rubber cap, snap-ring, fluid feed fitting and O-ring**

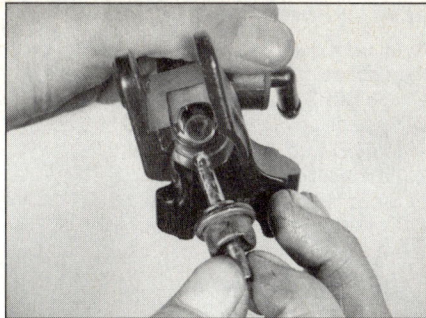

**20.7  Remove the rubber boot and pushrod**

**20.8a  Remove the snap-ring from the end of the master cylinder**

**7**  Remove the rubber boot and pushrod from the master cylinder **(see illustration)**.

**8**  Remove the snap-ring and washer, then dump out the piston complete with secondary cup, primary cup and spring **(see illustrations)**. If the piston won't come out, blow compressed air into the fluid line hole.

⚠️ *Warning: The piston may shoot out forcefully enough to cause injury. Point the open end of the master cylinder at a block of wood or a pile of rags inside a box and apply air pressure gradually. Never point the end of the cylinder at yourself, including your fingers.*

**9**  Remove the rubber grommet from the bore around the fluid feed fitting. Remove the snap-ring, lift the fitting out and remove the O-ring.

**10**  Thoroughly clean all of the components in clean brake fluid (don't use any type of petroleum-based solvent).

**11**  Check the piston and cylinder bore for wear, scratches and rust. If the piston shows these conditions, replace it and both rubber cups as a set. If the cylinder bore has any defects, replace the entire master cylinder.

**12**  Install the spring in the cylinder bore, wide end first.

**13**  Coat a new primary cup with brake fluid and install it in the cylinder, wide end first.

**14**  Coat the piston and secondary cup with brake fluid and install them in the cylinder.

**15**  Install the washer. Press the piston into the bore and install the snap-ring to hold it in place. Make sure the snap-ring seats securely in its groove.

**16**  Install the rubber boot and pushrod.

**17**  When you install the clutch lever, align the hole in the bushing with the pushrod. Tighten the lever pivot screw and its locknut to the torques listed in this Chapter's Specifications.

**18**  Install a new O-ring in the fluid feed fitting bore. Install the fitting, secure it with the snap-ring and push the rubber cap into position.

### Master cylinder installation

**19**  Installation is the reverse of the removal steps, with the following additions:
  a) Make sure the UP mark on the master cylinder clamp is upright **(see illustration)**.

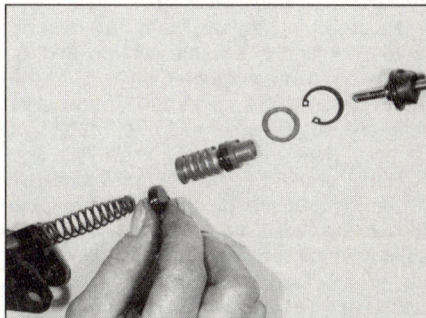

**20.8b  Remove the washer, piston, primary cup and spring**

  b) If you're working on a J or K model, rotate the master cylinder to place the clutch lever downward as far as possible, so it won't strike the fairing or the instrument cluster.

⚠️ *Warning: If the lever touches the fairing or instrument cluster while the bike is being operated, the clutch might disengage unexpectedly, leading to loss of control of the motorcycle.*

  c) If you're working on an L or M model, align the lower corner of the master cylinder front section with the punch mark in the handlebar **(see illustration 20.3a)**.
  d) Tighten the clamp bolts to the torque listed in this Chapter's Specifications. Tighten the upper clamp bolt first, then the lower one. If there is a small gap at

**20.20  Loosen the banjo bolt only if the slave cylinder will be removed completely**

  A  Mounting bolts     C  Bleed valve
  B  Banjo bolt               cap

**20.19  The UP mark on the master cylinder clamp must be upright**

  the bottom between the clamp and the master cylinder, don't try to close it by further tightening.
  e) Use a new sealing washer on each side of the banjo bolt fitting and tighten the banjo bolt to the torque listed in this Chapter's Specifications.
  f) Fill and bleed the master cylinder as described below. Operate the clutch and check for fluid leaks.

### Slave cylinder removal

**20**  If you're removing the slave cylinder for overhaul, loosen the banjo fitting bolt while the slave cylinder is still mounted on the engine **(see illustration)**. If you're just removing it for access to other components, leave the hydraulic line connected.

**21**  Remove the mounting bolt and take the slave cylinder off **(see illustration)**.

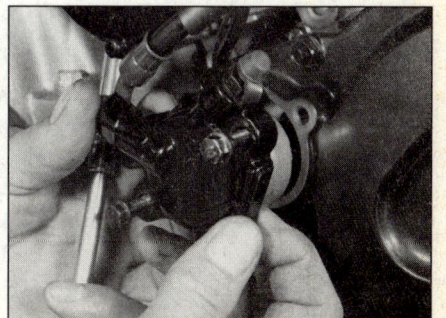

**20.21  Take the slave cylinder off the engine sprocket cover**

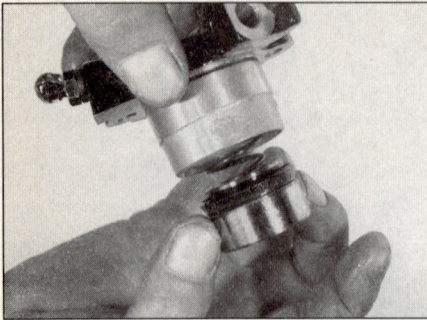

**20.24 Remove the piston and detach the spring**

**22** If you're removing the slave cylinder for overhaul, remove the banjo fitting bolt and detach the hydraulic line. Place the end of the line in a container to catch dripping fluid. *Caution: Brake fluid will damage paint. Wipe up any spills immediately and wash the area with soap and water.*
**23** If the hydraulic line is still connected, push the piston as far into the bore as it will go. Hold the piston in, slowly squeeze the clutch lever to the handlebar and tie the clutch lever in that position. Otherwise, the slave cylinder piston will fall out.

### Slave cylinder overhaul

**24** Let the pressure of the slave cylinder spring push the piston out of the cylinder, then remove the spring **(see illustration 20.5d and the accompanying illustration)**.
**25** Separate the spring from the piston.
**26** Thoroughly clean all parts in clean brake fluid (don't use any type of petroleum-based solvent).
**27** Check the cylinder bore and piston for wear, scratches and corrosion. If the piston shows these conditions, replace it and the seal as a set. If the cylinder bore has any defects, replace the entire slave cylinder. If the piston and bore are good, carefully remove the seal from the piston and install a new one with the lip facing into the bore.

### Slave cylinder installation

**28** Installation is the reverse of the removal procedure with the following additions:
  a) *Use new sealing washers on the clutch fluid line and tighten the banjo bolt to the torque listed in this Chapter's Specifications.*
  b) *Bleed the clutch (see below).*
  c) *Operate the clutch and check for fluid leaks.*

### System bleeding

**29** Support the bike securely upright.
**30** Remove the master cylinder reservoir cap, retainer and diaphragm. Top up the reservoir with fluid to the upper level line.
**31** Remove the cap from the bleed valve on the slave cylinder **(see illustration 20.20)**. Place a box wrench (ring spanner) over the bleed valve. Attach a rubber tube to the valve fitting and put the other end of the tube in a container. Pour enough clean brake fluid into the container to cover the end of the tube.
**32** Slowly squeeze the clutch lever several times. At the same time, tap on the clutch fluid line, starting at the bottom and working your way to the top. Stop when no more air bubbles can be seen rising from the bottom of the reservoir.
**33** Squeeze the clutch lever several times until you feel an increase in the effort required to pull the lever, then hold it in.
**34** With the lever held in, quickly open the bleed valve to let air and fluid escape, then close it.
**35** Repeat Steps 33 and 34 until there aren't any more bubbles in the fluid flowing into the container. **Note:** *Keep an eye on the fluid level in the reservoir. If it drops too low, air will be sucked into the line and the procedure will have to be done over.*
**36** Replenish the reservoir with fluid, then reinstall the diaphragm, retainer and cap.
**37** Disconnect the rubber tube from the valve fitting, clean up any brake fluid spills and put the cap back on the bleed valve.

### 21 External shift mechanism – removal, inspection and installation

### Shift pedal and rod

#### Removal

**1** Support the bike securely upright.
**2** Look for alignment marks on the end of the shift lever and shift shaft **(see illustration)**. If they aren't visible, make your own marks with a sharp punch.
**3** Remove the shift lever pinch bolt and slide the lever off the shaft **(see illustration)**.
**4** The shift pedal is removed together with the rider's footpeg on the left side of the bike (see Chapter 8).

#### Installation and adjustment

**5** Install the shift pedal (if removed) and tighten its bolt to the torque listed in Chapter 8 Specifications.
**6** Slip the shift lever onto the shaft, aligning the match marks. Install the pinch bolt and tighten it securely.
**7** Loosen the locknuts on the shift rod. **Note:** *The locknut closest to the knurled portion of the shift rod has left-hand threads. Turn it clockwise to loosen. Turn the rod so the shift levers are at right angles to the rod. If you're working on a J or K model, the shift pedal pad should be approximately 30 mm (1.2 inch) below the top of the footpeg (see illustration). If you're working on an L or M model, the center of the shift pedal should be even with the centerline of the shift rod (see illustration).*

### External shift linkage

#### Removal

**8** If you're working on a J or K model, remove the engine (see Section 5).
**9** If you're working on an L or M model,

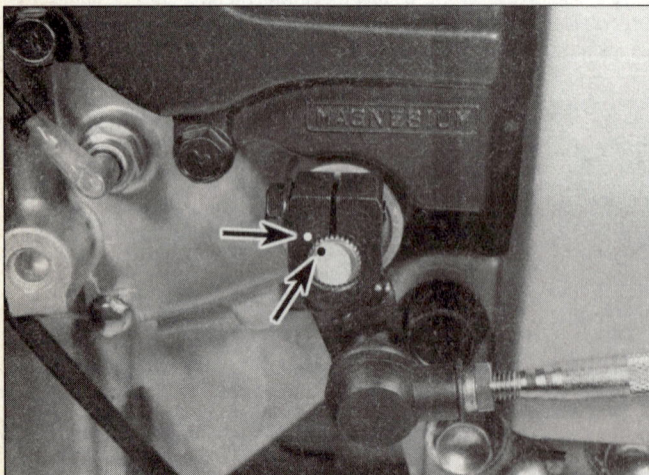

**21.2 Look for matchmarks on the shift lever and shaft (arrows)**

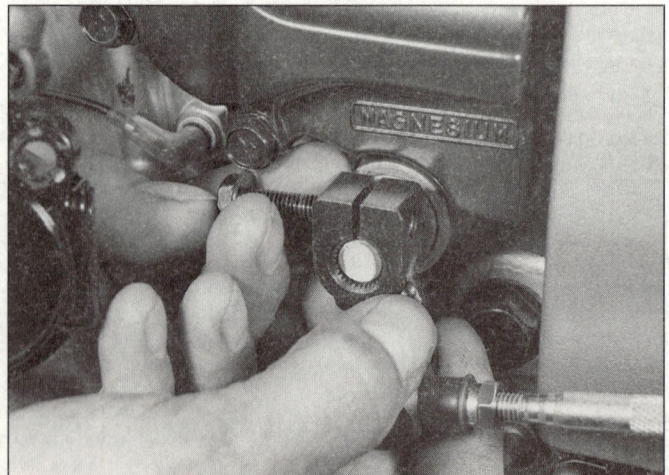

**21.3 Remove the bolt and slip the lever off the shaft**

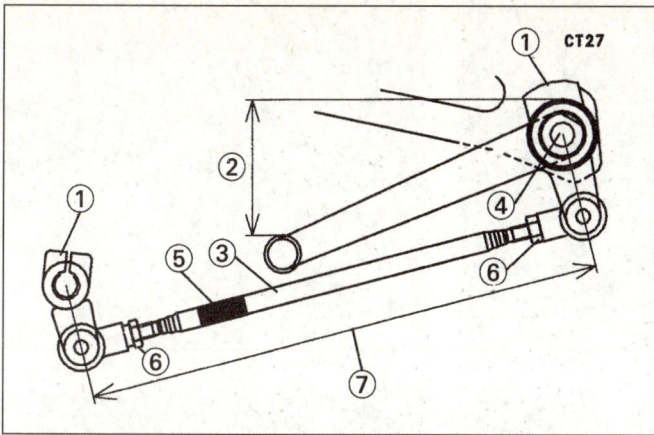

21.7a  Shift pedal details (J and K models)

1  Shift lever
2  Pedal height (measured from top of footpeg)
3  Shift rod

4  Footpeg
5  Knurled portion
6  Locknuts
7  Rod length

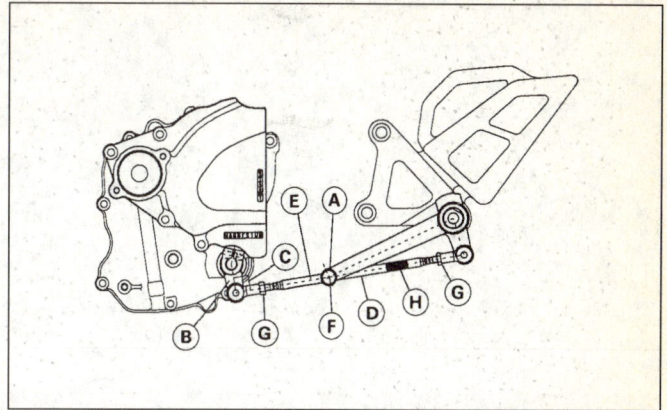

21.7b  Shift pedal details (L and M models)

A  Shift pedal pad
B  Shift lever
C  Shift lever at right angle to shift rod
D  Shift rod

E  Centerline of shift rod
F  Shift pedal pad intersects centerline of shift rod
G  Locknut
H  Knurled portion

remove the left lower fairing panel (see Chapter 8) and the engine sprocket (see Chapter 6).

**10** Remove the water pump (see Chapter 3).

**11** Position a drain pan under the shift mechanism cover. Remove the shift mechanism cover bolts and Phillips screws, then remove the cover, gasket and dowels **(see illustrations)**.

**12** If it's necessary to remove the shift drum cam, remove its Allen bolt and take it off the shift drum **(see illustrations)**. The external shift mechanism can be removed with the shift drum cam in place; to do this, pull the shift mechanism arm away from the shift drum so its points clear the shift drum cam **(see illustration)**.

**13** Pull the mechanism and shaft off **(see illustrations)**.

21.11a  Remove the cover bolts and screws (arrows) . . .

21.11b  . . . use an impact driver on the screws

21.11c  Take the cover off . . .

21.11d  . . . and remove the gasket . . .

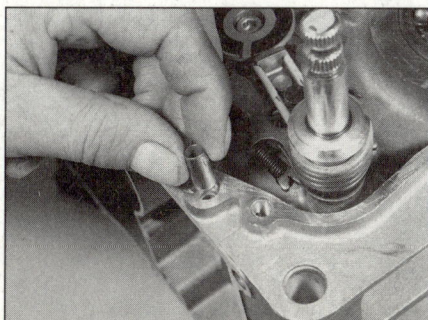

21.11e  . . . the lower dowel . . .

21.11f  . . . and the upper dowel (which has an O-ring)

**21.12a Remove the shift drum cam Allen bolt . . .**

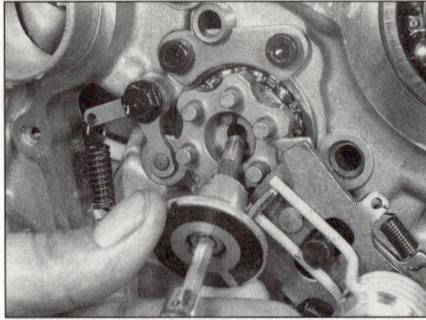

**21.12b . . . take off the shift drum cam . . .**

**21.12c . . . and remove its pin**

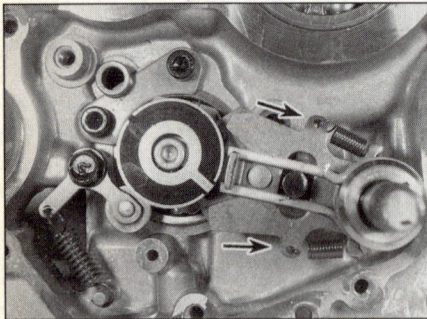

**21.12d Pull the shift arm in the direction of the arrows to clear the shift drum cam**

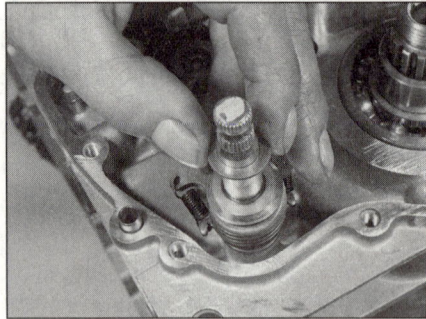

**21.13a Remove the washer . . .**

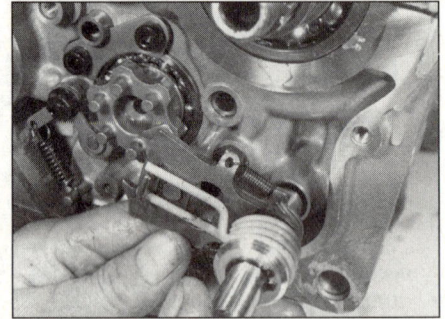

**21.13b . . . and take the external shift linkage out of the crankcase**

*Caution: Don't pull the shift rod out of the crankcase – the shift forks will fall out of position, and the crankcase will have to be separated to reinstall them.*

14 Unhook the spring from the gear positioning lever. Remove the lever bolt and take off the lever and the collar behind it **(see illustrations)**.

**Inspection**

15 Remove the snap-ring, bushing and return spring from the shift mechanism **(see illustration)**. Unhook the pawl springs, noting the direction they face **(see illustration)**.

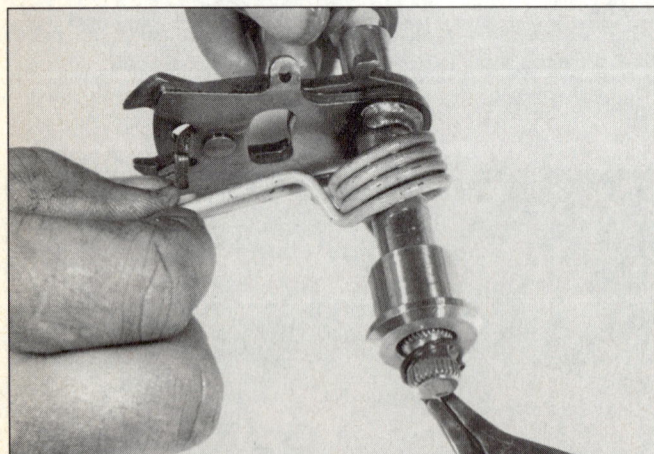

**21.14a Unhook the spring and remove the bolt (arrow) . . .**

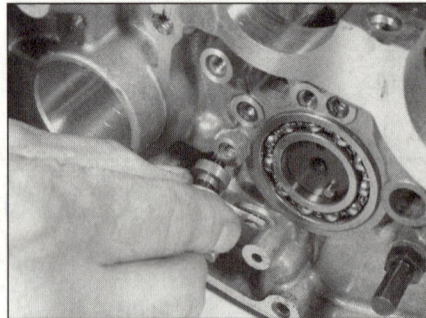

**21.14b . . . and remove the gear positioning lever and its collar**

**21.15a Remove the snap-ring, bushing and return spring from one end of the shift shaft, and the collar and shift arm from the other end**

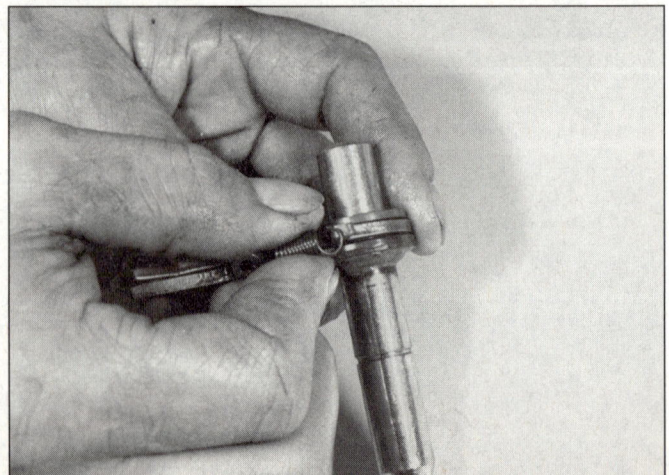

**21.15b The shift arm springs must be attached with the open ends facing in the proper direction**

**21.19 Check the return spring pin for looseness**

**21.20 Replace the bearing and seals in the cover if they're worn or damaged**

16 Check the shift shaft for bends and damage to the splines. If the shaft is bent, you can attempt to straighten it, but if the splines are damaged it will have to be replaced.
17 Check the condition of the return springs and the pawl spring. Replace them if they are cracked or distorted.
18 Check the shift mechanism arm and the overshift limiter for cracks, distortion and wear. If any of these conditions are found, replace the shift mechanism.
19 Make sure the return spring pin isn't loose **(see illustration)**. If it is, unscrew it, apply a non-hardening locking compound to the threads, then reinstall it and tighten it securely.
20 Check the condition of the seals and the shift pedal bearing in the cover **(see illustration)**. If the seals have been leaking or the bearing is worn or damaged, drive them out with a hammer and punch. New seals can be installed by driving them in with a socket.
*Caution: The bearing should be installed with a shouldered drift that fits the inside of the bearing exactly. If you don't have the proper tool, have the bearing replaced by a Kawasaki dealer.*

### Installation

21 Assemble the external shift mechanism. Install the two small pawl springs so their

hook ends will point outward (away from the engine) when the external shift mechanism is installed **(see illustration 21.12d)**.
22 If the shift drum cam was removed, install its dowel pin. Install the shift drum cam over the dowel pin, then install its Allen bolt and tighten it to the torque listed in this Chapter's Specifications.
23 Lubricate the inner end of the shift shaft and slide the external shift mechanism into place. Pull back the shift mechanism arm so its points clear the shift drum. Make sure the legs of the return spring are on either side of the pin.
24 Install the gear positioning lever and its return spring. Be sure to install the collar behind the lever with its narrow diameter facing away from the engine, so the lever will fit over it. Apply non-hardening thread locking agent to the threads of the bolt and tighten it to the torque listed in this Chapter's Specifications.
25 Apply high-temperature grease to the lips of the seals. Wrap the splines of the shift shaft with electrical tape, so the splines won't damage the seals as the cover is installed.
26 Apply a thin coat of RTV sealant to the cover mating areas on the crankcase, where the halves of the crankcase join, then install the gasket **(see illustration)**.

27 Make sure the rubber damper is in place inside the cover **(see illustration)**.
28 Make sure the washer is in place on the end of the external shift mechanism **(see illustration 21.13a)**. Install a new O-ring on the upper cover dowel and install both dowels in the crankcase **(see illustrations 21.11e and 21.11f)**.
29 Carefully guide the cover into place. Apply non-hardening thread locking agent to the threads of the upper front bolt and the two Phillips screws. Install the screws and bolts, tightening them evenly to the torque settings listed in this Chapter's Specifications (the bolts and screws have different torque settings).
30 The remainder of installation is the reverse of the removal steps.
31 Check the engine oil level and add some, if necessary (see 'Daily (pre-ride) checks').

## 22 Crankcase – disassembly and reassembly

1 To examine and repair or replace the crankshaft, connecting rods, bearings, transmission components and alternator shaft/starter motor clutch, the crankcase must be split into two parts.

### Disassembly

2 Remove the engine from the motorcycle (see Section 5).
3 Remove the carburetors (see Chapter 4).
4 Remove the water pump (see Chapter 3).
5 Remove the pickup coil (see Chapter 5).
6 Remove the clutch (see Section 19) and the alternator chain (see Section 27). Remove the cover from the external shift mechanism (see Section 21).
7 If you're going to remove the alternator shaft, remove the alternator (see Chapter 9) and the alternator coupling (see Section 27).

**21.26 Apply sealant to the crankcase seam areas (arrows), then install the gasket**

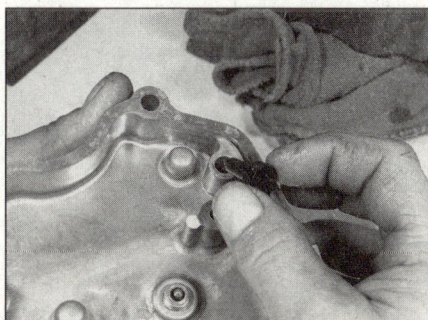
**21.27 Don't forget the rubber damper that goes in the cover**

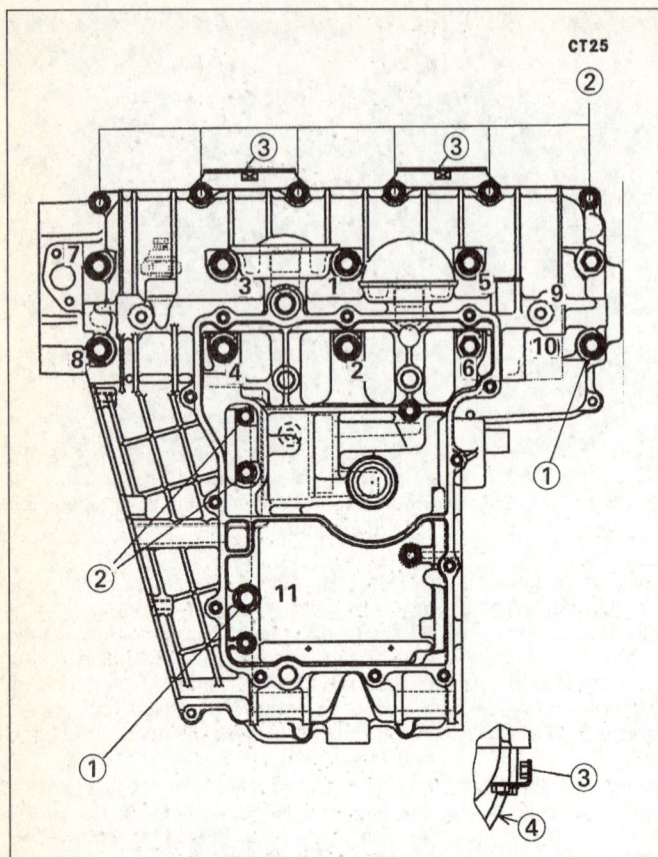

**22.11a  Crankcase lower bolts**

| 1  8 mm bolts | 3  Radiator bracket | 4  Lower crankcase |
| 2  6 mm bolts |     support brackets | |

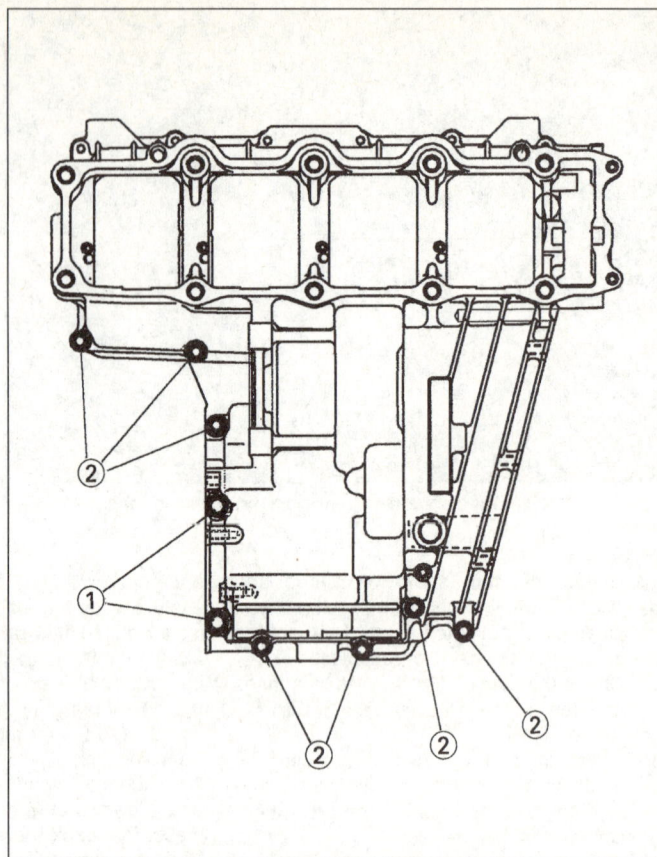

**22.11b  Crankcase upper bolts**

| 1  8 mm bolts | 2  6 mm bolts |

**8** If the crankcase is being separated to remove the crankshaft, remove the cylinder head, cylinder block and pistons (see Sections 10, 13 and 14).

**9** Remove the oil pan and engine sprocket cover (see Section 16 and Chapter 6).

**10** Check carefully to make sure there aren't any remaining components that attach the upper and lower halves of the crankcase together.

**11** Remove the 6 mm crankcase bolts **(see illustrations)**. Remove the remaining crankcase bolts in the reverse order of the tightening sequence. The bolt numbers are cast in the crankcase halves **(see illustration)**.

**12** Carefully pry the crankcase apart and lift the lower half off the upper half **(see illustration)**.

*Caution: Don't pry against the mating surfaces or they'll develop leaks.*

**22.11c  The crankcase bolt numbers are cast in the crankcase (arrow)**

**22.12  Lift the lower crankcase half off the upper half**

22.13a  There's a dowel at the cam chain end of the crankshaft . . .

22.13b  . . . and at the seal end . . .

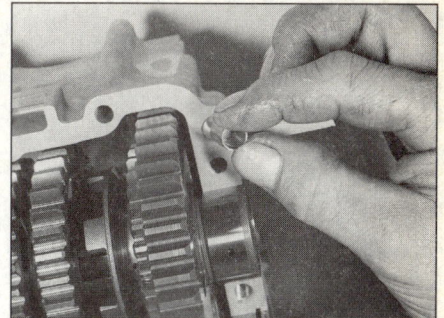

22.13c  . . . and one near the transmission output shaft

22.13d  Remove the oil nozzle and its O-ring

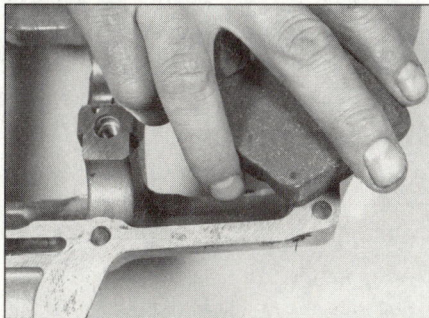

22.15  Remove all traces of sealant from the mating surfaces with a sharpening stone

22.16  Install the oil nozzle and a new O-ring

13  Remove the three crankcase dowels, the oil nozzle and its O-ring **(see illustrations)**.
14  Refer to Sections 23 through 32 for information on the internal components of the crankcase.

### Reassembly

15  Remove all traces of sealant from the crankcase mating surfaces with a sharpening stone or similar tool **(see illustration)**. Be careful not to let any fall into the case as this is done and be careful not to damage the mating surfaces.
16  Check to make sure the three dowel pins

are in place in their holes in the mating surface of the upper crankcase half **(see illustrations 22.13a, 22.13b and 22.13c)**. Install the oil nozzle with a new O-ring **(see illustration)**.
17  Make sure the cam chain is installed on the crankshaft sprocket.
18  Pour some engine oil over the transmission gears, the crankshaft main bearings and the shift drum. Don't get any oil on the crankcase mating surface.
19  Apply a thin, even bead of silicone sealant to the indicated areas of the crankcase mating surfaces **(see illustrations)**.
*Caution: Don't apply an excessive amount*

*of sealant, as it will ooze out when the case halves are assembled and may obstruct oil passages. Don't put sealant on the ends of the main bearing shells.*
20  Check the position of the shift drum – make sure it's in the neutral position (see Section 32).
21  Carefully place the lower crankcase half onto the upper crankcase half. While doing this, make sure the shift forks fit into their gear grooves (see Section 32). Make sure the crankshaft end seal is positioned properly when the case halves are mated **(see illustration)**.

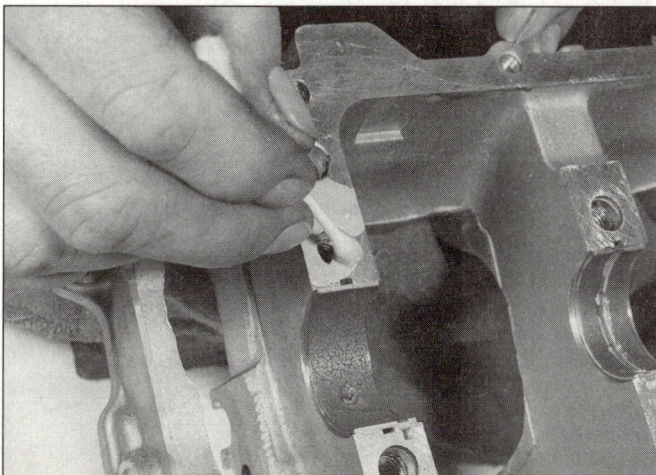

22.19a  Apply silicone sealant to the crankcase mating surfaces

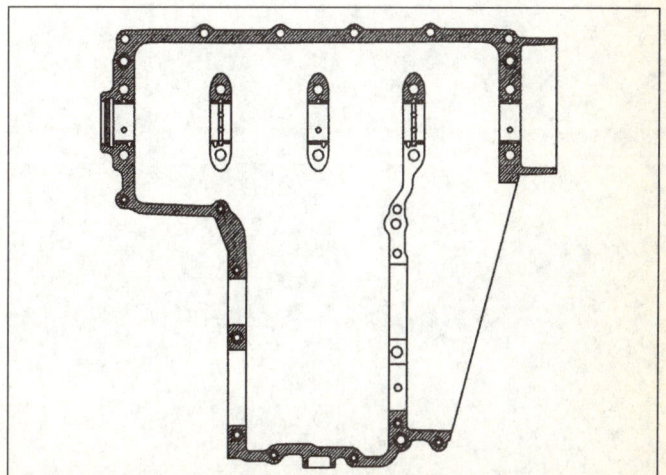

22.19b  Apply sealant to the shaded areas

22.21 The crankcase end seal should be installed like this when the cases are assembled

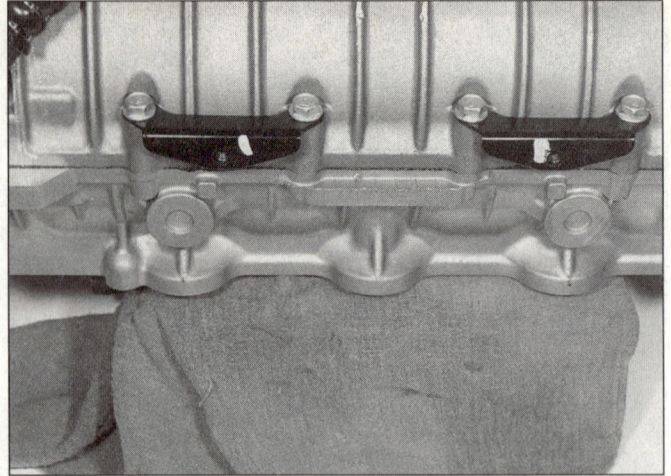

22.22 Four of the crankcase bolts also secure the support brackets for the radiator mounting bracket

22 Install the lower crankcase half bolts and tighten them so they are just snug. Note that four of the bolts secure the support brackets for the radiator mounting bracket **(see illustration)**.
23 Tighten the larger crankcase bolts (8 mm), in the numbered sequence cast in the crankcase, to half of the torque listed in this Chapter's Specifications. Then go through the sequence again, tightening the 8 mm bolts to the torque listed in this Chapter's Specifications.
24 After tightening the 8 mm bolts, tighten the 6 mm bolts to the torque listed in this Chapter's Specifications.
25 Turn the transmission mainshaft and output shaft to make sure they turn freely. Also make sure the crankshaft turns freely.
26 The remainder of installation is the reverse of removal.

## 23 Crankcase components – inspection and servicing

1 Separate the crankcase and remove the following:
a) *Crankshaft and main bearings*
b) *Transmission shafts*
c) *Shift drum and forks*
d) *Transmission oil pipe*
e) *Starter clutch*
f) *Breather assembly* **(see illustration)**
2 Clean the crankcase halves thoroughly with new solvent and dry them with compressed air. All oil passages should be blown out with compressed air and all traces of old gasket sealant should be removed from the mating surfaces.
*Caution: Be very careful not to nick or gouge the crankcase mating surfaces or*

*leaks will result. Check both crankcase sections very carefully for cracks and other damage.*
3 Check the ball bearings in the case **(see illustration)**. If they don't turn smoothly, drive them out with a bearing driver or a socket having an outside diameter slightly smaller than that of the bearing outer race. Before installing them, allow them to sit in the freezer overnight, and about fifteen-minutes before installation, place the case half in an oven, set to about 200°F, and allow it to heat up. The bearings are an interference fit, and this will ease installation.

⚠️ *Warning: Before heating the case, wash it thoroughly with soap and water so no explosive fumes are present. Also, don't use a flame to heat the case.*
4 If any damage is found that can't be repaired, replace the crankcase halves as a set.

23.1 Squeeze the hose clips and slide them down the hoses, then detach the breather assembly from the engine

23.3 Replace the ball bearings in the case if they're loose, rough or noisy

## 24 Main and connecting rod bearings – general note

1 Even though main and connecting rod bearings are generally replaced with new ones during the engine overhaul, the old bearings should be retained for close examination as they may reveal valuable information about the condition of the engine.

2 Bearing failure occurs mainly because of lack of lubrication, the presence of dirt or other foreign particles, overloading the engine and/or corrosion. Regardless of the cause of bearing failure, it must be corrected before the engine is reassembled to prevent it from happening again.

3 When examining the bearings, remove the main bearings from the case halves and the rod bearings from the connecting rods and caps and lay them out on a clean surface in the same general position as their location on the crankshaft journals. This will enable you to match any noted bearing problems with the corresponding side of the crankshaft journal.

4 Dirt and other foreign particles get into the engine in a variety of ways. It may be left in the engine during assembly or it may pass through filters or breathers. It may get into the oil and from there into the bearings. Metal chips from machining operations and normal engine wear are often present. Abrasives are sometimes left in engine components after reconditioning operations such as cylinder honing, especially when parts are not thoroughly cleaned using the proper cleaning methods. Whatever the source, these foreign objects often end up imbedded in the soft bearing material and are easily recognized. Large particles will not imbed in the bearing and will score or gouge the bearing and journal. The best prevention for this cause of bearing failure is to clean all parts thoroughly and keep everything spotlessly clean during engine reassembly. Frequent and regular oil and filter changes are also recommended.

5 Lack of lubrication or lubrication breakdown has a number of interrelated causes. Excessive heat (which thins the oil), overloading (which squeezes the oil from the bearing face) and oil leakage or throw off (from excessive bearing clearances, worn oil pump or high engine speeds) all contribute to lubrication breakdown. Blocked oil passages will also starve a bearing and destroy it. When lack of lubrication is the cause of bearing failure, the bearing material is wiped or extruded from the steel backing of the bearing. Temperatures may increase to the point where the steel backing and the journal turn blue from overheating.

6 Riding habits can have a definite effect on bearing life. Full throttle low speed operation, or lugging (laboring) the engine, puts very high loads on bearings, which tend to squeeze out the oil film. These loads cause the bearings to flex, which produces fine cracks in the

25.1a Lift the crankshaft out of the crankcase, together with the cam chain

bearing face (fatigue failure). Eventually the bearing material will loosen in pieces and tear away from the steel backing. Short trip driving leads to corrosion of bearings, as insufficient engine heat is produced to drive off the condensed water and corrosive gases produced. These products collect in the engine oil, forming acid and sludge. As the oil is carried to the engine bearings, the acid attacks and corrodes the bearing material.

7 Incorrect bearing installation during engine assembly will lead to bearing failure as well. Tight fitting bearings which leave insufficient bearing oil clearances result in oil starvation. Dirt or foreign particles trapped behind a bearing insert result in high spots on the bearing which lead to failure.

8 To avoid bearing problems, clean all parts thoroughly before reassembly, double check all bearing clearance measurements and lubricate the new bearings with engine assembly lube or moly-based grease during installation.

## 25 Crankshaft and main bearings – removal, inspection and installation

### Removal

1 Crankshaft removal is a simple matter of lifting it out of place once the crankcase has been separated (see illustration). Before removing the crankshaft check the endplay. Insert a feeler gauge between the crankshaft and the no. 2 crankcase main journal (see illustration). Compare your findings with this Chapter's Specifications. If the endplay is excessive, both case halves must be replaced.

2 The main bearing inserts can be removed from their saddles by pushing their centers to the side, then lifting them out (see illustration). Keep the bearing inserts in order. The main bearing oil clearance should be checked, however, before removing the inserts (see Step 8).

### Inspection

3 Mark and remove the connecting rods from the crankshaft (see Section 26).

25.1b Measure crankshaft end play with a feeler gauge between the no. 2 bearing web and the crankshaft

4 Clean the crankshaft with solvent, using a rifle-cleaning brush to scrub out the oil passages. If available, blow the crank dry with compressed air. Check the main and connecting rod journals for uneven wear, scoring and pits. Rub a penny across the journal several times – if a journal picks up copper from the penny, it's too rough. Replace the crankshaft.

5 Check the camshaft chain sprocket and the primary gear on the crankshaft for chipped teeth and other wear. If any undesirable conditions are found, replace the crankshaft. Check the chain as described in Section 29.

6 Check the rest of the crankshaft for cracks and other damage. It should be magnafluxed to reveal hidden cracks – a dealer service department or motorcycle machine shop will handle the procedure.

7 Set the crankshaft on V-blocks and check the runout with a dial indicator touching the center main journal, comparing your findings with this Chapter's Specifications. If the runout exceeds the limit, replace the crankshaft.

### Main bearing selection

8 To check the main bearing oil clearance, clean off the bearing inserts (and reinstall them, if they've been removed from the case) and lower the crankshaft into the upper half of the case. Cut five pieces of Plastigauge (type HPG-1) and lay them on the crankshaft main journals, parallel with journal axis (see illustration).

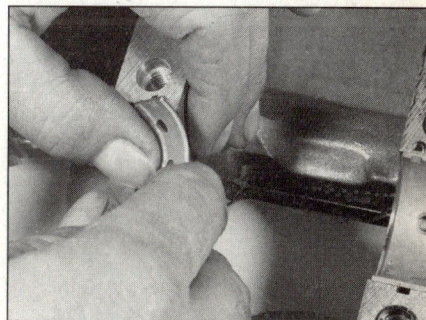
25.2 Push the centers of the bearing inserts sideways and rotate them out of their webs

**25.8  Lay a strip of Plastigauge along the bearing journal, parallel to the crankshaft centerline**

**25.9  Lay the lower crankcase half on the upper crankcase half and DO NOT rotate the crankshaft**

**9** Very carefully, guide the lower case half down onto the upper case half **(see illustration)**. Install the large (8 mm) bolts and tighten them, using the stages and sequence described in Section 22. Don't rotate the crankshaft!

**10** Now, remove the bolts and carefully lift the lower case half off. Compare the width of the crushed Plastigauge on each journal to the scale printed on the Plastigauge envelope to obtain the main bearing oil clearance **(see illustration)**. Write down your findings, then remove all traces of Plastigauge from the journals, using your fingernail or the edge of a credit card.

**25.10  Measure the width of the crushed Plastigauge with the scale on the envelope**

**11** If the oil clearance falls into the range listed in this Chapter's Specifications, no bearing replacement is required (provided they are in good shape). If the clearance is more than 0.036 mm (0.0014 inch), but less than the service limit of 0.066 mm (0.0025 inch), replace the bearing inserts with inserts that have blue paint marks **(see illustration)**, then check the oil clearance once again. Always replace all of the inserts at the same time.

**12** The clearance might be slightly greater than the standard clearance, but that doesn't matter, as long as it isn't greater than the maximum clearance or less than the minimum clearance.

**13** If the clearance is greater than the service limit listed in this Chapter's Specifications, measure the diameter of the crankshaft journals with a micrometer and compare your findings with this Chapter's Specifications. Also, by measuring the diameter at a number of points around each journal's circumference, you'll be able to determine whether or not the journal is out-of-round. Take the measurement at each end of the journal, near the crank throws, to determine if the journal is tapered.

**14** If any main bearing journal has worn down past the service limit, replace the crankshaft.

**15** If the diameters of the journals aren't less than the service limit but differ from the original markings on the crankshaft **(see illustration)**, apply new marks with a hammer and punch.

*If the journal measures between 31.984 and 31.992 mm (1.2592 and 1.2595 inch) don't make any marks on the crankshaft (there shouldn't be any marks there, anyway).*

*If the journal measures between 31.993 and 32.000 mm (1.2596 and 1.2598 inch) make a 1 mark on the crank in the area indicated (if it's not already there).*

**16** Remove the main bearing inserts and assemble the case halves (see Section 22). Using a telescoping gauge and a micrometer, measure the diameters of the main bearing bores, then compare the measurements with the marks on the upper case half **(see illustration)**.

*If the bores measure between 35.009 to 35.016 mm (1.3783 and 1.3785 inch), there shouldn't be any marks in the indicated areas.*

*If the bores measure between 35.000 and 35.008 mm (1.3779 and 1.3782 inch) there should be a O mark in the indicated areas.*

**17** Using the marks on the crankshaft and the

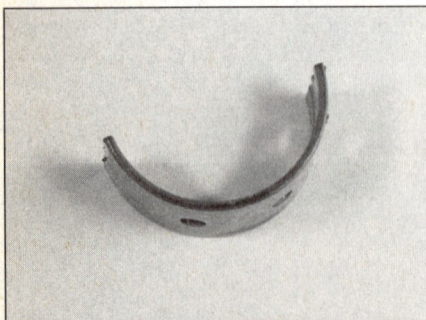

**25.11  Bearing thickness is indicated by a color code painted on the side of the bearing**

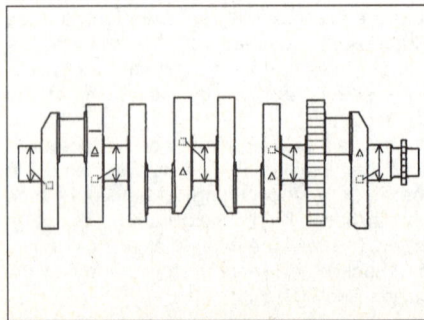

**25.15  Use the crankshaft marks (1 or none) . . .**

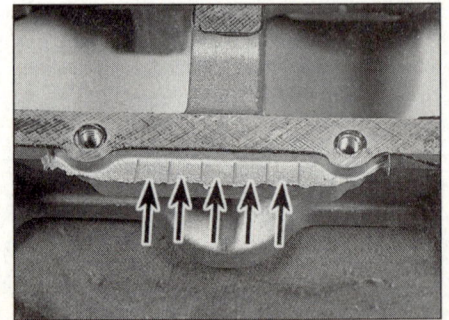

**25.16  . . . in conjunction with the crankcase marks (arrow) (O, or as shown here, no mark) . . .**

| Crankcase Main Bearing Bore Diameter Mark | Crankshaft Main Journal Diameter Mark | Crankshaft Bearing Insert* | | |
|---|---|---|---|---|
| | | Size Color | Part Number | Journal Nos. |
| O | 1 | Brown | 92028-1628 | 1, 3 ,5 |
| | | | 92028-1631 | 2, 4 |
| None | 1 | Black | 92028-1627 | 1, 3, 5 |
| O | None | | 92028-1630 | 2, 4 |
| None | None | Blue | 92028-1626 | 1, 3, 5 |
| | | | 92028-1629 | 2, 4 |

**25.17 . . . to determine the proper main bearing sizes**

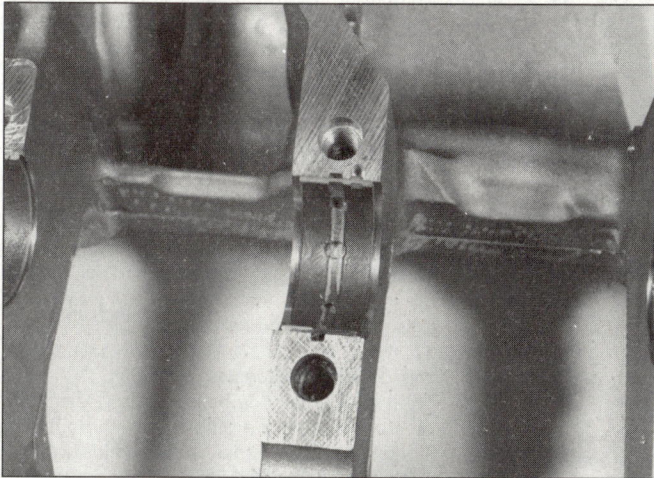

**25.18a Make sure the tabs on the bearing inserts fit into the notches in the webs**

**25.18b The no. 2 and no. 4 bearing inserts have oil grooves (arrows)**

marks on the case, determine the bearing sizes required by referring to the accomp-anying bearing selection chart **(see illustration)**.

### Installation

**18** Separate the case halves once again. Clean the bearing saddles in the case halves, then install the bearing inserts in all five of the webs in the case **(see illustration)**. The bearing inserts for journals 2 and 4 have oil grooves **(see illustration)**. When installing the bearings, use your hands only – don't tap them into place with a hammer.

**19** Lubricate the bearing inserts with engine assembly lube or moly-based grease.
**20** Install the connecting rods, if they were removed (see Section 26).
**21** Install a new crankshaft end seal at the opposite end of the crankcase to the cam chain sprocket **(see illustration)**.
**22** Install the cam chain on the crankshaft sprocket **(see illustration)**.
**23** Carefully lower the crankshaft into place.
**24** Assemble the case halves (see Section 22) and check to make sure the crank-shaft and the transmission shafts turn freely.

### 26 Connecting rods and bearings – removal, inspection and installation

### Removal

**1** Before removing the connecting rods from the crankshaft, measure the side clearance of each rod with a feeler gauge **(see illustration)**. If the clearance on any rod is greater than that listed in this Chapter's Specifications, that rod will have to be replaced with a new one.

**25.21 The ribs on the crankcase end seal must seat in the case groove**

**25.22 Be sure to drape the chain over the crankshaft sprocket**

**26.1 Check the connecting rod side clearance with a feeler gauge**

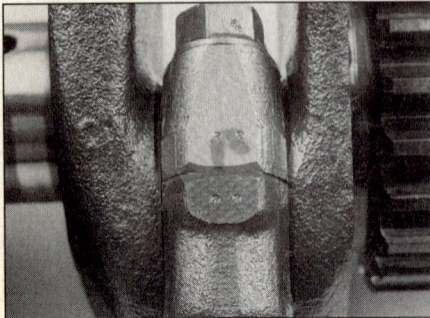

**26.2 Using a hammer and punch, make matching number marks on the connecting rod and its cap**

**2** Using a center punch, mark the position of each rod and cap, relative to its position on the crankshaft **(see illustration)**.

**3** Unscrew the bearing cap nuts **(see illustration)**, separate the cap from the rod, then detach the rod from the crankshaft. If the cap is stuck, tap on the ends of the rod bolts with a soft-faced hammer to free them.

**4** Separate the bearing inserts from the rods and caps, keeping them in order so they can be reinstalled in their original locations. Wash the parts in solvent and dry them with compressed air, if available.

## Inspection

**5** Check the connecting rods for cracks and other obvious damage. Lubricate the piston pin for each rod, install it in the proper rod and check for play **(see illustration)**. If it is loose, replace the connecting rod and/or the pin. **Note:** *Connecting rods are graded according to weight and installed so that the two rods at the left side of the engine (no. 1 and no. 2) and the two rods at the right side of the engine (no. 3 and no. 4) are the same weight grade. If a rod is replaced, make sure it matches the weight grade of the rod it's paired with.*

**6** Refer to Section 24 and examine the connecting rod bearing inserts. If they are scored, badly scuffed or appear to have been seized, new bearings must be installed. Always replace the bearings in the connecting rods as a set. If they are badly damaged, check the corresponding crankshaft journal.

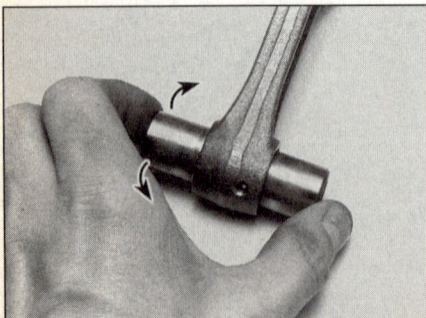

**26.5 Checking the piston pin and connecting rod bore for wear**

**26.3 Remove the bearing cap nuts**

**HAYNES HiNT** *Evidence of extreme heat, such as discoloration, indicates that lubrication failure has occurred. Be sure to thoroughly check the oil pump and pressure relief valve as well as all oil holes and passages before reassembling the engine.*

**7** Have the rods checked for twist and bending at a dealer service department or other motorcycle repair shop.

### Bearing selection

**8** If the bearings and journals appear to be in good condition, check the oil clearances as follows:

**9** Start with the rod for the number one cylinder. Wipe the bearing inserts and the connecting rod and cap clean, using a lint-free cloth.

**10** Install the bearing inserts in the connecting rod and cap. Make sure the tab on the bearing engages with the notch in the rod or cap.

**11** Wipe off the connecting rod journal with a lint-free cloth. Lay a strip of Plastigauge (type HPG-1) across the top of the journal, parallel with the journal axis **(see illustration 25.8)**.

**12** Position the connecting rod on the bottom of the journal, then install the rod cap and nuts. Tighten the nuts to the torque listed in this Chapter's Specifications, but don't allow the connecting rod to rotate at all.

**13** Unscrew the nuts and remove the

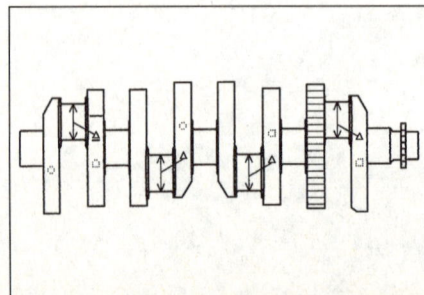

**26.18 The marks on the crank throws (arrows) should coincide with the diameters of the connecting rod journals**

connecting rod and cap from the journal, being very careful not to disturb the Plastigauge. Compare the width of the crushed Plastigauge to the scale printed in the Plastigauge envelope **(see illustration 25.10)** to determine the bearing oil clearance.

**14** If the clearance is within the range listed in this Chapter's Specifications and the bearings are in perfect condition, they can be reused. If the clearance is more than 0.066 mm (0.0026-inch) but less than the service limit of 0.10 mm (0.0039 inch), replace the bearing inserts with inserts that have blue paint marks, then check the oil clearance once again. Always replace all of the inserts at the same time.

**15** The clearance might be slightly greater than the standard clearance, but that doesn't matter, as long as it isn't greater than the maximum clearance or less than the minimum clearance.

**16** If the clearance is greater than the service limit listed in this Chapter's Specifications, measure the diameter of the connecting rod journal with a micrometer and compare your findings with this Chapter's Specifications. Also, by measuring the diameter at a number of points around the journal's circumference, you'll be able to determine whether or not the journal is out-of-round. Take the measurement at each end of the journal to determine if the journal is tapered.

**17** If any journal has worn down past the service limit, replace the crankshaft.

**18** If the diameter of the journal isn't less than the service limit but differs from the original markings on the crankshaft **(see illustration)** apply new marks with a hammer and punch.

> *If the journal measures between 33.984 and 33.992 mm (1.3379 and 1.3382 inch) (don't make any marks on the crank (there shouldn't be one there anyway).*
> *If the journal measures between 33.993 and 34.000 mm (1.3383 and 1.3385 inch), make a "0" mark on the crank in the area indicated (if not already there).*

**19** Remove the bearing inserts from the connecting rod and cap, then assemble the cap to the rod. Tighten the nuts to the torque listed in this Chapter's Specifications.

**20** Using a telescoping gauge and a micrometer, measure the inside diameter of the connecting rod **(see illustration)**. The mark on the connecting rod (if any) should coincide with the measurement, but if it doesn't, make a new mark **(see illustration)**.

> *If the inside diameter measures between 37.000 and 37.008 mm (1.4567 and 1.4570 inch), don't make any mark on the rod (there shouldn't be any there anyway).*
> *If the inside diameter measures between 37.009 and 37.016 mm (1.4570 and 1.4573 inch), make a 0 mark on the rod (it should already be there).*

**21** By referring to the accompanying chart **(see illustration)**, select the correct connecting rod bearing inserts.

**22** Repeat the bearing selection procedure for the remaining connecting rods.

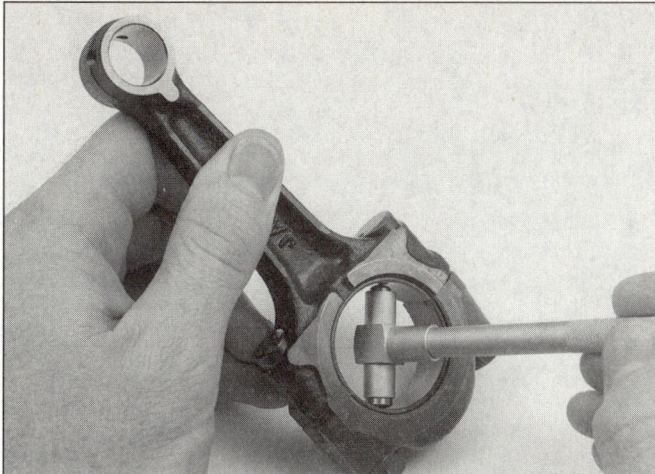

26.20a Measure the diameter of the connecting rod with a telescoping gauge then measure the gauge

26.20b The letter on the crank throw is a weight grade mark – the "O" mark (or lack of a mark, as shown here), next to the weight grade letter, in conjunction with the mark on the crank throw . . .

| Con-Rod Big End Bore Diameter Marking | Crankpin Diameter Mark | Bearing Insert | |
|---|---|---|---|
| | | Size Color | Part Number |
| None | O | Brown | 92028-1625 |
| None | None | Black | 92028-1624 |
| O | O | | |
| O | None | Blue | 92028-1623 |

26.21 . . . can be used, along with this chart, to determine the correct connecting rod bearing inserts to install

26.26 Assemble the connecting rods with the number marks pointing toward the primary gear end of the crankshaft

## Installation

23 The connecting rod bolts are designed to stretch in use. For this reason, they must be replaced with new bolts whenever they're removed.

24 Thoroughly clean the bolts, nut and connecting rods with high-flash point solvent, making sure to clean all of the preservative off of new parts.

25 Wipe off the bearing inserts and connecting rods and caps. Install the inserts into the rods and caps, using your hands only, making sure the tabs on the inserts engage with the notches in the rods and caps. When all the inserts are installed, lubricate them with engine assembly lube or moly-based grease. Don't get any lubricant on the mating surfaces of the rod or cap.

26 Assemble each connecting rod to its proper journal, making sure the previously applied matchmarks correspond to each other and the number casting points to the primary gear end of the crankshaft (see illustration). Also, the letter present at the rod/cap seam on one side of the connecting rod is a weight mark. If new rods are being installed and they don't all have the same letter on them, two rods with the same letter should be installed on one side of the crank, and the letters on the other two rods should match each other. This will minimize vibration.

27 Measure the length of each connecting rod bolt and write this number down. You'll need it later to determine bolt stretch.

28 When you're sure the rods are positioned correctly, apply a film of clean engine oil to the bolt and nut threads, under the heads of the bolts and to the surfaces of the nuts that contact the connecting rod. Tighten the nuts to the torque listed in this Chapter's Specifications.

29 Tighten the connecting rod nuts to the torque listed in this Chapter's Specifications. Note: *There are different torque settings for old and new nuts.*

30 After tightening, measure the length of the connecting rod bolts (see illustration). Calculate the difference between the original length (written down in Step 27) and the length after tightening. The difference is bolt stretch. If it's more than the value listed in this Chapter's Specifications, replace the bolts and nuts with new ones.

*Caution: Don't skip this step. Overstretched bolts may break while the engine is running, causing extensive engine damage.*

31 Turn the rods on the crankshaft. If any of them feel tight, tap on the bottom of the connecting rod caps with a hammer – this should relieve stress and free them up. If it doesn't, recheck the bearing clearance.

32 As a final step, recheck the connecting rod side clearances (see Step 1). If the clearances aren't correct, find out why before proceeding with engine assembly.

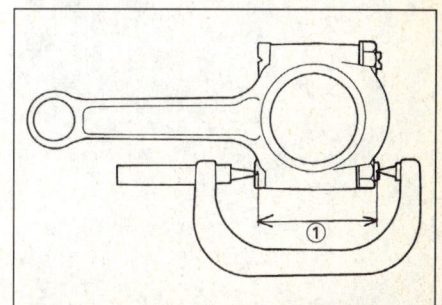

26.30 Measure connecting rod bolt length (1) after the bolts are tightened

27.3 Remove the lower tensioner's Allen bolts and take the tensioner off

27.4 Loosen the pivot bolt (A), set bolt (B) and locknut (C) and tighten the adjusting bolt (D) until the lower chain run is tight

## 27 Alternator chain –
adjustment, removal, inspection and installation

1 Remove the clutch (see Section 19).

### Adjustment

2 The alternator chain can be adjusted if it's noisy.

3 Unbolt the lower chain tensioner and remove it from the engine (see illustration).

4 Loosen the pivot bolt, set bolt and locknut on the upper chain tensioner (see illustration).

5 Turn the adjusting bolt on the upper chain tensioner counterclockwise (anti-clockwise) until the lower run of the chain is tight. Then tighten the locknut, the set bolt and the pivot bolt to the torques listed in this Chapter's Specifications.

6 Release the latch on the lower tensioner with a screwdriver and push the piston all the way in. Holding the piston in this position, install the lower tensioner (see illustration). Apply a non-permanent thread locking agent to the threads of the tensioner Allen bolts, install them and tighten them to the torque listed in this Chapter's Specifications.

### Removal

7 Remove the lower chain tensioner (see Step 3 above).

8 Unbolt the upper chain tensioner and remove it from the engine (see illustration).

9 Remove the alternator shaft oil pipe (see illustration).

10 Remove the snap-rings from the oil pump sprocket and alternator driven sprocket (see illustrations).

11 Slide the alternator drive sprocket, alternator driven sprocket and oil pump sprocket off their shafts, together with the chain (see illustration).

27.6 Hold the tensioner piston compressed and install the lower tensioner on the engine

27.8 Remove the set bolt and pivot bolt and take the upper tensioner off

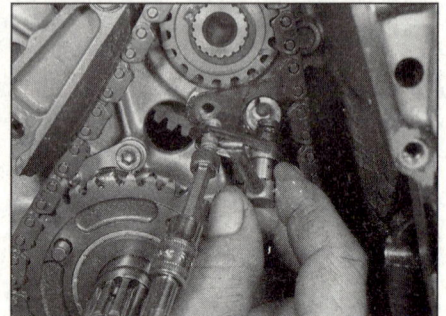

27.9 Remove the Allen bolt and take the oil pipe off

27.10a Remove the snap-ring from the oil pump sprocket . . .

27.10b . . . and the alternator driven sprocket . . .

27.11 . . . slide off the sprockets and chain . . .

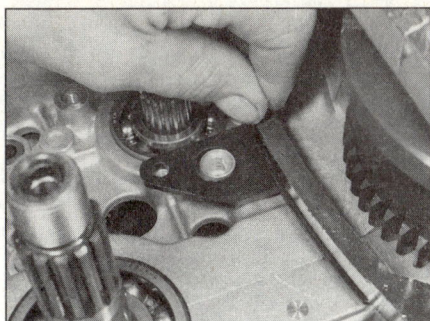
27.12 ... then remove the chain guide

27.13 Remove the thrust collar from behind the transmission mainshaft

27.26 The chain and sprockets should look like this when they're assembled

12 Take the chain guide off the oil pipe boss (see illustration).
13 Slide the thrust collar off the transmission shaft (see illustration).

### Inspection

14 Check the sprocket teeth for wear or damage. Check the drive pins in the alternator drive sprocket for looseness, wear or damage. Check the chain for wear, damage or loose pins. Replace the sprockets and chain as a set if any defects are found.
15 Check the oil pipe for clogging. Flush it with clean solvent and blow it out with compressed air, if available.
16 Check the tensioner guides and the chain guide for score marks or excessive wear. Replace the tensioners or the guide if defects are found.

28.4a Remove the bolt ...

17 Check the latch on the lower tensioner for wear or damage and replace the tensioner if defects are found.

### Installation

18 Install the alternator driven sprocket and secure it with a new snap-ring.
19 Install the chain guide on the oil pipe boss.
20 Install a new O-ring on the oil pipe and push the oil pipe into position in the engine. Apply non-permanent thread locking agent to the threads of the oil pipe bolt, then install the bolt and tighten it to the torque listed in this Chapter's Specifications.
21 Install the alternator chain over the driven sprocket.
22 Install the oil pump sprocket and secure it with a new snap-ring.
23 Install the thrust collar on the transmission shaft, then install the drive sprocket with its two pins outward (away from the engine).
24 Apply non-permanent thread locking agent to the threads of the upper chain tensioner pivot bolt and set bolt, then install the tensioner and tighten the bolts slightly.
25 Refer to Steps 4 through 6 above to adjust chain tension and install the lower tensioner.
26 Check to make sure all of the alternator chain components are installed correctly before installing the clutch (see illustration).

## 28 Alternator shaft and starter clutch – removal, inspection and installation

### Alternator coupling removal

1 The alternator coupling can be removed with the engine in the frame. To remove the alternator driveshaft or the starter clutch, the engine must be removed and the crankcases disassembled.
2 Remove the lower left fairing panel (see Chapter 8).
3 Remove the alternator and its damper (see Chapter 9).
4 Remove the alternator coupling bolt and washer and slide the coupling off the shaft splines (see illustrations). Note: To keep the engine from turning while you loosen the bolt, remove the pickup coil cover (see Chapter 5) and place a box wrench (ring spanner) over the flats of the timing rotor.
Caution: Don't try to keep the engine from turning by holding the timing rotor Allen bolt, or it may snap off.

### Alternator shaft and starter clutch removal

5 Perform Steps 1 through 4 above.
6 Remove the engine (see Section 5).
7 Separate the crankcase halves (Section 22).
8 Remove the transmission shafts and the transmission oil pipe (see Section 30).
9 Remove the bearing holder from the alternator shaft (see illustration).

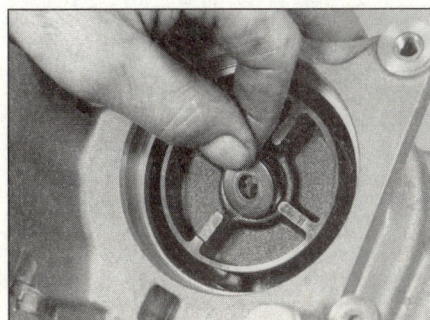
28.4b ... the washer ...

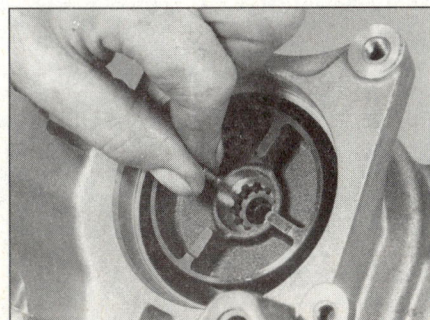
28.4c ... and the alternator coupling

28.9 Remove the Allen bolts and take off the bearing holder

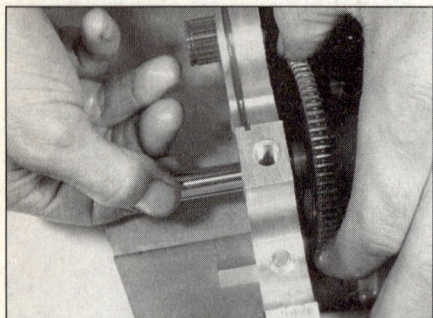

**28.10 Pull out the idler gear shaft and lift out the gear**

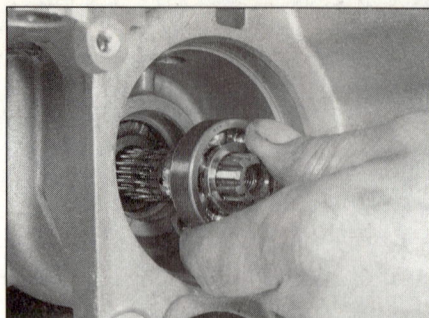

**28.11a Pull out the alternator driveshaft . . .**

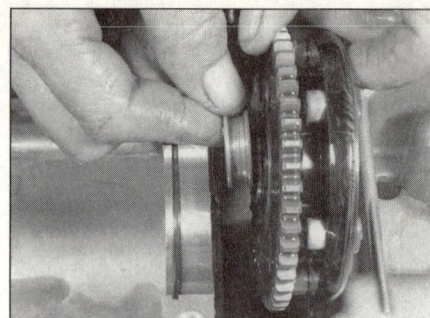

**28.11b . . . and lift out the starter clutch and spacer**

**28.12 Hold the starter clutch and try to rotate the gear**

**28.13a Remove the snap-ring . . .**

**28.13b . . . the thrust washer . . .**

**28.13c . . . the starter clutch gear . . .**

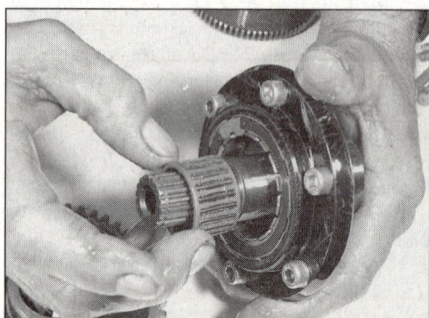

**28.13d . . . and the needle roller bearing . . .**

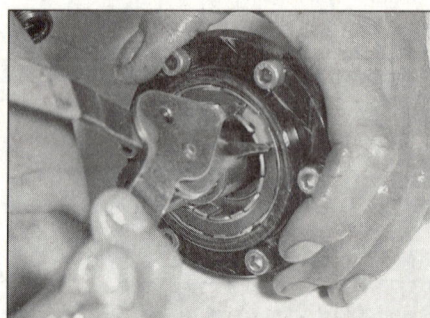

**28.13e . . . remove the large snap-ring from its groove . . .**

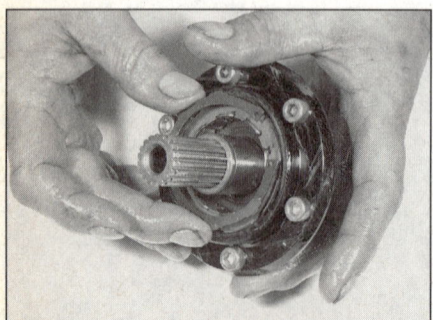

**28.13f . . . and take it off . . .**

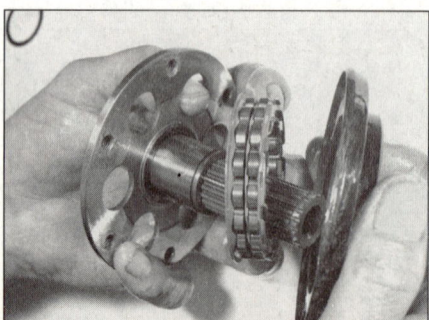

**28.13g . . . then remove the one-way clutch race and one-way clutch from the one-way clutch boss . . .**

**28.13h . . . and remove the thrust washer**

28.15 Check the idler gear for worn or damaged teeth and for looseness on its shaft

28.17a Check the alternator shaft bearings for looseness, roughness or noise . . .

28.17b . . . there's one at each end of the shaft

28.18a The spacer and starter clutch fit on the alternator shaft like this

28.18b Place the starter clutch in the engine . . .

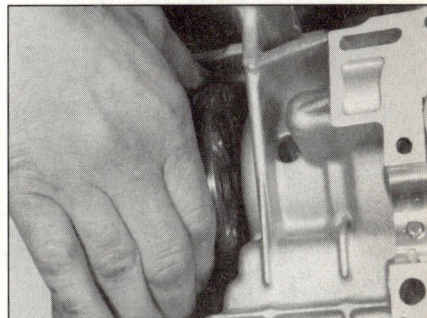

28.18c . . . and slide the alternator shaft partway in . . .

10  Pull out the idler gear shaft and lift out the idler gear **(see illustration)**.

11  Pull out the alternator shaft and bearing, then lift out the starter clutch and spacer **(see illustrations)**.

### Inspection

#### Starter motor clutch

12  Hold the starter motor clutch and attempt to turn the starter motor clutch gear back and forth **(see illustration)**. It should only turn in one direction.

13  If the starter motor clutch turns freely in both directions, or if it's locked up, disassemble it and inspect the components **(see illustrations)**.

14  Starter clutch assembly is the reverse of the disassembly steps. Lubricate all parts with clean engine oil during assembly. Apply non-permanent thread locking agent to the threads of the six Allen bolts and tighten them to the torque listed in this Chapter's Specifications. Make sure the snap-rings seat securely in their grooves.

#### Starter motor idler gear

15  Inspect the teeth on the starter motor idler gear for cracks and chips **(see illustration)**. Turn the idler gear on its shaft to make sure it spins freely. If the idler gear exhibits any undesirable conditions, replace it.

#### Alternator shaft and bearing

16  Check the splines on the shaft for wear or damage.

17  Turn the bearings and feel for tight spots and roughness **(see illustrations)**. If the condition of the bearings is in doubt or definitely bad, replace them. To replace the bearing mounted in the case, refer to Section 23.

### Installation

18  Place the starter clutch in the case and slide the alternator shaft in **(see illustrations)**.

19  Position the idle gear in the case. Lubricate its shaft with clean engine oil and slide it into the case and gear **(see illustrations)**.

20  The remainder of installation is the reverse of the removal procedure. Be sure to fill the cooling system with the proper coolant mixture and the crankcase with the recommended engine oil (see Chapter 1 and *Daily (pre-ride) checks*).

28.18d . . . then install the spacer and push the alternator shaft the rest of the way in

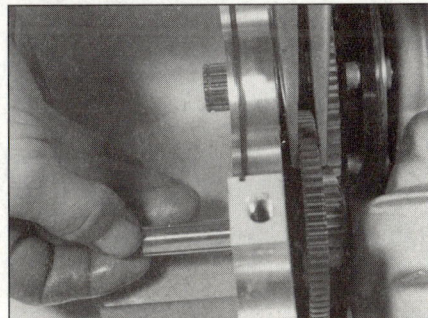

28.19a Position the idler gear in the case and install its shaft

28.19b The idler gear should mesh with the starter clutch like this when installed

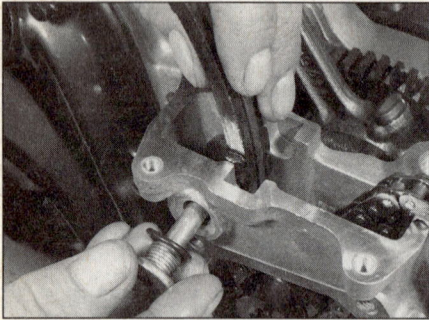

**29.3  Remove the Allen bolt and O-ring and lift the tensioner out (cylinder block removed for clarity)**

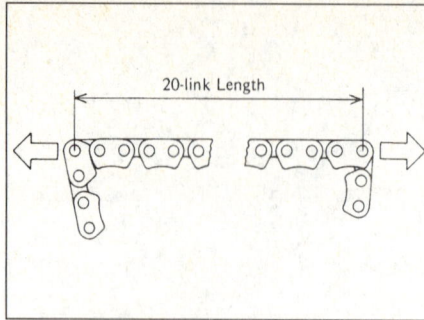

20-link Length

**29.10  When checking the cam chain, measure the length of 20 links**

9  Remove the chain from the crankshaft.
10  Pull the chain tight to eliminate all slack and measure the length of twenty links, pin-to-pin **(see illustration)**. Compare your findings to this Chapter's Specifications.
11  Also check the chain for binding and obvious damage.
12  If the twenty-link length is not as specified, or there is visible damage, replace the chain.
13  Installation is the reverse of the removal steps.

## 29 Camshaft chain and guides – removal, inspection and installation

### Chain guides

1  Remove the camshafts (see Section 9).
2  To remove the front (exhaust side) guide, lift it out of the engine **(see illustration 9.8a)**.
3  To remove the rear (intake side) chain guide, remove its Allen bolt and O-ring and lift the guide out **(see illustration)**.
4  Check the guides for deep grooves, cracking and other obvious damage, replacing them if necessary.
5  Installation is the reverse of the removal steps, with the following additions:
   a) Make sure the front guide fits securely in its slot **(see illustration 9.8b)**.
   b) Use a new O-ring on the rear guide's pivot bolt. Lubricate the O-ring with engine oil and tighten the bolt securely.

### Camshaft chain

6  Remove the engine (see Section 5).
7  Separate the crankcase halves (Section 22).
8  Remove the crankshaft (see Section 25).

## 30 Transmission shafts – removal and installation

### Removal

1  Remove the engine and clutch, then separate the case halves (see Sections 5, 19 and 22).
2  If necessary, remove the Allen bolt and slide the transmission oil pipe out of the case **(see illustrations)**.
3  The shafts can simply be lifted out of the upper half of the case **(see illustrations)**. If they are stuck, use a soft-faced hammer and gently tap on the bearings on the ends of the shafts to free them. The shaft nearest the rear of the case is the output shaft – the other shaft is the mainshaft.
4  Refer to Section 31 for information pertaining to transmission shaft service and Section 32 for information pertaining to the shift drum and forks.

### Installation

5  If the oil pipe was removed, slide it back into the case **(see illustration)**. Apply non-permanent thread locking agent to the threads of the oil pipe Allen bolt and tighten it to the torque listed in this Chapter's Specifications.

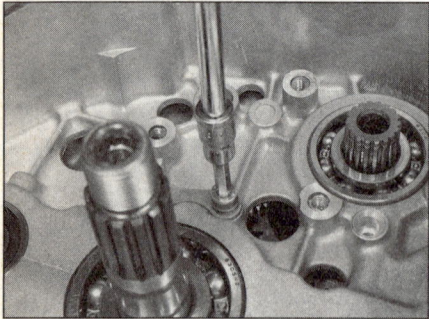

**30.2a  Remove the Allen bolt . . .**

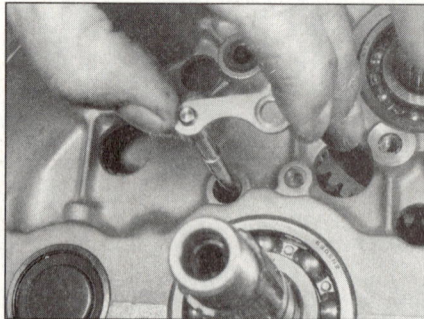

**30.2b  . . . and pull the oil pipe out of the crankcase**

**30.3a  Lift out the mainshaft . . .**

**30.3b  . . . and the output shaft**

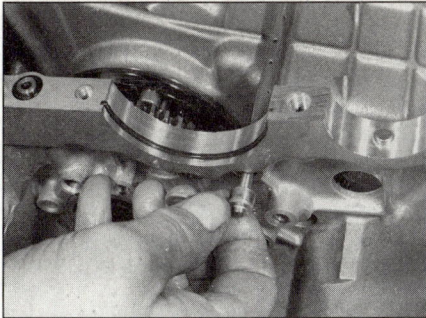
**30.5 Position the oil pipe in the crankcase and install its Allen bolt**

**30.6 The oil pipe should look like this when installed; make sure the set pins and half-circle bearing retainers (arrows) are in position**

**30.7 Engage the set pins with the holes in the bearing outer races**

**30.8 Place the gears in the neutral position like this so the shifting forks will engage the gear grooves (arrows)**

**6** Check to make sure the set pins and rings are present in the upper case half, where the shaft bearings seat **(see illustration)**.
**7** Carefully lower each shaft into place. The holes in the needle bearing outer races must engage with the set pins, and the grooves in the ball bearing outer races must engage with the set rings **(see illustration)**.
**8** Place the gears in the neutral position **(see illustration)**.
**9** The remainder of installation is the reverse of removal.

## 31 Transmission shafts –
disassembly, inspection and reassembly

**Note:** *When disassembling the transmission shafts, place the parts on a long rod or thread*

*a wire through them to keep them in order and facing the proper direction (see **Haynes Hint** in Part 2A).*
**1** Remove the shafts from the case (see Section 30).

### Mainshaft

#### Disassembly

**2** Remove the needle bearing outer race, then remove the snap-ring from the end of the shaft and slide the needle bearing off **(see illustrations)**.
**3** Remove the thrust washer and slide second gear off the shaft **(see illustrations)**.
**4** Remove sixth gear, its bushing and the toothed washer **(see illustrations)**.
**5** Remove the snap-ring **(see illustration)**.
**6** Remove the third/fourth gear cluster from

the shaft **(see illustration)**.
**7** Remove the next snap-ring, then slide the washer, fifth gear and its bushing off the shaft **(see illustrations)**.

**31.2a Remove the outer race . . .**

**31.2b . . . the snap-ring . . .**

**31.2c . . . and the needle roller bearing**

**31.3a Remove the thrust washer . . .**

**31.3b . . . and second gear**

**31.4a Remove sixth gear . . .**

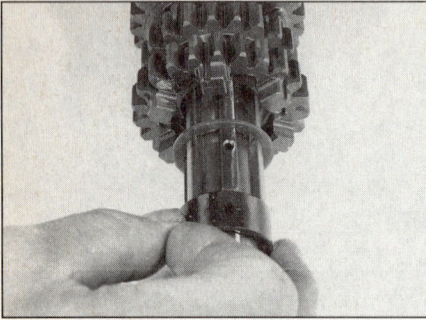
31.4b . . . its bushing . . .

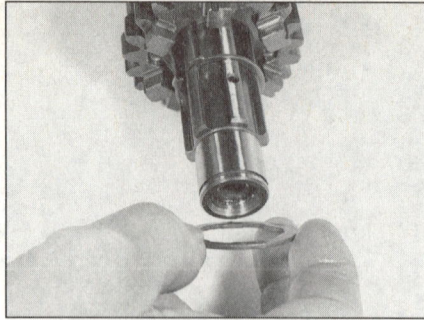
31.4c . . . and the toothed washer

31.5 Remove the snap-ring

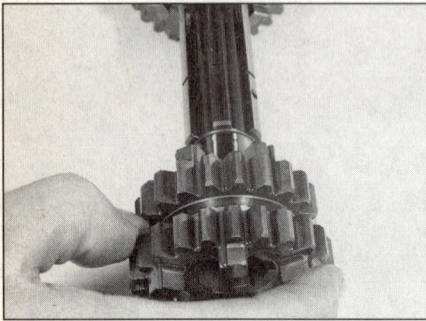
31.6 Slide off the third/fourth gear cluster

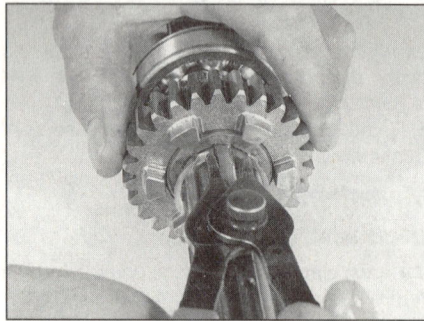
31.7a Remove the snap-ring . . .

31.7b . . . the toothed washer . . .

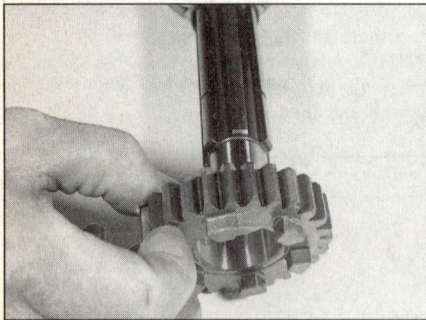
31.7c . . . fifth gear . . .

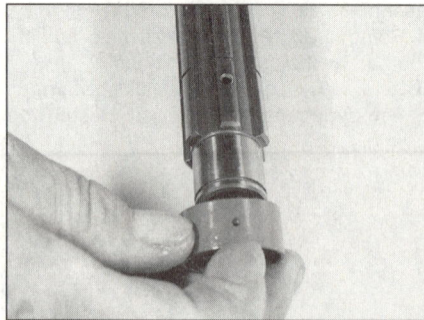
31.7d . . . and the bushing

## Inspection

8 Wash all of the components in clean solvent and dry them off. Rotate the ball bearing on the shaft, feeling for tightness, rough spots, excessive looseness and listening for noises (see illustration). If any of these conditions are found, replace the bearing. This will require the use of a hydraulic press or a bearing puller setup. If you don't have access to these tools, take the shaft and bearing to a Kawasaki dealer or other motorcycle repair shop and have them press the old bearing off the shaft and install the new one.

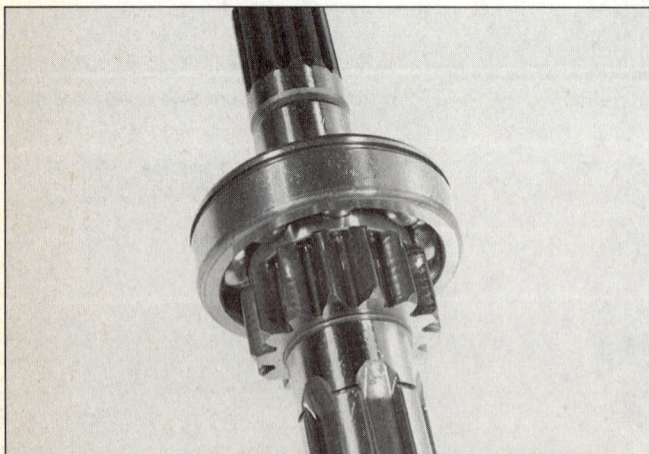
31.8 The ball bearing can be left on the shaft unless it needs to be replaced; the shaft must be replaced if first gear is worn or damaged

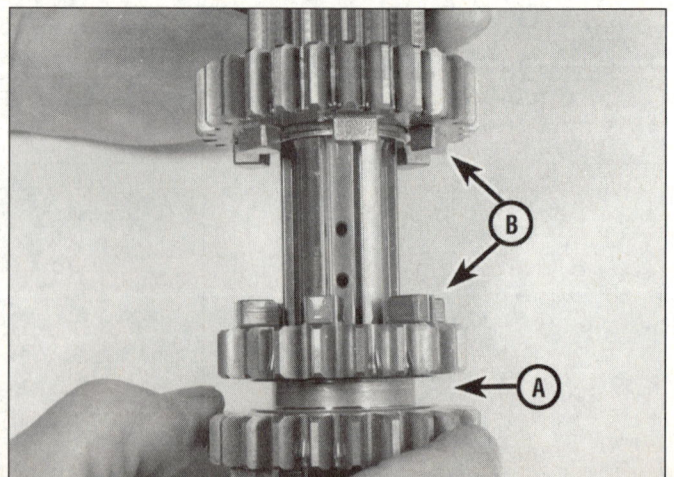
31.9 Measure the width of the shift fork groove (A) and check the dogs (B) for wear or damage

**31.11 Inspect the bushing (left arrow) in gears so equipped for wear – if the edges of the slots (right arrow) are rounded, replace the gear**

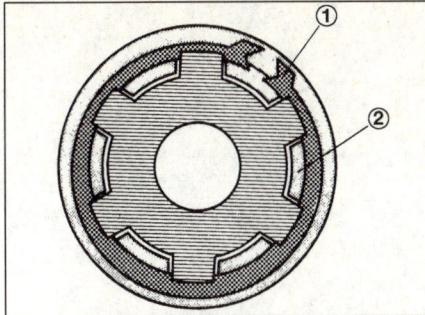

**31.13a Install the snap-rings so the opening in the ring is aligned with a spline groove**

*1 Snap-ring opening*      *2 Spline groove*

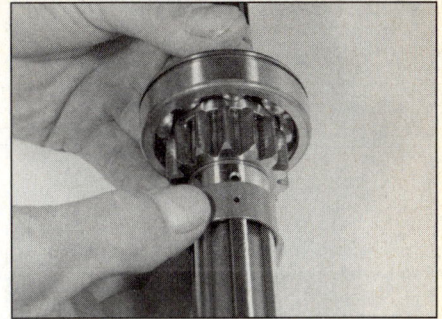

**31.13b Align the oil hole in the first gear bushing with the oil hole in the shaft . . .**

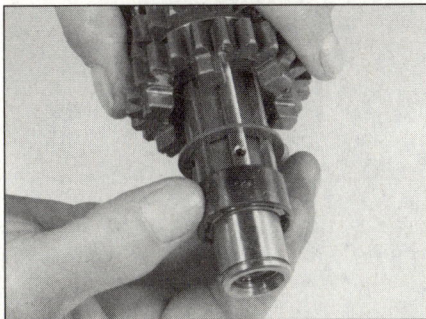

**31.13c . . . and do the same for the sixth gear bushing**

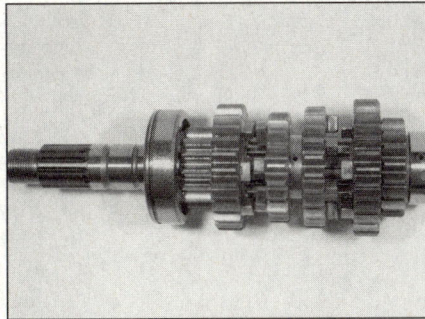

**31.13d The assembled mainshaft should look like this**

9 Measure the shift fork groove between third and fourth gears **(see illustration)**. If the groove width exceeds the figure listed in this Chapter's Specifications, replace the third/fourth gear assembly, and also check the third/fourth gear shift fork (see Section 32).

10 Check the gear teeth for cracking and other obvious damage. Check the bushing and surface in the inner diameter of fifth and sixth gears for scoring or heat discoloration. If either one is damaged, replace it.

11 Inspect the bushings and dog holes on gears so equipped for excessive wear or rounding off **(see illustration)**. Replace the

paired gears as a set if necessary.

12 Check the needle bearing and race for wear or heat discoloration and replace them if necessary.

## Reassembly

13 Reassembly is the basically the reverse of the disassembly procedure, but take note of the following points:

a) Always use new snap-rings and align the opening of the ring with a spline groove **(see illustration)**.

b) When installing the bushings for fifth and sixth gears to the shaft, align the oil hole in the shaft with the oil hole in the bushing **(see illustrations)**.

c) Lubricate the components with engine oil before assembling them.

d) After assembly, check the gears to make sure they're installed correctly **(see illustrations)**.

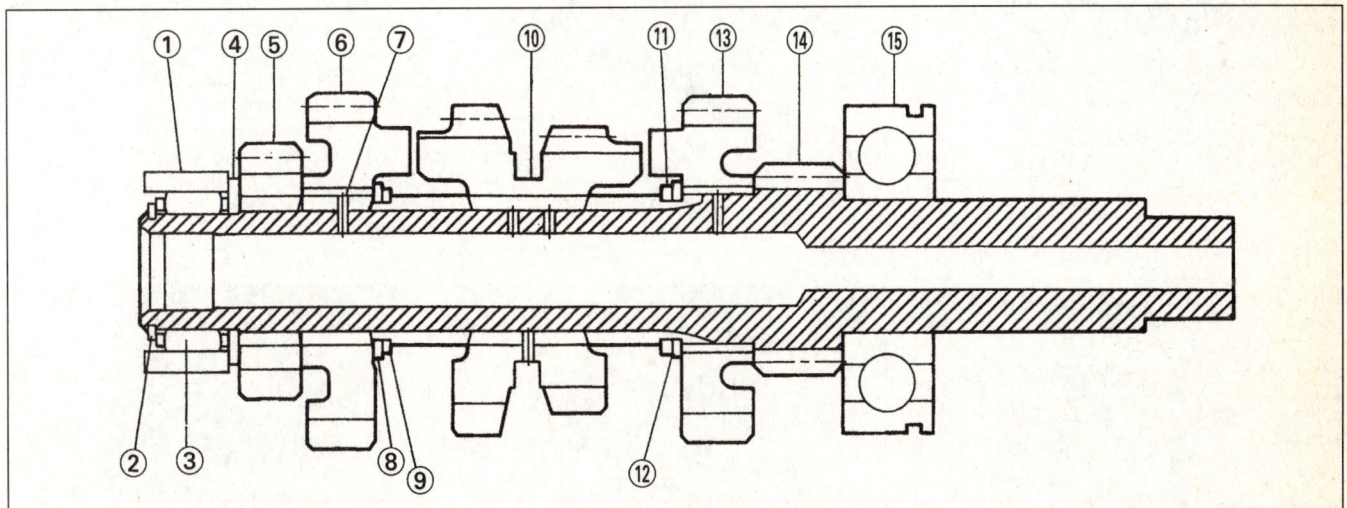

**31.13e Mainshaft details**

| | | | |
|---|---|---|---|
| 1 Bearing outer race | 5 Second gear | 9 Snap-ring | 12 Washer |
| 2 Snap-ring | 6 Sixth gear | 10 Third/fourth gear | 13 Fifth gear |
| 3 Needle bearing | 7 Bushing | cluster | 14 First gear/mainshaft |
| 4 Thrust washer | 8 Washer | 11 Snap-ring | 15 Ball bearing |

**31.14 Remove the bearing outer race . . .**

**31.15a . . . the needle roller bearing . . .**

**31.15b . . . the thrust washer . . .**

**31.15c . . . and first gear**

**31.16 Hold third gear (B) and spin the shaft rapidly with one hand while trying to slide fifth gear (A) off with the other**

## Output shaft

### Disassembly

**14** Remove the needle bearing outer race **(see illustration)**.

**15** Slide the needle bearing, washer and first gear off **(see illustrations)**.

**16** Remove fifth gear from the shaft. Fifth gear has three steel balls in it for the positive neutral finder mechanism. To remove the gear, grasp third gear and hold the shaft in a vertical position with one hand, and with the other hand, spin the shaft back and forth, holding onto fifth gear and pulling up **(see illustration)**.

**31.17a Remove the snap-ring . . .**

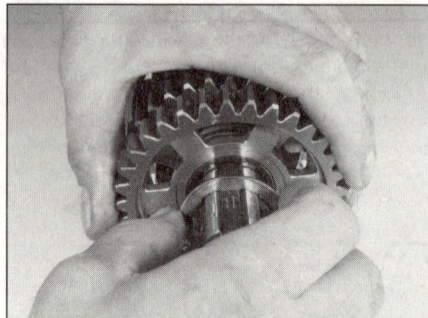

**31.17b . . . the toothed washer . . .**

**31.17c . . . third gear . . .**

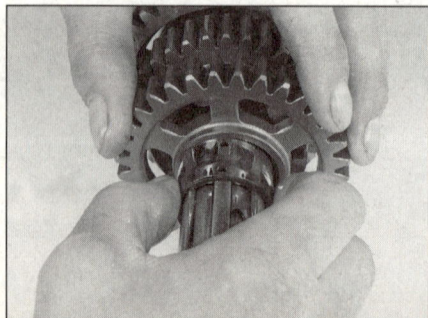

**31.17d . . . the third/fourth gear bushing . . .**

**31.17e . . . and fourth gear**

31.18a  Remove the toothed washer . . .

31.18b  . . . the snap-ring . . .

31.18c  . . . and sixth gear

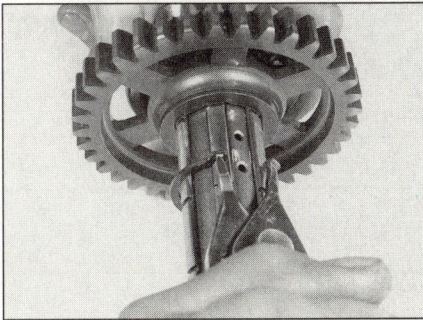

31.19a  Remove the snap-ring . . .

31.19b  . . . the toothed washer . . .

31.19c  . . . and second gear

*Caution: Don't pull the gear up too hard or fast – the balls will fly out of the gear.*

17  Remove the snap-ring, toothed washer, third gear, bushing and fourth gear from the shaft **(see illustrations)**.

31.20  The collar and ball bearing can be left on the shaft unless worn or damaged

18  Remove the toothed washer, snap-ring and sixth gear **(see illustrations)**.

19  Remove the next snap-ring, toothed washer, second gear and its bushing **(see illustrations)**.

20  The ball bearing and collar can be left on the shaft unless they need to be replaced **(see illustration)**.

### Inspection

21  Refer to Steps 8 through 12 for the inspection procedures. They are the same, except when checking the shift fork groove width you'll be checking it on fifth and sixth gears.

### Reassembly

22  Reassembly is the basically the reverse of the disassembly procedure, but take note of the following points:

a) *Always use new snap-rings and align the opening of the ring with a spline groove* **(see illustration 31.13a)**.

b) *When installing the bushings for second, third and fourth gears, align the oil hole in the bushing with the hole in the shaft* **(see illustrations)**.

c) *When installing fifth gear, don't use grease to hold the balls in place – to do so would impair the positive neutral finder mechanism. Just set the balls in their holes (the holes that they can't pass through), keep the gear in a vertical position and carefully set it on the shaft (engine oil will help keep them in place)* **(see illustrations)**. *The spline grooves that contain the holes with the balls must be aligned with the slots in the shaft spline grooves.*

d) *Lubricate the components with engine oil before assembling them.*

e) *After assembly, check the gears to make sure they're installed correctly* **(see illustrations)**.

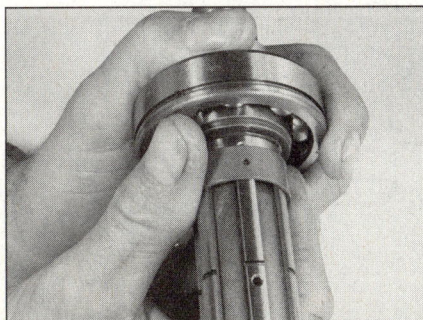

31.22a  Align the oil hole in the second gear bushing with the hole in the shaft . . .

31.22b  . . . and do the same with the holes for the third-fourth gear bushing

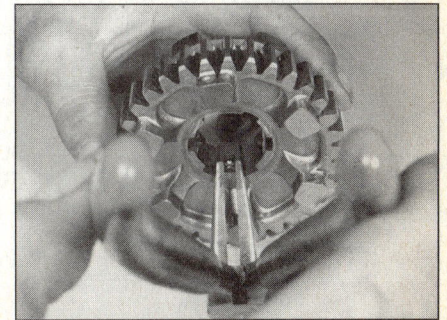

31.22c  Install the steel balls in the smaller holes in the gear . . .

**31.22d** . . . and align the steel balls with the shaft slots (arrow) when installing the gear

**31.22e** The assembled output shaft should look like this

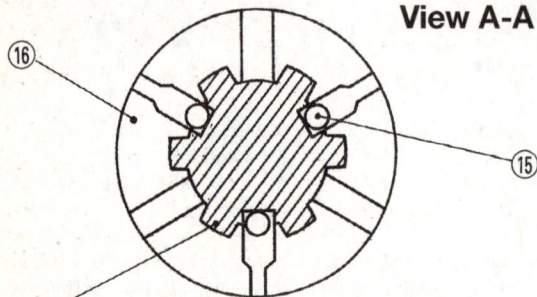

**View A-A**

**31.22f Output shaft details**

| | | |
|---|---|---|
| 1 Collar | 8 Snap-ring | 15 Steel ball |
| 2 Ball bearing | 9 Toothed washer | 16 Fifth gear |
| 3 Output shaft | 10 Fourth gear | 17 First gear |
| 4 Second gear | 11 Bushing | 18 Thrust washer |
| 5 Toothed washer | 12 Third gear | 19 Needle roller bearing |
| 6 Snap-ring | 13 Toothed washer | 20 Bearing outer race |
| 7 Sixth gear | 14 Snap-ring | |

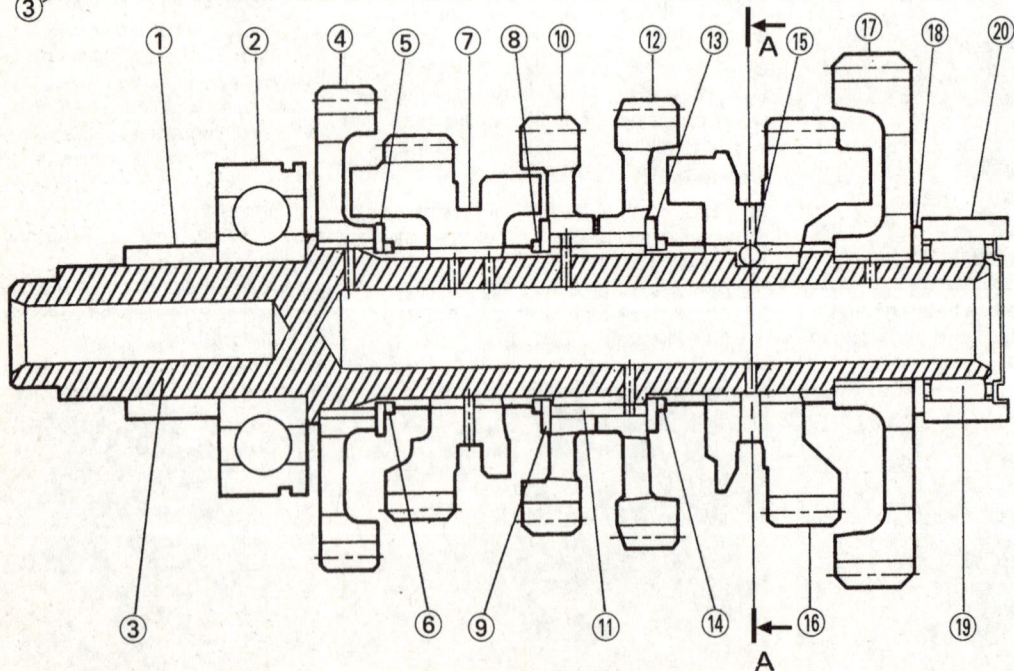

**32.4  Remove the Allen bolt (arrow) and slide the shift rod retainer out of the shift rod groove**

**32.5  Remove the Allen bolts (arrows) and slide the shift drum bearing holder out of the other shift rod groove**

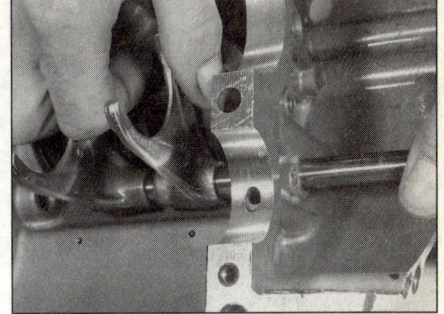

**32.6a  Support the shift forks and slide out the shift rod . . .**

**32.6b  . . . then support the remaining shift fork and slide out the other shift rod**

## 32  Shift drum and forks –
removal, inspection and installation

### Removal

**1**  Remove the engine and separate the crankcase halves (see Sections 5 and 22).
**2**  Remove the transmission shaft assemblies (see Section 30).
**3**  Remove the external shift mechanism (see Section 21).
**4**  Remove the shift rod retainer from the right side of the crankcase (see illustration).
**5**  Remove the shift drum bearing holder (which also acts as a shift rod retainer) from the left side of the crankcase (see illustration).
**6**  Slide out the shift rods and remove the forks (see illustrations). Reassemble the forks in their installed positions on the rods so

you will remember where they go (see illustration). The two shift rods are interchangeable but should be returned to their original locations if they will be reused.
**7**  Remove the dowel pin from the end of the shift drum (see illustration).
**8**  Slide the shift drum partway out, remove the bearing and slide the shift drum the rest of the way out (see illustrations).

### Inspection

**9**  Check the edges of the grooves in the drum for signs of excessive wear (see illustration). Measure the widths of the grooves and compare your findings to this Chapter's Specifications.
**10**  Check the shift forks for distortion and wear, especially at the fork ears (see illustration). Measure the thickness of the fork ears and compare your findings with this Chapter's Specifications. If they are discolored

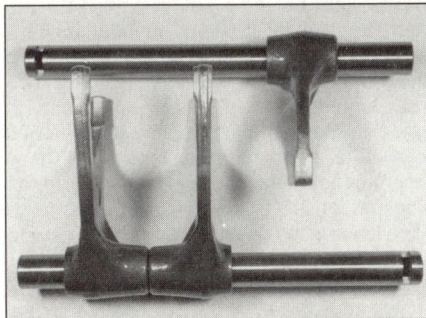

**32.6c  Install the shift forks back on their rods to ensure correct reassembly**

**32.7  Pull the pin out of the shift drum so it won't be lost**

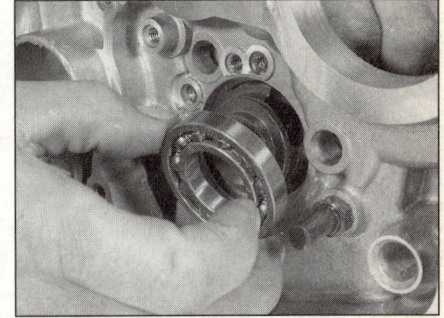

**32.8a  Push the shift drum partway out and remove the bearing . . .**

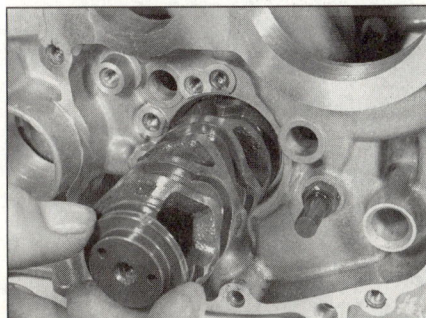

**32.8b  . . . then remove the shift drum from the case**

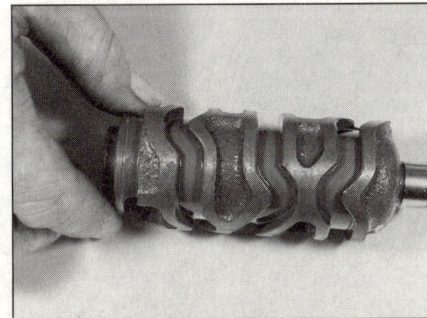

**32.9  Check the grooves in the shift drum for wear, especially at the points**

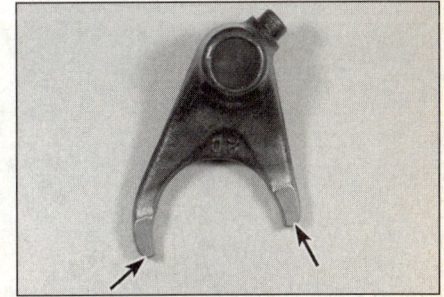

**32.10  Check the shift forks for wear and damage, especially at the fork ears (arrows)**

**32.13a Slide the shift drum partway in, lubricate its end (arrow) . . .**

**32.13b . . . then slide the shift drum the rest of the way in**

or severely worn they are probably bent. If damage or wear is evident, check the shift fork groove in the corresponding gear as well. Inspect the guide pins and the shaft bore for excessive wear and distortion and replace any defective parts with new ones.

**11** Check the shift fork shafts for evidence of wear, galling and other damage. Make sure the shift forks move smoothly on the shafts. If the shafts are worn or bent, replace them with new ones.

**12** Hold the inner race of the shift drum bearing with fingers and spin the outer race. Replace the bearing if it's rough, loose or noisy.

### Installation

**13** Lubricate the shift drum with clean engine oil and slide it into the case **(see illustrations)**.
**14** Install the shift drum bearing **(see illustrations)**, then install the dowel pin in the end of the shift drum **(see illustration 32.7)**.
**15** The remainder of installation is the reverse of removal, noting the following points:
  a) *Lubricate all parts with engine oil before installing them.*
  b) *Position the shift fork pins in the shift drum grooves. The pin on the shift fork with the shortest ears goes in the center groove of the shift drum. The longer ends of the two remaining shift forks face toward each other.*
  c) *Apply non-permanent thread locking agent to the threads of the shift rod retainer bolt and the shift drum bearing holder bolts. Tighten the bolts to the*

*torque listed in this Chapter's Specifications.*

## 33 Initial start-up after overhaul

**Note:** *Make sure the cooling system is checked carefully (especially the coolant level) before starting and running the engine.*
**1** Make sure the engine oil level is correct, then remove the spark plugs from the engine. Place the engine kill switch in the Off position and unplug the primary (low tension) wires from the coil.
**2** Turn on the key switch and crank the engine over with the starter until the oil pressure indicator light goes off (which indicates that oil pressure exists). Reinstall the spark plugs, connect the wires and turn the switch to On.
**3** Make sure there is fuel in the tank, then operate the choke.
**4** Start the engine and allow it to run at a moderately fast idle until it reaches operating temperature.

⚠️ **Warning: If the oil pressure indicator light doesn't go off, or it comes on while the engine is running, stop the engine immediately.**
**5** Check carefully for oil leaks and make sure the transmission and controls, especially the brakes, function properly before road testing the machine. Refer to Section 34 for the recommended break-in procedure.

**6** Upon completion of the road test, and after the engine has cooled down completely, recheck the valve clearances (see Chapter 1).

## 34 Recommended break-in procedure

**1** Any rebuilt engine needs time to break-in, even if parts have been installed in their original locations. For this reason, treat the machine gently for the first few miles to make sure oil has circulated throughout the engine and any new parts installed have started to seat.
**2** Even greater care is necessary if the engine has been rebored or a new crankshaft has been installed. In the case of a rebore, the engine will have to be broken in as if the machine were new. This means greater use of the transmission and a restraining hand on the throttle until at least 500 miles (800 km) have been covered. There's no point in keeping to any set speed limit – the main idea is to keep from lugging (laboring) the engine and to gradually increase performance until the 500 mile (800 km) mark is reached. These recommendations can be lessened to an extent when only a new crankshaft is installed. Experience is the best guide, since it's easy to tell when an engine is running freely.
**3** If a lubrication failure is suspected, stop the engine immediately and try to find the cause. If an engine is run without oil, even for a short period of time, irreparable damage will occur.

**32.14a Install the shift drum bearing**

**32.14b The shift rods and forks should look like this when assembled**

# Chapter 3
# Cooling system

## Contents

## Degrees of difficulty

| Easy, suitable for novice with little experience | Fairly easy, suitable for beginner with some experience | Fairly difficult, suitable for competent DIY mechanic | Difficult, suitable for experienced DIY mechanic | Very difficult, suitable for expert DIY or professional |
|---|---|---|---|---|

## Specifications

### General
Coolant type . . . . . . . . . . . . . . . . . . . . . . . . . . . . . . . . . . . . . . . . . .   see Chapter 1
Mixture ratio . . . . . . . . . . . . . . . . . . . . . . . . . . . . . . . . . . . . . . . . . .   see Chapter 1
Radiator cap pressure rating . . . . . . . . . . . . . . . . . . . . . . . . . . . . .   0.96 to 1.24 Bars (14 to 18 psi)

### Thermostat
Opening temperature . . . . . . . . . . . . . . . . . . . . . . . . . . . . . . . . . . .   80 to 84°C (176 to 183°F)
Fully open temperature . . . . . . . . . . . . . . . . . . . . . . . . . . . . . . . . . .   95°C (203°F)
Valve travel (when fully open) . . . . . . . . . . . . . . . . . . . . . . . . . . . .   Not less than 8 mm (5/16 inch)

### Torque specifications
Thermostatic fan switch-to-radiator
  H1 models . . . . . . . . . . . . . . . . . . . . . . . . . . . . . . . . . . . . . . . . . .   18 Nm (13 ft-lbs)
  H2 models . . . . . . . . . . . . . . . . . . . . . . . . . . . . . . . . . . . . . . . . . .   27 Nm (20 ft-lbs)
  J, K, L and M models . . . . . . . . . . . . . . . . . . . . . . . . . . . . . . . . .   18 Nm (13 ft-lbs)
Coolant temperature sending unit-to-thermostat housing
  H models . . . . . . . . . . . . . . . . . . . . . . . . . . . . . . . . . . . . . . . . . .   7.4 Nm (65 in-lbs)
  J, K, L and M models . . . . . . . . . . . . . . . . . . . . . . . . . . . . . . . . .   7.8 Nm (69 in-lbs)
Coolant passage bolts (J, K, L and M models) . . . . . . . . . . . . . . . . .   9.8 Nm (87 in-lbs)
Oil cooler hose union bolts (H1 models) . . . . . . . . . . . . . . . . . . . .   25 Nm (18 ft-lbs)
Oil cooler bolt (J, K, L and M models) . . . . . . . . . . . . . . . . . . . . . .   49 Nm (36 ft-lbs)

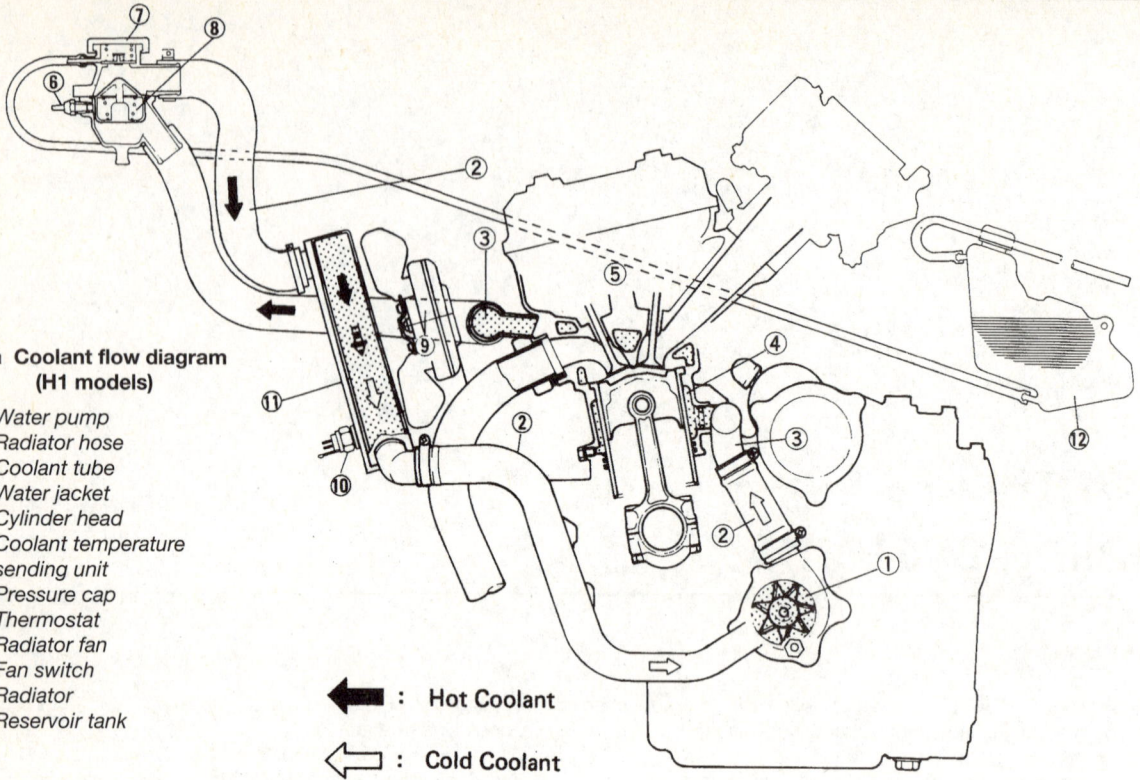

**1.1a  Coolant flow diagram
      (H1 models)**

1  Water pump
2  Radiator hose
3  Coolant tube
4  Water jacket
5  Cylinder head
6  Coolant temperature
   sending unit
7  Pressure cap
8  Thermostat
9  Radiator fan
10 Fan switch
11 Radiator
12 Reservoir tank

◀  :  Hot Coolant

⇦  :  Cold Coolant

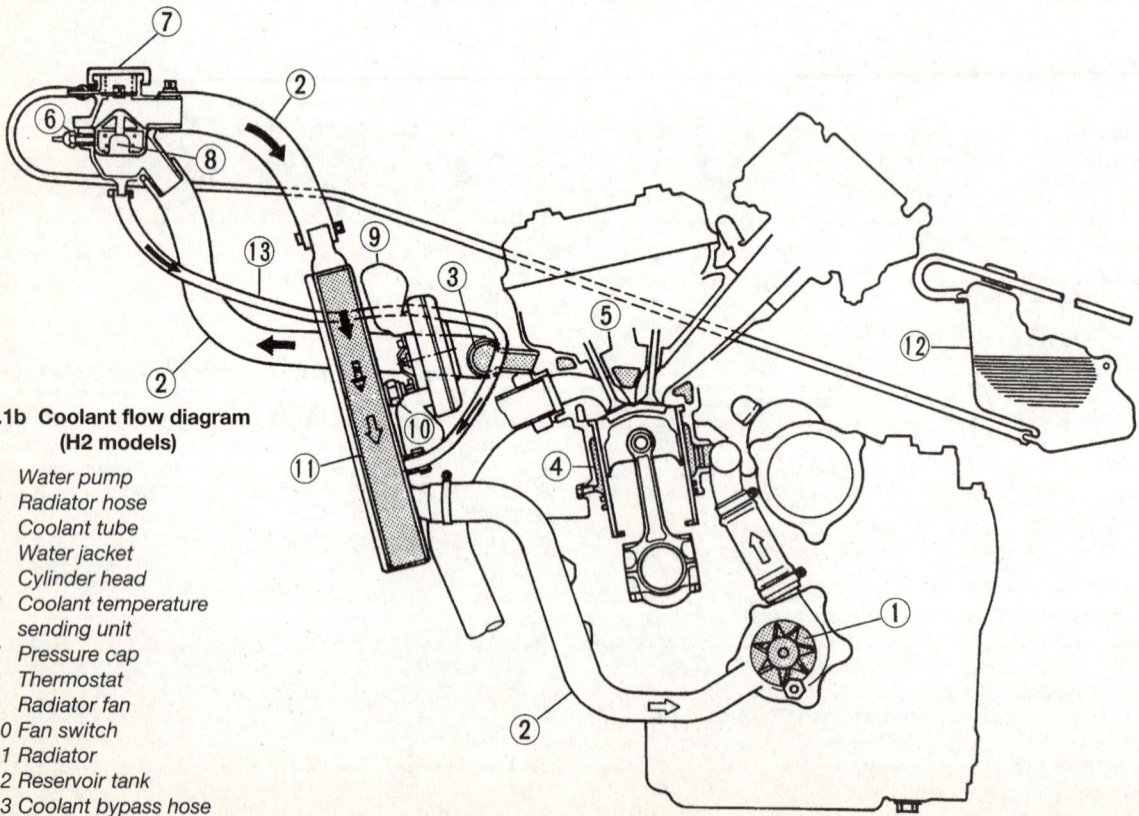

**1.1b  Coolant flow diagram
      (H2 models)**

1  Water pump
2  Radiator hose
3  Coolant tube
4  Water jacket
5  Cylinder head
6  Coolant temperature
   sending unit
7  Pressure cap
8  Thermostat
9  Radiator fan
10 Fan switch
11 Radiator
12 Reservoir tank
13 Coolant bypass hose

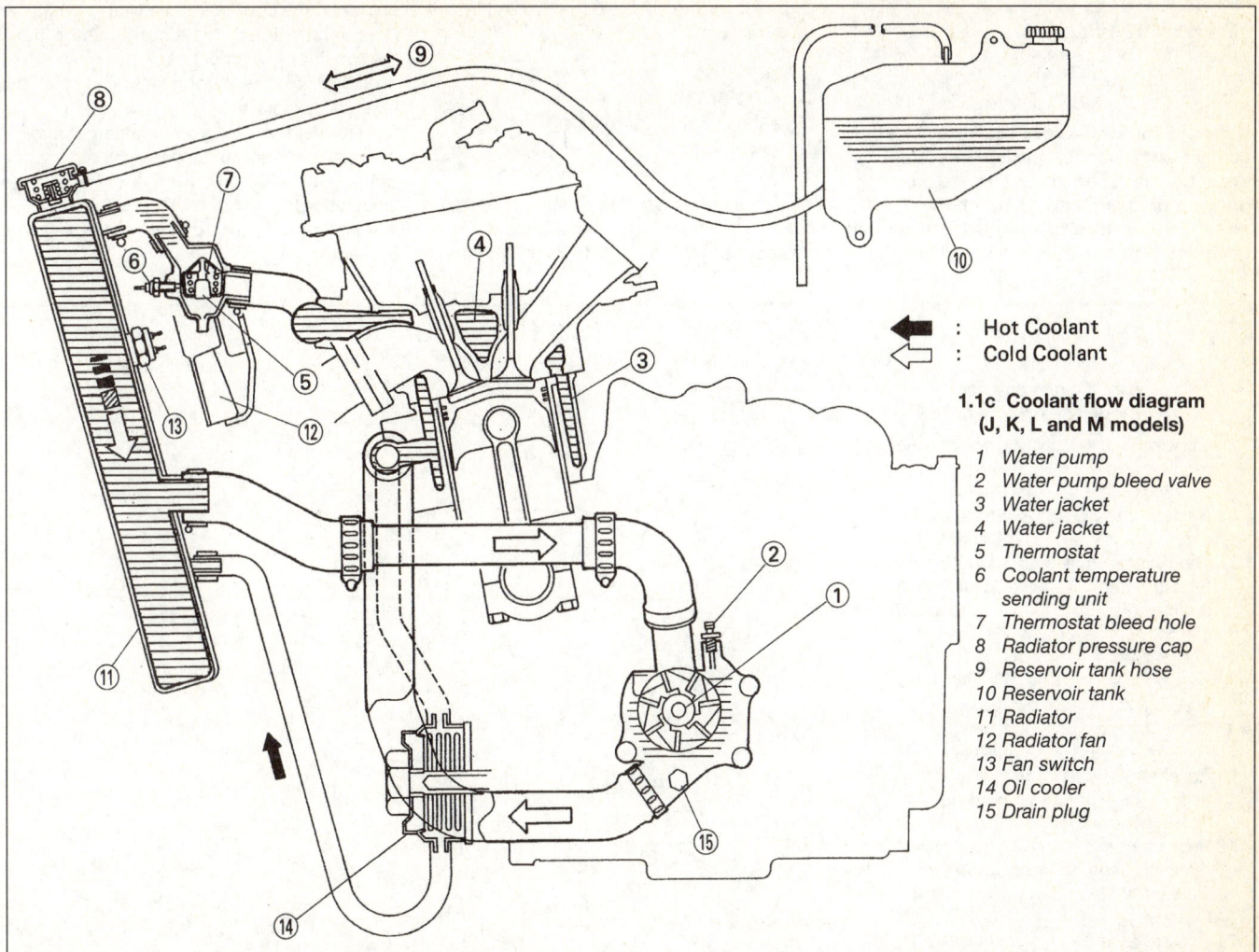

: Hot Coolant
: Cold Coolant

**1.1c Coolant flow diagram (J, K, L and M models)**

1 Water pump
2 Water pump bleed valve
3 Water jacket
4 Water jacket
5 Thermostat
6 Coolant temperature sending unit
7 Thermostat bleed hole
8 Radiator pressure cap
9 Reservoir tank hose
10 Reservoir tank
11 Radiator
12 Radiator fan
13 Fan switch
14 Oil cooler
15 Drain plug

## 1 General information

The models covered by this manual are equipped with a liquid cooling system which utilizes a water/antifreeze mixture to carry away excess heat produced during the combustion process **(see illustrations)**. The cylinders are surrounded by water jackets, through which the coolant is circulated by the water pump. The pump is mounted to the left side of the crankcase and is driven by the oil pump shaft. The coolant passes up through a flexible hose and a coolant pipe, which distributes it around the four cylinders. It flows through the water passages in the cylinder head, through another pipe (or hoses) and into the thermostat housing. The hot coolant then flows down into the radiator (which is mounted on the frame downtubes to take advantage of maximum air flow), where it is cooled by the passing air, through another hose and back to the water pump, where the cycle is repeated.

An electric fan, mounted behind the radiator and automatically controlled by a thermostatic switch, provides a flow of cooling air through the radiator when the motorcycle is not moving. Under certain conditions, the fan may come on even after the engine is stopped, and the ignition switch is off, and may run for several minutes.

The coolant temperature sending unit, threaded into the thermostat housing, senses the temperature of the coolant and controls the coolant temperature gauge on the instrument cluster.

The entire system is sealed and pressurized. The pressure is controlled by a valve which is part of the radiator cap. By pressurizing the coolant, the boiling point is raised, which prevents premature boiling of the coolant. An overflow hose, connected between the radiator and reservoir tank, directs coolant to the tank when the radiator cap valve is opened by excessive pressure. The coolant is automatically siphoned back to the radiator as the engine cools.

Many cooling system inspection and service procedures are considered part of routine maintenance and are included in Chapter 1.

⚠ *Warning 1: Do not allow antifreeze to come in contact with your skin or painted surfaces of the vehicle. Rinse off spills immediately with plenty of water. Antifreeze is highly toxic if ingested. Never leave antifreeze lying around in an open container or in puddles on the floor; children and pets are attracted by its sweet smell and may drink it. Check with local authorities about disposing of used antifreeze. Many communities have collection centers which will see that antifreeze is disposed of safely.*

⚠ *Warning 2: Don't remove the radiator cap when the engine and radiator are hot. Scalding hot coolant and steam may be blown out under pressure, which could cause serious injury. To open the radiator cap, place a thick rag, like a towel, over the radiator cap; slowly rotate the cap anti-clockwise to the first stop. This procedure allows any residual pressure to escape. When the steam has stopped escaping, press down on the cap while turning anti-clockwise and remove it.*

## 2  Radiator cap – check

If problems such as overheating and loss of coolant occur, check the entire system as described in Chapter 1. The radiator cap opening pressure should be checked by a dealer service department or service station equipped with the special tester required to do the job. If the cap is defective, replace it with a new one.

## 3  Coolant reservoir – removal and installation

1 Remove the right fairing panel (H models) or right side cover (all others) (see Chapter 8).
2 Disconnect the hose(s) from the reservoir and drain the coolant into a suitable container **(see illustrations)**. It's a good idea to mark the positions of the hoses so they aren't attached to the wrong fitting when the reservoir is installed.
3 Remove the reservoir retaining band (H models) or mounting fasteners (all others) and detach the reservoir from the frame **(see illustration)**.
4 Installation is the reverse of the removal procedure.

### 3.2a  Cooling system (H1 models) – exploded view

1 Coolant temperature sending unit
2 Radiator pressure cap
3 Thermostat cover
4 Pressure cap retainer
5 O-ring
6 Thermostat
7 Thermostat housing
8 Coolant hose
9 Overflow tube
10 Reservoir tank hose
11 Reservoir tank
12 Reservoir tank retaining band
13 Coolant hose
14 O-rings
15 Coolant tube
16 Coolant tube
17 Coolant hose
18 Fan
19 Radiator
20 Stone shield
21 Air baffle
22 Mounting damper
23 Collar
24 Mounting bolt
25 Fan switch
26 Clip
27 Coolant hose
28 Water pump cover
29 O-ring
30 Water pump housing
31 O-ring
32 Bleed valve
33 Carburetor warmer and coolant filters (UK models)

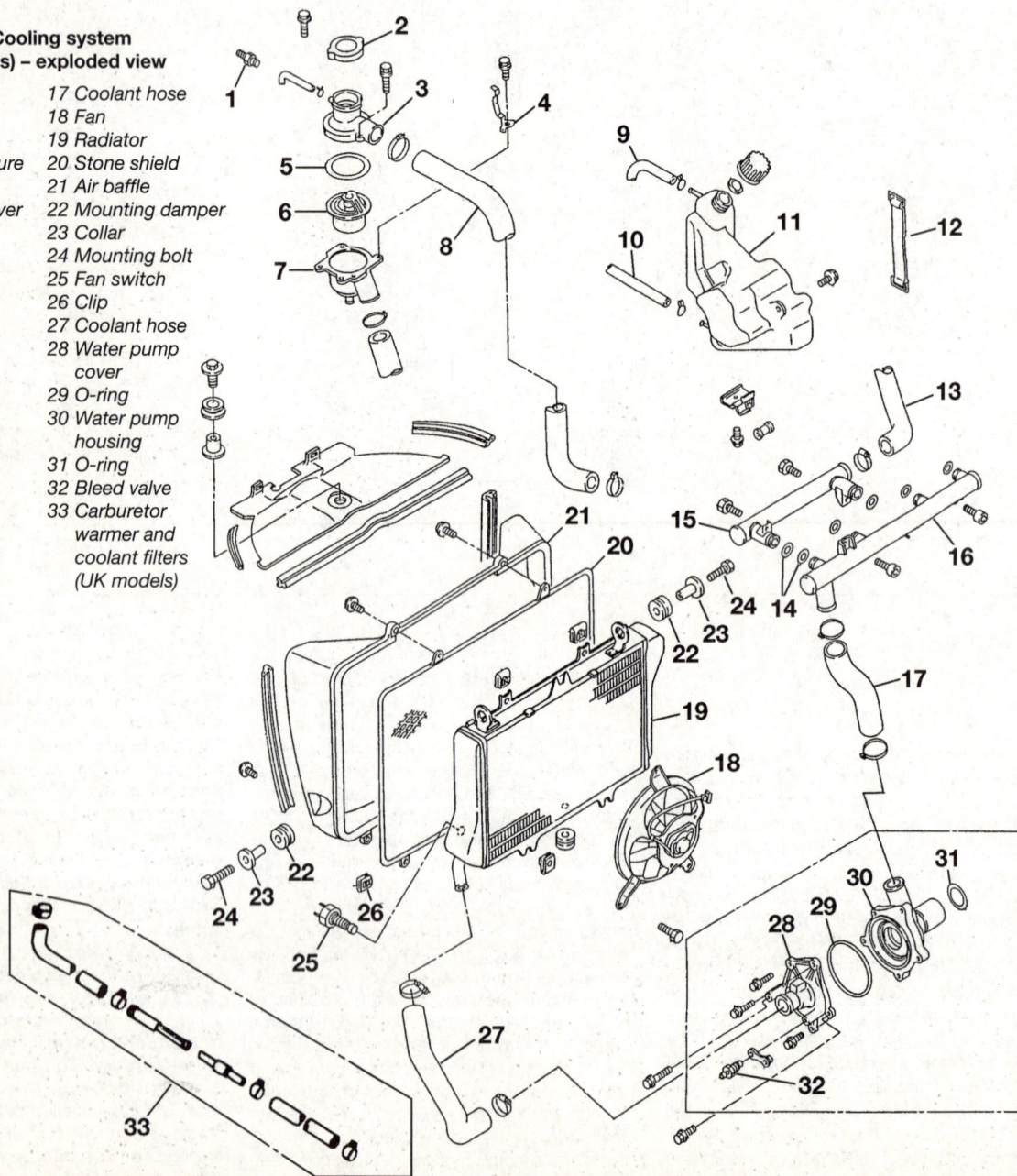

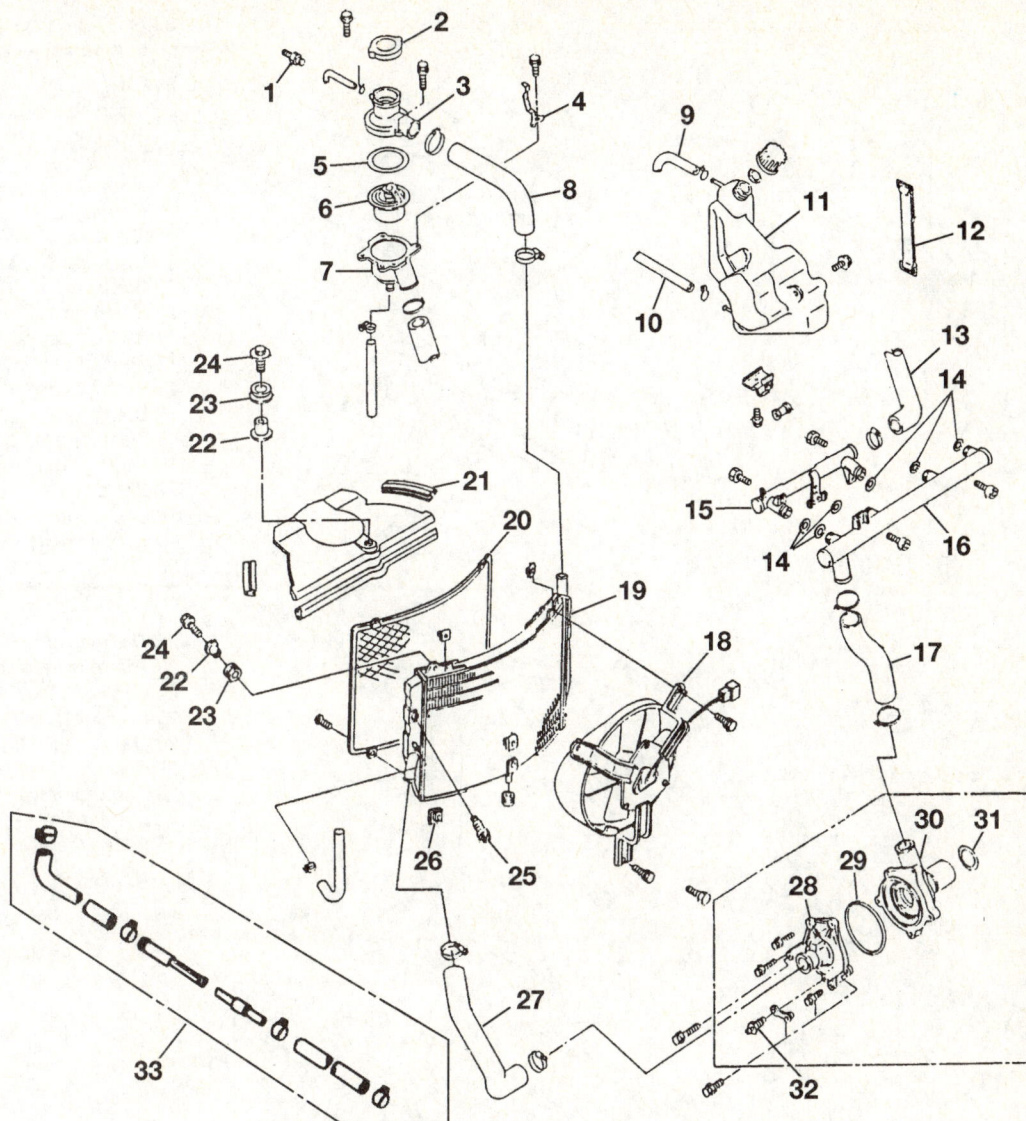

**3.2b  Cooling system
(H2 models) – exploded view**

| | | |
|---|---|---|
| 1 Coolant temperature sending unit | 12 Reservoir tank retainer | 24 Mounting bolt |
| 2 Pressure cap | 13 Coolant hose | 25 Fan switch |
| 3 Thermostat cover | 14 O-rings | 26 Clip |
| 4 Pressure cap retainer | 15 Coolant tube | 27 Coolant hose |
| 5 O-ring | 16 Coolant tube | 28 Water pump cover |
| 6 Thermostat | 17 Coolant hose | 29 O-ring |
| 7 Thermostat housing | 18 Fan | 30 Water pump housing |
| 8 Coolant hose | 19 Radiator | 31 O-ring |
| 9 Overflow tube | 20 Stone shield | 32 Bleed valve |
| 10 Reservoir tank hose | 21 Air baffle | 33 Carburetor warmer and |
| 11 Reservoir tank | 22 Mounting damper | coolant filters |
| | 23 Collar | (UK models) |

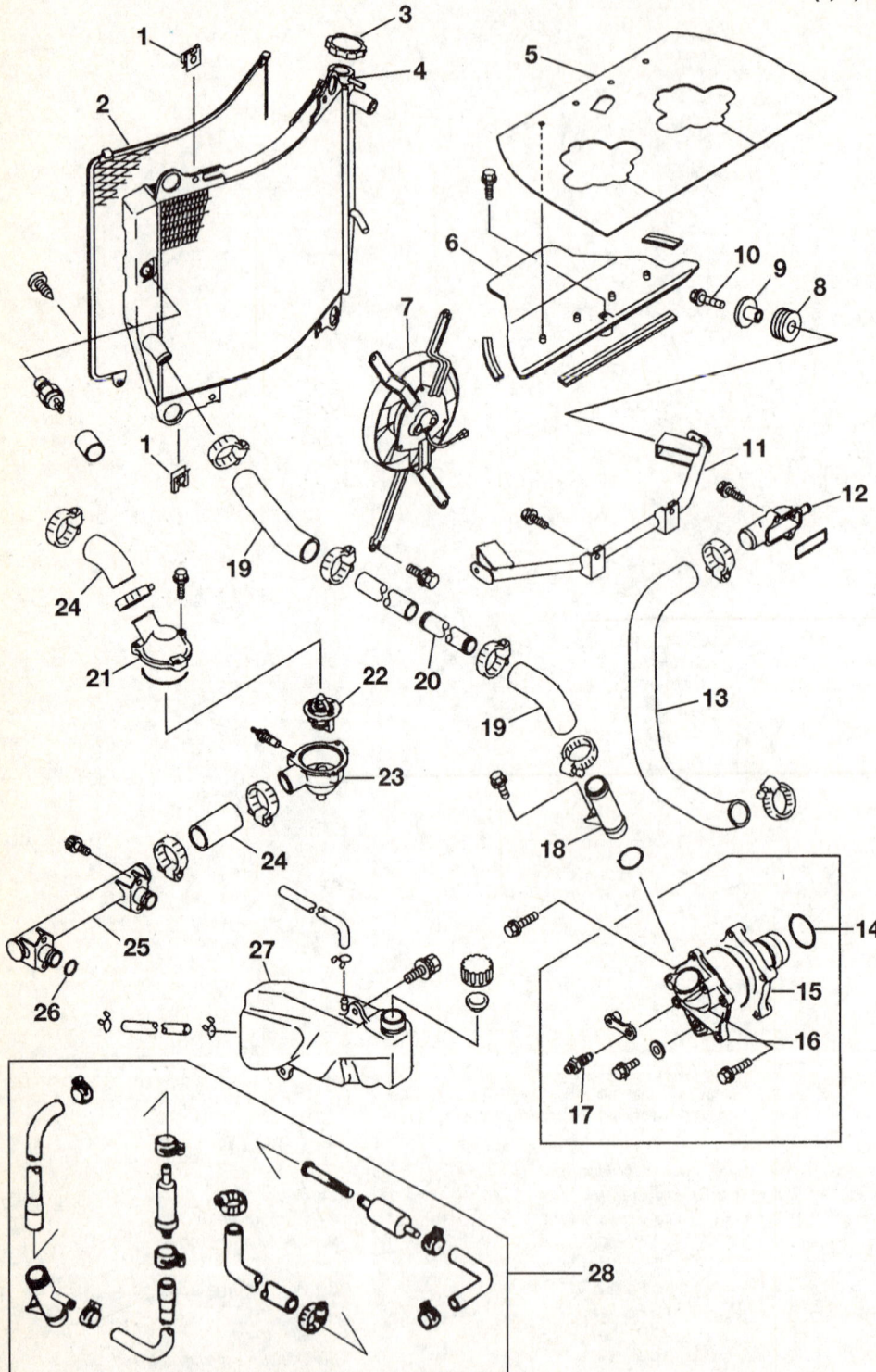

**3.2c  Cooling system
(J, K, L and M models) – exploded view**

1  Clip
2  Stone shield
3  Radiator pressure cap
4  Radiator
5  Air baffle
6  Air baffle
7  Fan
8  Mounting damper
9  Collar
10  Mounting bolt
11  Radiator bracket
12  Coolant passage
13  Coolant hose
14  O-ring
15  Water pump body
16  Water pump cover
17  Bleed valve
18  Coolant tube
19  Coolant hose
20  Coolant tube
21  Thermostat housing
    assembly
22  Thermostat
23  Thermostat housing
24  Coolant hose
25  Coolant tube
26  O-ring
27  Reservoir tank
28  Carburetor warmer and
    coolant filters
    (UK models)

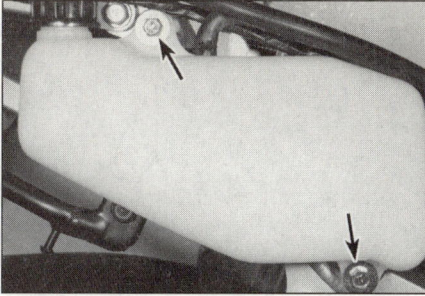

**3.3 Remove the mounting fasteners (arrows) from the top and bottom of the reservoir tank**

## 4  Cooling fan and thermostatic fan switch – check and replacement

### Check

**1** If the engine is overheating and the cooling fan isn't coming on, first remove the seat and check the main fuse and fan fuse (see Chapter 9). If a fuse is blown, check the fan circuit for a short to ground/earth (see the Wiring diagrams at the end of this book). If the fuses are good, remove the lower and upper fairings (see Chapter 8). Where necessary for access, remove the fuel tank (see Chapter 4).

**2** Follow the wiring harness from the fan motor to the connector and unplug it **(see illustration)**. Using two jumper wires, apply battery voltage to the terminals in the fan motor side of the electrical connector. If the fan doesn't work, replace the motor.

**3** If the fan does come on, the problem lies in the thermostatic fan switch or the wiring that connects the components. Remove the jumper wires and plug in the electrical connector to the fan. Unplug the two wires from the thermostatic fan switch and connect them together with a short jumper wire **(see illustration 1.1a, 1.1b or 1.1c for switch location)**. If the fan comes on, the circuit to the motor is okay, and the thermostatic fan switch is defective (see Step 10).

**4** If the fan still doesn't work, refer to the Wiring diagrams at the end of the book and check the fan circuit for broken wires or poor connections.

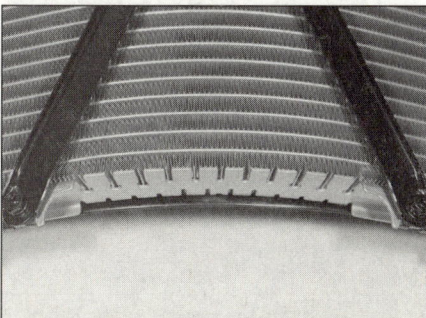

**4.2  Disconnect the fan connector (L model shown)**

### Replacement

#### Fan and motor

⚠ **Warning: The engine must be completely cool before beginning this procedure.**

**5** Disconnect the cable from the negative terminal of the battery.

**6** Remove the radiator (see Section 7).

**7** Remove the three bolts (H models) or four bolts (all others) securing the fan bracket to the radiator **(see illustrations)**, noting which bolt the ground/earth wire is attached to. Separate the fan and bracket from the radiator.

**8** The fan, motor and bracket aren't available separately. Replace them as an assembly if the fan doesn't run.

**9** Installation is the reverse of the removal procedure. Be sure to reinstall the ground/earth wire under the bracket-to-radiator bolt.

#### Thermostatic fan switch

⚠ **Warning: The engine must be completely cool before beginning this procedure.**

**10** Prepare the new switch by wrapping the threads with Teflon tape or by coating the threads with RTV sealant. Disconnect the wires from the old switch.

**11** Unscrew the switch from the radiator **(see illustration 1.1a, 1.1b or 1.1c for switch location)** and quickly install the new switch, tightening it to the torque listed in this Chapter's Specifications.

**12** Connect the wires to the switch.

**13** Check and, if necessary, add coolant to the system (see 'Daily (pre-ride) checks').

## 5  Coolant temperature gauge and sending unit – check and replacement

### Check

**1** If the engine has been overheating but the coolant temperature gauge hasn't been indicating a hotter than normal condition, begin with a check of the coolant level (see 'Daily (pre-ride) checks'). If it's low, add the recommended type of coolant and be sure to locate the source of the leak.

**2** Remove the seat and the fuel tank (see Chapter 4). Locate the coolant temperature sending unit, which is screwed into the thermostat housing. Unplug the electrical connector from the sending unit, turn the ignition key to the Run position (don't crank the engine over) and note the temperature gauge – it should read Cold.

**3** With the ignition key still in the On position, connect one end of a jumper wire to the sending unit wire and ground/earth the other end **(see illustration)**. The needle on the temperature gauge should swing over past the Hot mark.

*Caution: Don't ground/earth the wire any longer than necessary or the gauge may be damaged.*

**4** If the gauge passes both of these tests, but doesn't operate correctly under normal riding conditions, the temperature sending unit is defective and must be replaced.

**5** If the gauge didn't respond to the tests properly, either the wire to the gauge is bad or the gauge itself is defective.

### Replacement

#### Sending unit

⚠ **Warning: The engine must be completely cool before beginning this procedure.**

**6** Prepare the new sending unit by wrapping the threads with Teflon tape or by coating the threads with RTV sealant.

**7** Disconnect the wire from the sending unit. Unscrew the sending unit from the thermostat housing and quickly install the new unit, tightening it to the torque listed in this Chapter's Specifications.

**4.7a  Remove two bolts at the bottom of the fan bracket . . .**

**4.7b  . . . and two bolts at the top**

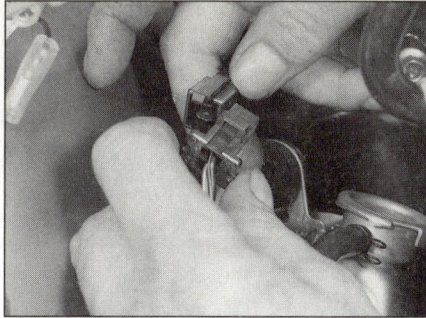

**5.3  Disconnect the temperature sending unit wire (arrow) and connect it to ground (earth)**

**6.4 Remove the thermostat housing bolts (arrows) . . .**

**6.5 . . . and lift off the cover to expose the thermostat; note the O-ring (arrow)**

**8** Reconnect the electrical connector to the sending unit. Check and, if necessary, add coolant to the system (see *'Daily (pre-ride) checks'*).

**Coolant temperature gauge**

**9** Refer to Chapter 9 for the coolant temperature gauge replacement procedure.

## 6  Thermostat –
removal, check and installation

⚠ *Warning: The engine must be completely cool before beginning this procedure.*

### Removal

**1** If the thermostat is functioning properly, the coolant temperature gauge should rise to the normal operating temperature quickly and

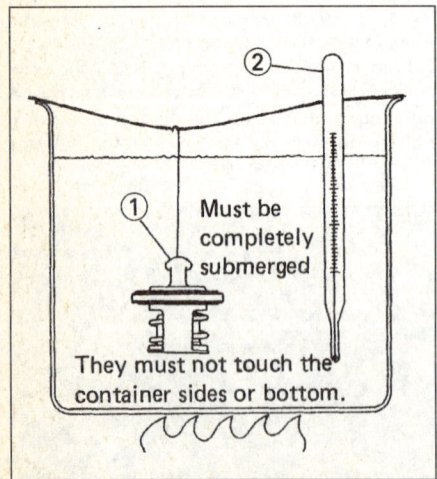

**6.7 Suspend the thermostat in water with a thermometer**

*1 Thermostat       2 Thermometer*

then stay there, only rising above the normal position occasionally when the engine gets unusually hot. If the engine does not reach normal operating temperature quickly, or if it overheats, the thermostat should be removed and checked, or replaced with a new one.
**2** Refer to Chapter 1 and drain the cooling system. Remove the upper fairing (H models) or lower fairing (J, K, L and M models) (see Chapter 8).
**3** If you're working on an H model, remove the pressure cap retainer and the two bolts securing the thermostat cover **(see illustration 3.2a or 3.2b)**. It's not necessary to disconnect the hoses from the cover.
**4** If you're working on a J, K, L and M model, remove the three bolts securing the thermostat cover **(see illustration)** and remove the cover. It's not necessary to disconnect the hose(s) from the cover, but note the location of the ground/earth wire terminal that fits over one of the housing bolts.
**5** On all models, withdraw the thermostat from the housing and remove the O-ring **(see illustration)**.

### Check

**6** Remove any coolant deposits, then visually check the thermostat for corrosion, cracks and other damage. If it was open when it was removed, it is defective. Check the O-ring for cracks and other damage.
**7** To check the thermostat operation, submerge it in a container of water along with a thermometer **(see illustration)**.
⚠ *Warning: Antifreeze is poisonous. Don't use a cooking pan. The thermostat should be suspended so it doesn't touch the container.*
**8** Gradually heat the water in the container with a hotplate or stove and check the temperature when the thermostat first starts to open.
**9** Continue heating the water and check the temperature when the thermostat is fully open.

**10** Lift the fully open thermostat out of the water and measure the distance the valve has opened.
**11** Compare the opening temperature, the fully open temperature and the valve travel to the values listed in this Chapter's Specifications.
**12** If these specifications are not met, or if the thermostat does not open while the water is heated, replace it with a new one.

### Installation

**13** Install the thermostat into the housing with the spring pointing into the housing.
**14** Install a new O-ring in the groove in the thermostat cover.
**15** Place the cover on the housing and install the bolts, tightening them securely. If you're working on a J, K, L and M model, be sure to place the ground/earth wire terminal over the proper bolt **(see illustration)**.
**16** The remainder of installation is the reverse of the removal procedure. Fill the cooling system with the recommended coolant (see *'Daily (pre-ride) checks'*). If you're working on an H model, install the pressure cap retainer.

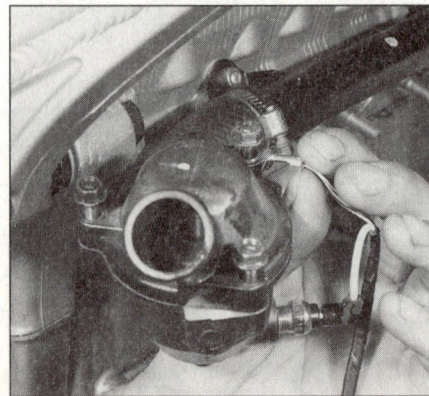

**6.15 Reinstall the ground (earth) wire connector when installing the thermostat cover (J, K, L and M models)**

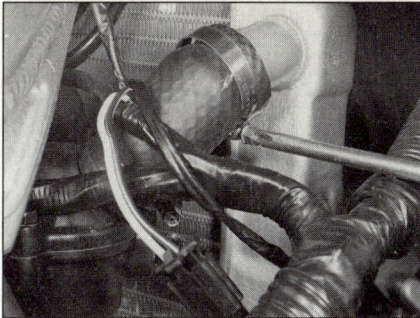

7.5a Remove the radiator hoses . . .

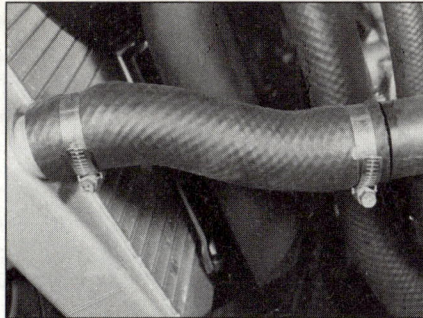

7.5b . . . (there's one on each side) . . .

7.5c . . . and the oil cooler hose on J, K, L and M models

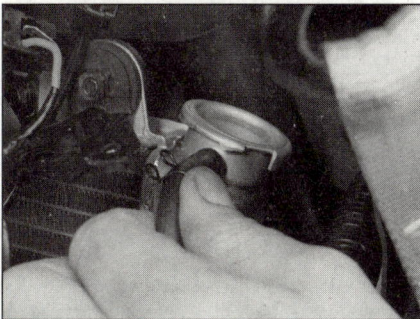

7.6 Slide the clip down the overflow hose and disconnect it from the fitting

7.8a Remove the radiator mounting bolts (arrow) – L model shown

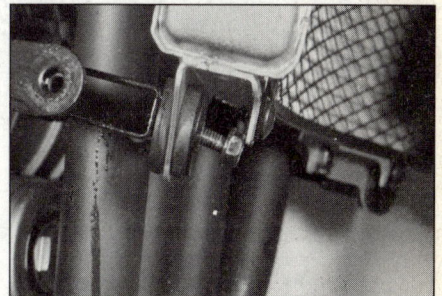

7.8b The lower bolts on some models may not come all the way out, but they can be unscrewed far enough for radiator removal

## 7  Radiator –
### removal and installation

⚠ **Warning: The engine must be completely cool before beginning this procedure.**

### Removal

1 Support the bike securely upright. Remove components as necessary for access to the battery and disconnect the cable from the negative terminal (see Chapter 9).

2 Remove the upper and lower fairings (H models) or the upper fairing (J, K, L and M models) (see Chapter 8).

3 Remove the fuel tank (see Chapter 4) and drain the coolant (see Chapter 1).

4 Unplug the electrical connectors for the fan motor and the thermostatic fan switch (see illustration 3.2a, 3.2b or 3.2c).

5 Loosen the hose clamps on both radiator hoses (see illustrations). On H2 models, loosen the clamp on the coolant bypass hose. If you're working on a J, K, L and M model, loosen the clamp on the oil cooler hose (see illustration). Detach the hoses.

6 If you're working on a J, K, L and M model, disconnect the overflow hose from the radiator filler neck (see illustration).

7 If you're working on an H model, remove the oil cooler (see Chapter 2).

8 Remove the radiator mounting bolts (see illustrations).

9 Lift the radiator out. Be sure to check the rubber mounting dampers and replace them if they're worn.

10 If the radiator is to be repaired, pressure checked or replaced, detach the cooling fan (see Section 4) and the stone shield (see illustrations).

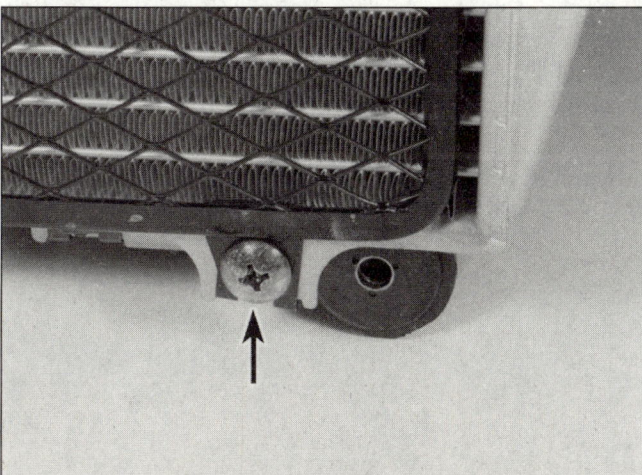

7.10a Remove the stone shield screws at the bottom of the radiator (arrow) . . .

7.10b . . . and lift the stone shield to disengage its upper hooks

*A Upper hook*          *B Bushing*          *C Washer*

**7.14 If the radiator bracket needs replacing, unbolt it from the brackets on the crankcase**

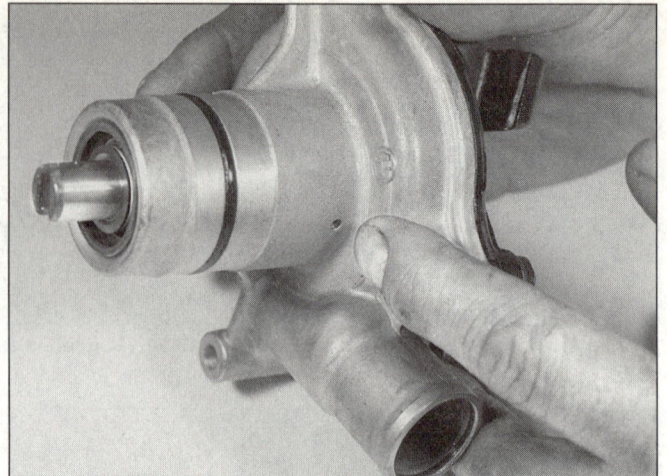

**8.5 Check the weep hole for signs of leakage (L model shown)**

**11** Carefully examine the radiator for evidence of leaks and damage. It is recommended that any necessary repairs be performed by a reputable radiator repair shop.
**12** If the radiator is clogged, or if large amounts of rust or scale have formed, the repair shop will also do a thorough cleaning job.
**13** Make sure the spaces between the cooling tubes and fins are clear. If necessary, use compressed air or running water to remove anything that may be clogging them.

> **HAYNES HiNT**
> *If the radiator fins are bent or flattened, straighten them very carefully with a small screwdriver.*

**14** If you're working on a J, K, L and M model, check the radiator mounting bracket for bending or cracks and replace it as necessary **(see illustration)**.

### Installation

**15** Installation is the reverse of the removal procedure, with the following additions:
a) *Be sure to replace the hoses if they are deteriorated.*
b) *Be sure the metal collars are in place in the rubber mounting dampers* **(see illustration 7.10b)**.

c) *Refill the cooling system with the recommended coolant (see 'Daily (pre-ride) checks').*

### 8 Water pump –
check, removal, inspection and installation

> ⚠ **Warning: The engine must be completely cool before beginning this procedure.**

**Note:** *The water pump on these models can't be overhauled – it must be replaced as a unit.*

### Check

**1** Visually check around the area of the water pump for coolant leaks. Try to determine if the leak is simply the result of a loose hose clamp or deteriorated hose.
**2** Support the bike securely upright.
**3** Remove the lower fairing (see Chapter 8).
**4** Drain the engine coolant following the procedure in Chapter 1.
**5** Check around the weep hole for signs of coolant leakage **(see illustration)**. If coolant has been leaking from the hole, the mechanical seal is damaged. It can't be replaced separately, so the entire pump must be replaced.

### Removal

#### H models

**6** Remove the left lower fairing panel (see Chapter 8).
**7** Drain the coolant (see Chapter 1).
**8** Remove the clutch release cylinder (see Chapter 2).
**9** Remove the engine sprocket cover (see Chapter 6).
**10** Loosen the coolant hose clamps at the water pump.
**11** Remove the shift pedal (see Chapter 2).
**12** Remove the water pump mounting bolts **(see illustration 3.2a or 3.2b)**. Pull the pump straight out to remove it, pulling it out of the coolant hoses as it's removed.

#### J, K, L and M models

**13** Remove the lower fairing panels (see Chapter 8).
**14** Drain the cooling system (see Chapter 1).
**15** Disconnect and remove the water pump upper hose **(see illustrations)**.
**16** Detach the brace, unbolt the coolant tube from the water pump cover and lift it out **(see illustrations)**.
**17** Loosen the clamp and disconnect the lower hose from the water pump **(see illustration)**.
**18** Remove the pump mounting bolts and lift the pump off **(see illustrations)**.

**8.15a Loosen the hose clamps . . .**

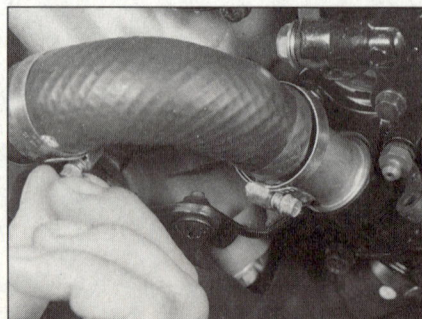

**8.15b . . . and remove the hose from the water pump coolant tube**

**8.16a Remove the screw that secures the brace to the air baffle . . .**

8.16b . . . unbolt the brace from the water pump cover . . .

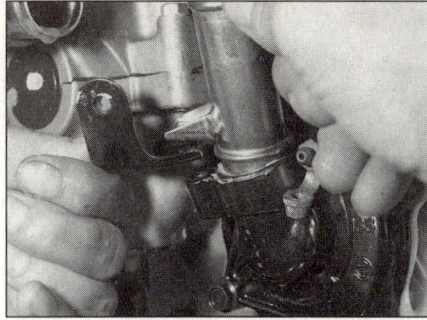
8.16c . . . and lift the coolant tube out of the water pump

8.17 Loosen the clamp and detach the lower hose from the water pump

8.18a Remove the pump mounting bolts (arrows); note the wire terminal on the lower bolt . . .

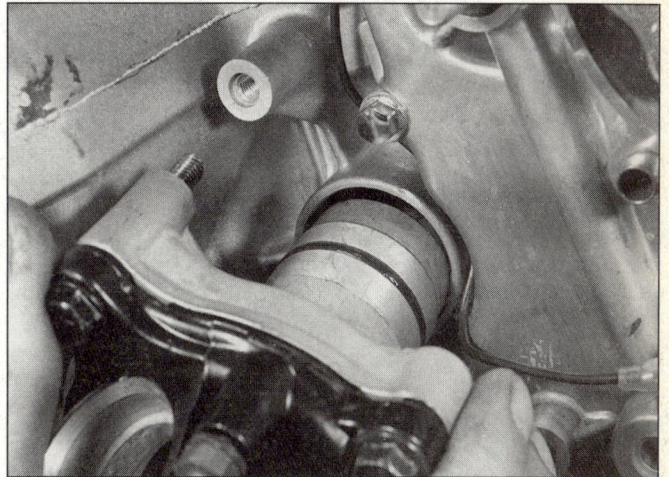
8.18b . . . and pull the water pump out of the engine

## Inspection

19 Unbolt the cover from the pump and lift it off.

20 Check the pump O-ring for brittleness or deterioration (see illustration). It's a good idea to replace the O-ring whenever the pump cover is removed.

21 Try to wiggle the water pump impeller back-and-forth and in-and-out (see illustration). If you can feel movement, the water pump must be replaced.

22 Check the impeller blades for corrosion. If they are heavily corroded, replace the water pump and flush the system thoroughly (it

would also be a good idea to check the internal condition of the radiator).

23 If you're working on a J, K, L and M model, check the O-ring on the coolant tube for brittleness or deterioration (see illustration). It's a good idea to replace it whenever the tube is removed.

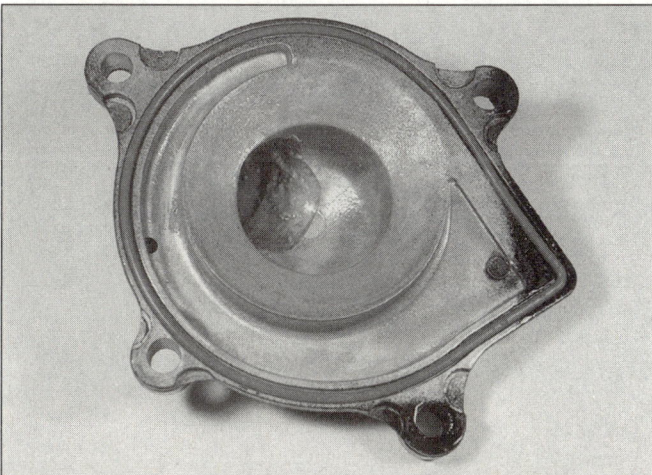
8.20 The cover O-ring should be replaced if there's any doubt about its condition

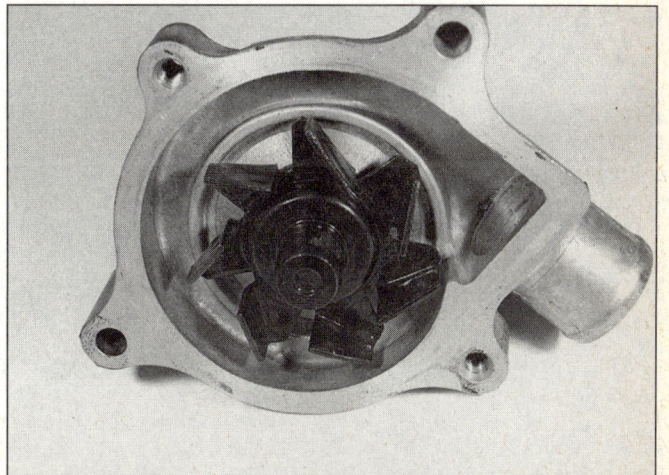
8.21 If the impeller is loose or corroded, replace the water pump

**8.23 The coolant tube O-ring should be replaced whenever the tube is removed from the pump cover**

**8.25a Align the slot in the pump shaft (arrow) . . .**

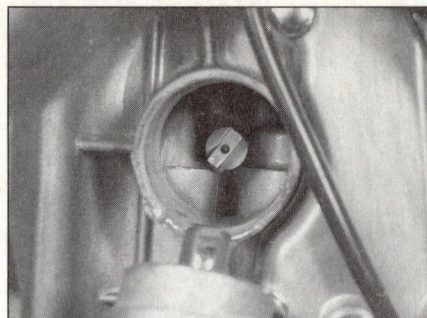

**8.25b . . . with the tooth on the shaft inside the engine**

## Installation

24 Lubricate the pump O-ring with a thin smear of multi-purpose grease.

25 Install the pump body before installing the cover. Turn the impeller so the slot in the pump shaft aligns with the drive tooth on the oil pump shaft inside the engine **(see illustrations)**.

26 Lubricate the cover O-ring with a thin smear of multi-purpose grease, then install the O-ring and the cover. Tighten the cover and mounting bolts evenly and securely, but don't overtighten and strip out the bolt holes.

27 The remainder of installation is the reverse of the removal steps.

28 Fill the cooling system with the recommended coolant (see 'Daily (pre-ride) checks').

29 Run the engine and check for coolant leaks.

### 9 Coolant pipe(s) and passage – removal and installation

⚠️ **Warning: The engine must be completely cool before beginning this procedure.**

1 Support the bike securely upright.

2 Remove the fairing panels as necessary for access (see Chapter 1).

3 Drain the engine coolant (see Chapter 1).

4 Where necessary, remove the radiator (see Section 7), carburetors or exhaust system (see Chapter 4).

## Coolant tubes

5 Remove the coolant tube-to-cylinder head bolts **(see illustration 3.2a, 3.2b or the accompanying illustration)** and separate the tube from the cylinder head.

6 Remove the O-rings from the ends of the tube and install new ones. If one of the ends doesn't have an O-ring, be sure to retrieve it from the engine.

7 Check the tube mounting holes in the engine for corrosion, and remove all traces of corrosion if any exists.

8 Install new O-rings on the tube ends, lubricating them with a little clean engine oil.

9 Install the ends of the coolant tube into the holes in the engine. Install the mounting bolts, tightening them securely.

10 Fill the cooling system with the recommended coolant (see Chapter 1) and check for leaks.

11 The remainder of installation is the reverse of the removal procedure.

## Coolant passage (J, K, L and M models)

12 Loosen the hose clamps and disconnect the coolant hoses from the passage **(see illustration)**.

13 Remove the mounting bolts and take the passage off the engine.

14 Clean all traces of old gasket and sealer from the passage.

15 Install the passage, using a new gasket, and tighten its bolts securely.

16 Fill the cooling system with the recommended coolant (see Chapter 1) and check for leaks.

17 The remainder of installation is the reverse of the removal procedure.

### 10 Oil cooler – removal and installation

## H models

**Note:** *Wait until the engine is cool before beginning this procedure.*

1 Support the bike securely upright and drain the engine oil (see Chapter 1).

2 Remove the lower fairing (see Chapter 8).

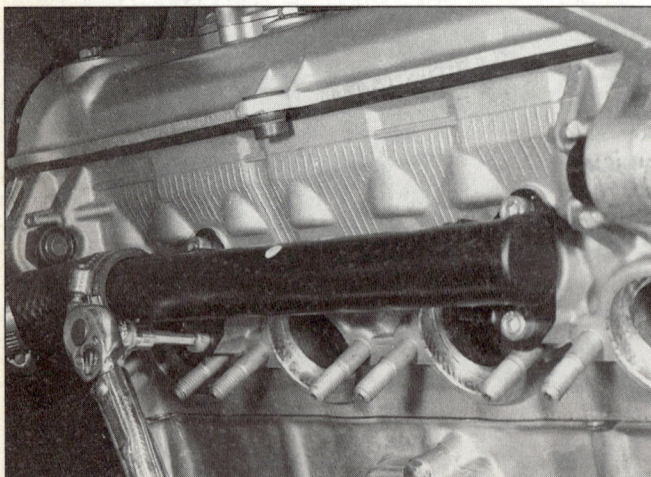

**9.5 Remove the Allen bolts to detach the coolant tube from the engine**

**9.12 Detach the coolant hoses and remove the mounting bolts (arrows)**

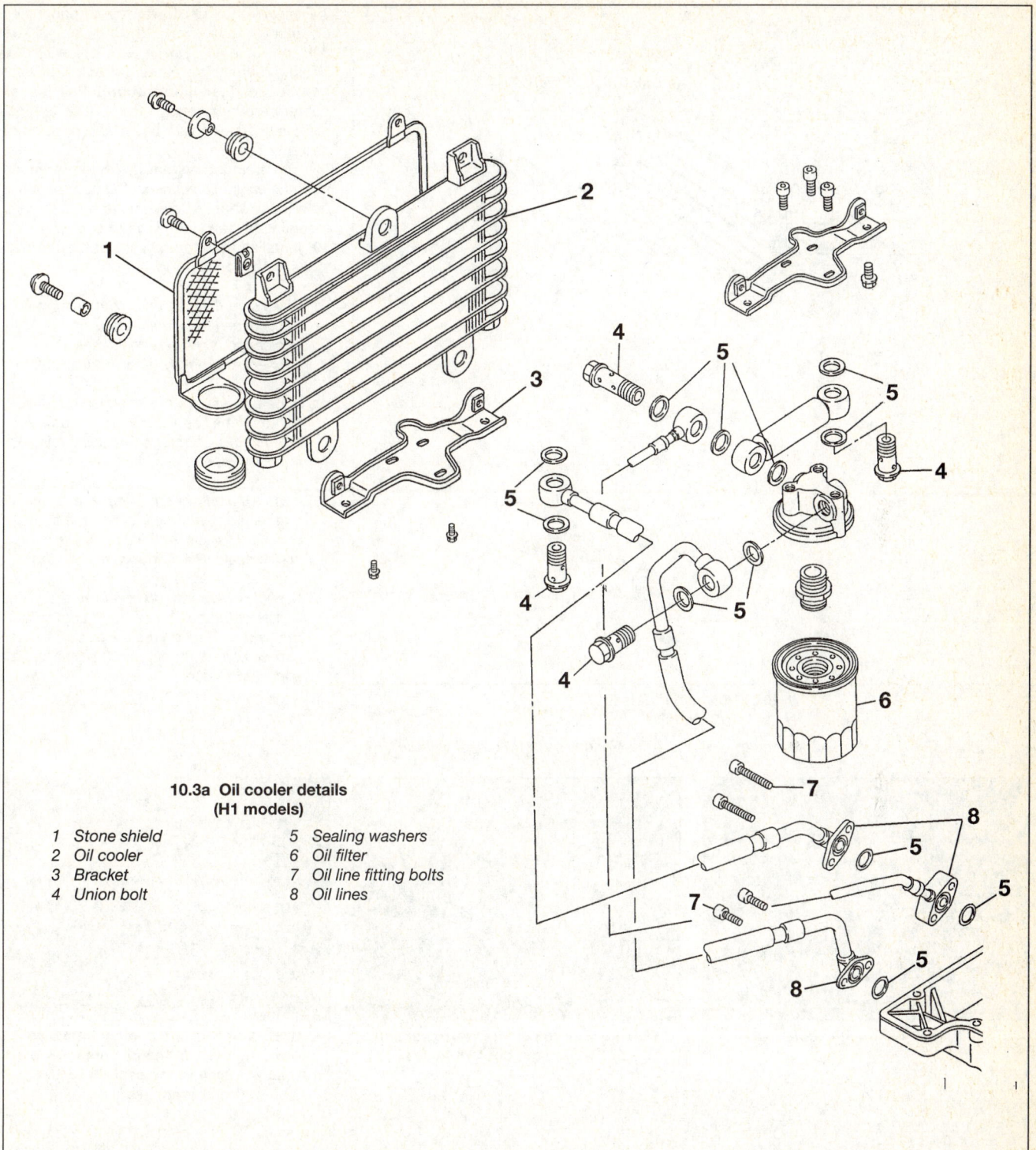

**10.3a  Oil cooler details**
**(H1 models)**

| | | | |
|---|---|---|---|
| 1 | Stone shield | 5 | Sealing washers |
| 2 | Oil cooler | 6 | Oil filter |
| 3 | Bracket | 7 | Oil line fitting bolts |
| 4 | Union bolt | 8 | Oil lines |

**3** Place a drain pan under the front of the crankcase and remove the oil cooler hose-to-crankcase union bolts **(see illustrations)**. Retrieve the sealing washers.
**4** Remove the oil cooler mounting bolts and take the oil cooler out.

**5** Installation is the reverse of removal. Be sure to use new sealing washers if the old ones were leaking or damaged, and fill the crankcase with the recommended type and amount of oil (see Chapter 1 and *Daily (pre-ride) checks*). If you're working on an H1

model, tighten the union bolts to the torque listed in this Chapter's Specifications.

## J, K, L and M models

**6** Support the bike securely upright. Drain the engine oil and coolant (see Chapter 1).

**7** Remove the right lower fairing panel (see Chapter 8).

**8** If the exhaust system is in place on the engine, remove the oil cooler bolt with a 27 mm socket **(see illustration)**. Pull the oil cooler forward away from the engine, disconnect the coolant hoses and take the oil cooler out.

**9** If the exhaust system has been removed from the engine, disconnect the coolant hoses **(see illustration)**, then remove the oil cooler mounting bolt and lift off the oil cooler.

**10** Installation is the reverse of the removal steps, with the following additions:

a) *Make sure the cast rib on the crankcase fits between the locating tabs on the oil cooler and make sure the hoses are connected properly* **(see illustration)**.

b) *If the exhaust system has been removed, tighten the oil cooler bolt to the torque listed in this Chapter's Specifications with a torque wrench.*

c) *If the exhaust system is in place on the engine, tighten the oil cooler bolt to the specified torque with a wrench and spring scale. Place the wrench on the bolt and place the spring scale on the wrench at a point 300 mm from the center of the bolt and at right angles to the wrench. Pull on the spring scale with a force of 17 kg to tighten the bolt to the specified torque*

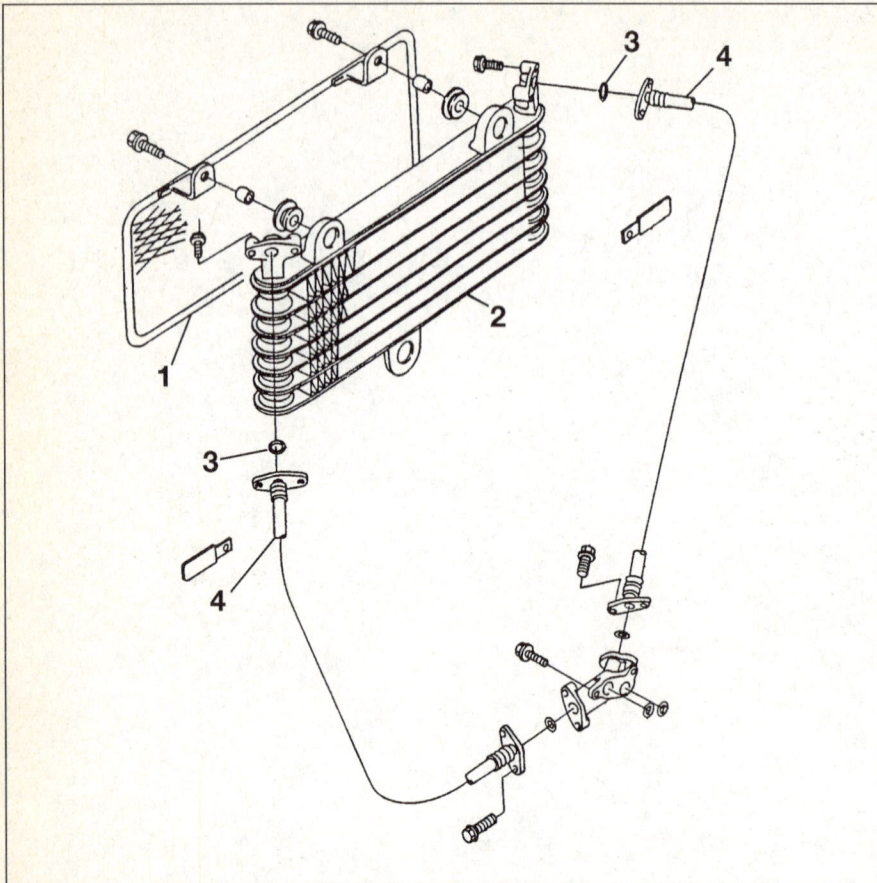

**10.3b Oil cooler details
(H2 models)**

| 1 Stone shield | 2 Oil cooler | 3 O-ring | 4 Oil line |

**10.8 Remove the oil cooler bolt**

**10.9 Loosen the clamps (arrows) and disconnect the coolant hoses**

**10.10 Connect the coolant hoses as shown; the tabs on the oil cooler (arrow) must fit around the cast rib on the crankcase**

# Chapter 4
# Fuel and exhaust systems

## Contents

## Degrees of difficulty

| Easy, suitable for novice with little experience | Fairly easy, suitable for beginner with some experience | Fairly difficult, suitable for competent DIY mechanic | Difficult, suitable for experienced DIY mechanic | Very difficult, suitable for expert DIY or professional |
|---|---|---|---|---|

## Specifications

### General

Fuel type . . . . . . . . . . . . . . . . . . . . . . . . . . . . . . . . . . . . . . . . . . . . . . . . . . Unleaded or low lead gasoline (petrol) subject to local regulations; minimum octane 91 RON (87 pump octane)

Fuel tank capacity
   H, J, K and L models . . . . . . . . . . . . . . . . . . . . . . . . . . . . . . . . . . . 18 liters (4.7 US gal, 4.0 Imp gal)
   M models . . . . . . . . . . . . . . . . . . . . . . . . . . . . . . . . . . . . . . . . . . . 17 liters (4.5 US gal, 3.7 Imp gal)

### Carburetor type

H1 models . . . . . . . . . . . . . . . . . . . . . . . . . . . . . . . . . . . . . . . . . . . . . Keihin CVK-D36 (four)
H2, J and L models . . . . . . . . . . . . . . . . . . . . . . . . . . . . . . . . . . . . . . . Keihin CVK-D38 (four)
K and M models . . . . . . . . . . . . . . . . . . . . . . . . . . . . . . . . . . . . . . . . . Keihin CVK-D39 (four)

### Jet sizes and settings (H1 model)

Main jet
   All except US high altitude models . . . . . . . . . . . . . . . . . . . . . . . . . . 130
   US high altitude models . . . . . . . . . . . . . . . . . . . . . . . . . . . . . . . . . 128
Main air jet . . . . . . . . . . . . . . . . . . . . . . . . . . . . . . . . . . . . . . . . . . . . . 100
Jet needle . . . . . . . . . . . . . . . . . . . . . . . . . . . . . . . . . . . . . . . . . . . . . . N52U
Pilot jet
   All except US high altitude models . . . . . . . . . . . . . . . . . . . . . . . . . . 38
   US high altitude models . . . . . . . . . . . . . . . . . . . . . . . . . . . . . . . . . 35
Pilot air jet . . . . . . . . . . . . . . . . . . . . . . . . . . . . . . . . . . . . . . . . . . . . . . 120
Pilot screw setting
   US models . . . . . . . . . . . . . . . . . . . . . . . . . . . . . . . . . . . . . . . . . . . Preset
   UK models . . . . . . . . . . . . . . . . . . . . . . . . . . . . . . . . . . . . . . . . . . . 1⅝ ± ¼ turns out
Choke (starter) jet . . . . . . . . . . . . . . . . . . . . . . . . . . . . . . . . . . . . . . . . 52
Float height . . . . . . . . . . . . . . . . . . . . . . . . . . . . . . . . . . . . . . . . . . . . . 13 ± 2 mm (0.512 ± 0.079 inch)
Fuel level . . . . . . . . . . . . . . . . . . . . . . . . . . . . . . . . . . . . . . . . . . . . . . . 4 to 6 mm (0.157 to 0.236 inches) below mark

## Jet sizes and settings (H2 model)

Main jet
    All except US high altitude models . . . . . . . . . . . . . . . . . . . . . . . . . . .  128
    US high altitude models . . . . . . . . . . . . . . . . . . . . . . . . . . . . . . . . . .  125
Main air jet . . . . . . . . . . . . . . . . . . . . . . . . . . . . . . . . . . . . . . . . . . . . . . .  100
Jet needle . . . . . . . . . . . . . . . . . . . . . . . . . . . . . . . . . . . . . . . . . . . . . . . . .  N76F
Pilot jet
    All except US high altitude models . . . . . . . . . . . . . . . . . . . . . . . . . . .  38
    US high altitude models . . . . . . . . . . . . . . . . . . . . . . . . . . . . . . . . . .  35
Pilot air jet
    All except California models . . . . . . . . . . . . . . . . . . . . . . . . . . . . . . . .  130
    California models . . . . . . . . . . . . . . . . . . . . . . . . . . . . . . . . . . . . . . .  135
Pilot screw setting
    US models . . . . . . . . . . . . . . . . . . . . . . . . . . . . . . . . . . . . . . . . . . .  Preset
    UK models . . . . . . . . . . . . . . . . . . . . . . . . . . . . . . . . . . . . . . . . . . .  1¾ ± ¼ turns out
Choke (starter) jet . . . . . . . . . . . . . . . . . . . . . . . . . . . . . . . . . . . . . . . . .  52
Float height . . . . . . . . . . . . . . . . . . . . . . . . . . . . . . . . . . . . . . . . . . . . . . .  13 ± 2 mm (0.512 ± 0.079 inch)
Fuel level . . . . . . . . . . . . . . . . . . . . . . . . . . . . . . . . . . . . . . . . . . . . . . . . .  4 to 6 mm (0.157 to 0.236 inches) below mark

## Jet sizes and settings (J model)

Main jet
    All except US high altitude models
        Cylinders 1 and 4 . . . . . . . . . . . . . . . . . . . . . . . . . . . . . . . . . . .  132
        Cylinders 2 and 3 . . . . . . . . . . . . . . . . . . . . . . . . . . . . . . . . . . .  138
    US high altitude models
        Cylinders 1 and 4 . . . . . . . . . . . . . . . . . . . . . . . . . . . . . . . . . . .  130
        Cylinders 2 and 3 . . . . . . . . . . . . . . . . . . . . . . . . . . . . . . . . . . .  135
Main air jet . . . . . . . . . . . . . . . . . . . . . . . . . . . . . . . . . . . . . . . . . . . . . . .  100
Jet needle
    Cylinders 1 and 4 . . . . . . . . . . . . . . . . . . . . . . . . . . . . . . . . . . . . . .  N96C
    Cylinders 2 and 3 . . . . . . . . . . . . . . . . . . . . . . . . . . . . . . . . . . . . . .  N96D
Pilot jet
    All except US high altitude models . . . . . . . . . . . . . . . . . . . . . . . . . . .  38
    US high altitude models . . . . . . . . . . . . . . . . . . . . . . . . . . . . . . . . . .  35
Pilot air jet . . . . . . . . . . . . . . . . . . . . . . . . . . . . . . . . . . . . . . . . . . . . . . .  130
Pilot screw setting
    US models . . . . . . . . . . . . . . . . . . . . . . . . . . . . . . . . . . . . . . . . . . .  Preset
    UK models . . . . . . . . . . . . . . . . . . . . . . . . . . . . . . . . . . . . . . . . . . .  2¼ turns out
Choke (starter) jet . . . . . . . . . . . . . . . . . . . . . . . . . . . . . . . . . . . . . . . . .  52
Float height . . . . . . . . . . . . . . . . . . . . . . . . . . . . . . . . . . . . . . . . . . . . . . .  13 ± 2 mm (0.512 ± 0.079 inch)
Fuel level . . . . . . . . . . . . . . . . . . . . . . . . . . . . . . . . . . . . . . . . . . . . . . . . .  4 to 6 mm (0.157 to 0.236 inches) below mark

## Jet sizes and settings (K model)

Main jet
    All except US high altitude models . . . . . . . . . . . . . . . . . . . . . . . . . . .  138
    US high altitude models . . . . . . . . . . . . . . . . . . . . . . . . . . . . . . . . . .  135
Main air jet . . . . . . . . . . . . . . . . . . . . . . . . . . . . . . . . . . . . . . . . . . . . . . .  100
Jet needle
    US except California models
        Cylinders 1 and 4 . . . . . . . . . . . . . . . . . . . . . . . . . . . . . . . . . . .  NOGC
        Cylinders 2 and 3 . . . . . . . . . . . . . . . . . . . . . . . . . . . . . . . . . . .  NOGD
    California models
        Cylinders 1 and 4 . . . . . . . . . . . . . . . . . . . . . . . . . . . . . . . . . . .  NOGA
        Cylinders 2 and 3 . . . . . . . . . . . . . . . . . . . . . . . . . . . . . . . . . . .  NOGB
    UK models
        Cylinders 1 and 4 . . . . . . . . . . . . . . . . . . . . . . . . . . . . . . . . . . .  OBEMP
        Cylinders 2 and 3 . . . . . . . . . . . . . . . . . . . . . . . . . . . . . . . . . . .  OBFMP
Pilot jet
    All except US high altitude models . . . . . . . . . . . . . . . . . . . . . . . . . . .  42
    US high altitude models . . . . . . . . . . . . . . . . . . . . . . . . . . . . . . . . . .  40
Pilot air jet . . . . . . . . . . . . . . . . . . . . . . . . . . . . . . . . . . . . . . . . . . . . . . .  120
Pilot screw setting
    US models . . . . . . . . . . . . . . . . . . . . . . . . . . . . . . . . . . . . . . . . . . .  Preset
    UK models . . . . . . . . . . . . . . . . . . . . . . . . . . . . . . . . . . . . . . . . . . .  1⅝ turns out
Choke (starter) jet . . . . . . . . . . . . . . . . . . . . . . . . . . . . . . . . . . . . . . . . .  55
Float height . . . . . . . . . . . . . . . . . . . . . . . . . . . . . . . . . . . . . . . . . . . . . . .  9 ± 2 mm (0.354 ± 0.079 inch)
Fuel level . . . . . . . . . . . . . . . . . . . . . . . . . . . . . . . . . . . . . . . . . . . . . . . . .  3 ± 1 mm (0.118 ± 0.039 inches) above mark

## Jet sizes and settings (L model)

Main jet
  All except US high altitude models
    Cylinders 1 and 4 ................................... 170
    Cylinders 2 and 3 ................................... 190
  US high altitude models
    Cylinders 1 and 4 ................................... 165
    Cylinders 2 and 3 ................................... 185
Main air jet ................................................ 90
Jet needle
  Cylinders 1 and 4 ..................................... N74E
  Cylinders 2 and 3 ..................................... N14I
Pilot jet
  All except US high altitude models ..................... 38
  US high altitude models .............................. 35
Pilot air jet ............................................... 120
Pilot screw setting
  US models ........................................... Preset
  UK models ........................................... 2¼ turns out
Choke (starter) jet ....................................... 52
Float height ............................................. 13 ± 2 mm (0.512 ± 0.079 inch)
Fuel level ............................................... 4 to 6 mm (0.157 to 0.236 inches) below mark

## Jet sizes and settings (M model)

Main jet
  All except US high altitude models
    Cylinders 1 and 4 ................................... 190
    Cylinders 2 and 3 ................................... 200
  US high altitude models
    Cylinders 1 and 4 ................................... 185
    Cylinders 2 and 3 ................................... 195
Main air jet
  Cylinders 1 and 4 ..................................... 70
  Cylinders 2 and 3 ..................................... 100
Jet needle
  Cylinders 1 and 4 ..................................... N1MA (clip in fourth groove from top)
  Cylinders 2 and 3 ..................................... N1HC (clip in fifth groove from top)
Pilot jet
  All except US high altitude models ..................... 42
  US high altitude models .............................. 40
Pilot air jet ............................................... 140
Pilot screw setting
  US models ........................................... Preset
  UK models ........................................... 2¼ turns out
Choke (starter) jet ....................................... 55
Float height ............................................. 9 ± 2 mm (0.354 ± 0.079 inch)
Fuel level ............................................... 3 ± 1 mm (0.118 ± 0.039 inches) above mark
Fuel pump pressure ..................................... 0.11 to 0.16 Bars (1.6 to 2.3 psi)

## 1  General information

The fuel system consists of the fuel tank, fuel tap, filter, electric fuel pump, carburetors and connecting lines, hoses and control cables.

The carburetors used on H, J and L models are four constant vacuum Keihins with butterfly-type throttle valves. For cold starting, an enrichment circuit is actuated by a cable and the choke lever mounted on the left handlebar.

The carburetors used on K and M models are four flat-slide Mikunis with accelerator pumps and fuel enrichers.

The exhaust system is a four-into-one design.

Many of the fuel system service procedures are considered routine maintenance items and for that reason are included in Chapter 1.

## 2  Fuel tank – removal and installation

**Warning: Gasoline (petrol) is extremely flammable, so take extra precautions when you work on any part of the fuel system. Don't smoke or allow open flames or bare light bulbs near the work area, and don't work in a garage where a natural gas-type appliance (such as a water heater or clothes dryer) is present. If you spill any fuel on your skin, rinse it off immediately with soap and water. When you perform any kind of work on the fuel system, wear safety glasses and have an extinguisher suitable for a class B type fire (flammable liquids) on hand.**

1 The fuel tank is held in place by bolts and rubber insulators, which fit through a flange projecting from the tank. On H models, the mounting flanges are bolted to the tank; on J, K, L and M models they're an integral part of the tank.

### Removal

#### H models

2 Remove the rear ducts, front seat and side covers (see Chapter 8).

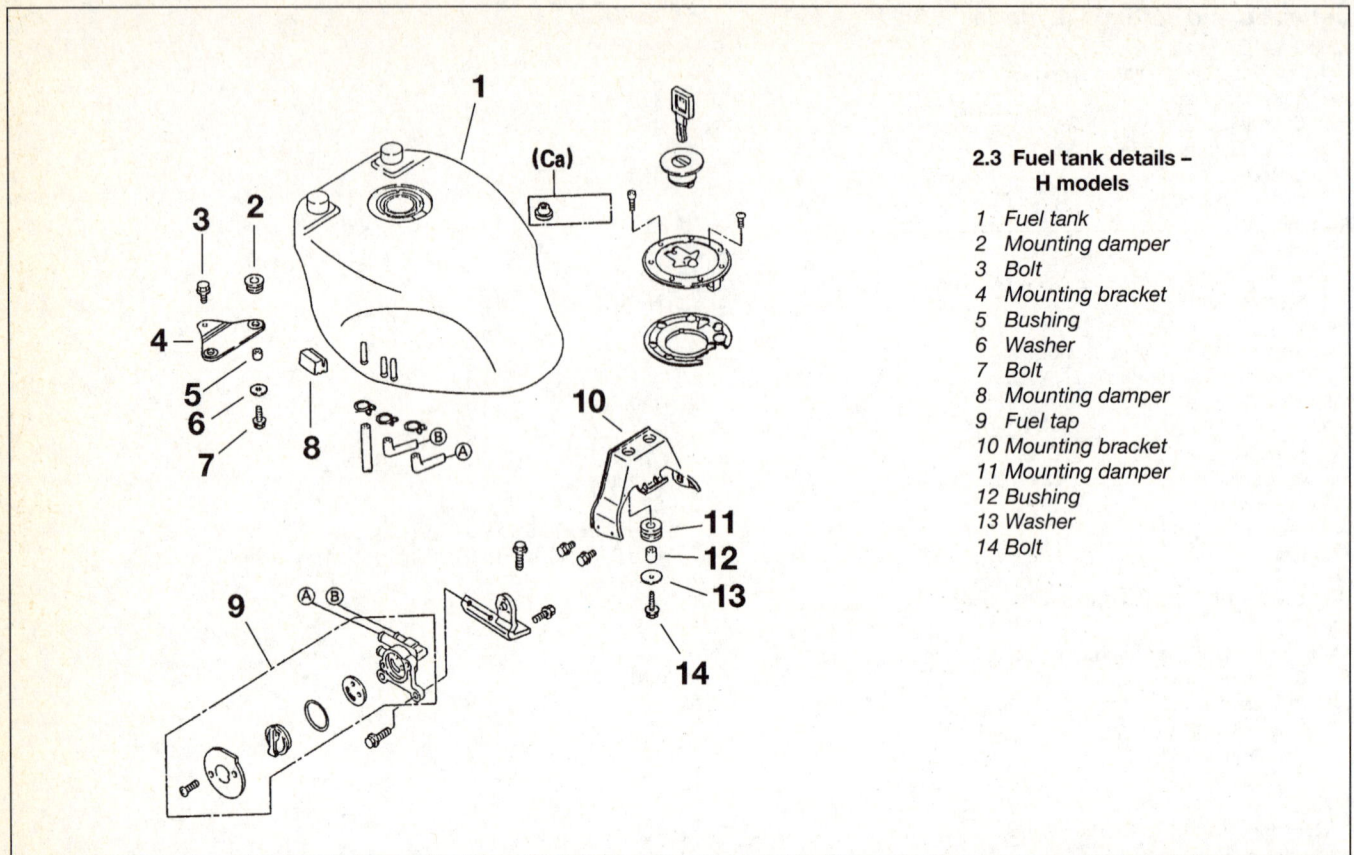

**2.3 Fuel tank details –**
**H models**

1 Fuel tank
2 Mounting damper
3 Bolt
4 Mounting bracket
5 Bushing
6 Washer
7 Bolt
8 Mounting damper
9 Fuel tap
10 Mounting bracket
11 Mounting damper
12 Bushing
13 Washer
14 Bolt

**3** Remove the fuel tank mounting bolts **(see illustration)**.
**4** Disconnect the fuel line between the fuel tap and carburetors. Detach the fuel tap from the frame, but leave the hoses that connect it to the fuel tank attached.
**5** If you're working on a California model, disconnect the evaporative emission hoses from their fittings on the tank (see Section 15).
**6** Lift the fuel tank off the motorcycle, together with the fuel tap.

**J, K, L and M models**

**7** If you're working on a J or L model, remove the side covers (see Chapter 8).
**8** If you're working on a K or M model, remove the seat (see Chapter 8).
**9** Disconnect the rear ends of the air ducts (see Chapter 8).
**10** Remove the mounting bolts at the front and rear of the fuel tank **(see illustrations)**. On L and M models, unbolt the tank mounting bracket from the frame **(see illustration)**.
**11** Disconnect the fuel tap line that runs to the carburetors **(see illustration)**. Remove the fuel tap mounting screws and spacers, but leave the fuel lines from the tap to the fuel tank connected **(see illustration)**.
**12** Lift the fuel tank off the motorcycle. If necessary, disconnect the fuel hoses (or evaporative emission system hoses on California models) from the tank **(see illustration)**.

## Installation

**13** Before installing the tank, check the condition of the rubber mounting dampers – if they're hardened, cracked, or show any other signs of deterioration, replace them.
**14** When installing the tank, reverse the removal procedure. Make sure the tank seats properly and does not pinch any control cables or wires.

## 3 Fuel tank – cleaning and repair

**1** All repairs to the fuel tank should be carried out by a professional who has experience in this critical and potentially dangerous work. Even after cleaning and flushing of the fuel system, explosive fumes can remain and ignite during repair of the tank.

**2** If the fuel tank is removed from the vehicle, it should not be placed in an area where sparks or open flames could ignite the fumes coming out of the tank. Be especially careful inside garages where a natural gas-type appliance is located, because the pilot light could cause an explosion.

## 4 Idle fuel/air mixture adjustment – general information

**1** Due to the increased emphasis on controlling motorcycle exhaust emissions, certain governmental regulations have been formulated which directly affect the carburetion of this machine. In order to comply with the regulations, the carburetors on some models have a metal sealing plug pressed into the hole over the pilot screw (which controls the idle fuel/air mixture) on each carburetor, so they can't be tampered with. These should only be removed in the event of a complete carburetor overhaul, and even then the screws should be returned to their original settings. The pilot screws on other models are accessible, but the use of an exhaust gas analyzer is the only accurate way to adjust the idle fuel/air mixture and be sure the machine doesn't exceed the emissions regulations.
**2** If the engine runs extremely rough at idle or continually stalls, and if a carburetor overhaul does not cure the problem, take the motorcycle to a Kawasaki dealer service department or other repair shop equipped with an exhaust gas analyzer. They will be able to properly adjust the idle fuel/air mixture to achieve a smooth idle and restore low speed performance.

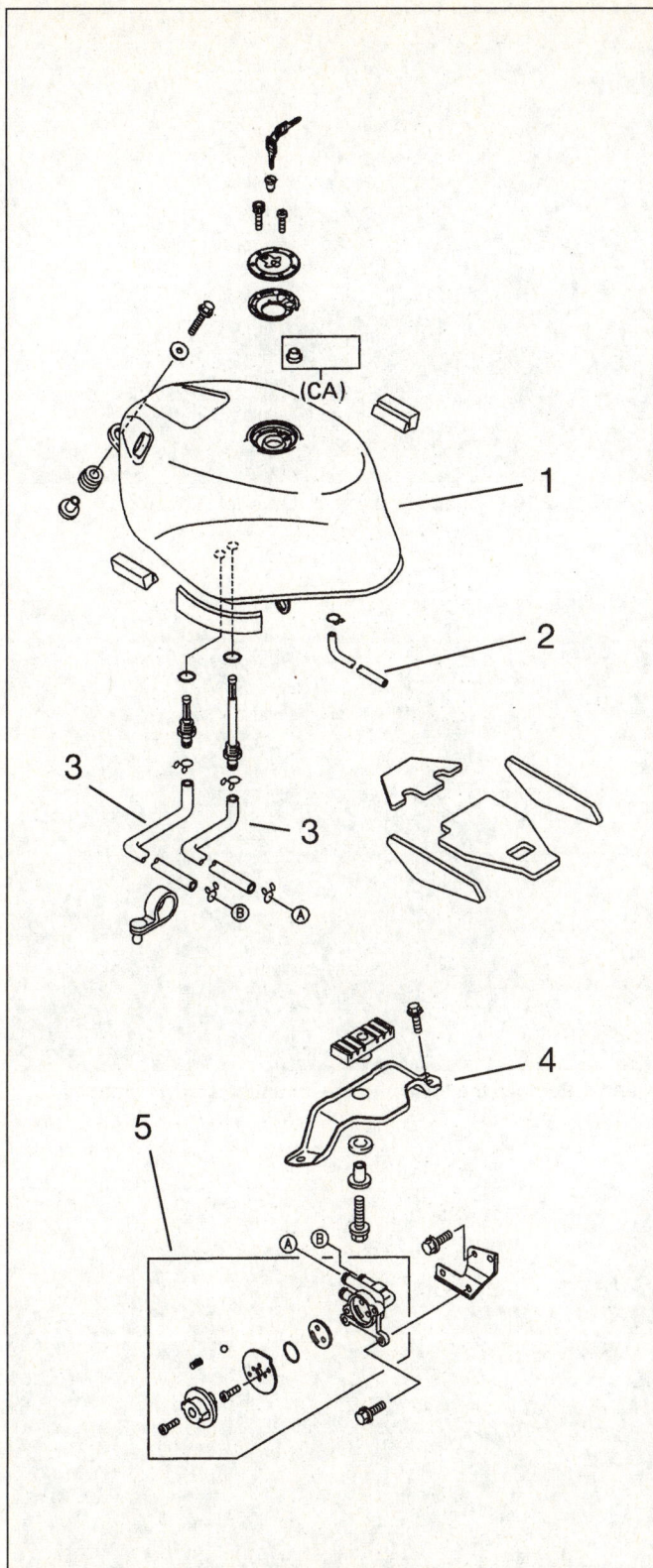

**2.10a Fuel tank details –
J and K models**

1  Fuel tank
2  Breather hose (except
   California models)
3  Fuel hoses
4  Mounting bracket
5  Fuel tap

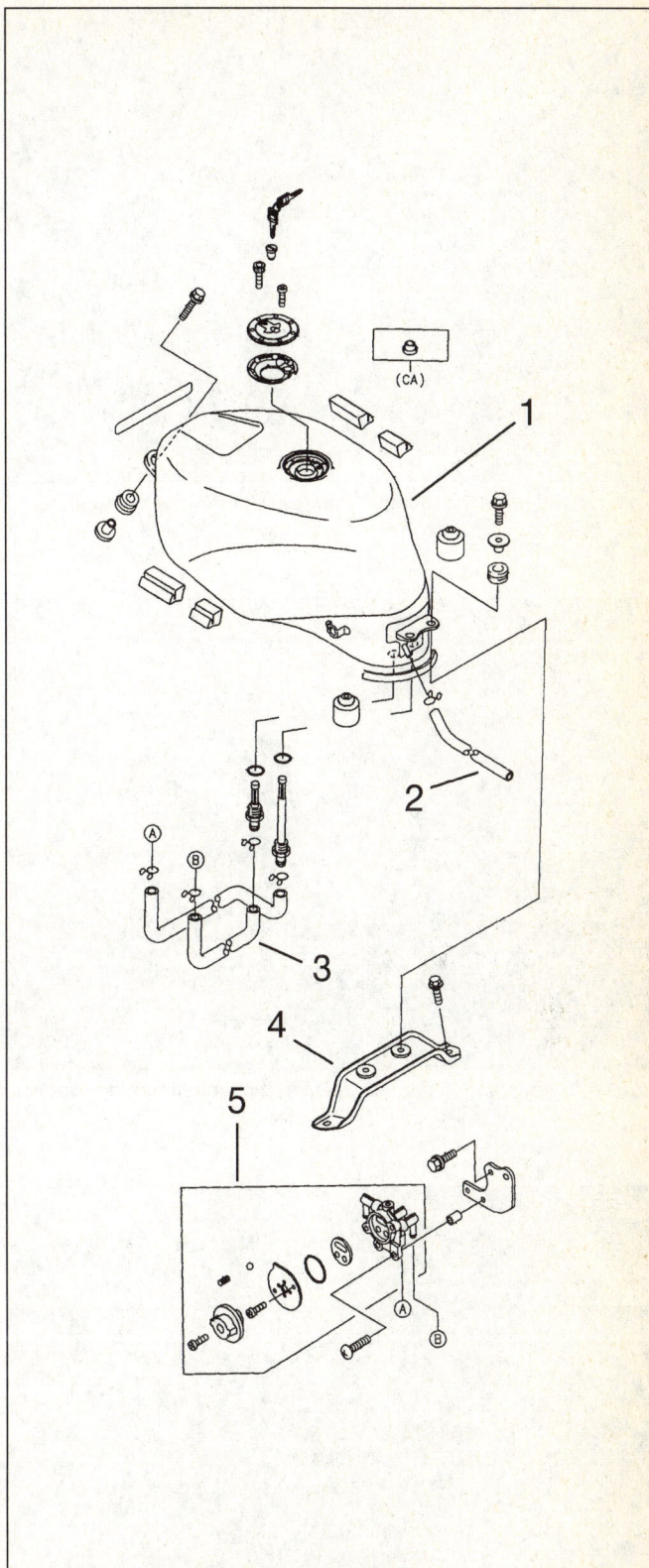

**2.10b Fuel tank details –
L and M models**

1  Fuel tank
2  Breather hose (except
   California models)
3  Fuel hoses
4  Mounting bracket
5  Fuel tap

2.10c  Remove the mounting bolts at the front of the tank . . .

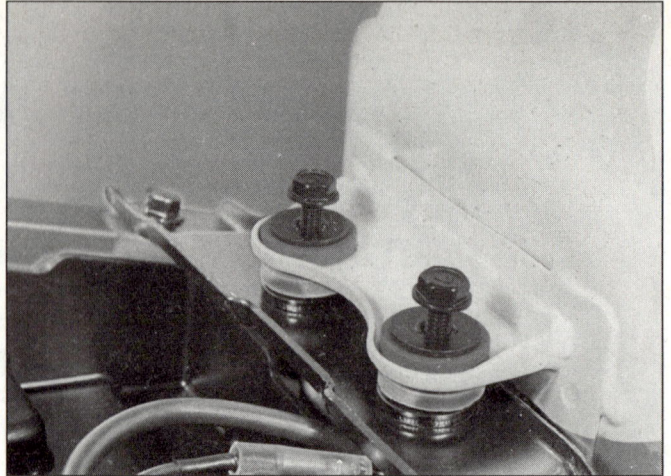

2.10d  . . . and at the rear (L model shown) . . .

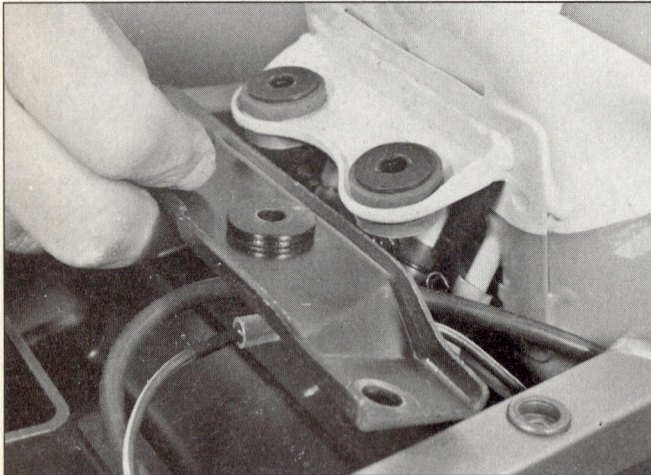

2.10e  . . . on L and M models, remove the tank mounting bracket

2.11a  Remove the hose and the mounting screws (arrows) . . .

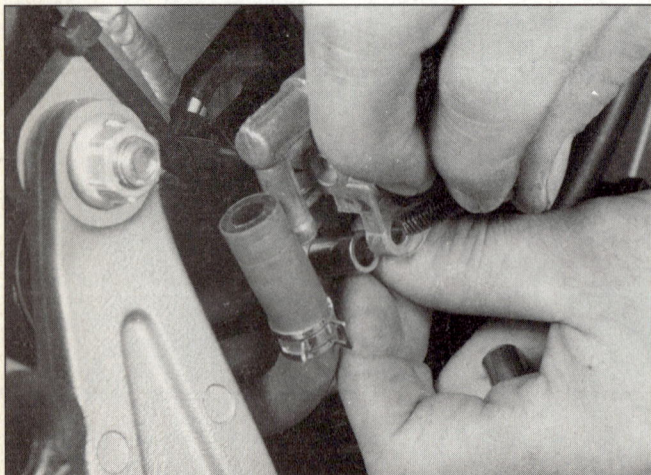

2.11b  . . . and the mounting screw spacers

2.12  Disconnect the tank hoses (L model shown)

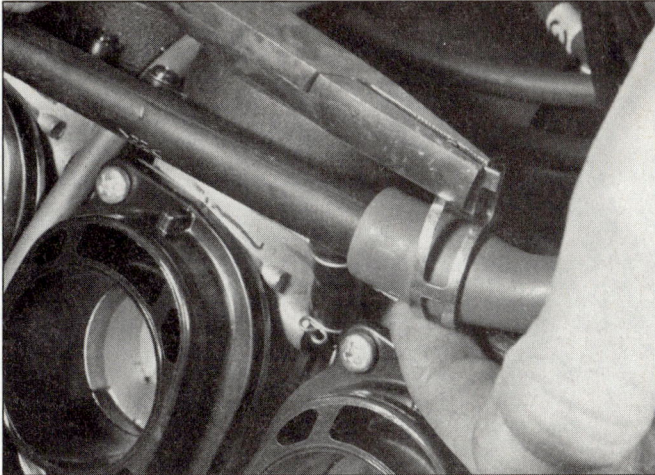

6.7a  Squeeze the clip and slide it down the vent hose . . .

6.7b  . . . and disconnect the hose from the carburetor fitting (arrow)

### 5  Carburetor overhaul – general information

**1** Poor engine performance, hesitation, hard starting, stalling, flooding and backfiring are all signs that major carburetor maintenance may be required.

**2** Keep in mind that many so-called carburetor problems are really not carburetor problems at all, but mechanical problems within the engine or ignition system malfunctions. Try to establish for certain that the carburetors are in need of maintenance before beginning a major overhaul.

**3** Check the fuel tap, the in-line fuel filter (and on J, K, L and M models also check the fuel tank filters), the fuel lines, the fuel tank cap vent (except California models), the intake manifold hose clamps, the vacuum hoses, the air filter element, the cylinder compression, the spark plugs, and the carburetor synchronization before assuming that a carburetor overhaul is required.

**4** Most carburetor problems are caused by dirt particles, varnish and other deposits which build up in and block the fuel and air passages. Also, in time, gaskets and O-rings shrink or deteriorate and cause fuel and air leaks which lead to poor performance.

**5** When the carburetor is overhauled, it is generally disassembled completely and the parts are cleaned thoroughly with a carburetor cleaning solvent and dried with filtered, unlubricated compressed air. The fuel and air passages are also blown through with compressed air to force out any dirt that may have been loosened but not removed by the solvent. Once the cleaning process is complete, the carburetor is reassembled using new gaskets, O-rings and, generally, a new inlet needle valve and seat.

**6** Before disassembling the carburetors, make sure you have a carburetor rebuild kit (which will include all necessary O-rings and other parts), some carburetor cleaner, a supply of rags, some means of blowing out the carburetor passages and a clean place to work.

> **HAYNES HiNT**  *It is recommended that only one carburetor be overhauled at a time to avoid mixing up parts.*

### 6  Carburetors – removal and installation

⚠ *Warning: Gasoline (petrol) is extremely flammable, so take extra precautions when you work on any part of the fuel system. Don't smoke or allow open flames or bare light bulbs near the work area, and don't work in a garage where a natural gas-type appliance (such as a water heater or clothes dryer) is present. If you spill any fuel on your skin, rinse it off immediately with soap and water. When you perform any kind of work on the fuel system, wear safety glasses and have a fire extinguisher suitable for a class B type fire (flammable liquids) on hand.*

### Removal

#### H, J and L models

**1** Remove the fuel tank (see Section 2).

**2** Remove the air filter housing (see Section 12).

**3** Disconnect the choke cable from the carburetor assembly (see Section 11).

**4** Loosen the throttle cables all the way and disconnect them from the throttle grip (see Section 10).

**5** If you're working on an H model, remove the idle speed screw extension.

**6** On all except L models, mark and disconnect the vent hoses from the carburetors.

**7** If you're working on an L model, squeeze the hose clamp and slide it down the vent hose, then disconnect the vent hose from the vent tube assembly (see illustrations).

**8** Disconnect the fuel line from the carburetor assembly (see illustration).

**9** Pull the carburetor assembly to the rear, clear of the intake manifold tubes (see illustration). Raise the assembly up far enough to disconnect the throttle cables from the throttle pulley, then remove the carburetors from the machine.

**10** After the carburetors have been removed, stuff clean rags into the intake manifold tubes to prevent the entry of dirt or other objects.

6.8  Disconnect the fuel line from the carburetor assembly

6.9  Pull the carburetor assembly clear of the intake manifold tubes (L model shown)

6.15 Disconnect the fuel enricher solenoid hoses (arrows)

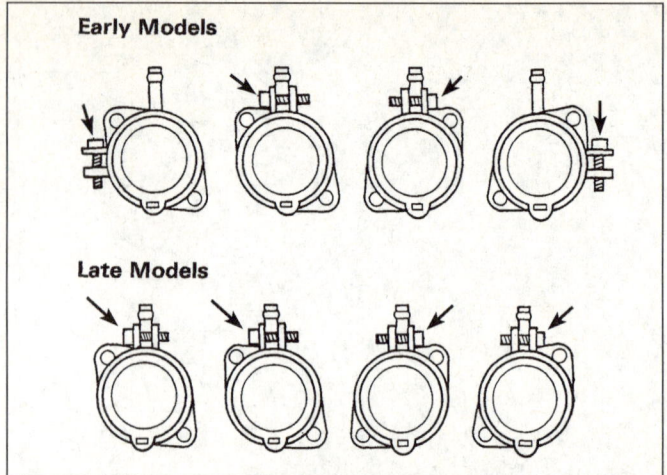

6.19a Intake joint screw positions –
H models

### K and M models

**11** Remove the fuel tank (see Section 2).

**12** Remove the air filter housing (see Section 12).

**13** Disconnect the choke cable from the carburetor assembly (see Section 11).

**14** Loosen the throttle cables all the way and disconnect them from the throttle grip (see Section 10).

**15** Mark and disconnect the fuel enricher solenoid hoses from the carburetors **(see illustration)**.

**16** Disconnect the fuel line from the carburetor assembly.

**17** Pull the carburetor assembly to the rear, clear of the intake manifold tubes. Raise the assembly up far enough to disconnect the throttle cables from the throttle pulley, then remove the carburetors from the machine.

**18** After the carburetors have been removed, stuff clean rags into the intake manifold tubes to prevent the entry of dirt or other objects.

### Installation

**19** Make sure the clamps on the intake manifolds are positioned correctly **(see illustrations)**.

**20** Position the assembly over the intake manifold tubes. Lightly lubricate the ends of the throttle cables with multi-purpose grease

and attach them to the throttle pulley. Make sure the accelerator and decelerator cables are in their proper positions.

**21** Tilt the front of the assembly down and insert the fronts of the carburetors into the intake manifold tubes. Push the assembly forward and tighten the clamps.

**22** Install the air filter housing (see Section 12).

**23** Make sure the ducts from the air filter housing are seated properly, then slide the spring bands into position.

**24** Connect the choke cable to the assembly and adjust it (see Section 11).

**25** Adjust the throttle grip freeplay (see Chapter 1).

**26** The remainder of installation is the reverse of the removal steps.

---

**7  Carburetors –**
disassembly, cleaning
and inspection

⚠️ *Warning: Gasoline (petrol) is extremely flammable, so take extra precautions when you work on any part of the fuel system. Don't smoke or allow open flames or bare light bulbs near the work area, and don't work in a garage where a natural gas-type appliance (such as a water heater or clothes dryer) is*

*present. If you spill any fuel on your skin, rinse it off immediately with soap and water. When you perform any kind of work on the fuel system, wear safety glasses and have a fire extinguisher suitable for class B fires (flammable liquids) on hand.*

### H, J and L models

#### Disassembly

**1** Remove the carburetors from the machine as described in Section 6. Set the assembly on a clean working surface. **Note:** *Unless the O-rings on the fuel and vent fittings between the carburetors are leaking, don't detach the carburetors from their mounting brackets. Also, work on one carburetor at a time to avoid getting parts mixed up.*

**2** If the carburetors must be separated from each other, remove the choke lever spring and choke lever by removing the three screws and six plastic washers (two washers per screw, one on each side of the lever) **(see illustrations)**, then remove the screws securing the mounting plate to the carburetors **(see illustrations)**. Mark the position of each carburetor and gently separate them, noting how the throttle linkage is connected **(see illustration)** and being careful not to lose any springs or fuel and vent fittings that are present between the carburetors **(see illustration)**.

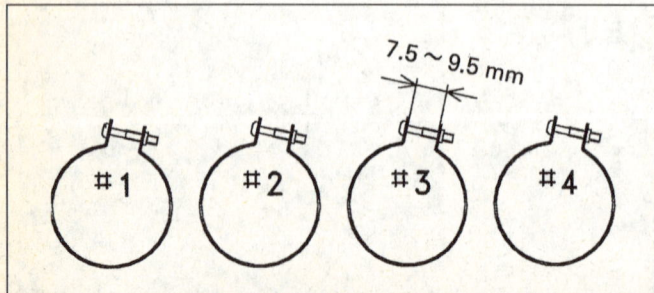

6.19b Intake joint screw positions –
J and L models

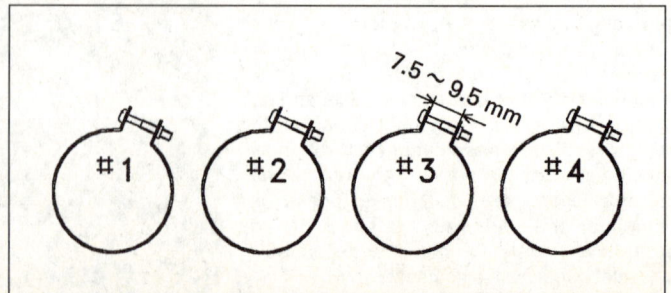

6.19c Intake joint screw positions –
K and M models

7.2a  Remove the choke shaft spring (arrow) . . .

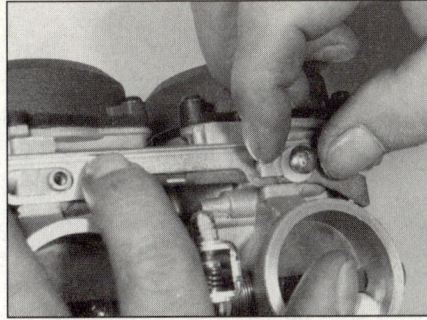

7.2b  . . . and the three choke shaft screws . . .

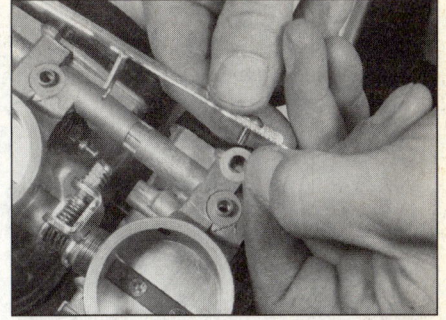

7.2c  . . . there are two plastic washers for each screw, one on each side of the choke shaft . . .

7.2d  . . . separate the choke shaft from the carburetors

7.2e  If the air cleaner intake boots must be separated from the carburetors, remove their screws

7.2f  The synchronizing screws and springs (arrows) should look like this

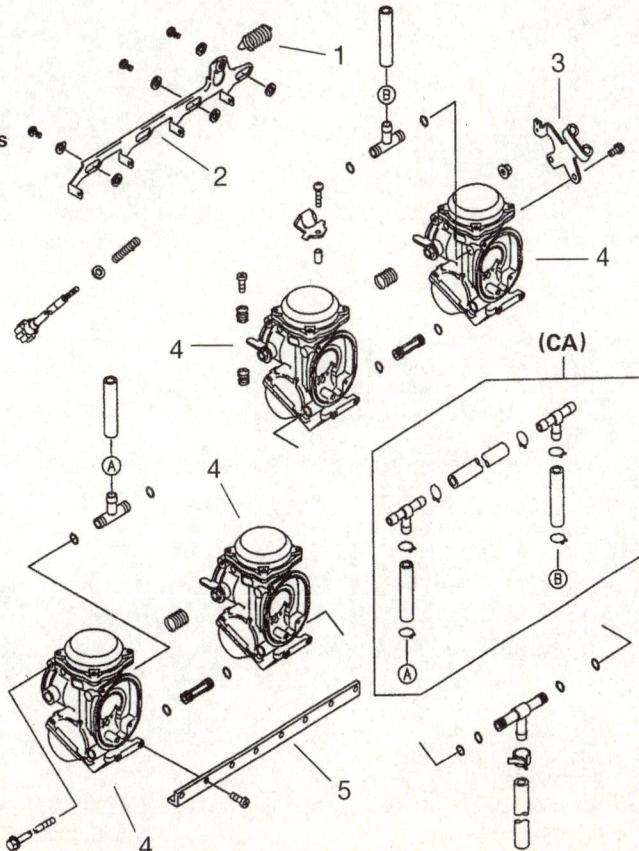

7.2g  Carburetor assembly details (H, J and L models)

1  Choke spring
2  Choke shaft
3  Throttle cable bracket
4  Carburetors
5  Mounting plate

(CA)

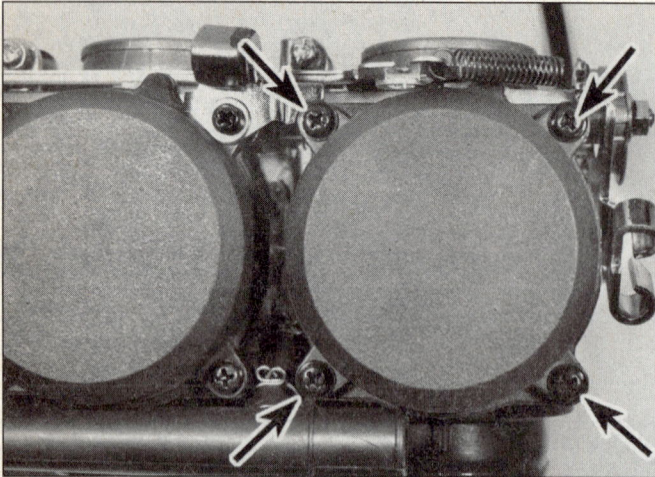

7.3a Remove the vacuum chamber cover screws (arrows) . . .

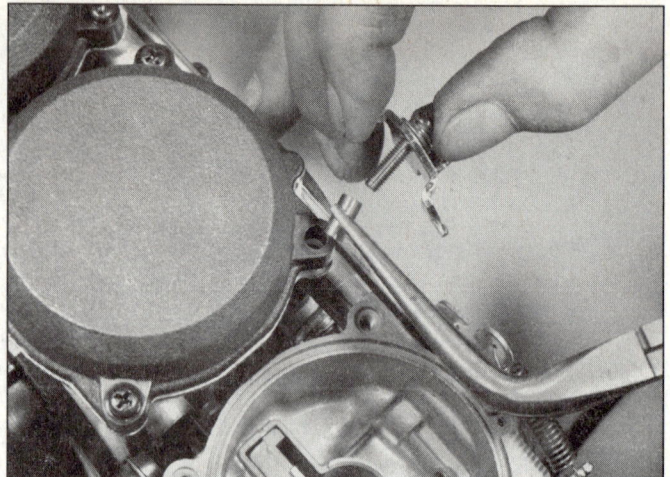

7.3b . . . the screw that secures the choke cable bracket has a dowel pin . . .

7.3c . . . lift off the cover and spring . . .

7.3d . . . and separate the diaphragm from its groove and the locating tab (arrow) from its notch

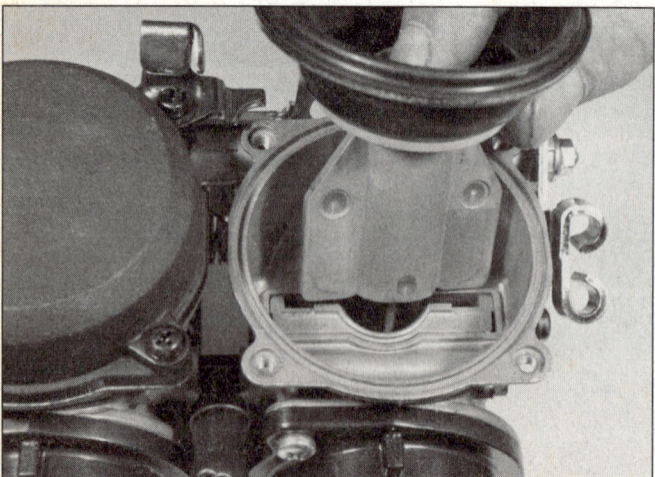

7.3e Lift the throttle piston out

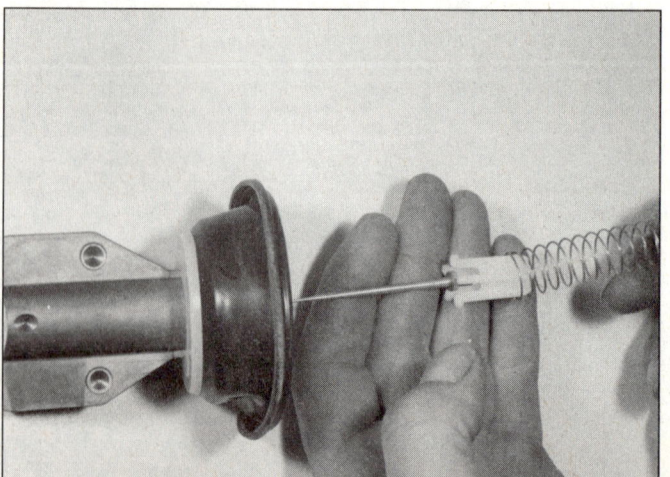

7.4a Separate the jet needle from the throttle piston

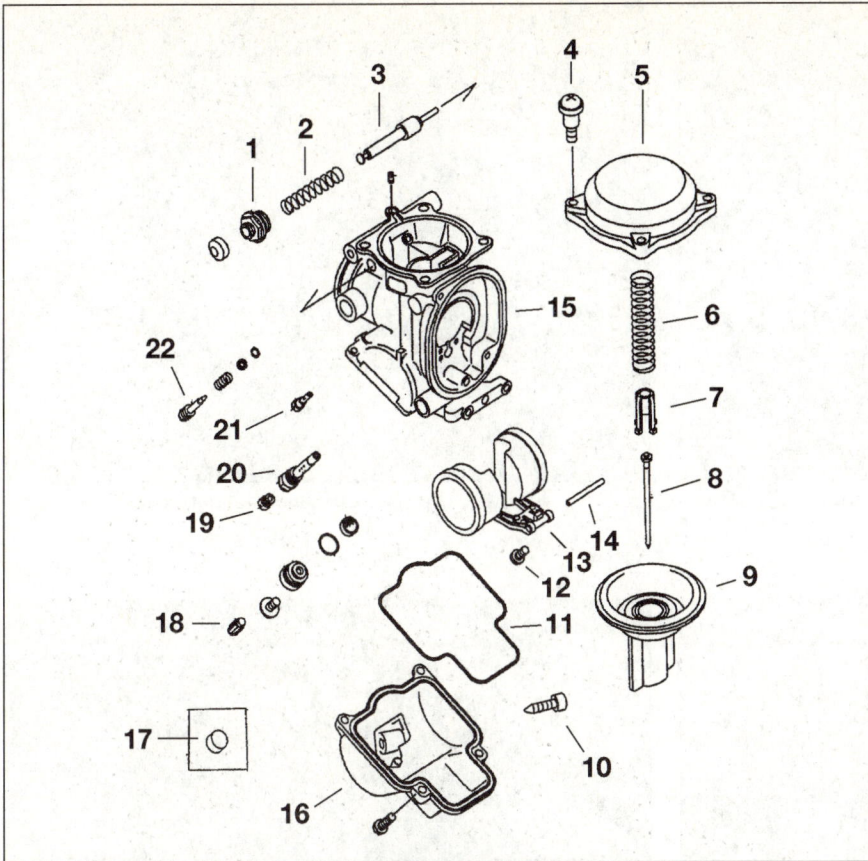
7.4b  Carburetor
(H, J and L models) – exploded view

| | | |
|---|---|---|
| 1  Choke plunger cap | 9  Throttle piston | 16 Float chamber |
| 2  Spring | 10 Float chamber drain | 17 Pilot screw plug |
| 3  Choke plunger | screw | (US models) |
| 4  Screw | 11 Float chamber gasket | 18 Needle valve and seat |
| 5  Vacuum chamber cover | 12 Retaining screw | 19 Main jet |
| 6  Vacuum piston spring | 13 Floats | 20 Needle jet holder |
| 7  Spring seat | 14 Float pivot pin | 21 Pilot jet |
| 8  Jet needle | 15 Carburetor body | 22 Pilot screw |

7.5a  Remove the float chamber cover screws . . .

7.5b  . . . and lift off the cover

7.6a  Loosen the screw shown

3  Remove the four screws securing the vacuum chamber cover to the carburetor body. Lift the cover off and remove the piston spring (see illustrations). Peel the diaphragm away from its groove in the carburetor body, being careful not to tear it (see illustration). Lift out the diaphragm/piston assembly (see illustration).
4  Remove the piston spring seat and separate the needle from the piston (see illustrations).
5  Remove the four screws retaining the float chamber to the carburetor body, then detach the chamber (see illustrations).

6  Push the float pivot pin out and detach the float (and fuel inlet valve needle) from the carburetor body (see illustrations). Detach the valve needle from the float. Remove the retaining screw and remove the needle valve seat (see illustration).

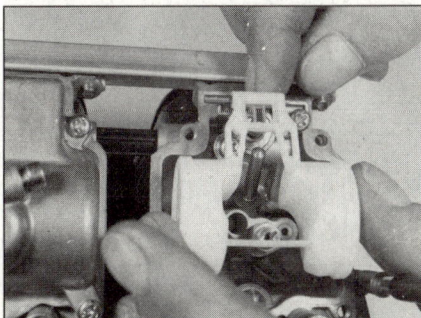
7.6b  . . . then push out the float pivot pin . . .

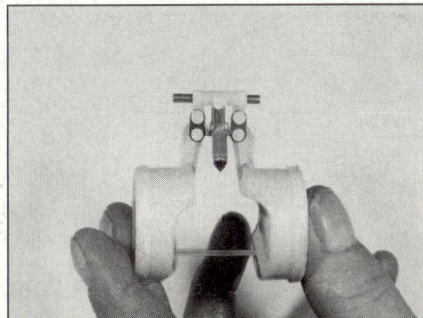
7.6c  . . . and lift out the float, together with the needle valve

7.6d  Remove the retaining screw and lift out the needle valve seat

7.7  Prevent the needle jet holder from turning and unscrew the main jet with a screwdriver

7.8  Unscrew the needle jet holder

7.9a  Unscrew the pilot jet . . .

7.9b  . . . and lift it out

7.10a  The pilot screw on US models is beneath a plug (arrow)

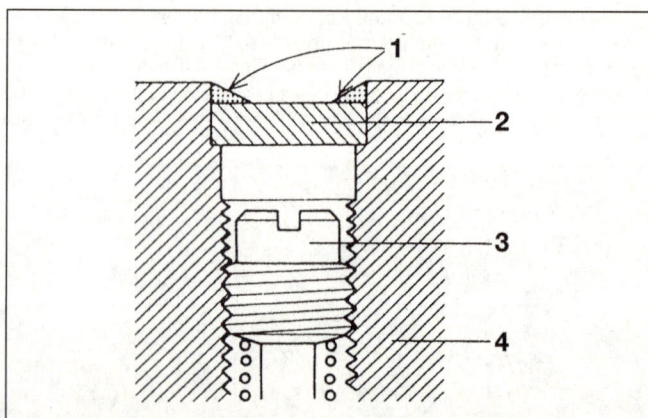

7.10b  On installation, apply bonding agent around the plug

1  Bonding agent     3  Pilot screw
2  Plug     4  Carburetor body

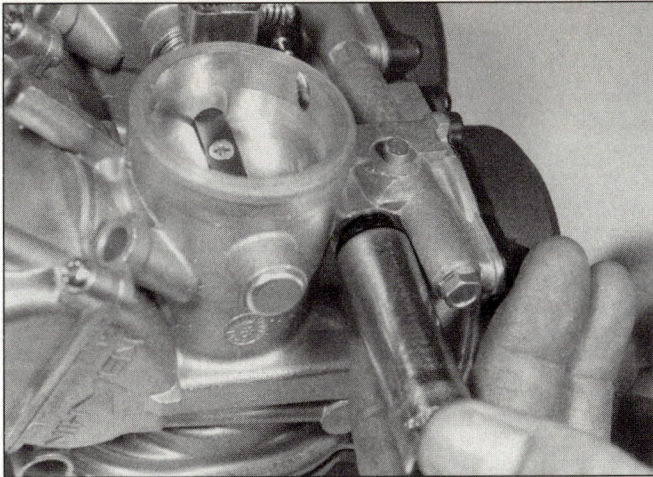

7.11a Unscrew the choke plunger cap with a socket . . .

7.11b . . . and take out the spring and plunger

7 Unscrew the main jet from the needle jet holder (see illustration).

8 Unscrew the needle jet holder/air bleed pipe (see illustration).

9 Using a small, flat-bladed screwdriver, remove the pilot jet (see illustrations).

10 The pilot (idle mixture) screw is located in the bottom of the carburetor body (see illustrations). On US models, this screw is hidden behind a plug which will have to be removed if the screw is to be taken out. To do this, punch a hole in the plug with an awl or a scribe, then pry it out. On all models, turn the pilot screw in, counting the number of turns until it bottoms lightly. Record that number for use when installing the screw. Now remove the pilot screw along with its spring, washer and O-ring.

11 The choke plunger can be removed by unscrewing the nut that retains it to the carburetor body (see illustrations) if the choke shaft has been removed (see Step 2).

## Cleaning

*Caution: Use only a carburetor cleaning solution that is safe for use with plastic parts (be sure to read the label on the container).*

12 Submerge the metal components in the carburetor cleaner for approximately thirty minutes (or longer, if the directions recommend it).

13 After the carburetor has soaked long enough for the cleaner to loosen and dissolve most of the varnish and other deposits, use a brush to remove the stubborn deposits. Rinse it again, then dry it with compressed air. Blow out all of the fuel and air passages in the main and upper body.

*Caution: Never clean the jets or passages with a piece of wire or a drill bit, as they will be enlarged, causing the fuel and air metering rates to be upset.*

## Inspection

14 Check the operation of the choke plunger. If it doesn't move smoothly, replace it, along with the return spring.

15 Check the tapered portion of the pilot screw for wear or damage (see illustration). Replace the pilot screw if necessary.

16 Check the carburetor body, float chamber and vacuum chamber cover for cracks, distorted sealing surfaces and other damage. If any defects are found, replace the faulty component, although replacement of the entire carburetor will probably be necessary (check with your parts supplier for the availability of separate components).

17 Check the jet needle for straightness by rolling it on a flat surface (such as a piece of glass). Replace it if it's bent or if the tip is worn.

18 Check the needle jet and replace the carburetor if it's worn or damaged (see illustration).

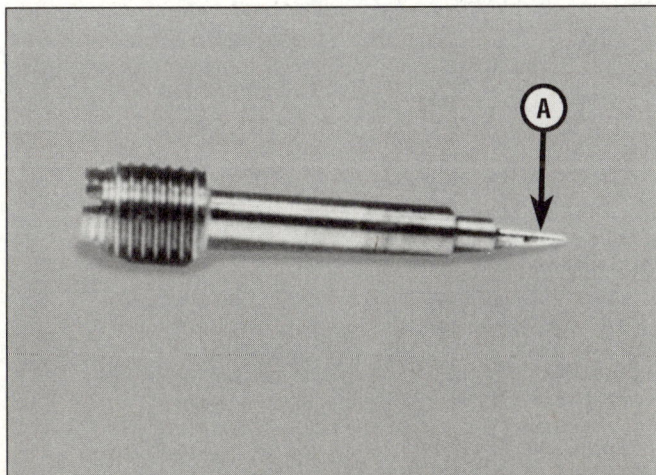

7.15 Check the tapered portion (A) of the pilot screw for wear or damage

7.18 Check the piston insert and the needle jet for wear or damage

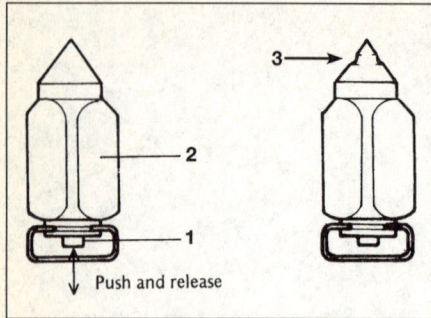

**7.19 Check the tip of the fuel inlet valve needle for grooves or scratches**

1  Rod
2  Valve needle
3  Groove in tip

**7.26a Remove the retaining screw . . .**

**7.26b . . . and lift out the jet needle**

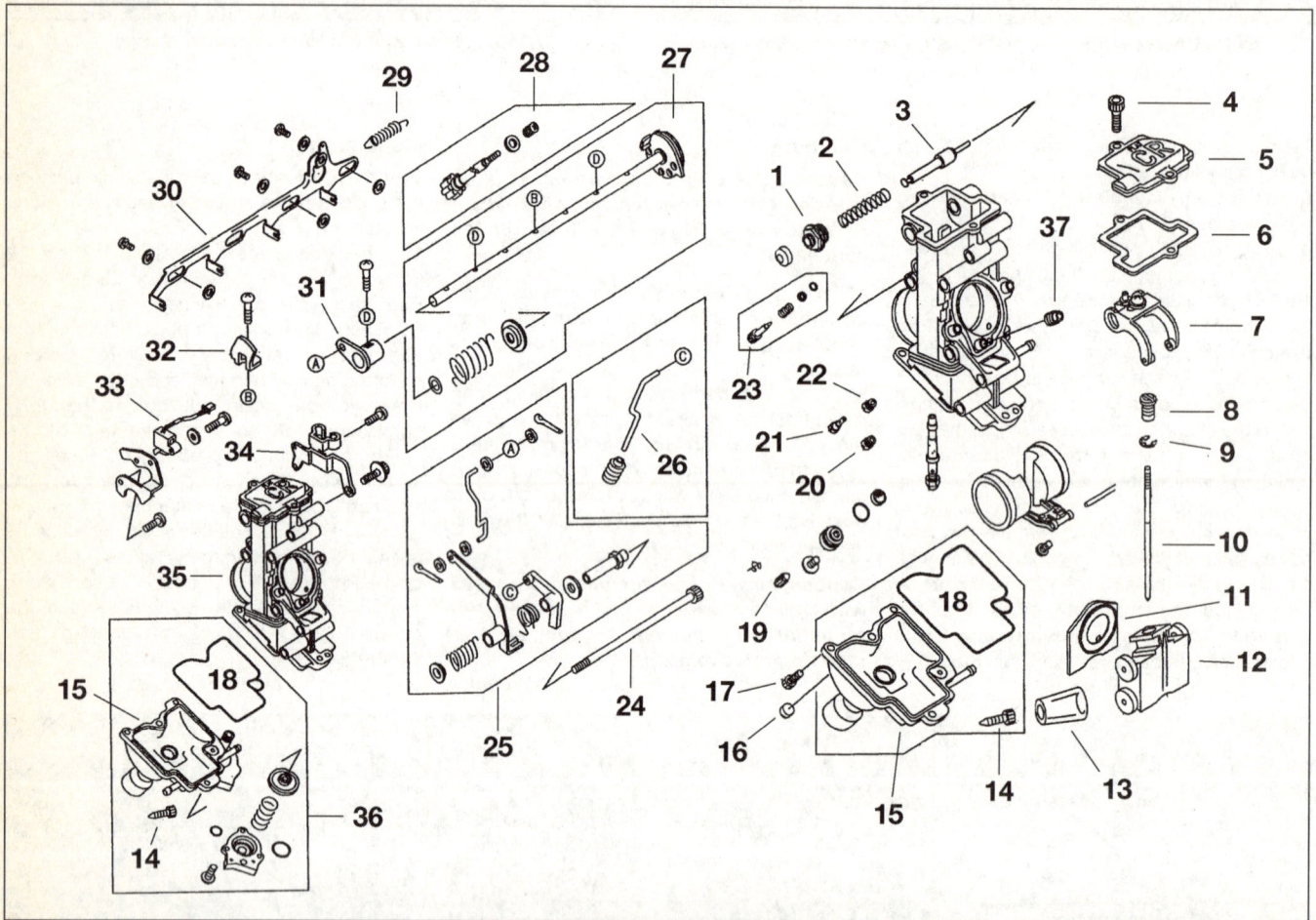

**7.26c Carburetors
(K and M models) – exploded view**

1  Choke plunger cap
2  Spring
3  Choke plunger
4  Cover screw
5  Mixing chamber cover
6  Cover gasket
7  Throttle valve arm
8  Needle screw
9  Jet needle clip
10  Jet needle
11  Throttle valve
12  Throttle valve
13  Drain screw cap
14  Drain screw
15  Float chamber cover
16  Pilot screw plug (US models)
17  Float chamber cover screw
18  Float chamber O-ring
19  Needle valve and seat
20  Main jet
21  Pilot jet
22  Choke (starter) jet
23  Pilot screw
24  Allen bolt
25  Accelerator link assembly
26  Accelerator pump rod
27  Accelerator shaft
28  Throttle stop screw
29  Choke shaft spring
30  Choke shaft
31  Accelerator link lever
32  Switch lever
33  Cruise cancel switch
34  Throttle cable bracket
35  Carburetor body
36  Accelerator pump
37  Air jet

7.27 Accelerator pump cover screws (A) and float chamber cover screws (B); pilot screw located in the recess in the cover (C)

7.32a The assembled carburetors should look like this

**19** Check the tip of the fuel inlet valve needle. If it has grooves or scratches in it, it must be replaced. Push in on the rod in the other end of the needle, then release it – if it doesn't spring back, replace the valve needle **(see illustration)**.

**20** Check the O-rings on the float chamber and the drain plug (in the float chamber). Replace them if they're damaged.

**21** Operate the throttle shaft to make sure the throttle butterfly valve opens and closes smoothly. If it doesn't, replace the carburetor.

**22** Check the floats for damage. This will usually be apparent by the presence of fuel inside one of the floats. If the floats are damaged, they must be replaced.

**23** Check the diaphragm for splits, holes and general deterioration. Holding it up to a light will help to reveal problems of this nature.

**24** Insert the vacuum piston in the carburetor body and see that it moves up-and-down smoothly. Check the surface of the piston for wear. If it's worn excessively or doesn't move smoothly in the bore, replace the carburetor.

### K and M models

#### Disassembly

**25** Refer to the carburetor synchronization procedure in Chapter 1 and remove the carburetor top cover.

**26** Unscrew the jet needle plug and lift out the jet needle **(see illustrations)**.

**27** On the bottom of the carburetor, remove the accelerator pump Allen bolts and take off the accelerator pump components **(see illustration)**.

**28** On the bottom of the carburetor, remove the float chamber cover Allen bolts and take off the float chamber cover **(see illustration 7.27)**.

**29** Remove the float pivot pin, float and needle valve. Remove the retaining screw and remove the needle valve seat.

**30** Unscrew the pilot jet, starter jet, main jet and main nozzle from the underside of the carburetor. Unscrew the air jet from its bore

just below the intake opening **(see illustration 7.26c)**.

**31** Perform Step 10 above to remove the pilot screw.

**32** If it's necessary to separate the carburetors, note carefully how the fuel fittings, vent fittings and throttle springs are installed, then remove the Allen bolts that hold the assembly together **(see illustrations)**. Carefully separate the carburetors from each other.

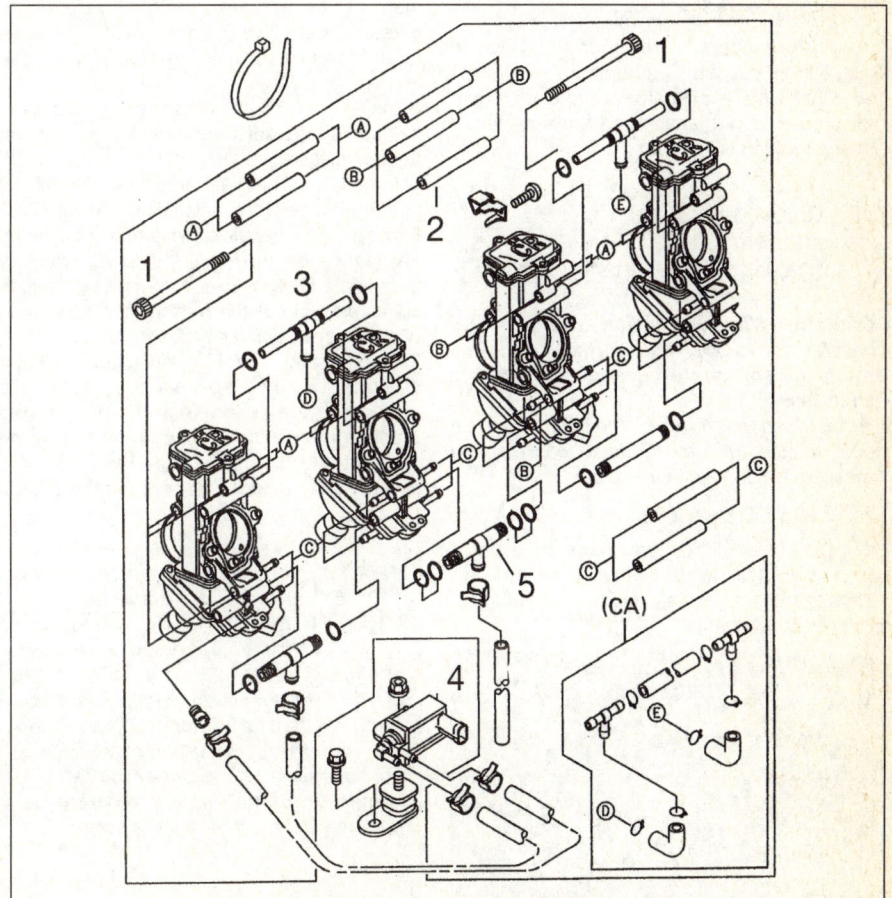

7.32b Carburetor assembly details
(K model shown, M model similar)

| | | |
|---|---|---|
| 1  Allen bolt | 3  Joint fittings | 5  Fuel fittings |
| 2  Collars | 4  Fuel enricher solenoid | |

**7.33a Disconnect the choke shaft spring (arrow) . . .**

**7.33b . . . and remove the three choke shaft screws (arrows)**

**33** To remove the choke shaft, disconnect its return spring and remove its mounting screws and washers **(see illustrations)**.

**34** To remove the choke plungers, refer to Step 11 above.

### Cleaning and inspection

**35** Perform Steps 12 through 22 above to clean and inspect the carburetors.

**36** Check the accelerator pump diaphragms for cracks or brittleness and replace them if their condition is in doubt.

## 8 Carburetors – reassembly and float height adjustment

*Caution: When installing the jets, be careful not to over-tighten them – they're made of soft material and can strip or shear easily.*

*Note: When reassembling the carburetors, be sure to use the new O-rings, gaskets and other parts supplied in the rebuild kit.*

### H, J and L models

**1** If the choke plunger was removed, install it in its bore, followed by its spring and nut. Tighten the nut securely and install the cap.

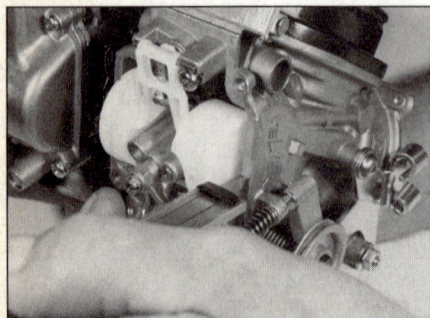

**8.9 Measure float height with a vernier caliper or similar tool**

**2** Install the pilot screw (if removed) along with its spring, washer and O-ring, turning it in until it seats lightly. Now, turn the screw out the number of turns that was previously recorded. If you're working on a US model, install a new metal plug in the hole over the screw. Apply a little bonding agent around the circumference of the plug after it has been seated.

**3** Install the pilot jet, tightening it securely.

**4** Install the needle jet holder/air bleed pipe, tightening it securely.

**5** Install the main jet into the needle jet holder/air bleed pipe, tightening it securely.

**6** Drop the jet needle down into its hole in the vacuum piston and install the spring seat over the needle. Make sure the spring seat doesn't cover the hole at the bottom of the vacuum piston – reposition it if necessary.

**7** Install the diaphragm/vacuum piston assembly into the carburetor body. Lower the spring into the piston. Seat the bead of the diaphragm into the groove in the top of the carburetor body, making sure the diaphragm isn't distorted or kinked **(see illustration 7.3d)**.

> **HAYNES HiNT** *If the diaphragm seems too large in diameter and doesn't want to seat in the groove, place the vacuum chamber cover over the carburetor diaphragm, insert your finger into the throat of the carburetor and push up on the vacuum piston. Push down gently on the vacuum chamber cover – it should drop into place, indicating the diaphragm has seated in its groove.*

**8** Install the vacuum chamber cover, tightening the screws securely. If you're working on the no. 3 carburetor, don't forget to install the dowel and choke cable bracket **(see illustration 7.3b)**.

**9** Invert the carburetor. Attach the fuel inlet valve needle to the float. Set the float into position in the carburetor, making sure the valve needle seats correctly. Install the float pivot pin. To check the float height, hold the carburetor so the float hangs down, then tilt it back until the valve needle is just seated (the rod in the end of the valve shouldn't be compressed). Measure the distance from the float chamber gasket surface to the top of the float **(see illustration)** and compare your measurement to the float height listed in this Chapter's Specifications. If it isn't as specified, carefully bend the tang that contacts the valve needle up or down until the float height is correct.

**10** Install the O-ring into the groove in the float chamber. Place the float chamber on the carburetor and install the screws, tightening them securely.

**11** If the carburetors were separated, install new O-rings on the fuel and vent fittings. Lubricate the O-rings on the fittings with a light film of oil and install them into their respective holes, making sure they seat completely **(see illustration 7.2g)**.

**12** Position the coil springs between the carburetors, gently push the carburetors together, then make sure the throttle linkages are correctly engaged. Check the fuel and vent fittings to make sure they engage properly also.

**13** Install the lower mounting plate and install the screws, but don't tighten them completely yet. Set the carburetors on a sheet of glass, then align them with a straightedge placed along the edges of the bores. When the centerlines of the carburetors are all in horizontal and vertical alignment, tighten the mounting plate screws securely.

**14** Install the choke lever, making sure it engages correctly with all the choke plungers **(see illustration)**. Position a plastic washer on each side of the choke lever, except on no. 2 carburetor **(see illustration)** and install the

**8.14a  Make sure the choke shaft engages all four plungers (arrows)**

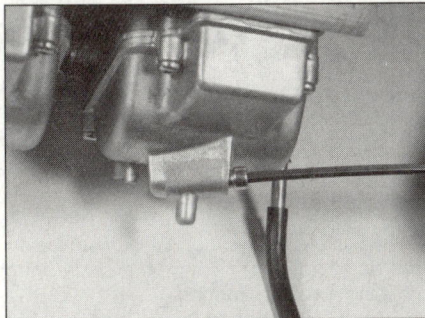

**8.14b  Be sure to reinstall the plastic washers on both sides of the shaft**

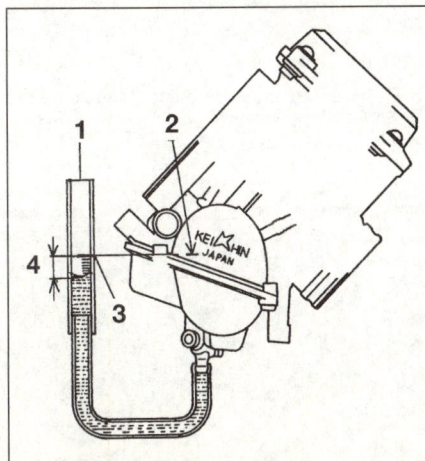

**8.16a  Throttle cable bracket details (L model shown)**

screws, tightening them securely. Install the lever return spring **(see illustration 7.2a)**, then make sure the choke mechanism operates smoothly.
**15** Install the throttle linkage springs **(see illustration 7.2f)**. Visually synchronize the throttle butterfly valves, turning the adjusting screws on the throttle linkage, if necessary, to equalize the clearance between the butterfly valve and throttle bore of each carburetor. Check to ensure the throttle operates smoothly.

**9.2a  Loosen the float chamber drain screw**

**9.2b  On H, J and L models, the fuel level measuring tool is attached like this . . .**

*1  Fuel level gauge
   (part no. 57001-
   1017)*
*2  Mark*
*3  Top line*
*4  Fuel level*

**16** If they were removed, install the throttle stop screw, throttle cable bracket and the air cleaner housing intake fittings **(see illustrations)**.

## K and M models

**17** Assembly is the reverse of the disassembly procedure. Refer to Step 9 above to set float height. Refer to Step 2 above to install the pilot screw.
**18** Install the carburetors (see Section 6), check the fuel level (see Section 9) and synchronize the carburetors (see Chapter 1).

---

### 9  Carburetors – fuel level adjustment

⚠️ *Warning: Gasoline (petrol) is extremely flammable, so take extra precautions when you work*

**9.2c  . . . on K and M models, it's attached like this**

*1  Fuel level gauge
   (part no. 57001-
   1017)*
*2  Mark*
*3  Top line*
*4  Fuel level*

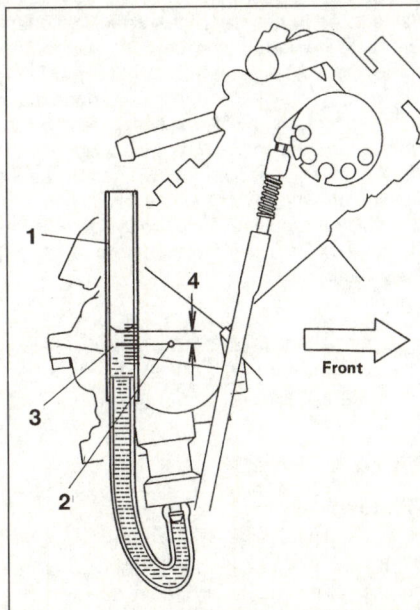

**8.16b  If the air cleaner joints were removed, install them like this**

*on any part of the fuel system. Don't smoke or allow open flames or bare light bulbs near the work area, and don't work in a garage where a natural gas-type appliance (such as a water heater or clothes dryer) is present. If you spill any fuel on your skin, rinse it off immediately with soap and water. When you perform any kind of work on the fuel system, wear safety glasses and have a fire extinguisher suitable for Class B fires (flammable liquids) on hand.*
**1** Remove the fuel tank (see Section 2) and the air filter housing (see Section 12). Connect an auxiliary fuel tank to the carburetors with a suitable length of hose, then support the motorcycle in an upright position.
**2** Attach Kawasaki service tool no. 57001-1017 to the drain fitting on the bottom of one of the carburetor float chambers (all four will be checked) **(see illustrations)**. This is a clear plastic tube graduated in millimeters. An alternative is to use a length of clear plastic tubing and an accurate ruler. Hold the graduated tube (or the free end of the clear plastic tube) against the carburetor body, as shown in the accompanying illustration. If the Kawasaki tool is being used, raise the zero mark to a point several millimeters above the bottom edge of the carburetor main body. If a piece of clear plastic tubing is being used, make a mark on the tubing at a point several millimeters above the bottom edge of the carburetor main body.
**3** Unscrew the drain screw at the bottom of the float chamber a couple of turns, then let

**10.3a Remove the screws from the front half of the throttle housing/switch assembly . . .**

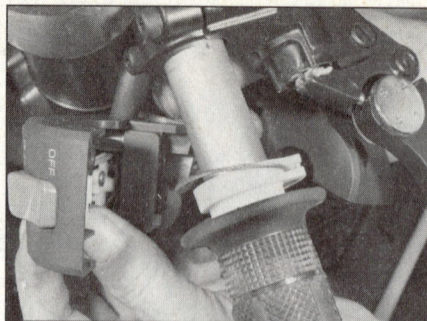

**10.3b . . . and take the rear half off the handlebar**

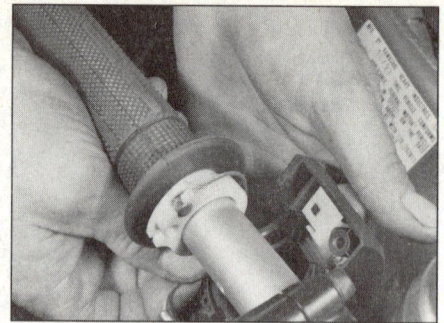

**10.4 Disconnect the cables from the throttle pulley**

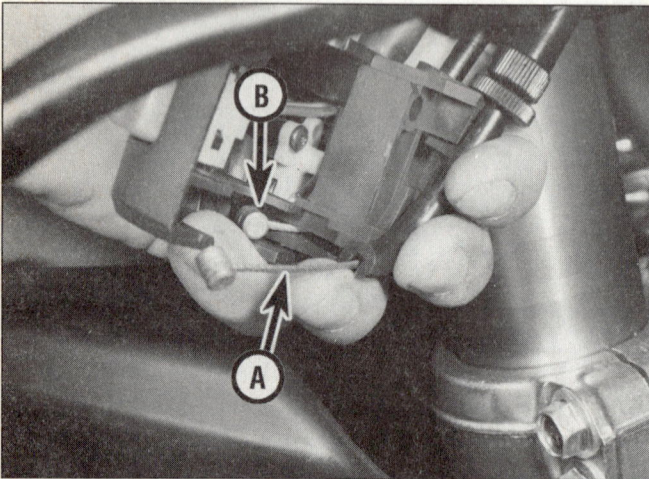

**10.5 Detach the accelerator cable (A) from its guide and the decelerator cable (B) from the housing**

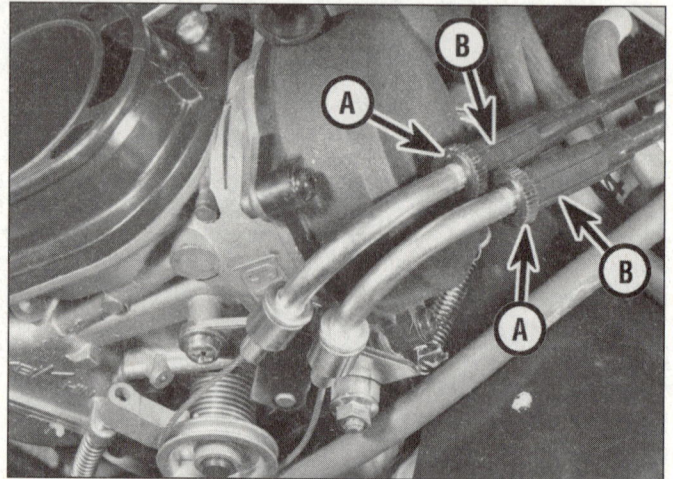

**10.6a Loosen the lockwheels (A) and turn the adjusters (B) to create slack in the cables**

fuel flow into the tube. Wait for the fuel level to stabilize, then slowly lower the tube until the zero mark is level with the fuel level mark on the carburetor body. **Note:** *Don't lower the zero mark below the bottom edge of the carburetor then bring it back up – the reading won't be accurate.*

4 Measure the distance between the mark and top of the fuel level in the tube or gauge. This distance is the fuel level – write it down on a piece of paper, screw in the drain screw, shut off the fuel flow, then move on to the next carburetor and check it the same way.

5 Compare your fuel level readings to the value listed in this Chapter's Specifications. If the fuel level in any carburetor is not correct, remove the float chamber and bend the tang up or down (see Section 8), as necessary, then recheck the fuel level. **Note:** *Bending the tang up increases the float height and lowers the fuel level – bending it down decreases the float height and raises the fuel level.*

6 After the fuel level for each carburetor has been adjusted, install the carburetor assembly (see Section 6).

---

**10 Throttle cables and grip –** removal, installation and adjustment

### Throttle cables

#### Removal

1 Remove the fuel tank (see Section 2).
2 Loosen the accelerator cable lockwheel and screw the cable adjuster in.
3 Remove the cable/switch housing screws **(see illustration)** and remove the front and rear halves of the housing **(see illustration)**.
4 Rotate the ends of the cables to align with the slots in the throttle grip pulley, then detach the cables from the pulley **(see illustration)**.

5 Detach the accelerator cable, its guide and the decelerator cable from the cable/switch housing **(see illustration)**.
6 Loosen the midline adjuster if necessary to create slack in the cables, then detach them from the throttle pulley **(see illustrations)**.
7 Remove the cables, noting how they are routed.

#### Installation

8 Route the cables into place. Make sure they don't interfere with any other components and aren't kinked or bent sharply.

**10.6b Lift the cable housings out of the guides and slide the cables through the slots . . .**

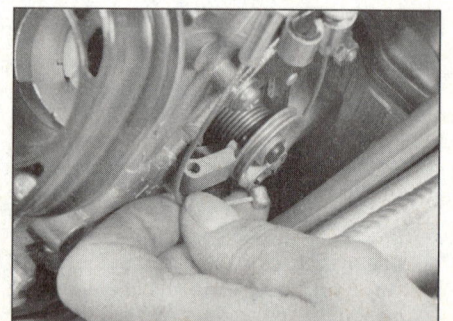

**10.6c . . . then rotate the cables to align them with the pulley slots and disengage them from the pulley**

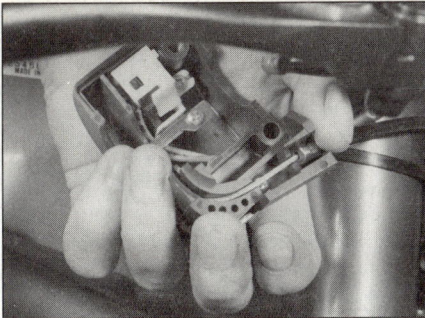

**10.12 Slide the decelerator cable into its slot in the housing**

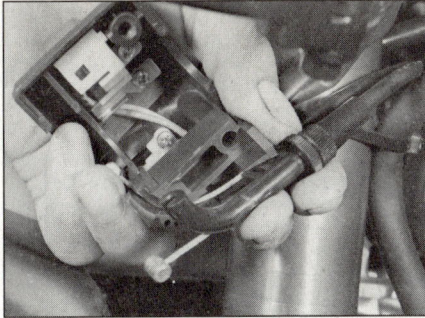

**10.13 Position the accelerator cable guide in the housing**

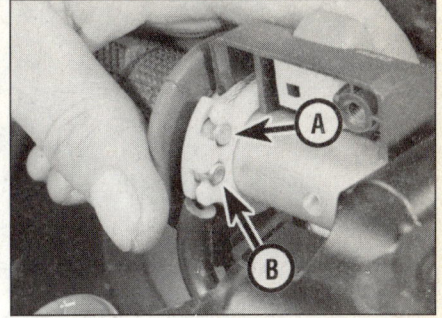

**10.14 Engage the decelerator cable (A) and accelerator cable (B) with the throttle pulley**

9 Lubricate the end of the accelerator cable with multi-purpose grease and connect it to the throttle pulley at the carburetor. Pass the inner cable through the slot in the bracket, then seat the cable housing in the bracket.

10 Repeat the previous step to connect the decelerator cable.

11 Route the decelerator cable around the backside of the handlebar and connect it to the rear hole in the throttle grip pulley.

12 Seat the decelerator cable in the throttle housing groove **(see illustration)**.

13 Push the accelerator cable guide into place, making sure the notched portion is correctly engaged with the housing **(see illustration)**.

14 From the front side of the handlebar, connect the accelerator cable to the forward hole in the throttle grip pulley. Connect the decelerator cable to the rearward hole **(see illustration)**.

15 Install the front half of the cable/switch housing, making sure the locating pin engages with the hole in the handlebar. If necessary, rotate the housing back and forth, until the locating pin drops into the hole and the housing halves mate together. Install the screws and tighten them securely.

**Adjustment**

16 Follow the procedure outlined in Chapter 1, *Throttle operation/grip freeplay – check and adjustment*, to adjust the cables.

17 Turn the handlebars back and forth to make sure the cables don't cause the steering to bind.

18 Install the fuel tank.

19 With the engine idling, turn the handlebars through their full travel (full left lock to full right lock) and note whether idle speed increases. If it does, the cables are routed incorrectly. Correct this dangerous condition before riding the bike.

### Throttle grip

#### Removal

20 Follow Steps 2 through 4 to detach the upper ends of the throttle cables from the throttle grip pulley.

21 Slide the throttle grip off the handlebar.

#### Installation

22 Clean the handlebar and apply a light coat of multi-purpose grease.

23 Push the throttle grip on.

24 Attach the cables following Steps 11 through 15, then adjust the cables following the procedure outlined in Chapter 1, *Throttle operation/grip freeplay – check and adjustment*.

---

**11 Choke cable –** removal, installation and adjustment

### Removal

1 Remove the seat and fuel tank (see Section 2).

2 Pull the choke cable casing away from its mounting bracket at the carburetor assembly, then pass the inner cable through the opening

in the bracket **(see illustration)**. Detach the cable end from the choke lever by the no. 4 carburetor.

3 Remove the two screws securing the choke cable/switch housing halves to the left handlebar **(see illustration)**. Pull the front half of the housing off and separate the choke cable from the lever **(see illustration)**.

4 Remove the cable, noting how it's routed.

### Installation

5 Route the cable into position. Connect the upper end of the cable to the choke lever. Make sure the cable guide seats properly in the housing. Place the housing up against the handlebar, making sure the pin in the housing fits into the hole in the handlebar. Install the screws, tightening them securely.

6 Connect the lower end of the cable to the choke lever. Pull back on the cable casing and connect it to the bracket on the no. 3 carburetor.

### Adjustment

7 Check the freeplay at the choke plunger lever on the carburetor assembly. It should move about two to three millimeters (1/8-inch).

8 If the freeplay isn't as specified, follow the cable to its midline adjuster (it resembles the throttle cable adjuster). Loosen the cable adjusting locknut and turn the adjusting nut in or out, as necessary, until the freeplay at the lever is correct. Tighten the locknut.

9 Install the fuel tank and all of the other components that were previously removed.

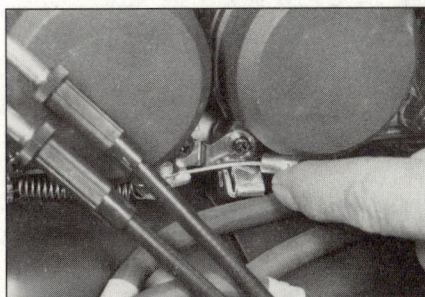

**11.2 Pull the cable free of the bracket and slide the cable through the slot in the bracket**

**11.3a Remove the choke cable/switch housing screws . . .**

**11.3b . . . separate the housing halves and disengage the cable from the lever**

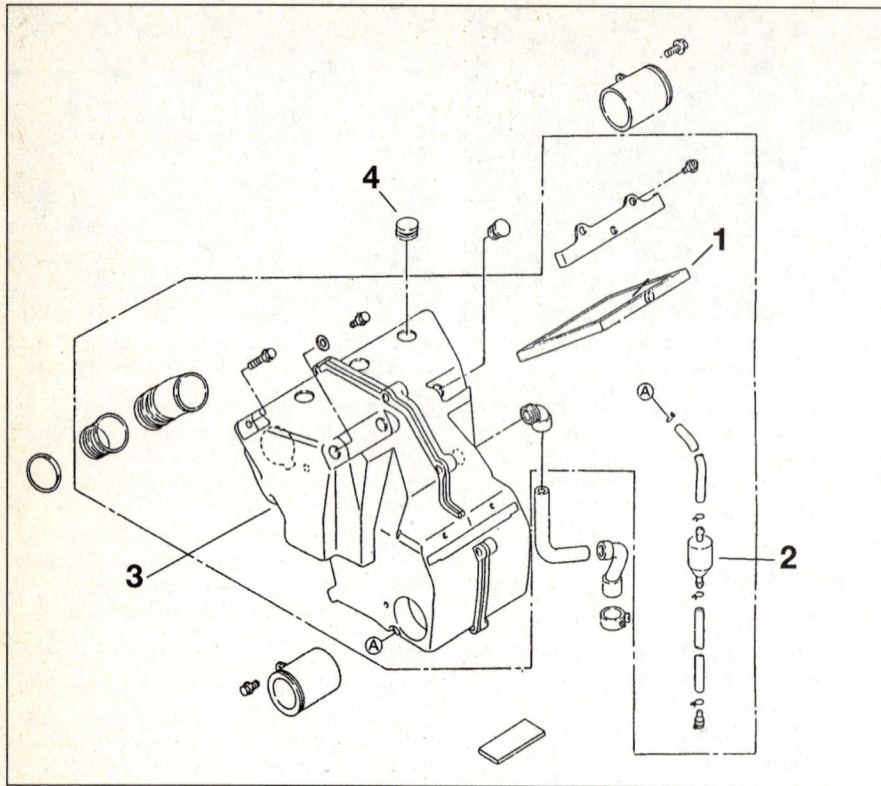

**12.2  Air filter housing details
(H models)**

| | | | |
|---|---|---|---|
| 1  Filter element | 2  Catch tank | 3  Air filter housing | 4  Plug |

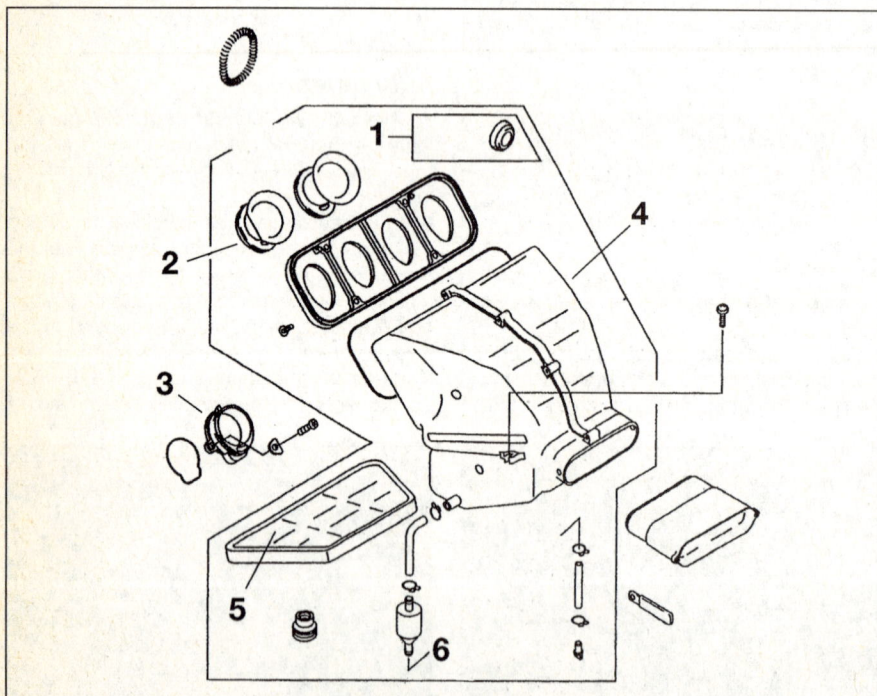

**12.7  Air filter housing details
(J and K models)**

| | | |
|---|---|---|
| 1  Plug (UK models) | 3  Carburetor intake | 4  Air filter housing | 6  Catch tank |
| 2  Intake fittings | manifold joints | 5  Filter element | |

## 12 Air filter housing –
### removal and installation

### H models

**1** Remove the fuel tank (see Section 2).
**2** Remove the inlet silencers on both sides of the air filter housing **(see illustration)**.
**3** Remove the plugs that cover the air filter housing bolts, then undo the bolts.
**4** Disconnect the catch tank drain hose and the crankcase breather hose, then lift the air filter housing off the motorcycle.
**5** Installation is the reverse of removal.

### J and K models

**6** Remove the fuel tank (see Section 2).
**7** Disconnect the catch tank drain hose **(see illustration)**.
**8** Remove the mounting screws (one at each rear corner of the housing) and lift the air filter housing off the motorcycle.
**9** Installation is the reverse of removal.

### L and M models

**10** Remove the side covers (see Chapter 8).
**11** Remove the fuel tank (see Section 2).
**12** Remove the mounting bolt at the front of the air filter housing **(see illustrations)**.
**13** Disconnect the hoses from the left side, the right side and the rear of the air box **(see illustrations)**.
**14** Lift the air cleaner filter off the motorcycle.
**15** Installation is the reverse of the removal steps. Apply a thin coat of engine oil to the air intake duct **(see illustration)** and the air intake fittings on the carburetors to ease installation.

## 13 Exhaust system –
### removal and installation

**1** Remove the lower fairing panels (see Chapter 8).
**2** Drain the coolant (see Chapter 1).

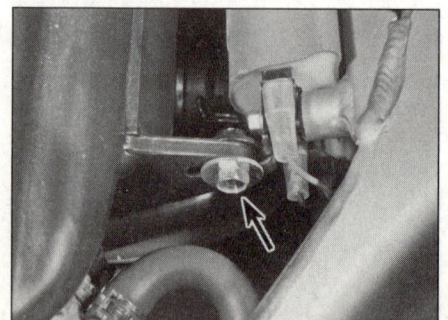

**12.12a  Remove the mounting bolt and washer (arrow)**

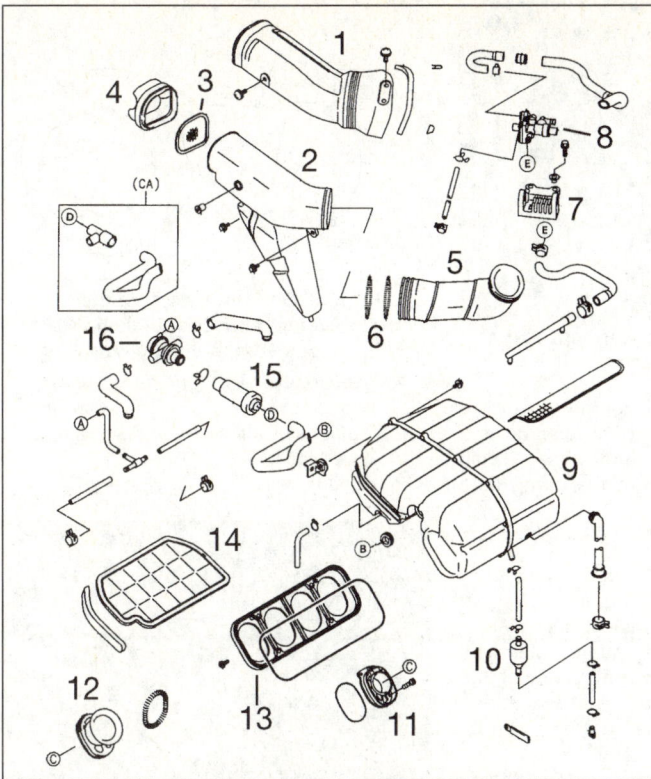

**12.12b  Air filter housing details
(L and M models)**

1   Air intake duct cover
2   Center air intake duct
3   Air intake screen
4   Front air intake duct
5   Rear air intake duct
6   Spring bands
7   Valve damper
8   Canister valve
    (California models)
9   Air filter housing
10  Catch tank
11  Intake fitting
12  Carburetor intake
    manifold joint
13  Air cleaner element
    holder
14  Air cleaner element
15  Intake silencer
16  Air switching valve
    (US models)

**12.13a  Disconnect the hoses from the left side (arrows) . . .**

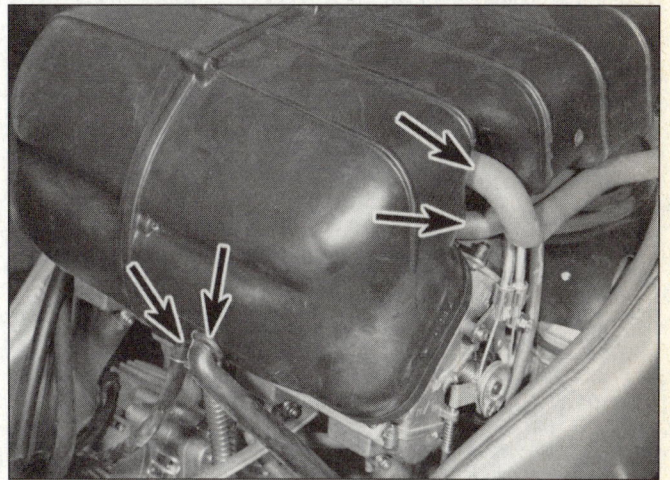

**12.13b  . . . and from the rear and right side (arrows)**

**3**  Remove the radiator (see Chapter 3).

**4**  Remove the oil cooler (see Chapter 3). If you're working on an H2 model, remove the oil cooler-to-radiator mounting bracket.

**5**  Remove the exhaust pipe holder nuts and washers and slide the holders off the mounting studs **(see illustrations)**. If the split keepers didn't come off with the holders, remove them from the pipes.

**6**  Remove the muffler mounting bolt at the footpeg bracket **(see illustration)**.

**7**  Pull the exhaust system forward, separate the pipes from the cylinder head and remove the system from the machine **(see illustration)**.

**12.15  Apply a thin coat of engine oil to the rim of the air intake duct**

**13.5a  Remove the mounting nuts and washers (arrows), noting how the holder halves fit together . . .**

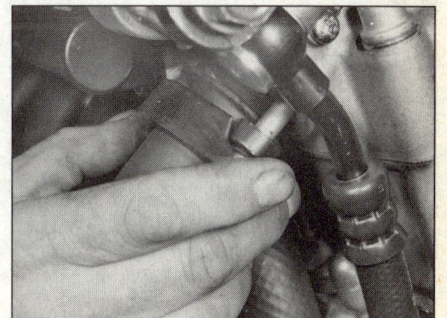

**13.5b  . . . and slide the holders off the studs**

**13.6 Unbolt the muffler/silencer**

**13.7 Pull the pipes forward out of the cylinder head**

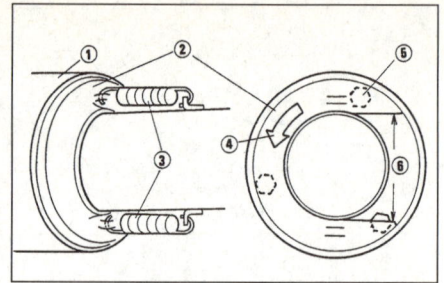
**13.8 On H models, turn the muffler/silencer cover anti-clockwise to expose the mounting nuts**

1  Muffler/silencer body
2  Muffler/silencer cover
3  Springs
4  Turn counter-clockwise (anti-clockwise)
5  Nuts
6  Opening

**8** To remove the muffler/silencer body on H models, remove the springs, turn the muffler cover anti-clockwise to remove it, then remove the muffler/silencer mounting nuts **(see illustration)**.
**9** To remove the muffler/silencer body on J, K, L and M models, remove the Allen bolts and take off the muffler/silencer body cover. Remove the nuts and separate the muffler/silencer body from the pipe.
**10** Installation is the reverse of removal, but be sure to install new gaskets at the muffler/silencer body (if it was removed) and at the cylinder head **(see illustration)**.

## 14 Fuel pump –
removal, testing and installation

### Removal

#### H models

**1** Remove the right side cover (see Chapter 8).
**2** Remove the air filter housing (See Section 12).

**3** Disconnect the fuel pump electrical connector. Squeeze the hose clamps and slide them along the fuel hoses, then disconnect the hoses from the pump **(see illustration)**.
**4** Slide the rubber mount off the bracket and lift the pump out.

#### J, K, L and M models

**5** Remove the side covers (see Chapter 8).
**6** Disconnect the fuel pump electrical connector **(see illustration)**. Locate the connector for the rear brake light switch and disconnect it as well, then unbolt the right footpeg bracket (see Chapter 8).
**7** Squeeze the hose clamps and slide them along the fuel hoses, then disconnect the hoses from the pump **(see illustration)**. Slide the rubber mount off the bracket and lift the pump out **(see illustration)**.

### Testing

#### Fuel pump

**8** Remove the fuel pump as described above. Connect a pressure gauge to the fuel pump and set it up so it pumps kerosene (paraffin) from a container and returns it to the same container **(see illustration)**.

**13.10 Use new gaskets at the cylinder head**

⚠ *Warning: Kerosene (paraffin) is used for this test because an open container of gasoline (petrol) is an extreme fire hazard.*
**9** Connect the fuel pump to the battery with two lengths of wire. The pump should run. If it doesn't, it's defective; replace it.

**14.3 Fuel pump details (H models)**

A  Front of motorcycle
B  IN mark
C  Inlet hose from fuel tank
D  Fuel filter arrow
E  Outlet hose to carburetors

**14.6 Disconnect the fuel pump connector (L model shown)**

14.7a Disconnect the hoses and slide the rubber mount off the bracket (L model shown)

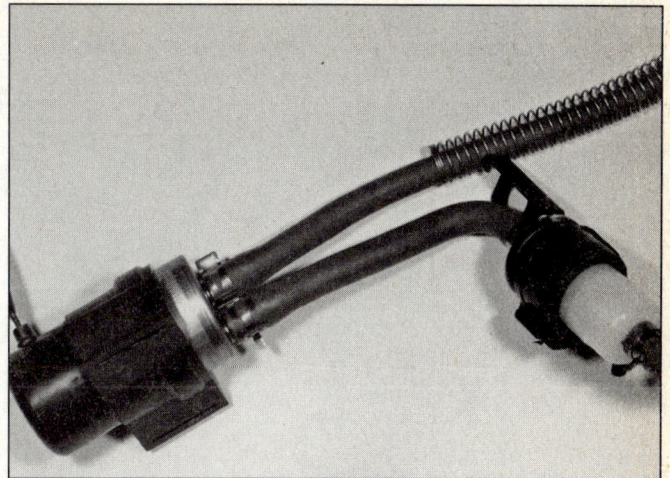

14.7b Fuel pump and filter details (L models)

**10** If the pump runs, the fuel pump relay is probably defective. Refer to Steps 12 and 13 below for relay testing procedures.

**11** With the pump running, pinch off the outlet hose and note the reading on the pressure gauge. Compare it to the value listed in this Chapter's Specifications. If it's not within the specified range, replace the fuel pump.

### Relay

**Note:** *Testing the relay requires a special Kawasaki tester (part no. 57001-983). Other ohmmeters may produce incorrect resistance readings. Ohmmeters with high-capacity*

*batteries will damage the relay. If you don't have the correct tester, have the relay tested by a Kawasaki dealer.*

**12** Locate the fuel pump relay. On H models, it's behind the flasher relay; remove the left side cover and seats for access (see Chapter 8). On J, K, L and M models, it's next to the rear brake fluid reservoir. The relay can be identified by its wire colors; red, black, black/blue and black/yellow.

**13** Disconnect the relay's electrical connector. Connect the tester to the relay terminals and note the resistance readings **(see illustration)**. If they're not as shown, replace the relay.

### Installation

**14** Installation is the reverse of the removal steps. Make sure the pump inlet and outlet hoses are connected to the proper fittings on the pump.

---

### 15 Evaporative emission control system – removal, inspection and installation

**Note:** *This procedure applies to California models only.*

**1** Remove the side cover(s) and seats as necessary for access to the system components (see Chapter 8).

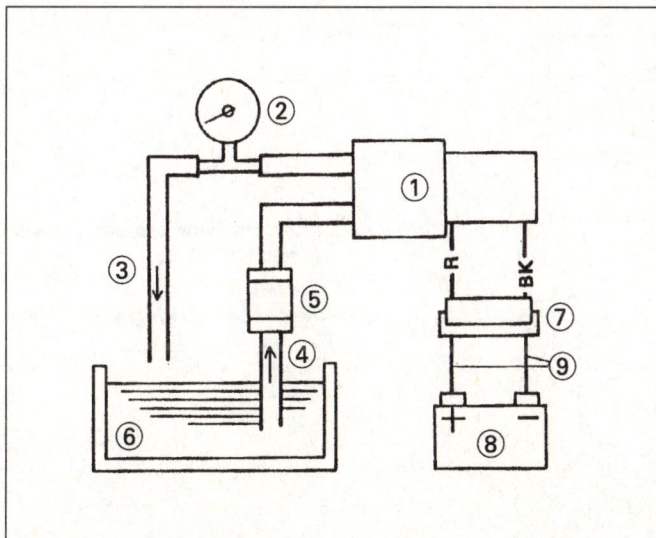

14.8 Fuel pump test setup

| 1 | Fuel pump | 6 | Kerosene (paraffin) |
| 2 | Pressure gauge | 7 | Two-pin connector |
| 3 | Outlet hose | 8 | Battery |
| 4 | Inlet hose | 9 | Lengths of wire |
| 5 | Fuel filter | | |

| Range | Tester (+) Lead Connection | | | |
|---|---|---|---|---|
| x 1 kΩ | 1 | 2 | 3 | 4 |
| † 1 | – | ∞ | ∞ | ∞ |
| 2 | ∞ | – | ∞ | ∞ |
| 3 | ∞ | 10 ~ 100 | – | ∞ |
| 4 | ∞ | 20 ~ 200 | 1 ~ 5 | – |

(†) : Tester (−) Lead Connection

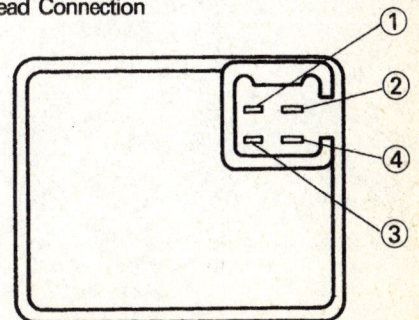

14.13 Fuel pump relay test

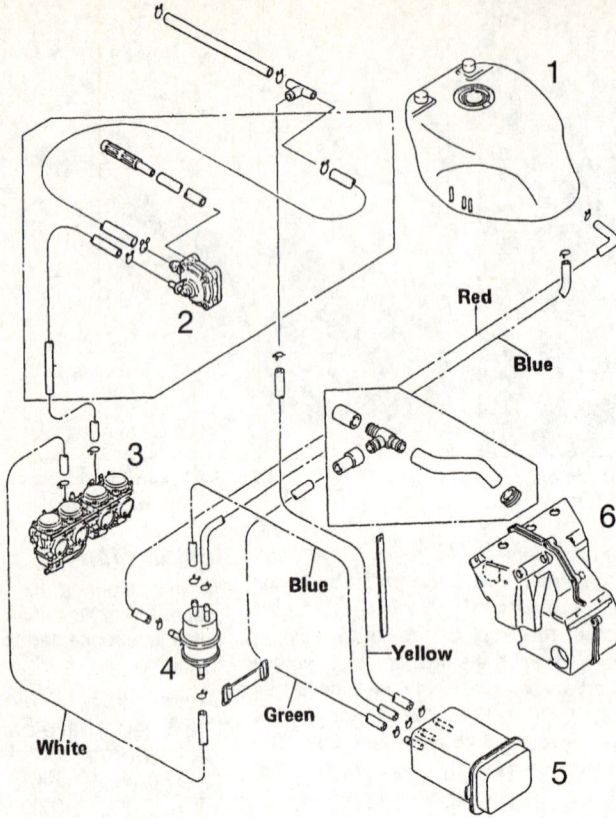

**15.2a Evaporative emission system details (H models)**

1 Fuel tank
2 Vacuum switching valve
3 Carburetor assembly
4 Separator
5 Canister
6 Air filter housing

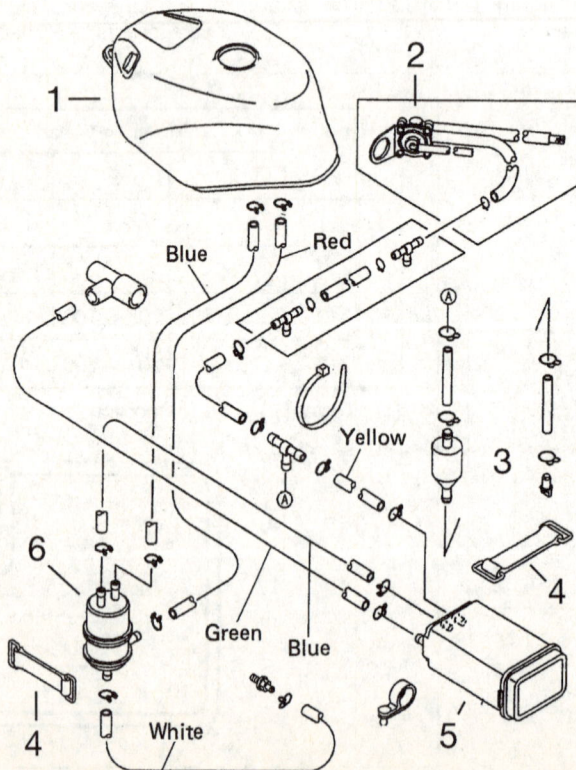

Red

Blue

Blue

Yellow

Green

White

**15.2b Evaporative emission system details (J and K models)**

1 Fuel tank
2 Vacuum switching valve
3 Catch tank
4 Retaining bands
5 Canister
6 Separator

Blue

Red

Yellow

Green

Blue

White

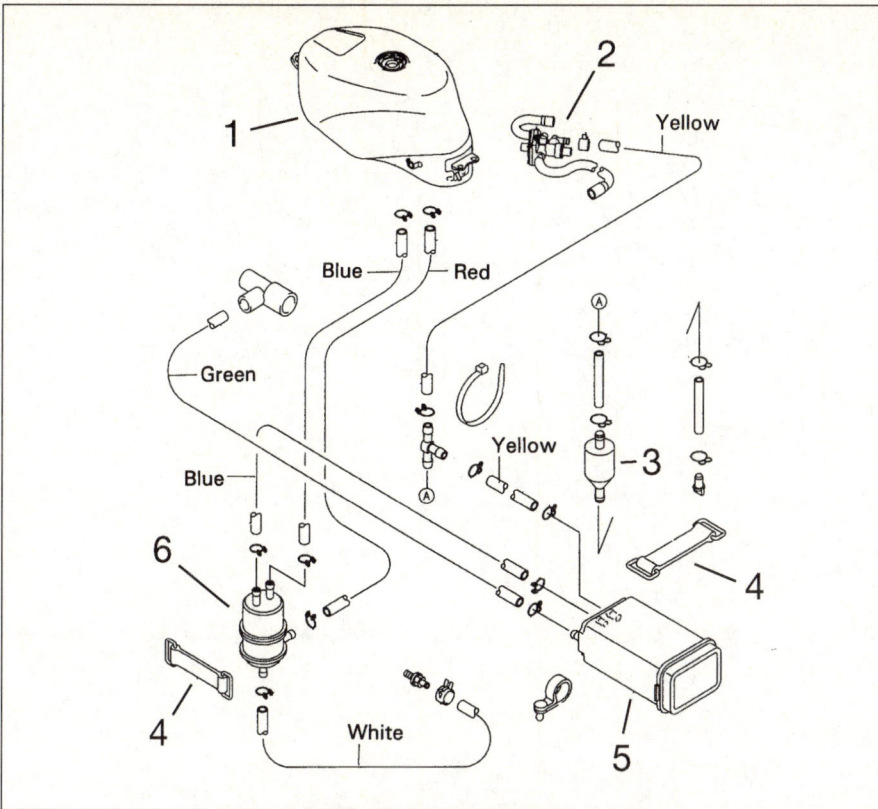

**15.2c Evaporative emission system details
(L and M models)**

| | | | | | |
|---|---|---|---|---|---|
| 1 | Fuel tank | 3 | Catch tank | 5 | Canister |
| 2 | Vacuum switching valve | 4 | Retaining bands | 6 | Separator |

2 Inspect the system hoses **(see illustrations)**. Replace any that are cracked, brittle or deteriorated.

3 Check the separator for obvious damage, such as cracks or broken fittings **(see illustration)**. If these problems are seen, replace the separator.

4 Disconnect the breather hose from the separator. Pour about 20 cc of gasoline (petrol) into the hose fitting on the separator, using an eye dropper or similar tool.

5 Disconnect the fuel return hose from the fuel tank fitting and place the disconnected hose end in a container level with the top of the fuel tank. Start the engine and let it idle. The fuel poured into the separator should come out of the hose into the container. If not, replace the separator.

6 Check the canister for obvious damage, such as cracks or broken hose fittings **(see illustration)**. Replace the canister if these are found. The canister is designed to last the life of the motorcycle, so it shouldn't fail from normal use.

7 Checking the vacuum valve requires special equipment and should be done by a Kawasaki dealer. To remove the valve, label and disconnect its hoses, remove its mounting bolts and lift it out. Installation is the reverse of the removal steps.

## 16 Air suction system – removal, inspection and installation

1 The air suction system is used on US and some European models to reduce exhaust emissions. Low-pressure pulses in the exhaust gas flow are used to draw air into the exhaust ports, where it combines with hot exhaust gas leaving the ports and reduces the amount of unburned hydrocarbons and carbon monoxide. Air is allowed to flow through the vacuum switching valve to the reed valves on the valve cover during cruising. When the throttle is closed, high vacuum closes the vacuum switching valve and cuts off air flow to prevent backfiring.

2 Reed valve inspection is covered in Chapter 1. Inspection of the vacuum switching valve requires special equipment and should be done by a Kawasaki dealer. If the engine backfires, the valve may be the cause.

3 To remove the vacuum switching valve, remove the fuel tank (see Section 2). Disconnect the vacuum hose and air hoses from the valve and lift it out **(see illustration)**. Installation is the reverse of the removal steps.

**15.3 Disconnect the hoses, remove the retaining band and lift the separator out (L model shown)**

**15.6 Check for colored dots on the canister that match the color codes on the hoses**

**16.3 Disconnect the hoses, remove the mounting bolts (arrows) and lift the vacuum switching valve out**

# Chapter 5
## Ignition system

## Contents

## Degrees of difficulty

| Easy, suitable for novice with little experience | Fairly easy, suitable for beginner with some experience | Fairly difficult, suitable for competent DIY mechanic | Difficult, suitable for experienced DIY mechanic | Very difficult, suitable for expert DIY or professional |
|---|---|---|---|---|

## Specifications

### Spark plug
Type and gap . . . . . . . . . . . . . . . . . . . . . . . . . . . . . . . . . . . . . . . see Chapter 1
Cap resistance
    H models . . . . . . . . . . . . . . . . . . . . . . . . . . . . . . . . . . . . . . . . Not specified
    J, K, L and M models . . . . . . . . . . . . . . . . . . . . . . . . . . . . . . . 3.75 to 6.25 K-ohms

### Ignition coil
Primary resistance . . . . . . . . . . . . . . . . . . . . . . . . . . . . . . . . . . . 1.8 to 2.8 ohms
Secondary resistance . . . . . . . . . . . . . . . . . . . . . . . . . . . . . . . . . 10 to 16 K-ohms
Arcing distance . . . . . . . . . . . . . . . . . . . . . . . . . . . . . . . . . . . . . . ¼ in (7 mm) or more

### Pick-up coil
Pick-up coil resistance
    H models . . . . . . . . . . . . . . . . . . . . . . . . . . . . . . . . . . . . . . . . 360 to 440 ohms
    J, K, L and M models . . . . . . . . . . . . . . . . . . . . . . . . . . . . . . . 375 to 565 ohms
Pick-up coil air gap (H models) . . . . . . . . . . . . . . . . . . . . . . . . . . 0.5 mm (0.20 inch)

### Ignition timing . . . . . . . . . . . . . . . . . . . . . . . . . . . . . . . . . . . . . Not adjustable

### Torque specifications
Pick-up coil cover bolts . . . . . . . . . . . . . . . . . . . . . . . . . . . . . . . . 9.8 Nm (87 inch-lbs)
Timing rotor . . . . . . . . . . . . . . . . . . . . . . . . . . . . . . . . . . . . . . . . 25 Nm (18 ft-lbs)

## 1 General information

This motorcycle is equipped with a battery operated, fully transistorized, breakerless ignition system. The system consists of the following components:

Pick-up coil
IC igniter unit
Battery and fuse
Ignition coils
Spark plugs
Engine stop (kill) and main (key) switches
Primary and secondary (HT) circuit wiring

The transistorized ignition system functions on the same principle as a DC ignition system with the pick-up unit and igniter performing the tasks previously associated with the breaker points and mechanical advance system. As a result, adjustment and maintenance of ignition components is eliminated (with the exception of spark plug replacement).

Because of their nature, the individual ignition system components can be checked but not repaired. If ignition system troubles occur, and the faulty component can be isolated, the only cure for the problem is to replace the part with a new one. Keep in mind that most electrical parts, once purchased, can't be returned. To avoid unnecessary expense, make very sure the faulty component has been positively identified before buying a replacement part.

2.3 Ground (earth) a spark plug to the engine and operate the starter – bright blue sparks should be visible

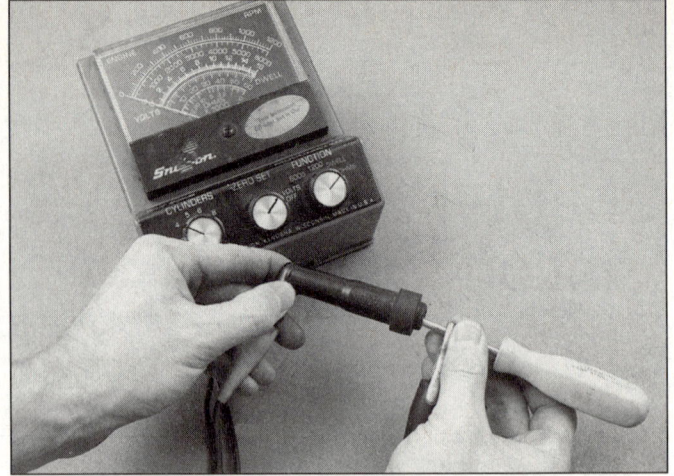

2.5 Unscrew the spark plug caps from the plug wires and measure their resistance with an ohmmeter

## 2 Ignition system – check

**Warning: Because of the very high voltage generated by the ignition system, extreme care should be taken when these checks are performed.**

1 If the ignition system is the suspected cause of poor engine performance or failure to start, a number of checks can be made to isolate the problem.

2 Make sure the ignition stop (kill) switch is in the Run or On position.

### Engine will not start

3 Refer to Chapter 1 and disconnect one of the spark plug wires. Connect the wire to a spare spark plug and lay the plug on the engine with the threads contacting the engine. If necessary, hold the spark plug with an insulated tool **(see illustration)**. Crank the engine over and make sure a well-defined, blue spark occurs between the spark plug electrodes.

**TOOL TiP**

A simple spark gap testing fixture can be made from a block of wood, two nails, a large alligator clip, a screw and a piece of wire

**Warning: Don't remove one of the spark plugs from the engine to perform this check – atomized fuel being pumped out of the open spark plug hole could ignite, causing severe injury!**

4 If no spark occurs, the following checks should be made:

5 Unscrew a spark plug cap from a plug wire and check the cap resistance with an ohmmeter **(see illustration)**. If the resistance is infinite, replace it with a new one. Repeat this check on the remaining plug caps.

6 Make sure all electrical connectors are clean and tight. Check all wires for shorts, opens and correct installation.

7 Check the battery voltage with a voltmeter. On models equipped with fillable batteries, check the specific gravity with a hydrometer (see Chapter 1). If the voltage is less than 12-volts or if the specific gravity is low, recharge the battery.

8 Check the ignition fuse and the fuse connections. If the fuse is blown, replace it with a new one; if the connections are loose or corroded, clean or repair them.

9 Refer to Section 3 and check the ignition coil primary and secondary resistance.

10 Refer to Section 4 and check the pick-up coil resistance.

11 If the preceding checks produce positive results but there is still no spark at the plug, remove the IC igniter and have it checked by a Kawasaki dealer service department or other repair shop equipped with the special tester required.

### Engine starts but misfires

12 If the engine starts but misfires, make the following checks before deciding that the ignition system is at fault.

13 The ignition system must be able to produce a spark across a seven millimeter (¼-inch) gap (minimum). A simple test fixture **(see Tool Tip)** can be constructed to make sure the minimum spark gap can be jumped.

Make sure the fixture electrodes are positioned seven millimeters apart.

14 Connect one of the spark plug wires to the protruding test fixture electrode, then attach the fixture's alligator clip to a good engine ground (earth) **(see illustration)**.

15 Crank the engine over (it will probably start and run on the remaining cylinders) and see if well-defined, blue sparks occur between the test fixture electrodes. If the minimum spark gap test is positive, the ignition coil for that cylinder (and its companion cylinder) is functioning properly. Repeat the check on one of the spark plug wires that is connected to the other coil. If the spark will not jump the gap during either test, or if it is weak (orange colored), refer to Steps 5 through 11 of this Section and perform the component checks described.

## 3 Ignition coils – check, removal and installation

### Check

1 In order to determine conclusively that the ignition coils are defective, they should be

2.14 Connect the tester – when the engine is cranked, sparks should jump the gap between the nails

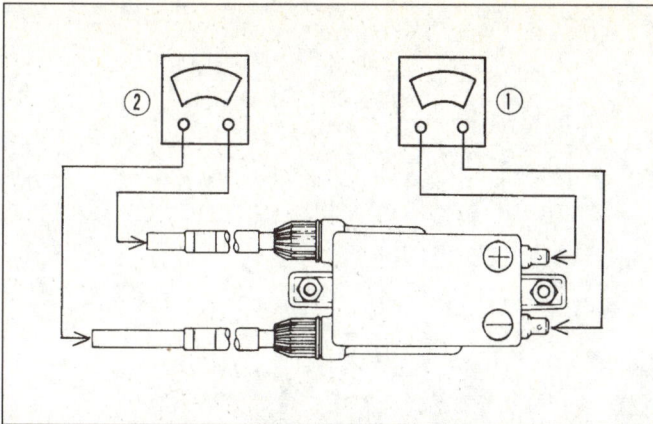

3.4 To check the resistance of the primary windings, connect the leads to the primary terminals (1); to check the resistance of the secondary (HT) windings, attach the ohmmeter to the spark plug wires (2)

3.9a It's a good idea to mark the ignition coils before removing them

tested by an authorized Kawasaki dealer service department which is equipped with the special electrical tester required for this check.
2 However, the coils can be checked visually (for cracks and other damage) and the primary and secondary coil resistances can be measured with an ohmmeter. If the coils are undamaged, and if the resistances are as specified, they are probably capable of proper operation.
3 To check the coils for physical damage, they must be removed (see Step 9). To check

3.9b On H through K models, the coil for cylinders 1 and 4 is mounted inside the left frame member . . .

the resistances, simply remove the fuel tank (see Chapter 4), unplug the primary circuit electrical connectors from the coil(s) and remove the spark plug wires from the plugs that are connected to the coil being checked. Mark the locations of all wires before disconnecting them. Unscrew the spark plug caps from the coil wires.
4 To check the coil primary resistance, attach one ohmmeter lead to one of the primary terminals and the other ohmmeter lead to the other primary terminal (see illustration).
5 Place the ohmmeter selector switch in the $\Omega$ x 1 position and compare the measured resistance to the value listed in this Chapter's Specifications.
6 If the coil primary resistance is as specified, check the coil secondary resistance by disconnecting the meter leads from the primary terminals and attaching them to the spark plug wire terminals (see illustration 3.4).
7 Place the ohmmeter selector switch in the K$\Omega$ position and compare the measured resistance to the values listed in this Chapter's Specifications.
8 If the resistances are not as specified, unscrew the spark plug wire retainers from the coil, detach the wires and check the

resistance again. If it is now within specifications, one or both of the wires are bad. If it's still not as specified, the coil is probably defective and should be replaced with a new one.

## Removal and installation

9 To remove the coils, refer to Chapter 4 and remove the fuel tank, then disconnect the spark plug wires from the plugs. After labeling them with tape to aid in reinstallation, unplug the coil primary circuit electrical connectors (see illustrations).
10 Support the coil with one hand and remove the coil mounting bolts, then lift the coil out.
11 Installation is the reverse of removal. If a new coil is being installed, unscrew the spark plug wire terminals from the coil, pull the wires out and transfer them to the new coil. Make sure the primary circuit electrical connectors are attached to the proper terminals. Just in case you forgot to mark the wires, the black and red wires connect to the no. 1 and 4 ignition coil (red to positive, black to negative) and the red and green wires attach to the no. 2 and 3 coil (red to positive, green to negative).

3.9c . . . and the coil for cylinders 2 and 3 is inside the right frame member

3.9d Both coils of L and M models are inside the front of the frame; the coil for cylinders 1 and 4 is on top . . .

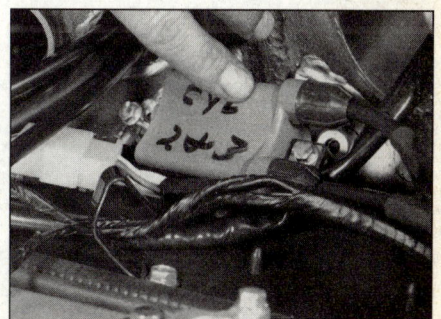

3.9e . . . and the coil for cylinders 2 and 3 is on the bottom

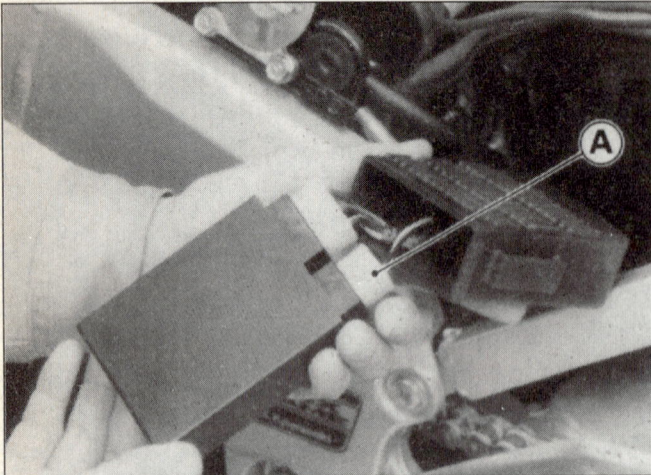

4.2 The igniter on H models is located in a holder on the left side of the motorcycle

A Pick-up coil connector

4.4 The igniter on J, K, L and M models is under the seats

## 4 Pick-up coil –
### check, removal and installation

### Check

#### H models

1 Remove the seats and the left side cover (see Chapter 8).
2 Locate the igniter on the left side of the motorcycle, just to the left of the battery. Pull it out of its holder and disconnect the pick-up coil connector (see illustration).

#### J, K, L and M models

3 Remove the seats (see Chapter 8). Where necessary for access, remove the air cleaner housing (see Chapter 4).
4 Disconnect the pick-up coil connector from the igniter (see illustration). The connector location varies according to model; refer to the Wiring diagrams at the end of the book to identify the connector.

#### All models

5 Connect an ohmmeter between the terminals in the pick-up coil connector and compare the resistance reading with the value listed in this Chapter's Specifications.
6 Set the ohmmeter on the highest resistance range. Measure the resistance between a good ground/earth and each terminal in the electrical connector. The meter should read infinity.
7 If the pick-up coil fails either of the above tests, it must be replaced.

### Removal

8 Remove the fairing right middle panel (see Chapter 8). Remove the bolts that secure the pick-up coil cover to the engine case (see illustration) and detach the cover from the engine.

9 Unscrew the pick-up coil mounting bolts and remove the pick-up coil (see illustrations). To remove the pick-up coil completely, disconnect its wiring from the igniter as described above, and free the wiring from any clamps or ties. On J, K, L and M models, it may be necessary to disconnect the oil pressure switch wire.

### Installation

#### H models

10 Bolt the pick-up coil loosely to the engine case. Measure the pick-up coil air gap with a feeler gauge (see illustration). Adjust the gap as necessary to the value listed in this Chapter's Specifications, then tighten the bolts securely.
11 Clean all traces of the old cover gasket from the cover and engine case.
12 Make sure the pick-up coil wiring harness grommet fits into the notch in the engine case.

4.8 Remove the cover bolts (arrows); the upper left bolt secures a wiring harness retainer and on L and M models, the lower two bolts secure a fairing bracket

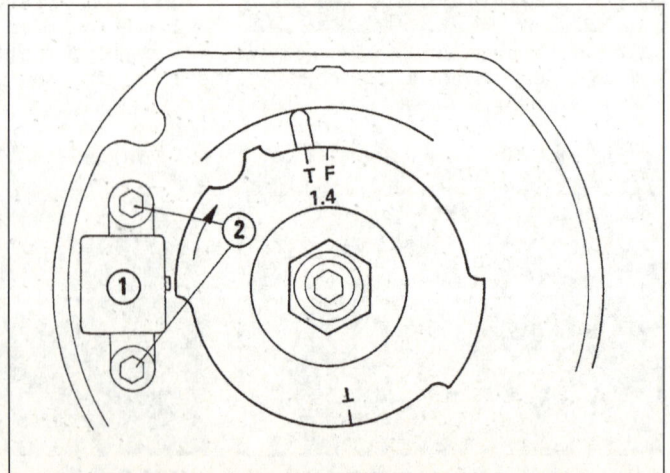

4.9a On H models, the pick-up coil (1) is secured by two bolts (2)

4.9b  On J, K, L and M models, remove the pick-up coil bolts (A); on installation, apply a coat of silicone sealant to the grommet surface (B) and the ends of the crankcase seams (C) . . .

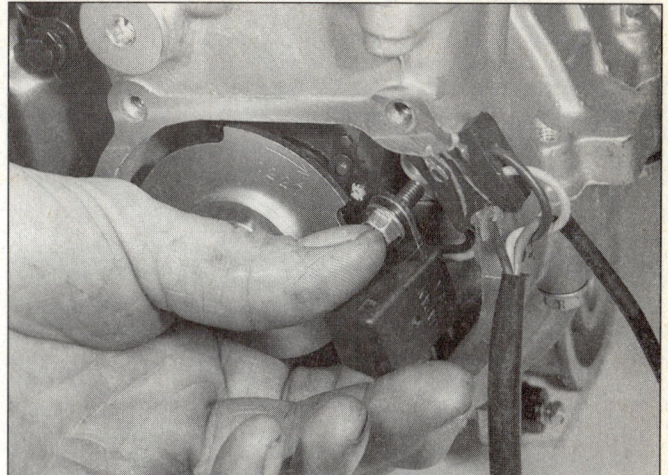

4.9c  . . . pull the pick-up coil out and detach the grommet from the crankcase

13  Install a new gasket and the pick-up coil cover, making sure the notch in the cover is positioned on the bottom. Apply non-permanent thread locking agent to the threads of the cover bolts and tighten them to the torque listed in this Chapter's Specifications.

### J, K, L and M models

14  Check the cover O-ring and replace it if it's brittle or deteriorated (see illustration).
15  Make sure the wiring harness grommet is in the notch in the engine case (see illustration 4.9c).
16  Apply a thin coat of sealant to the grommet surface and to the ends of the crankcase seam (see illustration 4.9b).
17  Apply non-permanent thread locking agent to the threads of the top two cover bolts. Install the cover and its O-ring, then

install all four bolts and tighten them evenly to the torque listed in this Chapter's Specifications. Note that one upper bolt secures a wiring harness retainer and on L and M models the two lower bolts secure a fairing bracket (see illustration 4.8).

### All models

18  Plug in the electrical connector and install all parts removed for access.

### 5  Timing rotor – removal and installation

1  Refer to Section 4 and remove the pick-up coil cover.
2  Hold the timing rotor with a box wrench (ring spanner) on its hex and remove the Allen bolt (see illustration).

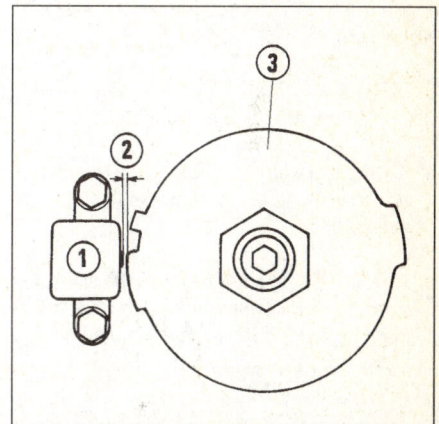

4.10  On H models, adjust the gap between the pick-up coil and timing rotor to specifications

1  Pick-up coil        3  Timing rotor
2  Gap

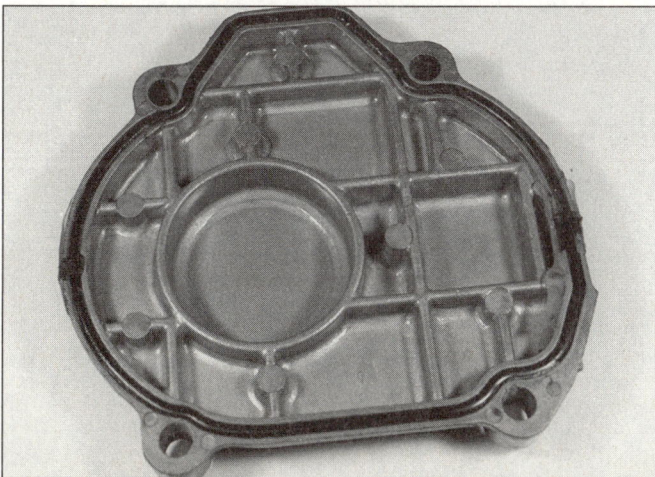

4.14  On J, K, L and M models, replace the cover O-ring if its condition is in any doubt

5.2  Hold the timing rotor with a box wrench and remove the Allen bolt

5.3  Take the bolt and timing rotor off

5.4  Align the notch in the rotor with the pin on the crankshaft (arrows)

3 Take the timing rotor off the crankshaft (see illustration).

4 Installation is the reverse of the removal steps, with the following additions:

a) Align the timing rotor notch with the pin on the end of the crankshaft (see illustration).

b) Hold the timing rotor hex with the same wrench used during removal and tighten the Allen bolt to the torque listed in this Chapter's Specifications.

## 6  IC igniter –
removal, check and installation

### Removal

#### H models

1 Remove the left side cover (see Chapter 8).
2 Locate the IC igniter in its holder to the left of the battery (see illustration 4.2). Pull it out and disconnect the electrical connectors.
3 Installation is the reverse of the removal steps.

#### J, K, L and M models

4 Remove the seats (see Chapter 8).
5 Disconnect the igniter electrical connectors (see illustration 4.4). Remove the mounting nuts and detach the igniter from its bracket.
6 Installation is the reverse of the removal steps.

### Check

7 A special tester is required to accurately measure the resistance values across the various terminals of the IC igniter. Take the unit to a Kawasaki dealer service department or other repair shop equipped with this tester.

# Chapter 6
## Steering, suspension and final drive

## Contents

## Degrees of difficulty

| Easy, suitable for novice with little experience | Fairly easy, suitable for beginner with some experience | Fairly difficult, suitable for competent DIY mechanic | Difficult, suitable for experienced DIY mechanic | Very difficult, suitable for expert DIY or professional |
|---|---|---|---|---|

## Specifications

**Front forks**

Fork spring length
  H models
    Standard . . . . . . . . . . . . . . . . . . . . . . . . . . . . . . . . . . . . . . . . 387.5 mm (15.256 inches)
    Minimum . . . . . . . . . . . . . . . . . . . . . . . . . . . . . . . . . . . . . . . . 380 mm (14.960 inches)
  J models
    Standard . . . . . . . . . . . . . . . . . . . . . . . . . . . . . . . . . . . . . . . . 291.6 mm (11.480 inches)
    Minimum . . . . . . . . . . . . . . . . . . . . . . . . . . . . . . . . . . . . . . . . 286 mm (11.260 inches)
  K and M models
    Standard . . . . . . . . . . . . . . . . . . . . . . . . . . . . . . . . . . . . . . . . 295.6 mm (11.638 inches)
    Minimum . . . . . . . . . . . . . . . . . . . . . . . . . . . . . . . . . . . . . . . . 289 mm (11.378 inches)
  L models
    Standard . . . . . . . . . . . . . . . . . . . . . . . . . . . . . . . . . . . . . . . . 300.6 mm (11.835 inches)
    Minimum . . . . . . . . . . . . . . . . . . . . . . . . . . . . . . . . . . . . . . . . 295 mm (11.614 inches)
Fork oil type . . . . . . . . . . . . . . . . . . . . . . . . . . . . . . . . . . . . . . . . SAE 5W fork oil
Fork oil level (fork fully compressed; spring removed)
  H models . . . . . . . . . . . . . . . . . . . . . . . . . . . . . . . . . . . . . . . . . 135 ± 2 mm (5.315 ± 0.079 inches)
  J models . . . . . . . . . . . . . . . . . . . . . . . . . . . . . . . . . . . . . . . . . 115 ± 2 mm (4.528 ± 0.0789 inches)
  K and M models . . . . . . . . . . . . . . . . . . . . . . . . . . . . . . . . . . . 112 ± 2 mm (4.409 ± 0.079 inches)
  L models . . . . . . . . . . . . . . . . . . . . . . . . . . . . . . . . . . . . . . . . . 91 ± 2 mm (3.583 ± 0.079 inches)

## Front forks (continued)

Fork oil capacity
  H and J models
    Dry fill (after overhaul) . . . . . . . . . . . . . . . . . . . . . . . . . . . . . . . . . 445 ± 4 cc
    Oil change . . . . . . . . . . . . . . . . . . . . . . . . . . . . . . . . . . . . . . . . . . approximately 380 cc
  K and M models
    Dry fill (after overhaul) . . . . . . . . . . . . . . . . . . . . . . . . . . . . . . . . . 380 ± 4 cc
    Oil change . . . . . . . . . . . . . . . . . . . . . . . . . . . . . . . . . . . . . . . . . . not specified
  L models
    Dry fill (after overhaul) . . . . . . . . . . . . . . . . . . . . . . . . . . . . . . . . . 400 ± 4 cc
    Oil change . . . . . . . . . . . . . . . . . . . . . . . . . . . . . . . . . . . . . . . . . . not specified
Fork protrusion from upper triple clamp
  J and K models . . . . . . . . . . . . . . . . . . . . . . . . . . . . . . . . . . . . . . . . 3.11 mm (0.122 inch)
  L models . . . . . . . . . . . . . . . . . . . . . . . . . . . . . . . . . . . . . . . . . . . . 5 mm (0.197 inch)
  M models . . . . . . . . . . . . . . . . . . . . . . . . . . . . . . . . . . . . . . . . . . . 10 mm (0.394 inch)

## Rear sprocket runout (maximum) . . . . . . . . . . . . . . . . . . . 0.5 mm (0.020 inch)

## Suspension adjustments

### H models

Front fork rebound damping
  Standard setting . . . . . . . . . . . . . . . . . . . . . . . . . . . . . . . . . . . . . . Fourth click
  Hardest setting . . . . . . . . . . . . . . . . . . . . . . . . . . . . . . . . . . . . . . . Fully clockwise
  Softest setting . . . . . . . . . . . . . . . . . . . . . . . . . . . . . . . . . . . . . . . Fully counterclockwise (anti-clockwise – thirteenth click)
Front fork spring preload (adjuster protrusion)
  Standard . . . . . . . . . . . . . . . . . . . . . . . . . . . . . . . . . . . . . . . . . . . 16 mm (0.630 inch) – 6½ spaces
  Hardest setting . . . . . . . . . . . . . . . . . . . . . . . . . . . . . . . . . . . . . . . 5 mm (0.192 inch) – 1 space
  Softest setting . . . . . . . . . . . . . . . . . . . . . . . . . . . . . . . . . . . . . . . 20 mm (0.787 inch) – 8½ spaces
Rear shock absorber rebound damping
  Standard setting . . . . . . . . . . . . . . . . . . . . . . . . . . . . . . . . . . . . . . II
  Softest setting . . . . . . . . . . . . . . . . . . . . . . . . . . . . . . . . . . . . . . . I
  Hardest setting . . . . . . . . . . . . . . . . . . . . . . . . . . . . . . . . . . . . . . IIII
Rear shock absorber spring preload
  Standard setting . . . . . . . . . . . . . . . . . . . . . . . . . . . . . . . . . . . . . . 10 mm 0.394 inch) less than free length
  Softest setting . . . . . . . . . . . . . . . . . . . . . . . . . . . . . . . . . . . . . . . 6 mm (0.236 inch) less than free length
  Hardest setting . . . . . . . . . . . . . . . . . . . . . . . . . . . . . . . . . . . . . . 26 mm (1.023 inch) less than free length

### J models

Front fork rebound damping
  Standard setting . . . . . . . . . . . . . . . . . . . . . . . . . . . . . . . . . . . . . . Seventh click
  Hardest setting . . . . . . . . . . . . . . . . . . . . . . . . . . . . . . . . . . . . . . . Fully clockwise
  Softest setting . . . . . . . . . . . . . . . . . . . . . . . . . . . . . . . . . . . . . . . Fully counterclockwise (anti-clockwise – thirteenth click)
Front fork spring preload (adjuster protrusion)
  Standard . . . . . . . . . . . . . . . . . . . . . . . . . . . . . . . . . . . . . . . . . . . 20 mm (0.787 inch) – 8½ spaces
  Hardest setting . . . . . . . . . . . . . . . . . . . . . . . . . . . . . . . . . . . . . . . 5 mm (0.192 inch) – 1 space
  Softest setting . . . . . . . . . . . . . . . . . . . . . . . . . . . . . . . . . . . . . . . 20 mm (0.787 inch) – 8½ spaces
Rear shock absorber rebound damping
  Standard setting . . . . . . . . . . . . . . . . . . . . . . . . . . . . . . . . . . . . . . 2
  Softest setting . . . . . . . . . . . . . . . . . . . . . . . . . . . . . . . . . . . . . . . 1
  Hardest setting . . . . . . . . . . . . . . . . . . . . . . . . . . . . . . . . . . . . . . 4
Rear shock absorber spring preload
  Standard setting . . . . . . . . . . . . . . . . . . . . . . . . . . . . . . . . . . . . . . 8 mm (0.315 inch) less than free length
  Softest setting . . . . . . . . . . . . . . . . . . . . . . . . . . . . . . . . . . . . . . . 8 mm (0.315 inch) less than free length
  Hardest setting . . . . . . . . . . . . . . . . . . . . . . . . . . . . . . . . . . . . . . 20 mm (0.787 inch) less than free length

### K models

Front fork rebound damping
  Standard setting . . . . . . . . . . . . . . . . . . . . . . . . . . . . . . . . . . . . . . Seventh click
  Hardest setting . . . . . . . . . . . . . . . . . . . . . . . . . . . . . . . . . . . . . . . Fully clockwise
  Softest setting . . . . . . . . . . . . . . . . . . . . . . . . . . . . . . . . . . . . . . . Fully counterclockwise (anti-clockwise – thirteenth click)
Front fork spring preload (adjuster protrusion)
  Standard . . . . . . . . . . . . . . . . . . . . . . . . . . . . . . . . . . . . . . . . . . . 20 mm (0.787 inch) – 8½ spaces
  Hardest setting . . . . . . . . . . . . . . . . . . . . . . . . . . . . . . . . . . . . . . . 5 mm (0.192 inch) – 1 space
  Softest setting . . . . . . . . . . . . . . . . . . . . . . . . . . . . . . . . . . . . . . . 20 mm (0.787 inch) – 8½ spaces
Front fork compression damping
  Standard . . . . . . . . . . . . . . . . . . . . . . . . . . . . . . . . . . . . . . . . . . . Fourth click
  Hardest setting . . . . . . . . . . . . . . . . . . . . . . . . . . . . . . . . . . . . . . . Fully clockwise
  Softest setting . . . . . . . . . . . . . . . . . . . . . . . . . . . . . . . . . . . . . . . Fully counterclockwise (anti-clockwise)

## Suspension adjustments (continued)

### K models (continued)

Rear shock absorber rebound damping
    Standard setting . . . . . . . . . . . . . . . . . . . . . . . . . . . . . . . . . . . . . 2
    Softest setting . . . . . . . . . . . . . . . . . . . . . . . . . . . . . . . . . . . . . . 1
    Hardest setting . . . . . . . . . . . . . . . . . . . . . . . . . . . . . . . . . . . . . 4

Rear shock absorber spring preload
    Standard setting . . . . . . . . . . . . . . . . . . . . . . . . . . . . . . . . . . . . 8 mm (0.315 inch) less than free length
    Softest setting . . . . . . . . . . . . . . . . . . . . . . . . . . . . . . . . . . . . . 8 mm (0.315 inch) less than free length
    Hardest setting . . . . . . . . . . . . . . . . . . . . . . . . . . . . . . . . . . . . 20 mm (0.787 inch) less than free length

Rear shock absorber compression damping
    Standard setting . . . . . . . . . . . . . . . . . . . . . . . . . . . . . . . . . . . . Seventh click
    Softest setting . . . . . . . . . . . . . . . . . . . . . . . . . . . . . . . . . . . . . Fully counterclockwise (anti-clockwise) – 25th click
    Hardest setting . . . . . . . . . . . . . . . . . . . . . . . . . . . . . . . . . . . . Fully clockwise

### L models

Front fork rebound damping
    Standard setting . . . . . . . . . . . . . . . . . . . . . . . . . . . . . . . . . . . . Seventh click
    Hardest setting . . . . . . . . . . . . . . . . . . . . . . . . . . . . . . . . . . . . Fully clockwise
    Softest setting . . . . . . . . . . . . . . . . . . . . . . . . . . . . . . . . . . . . . Fully counterclockwise (anti-clockwise – thirteenth click)

Front fork spring preload (adjuster protrusion)
    Standard . . . . . . . . . . . . . . . . . . . . . . . . . . . . . . . . . . . . . . . . 20 mm (0.787 inch) – 8½ spaces
    Hardest setting . . . . . . . . . . . . . . . . . . . . . . . . . . . . . . . . . . . . 5 mm (0.192 inch) – 1 space
    Softest setting . . . . . . . . . . . . . . . . . . . . . . . . . . . . . . . . . . . . . 20 mm (0.787 inch) – 8½ spaces

Rear shock absorber rebound damping
    Standard setting . . . . . . . . . . . . . . . . . . . . . . . . . . . . . . . . . . . . 2
    Softest setting . . . . . . . . . . . . . . . . . . . . . . . . . . . . . . . . . . . . . 1
    Hardest setting . . . . . . . . . . . . . . . . . . . . . . . . . . . . . . . . . . . . 4

Rear shock absorber spring preload
    Standard setting . . . . . . . . . . . . . . . . . . . . . . . . . . . . . . . . . . . . 8 mm (0.315 inch) less than free length
    Softest setting . . . . . . . . . . . . . . . . . . . . . . . . . . . . . . . . . . . . . 8 mm (0.315 inch) less than free length
    Hardest setting . . . . . . . . . . . . . . . . . . . . . . . . . . . . . . . . . . . . 20 mm (0.787 inch) less than free length

### M models

Front fork rebound damping
    Standard setting . . . . . . . . . . . . . . . . . . . . . . . . . . . . . . . . . . . . Seventh click
    Hardest setting . . . . . . . . . . . . . . . . . . . . . . . . . . . . . . . . . . . . Fully clockwise
    Softest setting . . . . . . . . . . . . . . . . . . . . . . . . . . . . . . . . . . . . . Fully counterclockwise (anti-clockwise – thirteenth click)

Front fork spring preload (adjuster protrusion)
    Standard . . . . . . . . . . . . . . . . . . . . . . . . . . . . . . . . . . . . . . . . 20 mm (0.787 inch) – 8½ spaces
    Hardest setting . . . . . . . . . . . . . . . . . . . . . . . . . . . . . . . . . . . . 5 mm (0.192 inch) – 1 space
    Softest setting . . . . . . . . . . . . . . . . . . . . . . . . . . . . . . . . . . . . . 20 mm (0.787 inch) – 8 spaces

Front fork compression damping
    Standard . . . . . . . . . . . . . . . . . . . . . . . . . . . . . . . . . . . . . . . . Fourth click
    Hardest setting . . . . . . . . . . . . . . . . . . . . . . . . . . . . . . . . . . . . Fully clockwise
    Softest setting . . . . . . . . . . . . . . . . . . . . . . . . . . . . . . . . . . . . . Fully counterclockwise (anti-clockwise)

Rear shock absorber rebound damping
    Standard setting . . . . . . . . . . . . . . . . . . . . . . . . . . . . . . . . . . . . 2
    Softest setting . . . . . . . . . . . . . . . . . . . . . . . . . . . . . . . . . . . . . 1
    Hardest setting . . . . . . . . . . . . . . . . . . . . . . . . . . . . . . . . . . . . 4

Rear shock absorber spring preload
    Standard setting . . . . . . . . . . . . . . . . . . . . . . . . . . . . . . . . . . . . 8 mm (0.315 inch) less than free length
    Softest setting . . . . . . . . . . . . . . . . . . . . . . . . . . . . . . . . . . . . . 8 mm (0.315 inch) less than free length
    Hardest setting . . . . . . . . . . . . . . . . . . . . . . . . . . . . . . . . . . . . 20 mm (0.787 inch) less than free length

Rear shock absorber compression damping
    Standard setting . . . . . . . . . . . . . . . . . . . . . . . . . . . . . . . . . . . . Seventh click
    Softest setting . . . . . . . . . . . . . . . . . . . . . . . . . . . . . . . . . . . . . Fully counterclockwise (anti-clockwise) – 25th click
    Hardest setting . . . . . . . . . . . . . . . . . . . . . . . . . . . . . . . . . . . . Fully clockwise

## Torque specifications

### Handlebars

Handlebar to bracket bolts*
    H models . . . . . . . . . . . . . . . . . . . . . . . . . . . . . . . . . . . . . . . . . 29 Nm (22 ft-lbs)
    J, K, L and M models . . . . . . . . . . . . . . . . . . . . . . . . . . . . . . . . 34 Nm (25 ft-lbs)
Handlebar bracket pinch bolts
    H1 models . . . . . . . . . . . . . . . . . . . . . . . . . . . . . . . . . . . . . . . . 29 Nm (22 ft-lbs)
    H2 models . . . . . . . . . . . . . . . . . . . . . . . . . . . . . . . . . . . . . . . . 20 Nm (14.5 ft-lbs)
    J, K, L and M models . . . . . . . . . . . . . . . . . . . . . . . . . . . . . . . . 23 Nm (16.5 ft-lbs)
Handlebar bracket to upper triple clamp bolts* . . . . . . . . . . . . . . . . 9.8 Nm (87 inch-lbs)

## Torque specifications (continued)

### Steering stem and triple clamps

Triple clamp pinch bolts (upper and lower)

| | |
|---|---|
| H models . . . . . . . . . . . . . . . . . . . . . . . . . . . . . . . . . . . . . . . . . . . . | 20 Nm (14.5 ft-lbs) |
| J, K, L and M models . . . . . . . . . . . . . . . . . . . . . . . . . . . . . . . . | 21 Nm (15 ft-lbs) |

Steering stem bearing locknut**

| | |
|---|---|
| Initial torque . . . . . . . . . . . . . . . . . . . . . . . . . . . . . . . . . . . . . . . . | 39 Nm (29 ft-lbs) |
| Final torque . . . . . . . . . . . . . . . . . . . . . . . . . . . . . . . . . . . . . . . . . | 4.9 Nm (43 inch-lbs) or hand tight |

Steering stem nut

| | |
|---|---|
| H models . . . . . . . . . . . . . . . . . . . . . . . . . . . . . . . . . . . . . . . . . . . . | 39 Nm (29 ft-lbs) |
| J, K, L and M models . . . . . . . . . . . . . . . . . . . . . . . . . . . . . . . . | 54 Nm (40 ft-lbs) |

### Front forks

H models

| | |
|---|---|
| Damper rod bolt* . . . . . . . . . . . . . . . . . . . . . . . . . . . . . . . . . . . . . | 39 Nm (29 ft-lbs) |
| Axle pinch bolts . . . . . . . . . . . . . . . . . . . . . . . . . . . . . . . . . . . . . . | 20 Nm (14.5 ft-lbs) |
| Piston rod nut . . . . . . . . . . . . . . . . . . . . . . . . . . . . . . . . . . . . . . . . | 13 Nm (113 inch-lbs) |
| Cap bolt . . . . . . . . . . . . . . . . . . . . . . . . . . . . . . . . . . . . . . . . . . . . | 23 Nm (16.5 ft-lbs) |

J, K, L and M models

| | |
|---|---|
| Damper rod bolt* . . . . . . . . . . . . . . . . . . . . . . . . . . . . . . . . . . . . . | 39 Nm (29 ft-lbs) |

Axle pinch bolts

| | |
|---|---|
| J and K models . . . . . . . . . . . . . . . . . . . . . . . . . . . . . . . . . . . . | 20 Nm (14.5 ft-lbs) |
| L and M models . . . . . . . . . . . . . . . . . . . . . . . . . . . . . . . . . . . | 21 Nm (15 ft-lbs) |
| Piston rod nut . . . . . . . . . . . . . . . . . . . . . . . . . . . . . . . . . . . . . . . . | 15 Nm (11 ft-lbs) |
| Cap bolt . . . . . . . . . . . . . . . . . . . . . . . . . . . . . . . . . . . . . . . . . . . . | 23 Nm (16.5 ft-lbs) |
| Compression damping adjuster to fork (K and M models only) . . . . . | 18 Nm (13 ft-lbs) |

### Rear suspension and swingarm

| | |
|---|---|
| Rear shock absorber and suspension linkage bolts/nuts . . . . . . . . . . . | 59 Nm (43 ft-lbs) |
| Swingarm pivot shaft holder bolts (H1 models only) . . . . . . . . . . . . . . | 20 Nm (14.5 ft-lbs) |
| Swingarm pivot shaft Allen head (H1 models only) . . . . . . . . . . . . . . | 9.8 Nm (87 inch-lbs) |

Swingarm pivot shaft nut

| | |
|---|---|
| H models . . . . . . . . . . . . . . . . . . . . . . . . . . . . . . . . . . . . . . . . . . . . | 88 Nm (65 ft-lbs) |
| J, K, L and M models . . . . . . . . . . . . . . . . . . . . . . . . . . . . . . . . | 145 Nm (110 ft-lbs) |

### Final drive

Engine sprocket cover bolts

| | |
|---|---|
| H models . . . . . . . . . . . . . . . . . . . . . . . . . . . . . . . . . . . . . . . . . . . . | 12 Nm (104 inch-lbs) |
| J, K, L and M models . . . . . . . . . . . . . . . . . . . . . . . . . . . . . . . . | 9.8 Nm (87 inch-lbs) |
| Engine sprocket nut . . . . . . . . . . . . . . . . . . . . . . . . . . . . . . . . . . . | 98 Nm (72 ft-lbs) |
| Rear sprocket-to-wheel coupling nuts . . . . . . . . . . . . . . . . . . . . . . . | 74 Nm (54 ft-lbs) |

*Apply non-permanent thread locking agent to the threads.
**Refer to text for adjustment procedure.

---

## 1  General information

The front forks are of the coil spring cartridge type. On H models, the forks are upright (inner fork tube on the top); on J, K, L and M models, they're inverted (inner fork tube on the bottom). The forks on all models have adjustments for rebound damping and spring preload. The forks on K and M models also have an adjustment for compression damping.

The rear suspension is Kawasaki's Uni-trak design, which consists of a single shock absorber, a rocker arm, two tie-rods and a square-section aluminum swingarm. The shock absorber on all models has adjustments for rebound damping and spring preload. On K and M models, compression damping is also adjustable. The rear shock on

K and M models uses a remote reservoir.

The final drive uses an endless chain (which means it doesn't have a master link). A rubber damper, often called a 'cush drive', is installed between the rear wheel coupling and the wheel.

## 2  Handlebars – removal, inspection and installation

1 The handlebars are individual assemblies made up of a tube and bracket (see illustration). The tube fits into the bracket, which slips over the fork tube. The upper triple clamp is installed on top of the brackets. If the handlebars must be removed for access to other components, such as the forks or the steering head, simply remove the bolts and slip the handlebar(s) off the fork tubes. It's not necessary to disconnect the cables, wires or

hoses, but it is a good idea to support the assembly with a piece of wire or rope, to avoid unnecessary strain on the cables, wires and the brake or clutch hose.
2 If the handlebars are to be removed completely, refer to Chapter 2 and Chapter 7 for the master cylinder removal procedures, Chapter 4 for the throttle grip removal procedure and Chapter 9 for the switch removal procedure.

### Removal

3 If you plan to remove the brackets and the handlebars, then separate them, it's a good idea to loosen the handlebar end bolts first (the brackets and forks will keep the handlebars from turning). **Note:** *On some early H models, the handlebar end bolts also act as the bracket pinch bolts. These must be loosened in order to remove the brackets from the fork tubes.*
4 Support the motorcycle securely upright.

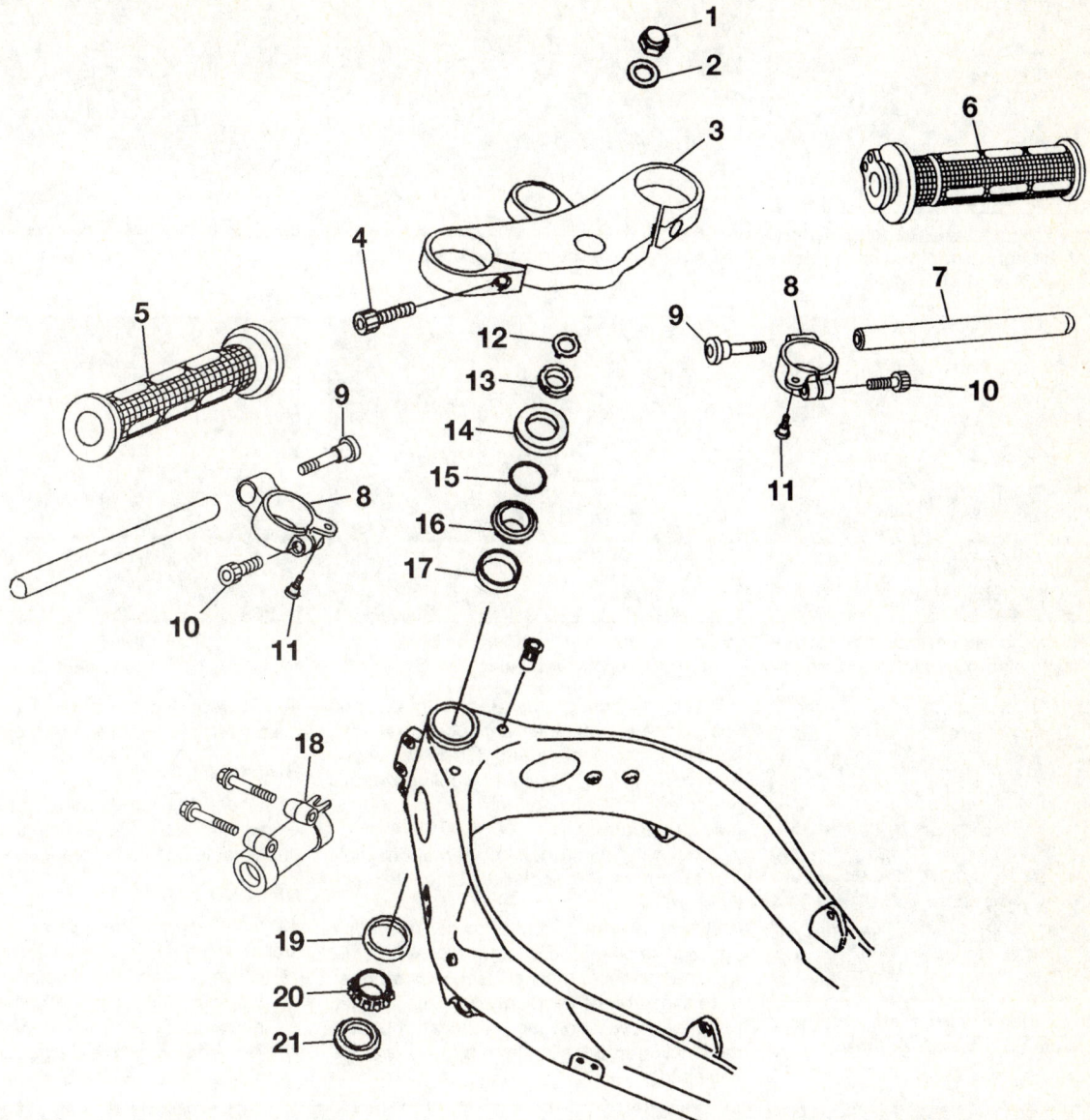

**2.1 Handlebar and steering head details**

1  Steering stem nut
2  Washer
3  Upper triple clamp
4  Triple clamp pinch bolt
5  Left handlebar and grip
6  Throttle grip
7  Right handlebar
8  Handlebar bracket
9  Handlebar end bolt
10 Handlebar bracket pinch
   bolt
11 Handlebar bracket to triple
   clamp bolt
12 Lockwasher
13 Locknut
14 Bearing cover
15 O-ring
16 Upper bearing
17 Upper bearing race
18 Brake hose union
19 Lower bearing race
20 Lower bearing
21 Steering stem grease seal
22 Steering stem/lower triple
   clamp
23 Triple clamp pinch bolt

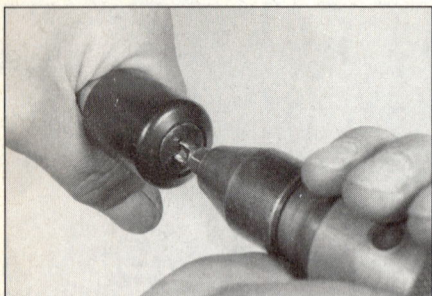

2.5 Remove the grip weight screw from the end of the handlebar with an impact driver (L and M models)

2.6a Remove the steering stem nut . . .

2.6b . . . and the washer

2.7 Loosen the handlebar pinch bolts and the upper triple clamp pinch bolts (arrows)

2.8a If necessary, remove the brake fluid reservoir bracket . . .

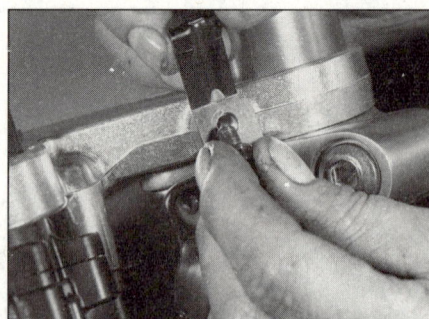

2.8b . . . and the clutch fluid reservoir bracket

2.8c Lift the handlebar brackets and upper triple clamp off the fork tubes

5 If you're working on an L or M model and you plan to remove the throttle grip or the left handlebar grip, loosen the Phillips screw and remove the grip end weight (see illustration).
6 Remove the steering stem nut and washer (see illustrations).
7 Loosen the pinch bolts on the upper triple clamp and the handlebar brackets (see illustration).
8 Detach the wiring harness guides and lift the upper triple clamp off the fork tubes, together with the handlebar brackets (see illustrations).
9 To separate the handlebar from the bracket, remove the end bolt (see illustration). Note the location of the alignment pin and notch (if equipped).

10 To separate the bracket from the triple clamp, remove the Allen bolt (see illustration).

## Inspection

11 Check the handlebars and brackets for cracks and distortion and replace them if any undesirable conditions are found.

## Installation

12 Installation is the reverse of the removal steps, with the following additions:
a) When installing the handlebars to the brackets, line up their alignment notch or pin correctly.
b) Tighten the bolts to the torque listed in this Chapter's Specifications.

2.9 Remove the handlebar bolt to separate it from the bracket; note the alignment pin and the notch it fits into

2.10 Remove the bolt and separate the handlebar bracket from the upper triple clamp

**3.9a  Lift the lockwasher off the steering stem . . .**

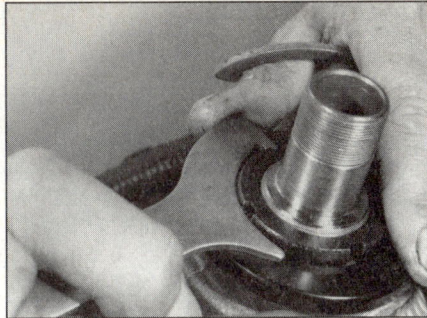

**3.9b  . . . undo the nut with a spanner wrench (C-spanner) . . .**

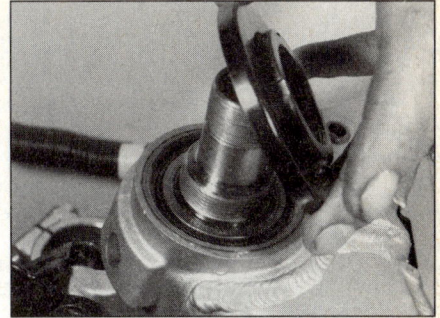

**3.9c  . . . and lift off the nut and bearing cover**

## 3  Steering stem and bearings – removal, inspection and installation

### Removal

1  If the steering head bearing check and adjustment (see Chapter 1) does not remedy excessive play or roughness in the steering head bearings, the entire front end must be disassembled and the bearings and races replaced with new ones.

### H models

2  Remove the fuel tank (see Chapter 4).
3  Remove the upper and lower fairings (see Chapter 8).

**3.14a  Driving out the lower bearing race with the special Kawasaki tool . . .**

### J, K, L and M models

4  Remove the upper fairing (see Chapter 8).
5  Disconnect the ignition main (key) switch electrical connector (see Chapter 9 if necessary).

### All models

6  Unbolt the brake hose union from the lower triple clamp. There is no need to disconnect the hydraulic hoses, but make sure no strain is placed on them.
7  Remove the front forks (see Section 4).
8  Refer to Section 2 and lift off the upper triple clamp.
9  Remove the lockwasher from the stem locknut (see illustration). Using a spanner wrench (C-spanner), remove the stem locknut and bearing cover (see illustrations) while supporting the steering head from the bottom.
10  Remove the steering stem and lower triple clamp assembly. If it's stuck, gently tap on the top of the steering stem with a plastic mallet or a hammer and a wood block.
11  Remove the upper bearing (see illustration).

### Inspection

12  Clean all the parts with solvent and dry them thoroughly, using compressed air, if available. Wipe the old grease out of the frame steering head and bearing races.
*Caution: If you do use compressed air, don't let the bearings spin as they're dried – it could ruin them.*
13  Examine the races in the steering head for cracks, dents, and pits. If even the slightest amount of wear or damage is evident, the races should be replaced with new ones.

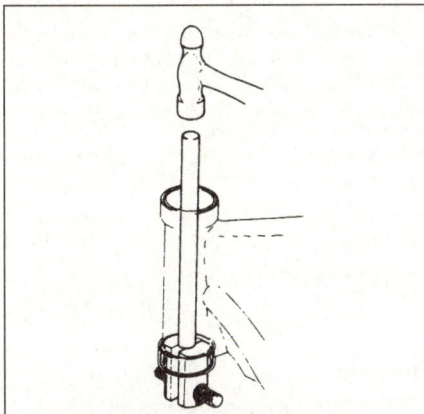

**3.11  Lift the bearing out of the steering head**

14  To remove the races, drive them out of the steering head with Kawasaki tool no. 57001-1107 or a hammer and drift punch (see illustrations). A slide hammer with the proper internal-jaw puller will also work. Since the races are an interference fit in the frame, installation will be easier if the new races are left overnight in a refrigerator. This will cause them to contract and slip into place in the frame with very little effort. When installing the races, use Kawasaki press shaft no. 57001-1075 and drivers no. 57001-1106 and 57001-1076 (see illustration), or tap them gently into place with a hammer and punch or a large socket. Do not strike the bearing surface or the race will be damaged.

**3.14d  Using the special Kawasaki tool to press the outer races into the frame**

1  Driver press shaft (tool no. 57001-1075)
2  Driver (tool no. 57001-1106)
3  Driver (tool no. 57001-1076)

**3.14b  . . . or by inserting a drift from above to drive out the lower race . . .**

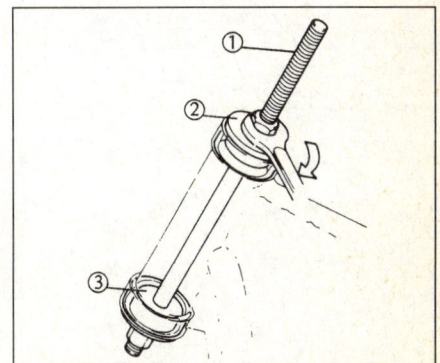

**3.14c  . . . then inserting the drift from below to drive out the upper race**

**3.16 Remove the lower bearing and grease seal from the steering stem only if they're to be replaced**

15 Check the bearings for wear. Look for cracks, dents, and pits in the races and flat spots on the bearings. Replace any defective parts with new ones. If a new bearing is required, replace both of them as a set.
16 Check the grease seal under the lower bearing and replace it with a new one if necessary **(see illustration)**.
17 To remove the lower bearing and grease seal from the steering stem, use a bearing puller (Kawasaki tool no. 57001-158), combined with adapter no. 57001-317 **(see illustration)**. A bearing puller, which can be rented, will also work. Don't remove this bearing unless it, or the grease seal underneath, must be replaced. Removal will damage the grease seal, so replace it whenever the bearing is removed.

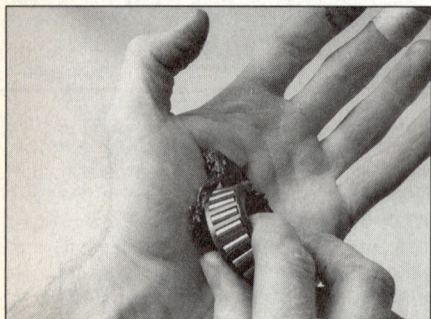
**3.20 Work the grease completely into the rollers**

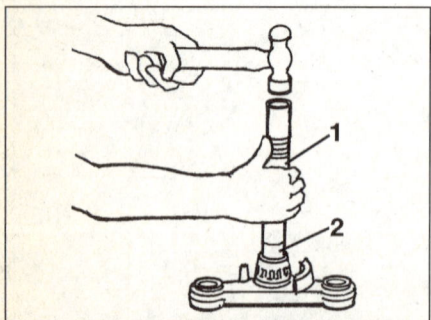
**3.21 Drive the lower bearing onto the steering stem with the special Kawasaki tools**

1  *Stem bearing driver (tool no. 57001-137)*
2  *Adapter (tool no. 57001-1074)*

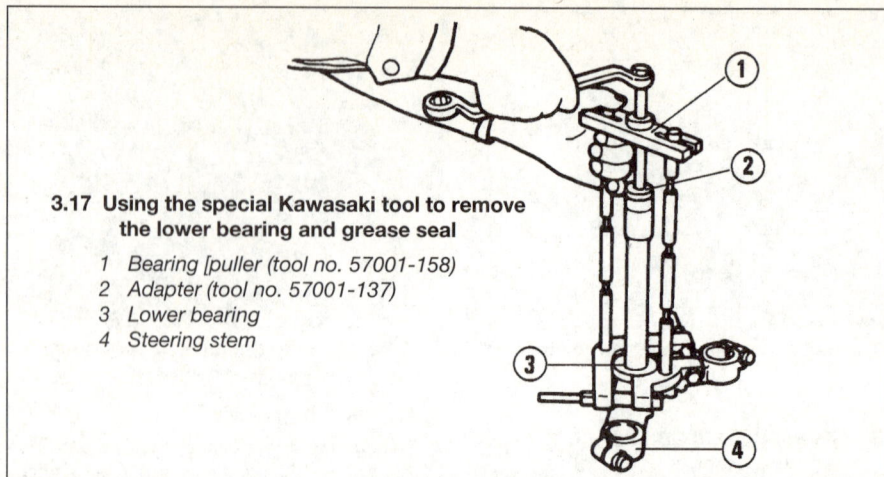
**3.17 Using the special Kawasaki tool to remove the lower bearing and grease seal**

1  *Bearing [puller (tool no. 57001-158)*
2  *Adapter (tool no. 57001-137)*
3  *Lower bearing*
4  *Steering stem*

18 Inspect the steering stem/lower triple clamp for cracks and other damage. Do not attempt to repair any steering components. Replace them with new parts if defects are found.
19 Check the O-ring under the stem locknut/stem cap assembly – if it's worn or deteriorated, replace it.
20 Pack the bearings with high-quality grease (preferably a moly-based grease) **(see illustration)**. Coat the outer races with grease also.
21 Install the grease seal and lower bearing onto the steering stem. Drive the lower bearing onto the steering stem using Kawasaki stem bearing driver no. 57001-137 and adapter no. 57001-1074 **(see illustration)**. Drive the bearing on until it's fully seated.

> **TOOL TiP** *If you don't have access to these tools, a section of pipe with a diameter the same as the inner race of the bearing can be used.*

### Installation

22 Insert the steering stem/lower triple clamp into the frame head. Install the upper bearing, O-ring, bearing cover and locknut **(see illustrations)**. Using the spanner wrench (C-spanner), tighten the locknut while moving the lower triple clamp back and forth. Continue to tighten the nut to the torque listed

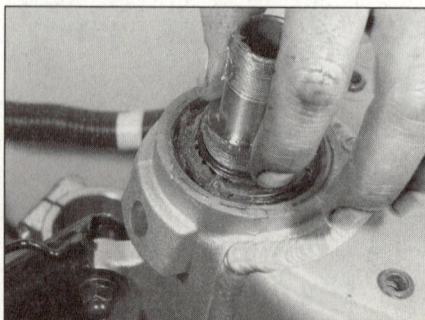
**3.22a Install the upper bearing and the O-ring**

in this Chapter's Specifications. To measure the torque, measure along the wrench to a distance of 180 mm (7.086 inches). Apply a force of 22.2 kg (48.84 lb) at right angles to the measured line **(see illustration)**.
23 Once the nut is tightened, back off the stem locknut just until it begins to turn easily. Now, tighten the locknut until it just becomes hard to turn, and tighten the locknut just a fraction of a turn from that point (not too much, though, or the steering will be too tight). The specified final torque is listed in this Chapter's Specifications. Make sure the steering head turns smoothly and that there's no play in the bearings.
24 Install the lockwasher, then install the upper triple clamp on the steering stem **(see illustration)**. Install the washer and nut, tightening the nut to the torque listed in this Chapter's Specifications.
25 The remainder of installation is the reverse of removal.

### 4  Forks – removal and installation

### Removal

1 Support the bike securely upright.
2 Remove the upper fairing and front fender (see Chapter 8).

**3.22b Install the bearing cover and tighten the locknut to the specified torque . . .**

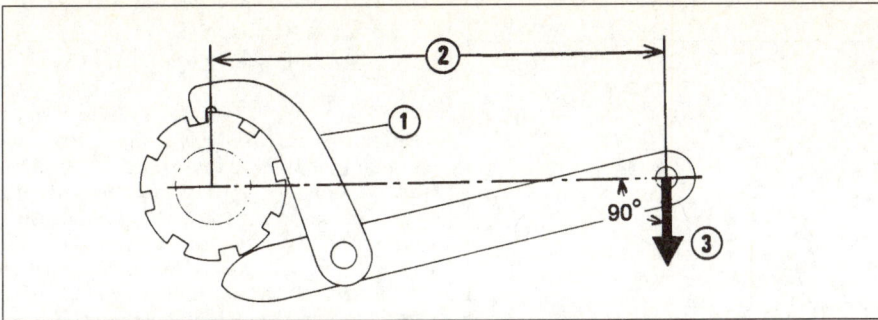

**3.22c . . . by measuring from the center of the steering stem along the tool handle and applying the amount of force described in the text**

1  Spanner wrench (C-spanner)
2  180 mm (7.086 inches)
3  Apply force in the direction of the arrow (at right angles to the measured line)

**3** If you're working on an L or M model, remove the air duct (see Chapter 4).
**4** Remove the front wheel (see Chapter 7).
**5** Remove any wiring harness clamps or straps from the fork tubes.
**6** If the forks will be disassembled after removal, loosen the fork cap bolt **(see illustration)**.
**7** Loosen the upper and lower triple clamp bolts **(see illustration 2.7 and the accompanying illustration)**, then slide the tubes down, using a twisting motion, and remove them from the motorcycle.

> **HAYNES HiNT** *If the fork legs are seized, spray the area with penetrating oil and allow time for it to soak in before trying again.*

**3.24 Install the lockwasher, making sure it engages the notches in the nut**

### Installation

**8** Slide each fork leg into the lower triple clamps.
**9** Slide the fork legs up, installing the tops of the tubes into the upper triple clamp.
**10** If you're working on an H model, position the fork legs so the top of the inner fork tube is flush with the upper triple clamp's top surface **(see illustration)**.
**11** If you're working on a J, K, L and M model, position the fork tube so it protrudes from the upper triple clamp the amount listed in this Chapter's Specifications **(see illustration)**.
**12** The remainder of installation is the reverse of the removal procedure. Be sure to tighten the triple clamp bolts to the torque listed in this Chapter's Specifications.
**13** Pump the front brake lever several times to bring the pads into contact with the discs.

**4.6  Loosen the cap bolt (arrow)**

**4.7  Loosen the lower triple clamp bolts (arrows)**

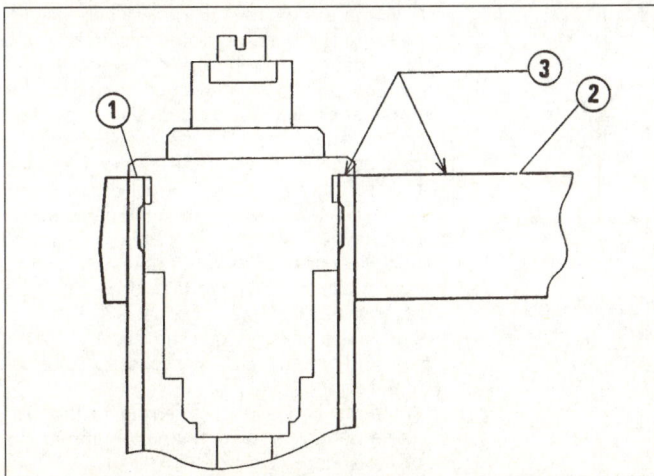

**4.10  The top edge of the fork tube should be flush with the upper surface of the upper triple clamp (H models)**

1  Top edge of fork tube
2  Upper triple clamp
3  Surfaces flush

**4.11  Measure fork protrusion with a vernier caliper; adjust rebound damping (A) and spring preload (B) with the adjusters on top of the fork (J through M models)**

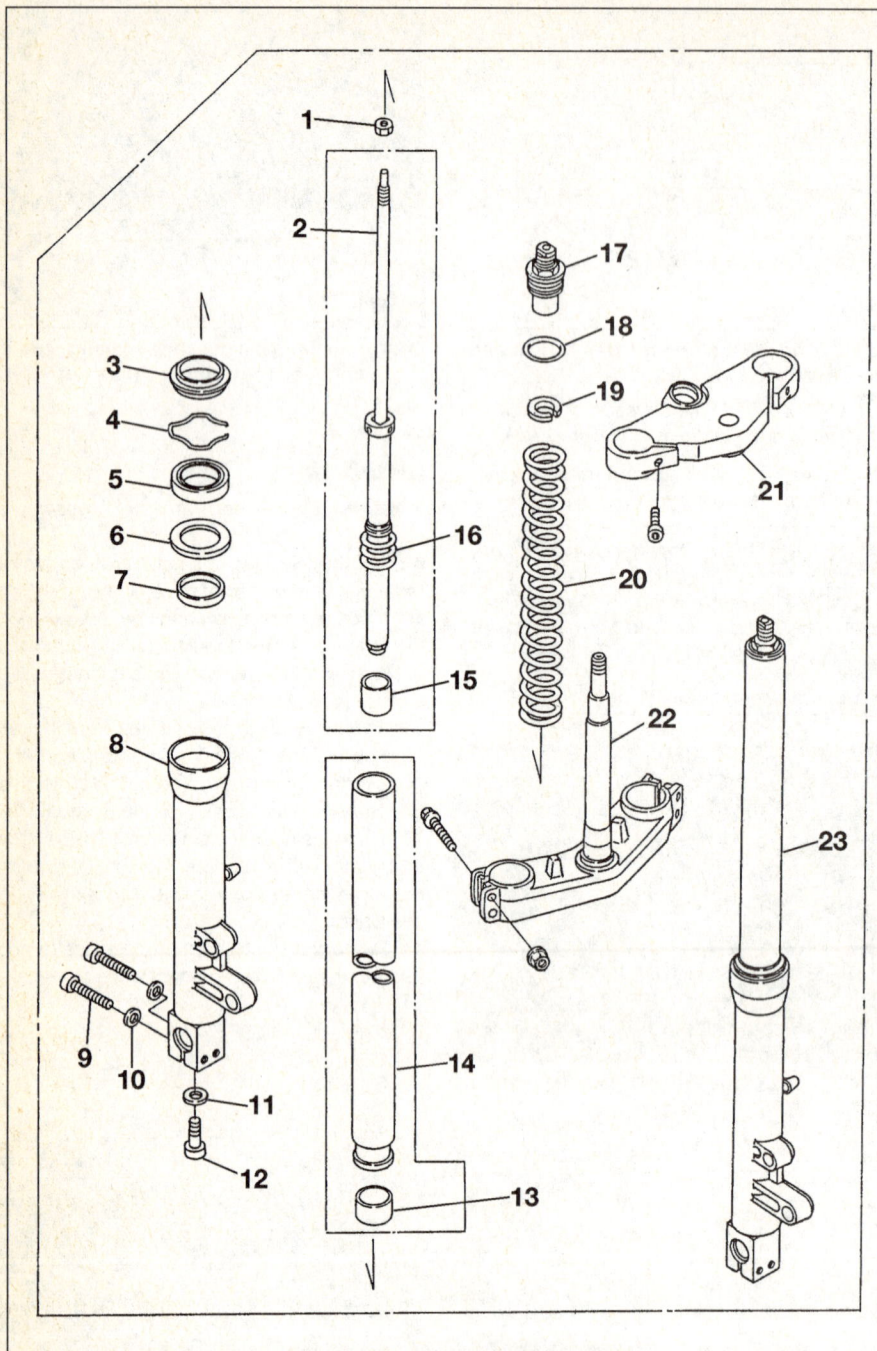

**5.2 Front fork (H models) –
exploded view**

| | |
|---|---|
| 1  Piston rod nut | 13  Inner tube bushing |
| 2  Cartridge | 14  Inner tube |
| 3  Dust seal | 15  Cylinder base |
| 4  Retaining ring | 16  Rebound spring |
| 5  Oil seal | 17  Cap bolt |
| 6  Washer | 18  O-ring |
| 7  Outer tube bushing | 19  Spring seat |
| 8  Outer tube | 20  Fork spring |
| 9  Axle pinch bolts | 21  Upper triple clamp |
| 10  Washers | 22  Steering stem/lower |
| 11  Sealing washer | triple clamp |
| 12  Damper rod bolt | 23  Fork assembly |

## 5  Forks (H models) –
### oil change and overhaul

**Note:** *Before overhauling the forks, read through the procedure, paying special attention to the steps involved in removing the damper rod bolt. If you don't have access to the special tools needed, you can loosen the damper rod bolt before the fork is disassembled, when the spring tension will help keep the damper rod from rotating.*

### Oil change

**1** Loosen the cap bolt and remove the fork leg (see Section 4).

**2** Unscrew the cap bolt from the fork **(see illustration)**. Invert the fork over a drain pan and pour out as much of the oil as will come (more will be poured out after partial disassembly).

**3** Compress the fork spring slightly and place an open-end wrench on the piston rod nut. Place another wrench on the flats of the damping adjuster. Hold the nut and turn the damping adjuster counterclockwise (anti-clockwise) until the damping adjuster and nut separate slightly. At this point, the damping adjuster can be unscrewed from the piston rod by hand.

**4** Remove the spring seat and fork spring. Label the top end of the spring so it can be reinstalled right way up.

**5** Invert the fork over the pan again and pump the piston rod up and down ten times or more to pump out the remaining oil.

**6** Turn the fork upright. Pour in fork oil of the amount and type listed in this Chapter's Specifications.

**7** Slowly pump the piston rod up and down ten times or more to purge any air bubbles from the inner fork tube, then allow several minutes for any remaining air bubbles to rise to the top of the oil.

**8** Compress the fork all the way and measure oil level. The best way to do this is with an oil level gauge, which can be made from a suction pump and a length of tubing **(see illustration 6.15)**. You can also use a tape measure or a rod with the specified oil level marked on it. Position the tubing in the fork tube so its end is at the correct distance from the top of the fork tube (this the same as fork oil level, listed in this Chapter's Specifications). Draw oil into the tube with the suction pump until it starts to draw air. At this point, the oil level will be correct.

**9** If you're using a tape measure or marked rod, add or remove oil as necessary to bring it to the correct level.

**10** Install the spring, referring to the label made during removal. Install the spring seat on the spring.

**11** Measure the depth of the damping adjuster inside the spring adjuster portion of the cap bolt **(see illustration 6.18)**. If necessary, turn it with a screwdriver so the depth is 25 mm (0.984 inch).

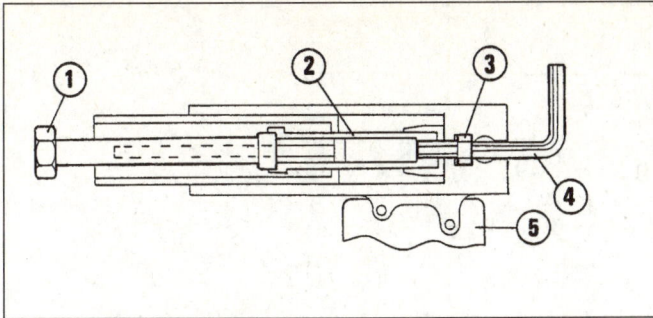

**5.17 The damper rod can be prevented from turning with a special Kawasaki tool**

1  *Damper rod holder*     3  *Damper rod bolt*
   *(Kawasaki tool no.*     4  *Allen wrench*
   *57001-1297)*         5  *Vise*
2  *Cartridge*

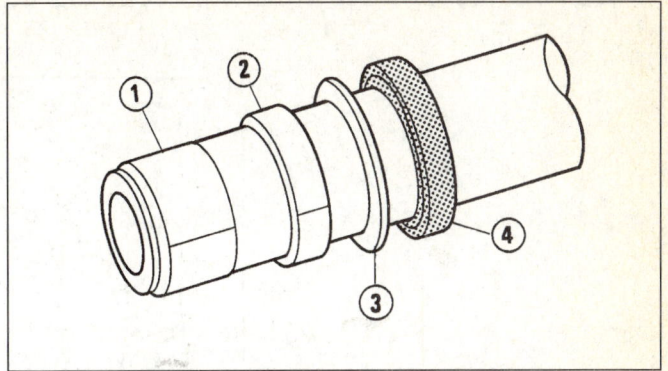

**5.21 The bushings will pop loose when the fork tubes are separated**

1  *Inner tube bushing*     3  *Washer*
2  *Outer tube bushing*     4  *Oil seal*

**12** Compress the spring and lodge the spacer guide against the nut as described in Step 4.
**13** Pull up the piston rod and thread the damping adjuster all the way onto it.
**14** Tighten the piston rod nut and damping adjuster against each other to the torque listed in this Chapter's Specifications **(see illustration 6.5a)**.
**15** Refer to Section 4 and install the fork leg. Change the oil in the other fork leg.

### Disassembly

**16** Perform Steps 1 through 5 above to disassemble the top end of the fork and drain the oil.
**17** Prevent the damper rod from turning using a holding handle and adapter which fit into the hex in the top of the cartridge **(see illustration)**. If you don't have the special tool, you can remove the damper rod bolt before removing the cap bolt; the tension of the fork spring should keep the damper rod from rotating. The usual practice in motorcycle shops is to use an air wrench, which spins the damper rod bolt suddenly enough to loosen it before the damper rod can rotate.
**18** Unscrew the damper rod bolt and remove its sealing washer.
**19** Pry the dust seal out of the outer fork tube.
**20** Pry the retaining ring out of its groove.
**21** Hold the outer tube and yank the inner tube away from it, repeatedly (like a slide-hammer) until the oil seal and bushings pop loose **(see illustration)**. Slide the seal and washer off the inner tube.
**22** Lift the cartridge out of the fork.
**23** Remove the axle pinch bolts and washers.

### Inspection

**24** Clean all parts in solvent and blow them dry with compressed air, if available. Check the inner and outer fork tubes, the guide bushings and the damper rod for score marks, scratches, flaking of the chrome and excessive or abnormal wear. Look for dents in

the tubes and replace them if any are found. Check the fork seal seat for nicks, gouges and scratches. If damage is evident, leaks will occur around the seal-to-outer tube junction. Replace worn or defective parts with new ones.
**25** Have the fork inner tube checked for runout at a dealer service department or other repair shop.
*Caution: If the fork tube is bent, it should not be straightened; replace it with a new one.*
**26** Measure the overall length of the fork spring and check it for cracks and other damage. Compare the length to the minimum length listed in this Chapter's Specifications. If it's defective or sagged, replace both fork springs with new ones. Never replace only one spring.
**27** Check the bushings for wear or damage **(see illustration 5.21)**. If their copper is worn or if they're scratched, replace them. Spread the smaller bushing and pull it off the inner tube, then slide off the larger bushing. When installing the new smaller bushing, spread it just enough to fit over the fork tube.

### Reassembly

**28** Assembly is the reverse of the disassembly procedure, with the following additions:

**6.2a  Unscrew the cap bolt**

a) *To seat the large bushing in the outer fork tube, place the old (used) bushing against the new bushing after the tubes are assembled. Make sure the split in the new bushing will face to one side of the motorcycle when installed, then tap against the washer with a fork seal driver (Kawasaki tool 57001-1340 or equivalent)* **(see illustrations 6.36b and 6.36c)**.
b) *Drive the oil seal in with the same tool used to install the bushing, then install the retaining ring and make sure it's securely seated in its groove.*
c) *Install the dust seal and make sure it's securely seated in the outer fork tube.*
d) *Use a new sealing washer on the damper rod bolt. Apply non-permanent thread locking agent to the threads and tighten the bolt to the torque listed in this Chapter's Specifications.*
e) *When you install the piston rod nut on the cartridge, thread it on finger-tight and make sure at least 12 mm of the threads are exposed above the nut.*

### 6  Forks (J, K, L and M models) – oil change and overhaul

**Note:** *Before overhauling the forks, read through the procedure, paying special attention to the steps involved in removing the damper rod bolt. If you don't have access to the special tools needed, you can loosen the damper rod bolt before the fork is disassembled, when the spring tension will help keep the damper rod from rotating.*

### Oil change

**1** Loosen the cap bolt and remove the fork leg (see Section 4).
**2** Unscrew the cap bolt from the fork **(see illustrations)**. Invert the fork over a drain pan and pour out as much of the oil as will come (more will be poured out after partial disassembly) **(see illustration)**.

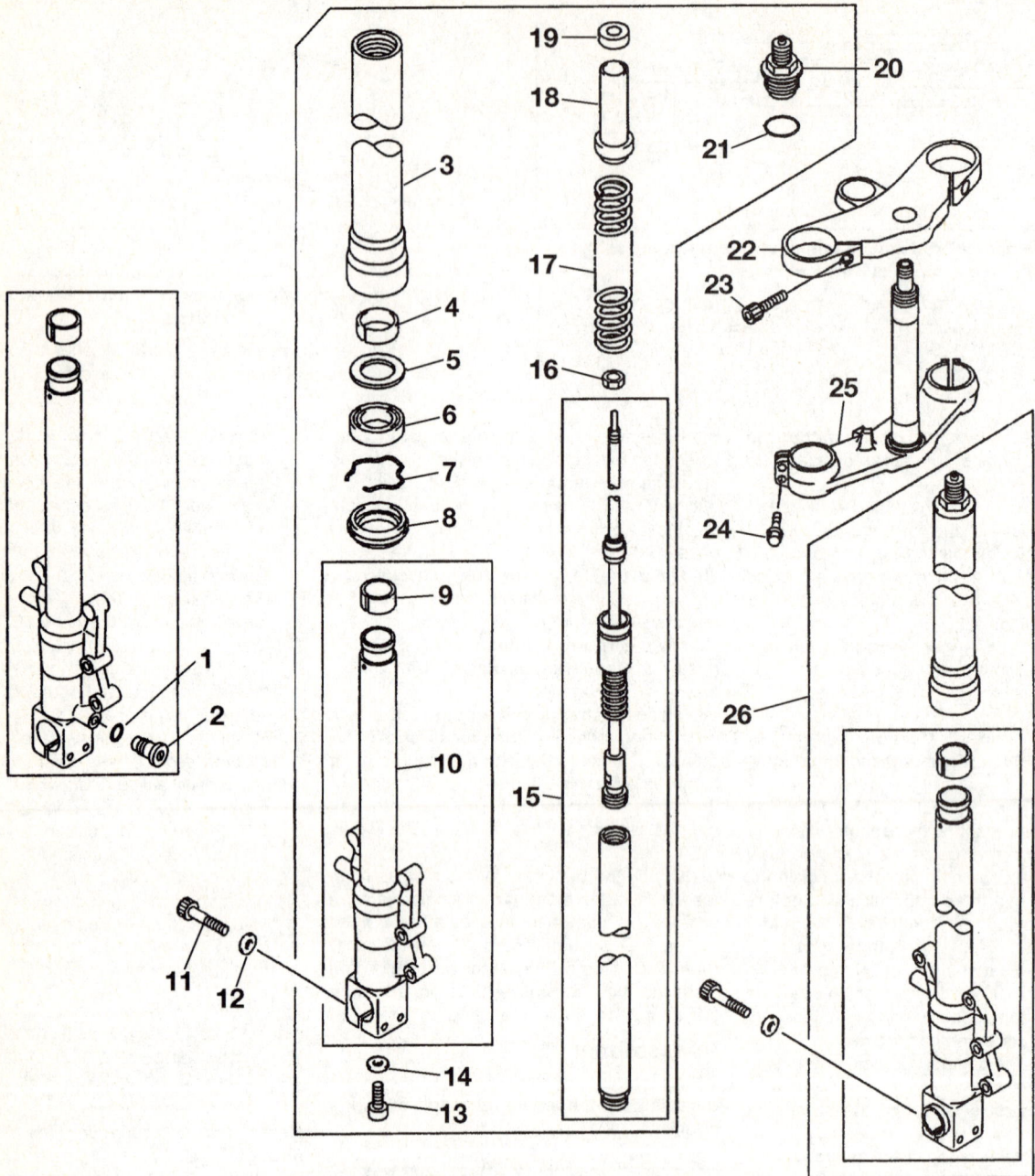

**6.2b Front fork**
**(J, K, L and M models) – exploded view**

| | | | |
|---|---|---|---|
| 1 O-ring | 7 Seal retainer | 14 Copper washer | 21 O-ring |
| 2 Compression damping adjuster (K and M models only) | 8 Dust seal | 15 Cartridge | 22 Upper triple clamp |
| | 9 Bushing | 16 Piston rod nut | 23 Pinch bolt |
| 3 Outer fork tube | 10 Inner fork tube | 17 Fork spring | 24 Pinch bolt |
| 4 Bushing | 11 Axle clamp bolt | 18 Spacer | 25 Steering stem/lower triple clamp |
| 5 Washer | 12 Washer | 19 Spacer guide | |
| 6 Oil seal | 13 Damper rod bolt | 20 Cap bolt | 26 Fork assembly |

6.2c  Pour the fork oil into a container

6.3  Unscrew the cap bolt from the preload and rebound damping adjusters to inspect the O-ring

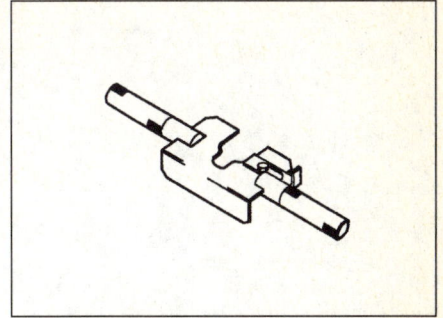

6.4a  Special Kawasaki tool used to compress the fork spring

6.4b  Press down on the pliers (arrows) . . .

6.4c  . . . and cock the spacer (A) sideways so the spacer guide (B) lodges against the piston rod nut (C)

**3** Unscrew the cap bolt from the rebound damping adjuster **(see illustration)**. Although this isn't absolutely necessary to change the fork oil, it is necessary to examine the O-ring on the damping adjuster.

**4** Kawasaki makes a special tool that's used to compress the fork spring by hand so the piston rod nut can be loosened **(see illustration)**. If you don't have the special tool,

place a pair of needle-nosed pliers over the lower lip of the spacer so the jaws rest on the spacer lip **(see illustration)**. Press down on both sides of the pliers (have an assistant help if necessary) until the spacer exposes the piston rod nut, then tilt the top of the spacer sideways and release the spring tension so the spacer guide lodges against the nut **(see illustration)**.

**5** Place an open-end wrench on the nut and another wrench on the flats of the preload adjuster **(see illustration)**. Hold the nut and turn the preload adjuster counterclockwise (anti-clockwise) until the preload adjuster and nut separate slightly. At this point, the preload adjuster can be unscrewed from the piston rod by hand **(see illustration)**.

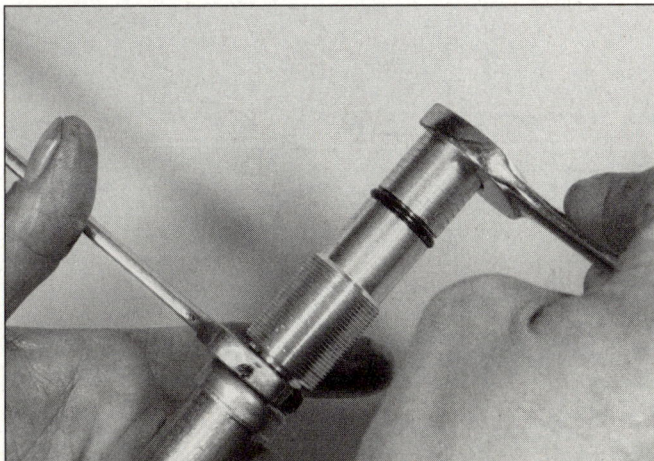

6.5a  Turn the piston rod nut and the preload adjuster away from each other . . .

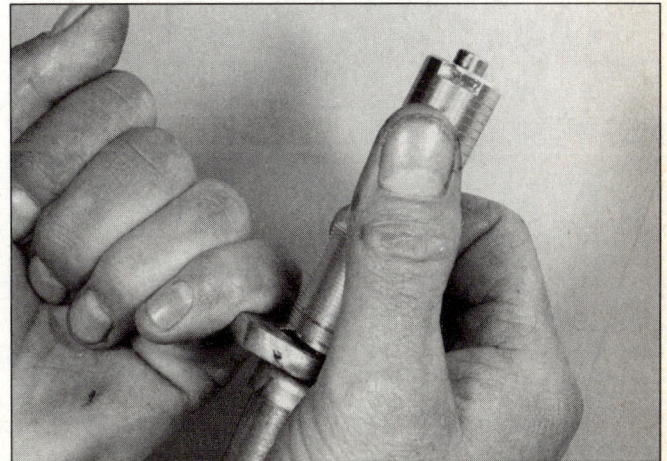

6.5b  . . . then unscrew the preload adjuster from the piston rod

**6.6a Take off the spacer and guide . . .**

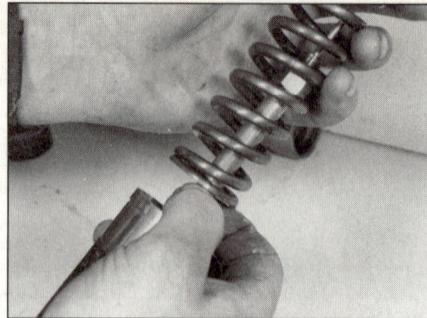

**6.6b . . . lift out the fork spring . . .**

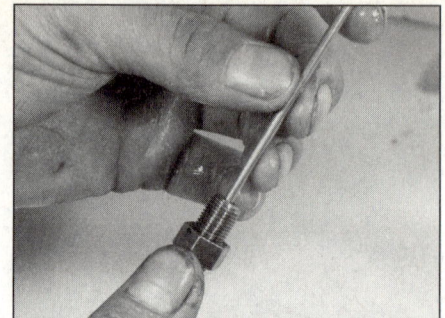

**6.6c . . . and lift the damping adjuster rod out of the piston rod**

**6** Remove the spacer, fork spring and damping adjuster rod **(see illustrations)**.

**7** Invert the fork over the pan again and pump the piston rod up and down ten times or more to pump out the remaining oil.

**8** Turn the fork upright and pull up the piston rod **(see illustration)**. Thread the nut by hand all the way down onto the piston rod and make sure at least 12 mm of threads are exposed above the nut.

**9** Insert the damping adjuster rod into the piston rod **(see illustration 6.6c)**.

**10** Push the piston rod all the way down into the fork and slide the outer fork tube all the way down the inner fork tube.

**11** Fill the fork to the top with the oil listed in this Chapter's Specifications (this is more than the specified amount).

**12** Slowly pump the piston rod up and down five times to purge any air bubbles from the inner fork tube.

**13** Extend and compress the fork tubes several times. While you're doing this, make sure fork oil level stays above the two holes in the top of the inner fork tube.

*Caution: During this step, don't extend the fork all the way or excess oil will be forced into the inner fork tube, raising oil level*

*approximately 30 mm (1.2 inch).*

**14** Let the fork stand upright for five minutes or more so any remaining air bubbles can rise to the top of the oil.

**15** Compress the fork all the way and measure oil level. The best way to do this is with an oil level gauge, which can be made from a suction pump and a length of tubing **(see illustration)**. Position the tubing in the fork tube so its end is at the correct distance from the top of the fork tube (this the same as fork oil level, listed in this Chapter's Specifications). Draw oil into the tube with the suction pump until it starts to draw air. At this point, the oil level will be correct.

**16** Install the spring with its closer-wound coils up. Install the spacer and spacer guide on the spring.

**17** Check the O-ring on the damping adjuster and replace it if its condition is in doubt.

**18** Measure the depth of the damping adjuster inside the spring adjuster portion of the cap bolt **(see illustration)**. If necessary, turn it with a screwdriver so the depth is 25 mm (0.984 inch).

**19** Compress the spring and lodge the spacer guide against the nut as described in Step 4.

**20** Pull up the piston rod and thread the damping adjuster all the way onto it.

**6.8 Hold the piston rod up and thread the nut all the way onto it**

**21** Tighten the piston rod nut and damping adjuster against each other to the torque listed in this Chapter's Specifications **(see illustration 6.5a)**.

**22** Install the cap bolt on the damping adjuster (if removed).

**23** Refer to Section 4 and install the fork leg. Change the oil in the other fork leg.

### Disassembly

**24** Perform Steps 1 through 7 above to disassemble the top end of the fork.

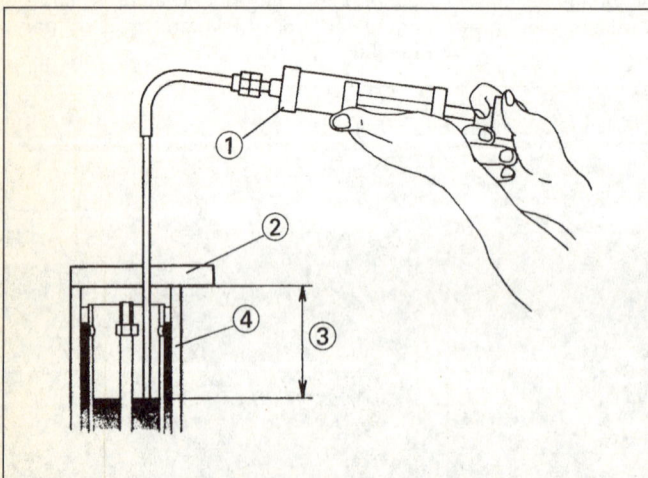

**6.15 Special Kawasaki tool used to measure fork oil level**

1   Oil level gauge (tool no. 57001-1290)
2   Stopper
3   Oil level
4   Outer fork tube

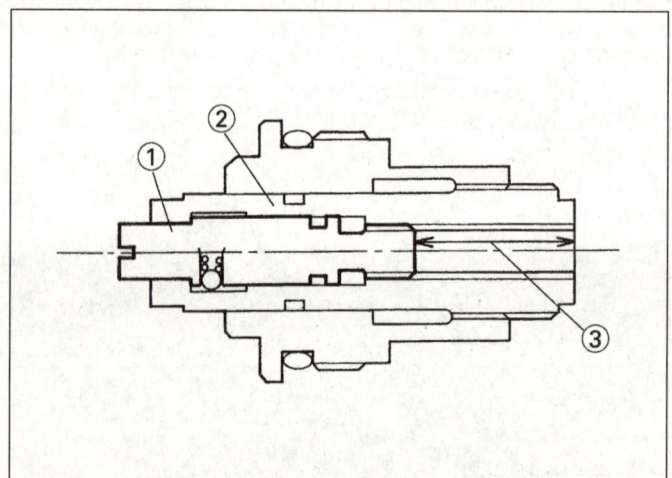

**6.18 Set the damping adjuster to the correct relationship with the preload adjuster**

1   Damping adjuster
2   Preload adjuster
3   Specified distance (see text)

**25** Pry the dust seal out of the outer fork tube **(see illustration)**.
**26** Pry the retaining ring out of its groove **(see illustration)**.
**27** Hold the outer tube and yank the inner tube away from it, repeatedly (like a slide-hammer) until the oil seal and bushing pop loose **(see illustration)**. Slide the seals and washer off the inner tube.
**28** Remove the axle pinch bolts and washers **(see illustration)**.
**29** Prevent the damper rod from turning using a holding handle and adapter which fit into the

hex in the top of the cartridge **(see illustrations)**. If you don't have the special tool, you can remove the damper rod bolt before removing the cap bolt; the tension of the fork spring should keep the damper rod from rotating. The usual practice in motorcycle shops is to use an air wrench, which spins the damper rod bolt suddenly enough to loosen it before the damper rod can rotate.
**30** Unscrew the damper rod bolt and remove its sealing washer **(see illustrations)**.
**31** Lift the cartridge out of the fork **(see illustration)**.

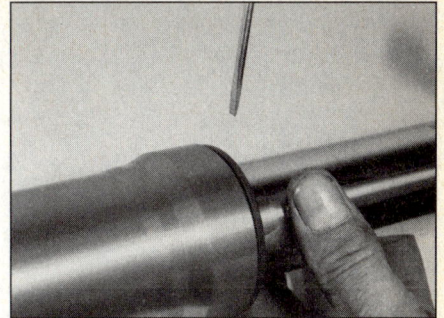

6.25  **Insert a screwdriver between the dust seal and fork tube and pry the seal out**

6.26  **Pry the oil seal retaining ring (arrow) out of its groove**

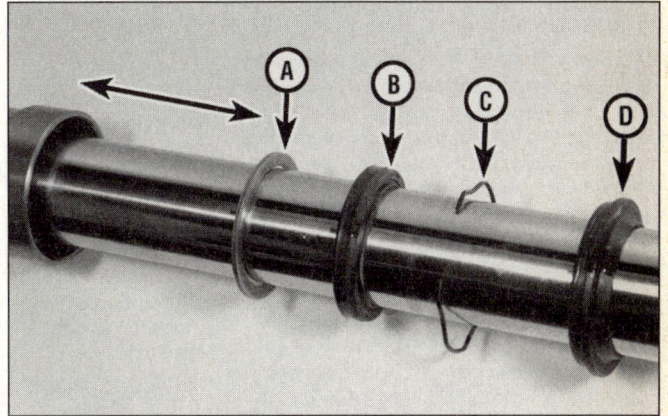

6.27  **Pull the tubes sharply apart until they separate – the oil seal, washer and bushing will come out**

A  *Washer*        C  *Retaining ring*
B  *Oil seal*       D  *Dust seal*

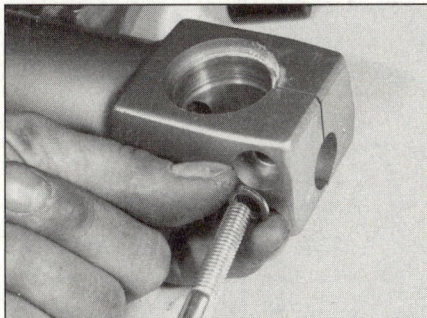

6.28  **Remove the axle pinch bolts and washers**

6.29a  **The special tool fits into the hex in the top of the cartridge (arrow)**

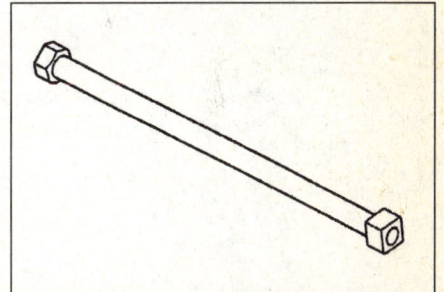

6.29b  **Special Kawasaki tool used to prevent the damper rod from turning while the bolt is loosened**

6.30a  **Insert an Allen wrench into the damper rod bolt . . .**

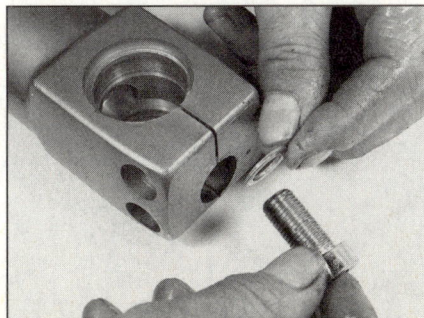

6.30b  **. . . then unscrew the bolt and remove its sealing washer**

6.31  **Lift the cartridge out of the fork**

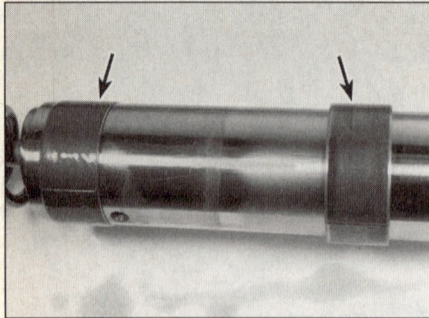

6.35a Check the bushings (arrows) for wear and damage

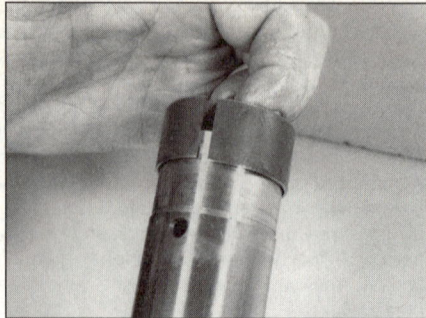

6.35b Pry the small bushing apart at the slit and take it off the fork tube . . .

6.35c . . . then slide the large bushing off

## Inspection

**32** Clean all parts in solvent and blow them dry with compressed air, if available. Check the inner and outer fork tubes, the guide bushings and the damper rod for score marks, scratches, flaking of the chrome and excessive or abnormal wear. Look for dents in the tubes and replace them if any are found. Check the fork seal seat for nicks, gouges and scratches. If damage is evident, leaks will occur around the seal-to-outer tube junction. Replace worn or defective parts with new ones.

**33** Have the fork inner tube checked for runout at a dealer service department or other repair shop.

*Caution: If the fork tube is bent, it should not be straightened; replace it with a new one.*

**34** Measure the overall length of the fork spring and check it for cracks and other damage. Compare the length to the minimum length listed in this Chapter's Specifications. If it's defective or sagged, replace both fork springs with new ones. Never replace only one spring.

**35** Check the bushings for wear or damage **(see illustration)**. If their copper is worn or if they're scratched, replace them. Spread the smaller bushing and pull it off the inner tube, then slide off the larger bushing **(see illustrations)**. When installing the new smaller bushing, spread it just enough to fit over the fork tube.

## Reassembly

**36** Assembly is the reverse of the disassembly procedure, with the following additions:

a) To seat the large bushing in the outer fork tube, place the washer against the bushing after the tubes are assembled. Tap against the washer with a fork seal driver (Kawasaki tool 57001-1340 or equivalent) **(see illustrations)**.

b) Wrap the end of the inner fork tube with tape or plastic wrap to protect the oil seal while it's installed **(see illustration)**. Drive the seal in with the same tool used to install the bushing, then install the retaining ring and make sure it's securely seated in its groove **(see illustrations)**.

c) Install the dust seal and make sure it's securely seated in the outer fork tube **(see illustration)**.

d) Use a new sealing washer on the damper rod bolt. Apply non-permanent thread locking agent to the threads and tighten the bolt to the torque listed in this Chapter's Specifications.

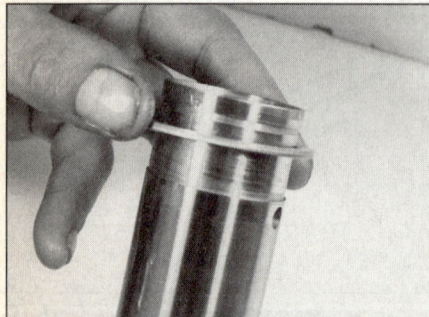

6.36a Install the washer . . .

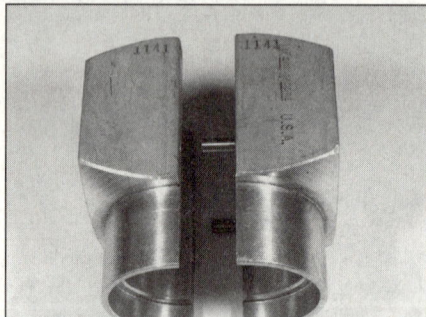

6.36b . . . and tap against it with a seal driver . . .

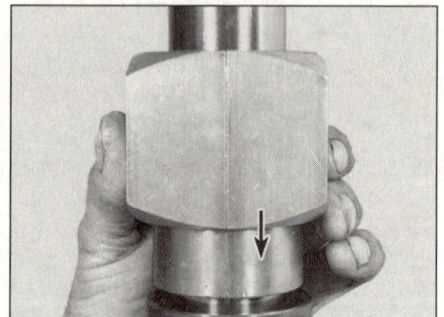

6.36c . . . tap down (arrow) gently and repeatedly to seat the bushing

6.36d Wrap the groove in the fork tube with tape to prevent the seal from being cut

6.36e Tap the seal into place until it seats in the fork tube (tool disassembled for clarity)

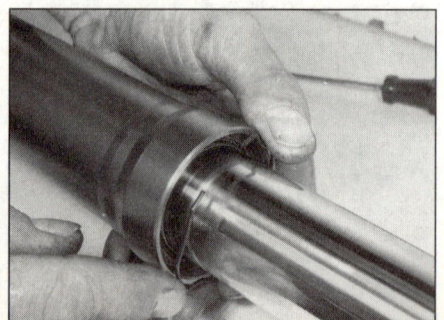

6.36f Compress the retaining ring and fit it securely into its groove

**6.36g  Seat the dust seal in the outer fork tube**

## 7  Rear shock absorber –
removal, disposal and installation

### Removal

**1**  Support the bike securely upright so there's no chance of it being knocked over during this procedure.

### H models

**2**  Remove the side covers and lower fairing panels (see Chapter 8).

**3**  Raise the rear wheel off the ground with a jack. Support the motorcycle securely so it can't be knocked over while it's jacked up.

**4**  Remove the nuts from all three bolts that secure the suspension rocker arm to the frame, the tie-rods and the shock absorber **(see illustration)**.

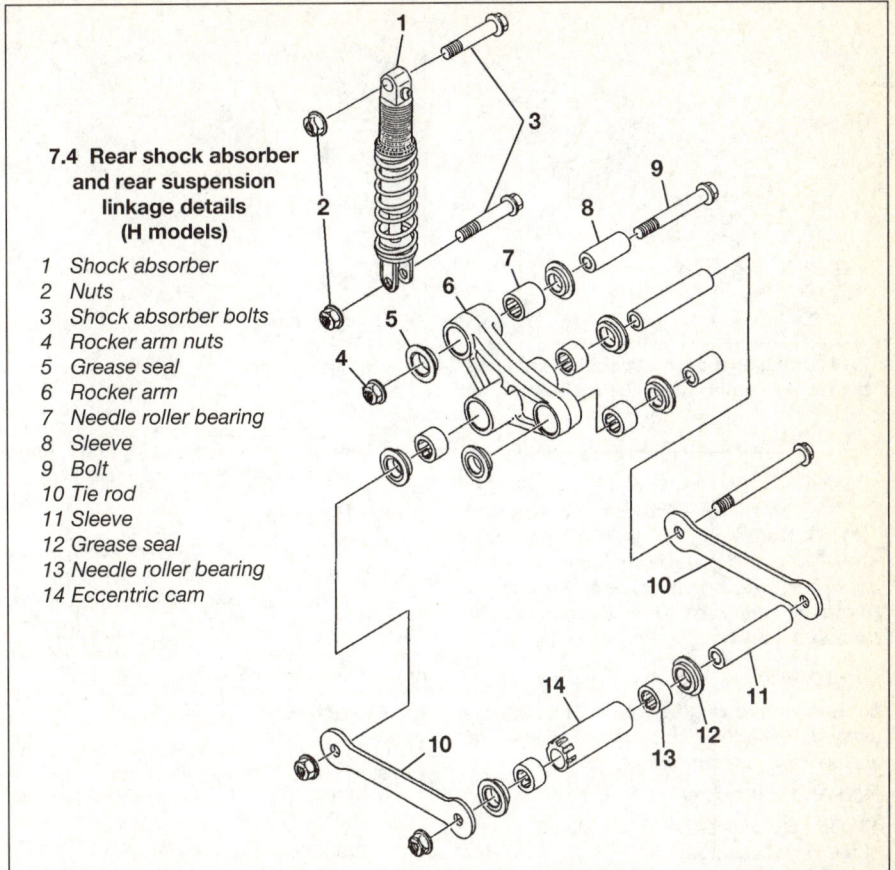

**7.4  Rear shock absorber and rear suspension linkage details (H models)**

1   Shock absorber
2   Nuts
3   Shock absorber bolts
4   Rocker arm nuts
5   Grease seal
6   Rocker arm
7   Needle roller bearing
8   Sleeve
9   Bolt
10  Tie rod
11  Sleeve
12  Grease seal
13  Needle roller bearing
14  Eccentric cam

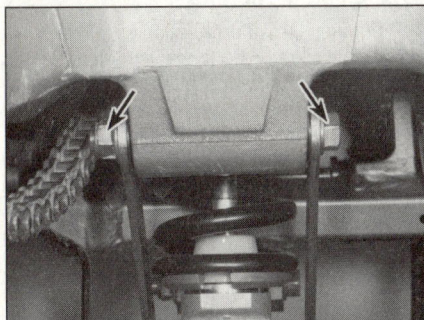

**7.9a  Remove the nut and bolt (arrows) and detach the upper ends of the tie-rods from the swingarm**

**5**  Support the muffler/silencer. Unbolt the rear footpeg bracket from the frame (see Chapter 8) – it also serves as the muffler/silencer bracket. Raise or lower the bracket as needed to provide removal clearance for the three bolts mentioned in Step 4 and slide the bolts out.

**6**  Remove the shock absorber upper bolt and lower the shock away from the motorcycle.

⚠ *Warning: The shock absorber contains high pressure nitrogen gas. If you're replacing the shock absorber, prepare it as described in the Disposal steps of this Section before throwing it away.*

**7**  Installation is the reverse of the removal

procedure. Tighten the shock absorber and tie-rod nuts to the torque values listed in this Chapter's Specifications.

### J, K, L and M models

**8**  If you're working on a K or M model, remove the pillion cover (see Chapter 8). Remove the fuel enricher solenoid (see Chapter 4), the fuel filter bracket and the clamping screws for the shock absorber gas reservoir.

**9**  Support the rear suspension rocker arm. Remove the bolt that secures the upper ends of the tie-rods to the swingarm and the bolts at the top and bottom of the shock absorber **(see illustrations)**. Lower the shock absorber and take it out from under the motorcycle.

**7.9b  Remove the nut (arrow) and bolt from the lower end of the shock absorber**

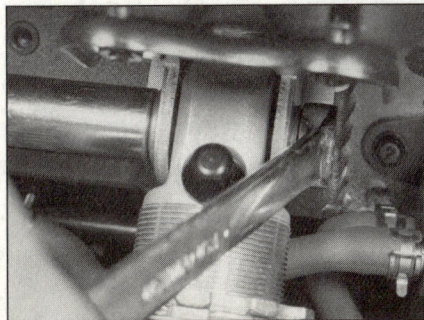

**7.9c  Hold the upper bolt with a wrench and remove the nut . . .**

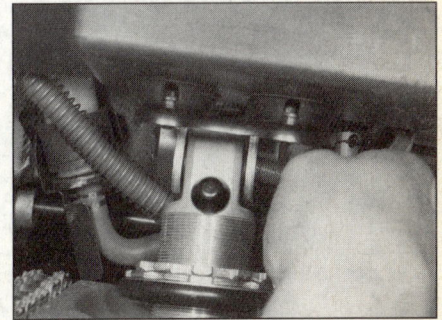

**7.9d  . . . then pull out the bolt and lower the shock absorber away from the bike**

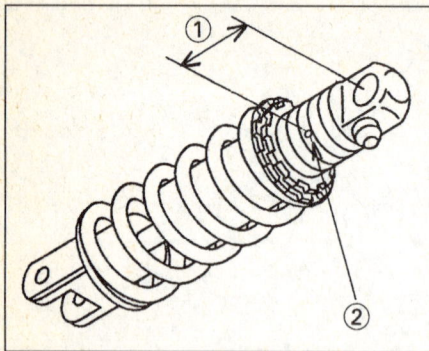

**7.14 Drill a hole at the specified distance below the centerline of the mounting bolt hole**

1  42 to 44 mm (1.65 to 1.73 inch)
2  Hole

⚠ **Warning: The shock absorber contains high pressure nitrogen gas. If you're replacing the shock absorber, prepare it as described in the Disposal steps of this Section before throwing it away.**

### Disposal

**10** Before throwing the old shock absorber away, release the gas pressure to avoid an accidental explosion.

#### H, J and L models

**11** Do not remove the plastic plug at the top of the shock absorber.
**12** Wear eye protection so you won't be injured by flying metal chips.
**13** Hold the shock absorber upright in its installed position so oil won't be blown out when drilling.
**14** Drill a hole at a point 42 to 44 mm (1.65 to 1.73 inch) below the centerline of the upper mounting bolt hole **(see illustration)**.

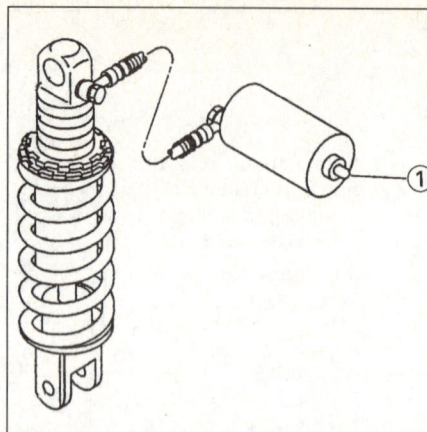

**7.15 Remove the valve cap (1) and release the gas through the valve**

#### K and M models

**15** Remove the cap from the valve on the reservoir **(see illustration)**.
**16** Point the valve away from yourself, then open the valve and release all gas from the reservoir. Remove the valve after the gas pressure has been released.

### Installation

**17** Installation is the reverse of the removal steps, with the following additions:
 a) Refer to Section 8 and lubricate the rocker arm needle bearings with molybdenum disulfide grease.
 b) Refer to Section 15 and adjust the damping and spring preload.
 c) Install the shock with the damping adjuster facing the right side of the bike (H models) or the rear (J, K, L and M models).
 d) Tighten the nuts and bolts to the torques listed in this Chapter's Specifications.

 e) On K and M models, install the gas reservoir with its hose against the tube of the rear frame section (so it won't rub against the seat while the bike is being ridden).

---

### 8  Rear suspension linkage – removal, check and installation

### Removal

**1** Support the bike securely upright so there's no chance of it being knocked over during this procedure.
**2** Raise the rear wheel off the ground with a jack. Support the motorcycle securely so it can't be knocked over while it's jacked up.

#### H models

**3** Remove the nuts from all three bolts that secure the suspension rocker arm to the frame, the tie-rods and the shock absorber **(see illustration 7.4)**.
**4** Support the muffler/silencer. Unbolt the rear footpeg bracket from the frame (see Chapter 8) – it also serves as the muffler/silencer bracket. Raise or lower the bracket as needed to provide removal clearance for the three bolts mentioned in Step 3 and slide the bolts out.
**5** Lower the rocker arm away from the motorcycle.
**6** Remove the bolt that secures the upper ends of the tie-rods to the swingarm **(see illustration 7.4)**. Take the tie-rods out.

#### J, K, L and M models

**7** Remove the muffler/silencer (see Chapter 4).
**8** Remove all three bolts that secure the rocker arm to the frame, shock absorber and tie-rods **(see illustrations)**. Lower the rocker arm away from the motorcycle **(see illustration)**.

**8.8a Remove the rocker arm nuts and bolt (lower arrows) and the tie-rod upper nut and bolt (upper arrow)**

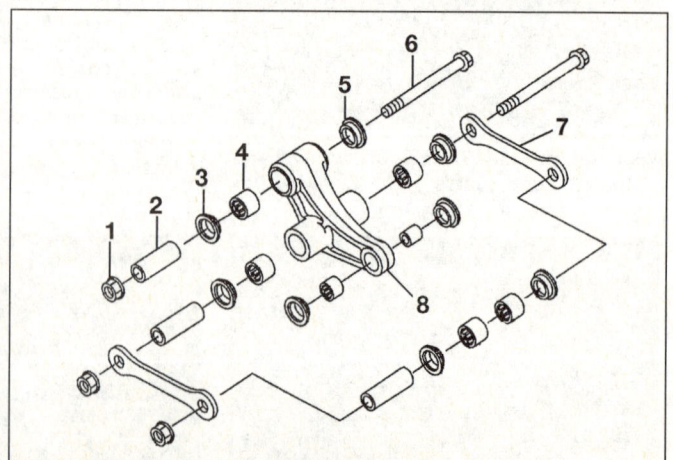

**8.8b Suspension linkage details (J, K, L and M models)**

| | | | |
|---|---|---|---|
| 1  Nut | 4  Needle roller | 6  Bolt |
| 2  Sleeve | bearing | 7  Tie-rod |
| 3  Grease seal | 5  Grease seal | 8  Rocker arm |

**9** Remove the bolt that secures the upper ends of the tie-rods to the swingarm **(see illustration 8.8a)**. Take the tie-rods out.

### Check

**10** Pry the grease seals off the rocker arm and push out the sleeves **(see illustrations)**.

**11** Check the sleeves and needle roller bearings **(see illustration)** for wear or damage. Replace the sleeves and bearings as a set if either is worn or damaged. Bearing removal requires a blind hole puller and slide hammer. Bearing installation requires a shouldered drift which fits the bearing exactly to prevent bearing damage (Kawasaki tool no. 57001-1058). If you don't have the correct tool, have the bearings replaced by a Kawasaki dealer or machine shop.

**12** Apply molybdenum disulfide grease to the bearings and sleeves and install each sleeve in its bearing. Press the grease seals into place around the sleeves.

**13** Inspect, and if necessary replace, the tie-rod bearings in the same manner as the

rocker arm bearings (they're located in the swingarm).

### Installation

**14** Installation is the reverse of removal. If you're working on an H model, align the center mark on the tie-rod eccentric cam with the split in the swingarm **(see illustration)**. Tighten the bolts and nuts to the torques listed in this Chapter's Specifications.

### 9 Swingarm bearings – check

**1** Refer to Chapter 7 and remove the rear wheel, then refer to Section 7 and remove the rear shock absorber.

**2** Grasp the rear of the swingarm with one hand and place your other hand at the junction of the swingarm and the frame. Try to move the rear of the swingarm from side-to-side. Any wear (play) in the bearings should be felt as movement between the swingarm and

8.8c  Take the rocker arm out

the frame at the front. The swingarm will actually be felt to move forward and backward at the front (not from side-to-side). If any play is noted, the bearings should be replaced with new ones (see Section 11).

**3** Next, move the swingarm up and down through its full travel. It should move freely, without any binding or rough spots. If it does not move freely, refer to Section 11 for servicing procedures.

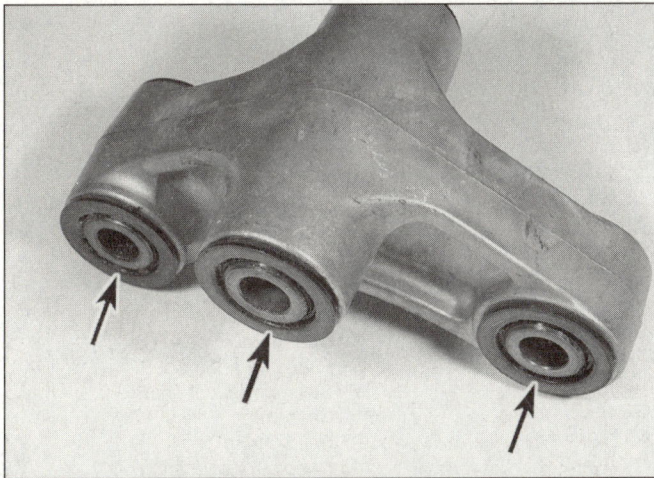
8.10a  Pry off the grease seals (arrows) . . .

8.10b  . . . and remove the sleeve from each bearing

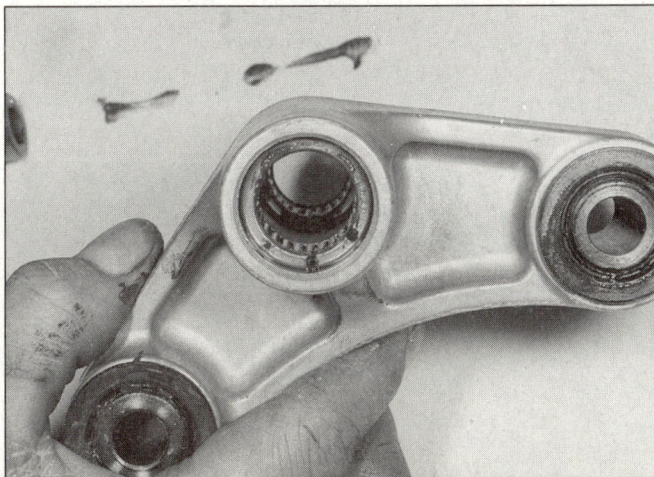
8.11  Check the bearings for wear and damage

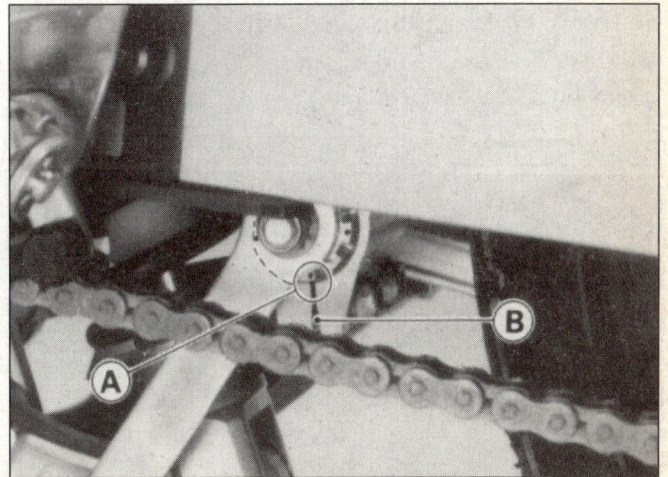
8.14  Align the center mark on the eccentric cam (A) with the split in the swingarm (B) (H models)

## 10 Swingarm –
### removal and installation

### Removal

**1** Support the bike securely upright so there's no chance of it being knocked over during this procedure.
**2** Raise the rear wheel off the ground with a jack. Support the motorcycle securely so it can't be knocked over while it's jacked up.
**3** Remove the rear wheel (see Chapter 7).

### H models

**4** Remove the lower fairing panels (see Chapter 8).

**5** Remove the shift pedal (see Chapter 2).
**6** Remove the left rider footpeg bracket (see Chapter 8).
**7** Remove the engine sprocket and slide the drive chain off the engine output shaft (see Section 12).
**8** Detach the rear brake hose from the swingarm.
**9** Remove the lower shock absorber bolt and the bolt that secures the lower ends of the tie-rods to the rocker arm **(see illustration 7.4)**.
**10** Remove the swingarm pivot shaft nut from the left side of the bike **(see illustrations)**. On the right side, remove the three bolts that secure the pivot shaft holder to the frame (H1 models only) or remove the pivot shaft cover (H2 models only).

**11** Support the swingarm and pull the pivot shaft out. Remove the swingarm. If necessary, remove the bolt and detach the tie-rods from the swingarm.

### J, K, L and M models

**12** Remove the rear brake torque link rear bolt and detach the rear brake hose from its clips and retainer (see Chapter 7).
**13** Remove the shock absorber (see Section 7).
**14** Remove the swingarm pivot shaft nut **(see illustrations)**.
**15** Support the swingarm and pull the pivot shaft out **(see illustration)**. Withdraw the swingarm to the rear of the motorcycle.
**16** To disengage the chain and remove the swingarm completely, remove the rear fender

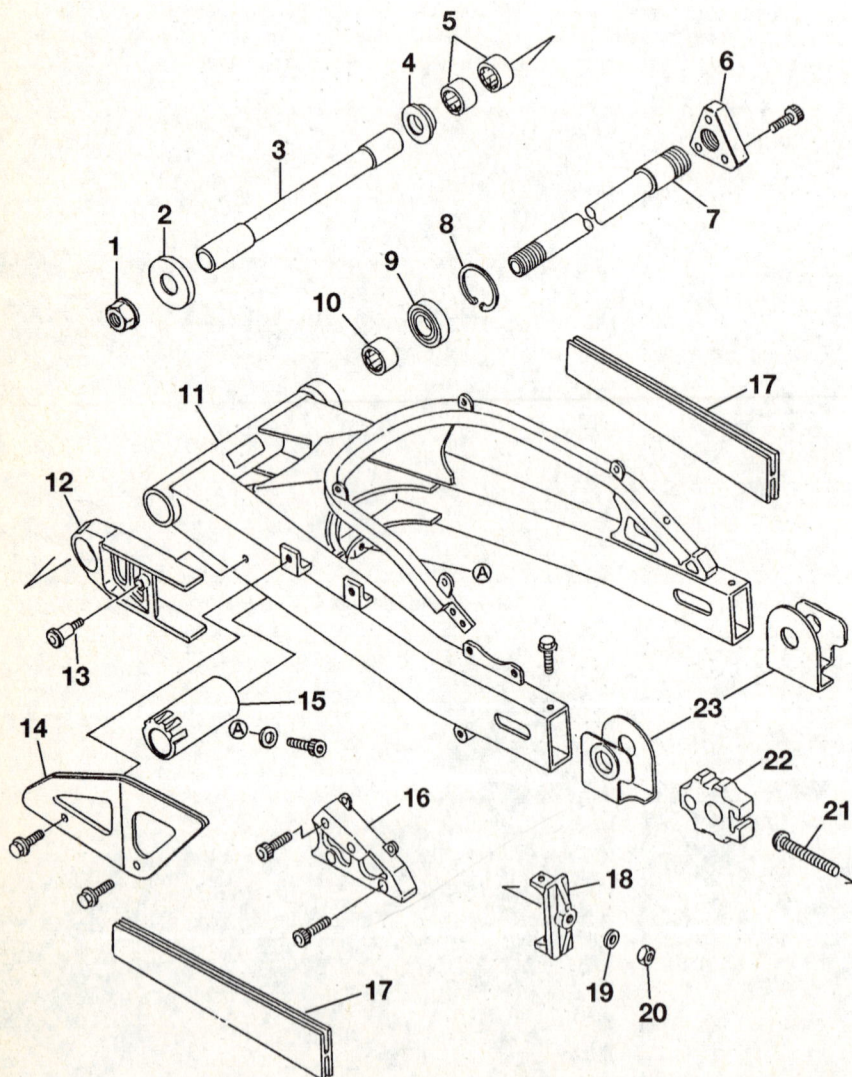

**10.10a Swingarm details (H1 models)**

1  Nut
2  Bearing cap
3  Pivot shaft bushing
4  Grease seal
5  Needle roller bearings
6  Pivot shaft holder
7  Swingarm pivot shaft
8  Snap-ring
9  Ball bearing
10 Needle roller bearing
11 Swingarm
12 Chain buffer
13 Chain buffer bolt
14 Chain guard
15 Eccentric cam
16 Chain guard
17 Chain guide
18 End cap
19 Washer
20 Chain adjuster locknut
21 Chain adjuster bolt
22 Chain adjuster
23 Axle holder

**10.10b  Swingarm details (H2 models)**

1  Nut
2  Bearing cap
3  Pivot shaft bushing
4  Grease seal
5  Needle roller bearings
6  Chain buffer
7  Eccentric cam
8  Chain adjuster locknut
9  Chain adjuster bolt
10  Chain adjuster
11  Swingarm
12  Pivot shaft cover
13  Swingarm pivot shaft
14  Collar
15  Grease seal
16  Snap-ring
17  Ball bearing
18  Needle roller bearing

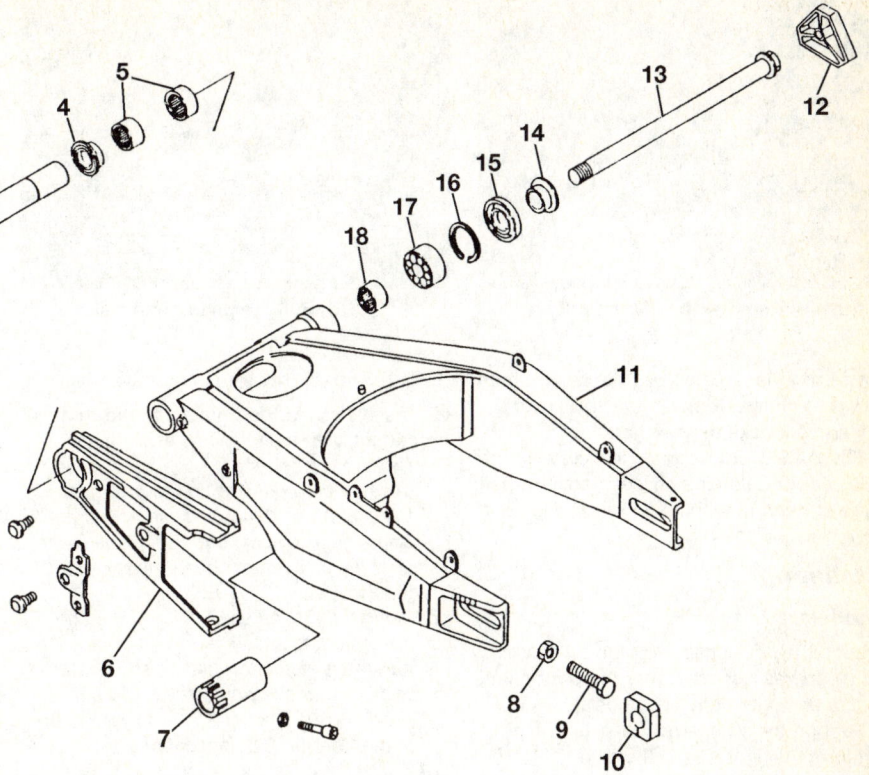

**10.14a  Swingarm details (J, K, L and M models)**

1  Nut
2  Collar
3  Pivot shaft bushing
4  Grease seal
5  Needle roller bearing
6  Swingarm
7  Needle roller bearing
8  Ball bearing
9  Snap-ring
10  Grease seal
11  Collar
12  Swing arm pivot shaft
13  Chain adjuster
14  Chain adjuster bolt
15  Locknut

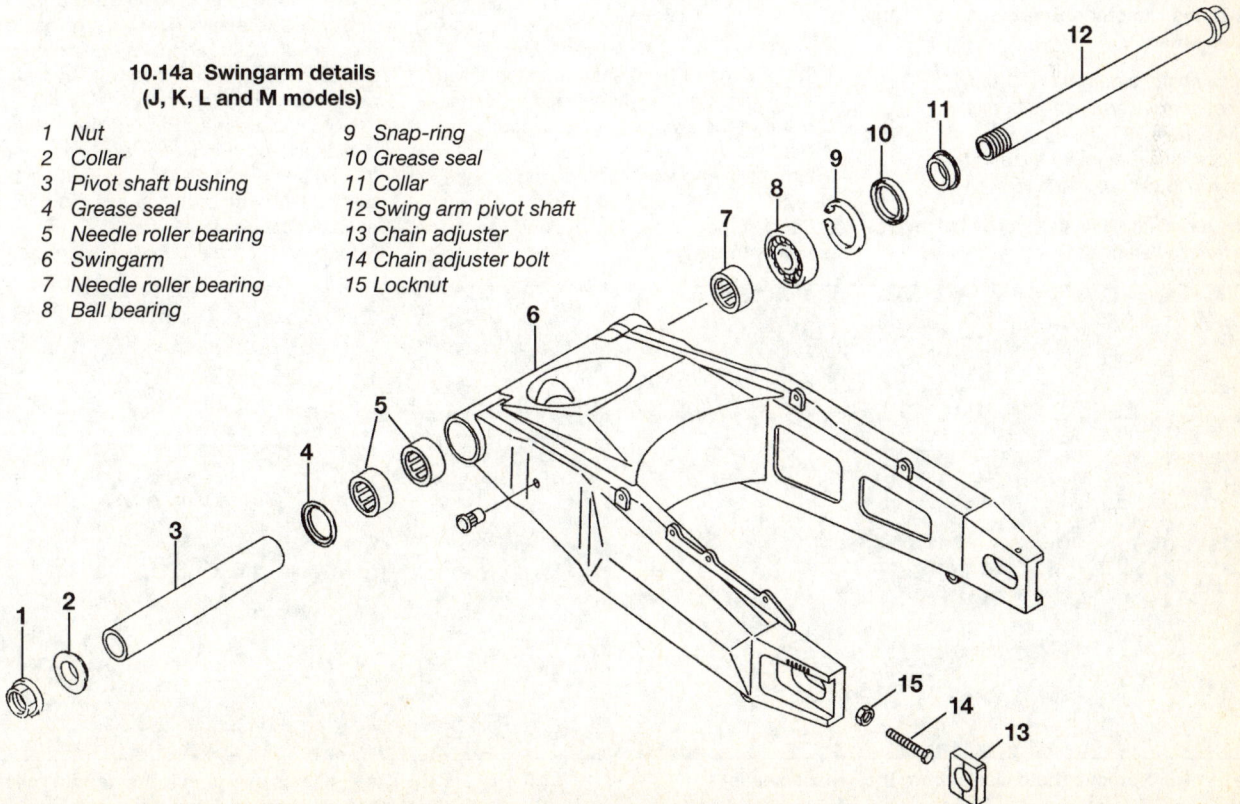

**10.14b Remove the pivot shaft nut**

**10.15 Pull the pivot shaft out**

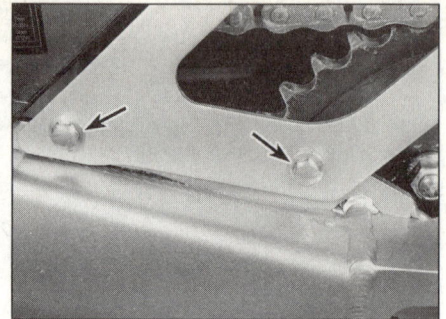

**10.16 Remove the chain guard bolts (arrows)**

front section (see Chapter 8) and the chain guard **(see illustration)**. Lift the chain over the swingarm and take the swingarm out.

17 Check the pivot bearings in the swingarm for dryness or deterioration (see Section 11). If they're in need of lubrication or replacement, refer to Section 11.

### Installation

#### H models

18 Loop the drive chain over the swingarm. If the holder was removed from the pivot shaft, thread it on finger-tight (H1 models).

19 Support the swingarm and install the pivot shaft from the right side of the bike.

20 Make sure the bearing grease seal and cap are in place (on the nut side of H1 models; on both sides of H2 models), then tighten the swingarm fasteners to the torque listed in this Chapter's Specifications, in the following order:

a) Pivot shaft holder bolts (H1 only)
b) Pivot shaft Allen head (H1 only)
c) Pivot shaft nut
d) Shock absorber lower nut and bolt
e) Tie rod lower nut and bolt

21 The remainder of installation is the reverse of the removal steps.

#### J, K, L and M models

22 Installation is the reverse of the removal procedure, with the following additions:

a) Be sure the bearing grease seals are in position before installing the pivot shaft.
b) Don't confuse the pivot shaft with the rear axle. The axle has a cotter pin hole in the end; the pivot shaft doesn't **(see illustration)**.
c) Tighten the pivot shaft nut and the shock absorber and suspension linkage nuts/bolts to the torque values listed in this Chapter's Specifications. Tighten the brake torque link bolt to the torque listed in the Chapter 7 Specifications.
d) Adjust the chain as described in Chapter 1.

---

### 11 Swingarm bearings – replacement

1 The swingarm pivot shaft rides on a sealed ball bearing and three needle roller bearings.
2 Remove the swingarm (see Section 10).
3 Remove the seal collar from the right side (H2 models) and the bearing cap from the left side (H1 and H2 models). Remove the seal collar from both sides (J, K, L and M models) **(see illustration)**.

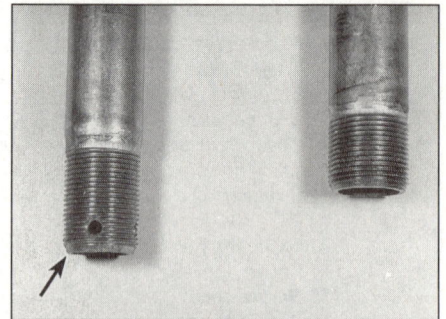

**10.22 The rear axle has a cotter pin hole in the end (arrow); the swingarm pivot shaft doesn't**

4 Slide the sleeve out **(see illustration 10.10a, 10.10b or 10.14b)**.
5 Pry out the seals **(see illustration)**.
6 Check the sleeve for wear or damage. If there's any doubt about its condition, replace the sleeve, the ball bearing and all three needle roller bearings as a set.
7 Rotate the center race of the ball bearing and check for roughness, looseness or play. If there's any doubt about its condition, replace the ball bearing, the sleeve and all three needle roller bearings as a set.
8 Inspect the needle roller bearings (use a

**11.3 Remove the collar(s) from the grease seal(s)**

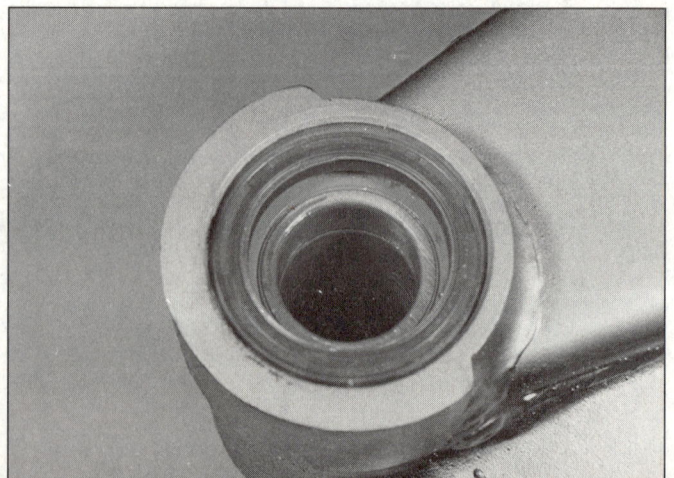

**11.5 Pry out the grease seals and check the needle roller bearings for wear and damage**

flashlight if necessary). If there was play in the swingarm, or if there's any doubt about the condition of any of the needle rollers bearings, replace the ball bearing, the sleeve and all three needle roller bearings as a set.

**9** Bearing removal requires a blind hole puller; installation of the needle roller bearings requires a shouldered drift of the exact same size as the bearings. If you don't have the correct tools, have the bearings replaced by a Kawasaki dealer. If you do the job yourself, remove the snap-ring before removing the ball bearing and install the ball bearing with its manufacturer's marks out.

**10** Lubricate the needle roller bearings and ball bearing with molybdenum disulfide grease. If the ball bearing is of the fully sealed type it will not require lubrication.

## 12 Drive chain –
removal, cleaning and installation

### *Removal*

**1** Remove the lower left fairing panel (see Chapter 8).

**12.4a Remove the engine sprocket cover bolts (749 cc engine shown) . . .**

**2** If you're working on an H model, remove the shift pedal (see Chapter 2) and the alternator bracket bolts (see Chapter 9).

**3** Remove the clutch slave cylinder (see Chapter 2).

**4** Remove the engine sprocket cover **(see illustrations)**.

**5** Bend back the tabs of the sprocket lockwasher **(see illustrations)**.

**6** If you're working on an H model, insert a rod between two sprocket teeth into the slot in the internal shift mechanism cover to hold the sprocket from turning, then loosen the nut

**12.4b . . . and take the cover off**

(see illustration).

**7** If you're working on a J, K, L and M model, shift the transmission into gear and have an assistant hold the brake on while you loosen the nut **(see illustration)**.

**8** Slide the sprocket off the output shaft and lift the chain off the sprocket **(see illustration)**.

**9** Detach the swingarm from the frame (see Section 10). Pull the swingarm back far enough to allow the chain to slip between the frame and the front of the swingarm **(see illustration)**.

**12.5a Bend back the lockwasher tabs (arrow) with a hammer and punch on a 748 cc engine . . .**

**12.5b . . . and on a 749 cc engine**

**12.6 On 748 cc engines, insert a rod through the sprocket to prevent the sprocket from turning**

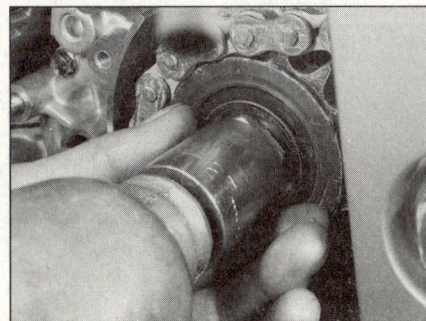

**12.7 Loosen the sprocket nut . . .**

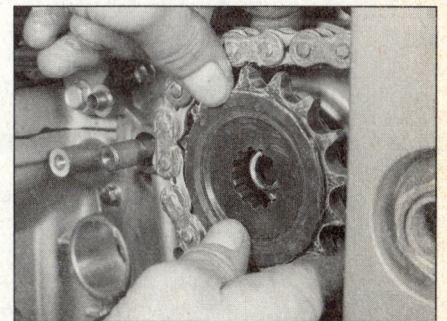

**12.8 . . . and take off the sprocket, then disengage it from the chain**

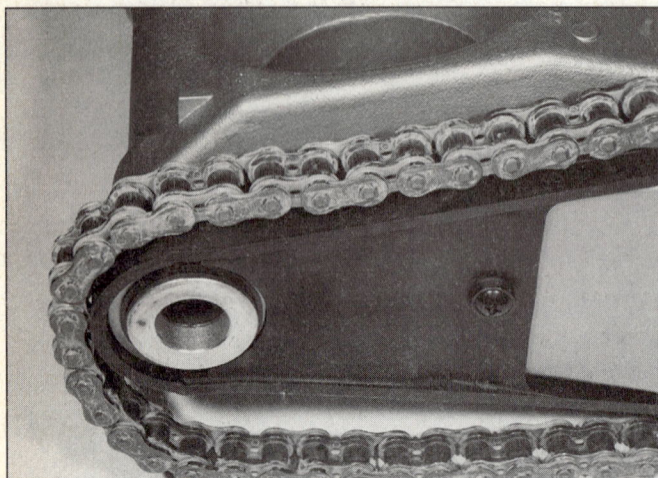
**12.9 Pass the chain over the swingarm to remove it**

**12.11 Make sure the cover dowels (arrow) are in position**

## Cleaning

**Caution: Don't use gasoline or other cleaning fluids for cleaning the drive chain.**
**10** Soak the chain in kerosene (paraffin) or diesel fuel for approximately five or six minutes. Remove the chain, wipe it off then blow dry it with compressed air immediately. The entire process shouldn't take longer than ten minutes – if it does, the O-rings in the chain rollers could be damaged.

## Installation

**11** Installation is the reverse of the removal procedure. Make sure the cover dowels are in position **(see illustration)**. Tighten the suspension fasteners, engine sprocket nut and the engine sprocket cover bolts to the torque values listed in this Chapter's Specifications. Tighten the rear axle nut to the torque listed in Chapter 7 Specifications.
**12** If you're working on an H model, install the shift pedal (see Chapter 2) and the alternator bracket (see Chapter 9); remember to set the alternator drivebelt tension.

**13** Lubricate the chain and check its freeplay following the procedure described in Chapter 1.

## 13 Sprockets – check and replacement

**1** Support the bike securely upright.
**2** Whenever the drive chain is inspected, the sprockets should be inspected also. If you are replacing the chain, replace the sprockets as well. Likewise, if the sprockets are in need of replacement, install a new chain also.
**3** Remove the engine sprocket cover following the procedure outlined in the previous Section.
**4** Attach a dial indicator to the swingarm, with the plunger of the indicator touching the rear sprocket near its outer diameter **(see illustration)**. Turn the wheel and measure the runout. If the runout exceeds the maximum runout listed in this Chapter's Specifications, replace the rear sprocket. As stated before,

it's a good idea to replace the chain and the sprockets as a set. However, if the components are relatively new or in good condition, but the sprocket is warped, you may be able to get away with just replacing the rear sprocket.
**5** Check the wear pattern on the sprockets (see Chapter 1). If the sprocket teeth are worn excessively, replace the chain and sprockets.
**6** To replace the rear sprocket, remove the rear wheel (see Chapter 7). Unscrew the nuts holding it to the wheel coupling and lift the sprocket off. Install the sprocket with the mark that indicates the number of teeth facing out and apply a non-hardening thread locking compound to the threads of the studs. Tighten the nuts to the torque listed in this Chapter's Specifications. Also, check the condition of the rubber damper under the rear wheel coupling (see Section 14).
**7** When installing the engine sprocket, engage it with the chain and slip it onto the output shaft. Install a new lockwasher, then tighten the nut to the torque listed in this

**13.4 Check the runout of the rear sprocket with a dial indicator**

**14.3 Lift the damper segments out of the wheel**

15.3 Fork compression damping is adjusted by turning a screw (arrow)

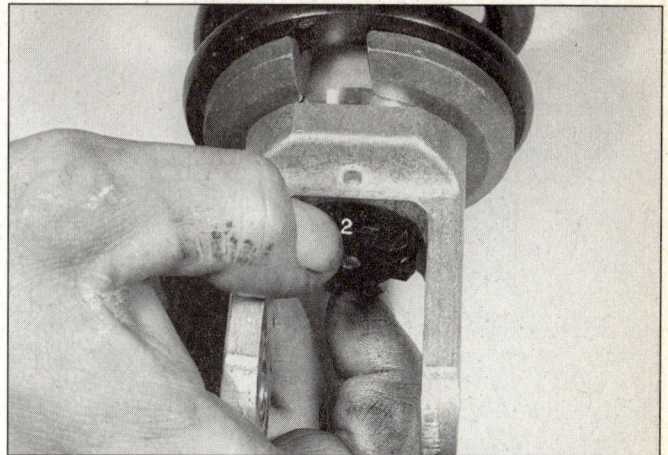

15.4 Rear shock absorber rebound damping is adjusted with a knob on the bottom of the shock (L model shown)

Chapter's Specifications and bend the lockwasher against the nut.

8 The remainder of installation is the reverse of the removal steps.

## 14 Rear wheel coupling/ rubber damper – check and replacement

1 Remove the rear wheel (see Chapter 7).

2 Lift the collar and rear sprocket/rear wheel coupling from the wheel (see Chapter 7).

3 Lift the rubber damper segments from the wheel (see illustration) and check them for cracks, hardening and general deterioration. Replace them as a set with new ones if necessary.

4 Checking and replacement procedures for the coupling bearing are similar to those described for the wheel bearings. Refer to Chapter 7.

5 Installation is the reverse of the removal procedure.

## 15 Suspension adjustments

### Front forks

⚠ Warning: Always adjust both front forks to the same settings or unstable handling may occur.

1 Set rebound damping with the adjuster on top of the fork (see illustration 4.11). Turn the adjuster all the way clockwise with a screwdriver, then back it out to the desired setting. Rebound damping settings are listed in this Chapter's Specifications.

2 Set spring preload by turning the adjuster with an open-end wrench on its flats (see illustration 4.11). To increase spring preload, turn the adjuster clockwise. To decrease it, turn it anti-clockwise. Preload settings are listed in this Chapter's Specifications.

3 Fork compression damping is adjustable on K and M models. Turn the adjuster all the way clockwise, then back it out to the desired setting (see illustration). Compression damping settings are listed in this Chapter's Specifications.

### Rear shock absorber

4 Set rebound damping with the adjuster on the bottom of the shock absorber (see illustration). On H models, you'll need to remove the cover for access to the adjuster. Suspension settings are listed in this Chapter's Specifications.

5 To set spring preload, refer to Section 7 and remove the shock absorber from the bike. Loosen the locknut, then the adjusting nut until the spring is completely extended (see illustration). Measure the spring free length, then tighten the adjusting nut to compress the spring to the desired length. Preload settings and spring lengths are listed in this Chapter's Specifications. Once preload is set, tighten the locknut against the adjusting nut.

6 Compression damping is adjustable on K and M models. Turn the adjuster all the way clockwise, then back it out to the desired setting (see illustration). Suspension settings are listed in this Chapter's Specifications.

15.5 Loosen the locknut with a special wrench then loosen the adjusting nut until the spring extends completely

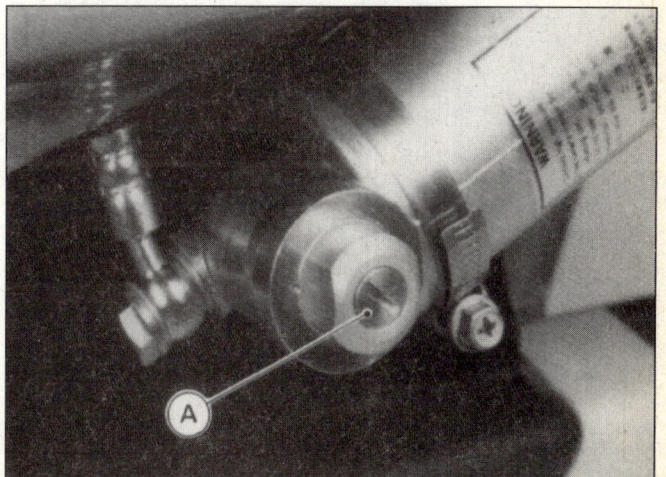

15.6 Rear shock absorber compression damping is adjusted by turning a screw on the gas reservoir (A)

# Chapter 7
# Brakes, wheels and tires

## Contents

## Degrees of difficulty

| Easy, suitable for novice with little experience | Fairly easy, suitable for beginner with some experience | Fairly difficult, suitable for competent DIY mechanic | Difficult, suitable for experienced DIY mechanic | Very difficult, suitable for expert DIY or professional |
|---|---|---|---|---|

## Specifications

### Brakes

| | |
|---|---|
| Brake fluid type . . . . . . . . . . . . . . . . . . . . . . . . . . . . . . . . . . . . . . . . | DOT 4 |
| Brake pad minimum thickness . . . . . . . . . . . . . . . . . . . . . . . . . . . . . . | See Chapter 1 |
| Brake pedal height . . . . . . . . . . . . . . . . . . . . . . . . . . . . . . . . . . . . . . . | See Chapter 1 |

Front disc thickness
  Standard
    H, J and K models . . . . . . . . . . . . . . . . . . . . . . . . . . . . . . . . . . . . 4.8 to 5.1 mm (0.189 to 0.200 inch)
    L and M models . . . . . . . . . . . . . . . . . . . . . . . . . . . . . . . . . . . . . 4.8 to 5.2 mm (0.189 to 0.204 inch)
  Minimum* . . . . . . . . . . . . . . . . . . . . . . . . . . . . . . . . . . . . . . . . . . . 4.5 mm (0.177 inch)
Rear disc thickness
  Standard . . . . . . . . . . . . . . . . . . . . . . . . . . . . . . . . . . . . . . . . . . . . 5.8 to 6.1 mm (0.228 to 0.240 inch)
  Minimum* . . . . . . . . . . . . . . . . . . . . . . . . . . . . . . . . . . . . . . . . . . . 5.5 mm (0.217 inch)
* Refer to marks stamped into the disc (they supersede information printed here)
Disc runout (maximum, front and rear, all models) . . . . . . . . . . . . . . . 0.3 mm (0.012 inch)

## Wheels and tires

Wheel runout
  Axial (side-to-side) . . . . . . . . . . . . . . . . . . . . . . . . . . . . . . . . . . . . . . . 0.5 mm (0.020 inch)
  Radial (out-of-round) . . . . . . . . . . . . . . . . . . . . . . . . . . . . . . . . . . . . . . 0.8 mm (0.031 inch)
  Rear axle runout . . . . . . . . . . . . . . . . . . . . . . . . . . . . . . . . . . . . . . . . . 0.2 mm (0.007 inch)
Tire pressures . . . . . . . . . . . . . . . . . . . . . . . . . . . . . . . . . . . . . . . . . . . . . . See Chapter 1
Tire tread depth . . . . . . . . . . . . . . . . . . . . . . . . . . . . . . . . . . . . . . . . . . . . See Chapter 1
Tire sizes
  H1 models
    Front . . . . . . . . . . . . . . . . . . . . . . . . . . . . . . . . . . . . . . . . . . . . . . . . . 120/70 VR17-V260
    Rear . . . . . . . . . . . . . . . . . . . . . . . . . . . . . . . . . . . . . . . . . . . . . . . . . 170/60 VR17-V260
  H2 models
    Front . . . . . . . . . . . . . . . . . . . . . . . . . . . . . . . . . . . . . . . . . . . . . . . . . 120/70 VR17-V260 or 120/70 ZR17
    Rear . . . . . . . . . . . . . . . . . . . . . . . . . . . . . . . . . . . . . . . . . . . . . . . . . 170/60 VR17-V260, 170/60 ZR17 or 180/55 ZR17
  J and K models
    Front . . . . . . . . . . . . . . . . . . . . . . . . . . . . . . . . . . . . . . . . . . . . . . . . . 120/70 VR17-V260 or 120/70 ZR17
    Rear . . . . . . . . . . . . . . . . . . . . . . . . . . . . . . . . . . . . . . . . . . . . . . . . . 180/55 VR17-V260 or 180/55 ZR17
  L and M models
    Front . . . . . . . . . . . . . . . . . . . . . . . . . . . . . . . . . . . . . . . . . . . . . . . . . 120/70 ZR17
    Rear . . . . . . . . . . . . . . . . . . . . . . . . . . . . . . . . . . . . . . . . . . . . . . . . . 180/55 ZR17

## Torque specifications

Front calipers
  Mounting bolts . . . . . . . . . . . . . . . . . . . . . . . . . . . . . . . . . . . . . . . . . . . 34 Nm (25 ft-lbs)
  Assembly bolts . . . . . . . . . . . . . . . . . . . . . . . . . . . . . . . . . . . . . . . . . . . 21 Nm (15 ft-lbs)
  Bleed valve . . . . . . . . . . . . . . . . . . . . . . . . . . . . . . . . . . . . . . . . . . . . . 7.8 Nm (69 inch-lbs)
  Pad cover screws . . . . . . . . . . . . . . . . . . . . . . . . . . . . . . . . . . . . . . . . 2.9 Nm (26 inch-lbs)
Rear caliper
  Mounting bolts . . . . . . . . . . . . . . . . . . . . . . . . . . . . . . . . . . . . . . . . . . . 25 Nm (18 ft-lbs)
  Assembly bolts . . . . . . . . . . . . . . . . . . . . . . . . . . . . . . . . . . . . . . . . . . . 32 Nm (24 ft-lbs)
  Bleed valves . . . . . . . . . . . . . . . . . . . . . . . . . . . . . . . . . . . . . . . . . . . . 7.8 Nm (69 inch-lbs)
Banjo fitting bolts . . . . . . . . . . . . . . . . . . . . . . . . . . . . . . . . . . . . . . . . . . 25 Nm (18 ft-lbs)
Brake disc-to-wheel bolts . . . . . . . . . . . . . . . . . . . . . . . . . . . . . . . . . . . 23 Nm (16.5 ft-lbs)
Front master cylinder
  Mounting bolts
    H models . . . . . . . . . . . . . . . . . . . . . . . . . . . . . . . . . . . . . . . . . . . . . 11 Nm (95 inch-lbs)
    J, K, L and M models . . . . . . . . . . . . . . . . . . . . . . . . . . . . . . . . . . . . 8.8 Nm (78 inch-lbs)
  Pivot screw . . . . . . . . . . . . . . . . . . . . . . . . . . . . . . . . . . . . . . . . . . . . . 1 Nm (9 inch-lbs)
  Pivot screw locknut . . . . . . . . . . . . . . . . . . . . . . . . . . . . . . . . . . . . . . . 5.9 Nm (52 inch-lbs)
  Brake light switch mounting screw . . . . . . . . . . . . . . . . . . . . . . . . . . . 1.2 Nm (10 inch-lbs)
  Reservoir retainer screw . . . . . . . . . . . . . . . . . . . . . . . . . . . . . . . . . . . 1.5 Nm (13 inch-lbs)
Rear master cylinder
  Mounting bolts
    H models . . . . . . . . . . . . . . . . . . . . . . . . . . . . . . . . . . . . . . . . . . . . . 25 Nm (18 ft-lbs)
    J, K, L and M models . . . . . . . . . . . . . . . . . . . . . . . . . . . . . . . . . . . . 23 Nm (16.5 ft-lbs)
  Reservoir mounting bolt(s) . . . . . . . . . . . . . . . . . . . . . . . . . . . . . . . . . 6.9 Nm (61 inch-lbs)
  Clevis locknut . . . . . . . . . . . . . . . . . . . . . . . . . . . . . . . . . . . . . . . . . . . 18 Nm (13 ft-lbs)
Torque link bolts and nuts . . . . . . . . . . . . . . . . . . . . . . . . . . . . . . . . . . . 25 Nm (18 ft-lbs)
Pedal pivot bolt . . . . . . . . . . . . . . . . . . . . . . . . . . . . . . . . . . . . . . . . . . . 25 Nm (18 ft-lbs)
Front axle and nut
  H models . . . . . . . . . . . . . . . . . . . . . . . . . . . . . . . . . . . . . . . . . . . . . . . 110 Nm (80 ft-lbs)
  J, K, L and M models . . . . . . . . . . . . . . . . . . . . . . . . . . . . . . . . . . . . . . 145 Nm (110 ft-lbs)
Front axle clamp bolts . . . . . . . . . . . . . . . . . . . . . . . . . . . . . . . . . . . . . . 20 Nm (14.5 ft-lbs)
Rear axle nut
  H models . . . . . . . . . . . . . . . . . . . . . . . . . . . . . . . . . . . . . . . . . . . . . . . 110 Nm (80 ft-lbs)
  J, K, L and M models . . . . . . . . . . . . . . . . . . . . . . . . . . . . . . . . . . . . . . 145 Nm (110 ft-lbs)

## 1  General information

The motorcycles covered by this manual are equipped with hydraulic disc brakes on the front and rear wheels. All models use a pair of dual-piston calipers on the front wheel and one dual-piston caliper on the rear wheel.

All models are equipped with cast aluminum wheels, which require very little maintenance and allow tubeless tires to be used.

*Caution: Disc brake components rarely require disassembly. Do not disassemble components unless absolutely necessary. If any hydraulic brake line connection in the system is loosened, the entire system should be disassembled, drained, cleaned and then properly filled and bled upon reassembly. Do not use solvents on internal brake components. Solvents will cause seals to swell and distort. Use only clean brake fluid or alcohol for cleaning. Use care when working with brake fluid as it can injure your eyes and it will damage painted surfaces and plastic parts.*

2.2a  Remove the banjo fitting bolt (A); the protrusion on the end of the hose rests against the stop (B)

2.2b  Separate the hose from the caliper; there's a sealing washer on each side of the fitting

2.3  Remove the caliper mounting bolts (arrows); the other Allen bolts are caliper assembly bolts only to be loosened during caliper overhaul

2.4a  Detach the hose from the clips on the swingarm . . .

## 2  Brake caliper – removal, overhaul and installation

**Warning: If a front caliper indicates the need for an overhaul (usually due to leaking fluid or sticky operation), BOTH front calipers should be overhauled and all old brake fluid flushed from the system. Also, the dust created by the brake system may contain asbestos, which is harmful to your health. Never blow it out with compressed air and don't inhale any of it. An approved filtering mask should be worn when working on the brakes. Do not, under any circumstances, use petroleum-based solvents to clean brake parts. Use brake cleaner or denatured alcohol only!**

**Note:** *If you are removing the caliper only to remove the front or rear wheel or to replace or inspect the rear brake pads, don't disconnect the hose from the caliper.*

### Removal

#### Front caliper

1  Support the bike securely upright. **Note:** *If you're planning to disassemble the caliper, read through the overhaul procedure, paying particular attention to the steps involved in removing the pistons with compressed air. If you don't have access to an air compressor, you can use the bike's hydraulic system to force the pistons out instead. To do this, remove the pads and pump the brake lever. If one piston comes out before the others, push it back into its bore and hold it in with a C-clamp (G-clamp) while pumping the brake lever to remove the remaining pistons.*
**Note:** *Remember, if you're just removing the caliper to remove the front wheel, ignore the following step.*
2  Disconnect the brake hose from the caliper. Remove the brake hose banjo fitting bolt and separate the hose from the caliper **(see illustrations)**. Discard the sealing washers. Plug the end of the hose or wrap a plastic bag tightly around it to prevent excessive fluid loss and contamination.

3  Unscrew the caliper mounting bolts **(see illustration)**. Lift off the caliper, being careful not to strain or twist the brake hose if it's still connected.

#### Rear caliper

4  If you're removing the rear caliper but leaving the brake hose connected, detach the brake hose from the clips on the swingarm and detach the brake hose retainer from the rear fender **(see illustrations)**.

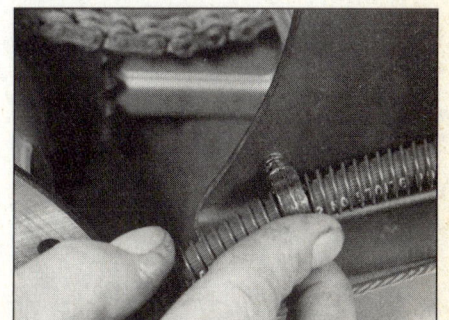

2.4b  . . . and (if equipped) detach the retainer from the fender

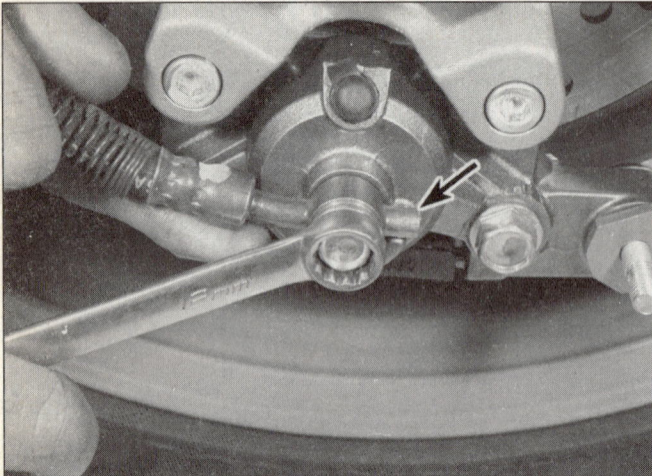

2.5a Remove the banjo fitting bolt; the protrusion on the end of the hose rests against the stop (arrow)

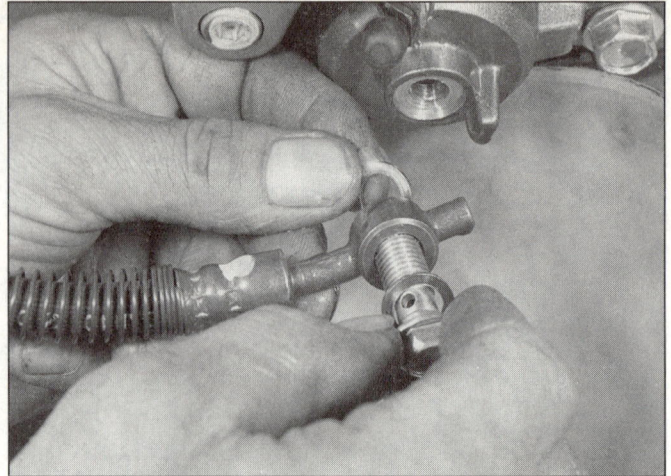

2.5b Separate the hose from the caliper; there's a sealing washer on each side of the fitting

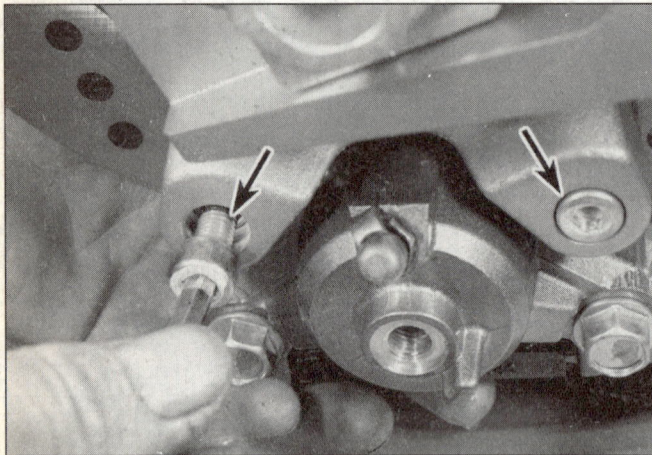

2.6a Remove the caliper mounting bolts (arrows); the two hex bolts are caliper assembly bolts only to be loosened during caliper overhaul

2.6b Lower the caliper away from the disc

**Note:** *If you're only removing the caliper to replace brake pads or remove the wheel, ignore the following step.*

5 Disconnect the brake hose from the caliper. Remove the brake hose banjo fitting bolt and separate the hose from the caliper **(see illustrations)**. Discard the sealing washers.

Plug the end of the hose or wrap a plastic bag tightly around it to prevent excessive fluid loss and contamination.

6 Unscrew the caliper mounting bolts **(see illustration)**. Lower the caliper, being careful not to strain or twist the brake hose if it's still connected **(see illustration)**.

7 Detach the caliper from the torque link and check the bushing for wear or damage **(see illustrations)**.

## Overhaul

### Front caliper

8 Remove the brake pads and anti-rattle spring from the caliper (see Section 3, if necessary). Clean the exterior of the caliper with denatured alcohol or brake system cleaner.

9 Remove the four Allen bolts that hold the caliper halves together and separate the caliper halves **(see illustrations)**.

10 Bolt a piece of wood at least 10 mm thick to the mating surface of each caliper half to prevent the pistons from flying out. The wood should block one of the fluid inlets. Use compressed air, directed into the other fluid inlet, to remove the piston(s) **(see illustrations)**. Use only enough air pressure to ease the piston(s) out of the bore. If a piston is blown out forcefully, even with the wood in place, it may be damaged.

2.7a Detach the rear end of the torque link from the caliper . . .

2.7b . . . and check the bushing for wear or damage

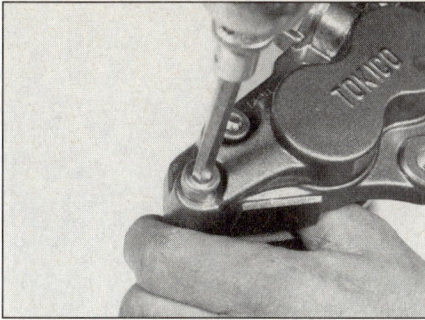

2.9a Remove the caliper assembly bolts . . .

2.9b . . . and separate the caliper halves

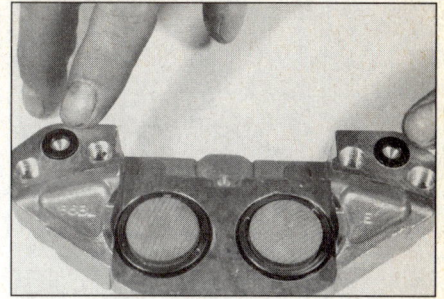

2.9c There's an O-ring at each of the fluid ports to be replaced whenever the caliper is disassembled

**Warning: Never place your fingers in front of the piston in an attempt to catch or protect it when applying compressed air, as serious injury could occur.**

**11** Using a wood or plastic tool, remove the piston seals **(see illustration)**. Metal tools may cause bore damage.

**12** Clean the pistons and the bores with denatured alcohol, clean brake fluid or brake system cleaner and blow dry them with filtered, unlubricated compressed air. Inspect the surfaces of the pistons for nicks and burrs and loss of plating. Check the caliper bores, too. If surface defects are present, the caliper must be replaced. If the caliper is in bad shape, the master cylinder should also be checked.

**13** Lubricate the piston seals with clean brake fluid and install them in their grooves in the caliper bore. Make sure they aren't twisted and seat completely.

**14** Lubricate the dust seals with clean brake fluid and install them in their grooves, making sure they seat correctly.

**15** Lubricate the pistons with clean brake fluid and install them into the caliper bores. Using your thumbs, push each piston all the way in **(see illustration)**, making sure it doesn't get cocked in the bore.

### Rear caliper

**16** Remove the brake pads from the caliper (see Section 3, if necessary). Clean the exterior of the caliper with denatured alcohol or brake system cleaner.

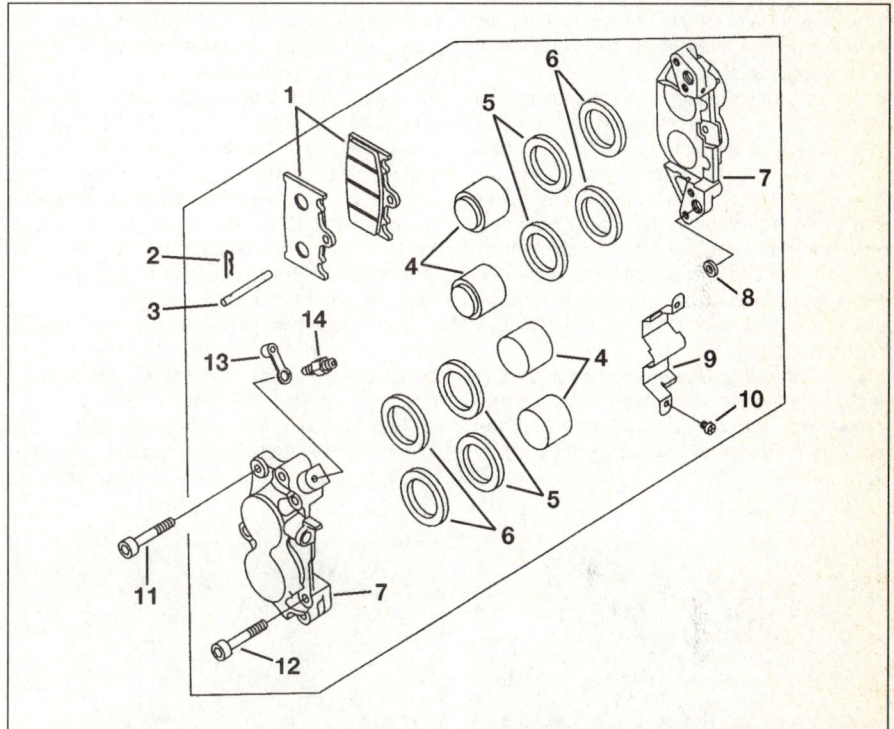

2.9d Front caliper – exploded view

1 Brake pads
2 Retaining pin clip
3 Retaining pin
4 Pistons
5 Dust seals
6 Piston seals
7 Caliper half
8 O-ring
9 Pad cover
10 Pad cover screw
11 Caliper mounting bolt
12 Caliper assembly bolt
13 Bleed valve cap
14 Bleed valve

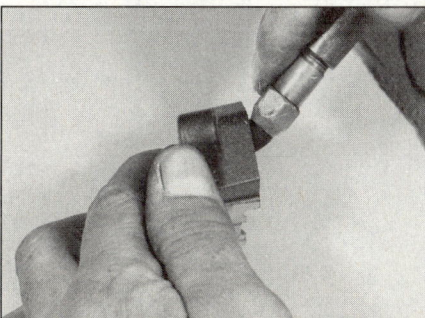

2.10a Keep your fingers out of the way of the piston and apply low-pressure compressed air to the fluid port

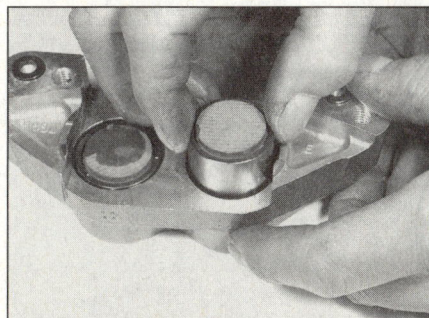

2.10b Take the pistons out of the bores

2.11 Remove the dust seals and piston seals with a wood or plastic tool

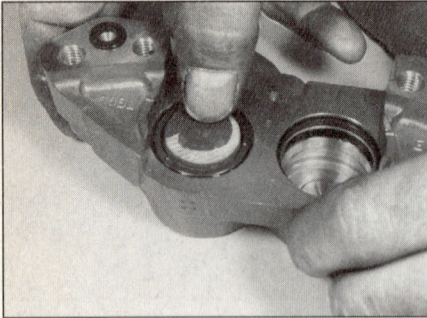

2.15 Push the pistons straight into the bores with the thumbs

2.17a Remove the caliper assembly bolts . . .

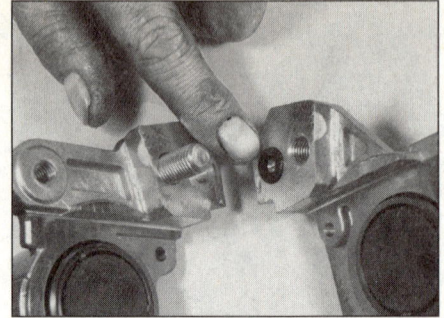

2.17b . . . and separate the caliper halves; replace the O-ring with a new one

**17** Remove the two bolts that hold the caliper halves together and separate the caliper halves **(see illustrations)**.

**18** To remove the pistons, place each caliper half face down on a workbench padded with rags. To remove the piston from the inner half, position it with the fluid port exposed over the edge of the bench. Blow air into the brake hose fitting (outer half) or fluid port (inner half) to ease the pistons out **(see illustrations)**. Use only enough air pressure to ease the piston(s) out of the bore. If a piston is blown out forcefully, even with the wood in place, it may be damaged.

⚠️ *Warning: Never place your fingers in front of the piston in an attempt to catch or protect it when applying compressed air, as serious injury could occur.*

**19** Using a wood or plastic tool, remove the piston seals **(see illustration)**. Metal tools may cause bore damage.

**20** Clean the pistons and the bores with denatured alcohol, clean brake fluid or brake system cleaner and blow dry them with filtered, unlubricated compressed air. Inspect the surfaces of the pistons for nicks and burrs and loss of plating. Check the caliper bores, too. If surface defects are present, the caliper must be replaced. If the caliper is in bad shape, the master cylinder should also be checked.

**21** Lubricate the piston seals with clean brake fluid and install them in their grooves in the caliper bore **(see illustration)**. Make sure they aren't twisted and seat completely.

**22** Lubricate the dust seals with clean brake fluid and install them in their grooves, making sure they seat correctly **(see illustration)**.

**23** Lubricate the pistons with clean brake fluid and install them into the caliper bores. Using your thumbs, push each piston all the way in **(see illustration)**, making sure it doesn't get cocked in the bore.

### Installation

**24** Installation is the reverse of the removal steps, with the following additions:

a) If you're installing a rear caliper, space the pads apart so the disc will fit between them *(see illustration)*.

b) Use new sealing washers on the brake hose fitting and position the protrusion on the fitting against the locating tab on the caliper.

c) Tighten the caliper mounting bolts and banjo fitting bolt to the torque listed in this Chapter's Specifications.

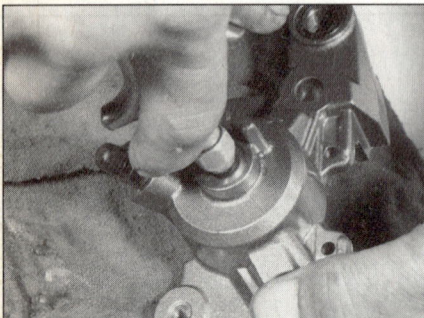

2.18a Apply low-pressure compressed air to the fluid port to ease the piston out

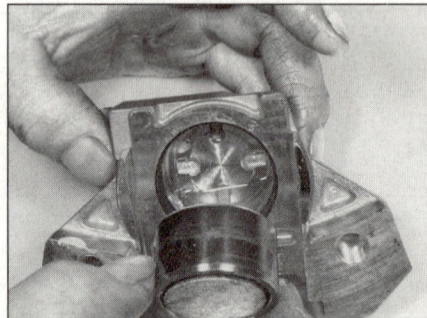

2.18b Remove the piston from the bore

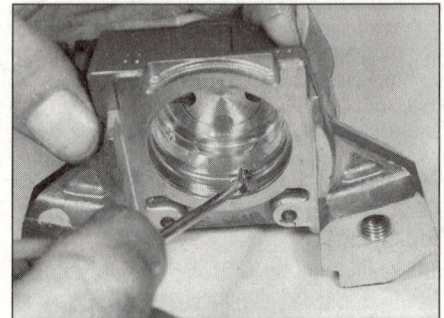

2.19 Remove the dust seals and piston seals with a wood or plastic tool

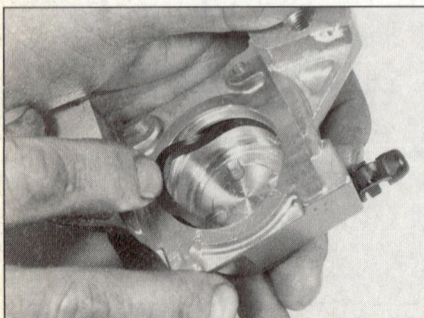

2.21 Insert the seals into the grooves . . .

2.22 . . . and make sure they seat completely

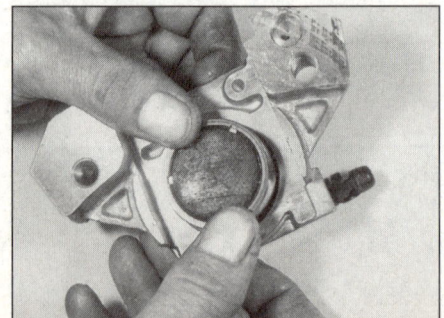

2.23 Push the piston straight into the bore with the thumbs

2.24  Leave enough space between the brake pads for the disc

d) On rear calipers, tighten the torque link bolt and nut to the torque listed in this Chapter's Specifications and install a new cotter pin.

**25** Fill the master cylinder with the recommended brake fluid (see Chapter 1) and bleed the system (see Section 10). Check for leaks.

**26** Check the operation of the brakes carefully before riding the motorcycle.

## 3  Brake pads – replacement

**Warning: When replacing the front brake pads always replace the pads in BOTH calipers – never just on one side. Also, the dust created by the brake system may contain asbestos, which is harmful to your health. Never blow it out with compressed air and don't inhale any of it. An approved filtering mask should be worn when working on the brakes.**

**1** Support the bike securely upright.

### Front caliper

**2** Remove the pad cover (see illustration).

**3** Remove the clip from the pad pin and withdraw the pin (see illustrations). Pull the pads out of the caliper opening.

**4** Refer to Chapter 1 and inspect the pads.

**5** Check the condition of the brake discs (see Section 4). If they're in need of machining or replacement, follow the procedure in that Section to remove them. If they are okay, deglaze them with sandpaper or emery cloth, using a swirling motion.

**6** Remove the cap from the master cylinder reservoir and siphon out some fluid. Push the pistons into the caliper as far as possible, while checking the master cylinder reservoir to make sure it doesn't overflow. If you can't depress the pistons with thumb pressure, try using a C-clamp (G-clamp). If the pistons stick, remove the caliper and overhaul it as described in Section 2.

**7** Install the new pads, the retaining pin and the clip. Install the pad cover (if equipped).

**8** Operate the brake lever several times to bring the pads into contact with the disc. Check the fluid level in the front brake fluid reservoir (see *Daily (pre-ride) checks*). Check the operation of the brakes carefully before riding the motorcycle.

### Rear caliper

**9** Remove the caliper (see Section 2).

**10** Remove the clip from one of the pad retaining pins and pull out the pin (see illustrations).

**11** Unhook the pad springs from the pads and remove them (see illustration).

**12** Remove the remaining clip and pin, then pull the pads out of the caliper (see illustrations).

3.2  Remove the screws and lift off the pad cover

3.3a  Pull out the clip (arrow) . . .

3.3b  . . . and the pad retaining pin

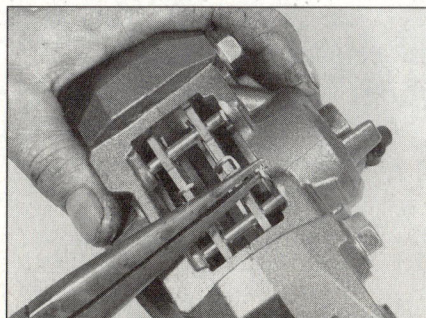
3.10a  Pull out one of the retaining pin clips . . .

3.10b  . . . and pull out the retaining pin

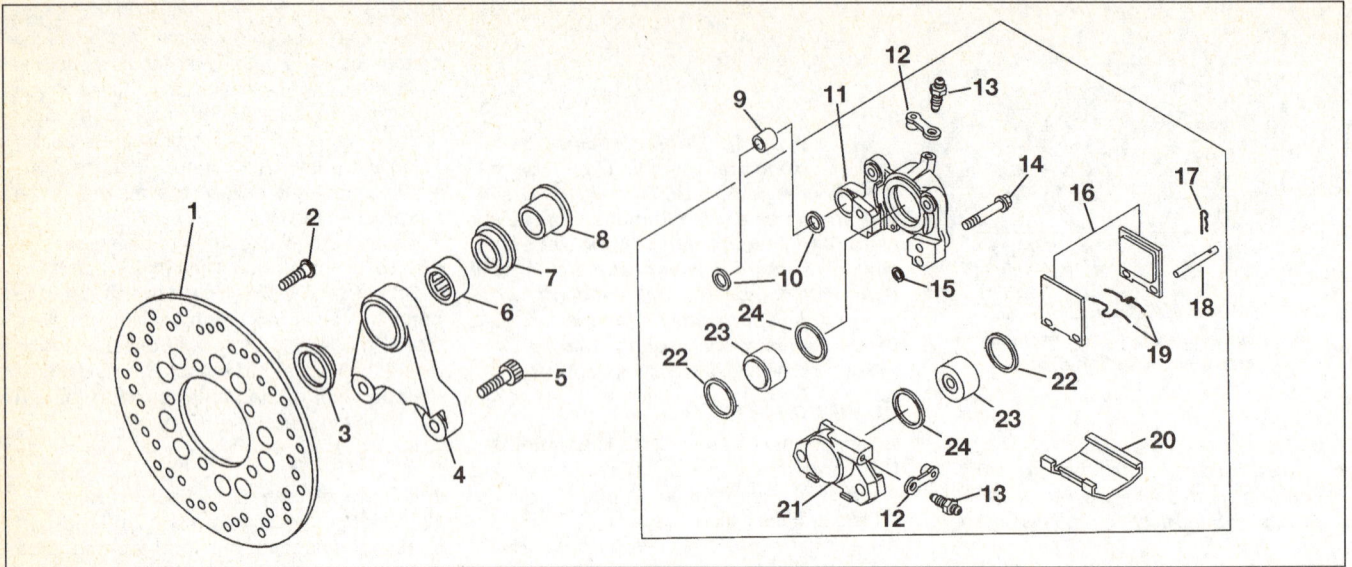

**3.10c Rear caliper – exploded view**

| | | | |
|---|---|---|---|
| 1 Brake disc | 7 Grease seal | 13 Bleed valve | 19 Pad springs |
| 2 Brake disc mounting bolt | 8 Collar | 14 Caliper assembly bolt | 20 Pad cover |
| 3 Grease seal | 9 Sleeve | 15 O-ring | 21 Caliper half |
| 4 Caliper bracket | 10 Grease seals | 16 Brake pads | 22 Dust seal |
| 5 Caliper mounting bolt | 11 Caliper half | 17 Retaining pin clip | 23 Piston |
| 6 Needle roller bearing | 12 Bleed valve cap | 18 Pad retaining pin | 24 Piston seal |

**13** Refer to Chapter 1 and inspect the pads.

**14** Check the condition of the brake discs (see Section 4). If they're in need of machining or replacement, follow the procedure in that Section to remove them. If they are okay, deglaze them with sandpaper or emery cloth, using a swirling motion.

**15** Remove the cap from the master cylinder reservoir and siphon out some fluid. Push the pistons into the caliper as far as possible, while checking the master cylinder reservoir to make sure it doesn't overflow. If you can't depress the pistons with thumb pressure, try using a C-clamp (G-clamp). If the pistons stick, remove the caliper and overhaul it as described in Section 2.

**16** Install the new pads, one retaining pin and its clip.

**17** Position the ends of the pad springs under the pin that's been installed and push them onto the pads (see illustration 3.11).

**18** Press down on the ends of the pad springs and install the other retaining pin on top of the springs. Install the clip to secure the pin (see illustration).

**19** Operate the brake pedal several times to bring the pads into contact with the disc. Check the fluid level in the rear brake fluid reservoir (see *Daily (pre-ride) checks*). Check the operation of the brakes carefully before riding the motorcycle.

**4 Brake disc(s) –** inspection, removal and installation

*Inspection*

**1** Support the bike securely upright. Place a jack beneath the bike and raise the wheel being checked off the ground. Be sure the bike is securely supported so it can't be knocked over.

**2** Visually inspect the surface of the disc(s) for score marks and other damage. Light scratches are normal after use and won't affect brake operation, but deep grooves and

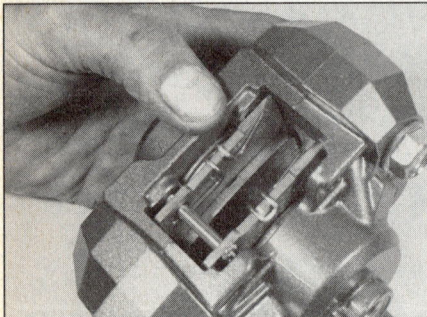

**3.11 Pull the springs off the pads**

**3.12a Pull the clip out of the other retaining pin . . .**

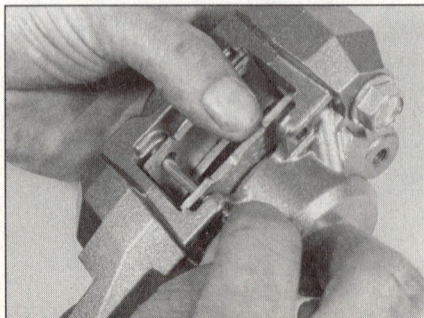

**3.12b . . . pull out the pin, then remove the pads**

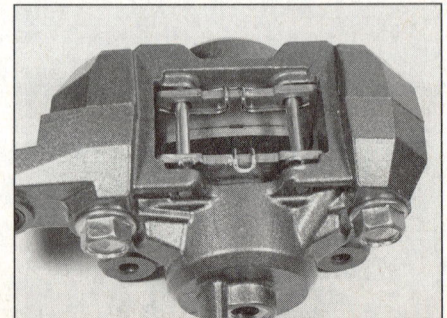

**3.18 The ends of the pad springs should fit under the retaining pins like this**

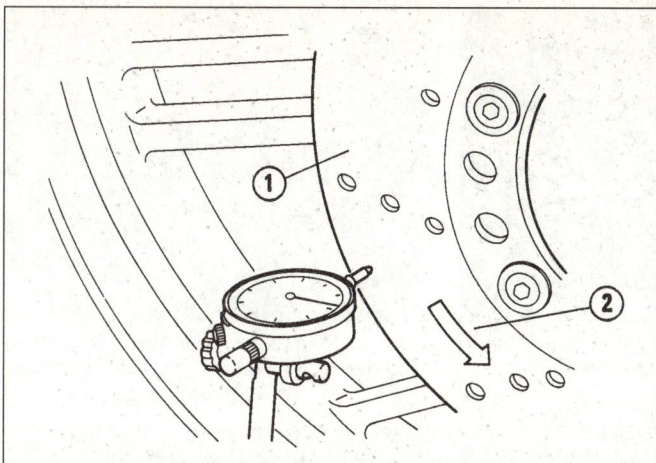

**4.3 Set up a dial indicator against the brake disc (1) and turn the disc in its normal direction of rotation (2) to measure runout**

**4.4 The minimum thickness is stamped into the disc**

*The arrow points in the forward rotating direction of the wheel*

heavy score marks will reduce braking efficiency and accelerate pad wear. If the discs are badly grooved they must be machined or replaced.

**3** To check disc runout, mount a dial indicator to a fork leg or the swingarm, with the plunger on the indicator touching the surface of the disc about _ inch from the outer edge **(see illustration)**. Slowly turn the wheel and watch the indicator needle, comparing your reading with the limit listed in this Chapter's Specifications. If the runout is greater than allowed, check the hub bearings for play (see Chapter 1). If the bearings are worn, replace them and repeat this check. If the disc runout is still excessive, it will have to be replaced.

**4** The disc must not be machined or allowed to wear down to a thickness less than the minimum allowable thickness, listed in this Chapter's Specifications. The thickness of the disc can be checked with a micrometer. If the thickness of the disc is less than the minimum allowable, it must be replaced. The minimum thickness is also stamped into the disc **(see illustration)**.

*Removal*

**5** Remove the wheel (see Section 13 for front wheel removal or Section 14 for rear wheel removal).
*Caution: Don't lay the wheel down and allow it to rest on one of the discs – the disc could become warped. Set the wheel on wood blocks so the disc doesn't support the weight of the wheel.*
**6** Mark the relationship of the disc to the wheel, so it can be installed in the same position. Remove the Allen head bolts that retain the disc to the wheel **(see illustration)**. Loosen the bolts a little at a time, in a criss-cross pattern, to avoid distorting the disc.
**7** Take note of any paper shims that may be present where the disc mates to the wheel. If there are any, mark their position and be sure to include them when installing the disc.

*Installation*

**8** Position the disc on the wheel, aligning the previously applied matchmarks (if you're reinstalling the original disc). Make sure the arrow (stamped on the disc) marking the

direction of rotation is pointing in the proper direction **(see illustration 4.4)**.
**9** Apply a non-hardening thread locking compound to the threads of the bolts. Install the bolts, tightening them a little at a time, in a criss-cross pattern, until the torque listed in this Chapter's Specifications is reached. Clean off all grease from the brake disc(s) using acetone or brake system cleaner.
**10** Install the wheel.
**11** Operate the brake lever or pedal several times to bring the pads into contact with the disc. Check the operation of the brakes carefully before riding the motorcycle.

**5 Rear brake torque link –** removal and installation

**1** Support the bike securely upright.
**2** At the rear end of the torque link, remove the rubber cap and cotter pin (if equipped), then unscrew the nut and remove the bolt **(see illustration)**.

**4.6 Remove the Allen bolts to detach the disc from the wheel**

**5.2 Remove the rubber cap and cotter pin (if equipped) then unscrew the nut and bolt**

**3** At the front end of the torque link, remove the clip from inside the frame member, then unscrew the Allen bolt **(see illustrations)**.

**4** Installation is the reverse of the removal steps. Tighten the bolts to the torque listed in this Chapter's Specifications.

## 6 Front brake master cylinder – removal, overhaul and installation

*Caution: Disassembly, overhaul and reassembly of the brake master cylinder must be done in a spotlessly clean work area to avoid contamination and possible failure of the brake hydraulic system components.*

**1** If the master cylinder is leaking fluid, or if the lever doesn't produce a firm feel when the brake is applied, and bleeding the brakes does not help, master cylinder overhaul is recommended. Before disassembling the master cylinder, read through the entire procedure and make sure that you have the

**5.3a At the front end of the torque link, remove the clip from the Allen bolt . . .**

**5.3b . . . then remove the Allen bolt (arrow) and the nut to detach the torque link from the frame**

correct rebuild kit. Also, you will need some new, clean brake fluid of the recommended type, some clean rags and internal snap-ring pliers. **Note:** *To prevent damage to the paint from spilled brake fluid, always cover the fuel tank when working on the master cylinder.*

### Removal

**2** If it's necessary to remove the reservoir,

remove the cap retainer and loosen the cap (this will be easier with the reservoir installed) **(see illustration)**. Disconnect the fluid supply hose, then remove the reservoir mounting screw and lift the reservoir off the motorcycle **(see illustrations)**.

**3** Place rags beneath the master cylinder to protect the paint in case of brake fluid spills. Remove the banjo fitting bolt **(see illustration)**

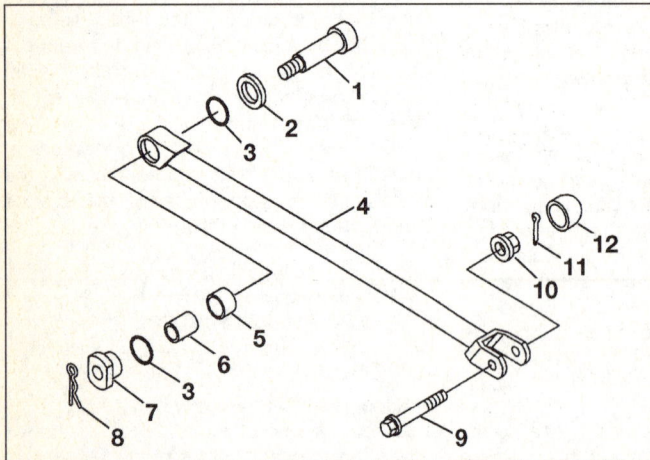
**5.3c Torque link details (L model shown; others similar)**

1 Allen bolt
2 Plain washer
3 O-ring
4 Torque link
5 Bushing
6 Sleeve
7 Nut
8 Clip
9 Bolt
10 Nut (self-locking on early models)
11 Cotter pin (late models only)
12 Rubber cap

**6.2a Remove the screw and take off the cap retainer . . .**

**6.2b . . . remove the reservoir mounting screw; disconnect the hose at the reservoir end (arrow) . . .**

**6.2c . . . or at the master cylinder end (arrow)**

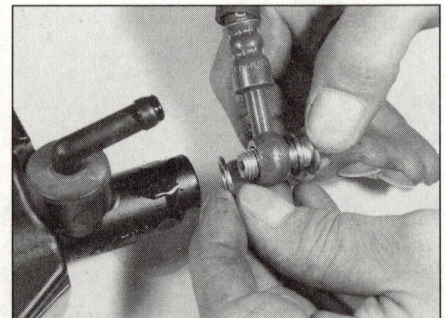
**6.3 Disconnect the banjo fitting from the master cylinder; there's a sealing washer on each side of the fitting**

**6.4 Remove the mounting bolts and separate the clamp from the cylinder body**

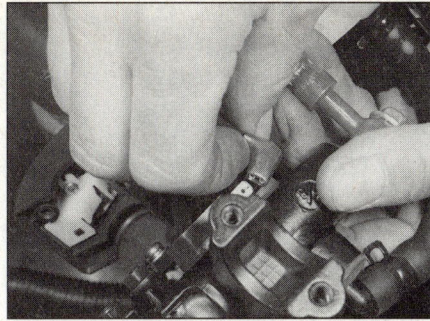

**6.5 Disconnect the brake light switch connectors from the switch**

**6.6a Remove the pivot screw locknut . . .**

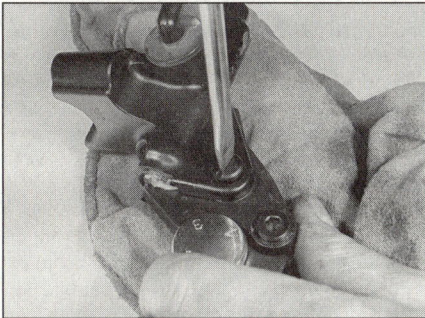

**6.6b . . . then undo the screw**

**6.7 Pry the rubber cover out of its groove, then remove the snap-ring, fluid fitting and O-ring**

**6.8a Remove the rubber boot . . .**

and separate the brake hose from the master cylinder. Wrap the end of the hose in a clean rag and suspend the hose in an upright position or bend it down carefully and place the open end in a clean container. The objective is to prevent excess loss of brake fluid, fluid spills and system contamination.

**4** Remove the master cylinder mounting bolts **(see illustration)** and separate the master cylinder from the handlebar.

**5** Disconnect the electrical connectors from the brake light switch **(see illustration)**.

### Overhaul

**6** Remove the locknut from the underside of the lever pivot screw, then remove the screw **(see illustrations)**.

**7** Lift the rubber cover from the feed hose fitting, then remove the snap-ring and take the

fitting and O-ring out of the master cylinder **(see illustration)**.

**8** Using snap-ring pliers, remove the snap-ring **(see illustrations)** and slide out the piston, the cup seals and the spring. Lay the parts out in the proper order to prevent confusion during reassembly.

**9** Clean all of the parts with brake system cleaner (available at auto parts stores), isopropyl alcohol or clean brake fluid.

*Caution: Do not, under any circum-*

*stances, use a petroleum-based solvent to clean brake parts. If compressed air is available, use it to dry the parts thoroughly (make sure it's filtered and unlubricated). Check the master cylinder bore for corrosion, scratches, nicks and score marks. If damage is evident, the master cylinder must be replaced with a new one. If the master cylinder is in poor condition, then the calipers should be checked as well.*

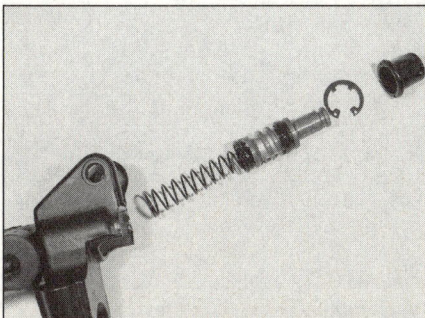

**6.8b . . . then the snap-ring, piston, cups and spring**

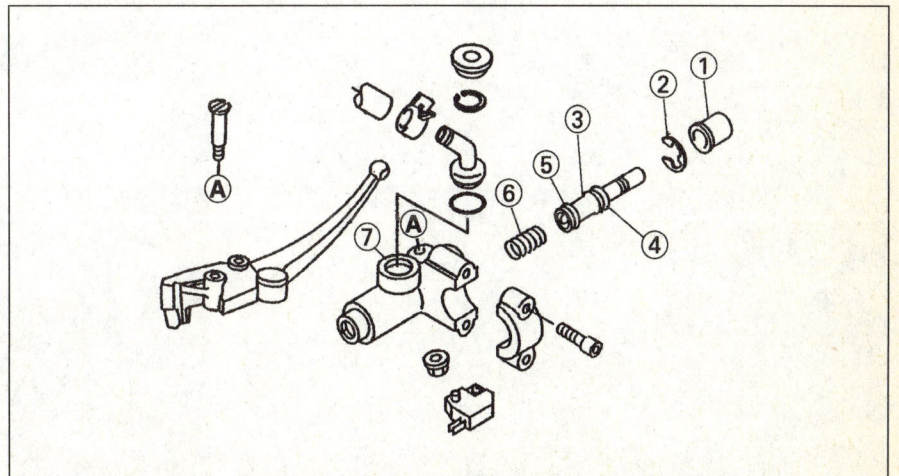

**6.8c Front master cylinder – exploded view**

| 1 Rubber boot | 4 Secondary cup | 6 Spring |
| 2 Snap-ring | 5 Primary cup | 7 Fluid port |
| 3 Piston | | |

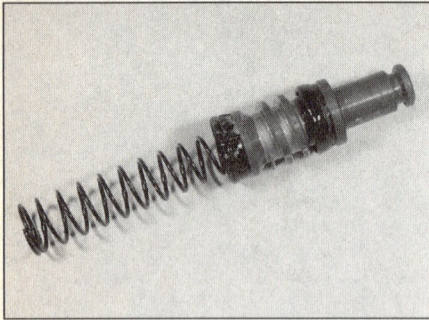

**6.10 Make sure the lips of the cups face in the proper direction**

**6.12 Make sure the snap-ring is securely seated in its groove**

**6.14 Align the lower left corner of the master cylinder's front mating surface with the punch mark on the handlebar**

**10** Remove the old cup seals from the piston and spring and install the new ones. Make sure the lips face away from the lever end of the piston **(see illustration)**. If a new piston is included in the rebuild kit, use it regardless of the condition of the old one.

**11** Before reassembling the master cylinder, soak the piston and the rubber cup seals in clean brake fluid for ten or fifteen minutes. Lubricate the master cylinder bore with clean brake fluid, then carefully insert the piston and related parts in the reverse order of disassembly. Make sure the lips on the cup seals do not turn inside out when they are slipped into the bore.

**12** Depress the piston, then install the snap-ring (make sure the snap-ring is properly seated in the groove with the sharp edge facing out) **(see illustration)**. Install the rubber dust boot (make sure the lip is seated properly in the piston groove).

**13** Install the brake lever and tighten the pivot bolt locknut.

## Installation

**14** Attach the master cylinder to the handlebar. On J and K models, align the upper gap between the master cylinder and clamp with the parting line between the halves of the throttle/switch housing. On L and M models, align the lower left corner of the master cylinder's front mating surface with the punch mark on the handlebar **(see illustration)**.

**15** Make sure the arrow and the word UP on the master cylinder clamp are pointing up, then tighten the bolts to the torque listed in this Chapter's Specifications. Tighten the top bolt fully, then tighten the lower bolt; don't try to close the gap at the lower bolt mating surface.

**16** Connect the brake hose to the master cylinder, using new sealing washers. Tighten the banjo fitting bolt to the torque listed in this Chapter's Specifications.

**17** If the reservoir was removed, reverse the removal steps to install it.

**18** Refer to Section 10 and bleed the air from the system.

## 7 Rear brake master cylinder – removal, overhaul and installation

*Caution: Disassembly, overhaul and reassembly of the brake master cylinder must be done in a spotlessly clean work* area *to avoid contamination and possible failure of the brake hydraulic system components.*

**1** If the master cylinder is leaking fluid, or if the pedal does not produce a firm feel when the brake is applied, and bleeding the brakes does not help, master cylinder overhaul is recommended.

**2** Before disassembling the master cylinder, read through the entire procedure and make sure that you have the correct rebuild kit. Also, you will need some new, clean brake fluid of the recommended type, some clean rags and internal snap-ring pliers.

## Removal

**3** Support the bike securely upright.

**4** Remove the cotter pin from the clevis pin on the master cylinder pushrod **(see illustration)**. Remove the clevis pin.

**5** Have a container and some rags ready to catch spilling brake fluid. Using a pair of pliers, slide the clamp up the fluid feed hose and detach the hose from the master cylinder **(see illustration)**. Direct the end of the hose into the container, unscrew the cap on the master cylinder reservoir and allow the fluid to drain.

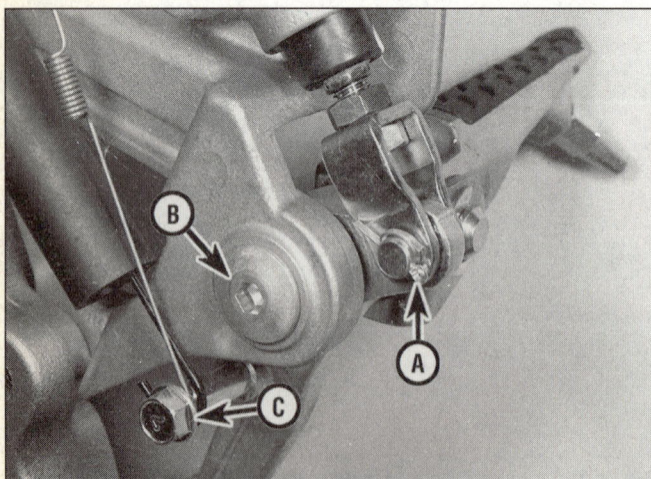

**7.4 Brake pedal installation details**

A  Cotter pin
B  Pedal pivot bolt
C  Pedal return and brake light switch springs

**7.5 Squeeze the clamp and slide it up the hose**

**7.6 Remove the banjo fitting bolt (upper arrow) and the mounting bolts (lower arrows) to detach the master cylinder**

6 Using a six-point box wrench (ring spanner), unscrew the banjo fitting bolt from the top of the master cylinder (see illustration). Discard the sealing washers on either side of the fitting.

7 Remove the two master cylinder mounting bolts and detach the cylinder from the bracket.

*Overhaul*

8 Using a pair of snap-ring pliers, remove the snap-ring from the fluid inlet fitting (see illustration) and detach the fitting from the master cylinder. Remove the O-ring from the bore.

9 Hold the clevis with a pair of pliers and loosen the locknut. Carefully remove the rubber dust boot from the pushrod.

10 Depress the pushrod and, using snap-ring pliers, remove the snap-ring. Slide out the piston, the cup seal and spring. Lay the parts out in the proper order to prevent confusion during reassembly.

11 Clean all of the parts with brake system cleaner (available at auto parts stores), isopropyl alcohol or clean brake fluid. *Caution: Do not, under any circumstances, use a petroleum-based solvent to clean brake parts. If compressed air is available, use it to dry the parts thoroughly (make sure it's filtered and unlubricated). Check the master cylinder bore for corrosion, scratches, nicks and score marks. If damage is evident, the master cylinder must be replaced with a new one. If the master cylinder is in poor condition, then the caliper should be checked as well.*

12 Remove the old cup seals from the piston and spring and install the new ones. Make sure the lips face away from the pushrod end of the piston. If a new piston is included in the rebuild kit, use it regardless of the condition of the old one.

13 Before reassembling the master cylinder, soak the piston and the rubber cup seals in clean brake fluid for ten or fifteen minutes. Lubricate the master cylinder bore with clean brake fluid, then carefully insert the parts in the reverse order of disassembly. Make sure the lips on the cup seals do not turn inside out when they are slipped into the bore.

**7.8 Rear master cylinder – exploded view**

1 Cap
2 Diaphragm retainer
3 Diaphragm
4 Reservoir mounting screw
5 Reservoir
6 Banjo fitting bolt
7 Sealing washers
8 Grommets (L and M models)
9 Hose clip assembly (L and M models)
10 Hose clip
11 Hose retainer
12 Master cylinder mounting bolt
13 Rear brake hose
14 Fluid feed fitting
15 Snap-ring
16 O-ring
17 Master cylinder body
18 Piston assembly
19 Snap-ring
20 Pushrod assembly
21 Clevis pin
22 Cotter pin
23 Clamp
24 Fluid feed hose

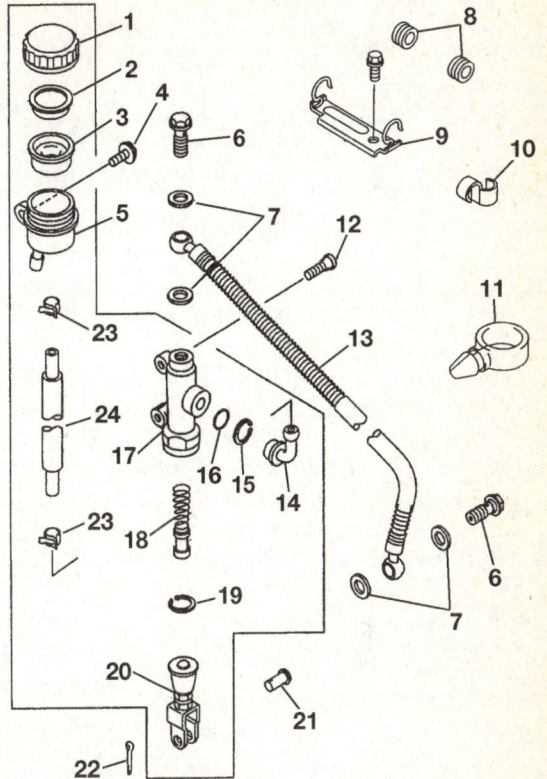

14 Lubricate the end of the pushrod with PBC (poly butyl cuprysil) grease, or silicone grease designed for brake applications, and install the pushrod and stop washer into the cylinder bore. Depress the pushrod, then install the snap-ring (make sure the snap-ring is properly seated in the groove with the sharp edge facing out). Install the rubber dust boot (make sure the lip is seated properly in the groove in the piston stop nut).

15 Install the clevis to the end of the pushrod, leaving about 3.5 to 5.5 mm of the end of the pushrod exposed past the adjusting nut, then tighten the locknut This will ensure the brake pedal will be positioned correctly.

16 Install the feed hose fitting, using a new O-ring. Install the snap-ring, making sure it seats properly in its groove.

*Installation*

17 Position the master cylinder on the frame and install the bolts, tightening them to the torque listed in this Chapter's Specifications.

18 Connect the banjo fitting to the top of the master cylinder, using new sealing washers on each side of the fitting. Tighten the banjo fitting bolt to the torque listed in this Chapter's Specifications.

19 Connect the fluid feed hose to the inlet fitting and install the hose clamp.

20 Connect the clevis to the brake pedal and secure the clevis pin with a new cotter pin.

21 Fill the fluid reservoir with new DOT 4

hydraulic fluid and bleed the system following the procedure in Section 10. Install the side cover.

22 Check the position of the brake pedal (see Chapter 1) and adjust it if necessary. Check the operation of the brakes carefully before riding the motorcycle.

**8  Brake pedal – removal and installation**

1 Support the bike securely upright so it can't be knocked over during this procedure.

2 Unhook the pedal return spring and the brake light switch spring from the pedal arm (see illustration 7.4).

3 Remove the pedal pivot bolt (it also secures the rider's footpeg on the right side) (see illustration 7.4 and the accompanying illustration). Remove the brake pedal from the motorcycle.

4 If it's necessary to remove the pedal bracket, remove the rear master cylinder (see Section 7), then unbolt the bracket from the motorcycle (see illustration).

5 Installation is the reverse of the removal steps, with the following additions:

a) Tighten the bracket bolts securely if they were removed. Tighten the pedal pivot bolt to the torque listed in this Chapter's Specifications.

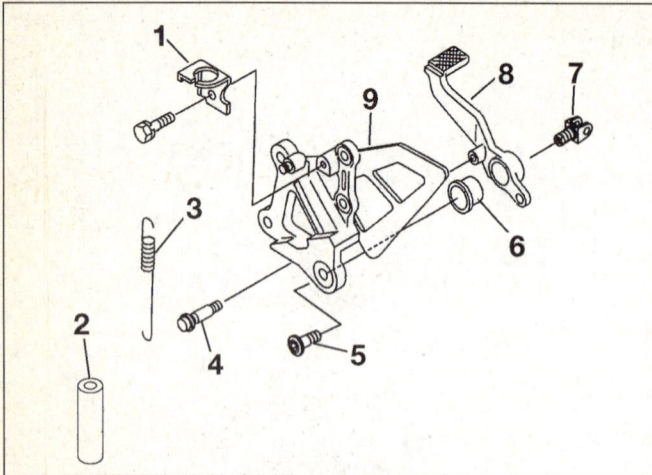

8.3 Brake pedal details

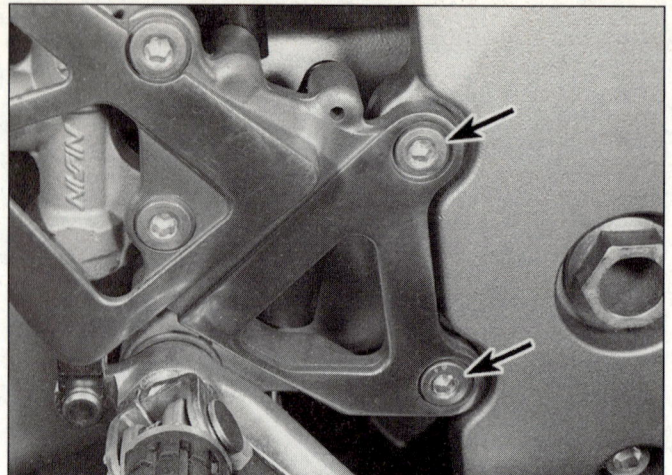

8.4 Remove the bracket bolts (arrows)

1  Brake light switch   3  Spring
   bracket             4  Pedal arm
2  Plastic tube (return 5  Pedal pivot
   spring cover)          bolt

6  Bushing
7  Footpeg pivot
8  Brake pedal
9  Pedal bracket

b) If the pedal spring was removed completely, attach its top end to the bracket (see illustration).
c) Refer to Chapter 1 and adjust brake pedal height.

## 9  Brake hoses and lines – inspection and replacement

### Inspection

1  Once a week, or if the motorcycle is used less frequently, before every ride, check the condition of the brake hoses.
2  Twist and flex the rubber hoses (see illustration) while looking for cracks, bulges and seeping fluid. Check extra carefully around the areas where the hoses connect with the banjo fittings, as these are common areas for hose failure.

### Replacement

3  All of the pressurized brake hoses have banjo fittings on each end of the hose (see illustration 7.8 and the accompanying illustrations). Fluid feed hoses, which connect the reservoirs to the master cylinders, are secured by spring clamps. Cover the surrounding area with plenty of rags and unscrew the banjo bolts on either end of the hose. Detach the hose from any clips that may be present and remove the hose.
4  Position the new hose, making sure it isn't twisted or otherwise strained, between the two components. Make sure the metal tube portion of the banjo fitting is located between the casting protrusions on the component it's connected to, if equipped. Install the banjo bolts, using new sealing washers on both sides of the fittings, and tighten them to the torque listed in this Chapter's Specifications.
5  Flush the old brake fluid from the system, refill the system with new DOT 4 hydraulic fluid and bleed the air from the system (see Section 10). Check the operation of the brakes carefully before riding the motorcycle.

## 10  Brake system bleeding

1  Bleeding the brake is simply the process of removing all the air bubbles from the brake fluid reservoir, the lines and the brake caliper. Bleeding is necessary whenever a brake system hydraulic connection is loosened, when a component or hose is replaced, or

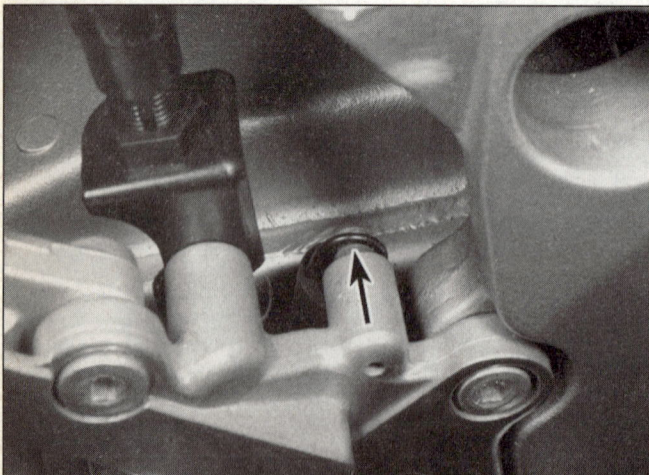

8.5 The upper end of the pedal return spring hooks to a post (arrow)

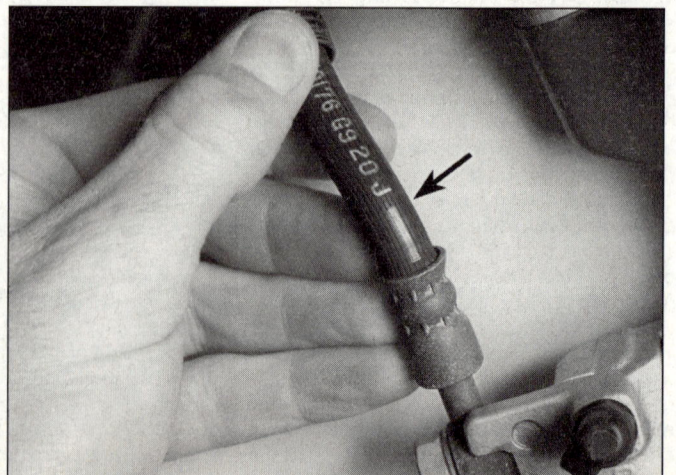

9.2 Flex the brake hoses and check for cracks, bulges and leaking fluid

**9.3a  Front brake hose details**

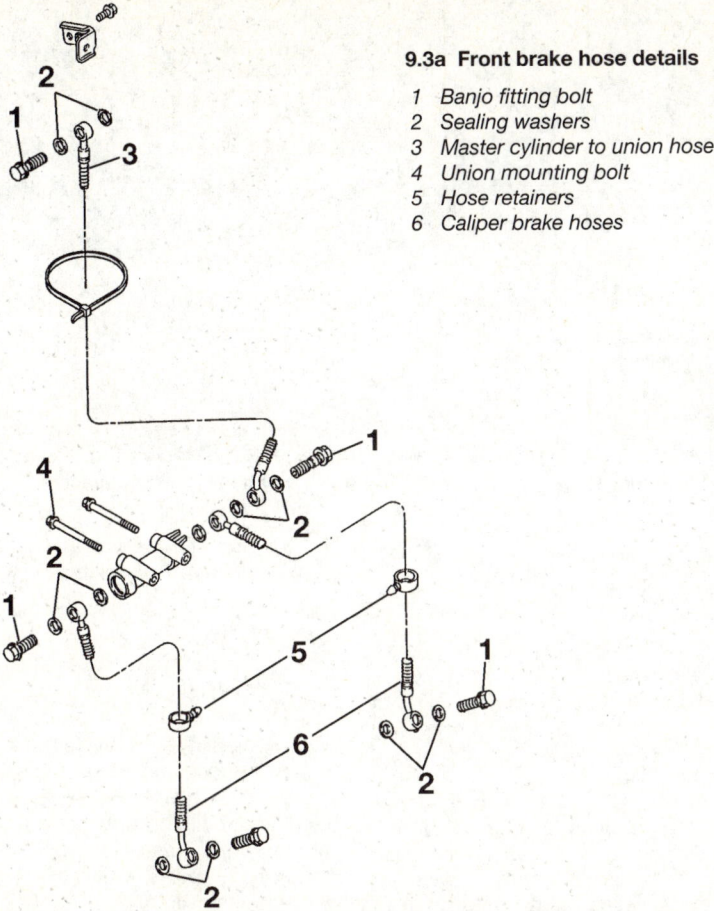

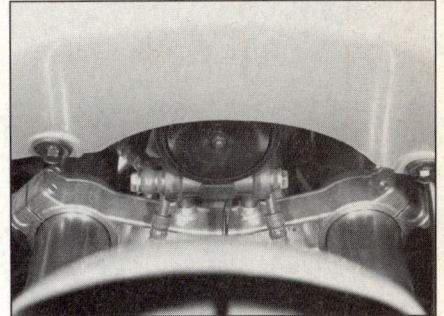

1  Banjo fitting bolt
2  Sealing washers
3  Master cylinder to union hose
4  Union mounting bolt
5  Hose retainers
6  Caliper brake hoses

**9.3b  Front brake hose union mounted on the lower triple clamp**

**9.3c  The neck of each hose fits into a notch in the hose union (arrow)**

mark during the bleeding process (see *Daily (pre-ride) checks*).

**7** Carefully pump the brake lever or pedal three or four times and hold it while opening the caliper bleed valve **(see illustrations)**. When the valve is opened, brake fluid will flow out of the caliper into the clear tubing and the lever will move toward the handlebar or the pedal will move down.

**8** Retighten the bleed valve, then release the brake lever or pedal gradually. Repeat the process until no air bubbles are visible in the brake fluid leaving the caliper and the lever or pedal is firm when applied. **Note:** *The rear calipers have two bleed valves – air must be bled from both, one after the other. Remember to add fluid to the reservoir as the level drops. Use only new, clean brake fluid of the recommended type. Never reuse the fluid lost during bleeding.*

when the master cylinder or caliper is overhauled. Leaks in the system may also allow air to enter, but leaking brake fluid will reveal their presence and warn you of the need for repair.

**2** To bleed the brake, you will need some new clean DOT 4 hydraulic brake fluid, a length of clear vinyl or plastic tubing, a small container partially filled with clean brake fluid, some rags and a wrench to fit the brake caliper bleed valve.

**3** Cover the fuel tank and other painted components to prevent damage in the event

that brake fluid is spilled.

**4** Remove the reservoir cap and slowly pump the brake lever or pedal a few times, until no air bubbles can be seen floating up from the holes at the bottom of the reservoir. Doing this bleeds the air from the master cylinder end of the line. Reinstall the reservoir cap.

**5** Attach one end of the clear vinyl or plastic tubing to the brake caliper bleed valve and submerge the other end in the brake fluid in the container **(see illustrations)**.

**6** Check the fluid level in the reservoir. Do not allow the fluid level to drop below the lower

**10.5a  Remove the dust cap (arrow) to expose the bleed valve**

**10.5b  On rear calipers, there's a bleed valve on the outer half (arrow) and one on the rearward side of the inner half**

**10.5c  Connect a plastic tube to the bleed valve and place the other end in a container**

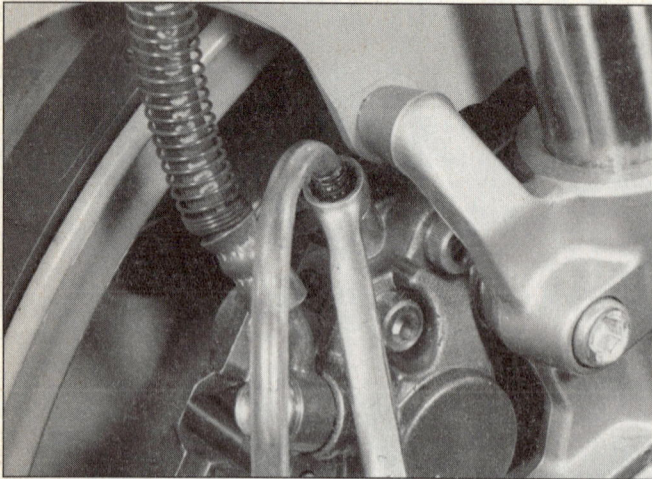

**10.7a Place a box wrench (ring spanner) on the bleed valve; this is a front caliper . . .**

**10.7b . . . and this is a rear caliper**

**9** If you're bleeding the front brakes, repeat this procedure to the other caliper. Be sure to check the fluid level in the master cylinder reservoir frequently.

**10** Replace the reservoir cap, wipe up any spilled brake fluid and check the entire system for leaks.

> **HAYNES HINT** *If bleeding is difficult, it may be necessary to let the brake fluid in the system stabilize for a few hours (it may be aerated). Repeat the bleeding procedure when the tiny bubbles in the system have settled out.*

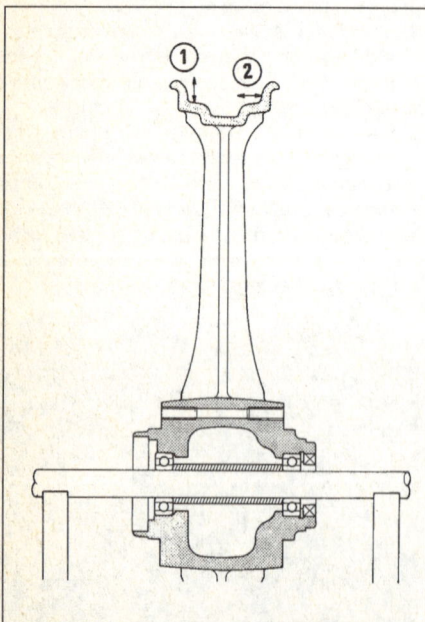

**11.2 Measure wheel runout with a dial indicator**

*1 Radial runout*   *2 Axial runout*

## 11 Wheels – inspection and repair

**1** Clean the wheels thoroughly to remove mud and dirt that may interfere with the inspection procedure or mask defects. Make a general check of the wheels and tires as described in Chapter 1.

**2** Support the motorcycle securely upright with the wheel to be checked in the air, then attach a dial indicator to the fork slider or the swingarm and position the stem against the side of the rim **(see illustration)**. Spin the wheel slowly and check the side-to-side (axial) runout of the rim, then compare your readings with the value listed in this Chapter's Specifications. In order to accurately check radial runout with the dial indicator, the wheel would have to be removed from the machine and the tire removed from the wheel. With the axle clamped in a vise, the wheel can be rotated to check the runout.

**3** An easier, though slightly less accurate, method is to attach a stiff wire pointer to the fork slider or the swingarm and position the end a fraction of an inch from the wheel (where the wheel and tire join). If the wheel is true, the distance from the pointer to the rim will be constant as the wheel is rotated. Repeat the procedure to check the runout of the rear wheel. **Note:** *If wheel runout is excessive, refer to the appropriate Section in this Chapter and check the wheel bearings very carefully before replacing the wheel.*

**4** The wheels should also be visually inspected for cracks, flat spots on the rim and other damage. Since tubeless tires are involved, look very closely for dents in the area where the tire bead contacts the rim. Dents in this area may prevent complete sealing of the tire against the rim, which leads to deflation of the tire over a period of time.

**5** If damage is evident, or if runout in either direction is excessive, the wheel will have to be replaced with a new one. Never attempt to repair a damaged cast aluminum wheel.

## 12 Wheels – alignment check

**1** Misalignment of the wheels, which may be due to a cocked rear wheel or a bent frame or triple clamps, can cause strange and possibly serious handling problems. If the frame or triple clamps are at fault, repair by a frame specialist or replacement with new parts are the only alternatives.

**2** To check the alignment you will need an assistant, a length of string or a perfectly straight piece of wood and a ruler. A plumb bob or other suitable weight will also be required.

**3** Support the motorcycle securely upright, then measure the width of both tires at their widest points. Subtract the smaller measurement from the larger measurement, then divide the difference by two. The result is the amount of offset that should exist between the front and rear tires on both sides.

**4** If a string is used, have your assistant hold one end of it about half way between the floor and the rear axle, touching the rear sidewall of the tire.

**5** Run the other end of the string forward and pull it tight so that it is roughly parallel to the floor. Slowly bring the string into contact with the front sidewall of the rear tire, then turn the front wheel until it is parallel with the string. Measure the distance from the front tire sidewall to the string **(see illustration)**.

**6** Repeat the procedure on the other side of the motorcycle. The distance from the front tire sidewall to the string should be equal on both sides.

**7** As was previously pointed out, a perfectly straight length of wood may be substituted for the string **(see illustration)**. The procedure is the same.

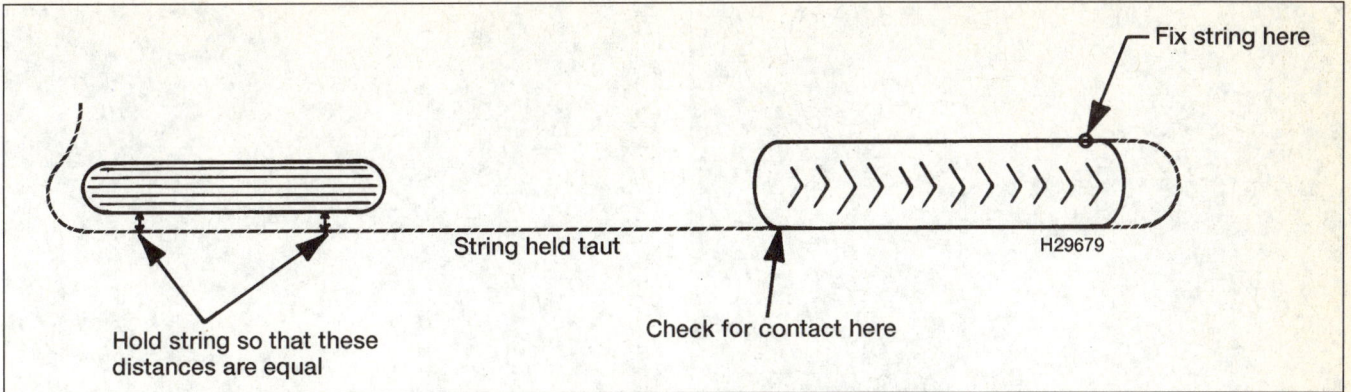

**12.5 Using the string method to check wheel alignment**

Fix string here

String held taut

H29679

Hold string so that these distances are equal

Check for contact here

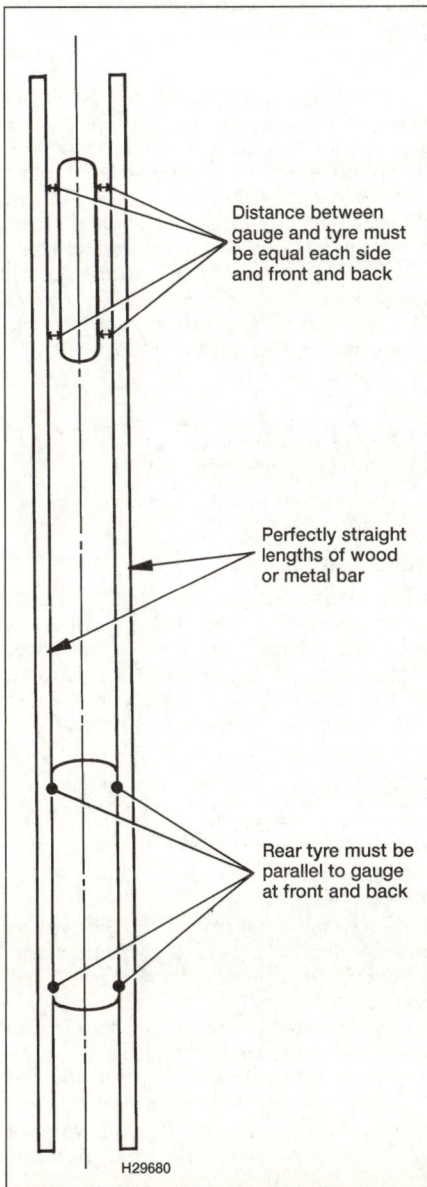

Distance between gauge and tyre must be equal each side and front and back

Perfectly straight lengths of wood or metal bar

Rear tyre must be parallel to gauge at front and back

H29680

**12.7 Using a straight-edge to check wheel alignment**

8 If the distance between the string and tire is greater on one side, or if the rear wheel appears to be cocked, refer to Chapter 6, *Swingarm bearings – check*, and make sure the swingarm is tight.

9 If the front-to-back alignment is correct, the wheels still may be out of alignment vertically.

10 Using the plumb bob, or other suitable weight, and a length of string, check the rear wheel to make sure it is vertical. To do this, hold the string against the tire upper sidewall and allow the weight to settle just off the floor. When the string touches both the upper and lower tire sidewalls and is perfectly straight, the wheel is vertical.

11 Once the rear wheel is vertical, check the

**13.5a Loosen the axle clamp bolts on the right fork leg (A); note the location of the spacer (B)**

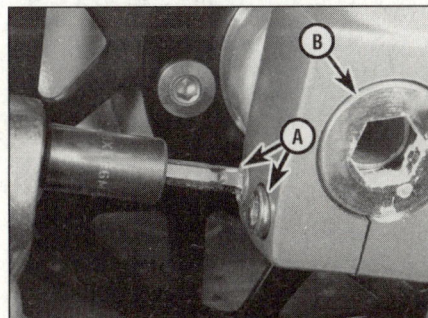

**13.5b Loosen the axle clamp bolts (A) on the left fork leg; hold the axle with an Allen wrench and remove the nut (B) . . .**

front wheel in the same manner. If both wheels are not perfectly vertical, the frame and/or major suspension components are bent.

## 13 Front wheel – removal and installation

### Removal

1 Remove the lower portion of the fairing (H models) or the front fender/mudguard (J, K, L and M models) (see Chapter 8).

2 Support the motorcycle securely upright, then raise the front wheel off the ground by placing a floor jack, with a wood block on the jack head, under the engine.

3 Disconnect the speedometer cable from the drive unit on the wheel (see Chapter 9).

4 Remove the brake calipers (see Section 2) and support them with pieces of wire. Don't disconnect the brake hoses from the calipers and don't let their weight hang on the brake hoses.

5 Loosen the axle clamp bolts on both forks, unscrew the axle nut and pull it out **(see illustrations)**.

6 Support the wheel, then pull out the axle **(see illustrations)** and carefully lower the wheel to the ground. Don't lose the spacer that fits into the right side of the hub.

**13.5c . . . with another Allen wrench . . .**

13.5d . . . pull out the nut . . .

13.6a . . . and the axle

**Caution: Don't lay the wheel down and allow it to rest on one of the discs – the disc could become warped. Set the wheel on wood blocks so the disc doesn't support the weight of the wheel. If the axle is corroded, remove the corrosion with fine emery cloth. Do not operate the front brake lever with the wheel removed.**

7 Check the condition of the wheel bearings (see Section 15).

### Installation

8 Installation is the reverse of removal. Apply a thin coat of grease to the seal lip, then slide the spacer into the right side of the hub. Position the speedometer drive unit in place in the left side of the hub, then slide the wheel into place. Make sure the notches in the speedometer drive housing line up with the lugs in the wheel.

9 Slip the axle into place, then tighten the axle nut to the torque listed in this Chapter's Specifications. Tighten the right side axle clamp bolts to the torque listed in this Chapter's Specifications.

10 Install the calipers. Apply the front brake, pump the forks up and down several times and check for binding and proper brake operation.

11 Install the fairing panels (H models) or the front fender (J, K, L and M models).

### 14 Rear wheel – removal and installation

#### Removal

1 Support the bike securely upright so it can't be knocked over when the rear wheel is removed. If necessary, remove the lower fairing panels (see Chapter 8) so you can place a jack beneath the motorcycle. On J, K, L and M models, it may also be necessary to remove the right lower fairing stay and the sidestand.

2 Loosen the chain adjusting bolt locknuts (see Chapter 1) and fully loosen both adjusting bolts.

3 Disconnect the rear end of the brake torque link (see Section 5).

4 Remove the cotter pin from the axle nut on the left side of the wheel **(see illustration)** and remove the plain washer (later models) and the nut.

5 Push the rear wheel as far forward as possible. Lift the top of the chain up off the rear sprocket and pull it to the left to disengage the chain from the sprocket.

⚠️ **Warning: Don't let your fingers slip between the chain and the sprocket.**

6 Support the wheel and slide the axle out **(see illustrations)**.

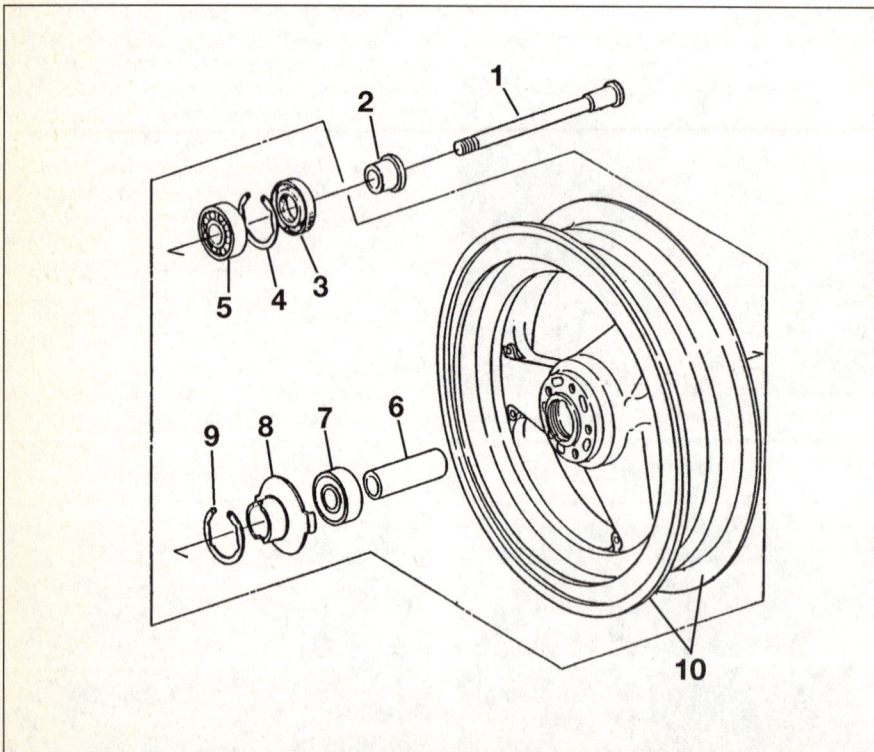

13.6b Front wheel details

| 1 | Axle | 4 | Snap-ring | 8 | Speedometer drive |
|---|------|---|-----------|---|-------------------|
| 2 | Spacer | 5 | Wheel bearing | 9 | Snap-ring |
| 3 | Grease seal | 6 | Spacer | 10 | Wheel |
| | | 7 | Wheel bearing | | |

**14.4 Rear wheel and coupling details**

1 Axle
2 Spacer
3 Grease seal
4 Snap-ring
5 Wheel bearing
6 Wheel
7 Spacer
8 Wheel bearing
9 Coupling damper (cush drive)
10 Coupling collar
11 Coupling
12 Sprocket stud
13 Coupling bearing
14 Snap-ring
15 Grease seal
16 Spacer
17 Washer
18 Axle nut
19 Cotter pin

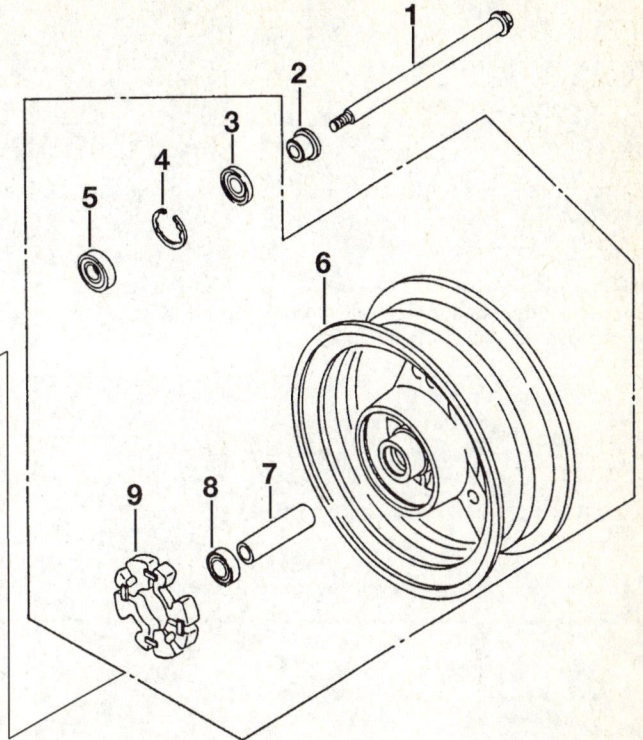

7 Lower the caliper bracket away from the disc, together with the caliper. Pull the wheel backward and remove it from the swingarm, being careful not to lose the spacers on either side of the hub (see illustrations).
Caution: Don't lay the wheel down and allow it to rest on the disc or the sprocket – they could become warped. Set the wheel on wood blocks so the disc or the sprocket doesn't support the weight of the wheel. Do not operate the brake pedal with the wheel removed.

8 Before installing the wheel, check the axle for straightness. If the axle is corroded, first remove the corrosion with fine emery cloth. Set the axle on V-blocks and check it for runout using a dial indicator (see illustration). If the axle exceeds the maximum allowable runout limit listed in this Chapter's Specifications, it must be replaced.

9 Check the condition of the wheel bearings (see Section 15).

*Installation*

10 Apply a thin coat of grease to the seal lips, then slide the spacers into their proper positions on the sides of the hub.

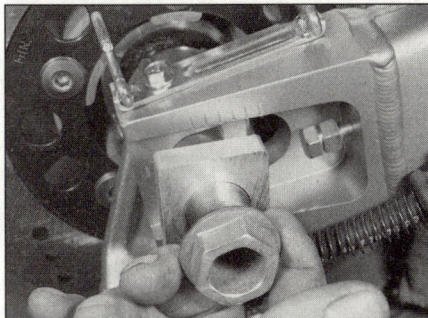

14.6a Grasp the axle and pull it out (note the location of the chain adjuster block) . . .

14.6b . . . with the wheel supported, pull the axle out

14.7a Remove the caliper bracket and the spacer from the right side . . .

14.7b . . . and remove the wheel rearwards

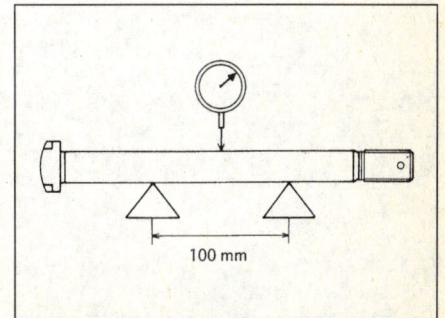

14.8 Check the axle for runout using a dial indicator and V-blocks

15.3a Lift out the speedometer drive unit and remove the snap-ring (arrows) . . .

15.3b . . . then remove the drive

15.4a Lift the spacer out of the wheel

15.4b A screwdriver can be used to pry out the seal if you don't have a removal tool like this one

15.4c Remove the snap-ring from the right side of the wheel

11 Slide the wheel into place, then slide the caliper bracket up until the pads fit over the disc and the caliper bracket hole lines up with the axle holes in the wheel and swingarm.

12 Pull the chain up over the sprocket, raise the wheel and install the axle, washer (later models) and axle nut. Don't tighten the axle nut at this time.

13 Adjust the chain slack (see Chapter 1) and tighten the adjuster locknuts.

14 Tighten the axle nut to the torque listed in this Chapter's Specifications. Install a new cotter pin, tightening the axle nut an additional amount, if necessary, to align the hole in the axle with the castellations on the nut.

15.5 Drive the bearings from the hub with a brass drift

15 Connect the torque link. Tighten the torque link nut to the torque listed in this Chapter's Specifications, then install a new cotter pin (and the plastic cap, if equipped).

16 Check the operation of the brake carefully before riding the motorcycle.

## 15 Wheel and rear coupling bearings – inspection and maintenance

### Front wheel bearings

1 Support the bike securely and remove the front wheel (see Section 13).

2 Set the wheel on blocks so as not to allow the weight of the wheel to rest on the brake disc.

3 From the left side of the wheel, remove the snap-ring securing the speedometer drive and remove the speedometer drive from the hub (see illustrations).

4 Remove the spacer and pry out the grease seal from the right side of the wheel (see illustrations). Remove the bearing snap-ring from beneath the grease seal (see illustration).

5 Using a metal rod (preferably a brass drift punch) inserted through the center of the hub

bearing, tap evenly around the inner race of the opposite bearing to drive it from the hub (see illustration). The bearing spacer will also come out.

6 Lay the wheel on its other side and remove the remaining bearing using the same technique.

7 If the bearings are open on one or both sides, clean them with a high flash-point solvent (one which won't leave any residue), blow them dry with compressed air (don't let the bearings spin as you dry them) and apply a few drops of oil to the bearing. On all bearings (open or sealed), hold the outer race of the bearing and rotate the inner race – if the

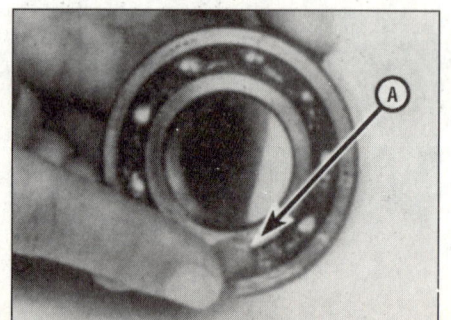

15.8 Press grease into the open side of the bearing (A) until it's full

**15.9  With the bearing in position, install the snap-ring and make sure it's securely seated in its groove**

**15.10  Install the speedometer drive and secure it with the snap-ring**

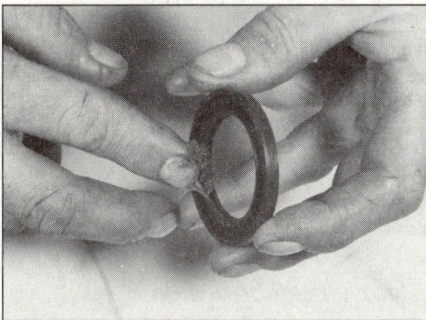

**15.11  Apply a coat of grease to the lip of the seal**

**15.15a  Remove the spacer . . .**

**15.15b  . . . and lift the coupling out of the wheel**

bearing doesn't turn smoothly, has rough spots or is noisy, replace it with a new one.

**8**  If an open bearing checks out okay and will be reused, wash it in solvent once again and dry it, then pack the bearing from the open side with high-quality bearing grease **(see illustration)**.

**9**  Thoroughly clean the hub area of the wheel. Install the right side bearing into its recess in the hub, with the marked or shielded side facing out. Using a bearing driver or a socket large enough to contact the outer race of the bearing, drive it in until the snap-ring groove is visible and install the snap-ring **(see illustration)**.

**10**  Turn the wheel over and install the bearing spacer and left side bearing, driving the bearing into place as described in Step 10, then install the speedometer drive and the snap-ring **(see illustration)**.

**11**  Coat the lip of a new grease seal with grease **(see illustration)**.

**12**  Install the grease seal on the right side of the wheel; it should go in with thumb pressure but if not, use a seal driver, large socket or a flat piece of wood to drive it into place.

**13**  Clean off all grease from the brake discs using acetone or brake system cleaner. Install the wheel (see Section 13).

### Rear coupling bearing

**14**  Refer to Section 14 and remove the rear wheel. Lay the wheel on its brake disc side, supported on blocks so its weight doesn't rest on the brake disc.

**15**  Lift off the spacer and rear wheel coupling **(see illustrations)**.

**16**  Pry out the grease seal and remove the snap-ring from the sprocket side of the coupling **(see illustrations)**.

**17**  Turn the wheel over and remove the coupling collar from the other side of the hub **(see illustration)**.

**15.16a  Pry out the grease seal . . .**

**15.16b  . . . and remove the snap-ring**

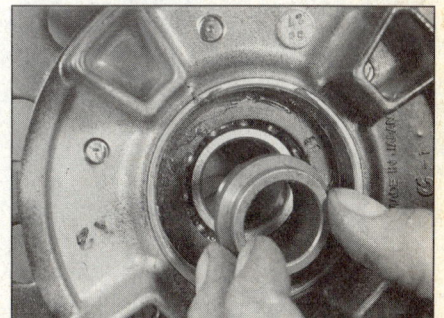

**15.17  Lift out the coupling collar**

**15.18 Drive out the coupling bearing**

**15.23 Press in the new grease seal**

**15.24 Pry out the grease seal**

**15.25 Remove the snap-ring**

22  Install the snap-ring to secure the bearing, making sure it fits securely in its groove. Install the collar on the other side of the coupling.

23  Coat the lip of a new grease seal with grease and install it on top of the snap-ring **(see illustration)**. It should go in with thumb pressure, but if not, tap it in with a hammer and socket, bearing driver or flat piece of wood. Install the coupling and spacer in the wheel and install the wheel (see Section 14).

### *Rear wheel bearings*

24  Pry out the grease seal on the brake disc side of the wheel **(see illustration)**.

25  Remove the snap-ring from beneath the grease seal with snap-ring pliers **(see illustration)**.

26  Using a metal rod (preferably a brass drift punch) inserted through the center of the hub bearing, tap evenly around the inner race of the opposite bearing to drive it from the hub **(see illustrations)**. The bearing spacer will also come out.

18  Drive the bearing out of the coupling with a bearing driver or drift punch **(see illustration)**.

19  If the bearings are open on one or both sides, clean the bearing with a high flash-point solvent (one which won't leave any residue), blow it dry with compressed air (don't let the bearing spin as you dry it) and apply a few drops of oil to the bearing. On all bearings (open or sealed), hold the outer race of the bearing and rotate the inner race – if the

bearing doesn't turn smoothly, has rough spots or is noisy, replace it with a new one.

20  If the bearing checks out okay and will be reused, wash it in solvent once again and dry it, then pack the bearing from the open side with high-quality bearing grease **(see illustration 15.8)**.

21  Drive the bearing into the coupling with a bearing driver or socket that bears against the outer race of the bearing.

**15.26a Insert a rod through the bearing on the brake disc side of the wheel and tap on it . . .**

**15.26b . . . to drive out the sprocket side bearing, then remove the spacer**

**27** Lay the wheel on its other side and remove the remaining bearing using the same technique **(see illustration)**.

**28** Clean the bearings with a high flash-point solvent (one which won't leave any residue) and blow them dry with compressed air (don't let the bearing spin as you dry them). Apply a few drops of oil to the bearing. Hold the outer race of the bearing and rotate the inner race – if the bearing doesn't turn smoothly, has rough spots or is noisy, replace it with a new one.

**29** If the bearing checks out okay and will be reused, wash it in solvent once again and dry it, then pack the bearing from the open side with high-quality bearing grease **(see illustration 15.8)**.

**30** Thoroughly clean the hub area of the wheel. Install the bearing into the recess in the hub, with the marked or shielded side facing out. Using a bearing driver or a socket large enough to contact the outer race of the

**15.27  Tap out the bearing on the brake disc side**

bearing, drive it in until the snap-ring groove is visible **(see illustration)**. Install the snap-ring **(see illustration 15.25)**.

**31** Turn the wheel over. Apply a coat of multi-purpose grease to the inside of the spacer **(see illustration)** and install it in the hub **(see illustration)**.

**15.30  Make sure the bearing is fully seated and the snap-ring groove is exposed**

**32** Pack the remaining bearing from the open side with grease **(see illustration)**, then install it in the hub **(see illustration)**, driving the bearing in with a socket or bearing driver large enough to contact the outer race of the bearing.

**33** Install a new grease seal. It should go in

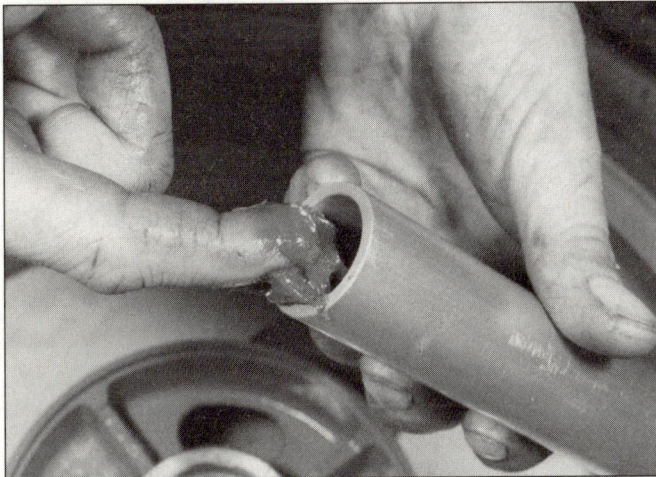

**15.31a  Apply a coat of grease to the inside of the spacer . . .**

**15.31b  . . . and install the spacer in the hub**

**15.32a  Pack the open side of the bearing with grease . . .**

**15.32b  . . . and install it in the hub**

located in the inside of the inner race (between the wheel and the coupling) **(see illustration 15.17)**.

**35** Clean off all grease from the brake discs using acetone or brake system cleaner. Install the wheel.

## 16 Tires –
general information and fitting

### General information

**1** The cast wheels fitted to all models are designed to take tubeless tires only.
**2** Refer to *'Daily (pre-ride) checks'* at the beginning of this manual for tire maintenance.

### Fitting new tires

**3** When selecting new tires, refer to the tire information label and the tire options listed in the owner's handbook. Ensure that front and rear tire types are compatible, the correct size and correct speed rating; if necessary seek advice from a Kawasaki dealer or tire fitting specialist **(see illustration)**.

**4** It is recommended that tires are fitted by a motorcycle tire specialist rather than attempted in the home workshop. This is particularly relevant in the case of tubeless tires because the force required to break the seal between the wheel rim and tire bead is substantial, and is usually beyond the capabilities of an individual working with normal tire levers. Additionally, the specialist will be able to balance the wheels after tire fitting.

**5** In the case of tubeless tires, note that punctured tires can in some cases be repaired. Kawasaki recommend that a repaired tire should not be used at speeds above 60 mph (100 kmh) for the first 24 hours after the repair, and thereafter not above 110 mph (180 kmh).

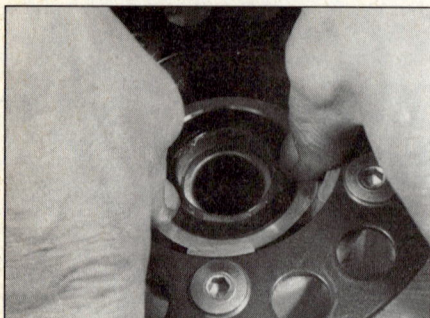

**15.33 Press the seal into the hub with your thumbs**

with thumb pressure **(see illustration)**, but if not, use a seal driver, large socket or a flat piece of wood to drive it into place.
**34** Press a little grease into the bearing in the rear wheel coupling. Install the coupling to the wheel, making sure the coupling collar is

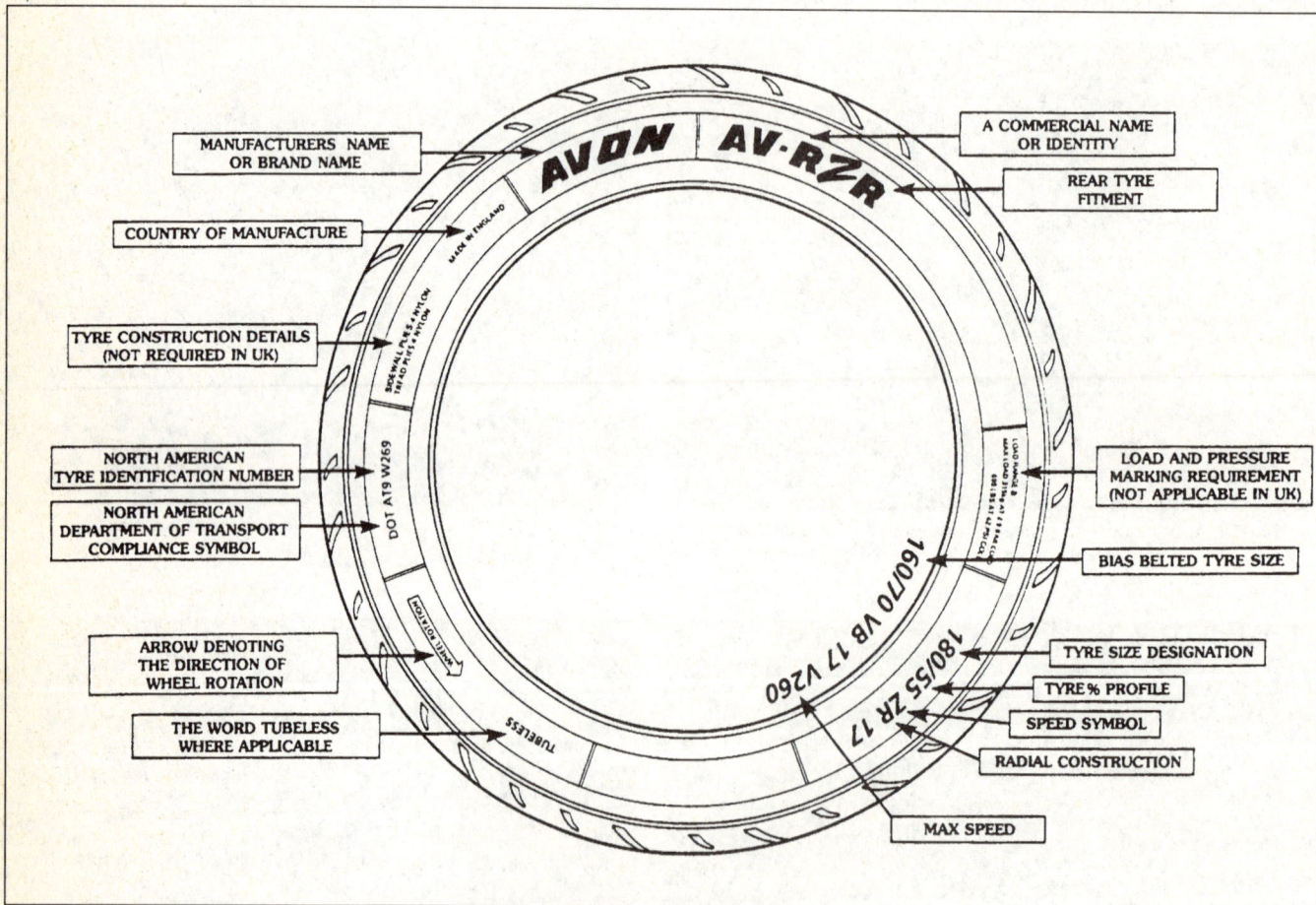

**16.3 Common tire sidewall markings**

# Chapter 8
# Fairing, bodywork and frame

## Contents

## Degrees of difficulty

| **Easy,** suitable for novice with little experience | 🔧 | **Fairly easy,** suitable for beginner with some experience | 🔧🔧 | **Fairly difficult,** suitable for competent DIY mechanic | 🔧🔧🔧 | **Difficult,** suitable for experienced DIY mechanic | 🔧🔧🔧🔧 | **Very difficult,** suitable for expert DIY or professional | 🔧🔧🔧🔧🔧 |

## Specifications

**Torque specifications**

| | |
|---|---|
| Frame downtube bolts (H models) . . . . . . . . . . . . . . . . . . . . . . . . . . . . . . | 44 Nm (33 ft-lbs) |
| Rear frame to main frame bolts . . . . . . . . . . . . . . . . . . . . . . . . . . . . . . . . | 44 Nm (33 ft-lbs) |
| Rear frame rear section to rear frame (J, K, L and M models) . . . . . . . . | 20 Nm (14.5 ft-lbs) |
| Sidestand bracket bolts . . . . . . . . . . . . . . . . . . . . . . . . . . . . . . . . . . . . . | 49 Nm (36 ft-lbs) |

---

### 1  General information

This Chapter covers the procedures necessary to remove and install the fairing and other body parts **(see illustrations)**.

Since many service and repair operations on these motorcycles require removal of the fairing and/or other body parts, the procedures are grouped here and referred to from other Chapters.

In the case of damage to the fairing or other body part, it is usually necessary to remove the broken component and replace it with a new (or used) one. The material that the fairing and other body parts is composed of doesn't lend itself to conventional repair techniques. There are, however, some shops that specialize in 'plastic welding', so it would be advantageous to check around first before throwing the damaged part away.

**Late Model**

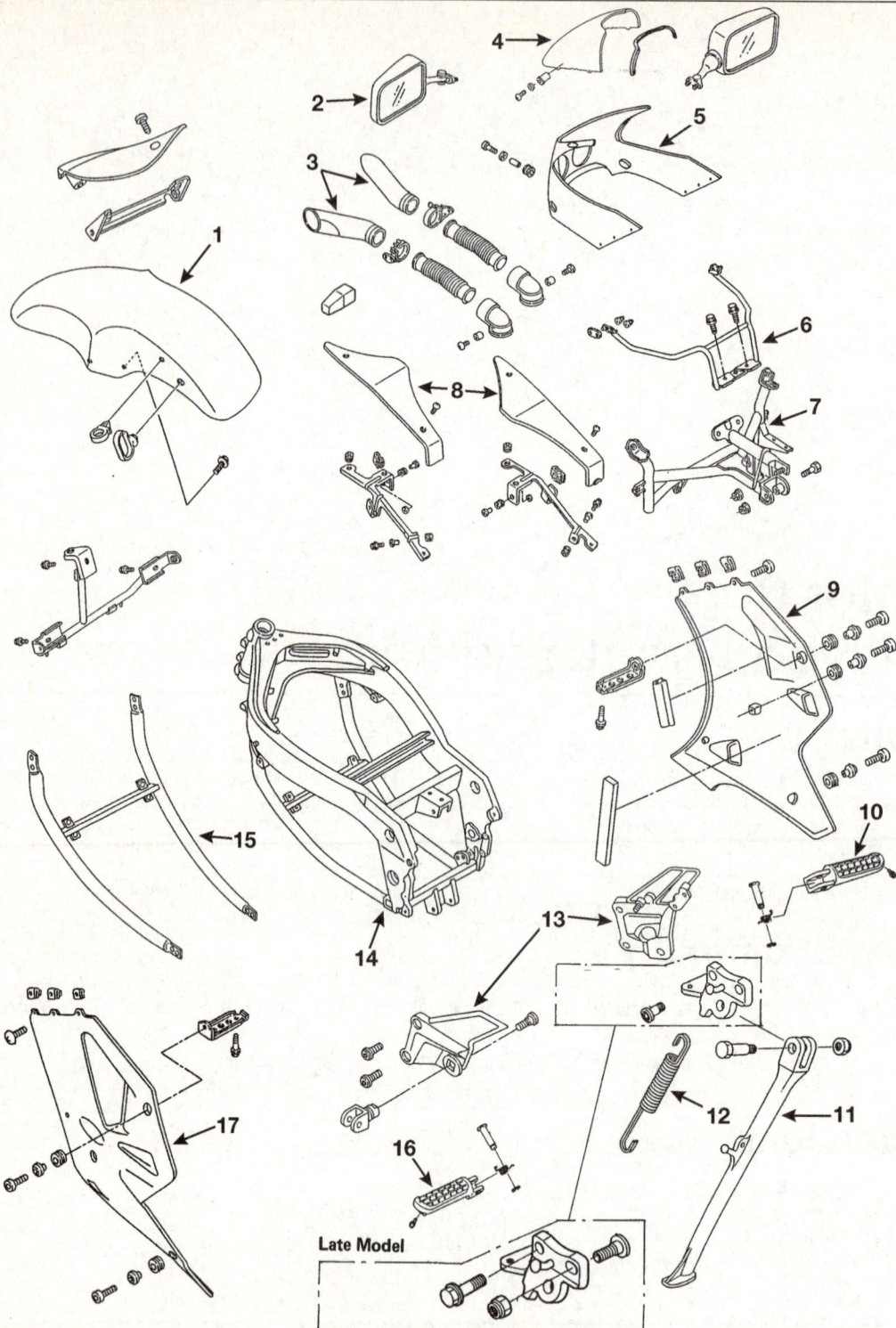

**1.1a Frame and bodywork details (H models) – part one of two**

| | | |
|---|---|---|
| 1 Front fender/ mudguard | 5 Upper fairing | 9 Right lower fairing panel | 14 Frame |
| 2 Mirror | 6 Fairing stay | 10 Rider's footpeg | 15 Frame downtubes |
| 3 Air ducts | 7 Fairing mount | 11 Sidestand | 16 Rider's footpeg |
| 4 Windshield | 8 Inner fairing panels | 12 Sidestand spring | 17 Left lower fairing panel |
| | | 13 Rider's footpeg brackets | |

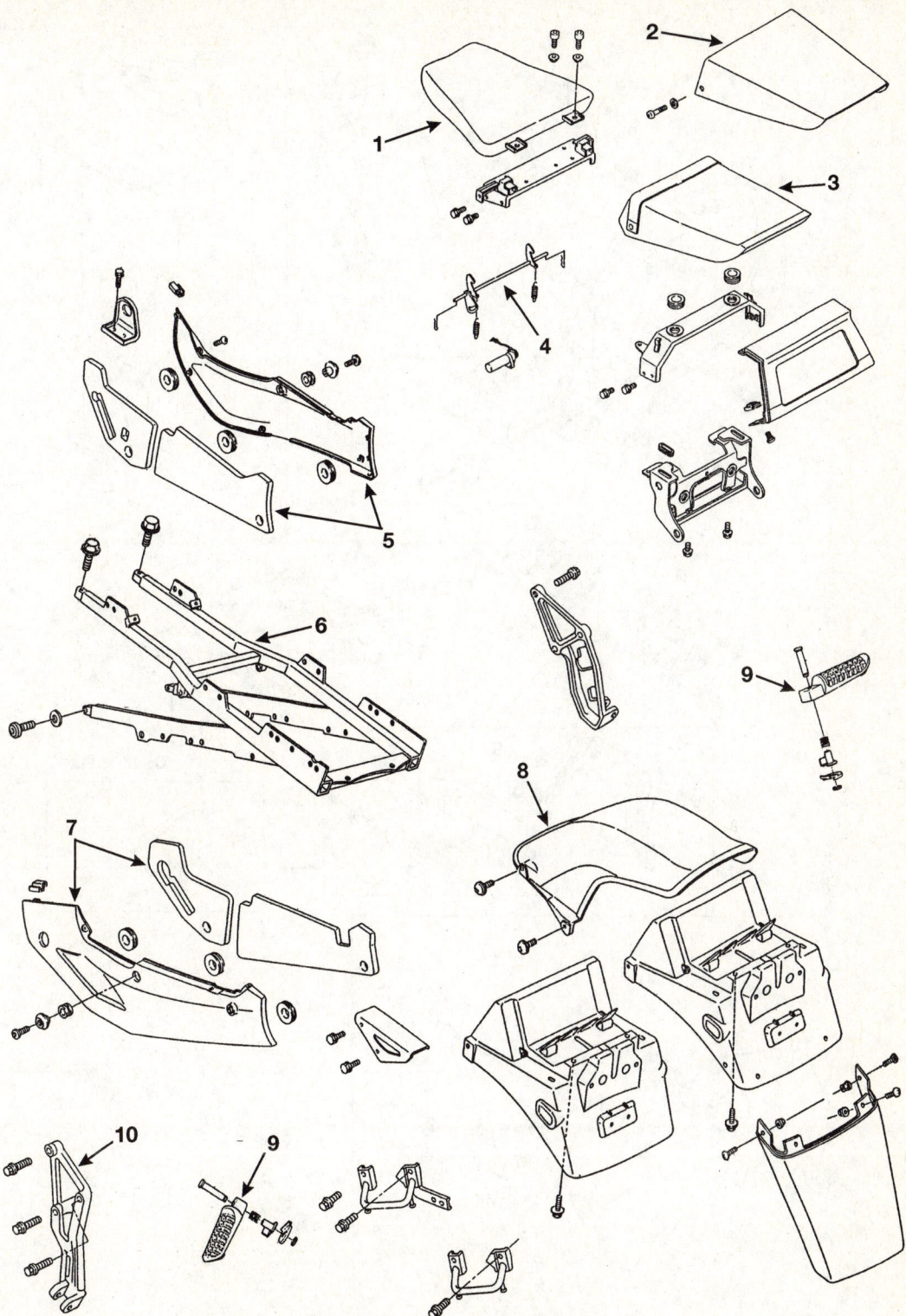

**1.1b  Frame and bodywork details (H models) – part two of two**

| | | | |
|---|---|---|---|
| 1 | Rider seat | 4 | Seat latch |
| 2 | Passenger seat cover | 5 | Right side cover |
| 3 | Passenger seat | 6 | Rear frame section |

| | | | |
|---|---|---|---|
| 7 | Left side cover | 9 | Passenger's footpeg |
| 8 | Rear fender/ mudguard | 10 | Passenger's footpeg bracket |

ZX750J

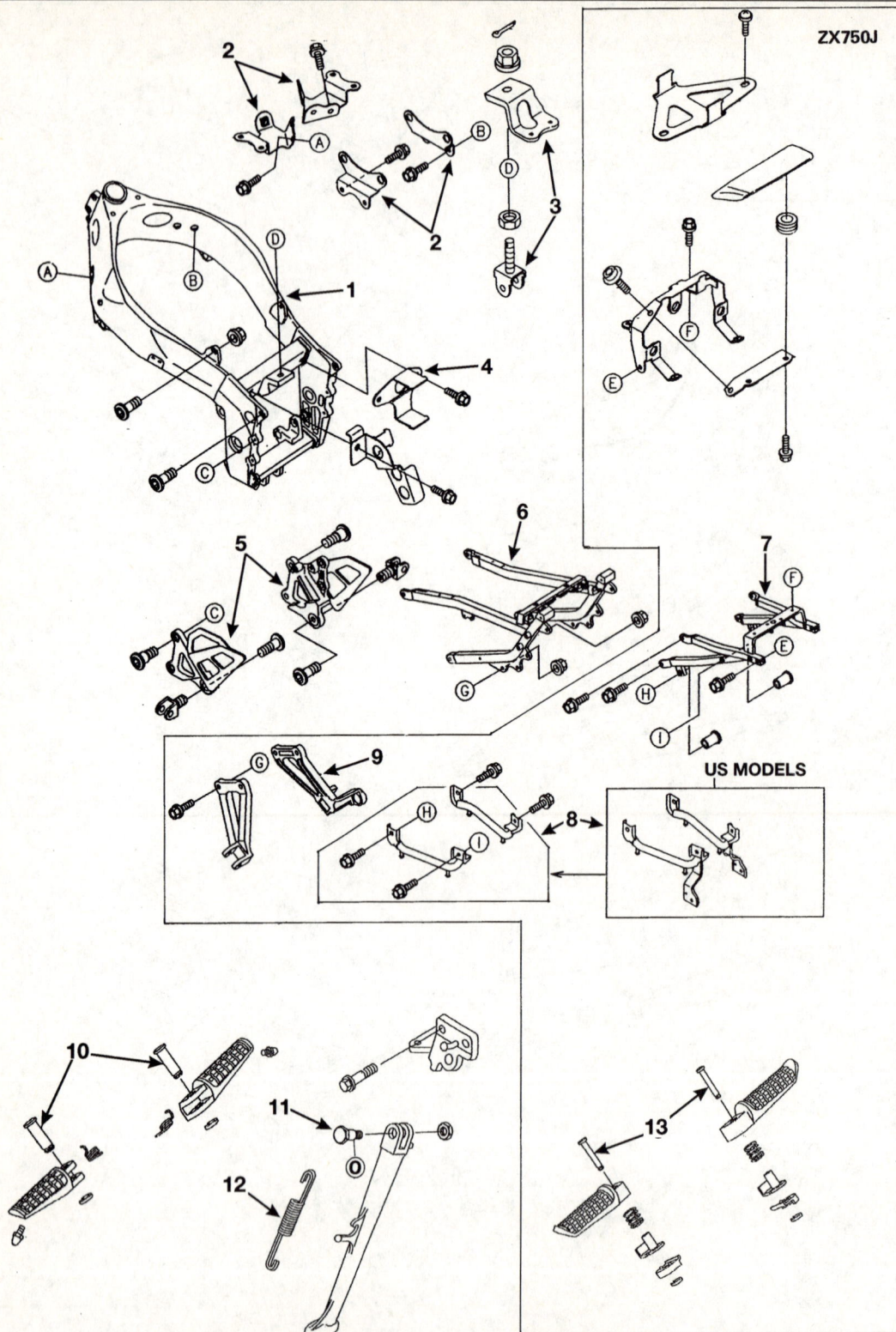

**1.1c Frame and bodywork details (J and K models) – part one of three**

1 Frame
2 Fairing brackets
3 Shock absorber bracket
4 Fuel pump bracket

5 Rider's footpeg brackets
6 Rear frame forward section
7 Rear frame rear section (J models)

8 Grab handle stays
9 Passenger's footpeg brackets
10 Rider's footpegs

11 Sidestand
12 Sidestand spring
13 Passenger footpegs

US MODELS

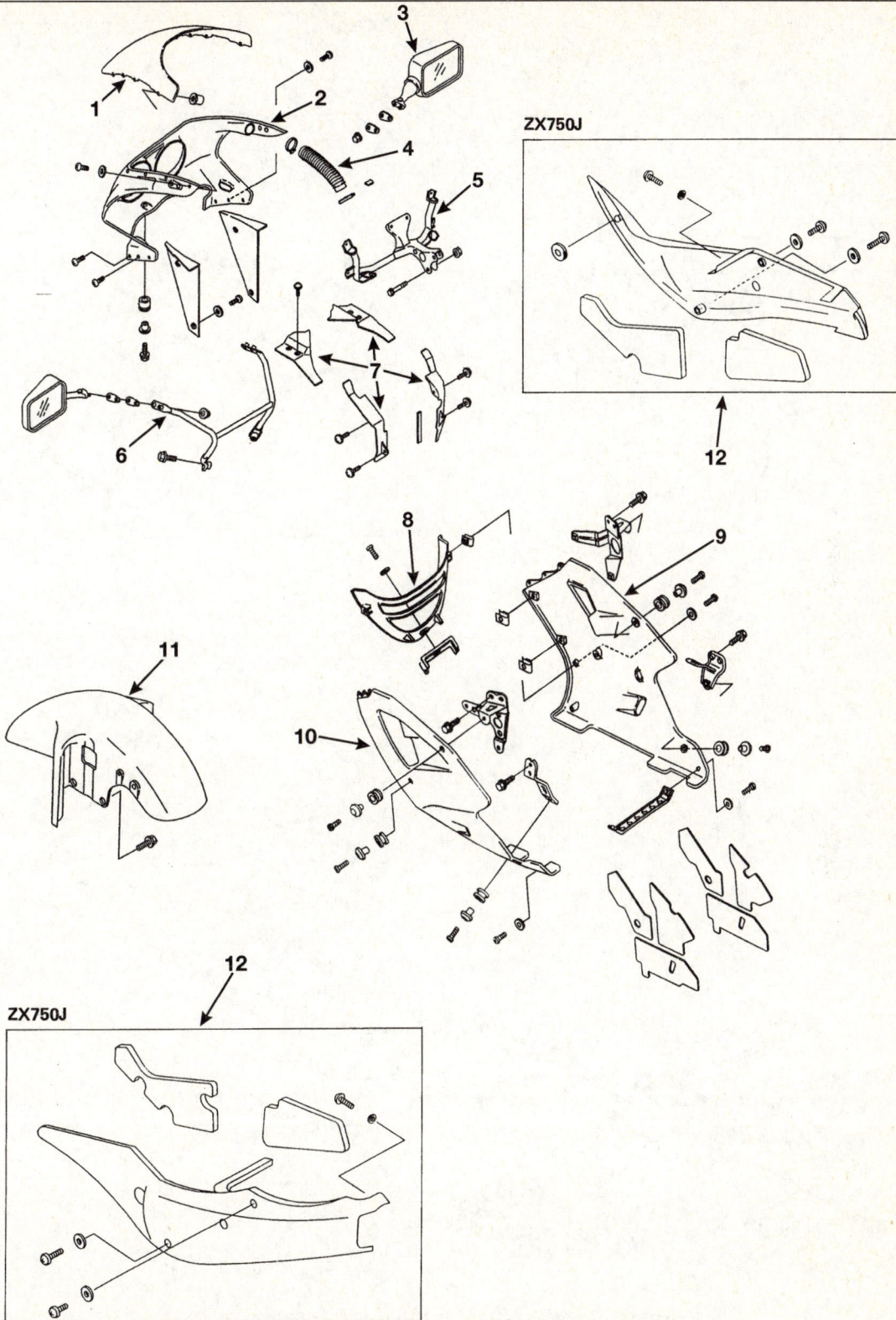

**1.1d  Frame and bodywork details (J and K models) – part two of three**

| | | | |
|---|---|---|---|
| 1 Windshield | 4 Air duct | 7 Inner fairing panels | 10 Lower left fairing panel |
| 2 Upper fairing | 5 Fairing mount | 8 Lower front fairing panel | 11 Front fender/mudguard |
| 3 Mirror | 6 Fairing stay | 9 Lower right fairing panel | 12 Side covers |

ZX750J

ZX750J

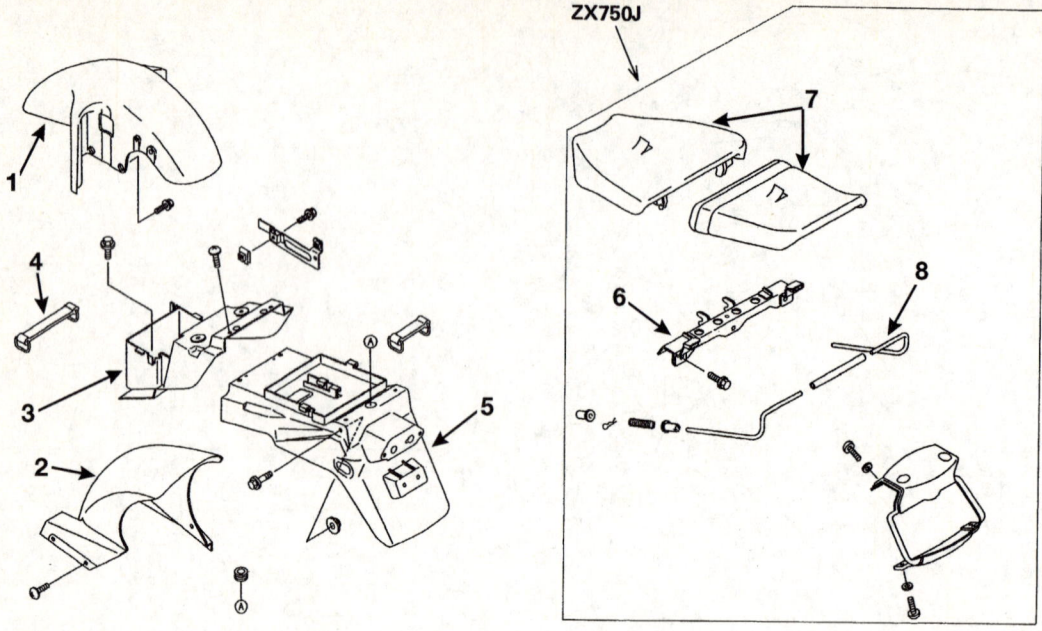

ZX750J

ZX750K

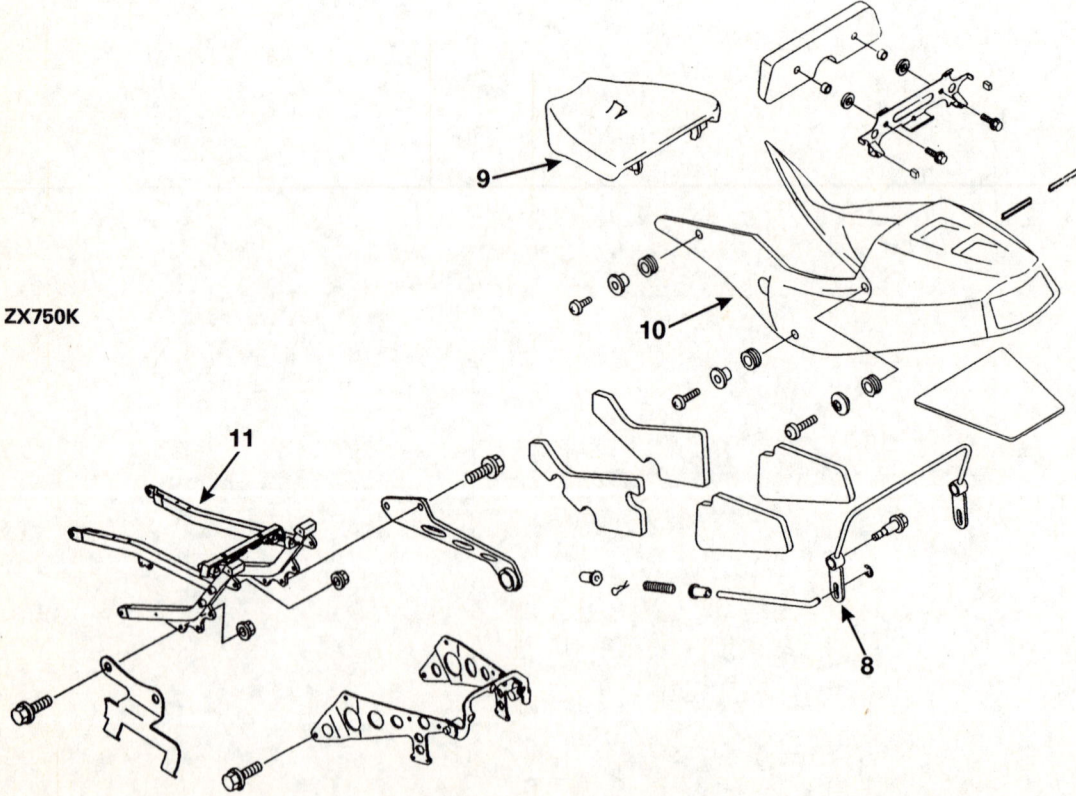

**1.1e  Frame and bodywork details (J and K models) – part three of three**

| | | | |
|---|---|---|---|
| 1 Front fender/<br>mudguard | 3 Battery box | 6 Bracket | 10 Side cover (K models) |
| | 4 Retaining strap | 7 Seats (J models) | 11 Rear frame rear section |
| 2 Rear fender/<br>mudguard front section | 5 Rear fender/mudguard rear<br>section | 8 Seat latch<br>9 Seat (K models) | (K models) |

ZX750L

US MODELS

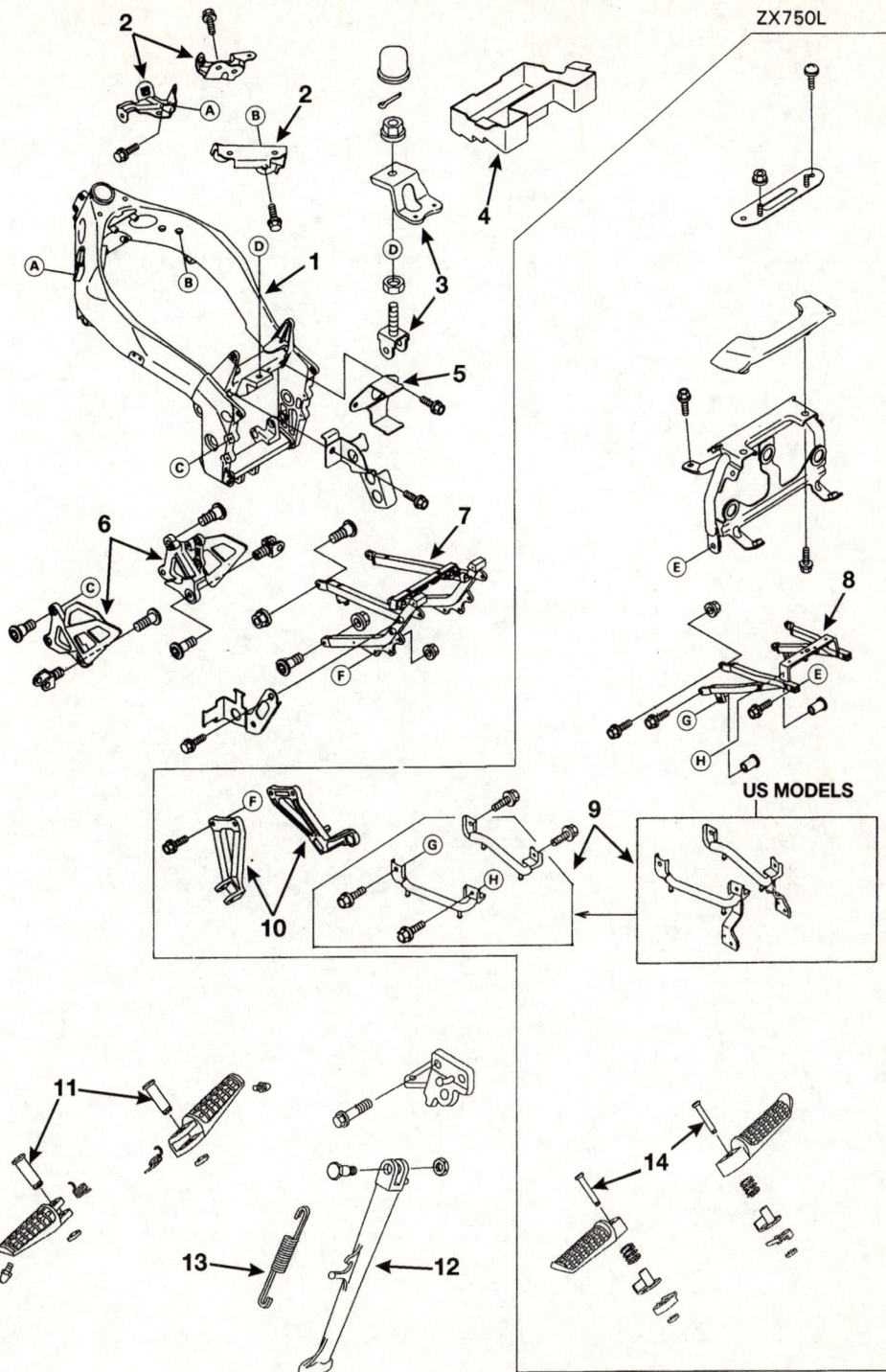

**1.1f  Frame and bodywork details (L and M models) – part one of three**

| | | | |
|---|---|---|---|
| 1  Frame | 4  Storage box | 8  Rear frame rear section (L models) | 11  Rider's footpegs |
| 2  Fairing brackets | 5  Fuel pump bracket | 9  Grab handle stays | 12  Sidestand |
| 3  Shock absorber and battery brackets | 6  Rider's footpeg brackets | 10  Passenger's footpeg brackets | 13  Sidestand spring |
| | 7  Rear frame front section | | 14  Passenger's footpegs |

ZX750L

ZX750L

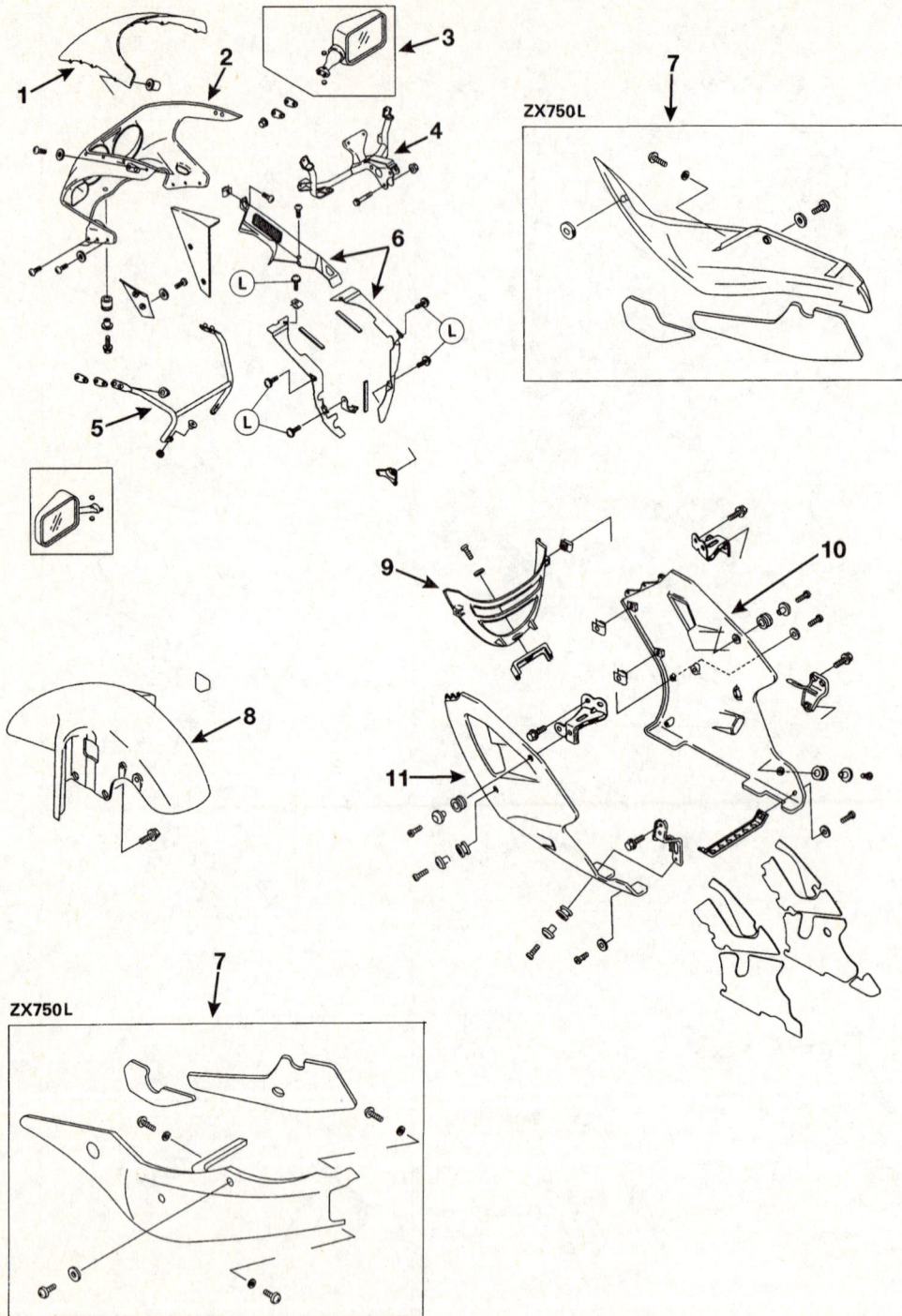

**1.1g Frame and bodywork details (L and M models) – part two of three**

| | | | |
|---|---|---|---|
| 1 Windshield | 4 Fairing mount | 7 Side cover (L models) | 10 Lower right fairing panel |
| 2 Upper fairing | 5 Fairing stay | 8 Front fender/mudguard | 11 Lower left fairing panel |
| 3 Mirror | 6 Inner fairing panels | 9 Lower front fairing panel | |

ZX750L

ZX750M

ZX750M

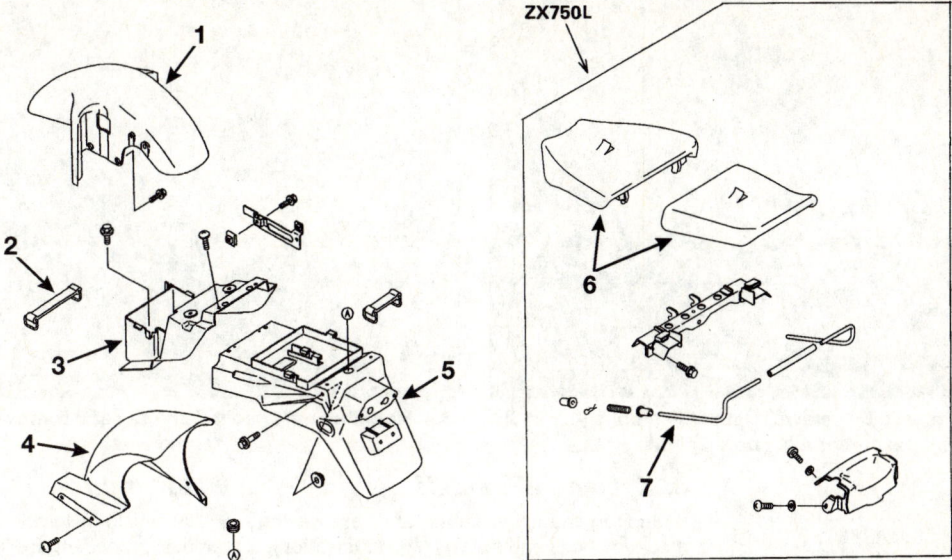

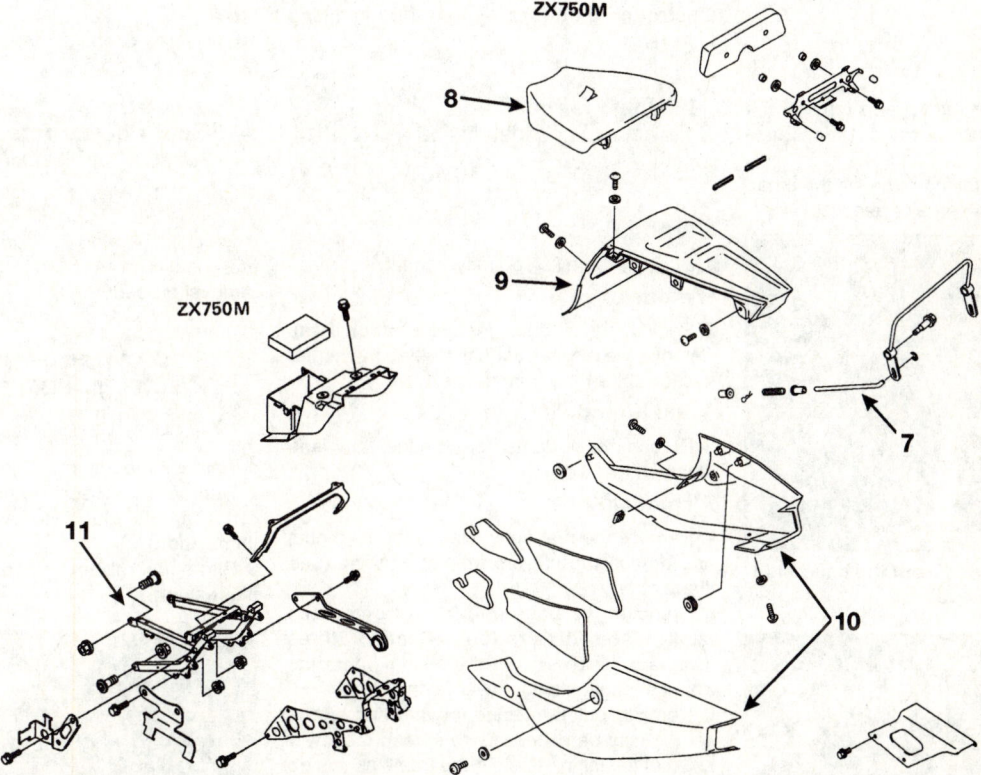

**1.1h  Frame and bodywork details (L and M models) – part three of three**

1  Front fender/mudguard
2  Retaining strap
3  Battery box
4  Rear fender/mudguard front section
5  Rear fender/mudguard rear section
6  Seats (L models)
7  Seat latch
8  Seat (M models)
9  Passenger seat cover (M models)
10  Side covers (M models)
11  Rear frame rear section (M models)

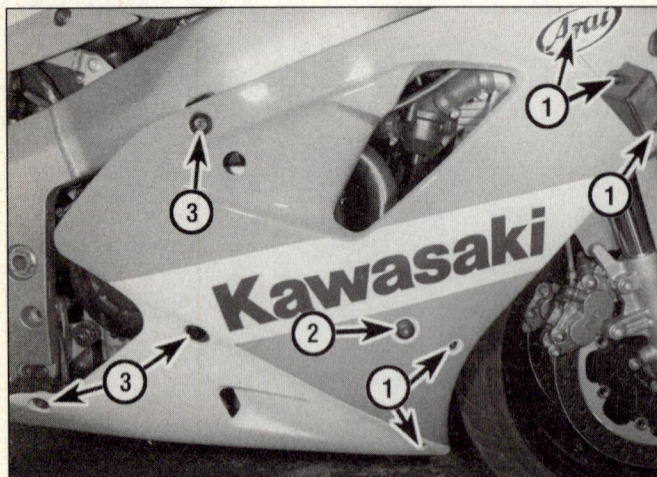

2.6 The fairing is secured by screws (1) and Allen bolts (2 and 3); install the center Allen bolt (2) first

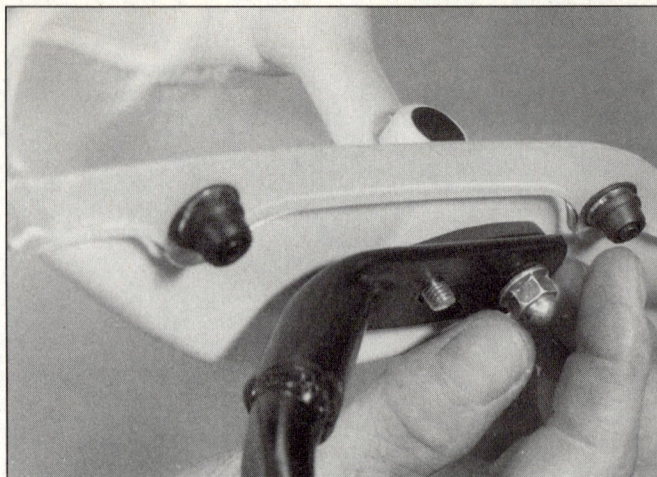

3.4 Hold the mirror so it won't fall and remove the mounting nuts and washers

## 2 Lower fairing – removal and installation

1 The lower fairing is made up of right and left panels.
2 Support the bike securely upright.

### H models

3 Support the lower fairing and remove the mounting bolts and screws (see illustration 1.1a).
4 Carefully take the fairing panel off the bike.
5 Installation is the reverse of removal.

3.5 Remove the bolts from the underside of the fairing

5.2 Remove the air duct cover screws (arrows)

### J, K, L and M models

6 Support the lower fairing and remove the mounting bolts and screws (see illustration).
7 Carefully take the fairing panel off the bike.
8 Installation is the reverse of removal. Tighten the forward Allen bolt first (see illustration 2.6), then tighten the remaining fasteners.

## 3 Upper fairing – removal and installation

### Removal

1 Support the bike securely upright.

### H models

2 Remove the air ducts (see illustration 1.1a). Remove the rear sections first, then the center sections, then the front section.

### J and K models

3 Detach the air ducts from the fuel tank (see illustration 1.1d).

### All models

4 Remove the rear view mirror mounting nuts and detach the mirrors from the fairing (see illustration).
5 Remove two bolts under the front of the fairing (see illustration). Remove three screws on each side of the fairing where the upper fairing joins the lower fairing panel.
6 Carefully pull the fairing forward and off the bike. It may be necessary to spread the lower sides of the fairing to clear the frame as you do this. Unplug the electrical connectors for the turn signals (refer to Chapter 9 if necessary).
7 If it's necessary to remove the windshield, remove the screws securing the windshield to the fairing.
8 Carefully separate the windshield from the fairing. If it sticks, don't attempt to pry it off – just keep applying steady pressure with your fingers.

### Installation

9 Windshield installation is the reverse of the removal procedure. Be sure each screw has a plastic washer under its head. Tighten the screws securely, but be careful not to overtighten them, as the windshield might crack.
10 Reverse Steps 1 through 6 to install the fairing on the motorcycle.

## 4 Rear view mirrors – removal and installation

Refer to Section 3, Step 4 – removing the rear view mirrors is part of the upper fairing removal procedure.

## 5 Air duct (L and M models) – removal and installation

1 Remove the upper fairing (see Section 3).
2 Remove three screws and take the cover off the air duct to frame joint (see illustration).
3 Slide the rubber joint off the duct (see illustration).

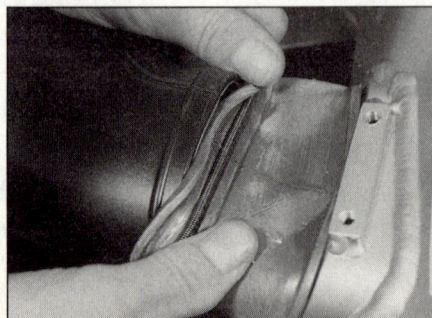

5.3 Pull back the spring band and separate the rubber joint from the duct

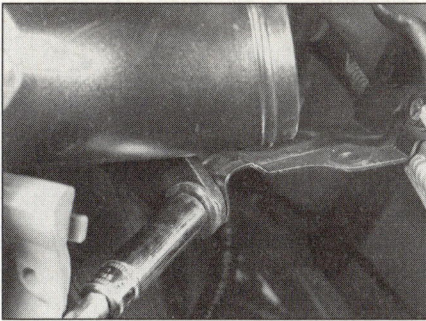

5.5a  Remove the duct bolts (rear bolt shown) . . .

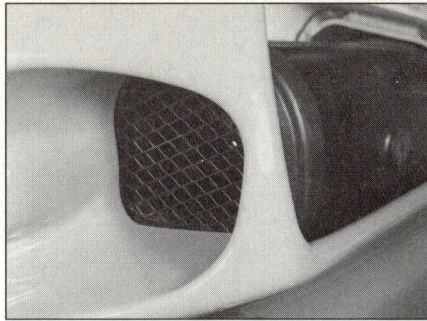

5.5b  . . . and pull the duct with its screen out of the fairing

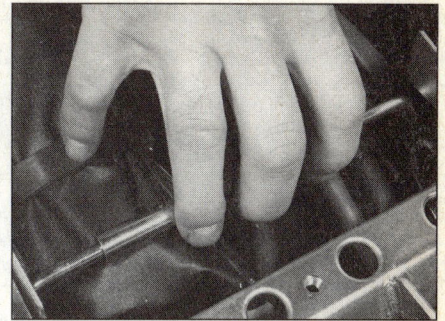

6.4  Pull back the latch to release the seat on J and L models

4  Remove the duct mounting bolts and lift it off.

5  If necessary, remove the front section of the duct and detach the screen from it (see illustrations).

6  Installation is the reverse of the removal steps.

## 6  Seats – removal and installation

### H models

1  Insert the ignition key in the seat lock. Press down on the rear seat, turn the key clockwise and lift out the rear seat, together with its cover.

2  Remove the two mounting bolts at the rear edge of the front seat, then pull it toward the rear of the motorcycle to remove it.

### J and L models

3  Insert the ignition key in the seat lock. Press down on the rear seat, turn the key counterclockwise (anti-clockwise) and lift out the rear seat.

4  Pull the seat latch lever to the rear (see illustration), then lift the rear end of the front seat and pull it backward to remove it.

### K and M models

5  Insert the ignition key in the seat lock. Turn the key clockwise and remove the pad at the back of the seat.

6  Reach inside the rear bodywork, pull the seat latch forward and lift the seat out.

### All models

7  Installation is the reverse of the removal steps. Make sure the seat lock or latch engages securely.

## 7  Side covers – removal and installation

### H models

1  Remove the seats (see Section 6).

2  Remove the side cover mounting screws and bolts (see illustration).

3  Carefully pull the side cover from the bike. There are lugs that fit into rubber bushings on

the frame, so apply a little extra force there (be careful not to break them off, though).

4  Installation is the reverse of the removal procedure.

### J and L models

5  Remove the seats (see Section 6).

6  Remove the screws (see illustration). Carefully pull the side cover from the bike. There's a lug that fit into a rubber bushing at the front of the cover, so apply a little extra force there (be careful not to break it off, though).

### K models

7  The side covers and rear bodywork are formed in one piece.

8  Remove the seat (see Section 6).

9  Remove the screws, Allen bolts and washers from each side of the motorcycle, then lift the side covers off (see illustration).

10  Installation is the reverse of the removal steps. Install the grommets on the leading edge of the opening (see illustration).

### M models

11  Remove the seats (see Section 6).

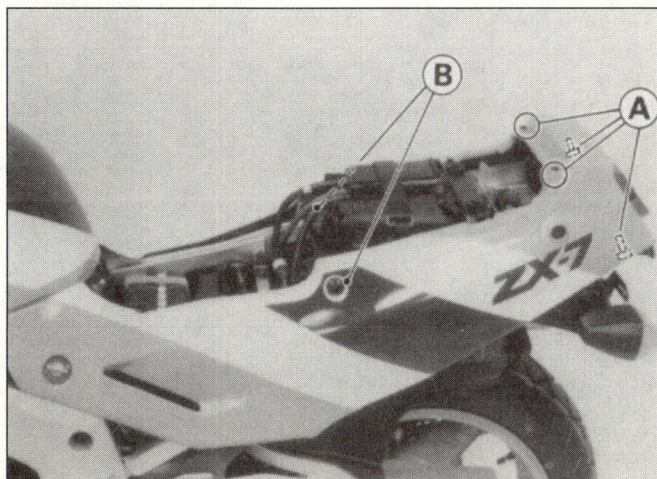

7.2  H model side cover fasteners

A  Screws          B  Bolts

7.6  J and L model side cover fasteners

A  White screws          C  Lug and grommet
B  Black screws

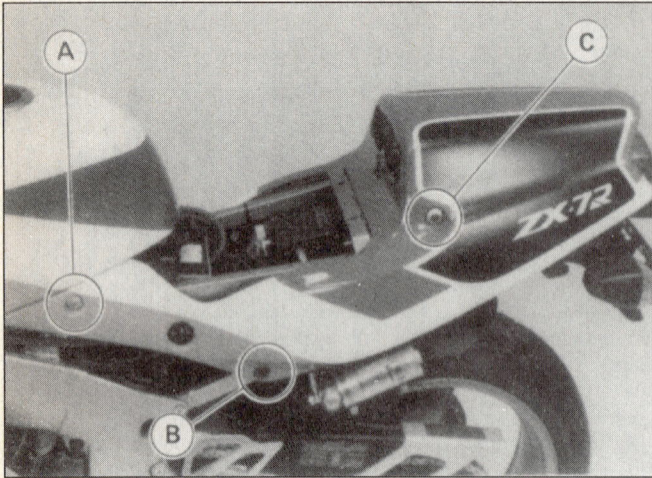

7.9 K model side cover fasteners

A  White screw            C  Allen bolt with thick washer
B  Allen bolt with flanged washer

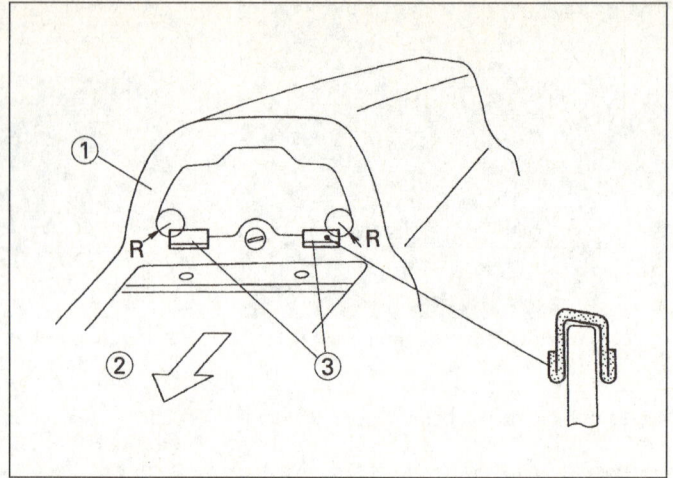

7.10  If the grommets were removed, install them like this

1  Seat cover               3  Grommets
2  Front of motorcycle

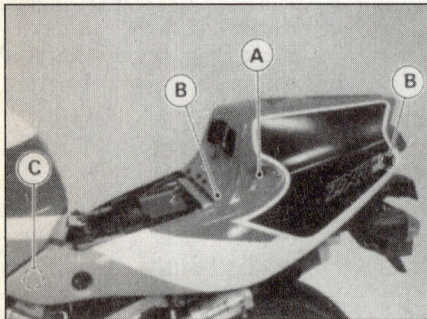

7.12 M model side cover fasteners

A  White screw          C  Lug and grommet
B  Black screws

**12** Remove the screws **(see illustration)**. Carefully pull the side cover from the bike. There's a lug that fits into a rubber bushing at the front of the cover, so apply a little extra

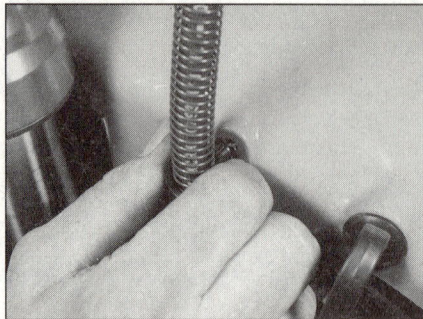

8.3  Detach the hose retainer from the fender/mudguard

force there (be careful not to break it off, though).
**13** The rear bodywork top section is retained to the frame's rear section by two screws at the front and two on each top edge.
**14** Installation is the reverse of the removal steps.

### 8  Front fender/mudguard – removal and installation

**1** Support the bike securely upright.
**2** Disconnect the speedometer cable from the speedometer drive.
**3** Detach the brake hose retainers (and speedometer cable retainer, if equipped) from the fender/mudguard **(see illustration)**.
**4** Remove the bolts from inside the fender/mudguard **(see illustration)**.
**5** Remove the fender/mudguard toward the front **(see illustration)**.
**6** Installation is the reverse of removal.

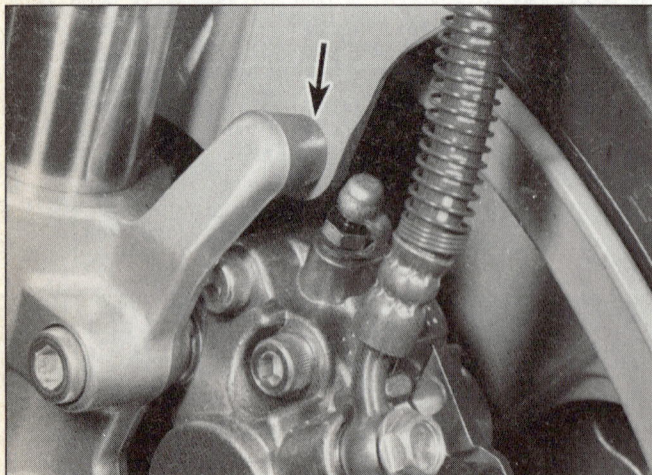

8.4  Remove two mounting bolts on each side (arrow) accessible from inside the fender/mudguard

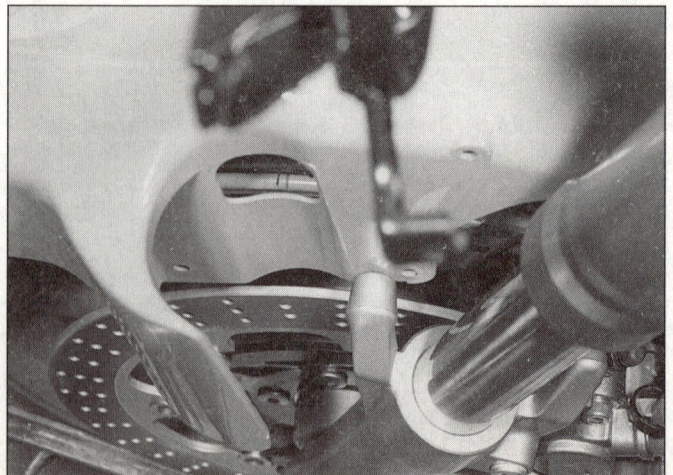

8.5  Rotate the fender/mudguard forward and lift it off

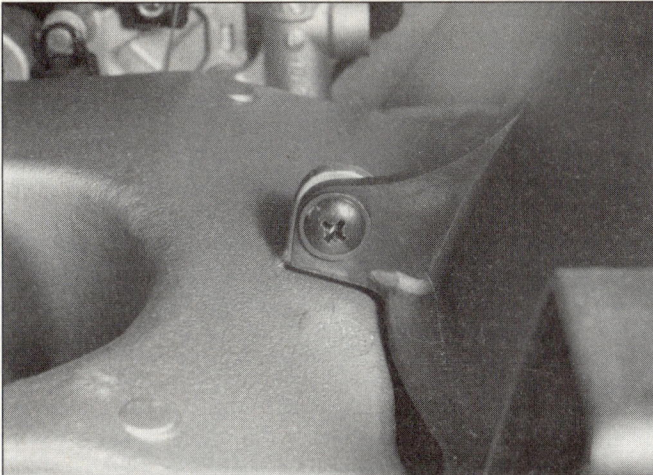

9.3a  On J, K, L and M models, the rear fender/mudguard is secured by one screw at the front . . .

9.3b  . . . and two screws on each side

## 9  Rear fender/mudguard – removal and installation

1  Support the bike securely upright.
2  Remove the rear wheel (see Chapter 7). Where necessary for access to the fender/mudguard fasteners, remove the swingarm (see Chapter 6).
3  Remove the mounting screws and detach the fender/mudguard from the swingarm **(see illustration 1.1b and the accompanying illustrations)**.
4  Installation is the reverse of the removal steps.

## 10  Footpegs and brackets – removal and installation

1  If it's only necessary to remove the footpeg, pry off the C-clip and remove the pivot pin and footpeg from the pivot **(see illustrations)**.
2  If it's necessary to remove the pivot from the bracket (rider footpegs only), remove the Allen bolt **(see illustration)**. The footpeg pivot on the right side also supports the brake pedal; refer to Chapter 7 for removal and installation procedures.
3  If it's necessary to remove the bracket for passenger footpegs or the rider's left footpeg, remove its mounting bolts and take it off the bike **(see illustrations 1.1a , 1.1b, 1.1c or 1.1f and the accompanying illustration)**.
4  To remove the bracket for the rider's right footpeg, refer to the rear brake master cylinder removal procedure in Chapter 7.

10.1a  Pry out the C-clip (arrow) (this is a front footpeg) . . .

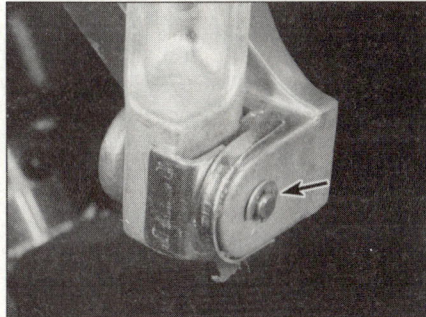

10.1b  . . . and this is a rear footpeg (arrow)

10.2  The front footpeg pivots are secured by Allen bolts (arrow)

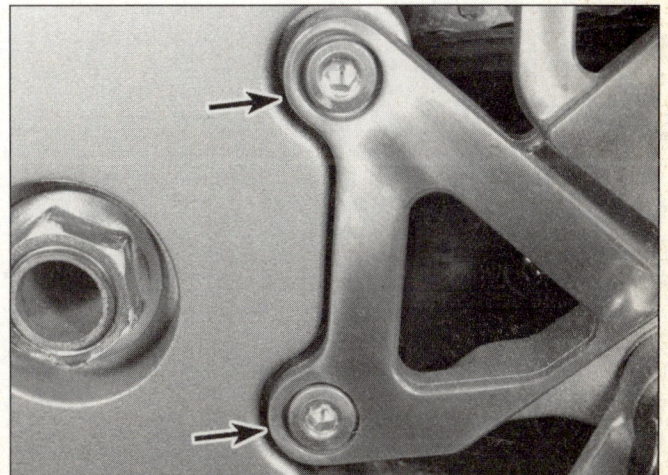

10.3  The left front footpeg bracket is secured by Allen bolts (arrows)

## 11 Sidestand maintenance

1 The sidestand pivots on a bolt attached to a bracket on the frame. Periodically, remove the pivot bolt and grease it thoroughly to avoid excessive wear.
2 Make sure the spring is in good condition and not overstretched **(see illustration)**. A broken or weak spring is an obvious safety hazard.

## 12 Frame –
general information,
inspection and repair

1 H models use a double-cradle frame with detachable downtubes. J, K, L and M models use a diamond-type frame without down-tubes. All models have a detachable rear frame section.
2 The frame shouldn't require attention unless accident damage has occurred. In most cases, frame replacement is the only satisfactory remedy for such damage. A few frame specialists have the jigs and other equipment necessary for straightening the frame to the required standard of accuracy, but even then there is no simple way of assessing to what extent the frame may have been overstressed.
3 After the machine has accumulated a lot of miles, the frame should be examined closely for signs of cracking or splitting at the welded joints. Corrosion can also cause weakness at these joints. Loose engine mount bolts can cause ovaling or fracturing to the engine mounting points. Minor damage can often be repaired by welding, depending on the nature and extent of the damage.

**11.2 The sidestand spring is installed like this**

4 Remember that a frame that is out of alignment will cause handling problems. If misalignment is suspected as the result of an accident, it will be necessary to strip the machine completely so the frame can be thoroughly checked.

# Chapter 9
## Electrical system

## Contents

## Degrees of difficulty

| Easy, suitable for novice with little experience | Fairly easy, suitable for beginner with some experience | Fairly difficult, suitable for competent DIY mechanic | Difficult, suitable for experienced DIY mechanic | Very difficult, suitable for expert DIY or professional |
|---|---|---|---|---|

## Specifications

**Battery**

Type
H1 models . . . . . . . . . . . . . . . . . . . . . . . . . . . . . . . . . . . . . . . . . . . . . 12 volt, 14 Ah (amp hours), fillable
H2 models . . . . . . . . . . . . . . . . . . . . . . . . . . . . . . . . . . . . . . . . . . . . . 12 volt, 12 Ah (amp hours), fillable
J and L models . . . . . . . . . . . . . . . . . . . . . . . . . . . . . . . . . . . . . . . . . 12 volt, 10 Ah (amp hours), maintenance free
K and M models . . . . . . . . . . . . . . . . . . . . . . . . . . . . . . . . . . . . . . . . 12 volt, 8 Ah (amp hours), maintenance free
Specific gravity (H models only) . . . . . . . . . . . . . . . . . . . . . . . . . . . . 1.280 at 20°C (68° F)
Terminal voltage (J, K, L and M models) . . . . . . . . . . . . . . . . . . . . . . 12.6 volts or more
Charging (J, K, L and M models)
    J and L models
        Standard charge . . . . . . . . . . . . . . . . . . . . . . . . . . . . . . . . . . . 1.2 amps for 5 to 10 hours*
        Quick charge . . . . . . . . . . . . . . . . . . . . . . . . . . . . . . . . . . . . . . 5.0 amps for one hour
    K and M models
        Standard charge . . . . . . . . . . . . . . . . . . . . . . . . . . . . . . . . . . . 0.9 amps for 5 to 10 hours*
        Quick charge . . . . . . . . . . . . . . . . . . . . . . . . . . . . . . . . . . . . . . 4.0 amps for one hour

*Refer to text for exact charging times.

## Bulbs

### US models

| | |
|---|---|
| Headlight | 2 x 60/55W |
| Tail/brake lights | 2 x 8/27W |
| License plate light | 8W |
| Turn signal lights | 4 x 23W |
| Instrument lights – H models | |
|   Neutral, oil pressure, high beam, turn signal warning | 3.4W |
|   Meter illumination | 3W |
| Instrument lights – J, K, L and M models | |
|   Neutral, oil pressure, high beam, turn signal warning | 3W |
|   Meter illumination | 3 x 1.7W, 2 x 2W |

### UK models

| | |
|---|---|
| Headlights | 2 x 60/55W |
| Parking light | 4W |
| Tail/brake lights | 2 x 5/21W |
| License plate light | 5W |
| Turn signal lights | 4 x 21W |
| Instrument lights – H models | |
|   Neutral, oil pressure, high beam, turn signal warning | 3.4W |
|   Meter illumination | 3W |
| Instrument lights – J, K, L and M models | |
|   Neutral, oil pressure, high beam, turn signal warning | 3W |
|   Meter illumination | 3 x 1.7W, 2 x 2W |

## Charging system

| | |
|---|---|
| Charging output voltage | |
|   H models | 14.5 volts at 4000 rpm |
|   J, K, L and M models | 14.2 to 14.8 volts at 4000 rpm |
| Stator coil resistance | One ohm or less |
| Rotor coil resistance | |
|   H models | 3 to 5 ohms |
|   J, K, L and M models | 2.3 to 3.5 ohms |
| Brush length | |
|   Standard | 10.5 mm (0.413 inch) |
|   Minimum | 4.5 mm (0.177 inch) |
| Slip ring diameter | |
|   Standard | 14.4 mm (0.567 inch) |
|   Minimum | 14.0 mm (0.551 inch) |

## Starter motor

| | |
|---|---|
| Brush length | |
|   Standard | 12 mm (0.472 inch) |
|   Minimum | 8.5 mm (0.335 inch) |
| Commutator diameter | |
|   Standard | 28 mm (1.102 inch) |
|   Minimum | 27 mm (1.063 inch) |

## Circuit fuse ratings

| | |
|---|---|
| Taillight | 10A |
| Accessory | 10A |
| Headlight relay | 10A |
| Fan fuse | 10A |
| Main fuse | 30A |
| Auxiliary headlight fuse | 20A |

## Torque specifications

| | |
|---|---|
| Alternator mounting bolts | |
|   H models | 39 Nm (29 ft-lbs) |
|   J, K, L and M models | 25 Nm (18 ft-lbs) |
| Starter mounting bolts | 9.8 Nm (87 inch-lbs) |
| Starter long screws (through-bolts) | 4.9 Nm (43 inch-lbs) |
| Starter terminal nut | 4.9 Nm (43 inch-lbs) |
| Neutral switch | 15 Nm (11 ft-lbs) |
| Oil pressure switch | 15 Nm (11 ft-lbs) |

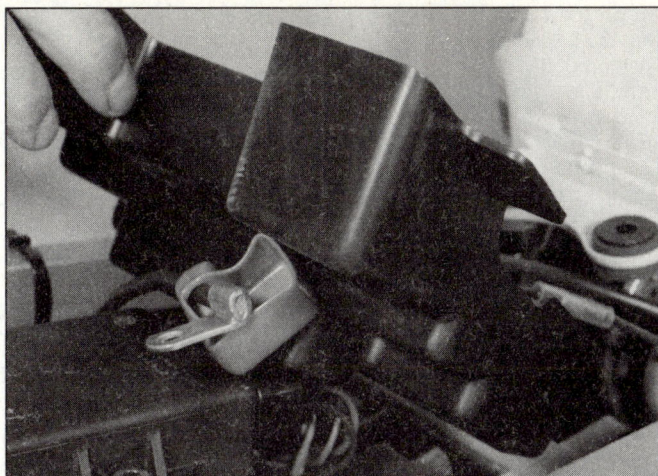
3.3 Lift out the storage box

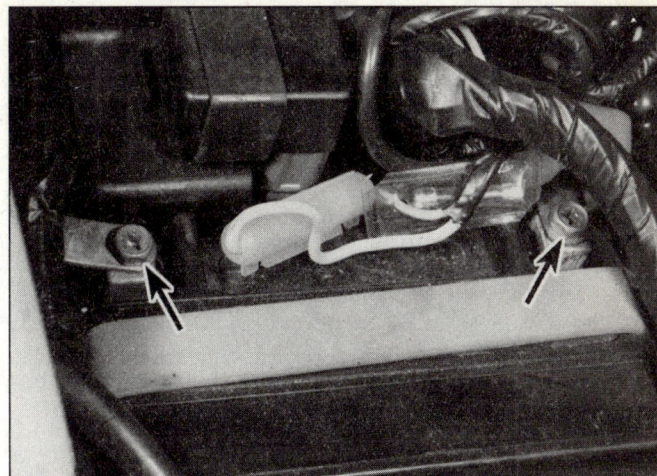
3.4 Disconnect the battery cables (arrows)

## 1  General information

The machines covered by this manual are equipped with a 12-volt electrical system. The components include a three-phase alternator with an integrated circuit regulator built in. The alternator is a separate unit, mounted on top of the engine. On H models, it's driven by a rubber belt. On J, K, L and M models, the alternator is driven by the alternator driveshaft through a rubber damper. The regulator maintains the charging system output within the specified range to prevent overcharging. The diodes convert the AC output of the alternator to DC current to power the lights and other components and to charge the battery.

An electric starter mounted to the engine case behind the bank of the cylinders is standard equipment. The starting system includes the motor, the battery, the solenoid, the starter circuit relay (part of the junction box) and the various wires and switches. If the engine kill switch and the main key switch are both in the On position, the circuit relay allows the starter motor to operate only if the transmission is in Neutral (Neutral switch on) or the clutch lever is pulled to the handlebar

(clutch switch on) and the sidestand is up (sidestand switch on).
**Note:** *Keep in mind that electrical parts, once purchased, can't be returned. To avoid unnecessary expense, make very sure the faulty component has been positively identified before buying a replacement part.*

## 2  Electrical fault finding

A typical electrical circuit consists of an electrical component, the switches, relays, etc. related to that component and the wiring and connectors that hook the component to both the battery and the frame. To aid in locating a problem in any electrical circuit, complete *Wiring diagrams* of each model are included at the end of this Chapter.

Before tackling any troublesome electrical circuit, first study the appropriate diagrams thoroughly to get a complete picture of what makes up that individual circuit. Trouble spots, for instance, can often be narrowed down by noting if other components related to that circuit are operating properly or not. If several components or circuits fail at one time, chances are the fault lies in the fuse or ground/earth connection, as several circuits often are routed through the same fuse and ground/earth connections.

Electrical problems often stem from simple causes, such as loose or corroded connections or a blown fuse. Prior to any electrical fault finding, always visually check the condition of the fuse, wires and connections in the problem circuit.

If testing instruments are going to be utilized, use the diagrams to plan where you will make the necessary connections in order to accurately pinpoint the trouble spot.

The basic tools needed for electrical fault finding include a test light or voltmeter, a continuity tester (which includes a bulb, battery and set of test leads) and a jumper

wire. For more extensive checks, a multimeter capable of measuring ohms, volts and amps will be required. Full details on the use of this equipment are given in *Fault Finding Equipment* in the Reference section at the end of this manual.

## 3  Battery –
inspection and maintenance

1  Most battery damage is caused by heat, vibration, and/or low electrolyte levels, so keep the battery securely mounted, check the electrolyte level frequently and make sure the charging system is functioning properly.
2  Refer to Chapter 1 for electrolyte level and specific gravity checking procedures on fillable batteries.
3  Remove the seats (see Chapter 8). If you're working on an L or M model, remove the storage box **(see illustration)**.
4  Disconnect the battery cables **(see illustration)**.

⚠ *Warning: Always disconnect the negative cable first and connect it last to prevent sparks which could cause the battery to explode.*

5  Unhook the battery retaining strap and lift the battery out **(see illustrations)**.

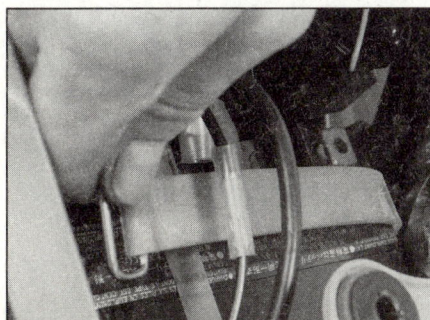
3.5a Unhook the retaining strap . . .

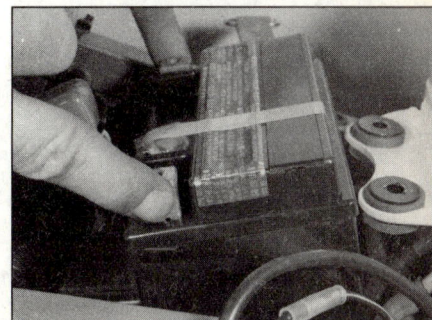
3.5b . . . and lift the battery out

**6** Check around the base inside of the battery for sediment, which is the result of sulfation caused by low electrolyte levels. These deposits will cause internal short circuits, which can quickly discharge the battery. Look for cracks in the case and replace the battery if either of these conditions is found.

**7** Check the battery terminals and cable ends for tightness and corrosion. If corrosion is evident, remove the cables from the battery and clean the terminals and cable ends with a wire brush or knife and emery paper. Reconnect the cables.

> **HAYNES HiNT** *Apply a thin coat of petroleum jelly to the battery connections to slow further corrosion.*

**8** The battery case should be kept clean to prevent current leakage, which can discharge the battery over a period of time (especially when it sits unused). Wash the outside of the case with a solution of baking soda and water. Do not get any baking soda solution in the battery cells. Rinse the battery thoroughly, then dry it.

**9** If acid has been spilled on the frame or battery box, neutralize it with the baking soda and water solution, dry it thoroughly, then touch up any damaged paint. Make sure the battery vent tube is directed away from the frame and is not kinked or pinched.

**10** If the motorcycle sits unused for long periods of time, disconnect the cables from the battery terminals. Refer to Section 4 and charge the battery approximately once every month.

**11** Reverse Steps 3 through 5 to install the battery.

## 4  Battery – charging

**1** If the machine sits idle for extended periods or if the charging system malfunctions, the battery can be charged from an external source. Charging procedures for the fillable batteries used on H models differ from the procedures for maintenance-free batteries, which are used on J, K, L and M models.

### Fillable batteries (H models)

**2** To properly charge the battery, you will need a charger of the correct rating, a hydrometer, a clean rag and a syringe for adding distilled water to the battery cells.

**3** The maximum charging rate for any battery is 1/10 of the rated amp/hour capacity. As an example, the maximum charging rate for the 14 amp/hour battery would be 1.4 amps. If the battery is charged at a higher rate, it could be damaged.

**4** Don't allow the battery to be subjected to a so-called quick charge (high rate of charge over a short period of time) unless you are prepared to buy a new battery. Quick charging can warp the plates in the battery or cause them to lose material. Either of these conditions can cause an internal short.

**5** When charging the battery, always remove it from the machine and be sure to check the electrolyte level before hooking up the charger. Add distilled water to any cells that are low.

**6** Loosen the cell caps, hook up the battery charger leads (red to positive, black to negative), cover the top of the battery with a clean rag, then, and only then, plug in the battery charger.

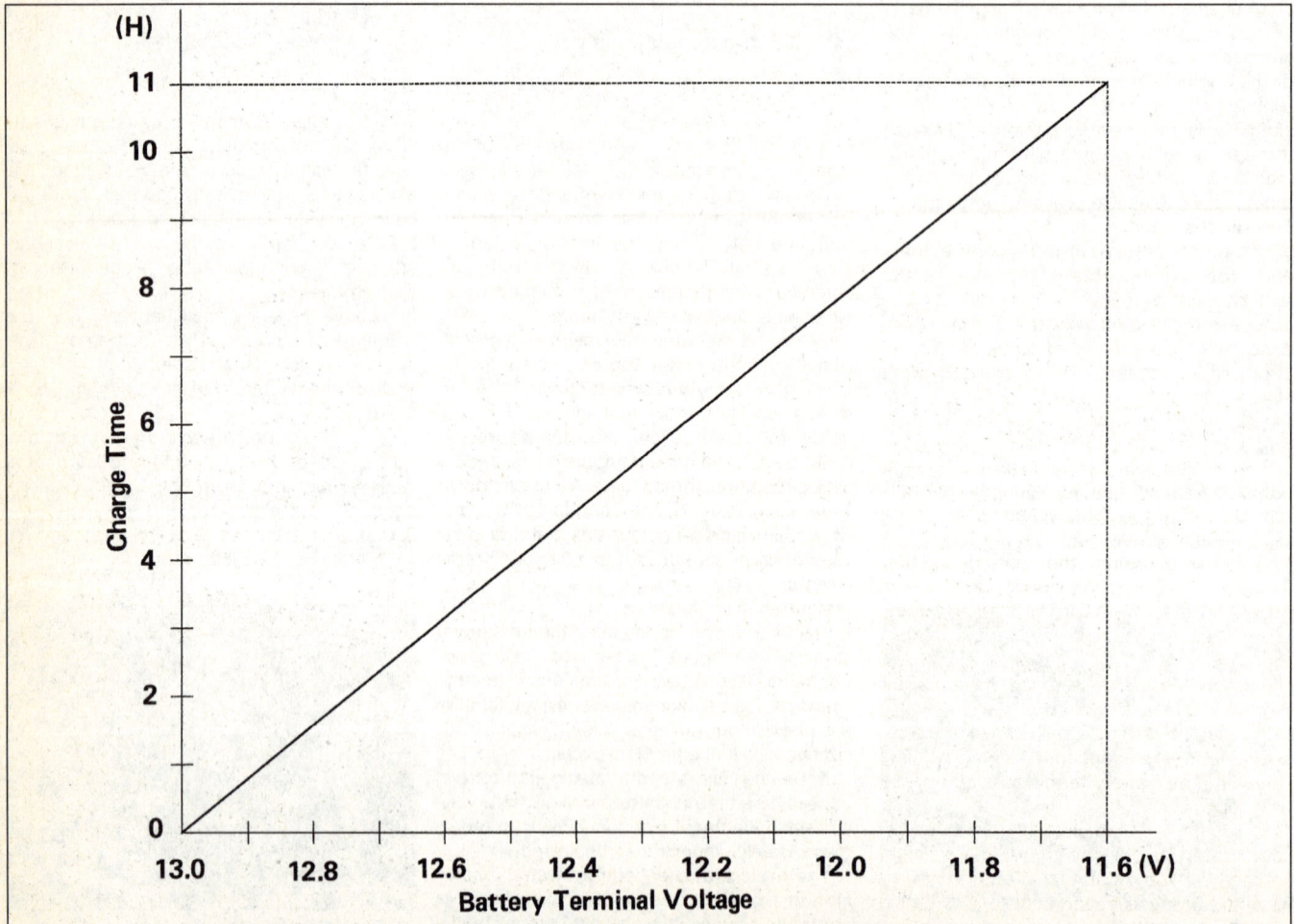

4.13  Battery charge time table (J, K, L and M models)

5.1a  Remove the fuse cover from the junction box . . .

5.1b  . . . to expose the fuses

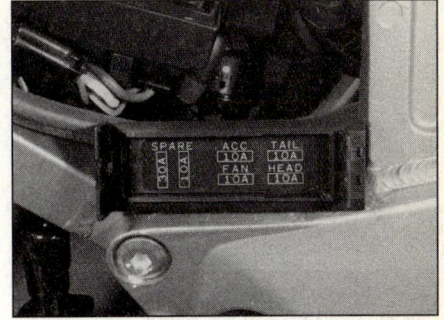
5.1c  Fuse ratings and functions are printed inside the cover

**Warning: The hydrogen gas escaping from a charging battery is explosive, so keep open flames and sparks well away from the area. Also, the electrolyte is extremely corrosive and will damage anything it comes in contact with.**

7 Allow the battery to charge until the specific gravity is as specified (refer to Chapter 1 for specific gravity checking procedures). The charger must be unplugged and disconnected from the battery when making specific gravity checks. If the battery overheats or gases excessively, the charging rate is too high. Either disconnect the charger or lower the charging rate to prevent damage to the battery.

8 If one or more of the cells do not show an increase in specific gravity after a long slow charge, or if the battery as a whole does not seem to want to take a charge, it's time for a new battery.

9 When the battery is fully charged, unplug the charger first, then disconnect the leads from the battery. Install the cell caps and wipe any electrolyte off the outside of the battery case.

### Maintenance-free batteries (J, K, L and M models)

10 Charging the maintenance-free battery used on these models requires a digital voltmeter and a variable-voltage charger with a built-in ammeter.

11 When charging the battery, always remove it from the machine and be sure to check the electrolyte level by looking through the translucent battery case before hooking up the charger. If the electrolyte level is low, the battery must be discarded; never remove the sealing plug to add water.

12 Disconnect the battery cables (negative cable first), then connect a digital voltmeter between the battery terminals and measure the voltage.

13 If terminal voltage is 12.6 volts or higher, the battery is fully charged. If it's lower, recharge the battery. Refer to the accompanying illustration and this Chapter's Specifications for charging rate and time (see illustration).

14 A quick charge can be used in an emergency, provided the maximum charge rates and times are not exceeded (exceeding the maximum rate or time may ruin the battery). A quick charge should always be followed as soon as possible by a charge at the standard rate and time.

15 Hook up the battery charger leads (positive lead to battery positive terminal, negative lead to battery negative terminal), then, and only then, plug in the battery charger.

**Warning: The hydrogen gas escaping from a charging battery is explosive, so keep open flames and sparks well away from the area. Also, the electrolyte is extremely corrosive and will damage anything it comes in contact with.**

16 Start charging at a high voltage setting (no more than 25 volts) and watch the ammeter for about 5 minutes. If the charging current doesn't increase, replace the battery with a new one.

17 When the charging current increases beyond the specified maximum, reduce the charging voltage to reduce the charging current to the rate listed in this Chapter's Specifications. Do this periodically as the battery charges.

18 Allow the battery to charge for the specified time listed in this Chapter's Specifications. If the battery overheats or gases excessively, the charging rate is too high. Either disconnect the charger or lower the charging rate to prevent damage to the battery.

19 After the specified time, unplug the charger first, then disconnect the leads from the battery.

20 Wait 30 minutes, then measure voltage between the battery terminals. If it's 12.6 volts or higher, the battery is fully charged. If it's between 12.0 and 12.6 volts, charge the battery again (refer to this Chapter's Specifications and illustration 4.13 for charge rate and time). If it's less than 12.0 volts, it's time for a new battery.

21 When the battery is fully charged, unplug the charger first, then disconnect the leads from the battery. Wipe off the outside of the battery case and install the battery in the bike.

## 5  Fuses – check and replacement

1 Most of the fuses are located under the seat, on the junction box (see illustrations). The fuses are protected by a plastic cover, which snaps into place. It contains fuses (and spares) which protect the main, headlight, taillight and accessory circuit wiring and components from damage caused by short circuits.

2 In addition to the junction box fuses, there's an individual 20-amp headlight fuse and a 30-amp main fuse. The 20-amp headlight fuse is located behind the left side cover (H2 models) or under the seat (all others). The main fuse is located on the starter relay (see illustration).

3 If you have a test light, the junction box fuses can be checked without removing them. Turn the ignition to the On position, connect one end of the test light to a good ground/earth, then probe each terminal on top of the fuse. If the fuse is good, there will be voltage available at both terminals. If the fuse is blown, there will only be voltage present at one of the terminals.

4 The fuses can be removed and checked visually. On all except the 20-amp headlight fuse, pull the fuse out with your fingertips or use a pair of needle-nose pliers. To remove the 20-amp headlight fuse, snap open its case and lift it out. A blown fuse is easily identified by a break in the element (see illustrations).

5.2  The main fuse is mounted on the starter relay (H2, J, K, L and M model location shown)

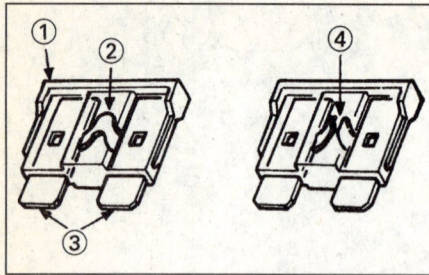

**5.4a  Blown fuses can be identified by the break in the element (this is a junction box fuse) . . .**

1  Fuse housing   3  Terminals
2  Element        4  Broken element

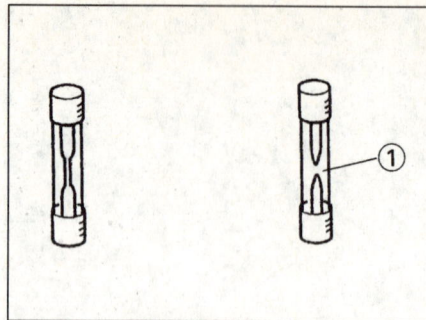

**5.4b  . . . and this is the 20-amp headlight fuse**

1  Broken element

### Fuse Circuit Inspection

| Meter Connection | Meter Reading (Ω) |
|---|---|
| 1 – 2 | 0 |
| *1 – 3B | 0 |
| 6 – 7 | 0 |
| 6 – 17 | 0 |
| 1 – 7 | ∞ |
| *8 – 17 | ∞ |

(*) : US, Canada Models only

**6.4a  Using an ohmmeter, check the continuity across the indicated pairs of terminals**

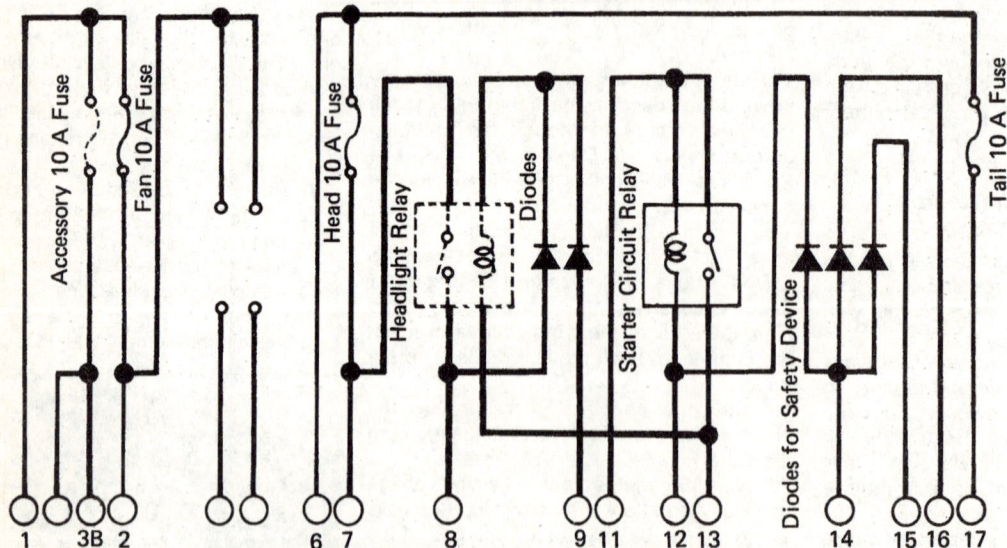

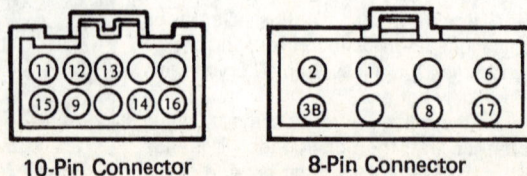

5 If a fuse blows, be sure to check the wiring harnesses very carefully for evidence of a short circuit. Look for bare wires and chafed, melted or burned insulation. If a fuse is replaced before the cause is located, the new fuse will blow immediately.

6 Never, under any circumstances, use a higher rated fuse or bridge the fuse block terminals, as damage to the electrical system could result.

7 Occasionally a fuse will blow or cause an open circuit for no obvious reason.

**HAYNES HiNT** *Corrosion of the fuse ends and fuse block terminals may occur and cause poor fuse contact. If this happens, remove the corrosion with a wire brush or emery paper, then spray the fuse end and terminals with electrical contact cleaner.*

### 6  Junction box – check

1 Aside from serving as the fuse block, the junction box also houses two relays – the starter circuit relay (not the starter solenoid) and (on US and Canadian models) the headlight relay. Neither of these relays is replaceable individually. If one of them fails, the junction box must be replaced.

**6.4b  Junction box circuit (H models)**

━ ━ ━  for US, Canada models

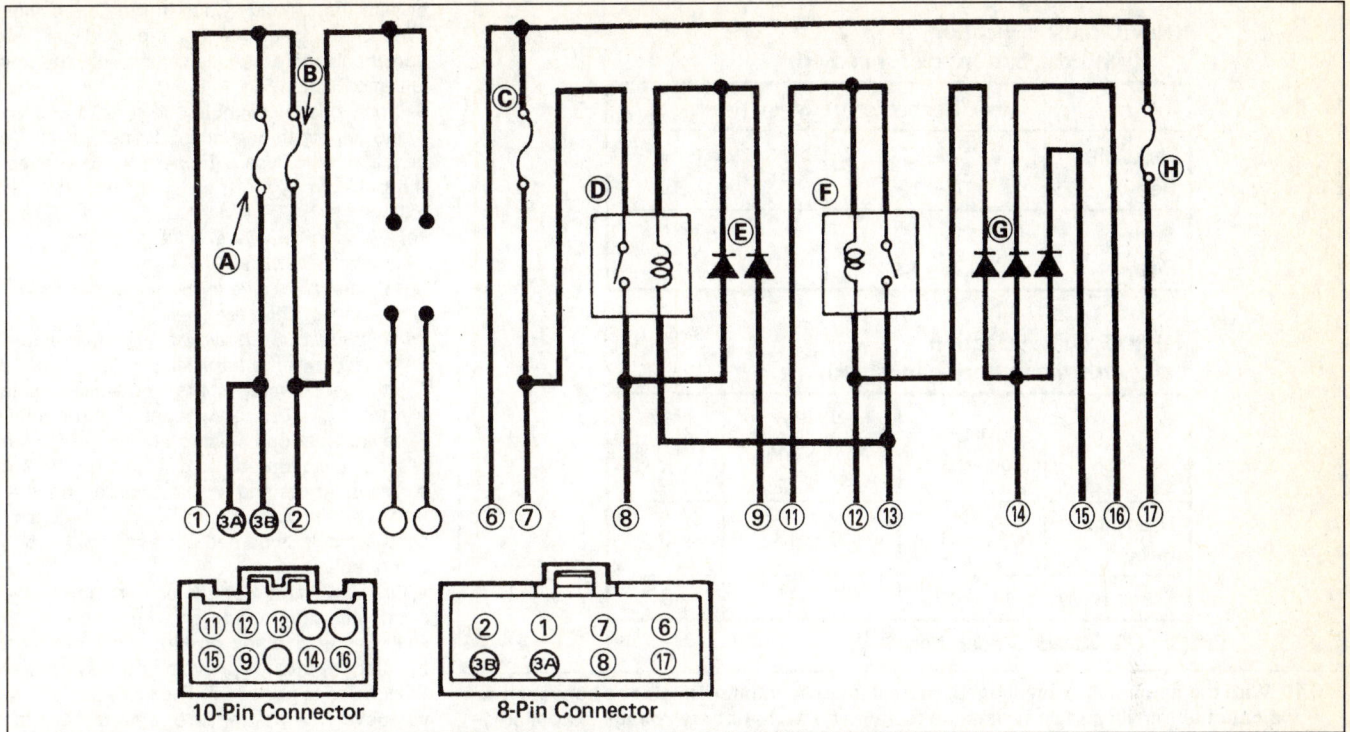

**6.4c Junction box circuit (J, K, L and M US and Canada models)**

A  10 amp accessory fuse          C  10 amp headlight fuse          E  Diodes                          G  Diodes for interlock circuit
B  10 amp fan fuse                D  Headlight relay                F  Starter circuit relay           H  10 amp taillight fuse

**2** In addition to the relay checks, the fuse circuits and diode circuits should be checked also, to rule out the possibility of an open circuit condition or blown diode within the junction block as the cause of an electrical problem.

### Fuse circuit check

**3** Remove the junction box by sliding it out of its holder. Unplug the electrical connectors from the box.

**4** If the terminals are dirty or bent, clean and straighten them. Using the accompanying tables as a guide, check the continuity across the indicated terminals with an ohmmeter – some should have no resistance and others should have infinite resistance **(see illustrations)**.

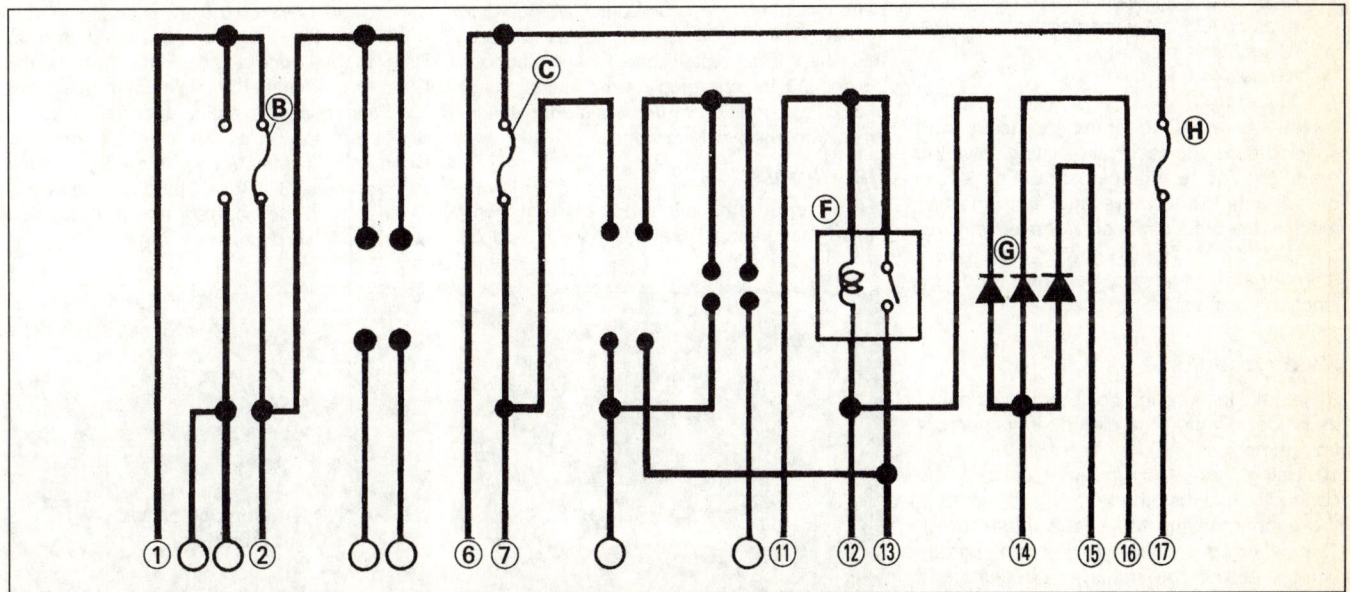

**6.4d Junction box circuit (J, K, L and M UK models)**

B  10 amp fan fuse               F  Starter circuit relay          H  10 amp taillight fuse
C  10 amp headlight fuse         G  Diodes for interlock circuit

### Relay Circuit Inspection
#### (with the battery disconnected)

|  | Meter Connection | Meter Reading (Ω) |
| --- | --- | --- |
| Headlight Relay | *7 – 8 | ∞ |
|  | *7 – 13 | ∞ |
| Starter Relay | 11 – 13 | ∞ |
|  | 12 – 13 | ∞ |

### Relay Circuit Inspection
#### (with the battery connected)

|  | Meter Connection | Battery Connection + – | Meter Reading (Ω) |
| --- | --- | --- | --- |
| Headlight Relay | *7 – 8 | *9 – 13 | 0 |
| Starter Relay | 11 – 13 | 11 – 12 | 0 |

(*) : US, Canada Models only

**6.10 With the junction box unplugged, there should be infinite resistance between the indicated terminals and no resistance (continuity) when battery voltage is applied**

**5** If the resistance values are not as specified, replace the junction box.

### Diode circuit check

**6** Remove the junction box by sliding it out of its holder. Unplug the electrical connectors from the box.
**7** Using an ohmmeter, check the resistance across the following pairs of terminals, then write down the readings. Here are the terminal pairs to be checked:

   13 and 8 (US and Canadian models only)
   13 and 9 (US and Canadian models only)
   12 and 14
   15 and 14
   16 and 14

**8** Now, reverse the ohmmeter leads and check the resistances again, writing down the readings. The resistances should be low in one direction and more than ten times as much in the other direction. If the readings for any pair of terminals are low or high in both directions, a diode is defective and the junction box must be replaced.

### Relay checks

**9** Remove the junction box by sliding it out of its holder. Unplug the electrical connectors from the box.
**10** Using an ohmmeter, check the conductivity across the terminals indicated in the accompanying table **(see illustration)**. Then, energize each relay by applying battery voltage across the indicated terminals and check the conductivity across the corresponding terminals shown on the table.
**11** If the junction box fails any of these tests, it must be replaced.

---

## 7 Lighting system – check

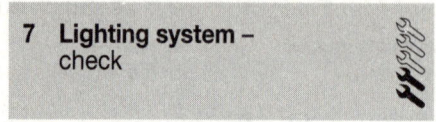

**1** The battery provides power for operation of the headlights, taillight, brake light, license plate light and instrument cluster lights. If none of the lights operate, always check battery voltage before proceeding. Low battery voltage indicates either a faulty battery, low battery electrolyte level or a defective charging system. Refer to Chapter 1 and Section 3 of this Chapter for battery checks and Sections 28 and 29 for charging system tests. Also, check the condition of the fuses and replace any blown fuses with new ones.

### Headlights

**2** If both of the headlight bulbs are out with the engine running (US and Canadian models)

**7.5a The high beam and low beam relays can be identified by their wire colors (this is L/M model low beam relay) . . .**

---

or with the lighting switch in the On position (UK models), check the 20 amp headlight fuse and the 10 amp headlight fuse with the key On (see Section 5).
**3** If only one headlight is out, refer to Section 8 and unplug the electrical connector for the headlight bulb. Use a jumper wire to connect the bulb directly to the battery terminals as follows:

   a) Black/yellow wire terminal to battery negative terminal
   b) White/yellow wire terminal (low beam) to battery positive terminal
   c) White/black wire terminal (high beam) to battery positive terminal.

   If the light comes on, the problem lies in the wiring or one of the switches in the circuit. Refer to Sections 14 and 20 for the switch testing procedures, and also the Wiring diagrams at the end of this Chapter. If either filament (high or low beam) of the bulb doesn't light, the bulb is burned out. Refer to Section 8 and replace it.
**4** On US and Canadian models, the headlights don't come on when the ignition switch is first turned on, but come on when the starter button is pressed and stay on until the ignition is turned off. The lights will go out whenever the starter is operated after the engine has stalled (this prevents excessive strain on the battery). This is accomplished by the headlight circuit relay in the junction box. To test the relay, refer to Section 6.

### High/low beam relay inspection

**5** If either the high beam or low beam headlights don't operate, test the high beam or low beam relay. On H models, the relays are at the front of the motorcycle below the instrument cluster. On J, K, L and M models, they're on the on the left side of the motorcycle, between the upper and lower rails of the rear frame section **(see illustrations)**. Refer to the Wiring diagrams at the end of this Chapter to identify the relays; high beam and low beam relays have different wire colors.
**6** Unplug the relay from its connector. Connect an ohmmeter, set to the R x 1 scale to terminals 3 and 4 **(see illustration)**. Connect a 12-volt battery (the motorcycle's battery will work) between terminals 1 and 2.

**7.5b . . . the L/M high beam relay (A) is mounted behind the turn signal relay (B)**

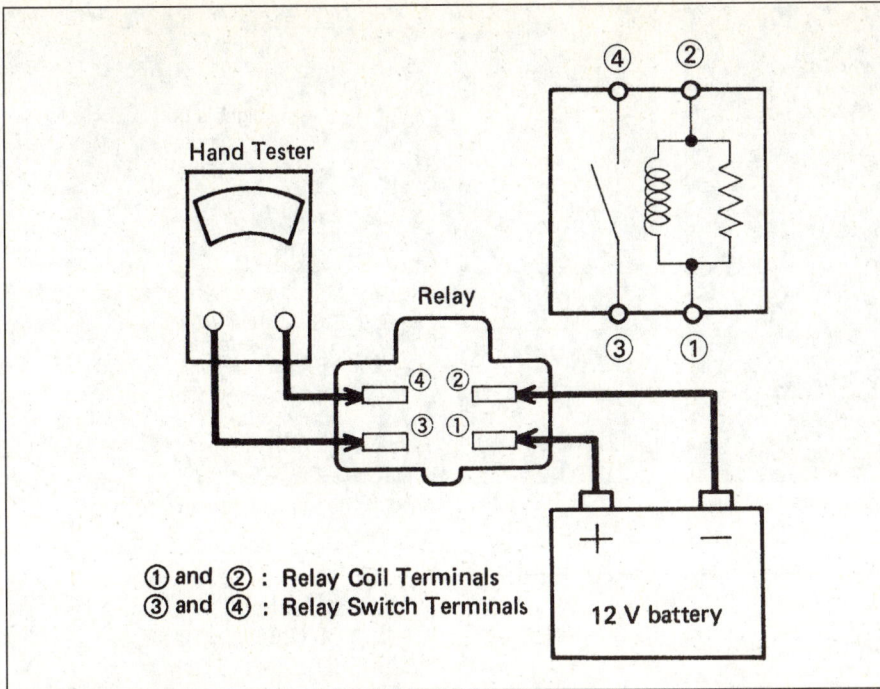

**7.6 Relay test details**

① and ② : Relay Coil Terminals
③ and ④ : Relay Switch Terminals

12 V battery

The ohmmeter should indicate zero ohms when the battery is connected and infinity when the battery is disconnected. If not, replace the relay.

## Taillight/license plate light

**7** If the taillight fails to work, check the bulbs and the bulb terminals first, then check for battery voltage at the power wire in the taillight. If voltage is present, check the ground/earth circuit for an open or poor connection.

**8** If no voltage is indicated, check the wiring between the taillight and the main (key) switch, then check the switch.

### Brake light

**9** See Section 14 for the brake light circuit checking procedure.

### Neutral indicator light

**10** If the neutral light fails to operate when the transmission is in Neutral, check the fuses and the bulb (see Section 17 for bulb removal procedures). If the bulb and fuses are in good condition, check for battery voltage at the wire attached to the neutral switch on the left side of the engine. If battery voltage is present, refer to Section 22 for the neutral switch check and replacement procedures.

**11** If no voltage is indicated, check the wiring between the junction box and the bulb, between the junction box and the switch and between the switch and the bulb for open circuits and poor connections.

### Oil pressure warning light

**12** See Section 18 for the oil pressure warning light circuit check.

## 8 Headlight bulb – replacement

**1** Where necessary for access, disconnect the upper end of the speedometer cable (see Section 15).

**2** Unplug the electrical connector from the headlight, then remove the dust cover (**see illustrations**).

**3** Lift up the retaining clip, swing it out of the way and remove the bulb holder (**see illustrations**).

**4** When installing the new bulb, reverse the removal procedure.

**HAYNES HINT** *Be sure not to touch the bulb with your fingers – oil from your skin will cause the bulb to overheat and fail prematurely. If you do touch the bulb, wipe it off with a clean rag dampened with rubbing alcohol.*

**5** UK models have a parking light bulb located beneath the headlight bulb. Pull the bulbholder out of the headlight unit for access to the bulb.

**8.2a Pull back the rubber cover . . .**

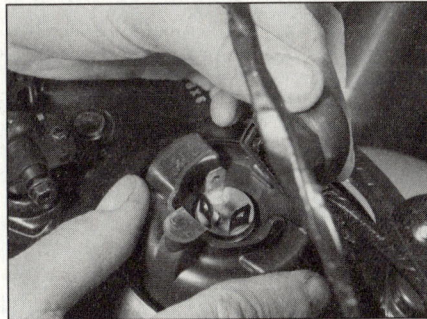

**8.2b . . . and unplug the wiring connector (headlight assembly removed for clarity)**

**8.2c Remove the dust cover . . .**

**8.3a . . . release the retainer . . .**

**8.3b . . . and lift the bulb holder out of the socket**

9.2a  Note the locations of any wiring harness retainers at the upper left corner . . .

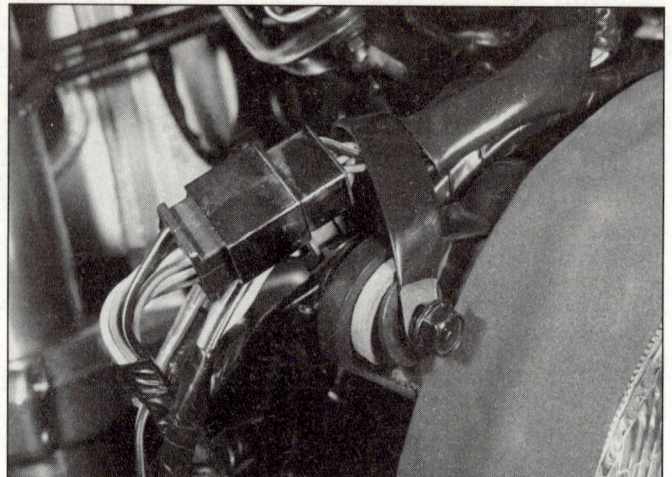
9.2b  . . . and at the upper right corner

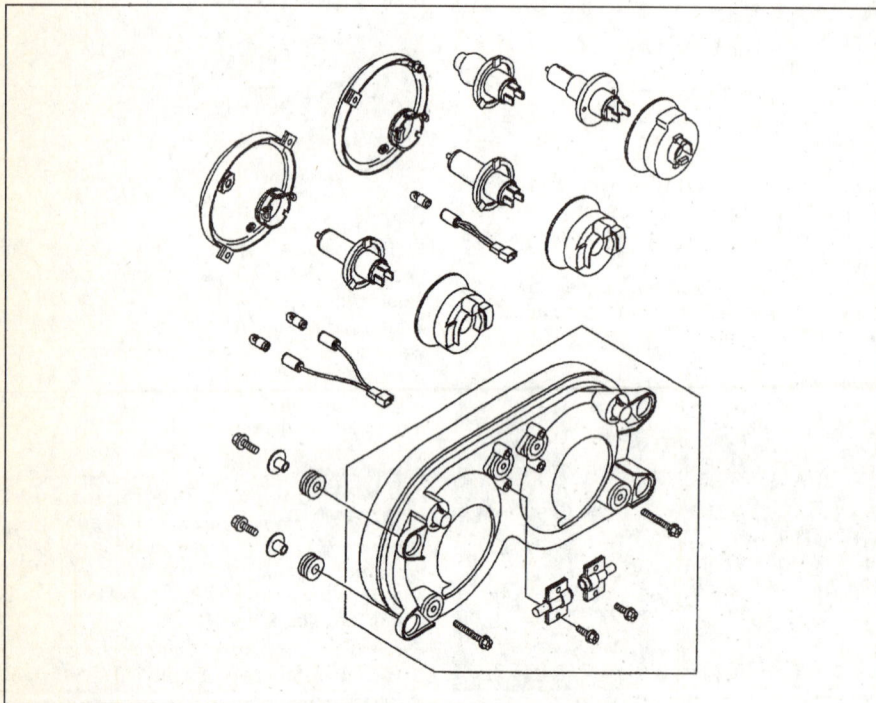
9.2c  Headlight assembly details

## 10  Headlight aim – check and adjustment

1  An improperly adjusted headlight may cause problems for oncoming traffic or provide poor, unsafe illumination of the road ahead. Before adjusting the headlight, be sure to consult with local traffic laws and regulations.
2  The headlight beam can be adjusted both vertically and horizontally. Before performing the adjustment, make sure the fuel tank is at least half full, and have an assistant sit on the seat.
3  Insert a Phillips screwdriver into the horizontal adjuster screw (see illustration), then turn the adjuster as necessary to center the beam.
4  To adjust the vertical position of the beam, insert the screwdriver into the vertical adjuster guide and turn the adjuster as necessary to raise or lower the beam.

## 11  Turn signal and taillight bulbs – replacement

### Turn signal bulbs

1  Bulb replacement for the turn signals is the same for the front and rear.

#### US models

2  Remove the screw that holds the lens/reflector assembly to the turn signal housing (see illustration). Pull out the lens/reflector assembly.
3  Remove the screws securing the lens to the reflector (see illustration).
4  Push the bulb in and turn it anti-clockwise to remove it. Check the socket terminals for corrosion and clean them if necessary. Line up the pins on the new bulb with the slots in the socket, push in and turn the bulb clockwise until it locks in place. Note: The

10.3  There's a horizontal adjuster for each headlight (arrow)

## 9  Headlight assembly – removal and installation

1  Remove the upper fairing (see Chapter 8). Unplug the headlight and on UK models also the parking light wiring connectors.
2  Remove the screws holding the headlight assembly to the fairing, noting the location of any wiring harness retainers (see illustrations). Take the assembly off the motorcycle.
3  Installation is the reverse of removal. Be sure to adjust the headlight aim (see Section 10).

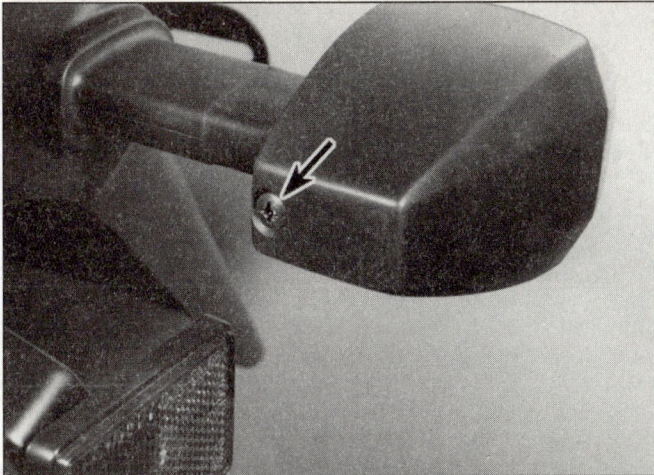

**11.2 Remove the screw (arrow) securing the lens/reflector assembly to the turn signal housing**

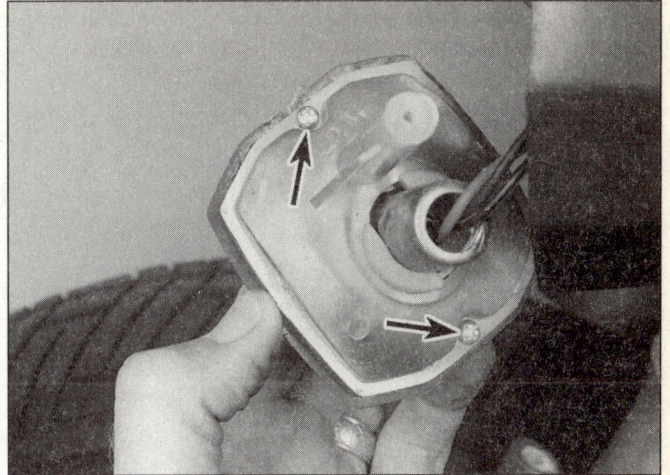

**11.3 Remove these screws (arrows) to detach the lens from the reflector**

pins on some bulbs are offset so they can only be installed one way.

> **HAYNES HINT** *It is a good idea to use a paper towel or dry cloth when handling the new bulb to prevent injury if the bulb should break and to increase bulb life.*

**5** Position the lens on the reflector and install the screws. Be careful not to overtighten them.
**6** Place the lens/reflector assembly into the housing and install the screw, tightening it securely.

### UK models

**7** Remove the lens securing screw and take off the lens. Push the bulb into its socket and turn anti-clockwise to remove. Install in the reverse order.

### Taillight bulbs

**8** If you're working on an H model, remove the passenger seat (see Chapter 8). If you're working on a J or L model, remove the side covers (see Chapter 8) and the passenger grip bracket **(see illustration)**.
**9** If you're working on a K or M model, remove the side covers/rear bodywork (see Chapter 8).
**10** Turn the bulb holders anti-clockwise **(see illustration)** until they stop, then pull straight out to remove them from the taillight housing. The bulbs can be removed from the holders by turning them anti-clockwise and pulling straight out.
**11** Check the socket terminals for corrosion and clean them if necessary. Line up the pins on the new bulb with the slots in the socket, push in and turn the bulb clockwise until it locks in place. **Note:** *The pins on the bulb are offset so it can only be installed one way.*

> **HAYNES HINT** *It is a good idea to use a paper towel or dry cloth when handling the new bulb to prevent injury if the bulb should break and to increase bulb life.*

**12** Make sure the rubber gaskets are in place and in good condition, then line up the smaller tab on the bulb holder with the smaller slot in the housing and push the bulb holder into the mounting hole. Turn it clockwise until it stops to lock it in place. **Note:** *The tabs and slots are two different sizes so the holders can only be installed one way.*
**13** Reinstall all components removed for access.

### License plate bulb

**14** Remove the lens securing screw and take off the lens. Push the bulb into its socket and turn anti-clockwise to remove. Install in the reverse order.

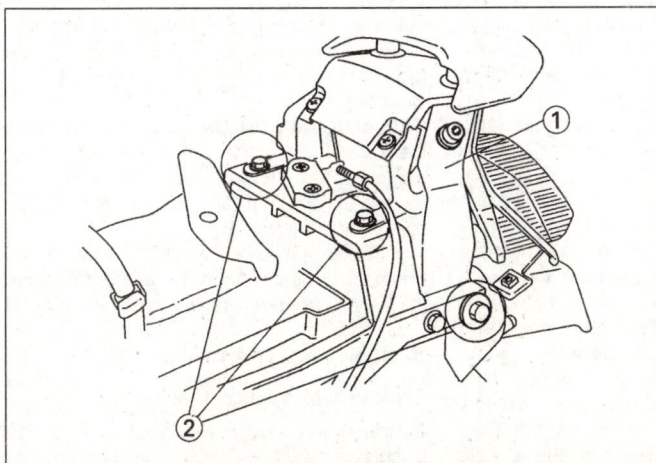

**11.8 Remove the grip bracket**

1 Grip bracket          2 Bolts

**11.10 Turn the bulbholders (arrows) anti-clockwise until they stop, then pull them out of the housing**

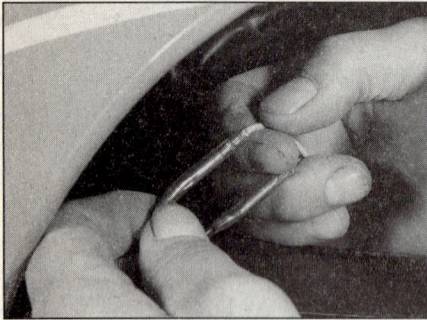

**12.2 Unplug the turn signal wiring connectors**

## 12 Turn signal assemblies – removal and installation

**1** The turn signal assemblies can be removed individually in the event of damage or failure.
**2** To remove a turn signal assembly, first follow the wiring harness from the turn signal to its electrical connectors **(see illustration)**. Mark the wires with pieces of numbered tape, then unplug the electrical connectors.
**3** Unscrew the nut that secures the turn signal assembly to the fairing or rear fender/mudguard **(see illustration)**.
**4** Detach the turn signal from the fairing or fender/mudguard. If you're installing a new turn signal, separate the stalk trim from the old stalk and transfer it to the new one.
**5** Installation is the reverse of the removal procedure.

## 13 Turn signal circuit – check

**1** The battery provides power for operation of the signal lights, so if they do not operate, always check the battery voltage (and specific gravity on fillable batteries) first. Low battery voltage indicates either a faulty battery, low electrolyte level or a defective charging

**13.3 On H models, the turn signal relay (A) and fuel pump relay (B) are mounted under the seat**

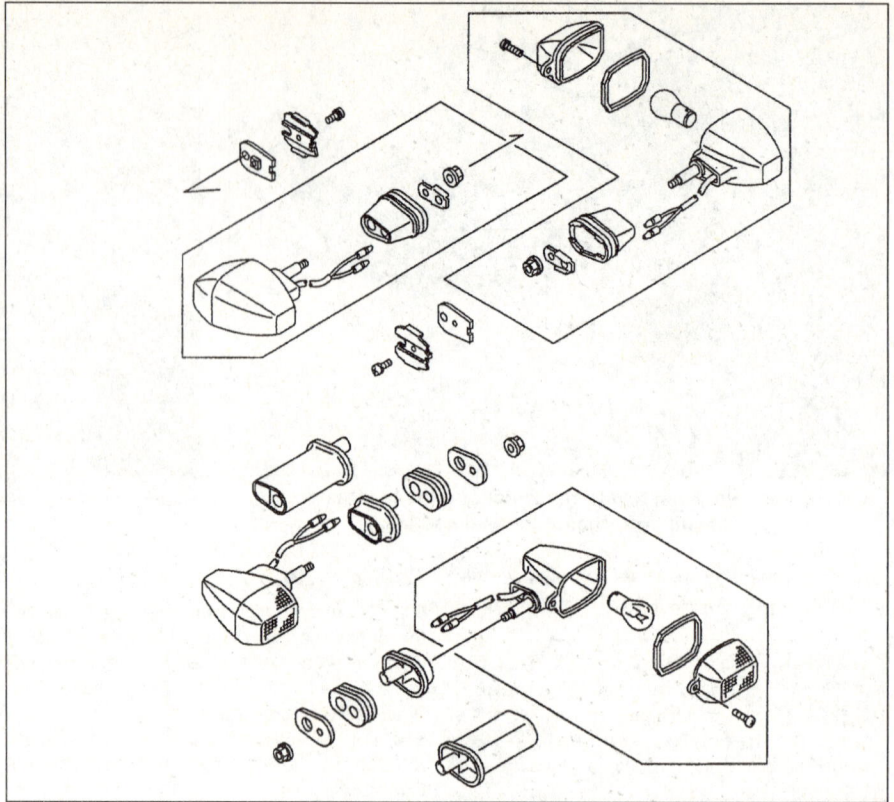

**12.3 Turn signal assembly details**

system. Refer to Chapter 1 and Section 3 of this Chapter for battery checks and Sections 28 and 29 for charging system tests. Also, check the fuses (see Section 5).
**2** Most turn signal problems are the result of a burned out bulb or corroded socket. This is especially true when the turn signals function properly in one direction, but fail to flash in the other direction. Check the bulbs and the sockets (see Section 11).
**3** If the bulbs and sockets check out okay, check for voltage at the turn signal relay's brown wire **(see illustration 7.5b or the accompanying illustration)** with the ignition On. If there's no voltage at the relay, follow the brown wire back to the junction box, checking for breaks and bad connections. Also check the Tail fuse on the junction box (see Section 5).
**4** If there's power at the turn signal relay, check for voltage at the orange wire terminal in the turn signal switch (this is the wire that connects the relay to the switch). If there's no power, check for breaks or bad connections in the orange wire (see the *Wiring diagrams* at the end of this Chapter). If the wire is good, the relay is probably defective.
**5** If there's power at the switch, refer to Section 20 and test the switch continuity. If the switch is good, refer to the *Wiring diagrams* and check for breaks or bad connections in the wiring from the switch to the individual turn signals.

## 14 Brake light switches – check and replacement

### Circuit check

**1** Before checking any electrical circuit, check the fuses (see Section 5).
**2** Using a test light connected to a good ground/earth, check for voltage to the brown wire at the brake light switch. If there's no voltage present, check the brown wire between the switch and the junction box (see the *Wiring diagrams* at the end of this Chapter).
**3** If voltage is available, touch the probe of the test light to the other terminal of the switch, then pull the brake lever or depress the brake pedal – if the test light doesn't light up, replace the switch.
**4** If the test light does light, check the wiring between the switch and the brake lights (see the *Wiring diagrams* at the end of this Chapter).

### Switch replacement
**Brake lever switch**

**5** Unplug the electrical connectors from the switch.
**6** Remove the mounting screw **(see illustration)** and detach the switch from the brake lever bracket/front master cylinder.

14.6 Remove the screw (arrow) to detach the brake light switch (master cylinder removed for clarity)

14.9 Rear brake light switch details

A Spring
B Plastic nut

C Bracket mounting bolt

7 Installation is the reverse of the removal procedure. The brake lever switch isn't adjustable.

### Brake pedal switch

8 Unplug the electrical connector in the switch harness.

15.3 Unscrew the speedometer cable nut from the speedometer

9 Disconnect the spring from the brake pedal switch (see illustration).
10 Hold the switch body from turning and rotate the adjuster nut until it clears the switch threads, then lift the switch out. To remove the nut from the bracket, compress its prongs and slip it out. To remove the switch bracket, unbolt it from the footpeg/brake pedal bracket.
11 Install the switch by reversing the removal procedure, then adjust the switch by following the procedure described in Chapter 1.

### 15 Instrument cluster and speedometer cable – removal and installation

1 Remove the upper fairing (see Chapter 8).
2 Remove the headlight assembly (see Section 9).
3 Detach the speedometer cable from the speedometer (see illustration).

4 Unplug the cluster harness electrical connectors.
5 Remove the instrument cluster mounting nuts and lift the cluster from the upper fairing mount (see illustrations).
*Caution: Always store the cluster with the gauges facing up or in a horizontal position – otherwise, the unit could be damaged.*
6 To complete removal of the speedometer cable, unscrew it from the drive unit on the front wheel and pull it out (see illustration).
7 Installation is the reverse of the removal procedure.

### 16 Meters and gauges – check and replacement

### *Temperature gauge*

1 Refer to Chapter 3 for the temperature gauge checking procedure.

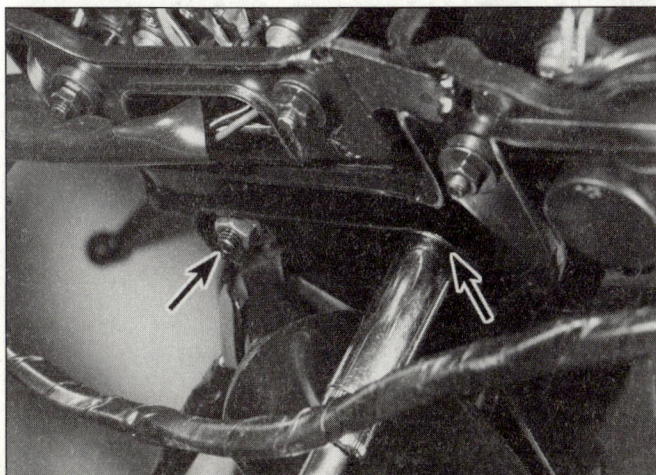

15.5a Remove the cluster mounting nuts (arrows) . . .

15.5b . . . and lift the cluster out

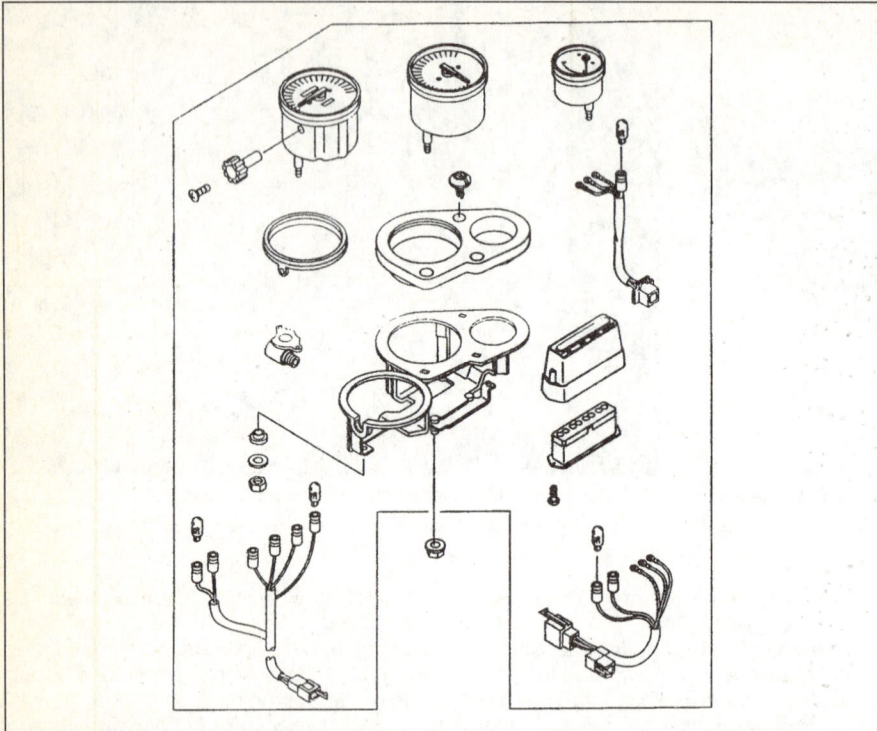

15.5c Instrument cluster details (J, K, L and M models); H models similar

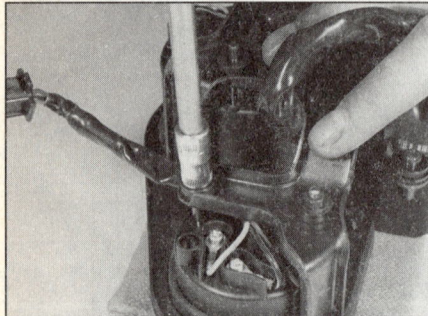

15.6 Unscrew the speedometer cable from the drive unit at the left front fork

locations, then remove their terminal screws and separate the wiring harness from the gauge (see illustration).

4 Installation is the reverse of the removal procedure.

### Tachometer

5 To check the tachometer, start by referring to the *Wiring diagrams* at the end of this Chapter and checking the tachometer circuit for breaks or bad connections.

6 If the wiring is good, disconnect the black lead from the ignition coil for cylinders 1 and 4 (refer to Chapter 5 for coil access procedures).

7 Connect a jumper wire to the battery positive terminal that's long enough to reach the disconnected black wire.

⚠️ Warning: Don't let the other end of the wire touch metal or sparks may occur. This could cause any fuel vapors or hydrogen gas from the battery to explode.

8 Turn the ignition key to On (but don't start the engine). Touch the end of the auxiliary wire to the disconnected black ignition coil wire. The tachometer needle should flick. If not, the tachometer is probably defective.

9 To replace the tachometer, remove its mounting nuts and grommets (see illustration). The wire colors of the three single-wire connectors should be molded in the gauge housing near the terminals. If they aren't, write down their locations, then remove their terminal screws and carefully pull the bulb sockets from the gauge (see illustration). Installation is the reverse of the removal procedure.

2 If it's necessary to replace the gauge, remove the instrument cluster (see Section 15).

3 Remove the gauge mounting nuts and grommets and separate the gauge from the cluster (see illustrations). The wire colors of the three single-wire connectors should be molded in the gauge housing near the terminals. If they aren't, write down their

16.3a Remove the temperature gauge mounting nuts . . .

16.3b . . . and the grommets

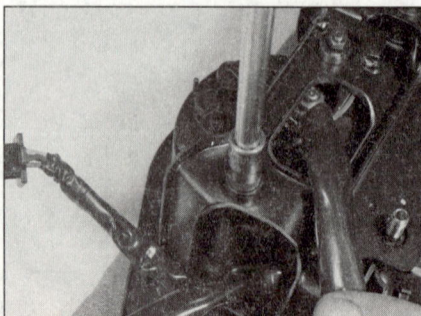

16.3c Remove the terminal screws and disconnect the wires

16.9a Remove the tachometer mounting nuts . . .

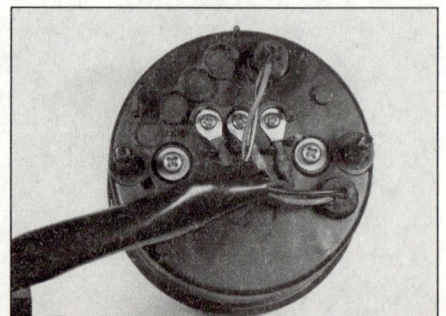

16.9b . . . remove the terminal screws and pull the bulb sockets out of the case

**16.11a** Remove the speedometer mounting nuts and detach the speedometer from the bracket . . .

**16.11b** . . . and pull out the bulb sockets

**16.12a** Remove the screws and lift off the gear housing . . .

**16.12b** . . . to expose the gear . . .

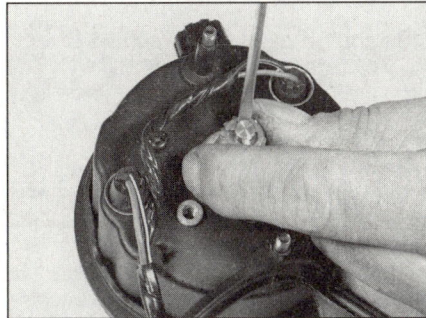

**16.12c** . . . pry off the C-clip . . .

**16.12d** . . . and lift the gear off the shaft

## Speedometer

**10** Special instruments are required to properly check the operation of the speedometer. Take the instrument to a Kawasaki dealer service department or other qualified repair shop for diagnosis.
**11** To remove the speedometer from the cluster, remove its mounting nuts and grommets and take it out (see illustration). Carefully separate the bulb sockets from the back of the gauge and remove the wiring harness (see illustration).
**12** To detach the gears from the speedometer, remove the drive gear mounting screws and lift it off to expose the driven gear (see illustrations). Pry the C-clip out of its groove and slide the driven gear off the shaft (see illustrations).

## 17 Instrument and warning light bulbs – replacement

**1** Some instrument cluster bulbs are accessible with the instrument cluster in position on the motorcycle (see illustration). To replace others, it will be necessary to remove the instrument cluster (see Section 15).
**2** To replace a cluster bulb, pull the appropriate rubber socket out of the back of the instrument cluster housing, then pull the bulb out of the socket (see illustration). If the socket contacts are dirty or corroded, they should be scraped clean and sprayed with electrical contact

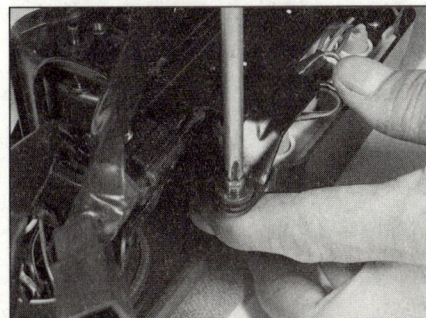

cleaner before new bulbs are installed.
**3** Carefully push the new bulb into position, then push the socket into the cluster housing.
**4** To replace a warning light bulb, remove the

instrument cluster and detach the warning light assembly (see illustration). Pull the bulb socket out of the assembly (see illustration), then pull the bulb out of the socket.

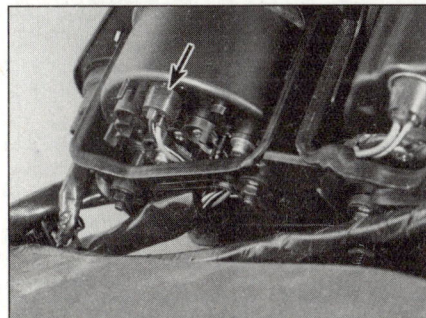

**17.1** Some instrument bulb sockets (arrow) can be removed without removing the instrument cluster

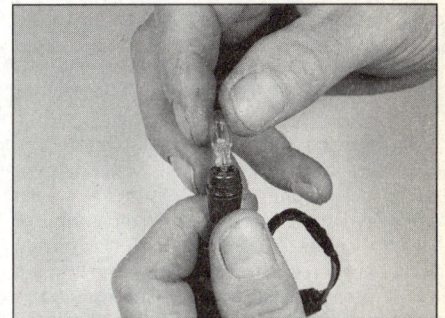

**17.2** Carefully work the bulb socket free of the case and pull the bulb out of the socket

**17.4a** Remove the screws and detach the warning light assembly . . .

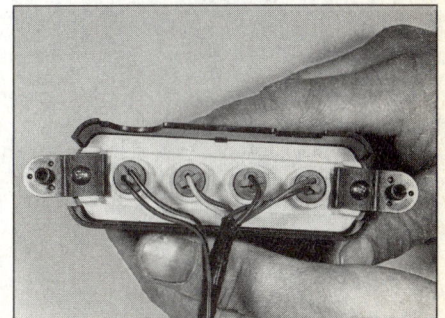

**17.4b** . . . then work the bulb socket(s) free of the assembly to expose the bulbs

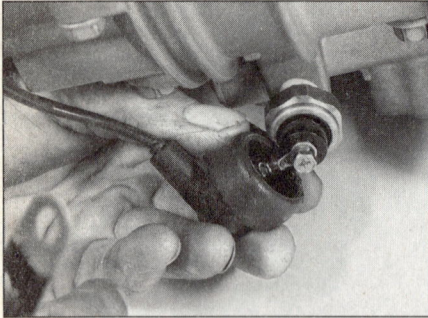

**18.2 Pull back the rubber boot, loosen the screw and disconnect the wire from the oil pressure switch**

| | BR | W | Y | BL | R |
|---|---|---|---|---|---|
| OFF, LOCK | | | | | |
| ON | ○━━━━━━○━━━━━━○ | | | ○━━━━━━○ | |
| P (Park) | | ○━━━━━━○━━━━━━━━━━━━━○ | | | |

**19.2 Check the continuity of the ignition switch in the different switch positions across the indicated terminals (yellow wire shown is used on H models; this wire is green/yellow on J, K, L and M models)**

## 18 Oil pressure sending unit – check and replacement

**1** If the oil pressure warning light fails to operate properly, check the oil level and make sure it is correct.

**2** If the oil level is correct, pull back the rubber cap and disconnect the wire from the oil pressure switch, which is located at the left rear of the oil pan (H1 models), right rear of the oil pan (H2 models) or on the front of the engine case (J, K, L and M models) **(see illustration)**. Turn the main switch On and ground/earth the end of the wire. If the light comes on, the oil pressure switch is defective and must be replaced with a new one (only after draining the engine oil).

**3** If the light does not come on, check the oil pressure warning light bulb, the wiring between the oil pressure sending unit and the light, and between the light and the junction box (see the *Wiring diagrams* at the end of this Chapter).

**4** To replace the sending unit, drain the engine oil (see Chapter 1) and unscrew the sending unit from the case. Wrap the threads of the new sending unit with Teflon tape or apply a thin coat of sealant to them, then screw the unit into its hole, tightening it to the torque listed in this Chapter's Specifications.

**5** Fill the crankcase with the recommended type and amount of oil (see Chapter 1 and *Daily (pre-ride) checks*) and check for leaks.

## 19 Ignition main (key) switch – check and replacement

### Check

**1** Follow the wiring harness from the ignition switch to the connector and unplug the connector.

**2** Using an ohmmeter, check the continuity of the terminal pairs indicated in the accompanying table **(see illustration)**. Continuity should exist between the terminals connected by a solid line when the switch is in the indicated position.

**3** If the switch fails any of the tests, replace it.

### Replacement

**4** The ignition switch is secured to the upper triple clamp by two self-shearing screws **(see illustration)**. These screws have heads that snap off when the screws are tightened.

**5** If you haven't already done so, unplug the switch electrical connector.

**6** If the lock cylinder works properly, unlock the steering. Refer to Chapter 6 and remove the upper triple clamp together with the ignition switch. Carefully drill holes through the centers of the self-shearing screws **(see illustration)**, then remove them with a screw extractor. Detach the switch from the upper triple clamp.

**7** If the lock cylinder doesn't work, you'll have to drill out the self-shearing screws from below.

**8** Hold the new switch in position and install the new self-shearing screws. Tighten the bolts until the heads break off.

**9** The remainder of installation is the reverse of the removal procedure.

## 20 Handlebar switches – check

**1** Generally speaking, the switches are reliable and trouble-free. Most troubles, when they do occur, are caused by dirty or corroded contacts, but wear and breakage of internal parts is a possibility that should not be overlooked. If breakage does occur, the entire switch and related wiring harness will have to be replaced with a new one, since individual parts are not usually available.

**2** The switches can be checked for continuity with an ohmmeter or a continuity test light. Always disconnect the battery negative cable, which will prevent the possibility of a short circuit, before making the checks.

**3** Trace the wiring harness of the switch in question and unplug the electrical connectors.

**4** Using the ohmmeter or test light, check for continuity between the terminals of the switch harness with the switch in the various positions. Refer to the continuity diagrams contained in the *Wiring diagrams* at the end of this Chapter. Continuity should exist between the terminals connected by a solid line when the switch is in the indicated position.

**5** If the continuity check indicates a problem exists, refer to Section 21, disassemble the switch and spray the switch contacts with electrical contact cleaner. If they are accessible, the contacts can be scraped clean with a knife or polished with crocus cloth. If switch components are damaged or broken, it will be obvious when the switch is disassembled.

## 21 Handlebar switches – removal and installation

**1** The handlebar switches are composed of two halves that clamp around the bars. They

**19.4 The ignition switch is secured by a pair of self-shearing screws (arrows)**

**19.6 Drill out the self-shearing screws (arrows) and remove them with a screw extractor**

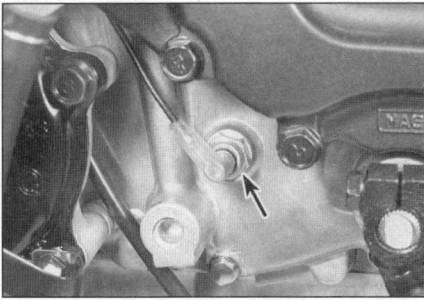

**22.1 The neutral switch is mounted on the left side of the engine (749 cc engine shown)**

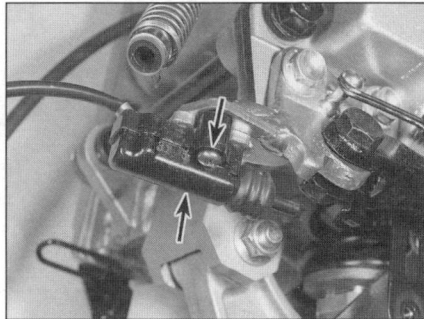

**23.6 The sidestand switch is secured by two screws (arrows)**

**24.2 Disconnect the wires from the horn terminals (arrows)**

are easily removed for cleaning or inspection by taking out the clamp screws and pulling the switch halves away from the handlebars.

2 To completely remove the switches, the electrical connectors in the wiring harness should be unplugged. The right side switch must be separated from the throttle cables, also.

3 When installing the switches, make sure the wiring harnesses are properly routed to avoid pinching or stretching the wires. Position the switches on the location pegs before tightening the screws.

## 22 Neutral switch – check and replacement

### Check

1 Disconnect the wire from the neutral switch **(see illustration)**. Connect one lead of an ohmmeter to a good ground/earth and the other lead to the post on the switch.
2 When the transmission is in neutral, the ohmmeter should read 0 ohms – in any other gear, the meter should read infinite resistance.
3 If the switch doesn't check out as described, replace it.

### Replacement

4 Where necessary, remove the lower left fairing panel for access (see Chapter 8).
5 Unscrew the neutral switch from the case.
6 Wrap the threads of the new switch with Teflon tape or apply a thin coat of RTV sealant to them. Install the switch in the case and tighten it to the torque listed in this Chapter's Specifications.

## 23 Sidestand switch – check and replacement

### Check

1 Where necessary for access, remove the lower left fairing panel (see Chapter 8).
2 Follow the wiring harness from the switch to the connector, then unplug the connector. Connect the leads of an ohmmeter between

the wire terminals in the switch side of the connector (not the harness side). With the sidestand in the up position, there should be continuity through the switch (0 ohms).
3 With the sidestand in the down position, the meter should indicate infinite resistance.
4 If the switch fails either of these tests, replace it.

### Replacement

5 Disconnect the switch electrical connector if you haven't already done so.
6 Unscrew the two Phillips head screws and remove the switch **(see illustration)**.
7 Installation is the reverse of the removal procedure.

## 24 Horn – check and replacement

### Check

1 Where necessary for access, remove the upper fairing.
2 Unplug the electrical connectors from the horn. Using two jumper wires, apply battery voltage directly to the terminals on the horn **(see illustration)**. If the horn sounds, check the switch (see Section 20) and the wiring between the switch and the horn (see the *Wiring diagrams* at the end of this Chapter).
3 If the horn doesn't sound, replace it.

### Replacement

4 Unbolt the horn bracket from the frame

(see illustration 24.2) and detach the electrical connectors.
5 Unbolt the horn from the bracket and transfer the bracket to the new horn.
6 Installation is the reverse of removal.

## 25 Starter relay – check and replacement

1 If you're working on an H1 model, remove the rider's seat (see Chapter 8).
2 If you're working on an H2, J, K, L and M model, remove the left side cover (see Chapter 8).
3 Disconnect the negative cable from the battery.
4 Pull back the plastic covers from the terminal nuts, remove the nuts and disconnect the starter relay cables **(see illustrations)**. Disconnect the remaining electrical connector.
5 Connect an ohmmeter between the terminals from which the nuts were removed **(see illustration)**. It should indicate infinite resistance.
6 Connect a 12-volt battery to the terminals from which the thin wires were disconnected (positive battery terminal to yellow/red; negative battery terminal to black/yellow). The motorcycle's battery can be used if it's fully charged. The ohmmeter should now indicate zero ohms.
7 If the relay doesn't perform as described, pull its rubber mount off the metal bracket and pull the relay out of the mount.

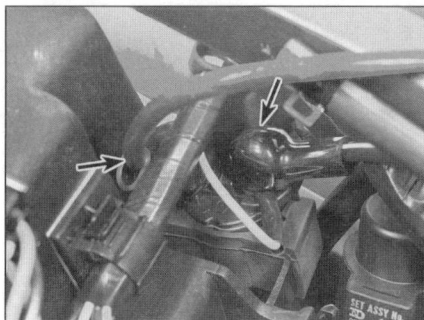

**25.4a Pull back the plastic covers . . .**

**25.4b . . . and remove the terminal nuts (arrows) to disconnect the cables**

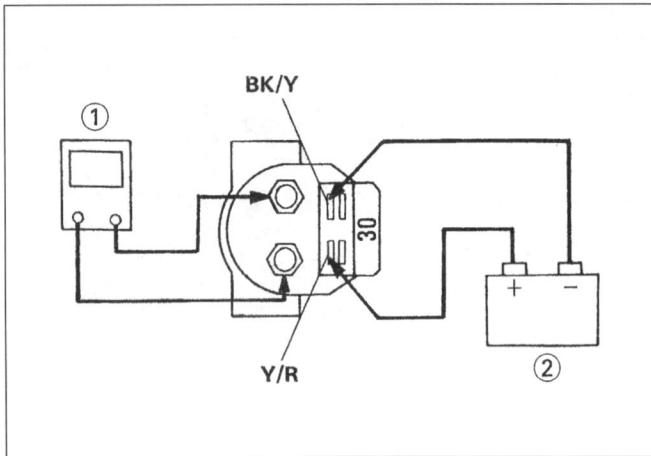

25.5 Connect an ohmmeter (1) to the cable terminals and a 12-volt battery (2) to the terminals for the yellow/red and black/yellow wires

26.7 Remove the starter mounting bolts (arrows) (749 cc engine shown)

8 Installation is the reverse of removal. Reconnect the negative battery cable after all the other electrical connections are made.

## 26 Starter motor – removal and installation

### Removal

#### H models

1 Remove the right lower fairing panel (see Chapter 8).
2 Remove the coolant reservoir tank (see Chapter 3).

#### J, K, L and M models

3 Remove the left lower fairing panel (see Chapter 8).
4 Remove the air cleaner housing (see Chapter 4).

#### All models

5 Disconnect the cable from the negative terminal of the battery.
6 Remove the nut retaining the starter cable to the starter.
7 Remove the starter mounting bolts (see illustration).

8 Lift the outer end of the starter up a little bit and slide the starter out of the engine case (see illustration).
*Caution: Don't drop or strike the starter or its magnets may be demagnetized, which will ruin it.*
9 Check the condition of the O-ring on the end of the starter and replace it if necessary.

### Installation

10 Remove any corrosion or dirt from the mounting lugs on the starter and the mounting points on the crankcase.
11 Apply a little engine oil to the O-ring and install the starter by reversing the removal procedure.

## 27 Starter motor – disassembly, inspection and reassembly

1 Remove the starter motor (see Section 26).

### Disassembly

2 Mark the position of the housing to each end cover. Remove the two long screws and detach the end cover from the pinion gear side (see illustrations).

3 At the other end of the starter, remove the brush end cover, together with the brushes (see illustration). Pull back the spring from the positive brush (the one that's attached to the terminal bolt), then slide that brush out of its holder in the brush plate and separate the brush plate from the end cover.
4 Pull the armature out of the housing. Note the location and number of any armature shims (see illustration).

### Inspection

5 The parts of the starter motor that most likely will require attention are the brushes. If one brush must be replaced, replace both of them. The brushes are replaced together with the terminal bolt and the brush plate. Brushes must be replaced if they are worn excessively, cracked, chipped, or otherwise damaged, or if one of them has a short or open.
6 Measure the length of the brushes and compare the results to the brush length listed in this Chapter's Specifications (see illustration). If either of the brushes is worn beyond the specified limits, replace them.
7 Connect an ohmmeter between the terminal bolt and the positive brush and check for continuity (see illustration). The meter should read close to 0 ohms. If it doesn't, the

26.8 Lift the starter and pull it out of the engine

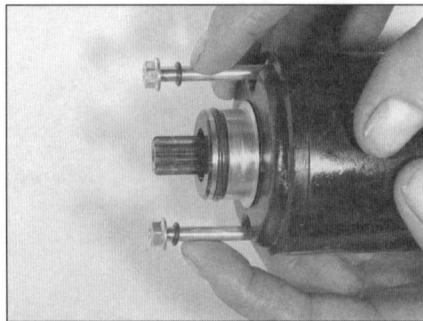

27.2a Remove the two long screws and their O-rings . . .

27.2b . . . and lift the cover off the pinion gear end of the starter

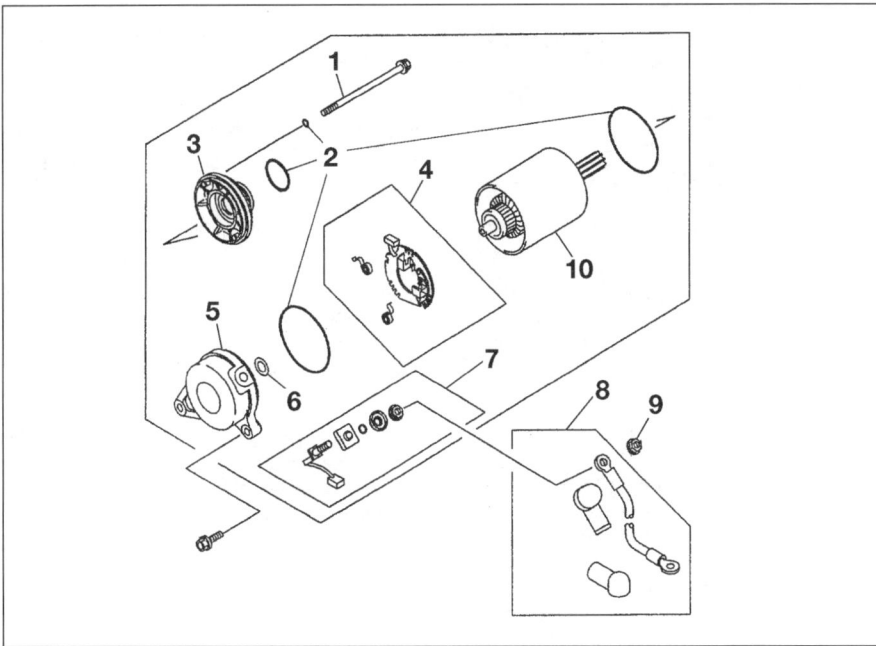

**27.2c  Starter – exploded view**

1  Long screw
2  O-rings
3  Pinion gear end cover
4  Brush plate and negative brush
5  Brush end cover
6  Armature shim
7  Terminal bolt and positive brush
8  Starter cable
9  Terminal nut
10 Armature and starter housing

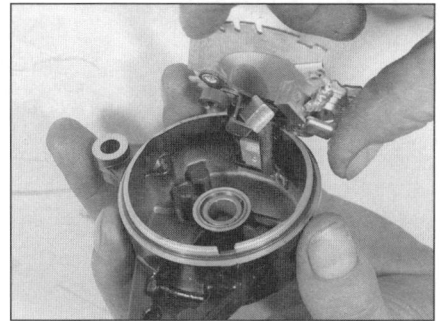

**27.3  Lift the brush plate out of the end cover**

**27.4  Note the location and number of any shims on the armature**

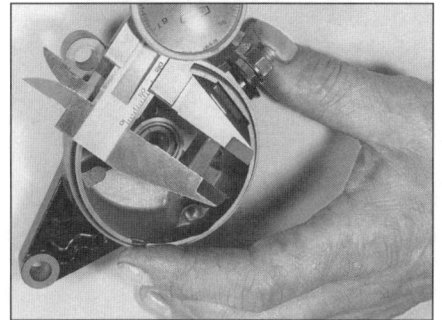

**27.6  Measure brush length and compare it to the Specifications**

brush lead has an open and the brushes must be replaced.

**8** Check for continuity between the negative brush and the brush plate **(see illustration)**. Again, the meter should read close to zero ohms. If it doesn't, replace the brushes.

**9** To detach the positive brush from the starter housing, remove the nut and push the terminal bolt through the housing.

**10** Inspect the commutator **(see illustration)** for scoring, scratches and discoloration. The commutator can be cleaned and polished with fine emery paper, but do not use sandpaper and do not remove copper from the commutator. After cleaning, clean out the grooves and wipe away any residue with a cloth soaked in an electrical system cleaner or denatured alcohol. Measure the commutator diameter with a micrometer or vernier caliper and compare it to the diameter listed in this Chapter's Specifications. If it is less than the service limit, the motor must be replaced with a new one.

**11** Using an ohmmeter or a continuity test

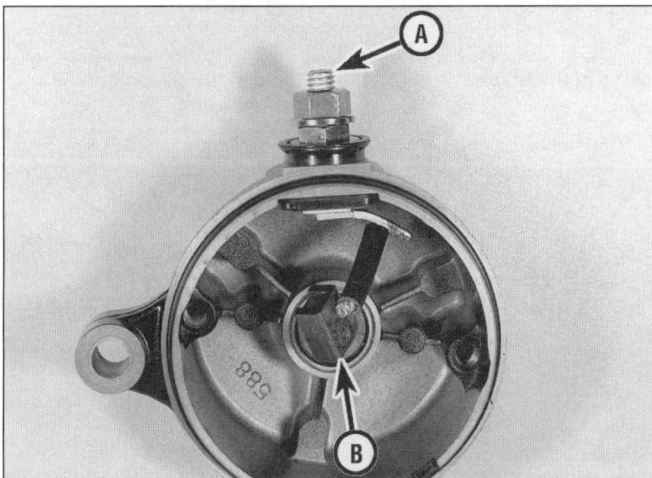

**27.7  There should be little or no resistance between the terminal bolt (A) and the positive brush (B)**

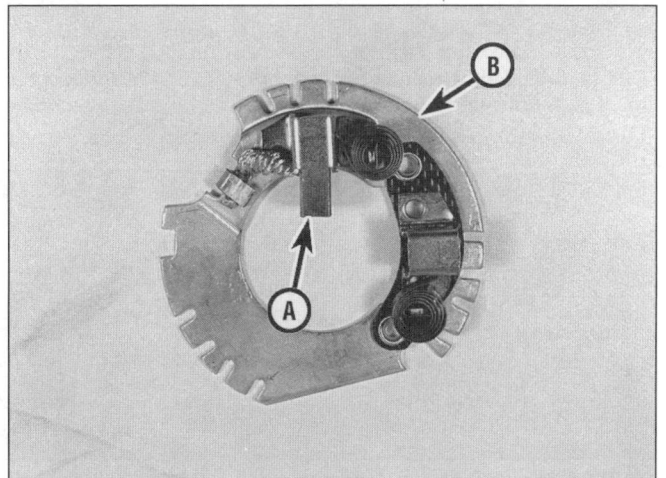

**27.8  There should be little or no resistance between the negative brush (A) and the brush holder (B)**

**27.10 Check the commutator for wear, cracks or burns**

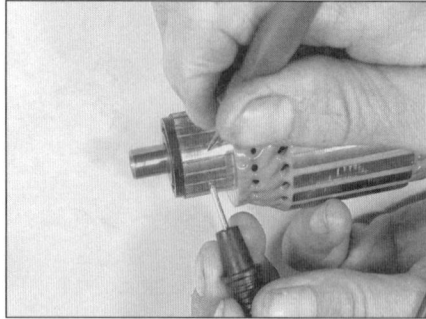

**27.11a Continuity should exist between pairs of commutator bars**

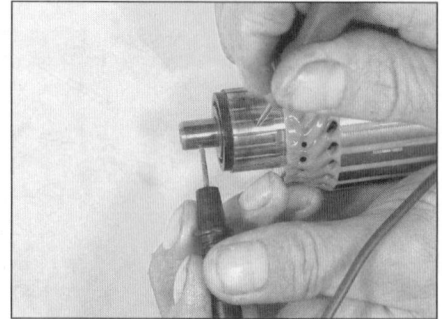

**27.11b There should be no continuity between the commutator bars and the armature shaft**

light, check for continuity between the commutator bars **(see illustration)**. Continuity should exist between each bar and all of the others. Also, check for continuity between the commutator bars and the armature shaft **(see illustration)**. There should be no continuity between the commutator and the shaft. If the checks indicate otherwise, the armature is defective.

**12** At the other end of the armature, check the bearing for roughness, looseness or loss of lubricant **(see illustration)**. Check with a motorcycle shop or Kawasaki dealer to see if the bearing can be replaced separately; if this isn't possible, replace the starter motor.
**13** Inspect the bushings in the end covers **(see illustrations)**. Replace the starter motor if the bushings are worn or damaged.
**14** Reposition the positive brush in the brush

plate and reinstall the brush plate in the end cover **(see illustration)**. Using the highest range on the ohmmeter, measure the resistance between the terminal bolt and the positive brush holder, the negative brush holders and the brush plate in turn. The reading should be infinite. If there is any reading at all, either the terminal bolt or the brush plate has a short. Replace the defective component.
**15** Check the starter pinion gear for worn, cracked, chipped and broken teeth. If the gear is damaged or worn, replace the starter motor.

### Reassembly

**16** Install the brush plate in the end cover **(see illustration 27.14)**. Make sure the terminal bolt nuts and washers are assembled

correctly **(see illustration 27.7)**. Tighten the terminal nut securely.
**17** Install any washers that were present on the end of the armature shaft **(see illustration 27.4)**.
**18** Push the brushes back into their holders against the pressure of the springs and slide the armature into place **(see illustrations)**.
**19** Align the matchmarks and install the housing on the brush end cover **(see illustrations)**. Align the match marks on the housing with the gear end cover, then install the cover **(see illustration)**.
**20** Make sure the O-rings are in place on the two long screws, then install and tighten them securely **(see illustrations)**. Make sure the O-ring is in place on the end of the starter before installing it in the engine.

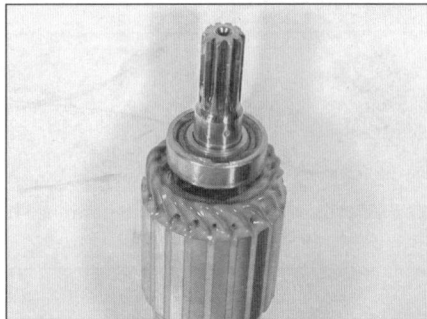

**27.12 Check the armature bearing for roughness, looseness or loss of lubricant**

**27.13a Inspect the bushing in the brush end housing . . .**

**27.13b . . . and in the pinion gear end housing**

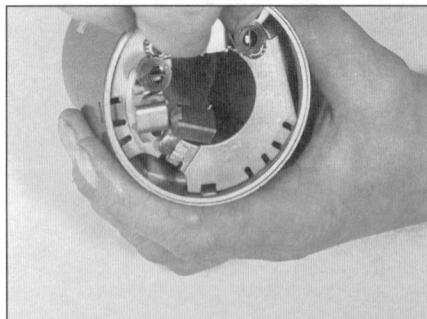

**27.14 There should be no continuity between the terminal bolt and the brush plate or either brush holder**

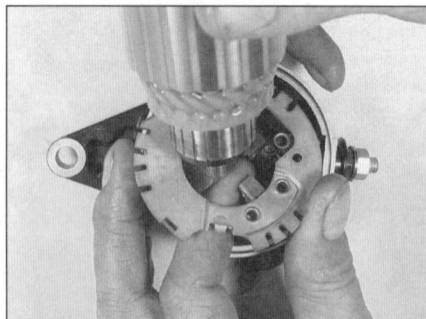

**27.18a Hold the brushes back with a finger and start the armature into the brush end cover. . .**

**27.18b . . . until the end of the armature shaft seats in the cover bushing**

27.19a Line up the notch in the brush plate with the mark on the brush end cover . . .

27.19b . . . and the mark on the starter housing

27.19c Line up the screw hole in the pinion gear end cover with the marks on the starter housing

## 28 Charging system testing – general information and precautions

1 If the performance of the charging system is suspect, the system as a whole should be checked first, followed by testing of the individual components (the alternator and the voltage regulator). **Note:** *Before beginning the checks, make sure the battery is fully charged and that all system connections are clean and tight.*

2 Checking the output of the charging system and the performance of the various components within the charging system requires the use of a voltmeter or a multimeter with a voltmeter function.

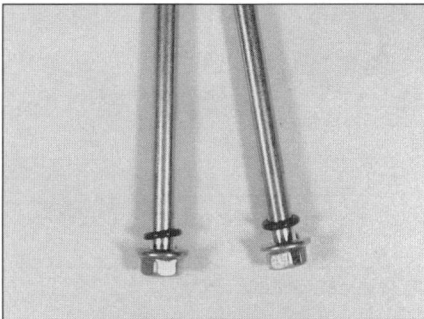

27.20a Make sure the O-rings are in position on the long screws . . .

27.20b . . . and install them in the starter; make sure the O-ring is in position on the end of the starter

3 When making the checks, follow the procedures carefully to prevent incorrect connections or short circuits, as irreparable damage to electrical system components may result if short circuits occur. Because of the special tools and expertise required, it is recommended that charging system tests beyond those described in this Chapter be left to a dealer service department or a reputable motorcycle repair shop.

## 29 Charging system – output test

*Caution: Never disconnect the battery cables from the battery while the engine is running. If the battery is disconnected, the alternator and regulator will be damaged.*

1 To check the charging system output, you will need a voltmeter or a multimeter with a voltmeter function.

2 The battery must be fully charged (charge it from an external source if necessary) and the engine must be at normal operating temperature to obtain an accurate reading.

3 Attach the positive (red) voltmeter lead to the positive (+) battery terminal and the negative (black) lead to the battery negative (-) terminal. The voltmeter selector switch (if equipped) must be in the 0 to 20 DC volt range.

4 Start the engine.

5 The charging system output should be

within the range listed in this Chapter's Specifications.

6 If the output is as specified, the alternator is functioning properly. If the charging system as a whole is not performing as it should, refer to Section 32 and check the alternator brushes.

7 Low voltage output may be the result of damaged windings in the alternator stator coils or wiring problems. Make sure all electrical connections are clean and tight, then refer to the following Sections to remove the alternator and inspect the brushes.

8 High voltage output (above the specified range) indicates a defective voltage regulator.

## 30 Alternator drivebelt (H models) – check and adjustment

1 The belt must be adjusted with the engine cold (at room temperature). The procedure requires a special tool and two torque wrenches, so it may be more practical to have a Kawasaki dealer adjust the belt for you.

2 Note which of the adjustment marks on the alternator scale aligns with the pointer **(see illustration)**. Write this down.

3 Remove the lower left fairing panel (see Chapter 8).

4 Remove the clutch slave cylinder (see Chapter 2).

5 Remove the alternator bracket, engine sprocket cover and the upper rear engine mounting bolt **(see illustration)**.

30.2 Note which of the marks on the scale (A) aligns with the pointer (B)

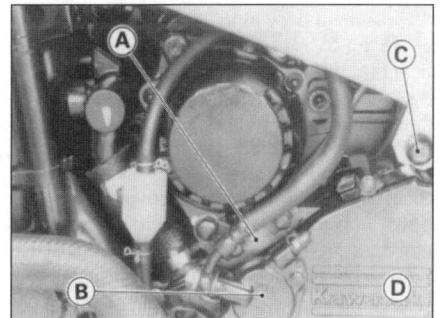

30.5 Remove the bracket (A), slave cylinder (B), engine mounting bolt (C) and engine sprocket cover (D)

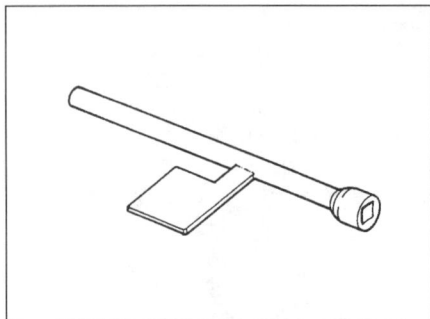

30.6a Alternator tension wrench

**6** Insert the alternator tension wrench (Kawasaki tool no.57001-1296) into the engine mounting bolt hole and place the tool lever against the alternator belt **(see illustrations)**.

**7** Loosen the alternator mounting bolts. Push the alternator down against the crankcase, then retighten the alternator mounting bolts with an Allen bolt bit. To get the correct torque for this stage of the procedure, turn the Allen bolt bit with a thumb and two fingers (don't use a ratchet handle).

**8** Attach a torque wrench to the special tool and apply 16 Nm (11.5 ft-lbs) torque.

**9** While holding the first torque wrench at the setting specified in Step 8, tighten the alternator mounting bolts to the torque listed in this Chapter's Specifications.

**10** Check the position of the pointer relative to the adjustment marks on the scale. It should be at the same position or higher. If it's lower, readjust the belt tension.

## 31 Alternator –
removed and installation

### Removal

#### H models

**1** Remove the fuel tank and carburetors (see Chapter 4). Remove the left lower fairing panel (see Chapter 8).

**2** Disconnect the cable from the battery negative terminal.

**3** Follow the wiring harness from the alternator to the connector under the left side cover and unplug the connector.

**4** Remove the alternator's upper cover **(see illustration 30.6b)**.

**5** Remove the clutch slave cylinder, alternator bracket and engine sprocket cover (see Section 30). Remove the alternator mounting bolts with a long Allen bolt bit. Remove the alternator from the left side of the motorcycle, disengaging its pulley from the drivebelt.

#### J, K, L and M models

**6** Remove the fuel tank. If you're working on a J or K model, remove the air filter housing (see Chapter 4).

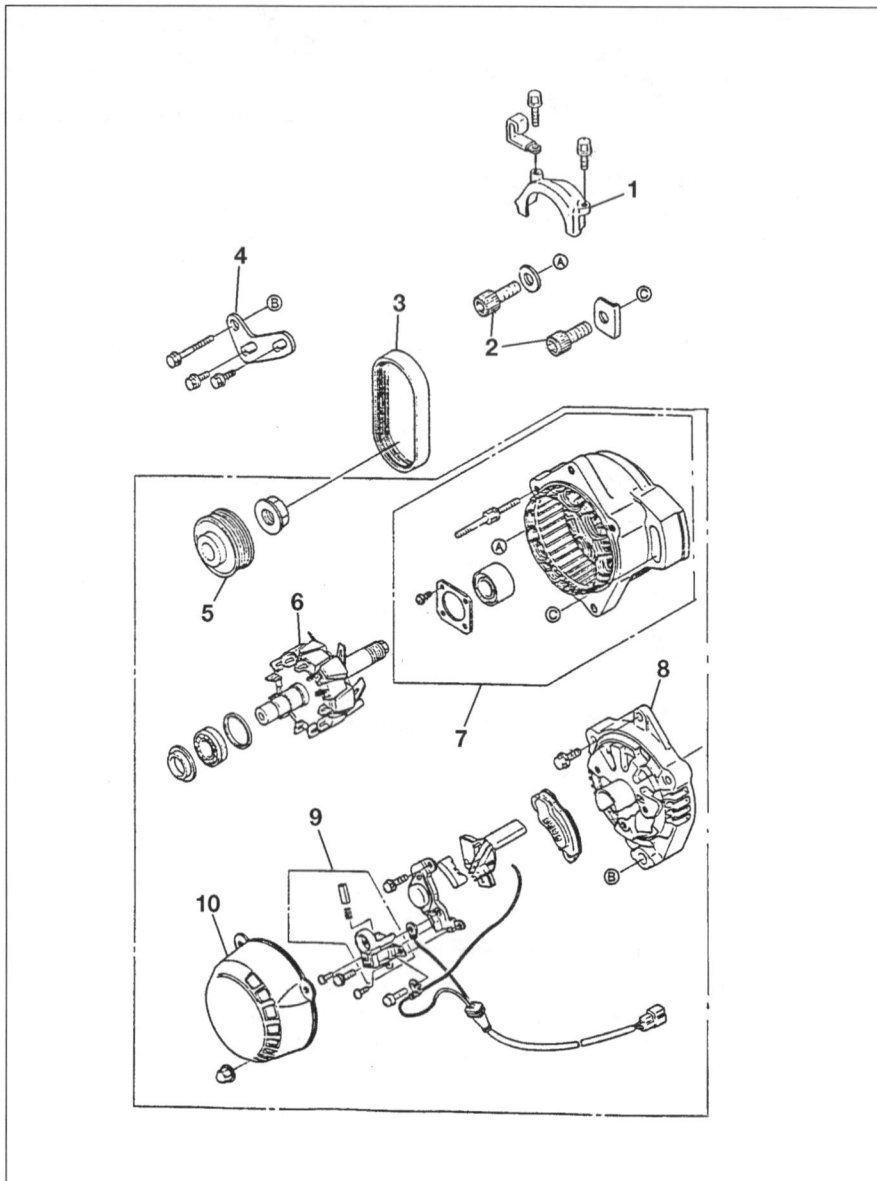

30.6b Alternator details (748 cc engine)

| | | |
|---|---|---|
| 1 Alternator cover | 4 Alternator adjusting bracket | 7 Stator |
| 2 Alternator mounting bolts | 5 Alternator pulley | 8 End housing |
| 3 Alternator drivebelt | 6 Rotor | 9 Brush assembly |
| | | 10 End cover |

**7** Remove the lower left fairing panel (see Chapter 8).

**8** Follow the wiring harness from the alternator to the connector and unplug the connector.

**9** Drain the cooling system (see Chapter 1). Remove the water pump coolant pipe and hose (see Chapter 3).

**10** Remove the clutch slave cylinder (see Chapter 2).

**11** Remove the alternator mounting bolts **(see illustrations)**. Pull the alternator out of the engine **(see illustration)** and remove it from the left side of the motorcycle.

**12** Check the rubber damper segments for wear or damage and replace them as necessary **(see illustration)**.

### Installation

**13** Installation is the reverse of the removal steps, with the following additions:
 a) On H models, ensure the drivebelt ribs fit over the alternator pulley grooves.
 b) Tighten the alternator mounting bolts to the torque listed in this Chapter's Specifications.
 c) If you're working on an H model, adjust the alternator belt (see Section 30).

31.11a Remove the alternator mounting bolts from the bottom (arrow) . . .

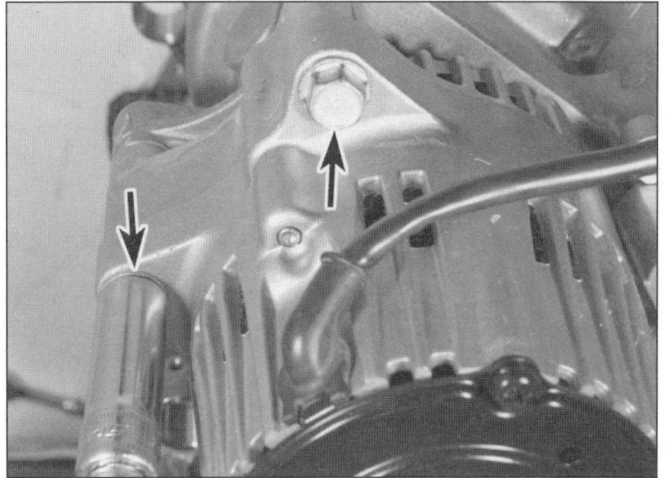

31.11b . . . and from the top (arrows) (749 cc engine shown)

31.11c On 749 cc engines, pull the alternator out of the engine

31.12 Replace the rubber damper segments if worn, brittle or deteriorated

## 32 Alternator brushes – inspection and replacement

1 This check, combined with the charging system output test described above, should diagnose most charging system problems. If the brushes are good but the alternator output is low, take the alternator to a dealer service department or other repair shop for further checks, or substitute a known good unit and recheck the charging system output.

2 Remove the end cover from the alternator (the alternator need not be removed from the motorcycle) (see illustration).

3 Remove the screws and lift out the brush assembly and regulator (see illustrations).

4 Press in on the brushes and make sure they move freely in and out of the holder (see illustration). Measure the length of the brushes and compare it to the value listed in this Chapter's Specifications. If the brushes are worn, replace them.

32.2 Remove the end cover from the alternator (749 cc shown)

32.3a Remove the screws (arrows) . . .

32.3b . . . and lift out the brush assembly (749 cc shown)

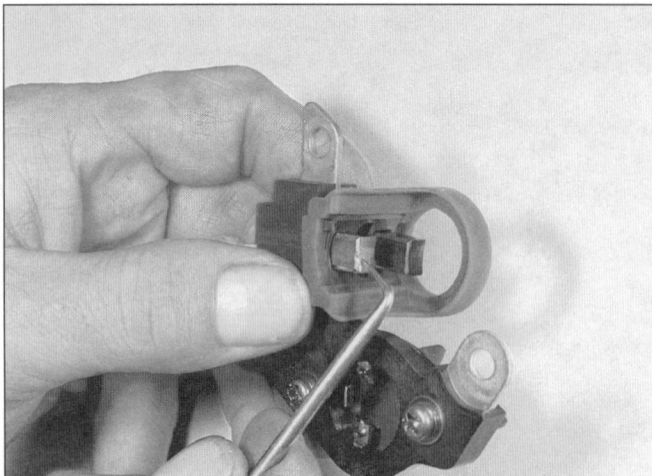

32.4 Push in on the brushes to make sure they move freely and return under spring pressure

32.5 Inspect the slip rings (arrows); the rotor must be replaced if worn or damaged

5 Inspect the slip rings (see illustration). If they're severely worn or damaged, the rotor must be replaced; take the alternator a Kawasaki dealer or reputable motorcycle shop.

6 Installation is the reverse of the removal steps.

## 33 Wiring diagrams

Prior to troubleshooting a circuit, check the fuses to make sure they're in good condition.

Make sure the battery is fully charged and check the cable connections.

When checking a circuit, make sure all connectors are clean, with no broken or loose terminals or wires. When unplugging a connector, don't pull on the wires – pull only on the connector housings themselves.

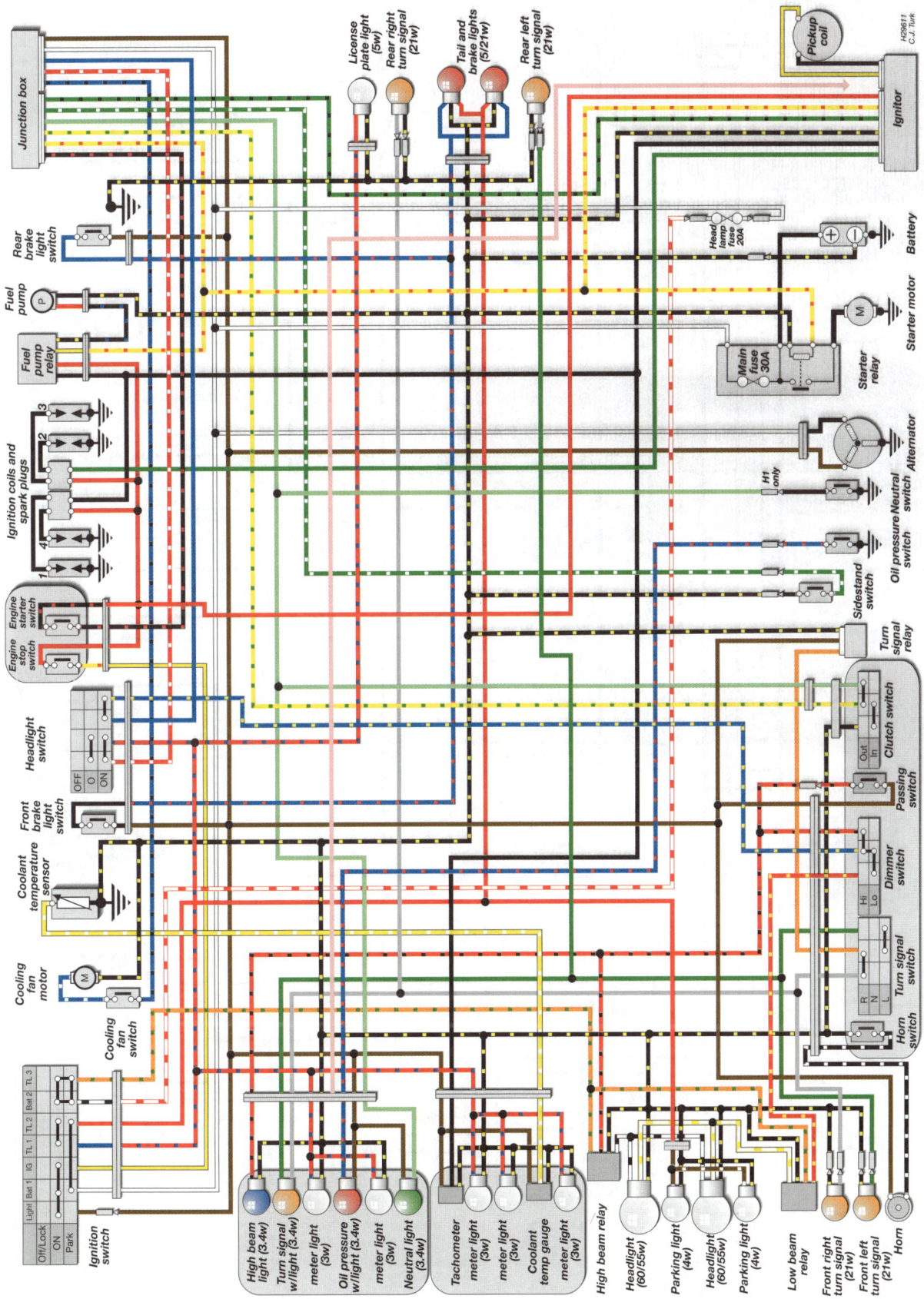

ZX750 H1 and H2 models (UK)

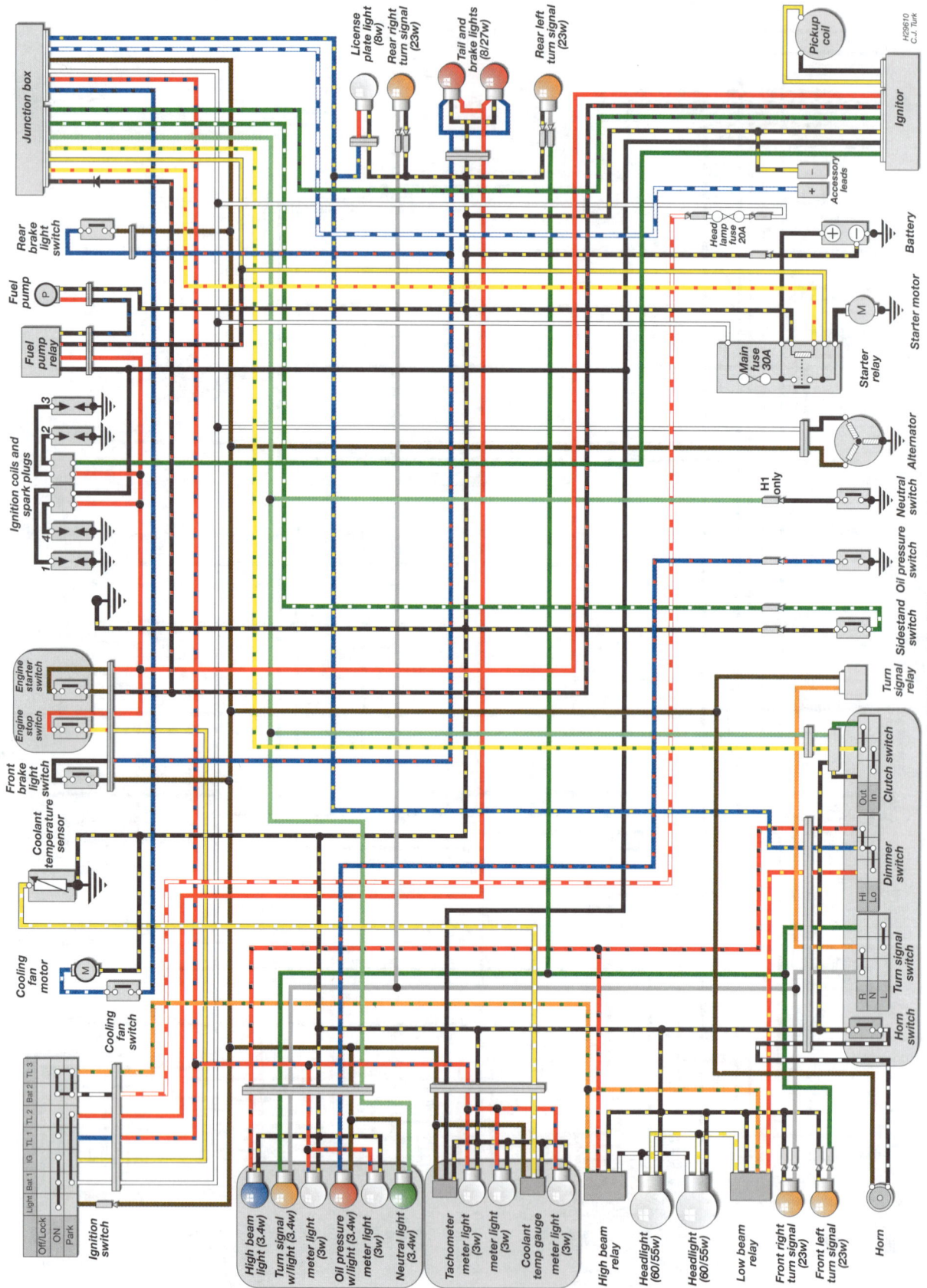

ZX750 H1 and H2 models (US and Canada)

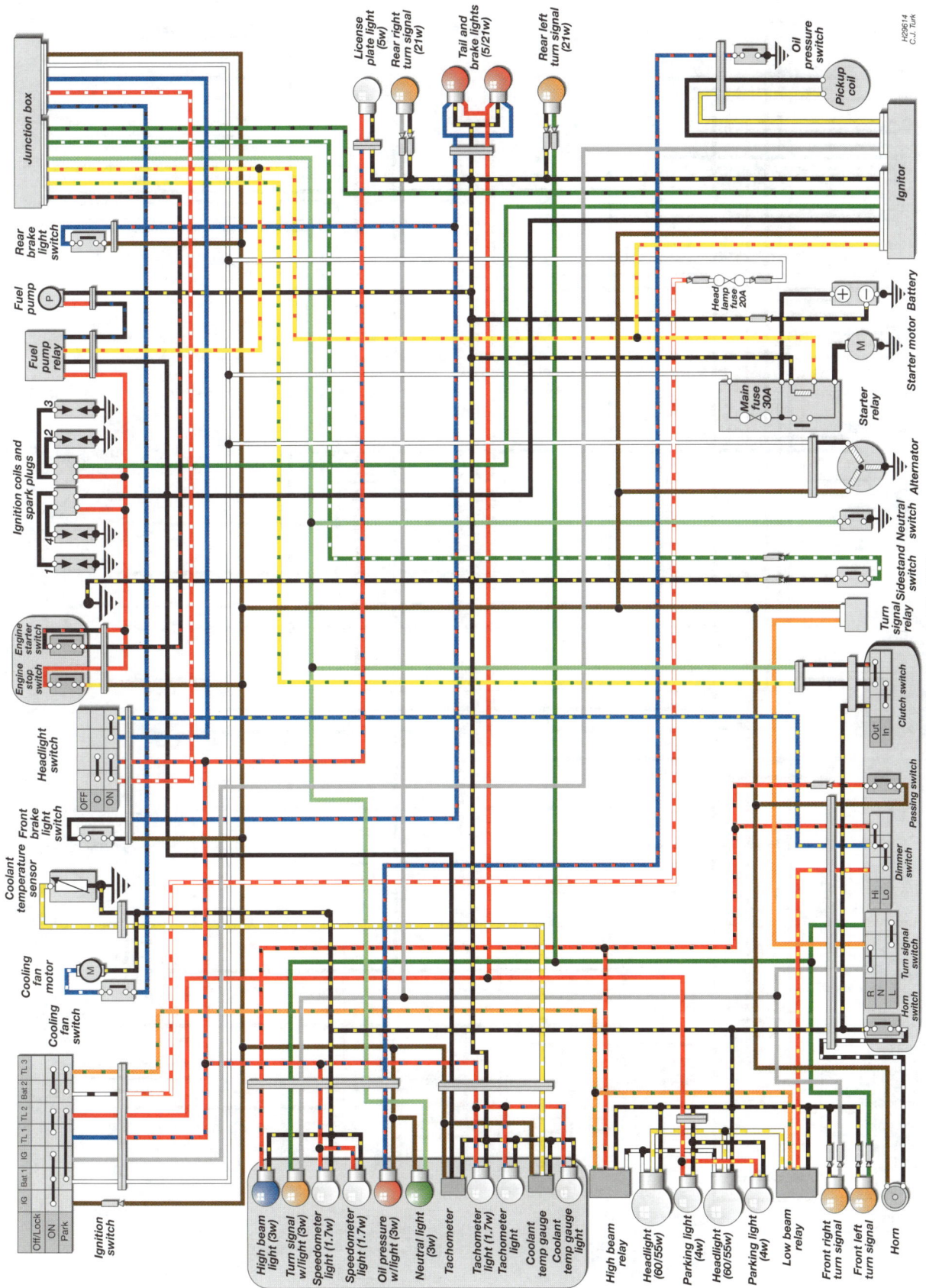

H29614
C.J. Turk

Junction box

License plate light (5w)

Rear right turn signal (21w)

Tail and brake lights (5/21w)

Rear left turn signal (21w)

Oil pressure switch

Pickup coil

Ignitor

Rear brake light switch

Fuel pump

Head lamp fuse 20A

Battery

Fuel pump relay

Main fuse 30A

Starter motor

Starter relay

Ignition coils and spark plugs

Alternator

Neutral switch

Sidestand switch

Turn signal relay

Engine starter switch

Engine stop switch

ZX750 J and L models (UK)

Clutch switch

Headlight switch

Passing switch

Front brake light switch

Coolant temperature sensor

Dimmer switch

Turn signal switch

Cooling fan motor

Horn switch

Cooling fan switch

Ignition switch

High beam light (3w)

Turn signal w/light (3w)

Speedometer light (1.7w)

Speedometer light (1.7w)

Oil pressure w/light (3w)

Neutral light (3w)

Tachometer

Tachometer light (1.7w)

Tachometer light

Coolant temp gauge

Coolant temp gauge light

High beam relay

Headlight (60/55w)

Parking light (4w)

Headlight (60/55w)

Parking light (4w)

Low beam relay

Front right turn signal

Front left turn signal

Horn

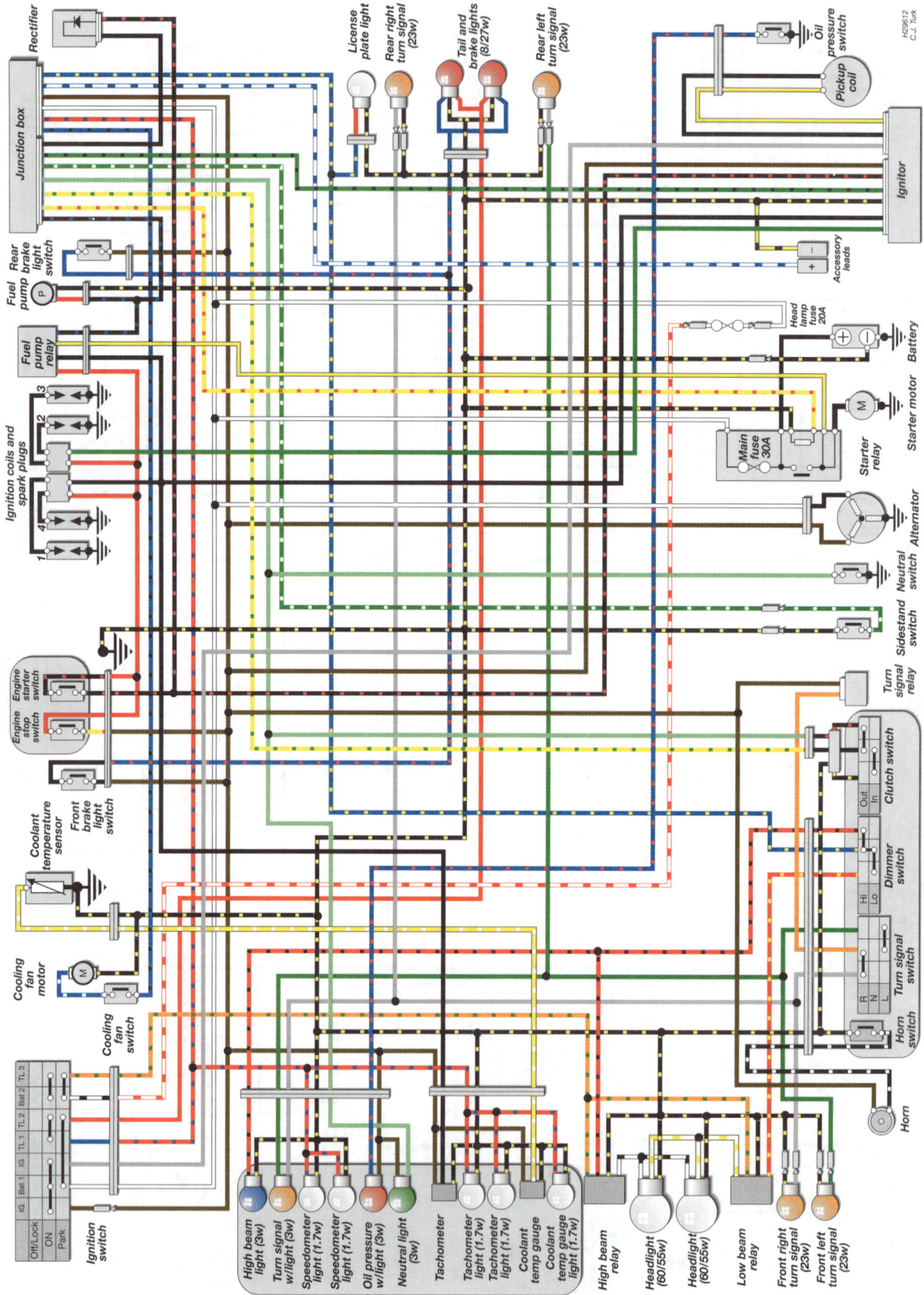

ZX750 J and L models (US and Canada)

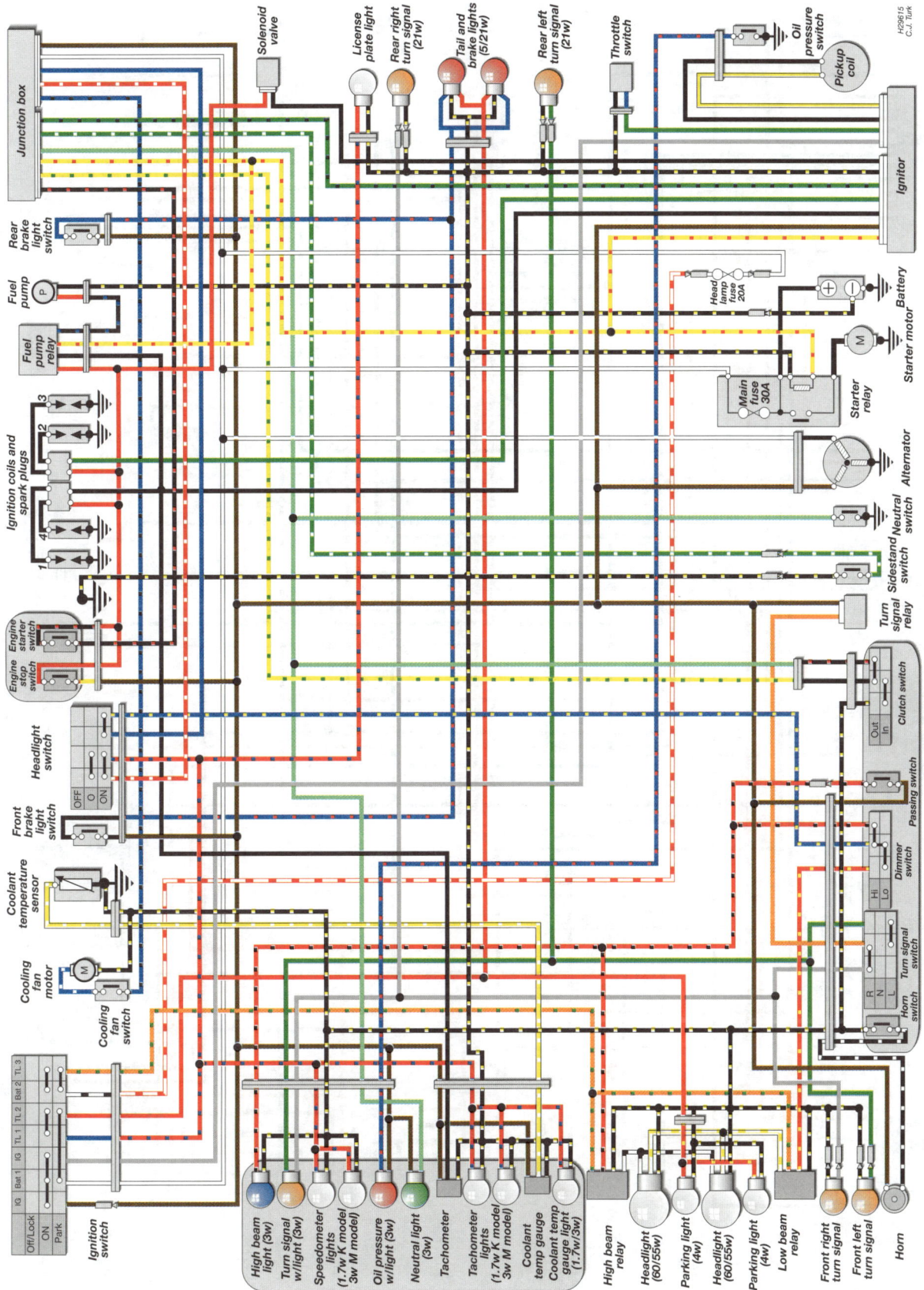

Solenoid valve

License plate light

Rear right turn signal (21w)

Tail and brake lights (5/21w)

Rear left turn signal (21w)

Throttle switch

Oil pressure switch

Pickup coil

H29615
C.J. Turk

Junction box

Ignitor

Rear brake light switch

Fuel pump

Head lamp fuse 20A

Battery

Fuel pump relay

Starter motor

Main fuse 30A

Starter relay

Ignition coils and spark plugs

Alternator

Neutral switch

Sidestand switch

Turn signal relay

Engine starter switch

Engine stop switch

Clutch switch

ZX750 K and M models (UK)

Passing switch

Headlight switch

Dimmer switch

Front brake light switch

Turn signal switch

Coolant temperature sensor

Horn switch

Cooling fan motor

Cooling fan switch

Ignition switch

High beam light (3w)

Turn signal w/light (3w)

Speedometer lights (1.7w K model 3w M model)

Oil pressure w/light (3w)

Neutral light (3w)

Tachometer

Tachometer lights (1.7w K model 3w M model)

Coolant temp gauge

Coolant temp gauge light (1.7w/3w)

High beam relay

Headlight (60/55w)

Parking light (4w)

Headlight (60/55w)

Parking light (4w)

Low beam relay

Front right turn signal

Front left turn signal

Horn

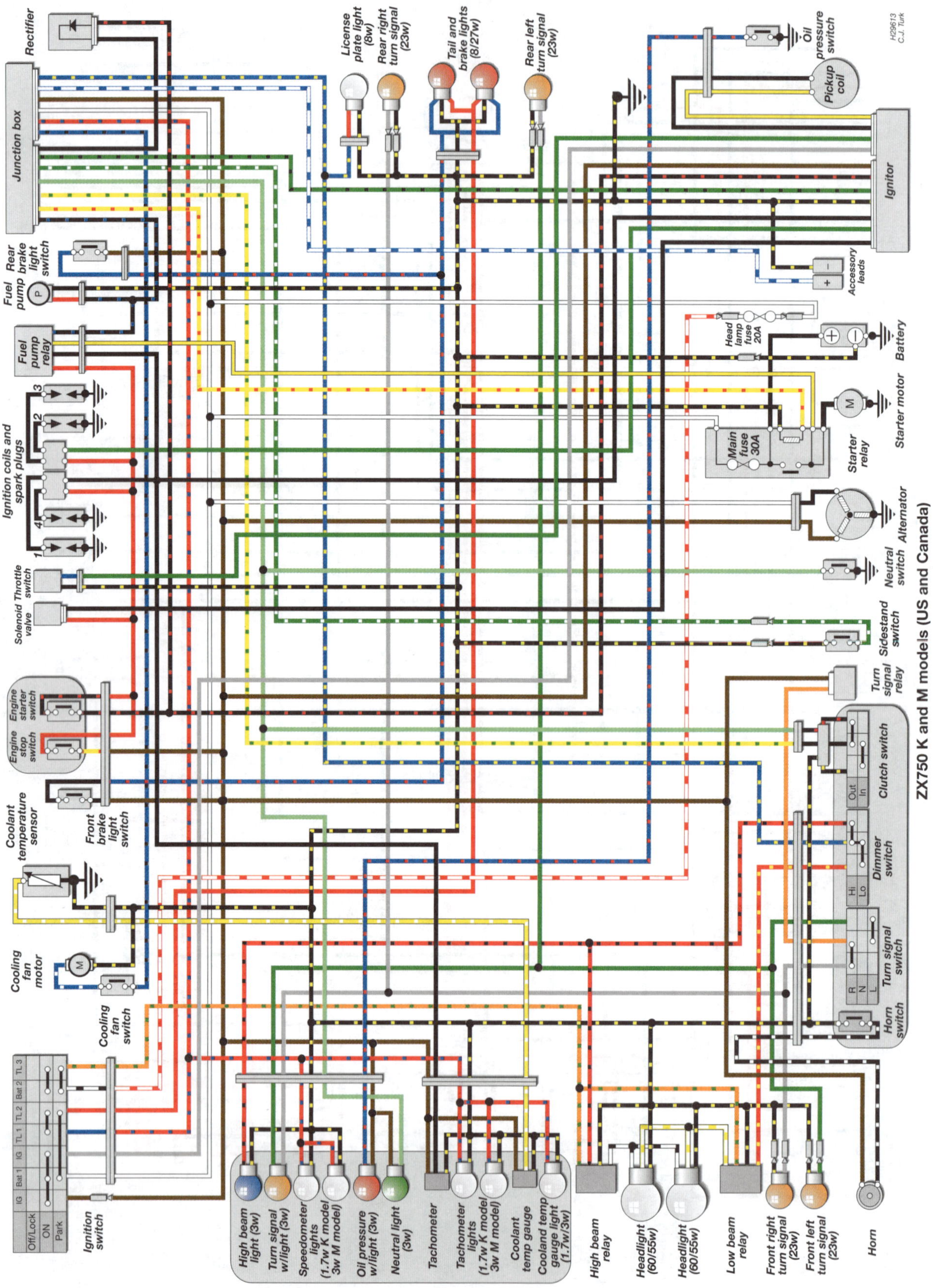
Rectifier
Junction box
Rear brake light switch
Fuel pump
P
Fuel pump relay
Ignition coils and spark plugs
1
2
3
4
Throttle switch
Solenoid valve
Engine starter switch
Engine stop switch
Coolant temperature sensor
Front brake light switch
Cooling fan motor
M
Cooling fan switch
IG
Bat 1
IG
TL 1
TL 2
Bat 2
TL 3
Off/Lock
ON
Park
Ignition switch
License plate light (8w)
Rear right turn signal (23w)
Tail and brake lights (8/27w)
Rear left turn signal (23w)
Oil pressure switch
Pickup coil
Ignitor
H29613
C.J. Turk
Accessory leads
Head lamp fuse 20A
Battery
Starter motor
Main fuse 30A
Starter relay
Alternator
Neutral switch
Sidestand switch
Turn signal relay
Out
In
Clutch switch
Hi
Lo
Dimmer switch
R
N
L
Turn signal switch
Horn switch
ZX750 K and M models (US and Canada)
High beam light (3w)
Turn signal w/light (3w)
Speedometer lights (1.7w K model 3w M model)
Oil pressure w/light (3w)
Neutral light (3w)
Tachometer
Tachometer lights (1.7w K model 3w M model)
Coolant temp gauge
Cooland temp gauge light (1.7w/3w)
High beam relay
Headlight (60/55w)
Headlight (60/55w)
Low beam relay
Front right turn signal (23w)
Front left turn signal (23w)
Horn

# Dimensions and weights

## Wheelbase
H models . . . . . . . . . . . . . . . . . . . . . . . . . . . .1455 mm (57.28 inches)
J and K models . . . . . . . . . . . . . . . . . . . . . . .1420 mm (55.90 inches)
L and M . . . . . . . . . . . . . . . . . . . . . . . . . . . . .1430 mm (56.29 inches)

## Overall length
H models . . . . . . . . . . . . . . . . . . . . . . . . . . .2090 mm (82.28 inches)
J, K, L and M models . . . . . . . . . . . . . . . . . . .2085 mm (82.08 inches)

## Overall width
H models . . . . . . . . . . . . . . . . . . . . . . . . . . . . .755 mm (29.72 inches)
J, K, L and M models . . . . . . . . . . . . . . . . . . . .730 mm (28.74 inches)

## Overall height
H models . . . . . . . . . . . . . . . . . . . . . . . . . . .1170 mm (46.06 inches)
J and K models . . . . . . . . . . . . . . . . . . . . . . .1120 mm (44.09 inches)
L and M models (US models) . . . . . . . . . . . . .1140 mm (44.88 inches)
L and M models (UK models) . . . . . . . . . . . . .1125 mm (44.29 inches)

## Seat height
H models . . . . . . . . . . . . . . . . . . . . . . . . . . . . .770 mm (30.31 inches)
J and K models . . . . . . . . . . . . . . . . . . . . . . . .780 mm (30.71 inches)
L and M models . . . . . . . . . . . . . . . . . . . . . . . .800 mm (31.49 inches)

## Weight (dry)
H1 (except California) . . . . . . . . . . . . . . . . . . . . . . . . .205 kg (451 lbs)
H1 (California) . . . . . . . . . . . . . . . . . . . . . . . . . . . . .205.5 kg (452 lbs)
H2 (except California) . . . . . . . . . . . . . . . . . . . . . . . .200 kg (440 lbs)
H2 (California) . . . . . . . . . . . . . . . . . . . . . . . . . . . . .200.5 kg (441 lbs)
J and K (except California) . . . . . . . . . . . . . . . . . . . .195 kg (429 lbs)
J and K (California) . . . . . . . . . . . . . . . . . . . . . . . . .195.5 kg (430 lbs)
L models (except California) . . . . . . . . . . . . . . . . . . .205 kg (451 lbs)
L models (California) . . . . . . . . . . . . . . . . . . . . . . . .205.5 kg (452 lbs)
M models (except California) . . . . . . . . . . . . . . . . . .200 kg (440 lbs)
M models (California) . . . . . . . . . . . . . . . . . . . . . . .200.5 kg (441 lbs)

## Buying tools

A toolkit is a fundamental requirement for servicing and repairing a motorcycle. Although there will be an initial expense in building up enough tools for servicing, this will soon be offset by the savings made by doing the job yourself. As experience and confidence grow, additional tools can be added to enable the repair and overhaul of the motorcycle. Many of the specialist tools are expensive and not often used so it may be preferable to hire them, or for a group of friends or motorcycle club to join in the purchase.

As a rule, it is better to buy more expensive, good quality tools. Cheaper tools are likely to wear out faster and need to be renewed more often, nullifying the original saving.

**Warning: To avoid the risk of a poor quality tool breaking in use, causing injury or damage to the component being worked on, always aim to purchase tools which meet the relevant national safety standards.**

The following lists of tools do not represent the manufacturer's service tools, but serve as a guide to help the owner decide which tools are needed for this level of work. In addition, items such as an electric drill, hacksaw, files, hammers, soldering iron and a workbench equipped with a vice, may be needed. Although not classed as tools, a selection of bolts, screws, nuts, washers and pieces of tubing always come in useful.

For more information about tools, refer to the Haynes *Motorcycle Workshop Practice Manual* (Bk. No. 1454).

## Manufacturer's service tools

Inevitably certain tasks require the use of a service tool. Where possible an alternative tool or method of approach is recommended, but sometimes there is no option if personal injury or damage to the component is to be avoided. Where required, service tools are referred to in the relevant procedure.

Service tools can usually only be purchased from a motorcycle dealer and are identified by a part number. Some of the commonly-used tools, such as rotor pullers, are available in aftermarket form from mail-order motorcycle tool and accessory suppliers.

# Maintenance and minor repair tools

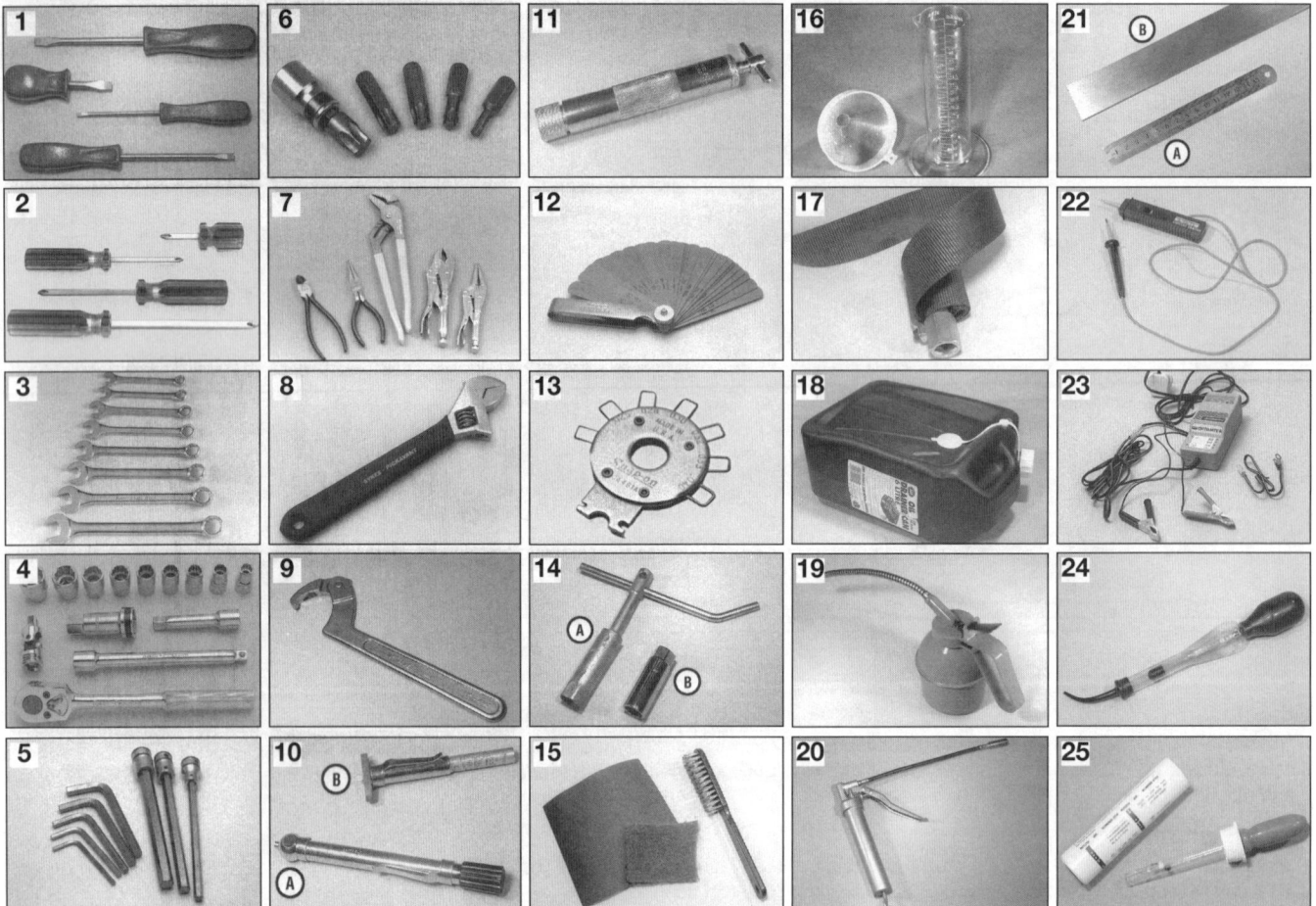

1 Set of flat-bladed screwdrivers
2 Set of Phillips head screwdrivers
3 Combination open-end & ring spanners
4 Socket set (3/8 inch or 1/2 inch drive)
5 Set of Allen keys or bits
6 Set of Torx keys or bits
7 Pliers and self-locking grips (Mole grips)
8 Adjustable spanner
9 C-spanner (ideally adjustable type)
10 Tyre pressure gauge (A) & tread depth gauge (B)
11 Cable pressure oiler
12 Feeler gauges
13 Spark plug gap measuring and adjusting tool
14 Spark plug spanner (A) or deep plug socket (B)
15 Wire brush and emery paper
16 Funnel and measuring vessel
17 Strap wrench, chain wrench or oil filter removal tool
18 Oil drainer can or tray
19 Pump type oil can
20 Grease gun
21 Steel rule (A) and straight-edge (B)
22 Continuity tester
23 Battery charger
24 Hydrometer (for battery specific gravity check)
25 Anti-freeze tester (for liquid-cooled engines)

## Repair and overhaul tools

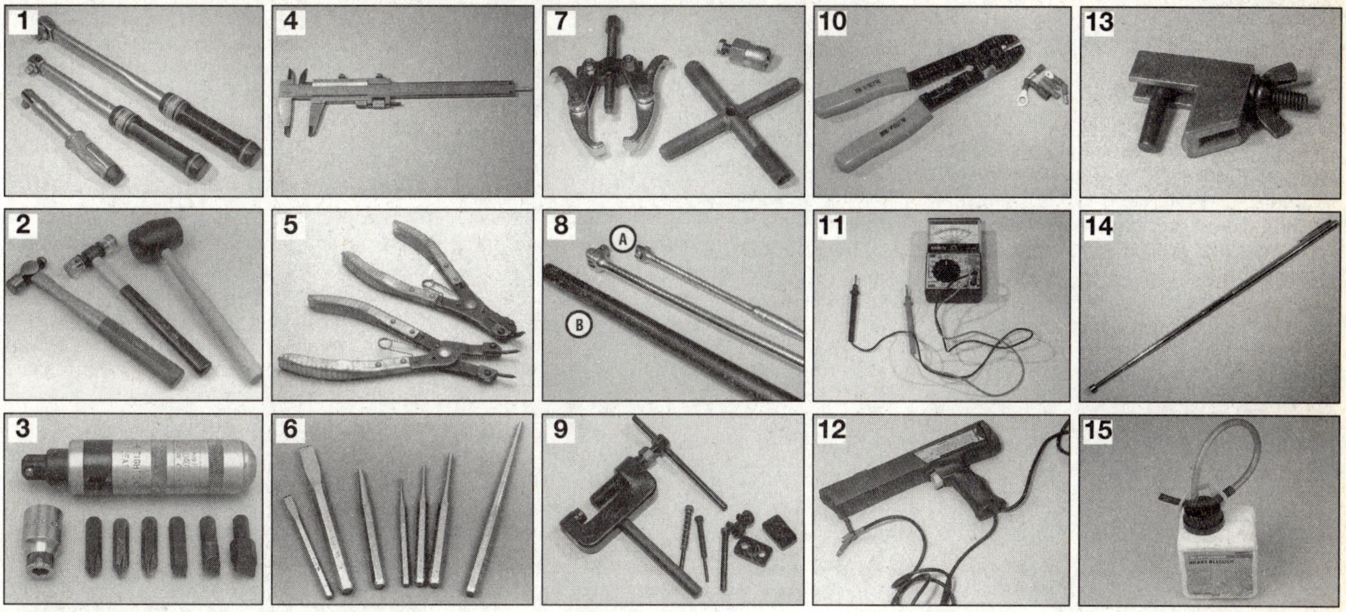

1 Torque wrench
  (small and mid-ranges)
2 Conventional, plastic or
  soft-faced hammers
3 Impact driver set
4 Vernier gauge
5 Circlip pliers (internal and
  external, or combination)
6 Set of punches
  and cold chisels
7 Selection of pullers
8 Breaker bars (A)
  and length of tubing (B)
9 Chain breaking/
  riveting tool
10 Wire crimper tool
11 Multimeter (measures
   amps, volts and ohms)
12 Stroboscope (for
   dynamic timing checks)
13 Hose clamp
   (wingnut type shown)
14 Magnetic arm
   (telescopic type shown)
15 One-man brake/clutch
   bleeder kit

## Specialist tools

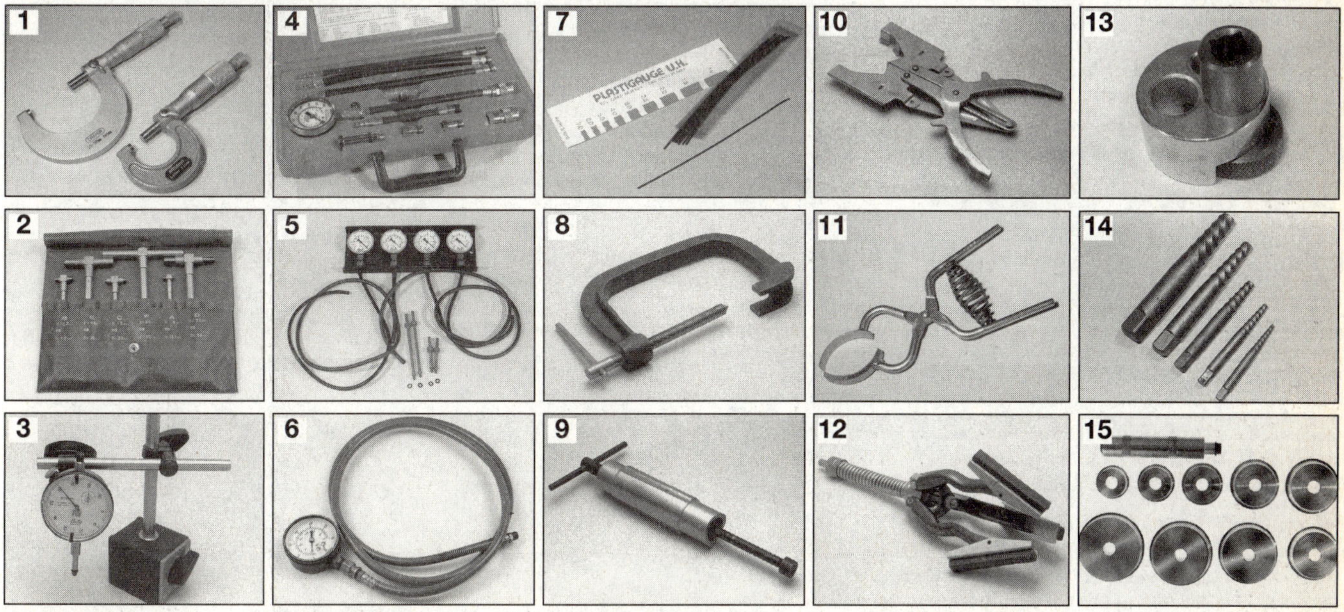

1 Micrometer
  (external type)
2 Telescoping gauges or
  small-hole gauges
3 Dial gauge
4 Cylinder
  compression gauge
5 Vacuum gauges (shown)
  or manometer
6 Oil pressure gauge
7 Plastigauge kit
8 Valve spring compressor
  (4-stroke engines)
9 Piston pin drawbolt tool
10 Piston ring removal and
   installation tool
11 Piston ring clamp
12 Cylinder bore hone
   (stone type shown)
13 Stud extractor
14 Screw extractor set
15 Bearing driver set

# Tools and Workshop Tips

## 1 Workshop equipment and facilities

### The workbench

● Work is made much easier by raising the bike up on a ramp - components are much more accessible if raised to waist level. The hydraulic or pneumatic types seen in the dealer's workshop are a sound investment if you undertake a lot of repairs or overhauls **(see illustration 1.1)**.

**1.1 Hydraulic motorcycle ramp**

● If raised off ground level, the bike must be supported on the ramp to avoid it falling. Most ramps incorporate a front wheel locating clamp which can be adjusted to suit different diameter wheels. When tightening the clamp, take care not to mark the wheel rim or damage the tyre - use wood blocks on each side to prevent this.
● Secure the bike to the ramp using tie-downs **(see illustration 1.2)**. If the bike has only a sidestand, and hence leans at a dangerous angle when raised, support the bike on an auxiliary stand.

**1.2 Tie-downs are used around the passenger footrests to secure the bike**

● Auxiliary (paddock) stands are widely available from mail order companies or motorcycle dealers and attach either to the wheel axle or swingarm pivot **(see illustration 1.3)**. If the motorcycle has a centrestand, you can support it under the crankcase to prevent it toppling whilst either wheel is removed **(see illustration 1.4)**.

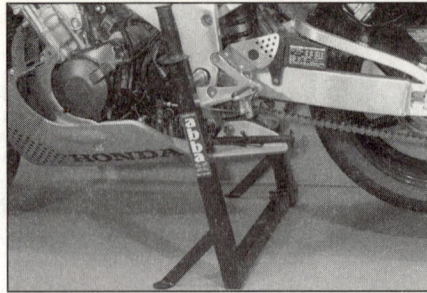

**1.3 This auxiliary stand attaches to the swingarm pivot**

**1.4 Always use a block of wood between the engine and jack head when supporting the engine in this way**

### Fumes and fire

● Refer to the Safety first! page at the beginning of the manual for full details. Make sure your workshop is equipped with a fire extinguisher suitable for fuel-related fires (Class B fire - flammable liquids) - it is not sufficient to have a water-filled extinguisher.
● Always ensure adequate ventilation is available. Unless an exhaust gas extraction system is available for use, ensure that the engine is run outside of the workshop.
● If working on the fuel system, make sure the workshop is ventilated to avoid a build-up of fumes. This applies equally to fume build-up when charging a battery. Do not smoke or allow anyone else to smoke in the workshop.

### Fluids

● If you need to drain fuel from the tank, store it in an approved container marked as suitable for the storage of petrol (gasoline) **(see illustration 1.5)**. Do not store fuel in glass jars or bottles.

**1.5 Use an approved can only for storing petrol (gasoline)**

● Use proprietary engine degreasers or solvents which have a high flash-point, such as paraffin (kerosene), for cleaning off oil, grease and dirt - never use petrol (gasoline) for cleaning. Wear rubber gloves when handling solvent and engine degreaser. The fumes from certain solvents can be dangerous - always work in a well-ventilated area.

### Dust, eye and hand protection

● Protect your lungs from inhalation of dust particles by wearing a filtering mask over the nose and mouth. Many frictional materials still contain asbestos which is dangerous to your health. Protect your eyes from spouts of liquid and sprung components by wearing a pair of protective goggles **(see illustration 1.6)**.

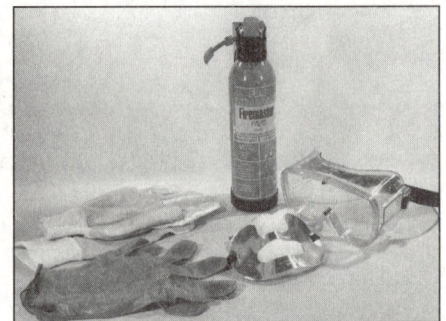

**1.6 A fire extinguisher, goggles, mask and protective gloves should be at hand in the workshop**

● Protect your hands from contact with solvents, fuel and oils by wearing rubber gloves. Alternatively apply a barrier cream to your hands before starting work. If handling hot components or fluids, wear suitable gloves to protect your hands from scalding and burns.

### What to do with old fluids

● Old cleaning solvent, fuel, coolant and oils should not be poured down domestic drains or onto the ground. Package the fluid up in old oil containers, label it accordingly, and take it to a garage or disposal facility. Contact your local authority for location of such sites or ring the oil care hotline.

**OIL CARE**

**FOLLOW THE CODE**

**OIL BANK LINE**
**0800 66 33 66**

*Note: It is antisocial and illegal to dump oil down the drain. To find the location of your local oil recycling bank, call this number free.*

*In the USA, note that any oil supplier must accept used oil for recycling.*

## 2 Fasteners -
screws, bolts and nuts

### Fastener types and applications

#### Bolts and screws

● Fastener head types are either of hexagonal, Torx or splined design, with internal and external versions of each type (see illustrations 2.1 and 2.2); splined head fasteners are not in common use on motorcycles. The conventional slotted or Phillips head design is used for certain screws. Bolt or screw length is always measured from the underside of the head to the end of the item (see illustration 2.11).

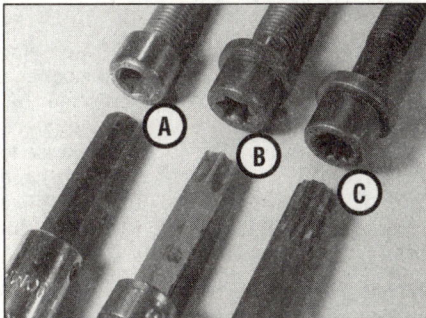

**2.1 Internal hexagon/Allen (A), Torx (B) and splined (C) fasteners, with corresponding bits**

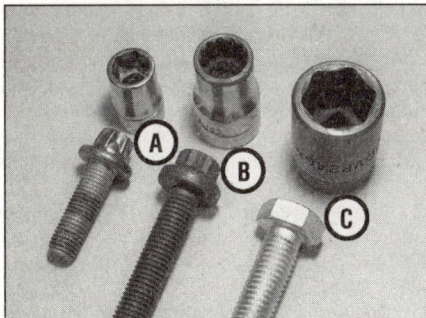

**2.2 External Torx (A), splined (B) and hexagon (C) fasteners, with corresponding sockets**

● Certain fasteners on the motorcycle have a tensile marking on their heads, the higher the marking the stronger the fastener. High tensile fasteners generally carry a 10 or higher marking. Never replace a high tensile fastener with one of a lower tensile strength.

#### Washers (see illustration 2.3)

● Plain washers are used between a fastener head and a component to prevent damage to the component or to spread the load when torque is applied. Plain washers can also be used as spacers or shims in certain assemblies. Copper or aluminium plain washers are often used as sealing washers on drain plugs.

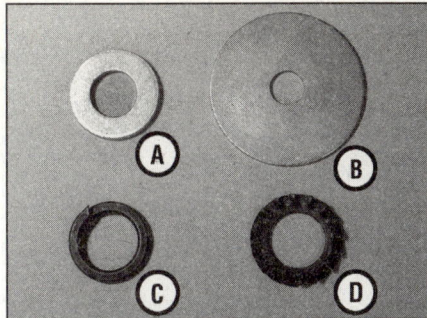

**2.3 Plain washer (A), penny washer (B), spring washer (C) and serrated washer (D)**

● The split-ring spring washer works by applying axial tension between the fastener head and component. If flattened, it is fatigued and must be renewed. If a plain (flat) washer is used on the fastener, position the spring washer between the fastener and the plain washer.

● Serrated star type washers dig into the fastener and component faces, preventing loosening. They are often used on electrical earth (ground) connections to the frame.

● Cone type washers (sometimes called Belleville) are conical and when tightened apply axial tension between the fastener head and component. They must be installed with the dished side against the component and often carry an OUTSIDE marking on their outer face. If flattened, they are fatigued and must be renewed.

● Tab washers are used to lock plain nuts or bolts on a shaft. A portion of the tab washer is bent up hard against one flat of the nut or bolt to prevent it loosening. Due to the tab washer being deformed in use, a new tab washer should be used every time it is disturbed.

● Wave washers are used to take up endfloat on a shaft. They provide light springing and prevent excessive side-to-side play of a component. Can be found on rocker arm shafts.

#### Nuts and split pins

● Conventional plain nuts are usually six-sided (see illustration 2.4). They are sized by thread diameter and pitch. High tensile nuts carry a number on one end to denote their tensile strength.

**2.4 Plain nut (A), shouldered locknut (B), nylon insert nut (C) and castellated nut (D)**

● Self-locking nuts either have a nylon insert, or two spring metal tabs, or a shoulder which is staked into a groove in the shaft - their advantage over conventional plain nuts is a resistance to loosening due to vibration. The nylon insert type can be used a number of times, but must be renewed when the friction of the nylon insert is reduced, ie when the nut spins freely on the shaft. The spring tab type can be reused unless the tabs are damaged. The shouldered type must be renewed every time it is disturbed.

● Split pins (cotter pins) are used to lock a castellated nut to a shaft or to prevent slackening of a plain nut. Common applications are wheel axles and brake torque arms. Because the split pin arms are deformed to lock around the nut a new split pin must always be used on installation - always fit the correct size split pin which will fit snugly in the shaft hole. Make sure the split pin arms are correctly located around the nut (see illustrations 2.5 and 2.6).

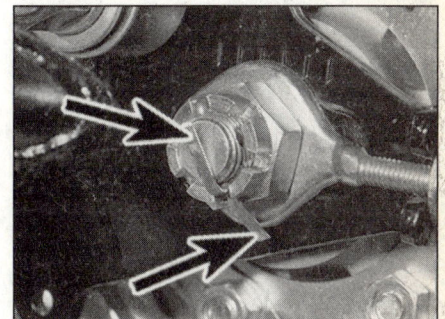

**2.5 Bend split pin (cotter pin) arms as shown (arrows) to secure a castellated nut**

**2.6 Bend split pin (cotter pin) arms as shown to secure a plain nut**

*Caution: If the castellated nut slots do not align with the shaft hole after tightening to the torque setting, tighten the nut until the next slot aligns with the hole - never slacken the nut to align its slot.*

● R-pins (shaped like the letter R), or slip pins as they are sometimes called, are sprung and can be reused if they are otherwise in good condition. Always install R-pins with their closed end facing forwards (see illustration 2.7).

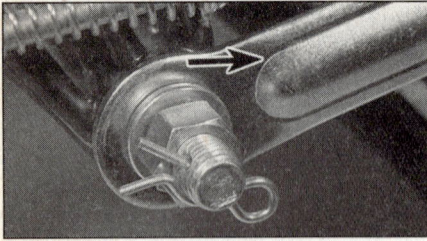
**2.7 Correct fitting of R-pin. Arrow indicates forward direction**

### Circlips (see illustration 2.8)

● Circlips (sometimes called snap-rings) are used to retain components on a shaft or in a housing and have corresponding external or internal ears to permit removal. Parallel-sided (machined) circlips can be installed either way round in their groove, whereas stamped circlips (which have a chamfered edge on one face) must be installed with the chamfer facing away from the direction of thrust load **(see illustration 2.9)**.

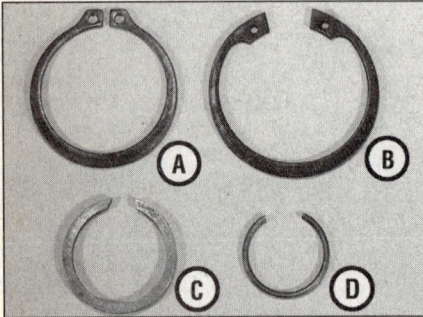
**2.8 External stamped circlip (A), internal stamped circlip (B), machined circlip (C) and wire circlip (D)**

● Always use circlip pliers to remove and install circlips; expand or compress them just enough to remove them. After installation, rotate the circlip in its groove to ensure it is securely seated. If installing a circlip on a splined shaft, always align its opening with a shaft channel to ensure the circlip ends are well supported and unlikely to catch **(see illustration 2.10)**.

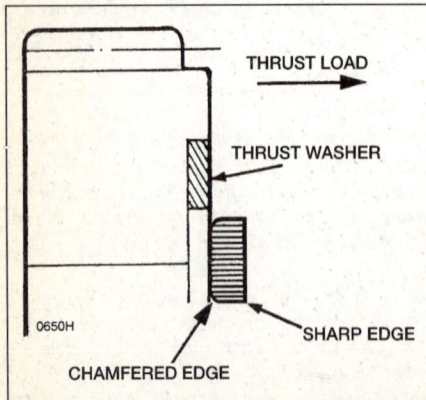
**2.9 Correct fitting of a stamped circlip**

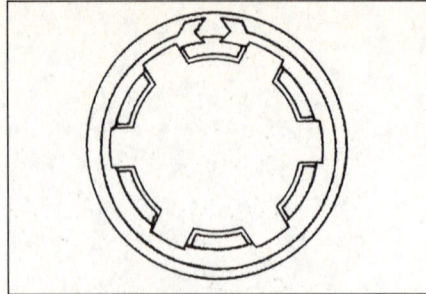
**2.10 Align circlip opening with shaft channel**

● Circlips can wear due to the thrust of components and become loose in their grooves, with the subsequent danger of becoming dislodged in operation. For this reason, renewal is advised every time a circlip is disturbed.

● Wire circlips are commonly used as piston pin retaining clips. If a removal tang is provided, long-nosed pliers can be used to dislodge them, otherwise careful use of a small flat-bladed screwdriver is necessary. Wire circlips should be renewed every time they are disturbed.

### Thread diameter and pitch

● Diameter of a male thread (screw, bolt or stud) is the outside diameter of the threaded portion **(see illustration 2.11)**. Most motorcycle manufacturers use the ISO (International Standards Organisation) metric system expressed in millimetres, eg M6 refers to a 6 mm diameter thread. Sizing is the same for nuts, except that the thread diameter is measured across the valleys of the nut.

● Pitch is the distance between the peaks of the thread **(see illustration 2.11)**. It is expressed in millimetres, thus a common bolt size may be expressed as 6.0 x 1.0 mm (6 mm thread diameter and 1 mm pitch). Generally pitch increases in proportion to thread diameter, although there are always exceptions.

● Thread diameter and pitch are related for conventional fastener applications and the following table can be used as a guide. Additionally, the AF (Across Flats), spanner or socket size dimension of the bolt or nut **(see illustration 2.11)** is linked to thread and pitch specification. Thread pitch can be measured with a thread gauge **(see illustration 2.12)**.

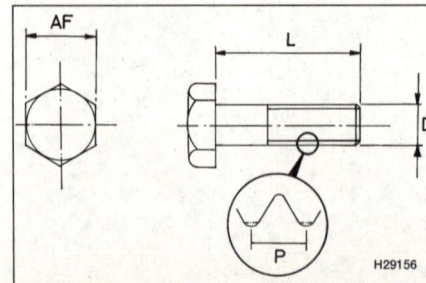
**2.11 Fastener length (L), thread diameter (D), thread pitch (P) and head size (AF)**

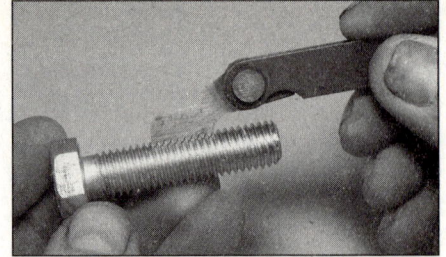
**2.12 Using a thread gauge to measure pitch**

| AF size | Thread diameter x pitch (mm) |
|---------|------------------------------|
| 8 mm    | M5 x 0.8                     |
| 8 mm    | M6 x 1.0                     |
| 10 mm   | M6 x 1.0                     |
| 12 mm   | M8 x 1.25                    |
| 14 mm   | M10 x 1.25                   |
| 17 mm   | M12 x 1.25                   |

● The threads of most fasteners are of the right-hand type, ie they are turned clockwise to tighten and anti-clockwise to loosen. The reverse situation applies to left-hand thread fasteners, which are turned anti-clockwise to tighten and clockwise to loosen. Left-hand threads are used where rotation of a component might loosen a conventional right-hand thread fastener.

### Seized fasteners

● Corrosion of external fasteners due to water or reaction between two dissimilar metals can occur over a period of time. It will build up sooner in wet conditions or in countries where salt is used on the roads during the winter. If a fastener is severely corroded it is likely that normal methods of removal will fail and result in its head being ruined. When you attempt removal, the fastener thread should be heard to crack free and unscrew easily - if it doesn't, stop there before damaging something.

● A smart tap on the head of the fastener will often succeed in breaking free corrosion which has occurred in the threads **(see illustration 2.13)**.

● An aerosol penetrating fluid (such as WD-40) applied the night beforehand may work its way down into the thread and ease removal. Depending on the location, you may be able to make up a Plasticine well around the fastener head and fill it with penetrating fluid.

**2.13 A sharp tap on the head of a fastener will often break free a corroded thread**

● If you are working on an engine internal component, corrosion will most likely not be a problem due to the well lubricated environment. However, components can be very tight and an impact driver is a useful tool in freeing them (see illustration 2.14).

**2.14 Using an impact driver to free a fastener**

● Where corrosion has occurred between dissimilar metals (eg steel and aluminium alloy), the application of heat to the fastener head will create a disproportionate expansion rate between the two metals and break the seizure caused by the corrosion. Whether heat can be applied depends on the location of the fastener - any surrounding components likely to be damaged must first be removed (see illustration 2.15). Heat can be applied using a paint stripper heat gun or clothes iron, or by immersing the component in boiling water - wear protective gloves to prevent scalding or burns to the hands.

**2.15 Using heat to free a seized fastener**

● As a last resort, it is possible to use a hammer and cold chisel to work the fastener head unscrewed (see illustration 2.16). This will damage the fastener, but more importantly extreme care must be taken not to damage the surrounding component.

*Caution: Remember that the component being secured is generally of more value than the bolt, nut or screw - when the fastener is freed, do not unscrew it with force, instead work the fastener back and forth when resistance is felt to prevent thread damage.*

**2.16 Using a hammer and chisel to free a seized fastener**

## Broken fasteners and damaged heads

● If the shank of a broken bolt or screw is accessible you can grip it with self-locking grips. The knurled wheel type stud extractor tool or self-gripping stud puller tool is particularly useful for removing the long studs which screw into the cylinder mouth surface of the crankcase or bolts and screws from which the head has broken off (see illustration 2.17). Studs can also be removed by locking two nuts together on the threaded end of the stud and using a spanner on the lower nut (see illustration 2.18).

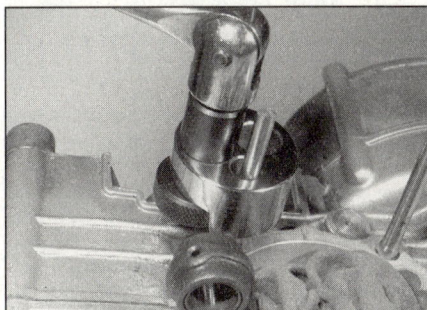

**2.17 Using a stud extractor tool to remove a broken crankcase stud**

**2.18 Two nuts can be locked together to unscrew a stud from a component**

● A bolt or screw which has broken off below or level with the casing must be extracted using a screw extractor set. Centre punch the fastener to centralise the drill bit, then drill a hole in the fastener (see illustration 2.19). Select a drill bit which is

**2.19 When using a screw extractor, first drill a hole in the fastener . . .**

approximately half to three-quarters the diameter of the fastener and drill to a depth which will accommodate the extractor. Use the largest size extractor possible, but avoid leaving too small a wall thickness otherwise the extractor will merely force the fastener walls outwards wedging it in the casing thread.

● If a spiral type extractor is used, thread it anti-clockwise into the fastener. As it is screwed in, it will grip the fastener and unscrew it from the casing (see illustration 2.20).

**2.20 . . . then thread the extractor anti-clockwise into the fastener**

● If a taper type extractor is used, tap it into the fastener so that it is firmly wedged in place. Unscrew the extractor (anti-clockwise) to draw the fastener out.

⚠ *Warning: Stud extractors are very hard and may break off in the fastener if care is not taken - ask an engineer about spark erosion if this happens.*

● Alternatively, the broken bolt/screw can be drilled out and the hole retapped for an oversize bolt/screw or a diamond-section thread insert. It is essential that the drilling is carried out squarely and to the correct depth, otherwise the casing may be ruined - if in doubt, entrust the work to an engineer.

● Bolts and nuts with rounded corners cause the correct size spanner or socket to slip when force is applied. Of the types of spanner/socket available always use a six-point type rather than an eight or twelve-point type - better grip

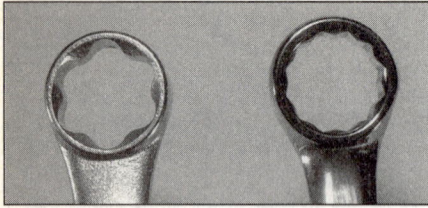

**2.21 Comparison of surface drive ring spanner (left) with 12-point type (right)**

is obtained. Surface drive spanners grip the middle of the hex flats, rather than the corners, and are thus good in cases of damaged heads **(see illustration 2.21)**.

● Slotted-head or Phillips-head screws are often damaged by the use of the wrong size screwdriver. Allen-head and Torx-head screws are much less likely to sustain damage. If enough of the screw head is exposed you can use a hacksaw to cut a slot in its head and then use a conventional flat-bladed screwdriver to remove it. Alternatively use a hammer and cold chisel to tap the head of the fastener round to slacken it. Always replace damaged fasteners with new ones, preferably Torx or Allen-head type.

*A dab of valve grinding compound between the screw head and screwdriver tip will often give a good grip.*

## Thread repair

● Threads (particularly those in aluminium alloy components) can be damaged by overtightening, being assembled with dirt in the threads, or from a component working loose and vibrating. Eventually the thread will fail completely, and it will be impossible to tighten the fastener.

● If a thread is damaged or clogged with old locking compound it can be renovated with a thread repair tool (thread chaser) **(see illustrations 2.22 and 2.23)**; special thread

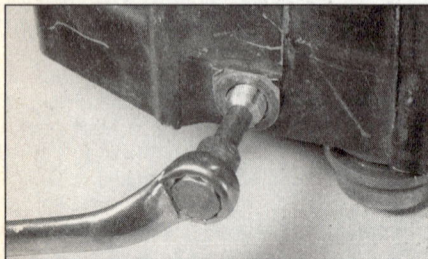

**2.22 A thread repair tool being used to correct an internal thread**

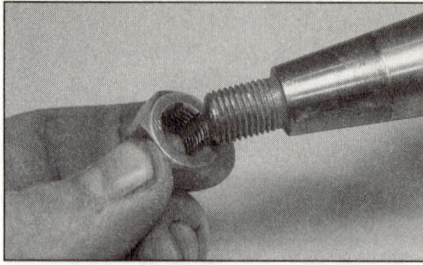

**2.23 A thread repair tool being used to correct an external thread**

chasers are available for spark plug hole threads. The tool will not cut a new thread, but clean and true the original thread. Make sure that you use the correct diameter and pitch tool. Similarly, external threads can be cleaned up with a die or a thread restorer file **(see illustration 2.24)**.

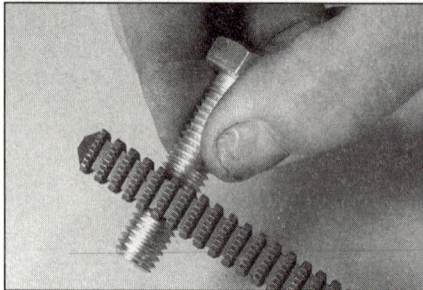

**2.24 Using a thread restorer file**

● It is possible to drill out the old thread and retap the component to the next thread size. This will work where there is enough surrounding material and a new bolt or screw can be obtained. Sometimes, however, this is not possible - such as where the bolt/screw passes through another component which must also be suitably modified, also in cases where a spark plug or oil drain plug cannot be obtained in a larger diameter thread size.

● The diamond-section thread insert (often known by its popular trade name of Heli-Coil) is a simple and effective method of renewing the thread and retaining the original size. A kit can be purchased which contains the tap, insert and installing tool **(see illustration 2.25)**. Drill out the damaged thread with the size drill specified **(see illustration 2.26)**. Carefully retap the thread **(see illustration 2.27)**. Install the

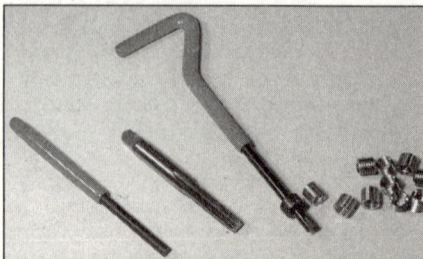

**2.25 Obtain a thread insert kit to suit the thread diameter and pitch required**

**2.26 To install a thread insert, first drill out the original thread . . .**

**2.27 . . . tap a new thread . . .**

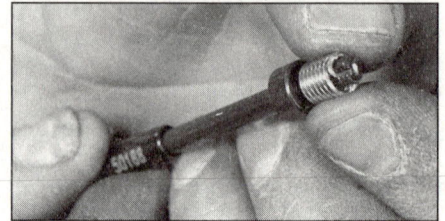

**2.28 . . . fit insert on the installing tool . . .**

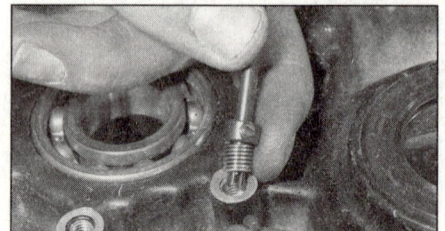

**2.29 . . . and thread into the component . . .**

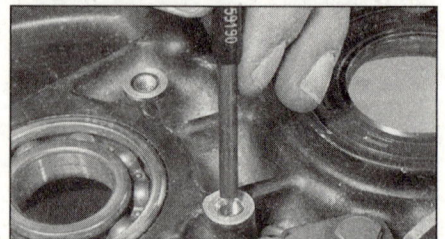

**2.30 . . . break off the tang when complete**

insert on the installing tool and thread it slowly into place using a light downward pressure **(see illustrations 2.28 and 2.29)**. When positioned between a 1/4 and 1/2 turn below the surface withdraw the installing tool and use the break-off tool to press down on the tang, breaking it off **(see illustration 2.30)**.

● There are epoxy thread repair kits on the market which can rebuild stripped internal threads, although this repair should not be used on high load-bearing components.

## Thread locking and sealing compounds

● Locking compounds are used in locations where the fastener is prone to loosening due to vibration or on important safety-related items which might cause loss of control of the motorcycle if they fail. It is also used where important fasteners cannot be secured by other means such as lockwashers or split pins.

● Before applying locking compound, make sure that the threads (internal and external) are clean and dry with all old compound removed. Select a compound to suit the component being secured - a non-permanent general locking and sealing type is suitable for most applications, but a high strength type is needed for permanent fixing of studs in castings. Apply a drop or two of the compound to the first few threads of the fastener, then thread it into place and tighten to the specified torque. Do not apply excessive thread locking compound otherwise the thread may be damaged on subsequent removal.

● Certain fasteners are impregnated with a dry film type coating of locking compound on their threads. Always renew this type of fastener if disturbed.

● Anti-seize compounds, such as copper-based greases, can be applied to protect threads from seizure due to extreme heat and corrosion. A common instance is spark plug threads and exhaust system fasteners.

## 3 Measuring tools and gauges

## Feeler gauges

● Feeler gauges (or blades) are used for measuring small gaps and clearances **(see illustration 3.1)**. They can also be used to measure endfloat (sideplay) of a component on a shaft where access is not possible with a dial gauge.

● Feeler gauge sets should be treated with care and not bent or damaged. They are etched with their size on one face. Keep them clean and very lightly oiled to prevent corrosion build-up.

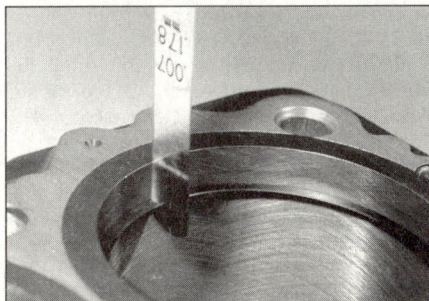

**3.1 Feeler gauges are used for measuring small gaps and clearances - thickness is marked on one face of gauge**

● When measuring a clearance, select a gauge which is a light sliding fit between the two components. You may need to use two gauges together to measure the clearance accurately.

## Micrometers

● A micrometer is a precision tool capable of measuring to 0.01 or 0.001 of a millimetre. It should always be stored in its case and not in the general toolbox. It must be kept clean and never dropped, otherwise its frame or measuring anvils could be distorted resulting in inaccurate readings.

● External micrometers are used for measuring outside diameters of components and have many more applications than internal micrometers. Micrometers are available in different size ranges, eg 0 to 25 mm, 25 to 50 mm, and upwards in 25 mm steps; some large micrometers have interchangeable anvils to allow a range of measurements to be taken. Generally the largest precision measurement you are likely to take on a motorcycle is the piston diameter.

● Internal micrometers (or bore micrometers) are used for measuring inside diameters, such as valve guides and cylinder bores. Telescoping gauges and small hole gauges are used in conjunction with an external micro-meter, whereas the more expensive internal micrometers have their own measuring device.

### External micrometer

**Note:** *The conventional analogue type instrument is described. Although much easier to read, digital micrometers are considerably more expensive.*

● Always check the calibration of the micrometer before use. With the anvils closed (0 to 25 mm type) or set over a test gauge (for

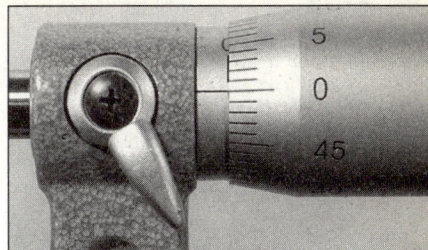

**3.2 Check micrometer calibration before use**

the larger types) the scale should read zero **(see illustration 3.2)**; make sure that the anvils (and test piece) are clean first. Any discrepancy can be adjusted by referring to the instructions supplied with the tool. Remember that the micrometer is a precision measuring tool - don't force the anvils closed, use the ratchet (4) on the end of the micrometer to close it. In this way, a measured force is always applied.

● To use, first make sure that the item being measured is clean. Place the anvil of the micrometer (1) against the item and use the thimble (2) to bring the spindle (3) lightly into contact with the other side of the item **(see illustration 3.3)**. Don't tighten the thimble down because this will damage the micrometer - instead use the ratchet (4) on the end of the micrometer. The ratchet mechanism applies a measured force preventing damage to the instrument.

● The micrometer is read by referring to the linear scale on the sleeve and the annular scale on the thimble. Read off the sleeve first to obtain the base measurement, then add the fine measurement from the thimble to obtain the overall reading. The linear scale on the sleeve represents the measuring range of the micrometer (eg 0 to 25 mm). The annular scale

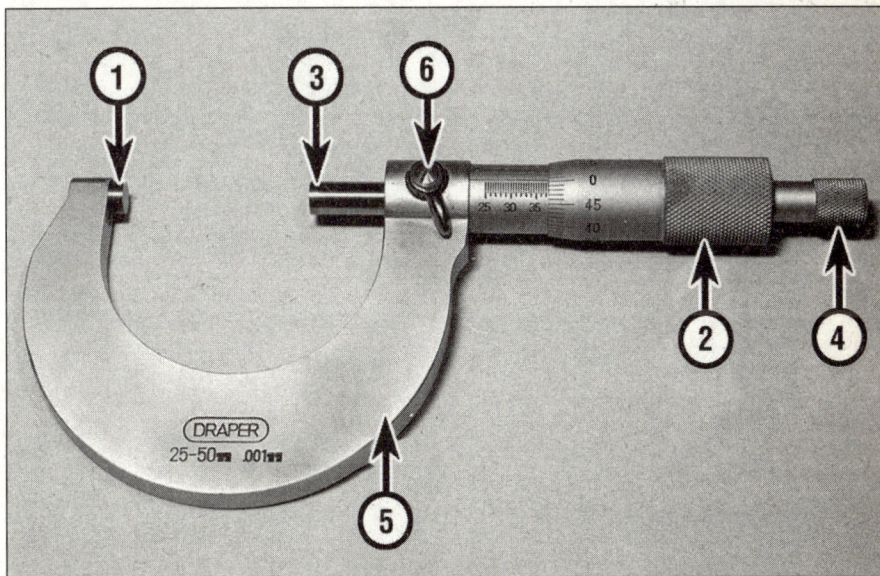

**3.3 Micrometer component parts**

| | | |
|---|---|---|
| 1 Anvil | 3 Spindle | 5 Frame |
| 2 Thimble | 4 Ratchet | 6 Locking lever |

on the thimble will be in graduations of 0.01 mm (or as marked on the frame) - one full revolution of the thimble will move 0.5 mm on the linear scale. Take the reading where the datum line on the sleeve intersects the thimble's scale. Always position the eye directly above the scale otherwise an inaccurate reading will result.

In the example shown the item measures 2.95 mm (see illustration 3.4):

| | |
|---|---|
| Linear scale | 2.00 mm |
| Linear scale | 0.50 mm |
| Annular scale | 0.45 mm |
| Total figure | **2.95 mm** |

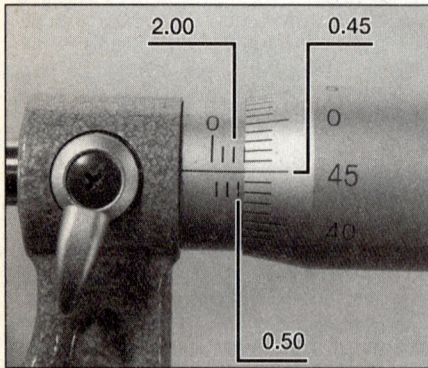

**3.4 Micrometer reading of 2.95 mm**

Most micrometers have a locking lever (6) on the frame to hold the setting in place, allowing the item to be removed from the micrometer.
● Some micrometers have a vernier scale on their sleeve, providing an even finer measurement to be taken, in 0.001 increments of a millimetre. Take the sleeve and thimble measurement as described above, then check which graduation on the vernier scale aligns with that of the annular scale on the thimble **Note**: *The eye must be perpendicular to the scale when taking the vernier reading - if necessary rotate the body of the micrometer to ensure this.* Multiply the vernier scale figure by 0.001 and add it to the base and fine measurement figures.

In the example shown the item measures 46.994 mm (see illustrations 3.5 and 3.6):

| | |
|---|---|
| Linear scale (base) | 46.000 mm |
| Linear scale (base) | 00.500 mm |
| Annular scale (fine) | 00.490 mm |
| Vernier scale | 00.004 mm |
| Total figure | **46.994 mm** |

### Internal micrometer

● Internal micrometers are available for measuring bore diameters, but are expensive and unlikely to be available for home use. It is suggested that a set of telescoping gauges and small hole gauges, both of which must be used with an external micrometer, will suffice for taking internal measurements on a motorcycle.

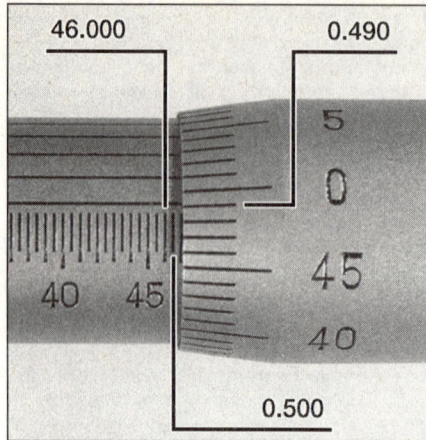

**3.5 Micrometer reading of 46.99 mm on linear and annular scales . . .**

**3.6 . . . and 0.004 mm on vernier scale**

● Telescoping gauges can be used to measure internal diameters of components. Select a gauge with the correct size range, make sure its ends are clean and insert it into the bore. Expand the gauge, then lock its position and withdraw it from the bore (see illustration 3.7). Measure across the gauge ends with a micrometer (see illustration 3.8).

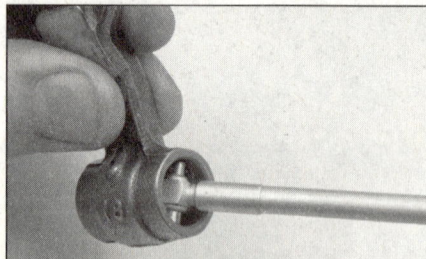

**3.7 Expand the telescoping gauge in the bore, lock its position . . .**

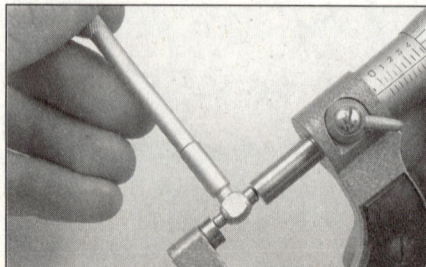

**3.8 . . . then measure the gauge with a micrometer**

**3.9 Expand the small hole gauge in the bore, lock its position . . .**

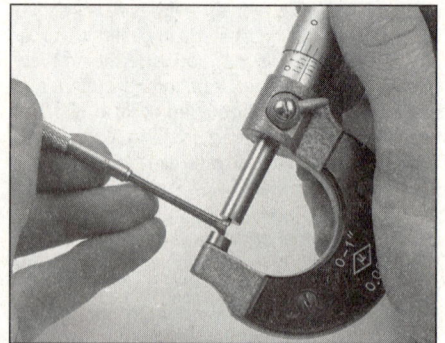

**3.10 . . . then measure the gauge with a micrometer**

● Very small diameter bores (such as valve guides) are measured with a small hole gauge. Once adjusted to a slip-fit inside the component, its position is locked and the gauge withdrawn for measurement with a micrometer (see illustrations 3.9 and 3.10).

### Vernier caliper

**Note**: *The conventional linear and dial gauge type instruments are described. Digital types are easier to read, but are far more expensive.*
● The vernier caliper does not provide the precision of a micrometer, but is versatile in being able to measure internal and external diameters. Some types also incorporate a depth gauge. It is ideal for measuring clutch plate friction material and spring free lengths.
● To use the conventional linear scale vernier, slacken off the vernier clamp screws (1) and set its jaws over (2), or inside (3), the item to be measured (see illustration 3.11). Slide the jaw into contact, using the thumb-wheel (4) for fine movement of the sliding scale (5) then tighten the clamp screws (1). Read off the main scale (6) where the zero on the sliding scale (5) intersects it, taking the whole number to the left of the zero; this provides the base measurement. View along the sliding scale and select the division which lines up exactly with any of the divisions on the main scale, noting that the divisions usually represents 0.02 of a millimetre. Add this fine measurement to the base measurement to obtain the total reading.

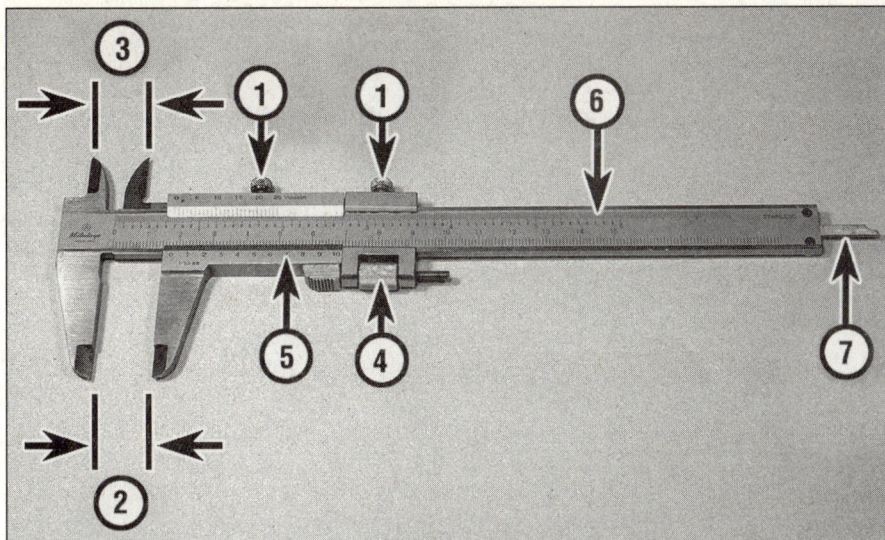

**3.11 Vernier component parts (linear gauge)**

| 1 | Clamp screws | 3 | Internal jaws | 5 | Sliding scale | 7 | Depth gauge |
|---|---|---|---|---|---|---|---|
| 2 | External jaws | 4 | Thumbwheel | 6 | Main scale | | |

In the example shown the item measures 55.92 mm **(see illustration 3.12)**:

| Base measurement | 55.00 mm |
|---|---|
| Fine measurement | 00.92 mm |
| **Total figure** | **55.92 mm** |

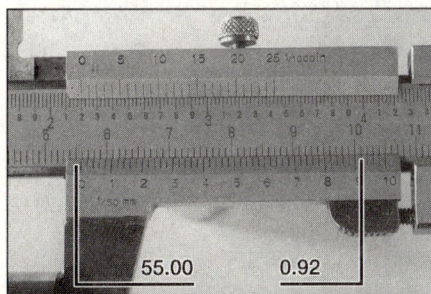

**3.12 Vernier gauge reading of 55.92 mm**

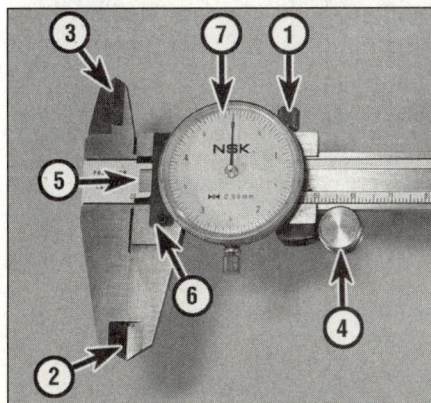

**3.13 Vernier component parts (dial gauge)**

| 1 | Clamp screw | 5 | Main scale |
|---|---|---|---|
| 2 | External jaws | 6 | Sliding scale |
| 3 | Internal jaws | 7 | Dial gauge |
| 4 | Thumbwheel | | |

● Some vernier calipers are equipped with a dial gauge for fine measurement. Before use, check that the jaws are clean, then close them fully and check that the dial gauge reads zero. If necessary adjust the gauge ring accordingly. Slacken the vernier clamp screw (1) and set its jaws over (2), or inside (3), the item to be measured **(see illustration 3.13)**. Slide the jaws into contact, using the thumbwheel (4) for fine movement. Read off the main scale (5) where the edge of the sliding scale (6) intersects it, taking the whole number to the left of the zero; this provides the base measurement. Read off the needle position on the dial gauge (7) scale to provide the fine measurement; each division represents 0.05 of a millimetre. Add this fine measurement to the base measurement to obtain the total reading.

In the example shown the item measures 55.95 mm **(see illustration 3.14)**:

| Base measurement | 55.00 mm |
|---|---|
| Fine measurement | 00.95 mm |
| **Total figure** | **55.95 mm** |

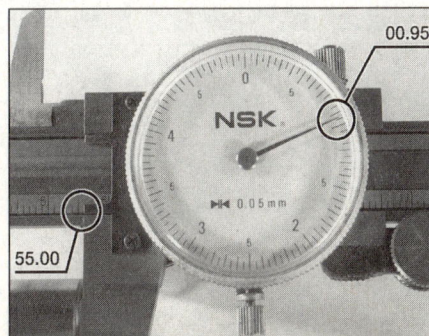

**3.14 Vernier gauge reading of 55.95 mm**

## Plastigauge

● Plastigauge is a plastic material which can be compressed between two surfaces to measure the oil clearance between them. The width of the compressed Plastigauge is measured against a calibrated scale to determine the clearance.

● Common uses of Plastigauge are for measuring the clearance between crankshaft journal and main bearing inserts, between crankshaft journal and big-end bearing inserts, and between camshaft and bearing surfaces. The following example describes big-end oil clearance measurement.

● Handle the Plastigauge material carefully to prevent distortion. Using a sharp knife, cut a length which corresponds with the width of the bearing being measured and place it carefully across the journal so that it is parallel with the shaft **(see illustration 3.15)**. Carefully install both bearing shells and the connecting rod. Without rotating the rod on the journal tighten its bolts or nuts (as applicable) to the specified torque. The connecting rod and bearings are then disassembled and the crushed Plastigauge examined.

**3.15 Plastigauge placed across shaft journal**

● Using the scale provided in the Plastigauge kit, measure the width of the material to determine the oil clearance **(see illustration 3.16)**. Always remove all traces of Plastigauge after use using your fingernails.
*Caution: Arriving at the correct clearance demands that the assembly is torqued correctly, according to the settings and sequence (where applicable) provided by the motorcycle manufacturer.*

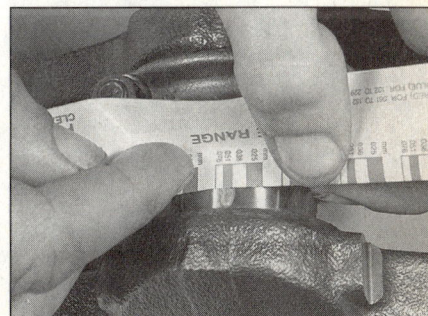

**3.16 Measuring the width of the crushed Plastigauge**

## Dial gauge or DTI (Dial Test Indicator)

● A dial gauge can be used to accurately measure small amounts of movement. Typical uses are measuring shaft runout or shaft endfloat (sideplay) and setting piston position for ignition timing on two-strokes. A dial gauge set usually comes with a range of different probes and adapters and mounting equipment.

● The gauge needle must point to zero when at rest. Rotate the ring around its periphery to zero the gauge.

● Check that the gauge is capable of reading the extent of movement in the work. Most gauges have a small dial set in the face which records whole millimetres of movement as well as the fine scale around the face periphery which is calibrated in 0.01 mm divisions. Read off the small dial first to obtain the base measurement, then add the measurement from the fine scale to obtain the total reading.

In the example shown the gauge reads 1.48 mm (see illustration 3.17):

| | |
|---|---|
| Base measurement | 1.00 mm |
| Fine measurement | 0.48 mm |
| Total figure | **1.48 mm** |

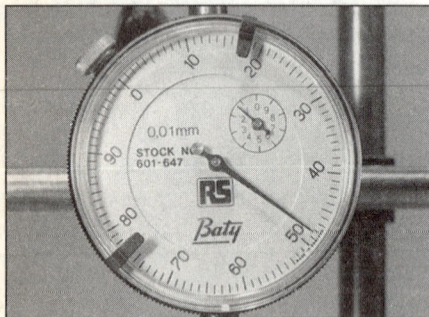

**3.17 Dial gauge reading of 1.48 mm**

● If measuring shaft runout, the shaft must be supported in vee-blocks and the gauge mounted on a stand perpendicular to the shaft. Rest the tip of the gauge against the centre of the shaft and rotate the shaft slowly whilst watching the gauge reading **(see illustration 3.18)**. Take several measurements along the length of the shaft and record the

**3.18 Using a dial gauge to measure shaft runout**

maximum gauge reading as the amount of runout in the shaft. **Note:** *The reading obtained will be total runout at that point - some manufacturers specify that the runout figure is halved to compare with their specified runout limit.*

● Endfloat (sideplay) measurement requires that the gauge is mounted securely to the surrounding component with its probe touching the end of the shaft. Using hand pressure, push and pull on the shaft noting the maximum endfloat recorded on the gauge **(see illustration 3.19)**.

**3.19 Using a dial gauge to measure shaft endfloat**

● A dial gauge with suitable adapters can be used to determine piston position BTDC on two-stroke engines for the purposes of ignition timing. The gauge, adapter and suitable length probe are installed in the place of the spark plug and the gauge zeroed at TDC. If the piston position is specified as 1.14 mm BTDC, rotate the engine back to 2.00 mm BTDC, then slowly forwards to 1.14 mm BTDC.

## Cylinder compression gauges

● A compression gauge is used for measuring cylinder compression. Either the rubber-cone type or the threaded adapter type can be used. The latter is preferred to ensure a perfect seal against the cylinder head. A 0 to 300 psi (0 to 20 Bar) type gauge (for petrol/gasoline engines) will be suitable for motorcycles.

● The spark plug is removed and the gauge either held hard against the cylinder head (cone type) or the gauge adapter screwed into the cylinder head (threaded type) **(see illustration 3.20)**. Cylinder compression is measured with the engine turning over, but not running - carry out the compression test as described in

**3.20 Using a rubber-cone type cylinder compression gauge**

*Fault Finding Equipment.* The gauge will hold the reading until manually released.

## Oil pressure gauge

● An oil pressure gauge is used for measuring engine oil pressure. Most gauges come with a set of adapters to fit the thread of the take-off point **(see illustration 3.21)**. If the take-off point specified by the motorcycle manufacturer is an external oil pipe union, make sure that the specified replacement union is used to prevent oil starvation.

**3.21 Oil pressure gauge and take-off point adapter (arrow)**

● Oil pressure is measured with the engine running (at a specific rpm) and often the manufacturer will specify pressure limits for a cold and hot engine.

## Straight-edge and surface plate

● If checking the gasket face of a component for warpage, place a steel rule or precision straight-edge across the gasket face and measure any gap between the straight-edge and component with feeler gauges **(see illustration 3.22)**. Check diagonally across the component and between mounting holes **(see illustration 3.23)**.

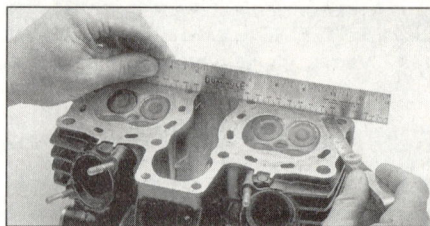

**3.22 Use a straight-edge and feeler gauges to check for warpage**

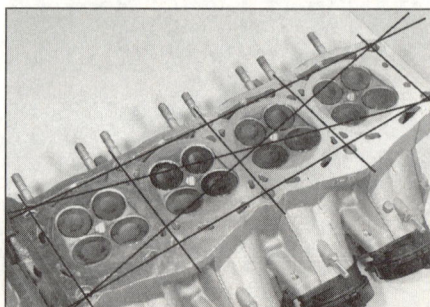

**3.23 Check for warpage in these directions**

● Checking individual components for warpage, such as clutch plain (metal) plates, requires a perfectly flat plate or piece or plate glass and feeler gauges.

## 4  Torque and leverage

### What is torque?

● Torque describes the twisting force about a shaft. The amount of torque applied is determined by the distance from the centre of the shaft to the end of the lever and the amount of force being applied to the end of the lever; distance multiplied by force equals torque.

● The manufacturer applies a measured torque to a bolt or nut to ensure that it will not slacken in use and to hold two components securely together without movement in the joint. The actual torque setting depends on the thread size, bolt or nut material and the composition of the components being held.

● Too little torque may cause the fastener to loosen due to vibration, whereas too much torque will distort the joint faces of the component or cause the fastener to shear off. Always stick to the specified torque setting.

### Using a torque wrench

● Check the calibration of the torque wrench and make sure it has a suitable range for the job. Torque wrenches are available in Nm (Newton-metres), kgf m (kilograms-force metre), lbf ft (pounds-feet), lbf in (inch-pounds). Do not confuse lbf ft with lbf in.

● Adjust the tool to the desired torque on the scale (see illustration 4.1). If your torque wrench is not calibrated in the units specified, carefully convert the figure (see *Conversion Factors*). A manufacturer sometimes gives a torque setting as a range (8 to 10 Nm) rather than a single figure - in this case set the tool midway between the two settings. The same torque may be expressed as 9 Nm ± 1 Nm. Some torque wrenches have a method of locking the setting so that it isn't inadvertently altered during use.

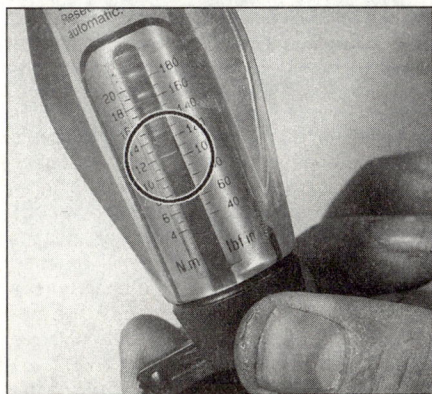

**4.1 Set the torque wrench index mark to the setting required, in this case 12 Nm**

● Install the bolts/nuts in their correct location and secure them lightly. Their threads must be clean and free of any old locking compound. Unless specified the threads and flange should be dry - oiled threads are necessary in certain circumstances and the manufacturer will take this into account in the specified torque figure. Similarly, the manufacturer may also specify the application of thread-locking compound.

● Tighten the fasteners in the specified sequence until the torque wrench clicks, indicating that the torque setting has been reached. Apply the torque again to double-check the setting. Where different thread diameter fasteners secure the component, as a rule tighten the larger diameter ones first.

● When the torque wrench has been finished with, release the lock (where applicable) and fully back off its setting to zero - do not leave the torque wrench tensioned. Also, do not use a torque wrench for slackening a fastener.

### Angle-tightening

● Manufacturers often specify a figure in degrees for final tightening of a fastener. This usually follows tightening to a specific torque setting.

● A degree disc can be set and attached to the socket (see illustration 4.2) or a protractor can be used to mark the angle of movement on the bolt/nut head and the surrounding casting (see illustration 4.3).

**4.2 Angle tightening can be accomplished with a torque-angle gauge . . .**

**4.3 . . . or by marking the angle on the surrounding component**

### Loosening sequences

● Where more than one bolt/nut secures a component, loosen each fastener evenly a little at a time. In this way, not all the stress of the joint is held by one fastener and the components are not likely to distort.

● If a tightening sequence is provided, work in the REVERSE of this, but if not, work from the outside in, in a criss-cross sequence (see illustration 4.4).

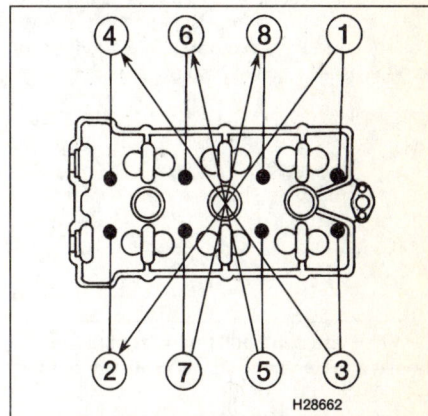

**4.4  When slackening, work from the outside inwards**

### Tightening sequences

● If a component is held by more than one fastener it is important that the retaining bolts/nuts are tightened evenly to prevent uneven stress build-up and distortion of sealing faces. This is especially important on high-compression joints such as the cylinder head.

● A sequence is usually provided by the manufacturer, either in a diagram or actually marked in the casting. If not, always start in the centre and work outwards in a criss-cross pattern (see illustration 4.5). Start off by securing all bolts/nuts finger-tight, then set the torque wrench and tighten each fastener by a small amount in sequence until the final torque is reached. By following this practice,

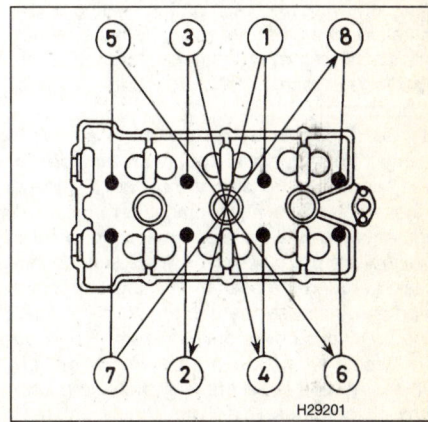

**4.5  When tightening, work from the inside outwards**

the joint will be held evenly and will not be distorted. Important joints, such as the cylinder head and big-end fasteners often have two- or three-stage torque settings.

### Applying leverage

● Use tools at the correct angle. Position a socket wrench or spanner on the bolt/nut so that you pull it towards you when loosening. If this can't be done, push the spanner without curling your fingers around it (see illustration 4.6) - the spanner may slip or the fastener loosen suddenly, resulting in your fingers being crushed against a component.

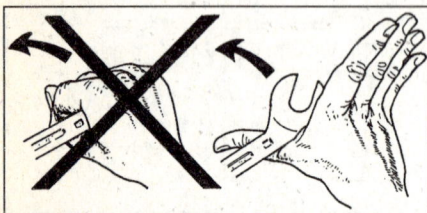

**4.6 If you can't pull on the spanner to loosen a fastener, push with your hand open**

● Additional leverage is gained by extending the length of the lever. The best way to do this is to use a breaker bar instead of the regular length tool, or to slip a length of tubing over the end of the spanner or socket wrench.
● If additional leverage will not work, the fastener head is either damaged or firmly corroded in place (see Fasteners).

## 5 Bearings

### Bearing removal and installation

#### Drivers and sockets

● Before removing a bearing, always inspect the casing to see which way it must be driven out - some casings will have retaining plates or a cast step. Also check for any identifying markings on the bearing and if installed to a certain depth, measure this at this stage. Some roller bearings are sealed on one side - take note of the original fitted position.
● Bearings can be driven out of a casing using a bearing driver tool (with the correct size head) or a socket of the correct diameter. Select the driver head or socket so that it contacts the outer race of the bearing, not the balls/rollers or inner race. Always support the casing around the bearing housing with wood blocks, otherwise there is a risk of fracture. The bearing is driven out with a few blows on the driver or socket from a heavy mallet. Unless access is severely restricted (as with wheel bearings), a pin-punch is not recommended unless it is moved around the bearing to keep it square in its housing.

● The same equipment can be used to install bearings. Make sure the bearing housing is supported on wood blocks and line up the bearing in its housing. Fit the bearing as noted on removal - generally they are installed with their marked side facing outwards. Tap the bearing squarely into its housing using a driver or socket which bears only on the bearing's outer race - contact with the bearing balls/rollers or inner race will destroy it (see illustrations 5.1 and 5.2).
● Check that the bearing inner race and balls/rollers rotate freely.

**5.1 Using a bearing driver against the bearing's outer race**

**5.2 Using a large socket against the bearing's outer race**

#### Pullers and slide-hammers

● Where a bearing is pressed on a shaft a puller will be required to extract it (see illustration 5.3). Make sure that the puller clamp or legs fit securely behind the bearing and are unlikely to slip out. If pulling a bearing

**5.3 This bearing puller clamps behind the bearing and pressure is applied to the shaft end to draw the bearing off**

off a gear shaft for example, you may have to locate the puller behind a gear pinion if there is no access to the race and draw the gear pinion off the shaft as well (see illustration 5.4). *Caution: Ensure that the puller's centre bolt locates securely against the end of the shaft and will not slip when pressure is applied. Also ensure that puller does not damage the shaft end.*

**5.4 Where no access is available to the rear of the bearing, it is sometimes possible to draw off the adjacent component**

● Operate the puller so that its centre bolt exerts pressure on the shaft end and draws the bearing off the shaft.
● When installing the bearing on the shaft, tap only on the bearing's inner race - contact with the balls/rollers or outer race with destroy the bearing. Use a socket or length of tubing as a drift which fits over the shaft end (see illustration 5.5).

**5.5 When installing a bearing on a shaft use a piece of tubing which bears only on the bearing's inner race**

● Where a bearing locates in a blind hole in a casing, it cannot be driven or pulled out as described above. A slide-hammer with knife-edged bearing puller attachment will be required. The puller attachment passes through the bearing and when tightened expands to fit firmly behind the bearing (see illustration 5.6). By operating the slide-hammer part of the tool the bearing is jarred out of its housing (see illustration 5.7).
● It is possible, if the bearing is of reasonable weight, for it to drop out of its housing if the casing is heated as described below. If this

**5.6 Expand the bearing puller so that it locks behind the bearing . . .**

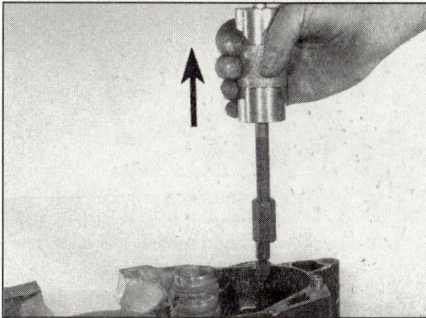

**5.7 . . . attach the slide hammer to the bearing puller**

method is attempted, first prepare a work surface which will enable the casing to be tapped face down to help dislodge the bearing - a wood surface is ideal since it will not damage the casing's gasket surface. Wearing protective gloves, tap the heated casing several times against the work surface to dislodge the bearing under its own weight **(see illustration 5.8)**.

**5.8 Tapping a casing face down on wood blocks can often dislodge a bearing**

● Bearings can be installed in blind holes using the driver or socket method described above.

## Drawbolts

● Where a bearing or bush is set in the eye of a component, such as a suspension linkage arm or connecting rod small-end, removal by drift may damage the component. Furthermore, a rubber bushing in a shock absorber eye cannot successfully be driven out of position. If access is available to a engineering press, the task is straightforward. If not, a drawbolt can be fabricated to extract the bearing or bush.

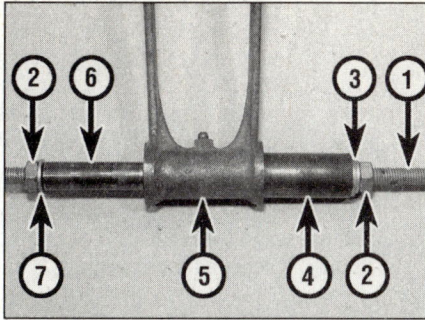

**5.9 Drawbolt component parts assembled on a suspension arm**

1  Bolt or length of threaded bar
2  Nuts
3  Washer (external diameter greater than tubing internal diameter)
4  Tubing (internal diameter sufficient to accommodate bearing)
5  Suspension arm with bearing
6  Tubing (external diameter slightly smaller than bearing)
7  Washer (external diameter slightly smaller than bearing)

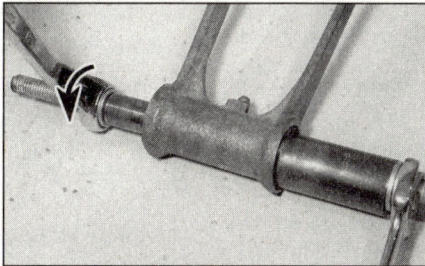

**5.10 Drawing the bearing out of the suspension arm**

● To extract the bearing/bush you will need a long bolt with nut (or piece of threaded bar with two nuts), a piece of tubing which has an internal diameter larger than the bearing/bush, another piece of tubing which has an external diameter slightly smaller than the bearing/bush, and a selection of washers **(see illustrations 5.9 and 5.10)**. Note that the pieces of tubing must be of the same length, or longer, than the bearing/bush.
● The same kit (without the pieces of tubing) can be used to draw the new bearing/bush back into place **(see illustration 5.11)**.

**5.11 Installing a new bearing (1) in the suspension arm**

## Temperature change

● If the bearing's outer race is a tight fit in the casing, the aluminium casing can be heated to release its grip on the bearing. Aluminium will expand at a greater rate than the steel bearing outer race. There are several ways to do this, but avoid any localised extreme heat (such as a blow torch) - aluminium alloy has a low melting point.
● Approved methods of heating a casing are using a domestic oven (heated to 100°C) or immersing the casing in boiling water **(see illustration 5.12)**. Low temperature range localised heat sources such as a paint stripper heat gun or clothes iron can also be used **(see illustration 5.13)**. Alternatively, soak a rag in boiling water, wring it out and wrap it around the bearing housing.

⚠ **Warning: All of these methods require care in use to prevent scalding and burns to the hands. Wear protective gloves when handling hot components.**

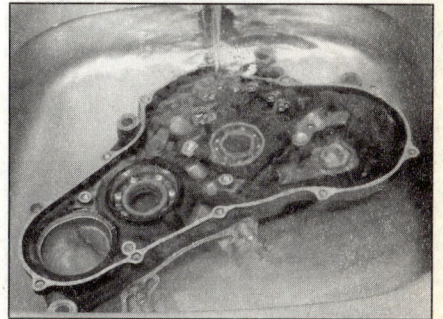

**5.12 A casing can be immersed in a sink of boiling water to aid bearing removal**

**5.13 Using a localised heat source to aid bearing removal**

● If heating the whole casing note that plastic components, such as the neutral switch, may suffer - remove them beforehand.
● After heating, remove the bearing as described above. You may find that the expansion is sufficient for the bearing to fall out of the casing under its own weight or with a light tap on the driver or socket.
● If necessary, the casing can be heated to aid bearing installation, and this is sometimes the recommended procedure if the motorcycle manufacturer has designed the housing and bearing fit with this intention.

● Installation of bearings can be eased by placing them in a freezer the night before installation. The steel bearing will contract slightly, allowing easy insertion in its housing. This is often useful when installing steering head outer races in the frame.

## Bearing types and markings

● Plain shell bearings, ball bearings, needle roller bearings and tapered roller bearings will all be found on motorcycles (see illustrations 5.14 and 5.15). The ball and roller types are usually caged between an inner and outer race, but uncaged variations may be found.

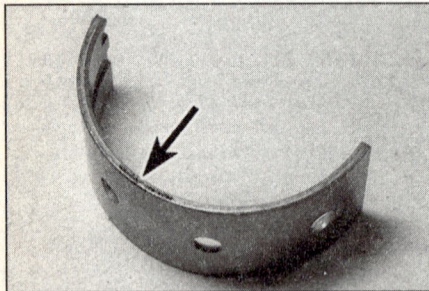

**5.14 Shell bearings are either plain or grooved. They are usually identified by colour code (arrow)**

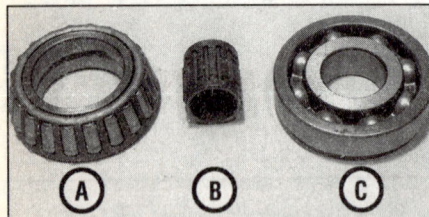

**5.15 Tapered roller bearing (A), needle roller bearing (B) and ball journal bearing (C)**

● Shell bearings (often called inserts) are usually found at the crankshaft main and connecting rod big-end where they are good at coping with high loads. They are made of a phosphor-bronze material and are impregnated with self-lubricating properties.

● Ball bearings and needle roller bearings consist of a steel inner and outer race with the balls or rollers between the races. They require constant lubrication by oil or grease and are good at coping with axial loads. Taper roller bearings consist of rollers set in a tapered cage set on the inner race; the outer race is separate. They are good at coping with axial loads and prevent movement along the shaft - a typical application is in the steering head.

● Bearing manufacturers produce bearings to ISO size standards and stamp one face of the bearing to indicate its internal and external diameter, load capacity and type (see illustration 5.16).

● Metal bushes are usually of phosphor-bronze material. Rubber bushes are used in suspension mounting eyes. Fibre bushes have also been used in suspension pivots.

**5.16 Typical bearing marking**

## Bearing fault finding

● If a bearing outer race has spun in its housing, the housing material will be damaged. You can use a bearing locking compound to bond the outer race in place if damage is not too severe.

● Shell bearings will fail due to damage of their working surface, as a result of lack of lubrication, corrosion or abrasive particles in the oil (see illustration 5.17). Small particles of dirt in the oil may embed in the bearing material whereas larger particles will score the bearing and shaft journal. If a number of short journeys are made, insufficient heat will be generated to drive off condensation which has built up on the bearings.

**5.17 Typical bearing failures**

● Ball and roller bearings will fail due to lack of lubrication or damage to the balls or rollers. Tapered-roller bearings can be damaged by overloading them. Unless the bearing is sealed on both sides, wash it in paraffin (kerosene) to remove all old grease then allow it to dry. Make a visual inspection looking to dented balls or rollers, damaged cages and worn or pitted races (see illustration 5.18).

● A ball bearing can be checked for wear by listening to it when spun. Apply a film of light oil to the bearing and hold it close to the ear - hold the outer race with one hand and spin the inner

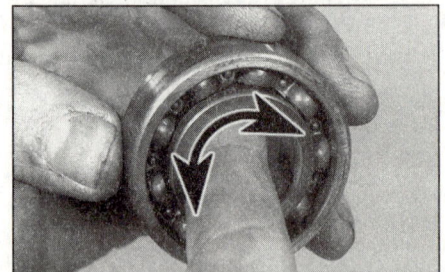

**5.18 Example of ball journal bearing with damaged balls and cages**

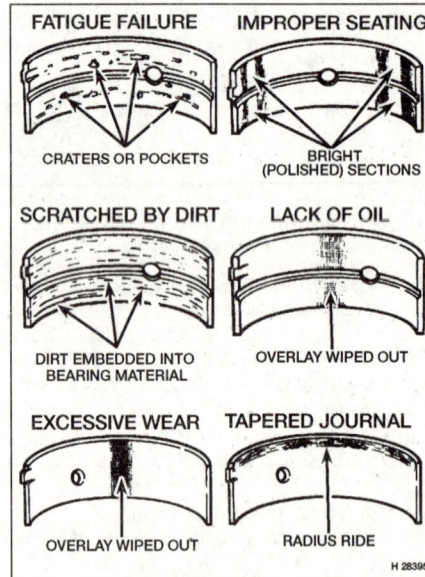

**5.19 Hold outer race and listen to inner race when spun**

race with the other hand (see illustration 5.19). The bearing should be almost silent when spun; if it grates or rattles it is worn.

## 6  Oil seals

## Oil seal removal and installation

● Oil seals should be renewed every time a component is dismantled. This is because the seal lips will become set to the sealing surface and will not necessarily reseal.

● Oil seals can be prised out of position using a large flat-bladed screwdriver (see illustration 6.1). In the case of crankcase seals, check first that the seal is not lipped on the inside, preventing its removal with the crankcases joined.

**6.1 Prise out oil seals with a large flat-bladed screwdriver**

● New seals are usually installed with their marked face (containing the seal reference code) outwards and the spring side towards the fluid being retained. In certain cases, such as a two-stroke engine crankshaft seal, a double lipped seal may be used due to there being fluid or gas on each side of the joint.

● Use a bearing driver or socket which bears only on the outer hard edge of the seal to install it in the casing - tapping on the inner edge will damage the sealing lip.

## Oil seal types and markings

● Oil seals are usually of the single-lipped type. Double-lipped seals are found where a liquid or gas is on both sides of the joint.
● Oil seals can harden and lose their sealing ability if the motorcycle has been in storage for a long period - renewal is the only solution.
● Oil seal manufacturers also conform to the ISO markings for seal size - these are moulded into the outer face of the seal (see illustration 6.2).

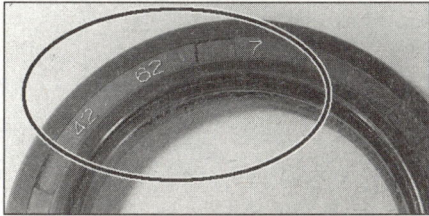

6.2 These oil seal markings indicate inside diameter, outside diameter and seal thickness

## 7 Gaskets and sealants

### Types of gasket and sealant

● Gaskets are used to seal the mating surfaces between components and keep lubricants, fluids, vacuum or pressure contained within the assembly. Aluminium gaskets are sometimes found at the cylinder joints, but most gaskets are paper-based. If the mating surfaces of the components being joined are undamaged the gasket can be installed dry, although a dab of sealant or grease will be useful to hold it in place during assembly.
● RTV (Room Temperature Vulcanising) silicone rubber sealants cure when exposed to moisture in the atmosphere. These sealants are good at filling pits or irregular gasket faces, but will tend to be forced out of the joint under very high torque. They can be used to replace a paper gasket, but first make sure that the width of the paper gasket is not essential to the shimming of internal components. RTV sealants should not be used on components containing petrol (gasoline).
● Non-hardening, semi-hardening and hard setting liquid gasket compounds can be used with a gasket or between a metal-to-metal joint. Select the sealant to suit the application: universal non-hardening sealant can be used on virtually all joints; semi-hardening on joint faces which are rough or damaged; hard setting sealant on joints which require a permanent bond and are subjected to high temperature and pressure. Note: Check first if

the paper gasket has a bead of sealant impregnated in its surface before applying additional sealant.
● When choosing a sealant, make sure it is suitable for the application, particularly if being applied in a high-temperature area or in the vicinity of fuel. Certain manufacturers produce sealants in either clear, silver or black colours to match the finish of the engine. This has a particular application on motorcycles where much of the engine is exposed.
● Do not over-apply sealant. That which is squeezed out on the outside of the joint can be wiped off, whereas an excess of sealant on the inside can break off and clog oilways.

### Breaking a sealed joint

● Age, heat, pressure and the use of hard setting sealant can cause two components to stick together so tightly that they are difficult to separate using finger pressure alone. Do not resort to using levers unless there is a pry point provided for this purpose (see illustration 7.1) or else the gasket surfaces will be damaged.
● Use a soft-faced hammer (see illustration 7.2) or a wood block and conventional hammer to strike the component near the mating surface. Avoid hammering against cast extremities since they may break off. If this method fails, try using a wood wedge between the two components.
Caution: If the joint will not separate, double-check that you have removed all the fasteners.

7.1 If a pry point is provided, apply gently pressure with a flat-bladed screwdriver

7.2 Tap around the joint with a soft-faced mallet if necessary - don't strike cooling fins

### Removal of old gasket and sealant

● Paper gaskets will most likely come away complete, leaving only a few traces stuck on

Most components have one or two hollow locating dowels between the two gasket faces. If a dowel cannot be removed, do not resort to gripping it with pliers - it will almost certainly be distorted. Install a close-fitting socket or Phillips screwdriver into the dowel and then grip the outer edge of the dowel to free it.

the sealing faces of the components. It is imperative that all traces are removed to ensure correct sealing of the new gasket.
● Very carefully scrape all traces of gasket away making sure that the sealing surfaces are not gouged or scored by the scraper (see illustrations 7.3, 7.4 and 7.5). Stubborn deposits can be removed by spraying with an aerosol gasket remover. Final preparation of

7.3 Paper gaskets can be scraped off with a gasket scraper tool . . .

7.4 . . . a knife blade . . .

7.5 . . . or a household scraper

7.6 Fine abrasive paper is wrapped around a flat file to clean up the gasket face

7.7 A kitchen scourer can be used on stubborn deposits

the gasket surface can be made with very fine abrasive paper or a plastic kitchen scourer **(see illustrations 7.6 and 7.7)**.

● Old sealant can be scraped or peeled off components, depending on the type originally used. Note that gasket removal compounds are available to avoid scraping the components clean; make sure the gasket remover suits the type of sealant used.

## 8  Chains

### Breaking and joining final drive chains

● Drive chains for all but small bikes are continuous and do not have a clip-type connecting link. The chain must be broken using a chain breaker tool and the new chain securely riveted together using a new soft rivet-type link. Never use a clip-type connecting link instead of a rivet-type link, except in an emergency. Various chain breaking and riveting tools are available, either as separate tools or combined as illustrated in the accompanying photographs - read the instructions supplied with the tool carefully.

⚠️ *Warning: The need to rivet the new link pins correctly cannot be overstressed - loss of control of the motorcycle is very likely to result if the chain breaks in use.*

● Rotate the chain and look for the soft link. The soft link pins look like they have been deeply centre-punched instead of peened over

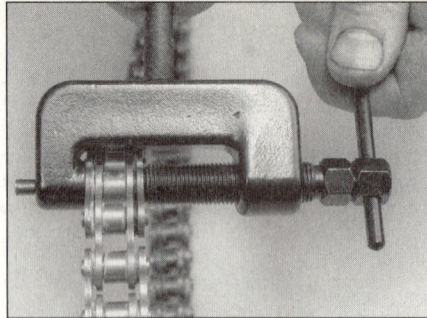

8.1 Tighten the chain breaker to push the pin out of the link . . .

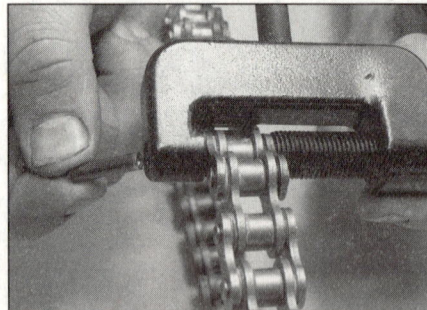

8.2 . . . withdraw the pin, remove the tool . . .

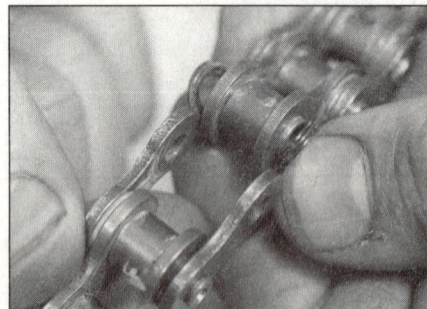

8.3 . . . and separate the chain link

like all the other pins **(see illustration 8.9)** and its sideplate may be a different colour. Position the soft link midway between the sprockets and assemble the chain breaker tool over one of the soft link pins **(see illustration 8.1)**. Operate the tool to push the pin out through the chain **(see illustration 8.2)**. On an O-ring chain, remove the O-rings **(see illustration 8.3)**. Carry out the same procedure on the other soft link pin.

*Caution: Certain soft link pins (particularly on the larger chains) may require their ends to be filed or ground off before they can be pressed out using the tool.*

● Check that you have the correct size and strength (standard or heavy duty) new soft link - do not reuse the old link. Look for the size marking on the chain sideplates **(see illustration 8.10)**.

● Position the chain ends so that they are engaged over the rear sprocket. On an O-ring chain, install a new O-ring over each pin of the link and insert the link through the two chain

8.4 Insert the new soft link, with O-rings, through the chain ends . . .

8.5 . . . install the O-rings over the pin ends . . .

8.6 . . . followed by the sideplate

ends **(see illustration 8.4)**. Install a new O-ring over the end of each pin, followed by the sideplate (with the chain manufacturer's marking facing outwards) **(see illustrations 8.5 and 8.6)**. On an unsealed chain, insert the link through the two chain ends, then install the sideplate with the chain manufacturer's marking facing outwards.

● Note that it may not be possible to install the sideplate using finger pressure alone. If using a joining tool, assemble it so that the plates of the tool clamp the link and press the sideplate over the pins **(see illustration 8.7)**. Otherwise, use two small sockets placed over

8.7 Push the sideplate into position using a clamp

**8.8 Assemble the chain riveting tool over one pin at a time and tighten it fully**

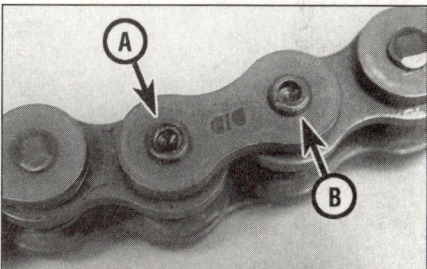

**8.9 Pin end correctly riveted (A), pin end unriveted (B)**

the rivet ends and two pieces of the wood between a G-clamp. Operate the clamp to press the sideplate over the pins.

● Assemble the joining tool over one pin (following the maker's instructions) and tighten the tool down to spread the pin end securely **(see illustrations 8.8 and 8.9)**. Do the same on the other pin.

⚠ *Warning: Check that the pin ends are secure and that there is no danger of the sideplate coming loose. If the pin ends are cracked the soft link must be renewed.*

### Final drive chain sizing

● Chains are sized using a three digit number, followed by a suffix to denote the chain type **(see illustration 8.10)**. Chain type is either standard or heavy duty (thicker sideplates), and also unsealed or O-ring/X-ring type.

● The first digit of the number relates to the pitch of the chain, ie the distance from the centre of one pin to the centre of the next pin **(see illustration 8.11)**. Pitch is expressed in eighths of an inch, as follows:

**8.10 Typical chain size and type marking**

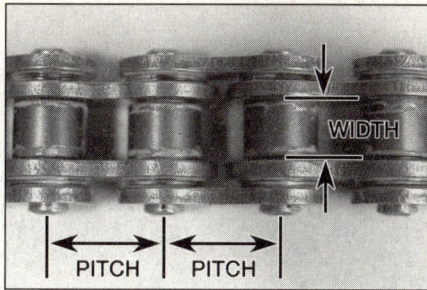

**8.11 Chain dimensions**

*Sizes commencing with a 4 (eg 428) have a pitch of 1/2 inch (12.7 mm)*

*Sizes commencing with a 5 (eg 520) have a pitch of 5/8 inch (15.9 mm)*

*Sizes commencing with a 6 (eg 630) have a pitch of 3/4 inch (19.1 mm)*

● The second and third digits of the chain size relate to the width of the rollers, again in imperial units, eg the 525 shown has 5/16 inch (7.94 mm) rollers **(see illustration 8.11)**.

## 9  Hoses

### Clamping to prevent flow

● Small-bore flexible hoses can be clamped to prevent fluid flow whilst a component is worked on. Whichever method is used, ensure that the hose material is not permanently distorted or damaged by the clamp.

a) A brake hose clamp available from auto accessory shops **(see illustration 9.1)**.
b) A wingnut type hose clamp **(see illustration 9.2)**.

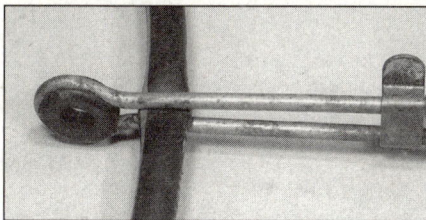

**9.1 Hoses can be clamped with an automotive brake hose clamp . . .**

**9.2 . . . a wingnut type hose clamp . . .**

c) Two sockets placed each side of the hose and held with straight-jawed self-locking grips **(see illustration 9.3)**.
d) Thick card each side of the hose held between straight-jawed self-locking grips **(see illustration 9.4)**.

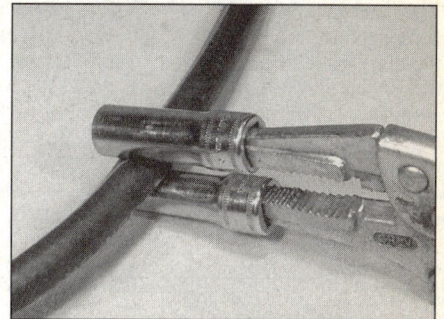

**9.3 . . . two sockets and a pair of self-locking grips . . .**

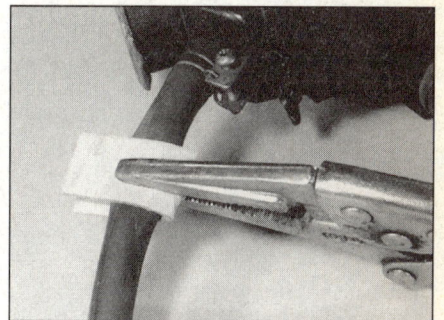

**9.4 . . . or thick card and self-locking grips**

### Freeing and fitting hoses

● Always make sure the hose clamp is moved well clear of the hose end. Grip the hose with your hand and rotate it whilst pulling it off the union. If the hose has hardened due to age and will not move, slit it with a sharp knife and peel its ends off the union **(see illustration 9.5)**.

● Resist the temptation to use grease or soap on the unions to aid installation; although it helps the hose slip over the union it will equally aid the escape of fluid from the joint. It is preferable to soften the hose ends in hot water and wet the inside surface of the hose with water or a fluid which will evaporate.

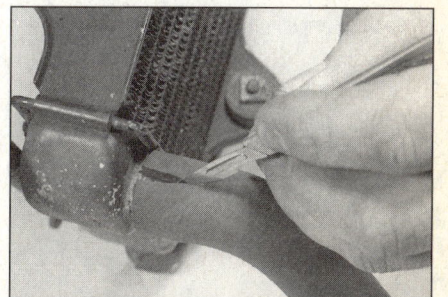

**9.5 Cutting a coolant hose free with a sharp knife**

# Conversion Factors

## Length (distance)

| | | | | | |
|---|---|---|---|---|---|
| Inches (in) | x 25.4 | = Millimetres (mm) | x 0.0394 | = | Inches (in) |
| Feet (ft) | x 0.305 | = Metres (m) | x 3.281 | = | Feet (ft) |
| Miles | x 1.609 | = Kilometres (km) | x 0.621 | = | Miles |

## Volume (capacity)

| | | | | | |
|---|---|---|---|---|---|
| Cubic inches (cu in; in³) | x 16.387 | = Cubic centimetres (cc; cm³) | x 0.061 | = | Cubic inches (cu in; in³) |
| Imperial pints (Imp pt) | x 0.568 | = Litres (l) | x 1.76 | = | Imperial pints (Imp pt) |
| Imperial quarts (Imp qt) | x 1.137 | = Litres (l) | x 0.88 | = | Imperial quarts (Imp qt) |
| Imperial quarts (Imp qt) | x 1.201 | = US quarts (US qt) | x 0.833 | = | Imperial quarts (Imp qt) |
| US quarts (US qt) | x 0.946 | = Litres (l) | x 1.057 | = | US quarts (US qt) |
| Imperial gallons (Imp gal) | x 4.546 | = Litres (l) | x 0.22 | = | Imperial gallons (Imp gal) |
| Imperial gallons (Imp gal) | x 1.201 | = US gallons (US gal) | x 0.833 | = | Imperial gallons (Imp gal) |
| US gallons (US gal) | x 3.785 | = Litres (l) | x 0.264 | = | US gallons (US gal) |

## Mass (weight)

| | | | | | |
|---|---|---|---|---|---|
| Ounces (oz) | x 28.35 | = Grams (g) | x 0.035 | = | Ounces (oz) |
| Pounds (lb) | x 0.454 | = Kilograms (kg) | x 2.205 | = | Pounds (lb) |

## Force

| | | | | | |
|---|---|---|---|---|---|
| Ounces-force (ozf; oz) | x 0.278 | = Newtons (N) | x 3.6 | = | Ounces-force (ozf; oz) |
| Pounds-force (lbf; lb) | x 4.448 | = Newtons (N) | x 0.225 | = | Pounds-force (lbf; lb) |
| Newtons (N) | x 0.1 | = Kilograms-force (kgf; kg) | x 9.81 | = | Newtons (N) |

## Pressure

| | | | | | |
|---|---|---|---|---|---|
| Pounds-force per square inch (psi; lbf/in²; lb/in²) | x 0.070 | = Kilograms-force per square centimetre (kgf/cm²; kg/cm²) | x 14.223 | = | Pounds-force per square inch (psi; lbf/in²; lb/in²) |
| Pounds-force per square inch (psi; lbf/in²; lb/in²) | x 0.068 | = Atmospheres (atm) | x 14.696 | = | Pounds-force per square inch (psi; lbf/in²; lb/in²) |
| Pounds-force per square inch (psi; lbf/in²; lb/in²) | x 0.069 | = Bars | x 14.5 | = | Pounds-force per square inch (psi; lbf/in²; lb/in²) |
| Pounds-force per square inch (psi; lbf/in²; lb/in²) | x 6.895 | = Kilopascals (kPa) | x 0.145 | = | Pounds-force per square inch (psi; lbf/in²; lb/in²) |
| Kilopascals (kPa) | x 0.01 | = Kilograms-force per square centimetre (kgf/cm²; kg/cm²) | x 98.1 | = | Kilopascals (kPa) |
| Millibar (mbar) | x 100 | = Pascals (Pa) | x 0.01 | = | Millibar (mbar) |
| Millibar (mbar) | x 0.0145 | = Pounds-force per square inch (psi; lbf/in²; lb/in²) | x 68.947 | = | Millibar (mbar) |
| Millibar (mbar) | x 0.75 | = Millimetres of mercury (mmHg) | x 1.333 | = | Millibar (mbar) |
| Millibar (mbar) | x 0.401 | = Inches of water (inH₂O) | x 2.491 | = | Millibar (mbar) |
| Millimetres of mercury (mmHg) | x 0.535 | = Inches of water (inH₂O) | x 1.868 | = | Millimetres of mercury (mmHg) |
| Inches of water (inH₂O) | x 0.036 | = Pounds-force per square inch (psi; lbf/in²; lb/in²) | x 27.68 | = | Inches of water (inH₂O) |

## Torque (moment of force)

| | | | | | |
|---|---|---|---|---|---|
| Pounds-force inches (lbf in; lb in) | x 1.152 | = Kilograms-force centimetre (kgf cm; kg cm) | x 0.868 | = | Pounds-force inches (lbf in; lb in) |
| Pounds-force inches (lbf in; lb in) | x 0.113 | = Newton metres (Nm) | x 8.85 | = | Pounds-force inches (lbf in; lb in) |
| Pounds-force inches (lbf in; lb in) | x 0.083 | = Pounds-force feet (lbf ft; lb ft) | x 12 | = | Pounds-force inches (lbf in; lb in) |
| Pounds-force feet (lbf ft; lb ft) | x 0.138 | = Kilograms-force metres (kgf m; kg m) | x 7.233 | = | Pounds-force feet (lbf ft; lb ft) |
| Pounds-force feet (lbf ft; lb ft) | x 1.356 | = Newton metres (Nm) | x 0.738 | = | Pounds-force feet (lbf ft; lb ft) |
| Newton metres (Nm) | x 0.102 | = Kilograms-force metres (kgf m; kg m) | x 9.804 | = | Newton metres (Nm) |

## Power

| | | | | | |
|---|---|---|---|---|---|
| Horsepower (hp) | x 745.7 | = Watts (W) | x 0.0013 | = | Horsepower (hp) |

## Velocity (speed)

| | | | | | |
|---|---|---|---|---|---|
| Miles per hour (miles/hr; mph) | x 1.609 | = Kilometres per hour (km/hr; kph) | x 0.621 | = | Miles per hour (miles/hr; mph) |

## Fuel consumption*

| | | | | | |
|---|---|---|---|---|---|
| Miles per gallon (mpg) | x 0.354 | = Kilometres per litre (km/l) | x 2.825 | = | Miles per gallon (mpg) |

## Temperature

Degrees Fahrenheit = (°C x 1.8) + 32

Degrees Celsius (Degrees Centigrade; °C) = (°F - 32) x 0.56

*It is common practice to convert from miles per gallon (mpg) to litres/100 kilometres (l/100km), where mpg x l/100 km = 282*

A number of chemicals and lubricants are available for use in motorcycle maintenance and repair. They include a wide variety of products ranging from cleaning solvents and degreasers to lubricants and protective sprays for rubber, plastic and vinyl.

● **Contact point/spark plug cleaner** is a solvent used to clean oily film and dirt from points, grime from electrical connectors and oil deposits from spark plugs. It is oil free and leaves no residue. It can also be used to remove gum and varnish from carburettor jets and other orifices.

● **Carburettor cleaner** is similar to contact point/spark plug cleaner but it usually has a stronger solvent and may leave a slight oily reside. It is not recommended for cleaning electrical components or connections.

● **Brake system cleaner** is used to remove grease or brake fluid from brake system components (where clean surfaces are absolutely necessary and petroleum-based solvents cannot be used); it also leaves no residue.

● **Silicone-based lubricants** are used to protect rubber parts such as hoses and grommets, and are used as lubricants for hinges and locks.

● **Multi-purpose grease** is an all purpose lubricant used wherever grease is more practical than a liquid lubricant such as oil. Some multi-purpose grease is coloured white and specially formulated to be more resistant to water than ordinary grease.

● **Gear oil** (sometimes called gear lube) is a specially designed oil used in transmissions and final drive units, as well as other areas where high friction, high temperature lubrication is required. It is available in a number of viscosities (weights) for various applications.

● **Motor oil**, of course, is the lubricant specially formulated for use in the engine. It normally contains a wide variety of additives to prevent corrosion and reduce foaming and wear. Motor oil comes in various weights (viscosity ratings) of from 5 to 80. The recommended weight of the oil depends on the seasonal temperature and the demands on the engine. Light oil is used in cold climates and under light load conditions; heavy oil is used in hot climates and where high loads are encountered. Multi-viscosity oils are designed to have characteristics of both light and heavy oils and are available in a number of weights from 5W-20 to 20W-50.

● **Petrol additives** perform several functions, depending on their chemical makeup. They usually contain solvents that help dissolve gum and varnish that build up on carburettor and inlet parts. They also serve to break down carbon deposits that form on the inside surfaces of the combustion chambers. Some additives contain upper cylinder lubricants for valves and piston rings.

● **Brake and clutch fluid** is a specially formulated hydraulic fluid that can withstand the heat and pressure encountered in brake/clutch systems. Care must be taken that this fluid does not come in contact with painted surfaces or plastics. An opened container should always be resealed to prevent contamination by water or dirt.

● **Chain lubricants** are formulated especially for use on motorcycle final drive chains. A good chain lube should adhere well and have good penetrating qualities to be effective as a lubricant inside the chain and on the side plates, pins and rollers. Most chain lubes are either the foaming type or quick drying type and are usually marketed as sprays. Take care to use a lubricant marked as being suitable for O-ring chains.

● **Degreasers** are heavy duty solvents used to remove grease and grime that may accumulate on engine and frame components. They can be sprayed or brushed on and, depending on the type, are rinsed with either water or solvent.

● **Solvents** are used alone or in combination with degreasers to clean parts and assemblies during repair and overhaul. The home mechanic should use only solvents that are non-flammable and that do not produce irritating fumes.

● **Gasket sealing compounds** may be used in conjunction with gaskets, to improve their sealing capabilities, or alone, to seal metal-to-metal joints. Many gasket sealers can withstand extreme heat, some are impervious to petrol and lubricants, while others are capable of filling and sealing large cavities. Depending on the intended use, gasket sealers either dry hard or stay relatively soft and pliable. They are usually applied by hand, with a brush, or are sprayed on the gasket sealing surfaces.

● **Thread locking compound** is an adhesive locking compound that prevents threaded fasteners from loosening because of vibration. It is available in a variety of types for different applications.

● **Moisture dispersants** are usually sprays that can be used to dry out electrical components such as the fuse block and wiring connectors. Some types can also be used as treatment for rubber and as a lubricant for hinges, cables and locks.

● **Waxes and polishes** are used to help protect painted and plated surfaces from the weather. Different types of paint may require the use of different types of wax polish. Some polishes utilise a chemical or abrasive cleaner to help remove the top layer of oxidised (dull) paint on older vehicles. In recent years, many non-wax polishes (that contain a wide variety of chemicals such as polymers and silicones) have been introduced. These non-wax polishes are usually easier to apply and last longer than conventional waxes and polishes.

## About the MOT Test

In the UK, all vehicles more than three years old are subject to an annual test to ensure that they meet minimum safety requirements. A current test certificate must be issued before a machine can be used on public roads, and is required before a road fund licence can be issued. Riding without a current test certificate will also invalidate your insurance.

For most owners, the MOT test is an annual cause for anxiety, and this is largely due to owners not being sure what needs to be checked prior to submitting the motorcycle for testing. The simple answer is that a fully roadworthy motorcycle will have no difficulty in passing the test.

This is a guide to getting your motorcycle through the MOT test. Obviously it will not be possible to examine the motorcycle to the same standard as the professional MOT

tester, particularly in view of the equipment required for some of the checks. However, working through the following procedures will enable you to identify any problem areas before submitting the motorcycle for the test.

It has only been possible to summarise the test requirements here, based on the regulations in force at the time of printing. Test standards are becoming increasingly stringent, although there are some exemptions for older vehicles. More information about the MOT test can be obtained from the HMSO publications, *How Safe is your Motorcycle* and *The MOT Inspection Manual for Motorcycle Testing*.

Many of the checks require that one of the wheels is raised off the ground. If the motorcycle doesn't have a centre stand, note that an auxiliary stand will be required. Additionally, the help of an assistant may prove useful.

Certain exceptions apply to machines under 50 cc, machines without a lighting system, and Classic bikes - if in doubt about any of the requirements listed below seek confirmation from an MOT tester prior to submitting the motorcycle for the test.

Check that the frame number is clearly visible.

> **HAYNES HiNT** *If a component is in borderline condition, the tester has discretion in deciding whether to pass or fail it. If the motorcycle presented is clean and evidently well cared for, the tester may be more inclined to pass a borderline component than if the motorcycle is scruffy and apparently neglected.*

# Electrical System

## Lights, turn signals, horn and reflector

✔ With the ignition on, check the operation of the following electrical components. **Note:** *The electrical components on certain small-capacity machines are powered by the generator, requiring that the engine is run for this check.*

a) *Headlight and tail light. Check that both illuminate in the low and high beam switch positions.*
b) *Position lights. Check that the front position (or sidelight) and tail light illuminate in this switch position.*
c) *Turn signals. Check that all flash at the correct rate, and that the warning light(s) function correctly. Check that the turn signal switch works correctly.*
c) *Hazard warning system (where fitted). Check that all four turn signals flash in this switch position.*
d) *Brake stop light. Check that the light comes on when the front and rear brakes are independently applied. Models first used on or after 1st April 1986 must have a brake light switch on each brake.*
e) *Horn. Check that the sound is continuous and of reasonable volume.*

✔ Check that there is a red reflector on the rear of the machine, either mounted separately or as part of the tail light lens.
✔ Check the condition of the headlight, tail light and turn signal lenses.

## Headlight beam height

✔ The MOT tester will perform a headlight beam height check using specialised beam setting equipment (see illustration 1). This equipment will not be available to the home mechanic, but if you suspect that the headlight is incorrectly set or may have been maladjusted in the past, you can perform a rough test as follows.
✔ Position the bike in a straight line facing a brick wall. The bike must be off its stand, upright and with a rider seated. Measure the height from the ground to the centre of the headlight and mark a horizontal line on the wall at this height. Position the motorcycle 3.8 metres from the wall and draw a vertical

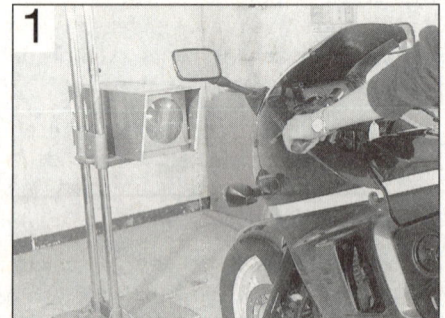

**Headlight beam height checking equipment**

line up the wall central to the centreline of the motorcycle. Switch to dipped beam and check that the beam pattern falls slightly lower than the horizontal line and to the left of the vertical line (see illustration 2).

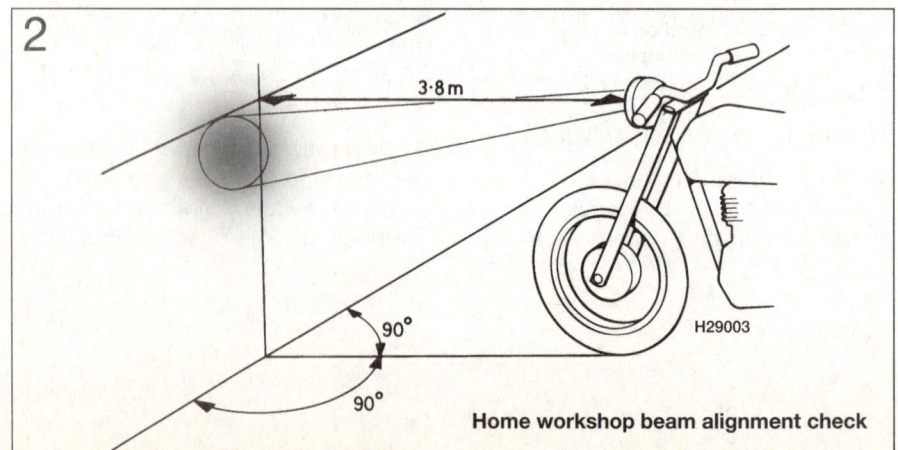

**Home workshop beam alignment check**

# Exhaust System and Final Drive

## Exhaust

✔ Check that the exhaust mountings are secure and that the system does not foul any of the rear suspension components.

✔ Start the motorcycle. When the revs are increased, check that the exhaust is neither holed nor leaking from any of its joints. On a linked system, check that the collector box is not leaking due to corrosion.

✔ Note that the exhaust decibel level ("loudness" of the exhaust) is assessed at the discretion of the tester. If the motorcycle was first used on or after 1st January 1985 the silencer must carry the BSAU 193 stamp, or a marking relating to its make and model, or be of OE (original equipment) manufacture. If the silencer is marked NOT FOR ROAD USE, RACING USE ONLY or similar, it will fail the MOT.

## Final drive

✔ On chain or belt drive machines, check that the chain/belt is in good condition and does not have excessive slack. Also check that the sprocket is securely mounted on the rear wheel hub. Check that the chain/belt guard is in place.

✔ On shaft drive bikes, check for oil leaking from the drive unit and fouling the rear tyre.

# Steering and Suspension

## Steering

✔ With the front wheel raised off the ground, rotate the steering from lock to lock. The handlebar or switches must not contact the fuel tank or be close enough to trap the rider's hand. Problems can be caused by damaged lock stops on the lower yoke and frame, or by the fitting of non-standard handlebars.

✔ When performing the lock to lock check, also ensure that the steering moves freely without drag or notchiness. Steering movement can be impaired by poorly routed cables, or by overtight head bearings or worn bearings. The tester will perform a check of the steering head bearing lower race by mounting the front wheel on a surface plate, then performing a lock to lock check with the weight of the machine on the lower bearing (see illustration 3).

✔ Grasp the fork sliders (lower legs) and attempt to push and pull on the forks (see illustration 4). Any play in the steering head bearings will be felt. Note that in extreme cases, wear of the front fork bushes can be misinterpreted for head bearing play.

✔ Check that the handlebars are securely mounted.

✔ Check that the handlebar grip rubbers are secure. They should by bonded to the bar left end and to the throttle cable pulley on the right end.

## Front suspension

✔ With the motorcycle off the stand, hold the front brake on and pump the front forks up and down (see illustration 5). Check that they are adequately damped.

✔ Inspect the area above and around the front fork oil seals (see illustration 6). There should be no sign of oil on the fork tube (stanchion) nor leaking down the slider (lower leg). On models so equipped, check that there is no oil leaking from the anti-dive units.

✔ On models with swingarm front suspension, check that there is no freeplay in the linkage when moved from side to side.

## Rear suspension

✔ With the motorcycle off the stand and an assistant supporting the motorcycle by its handlebars, bounce the rear suspension (see illustration 7). Check that the suspension components do not foul on any of the cycle parts and check that the shock absorber(s) provide adequate damping.

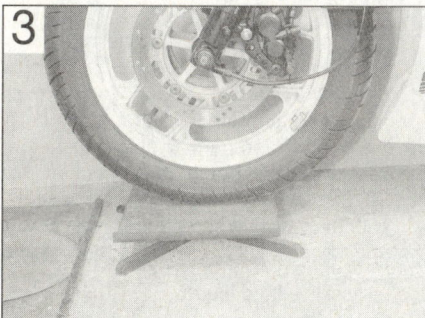

Front wheel mounted on a surface plate for steering head bearing lower race check

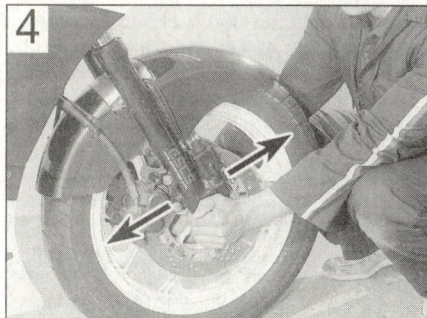

Checking the steering head bearings for freeplay

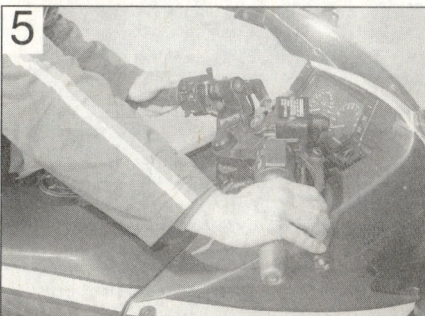

Hold the front brake on and pump the front forks up and down to check operation

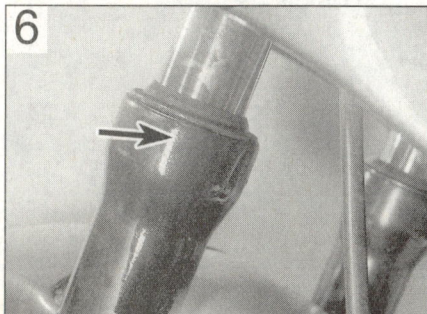

Inspect the area around the fork dust seal for oil leakage (arrow)

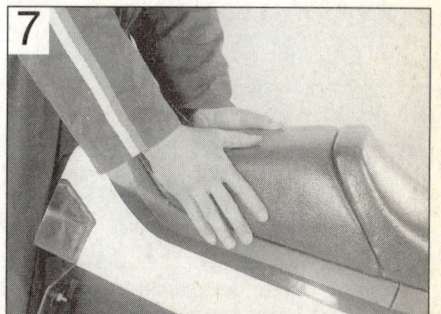

Bounce the rear of the motorcycle to check rear suspension operation

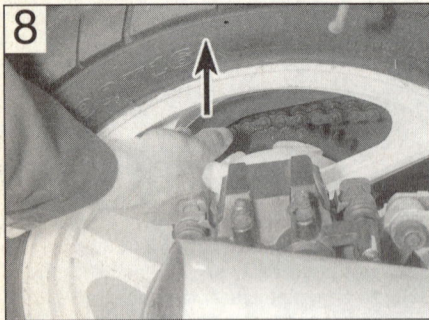

Checking for rear suspension linkage play

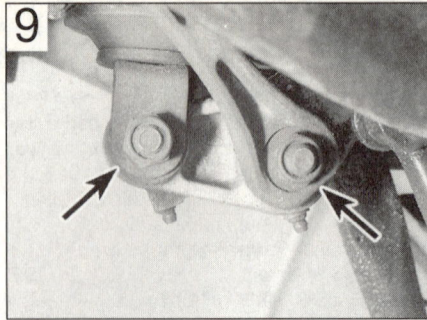

Worn suspension linkage pivots (arrows) are usually the cause of play in the rear suspension

Grasp the swingarm at the ends to check for play in its pivot bearings

✔ Visually inspect the shock absorber(s) and check that there is no sign of oil leakage from its damper. This is somewhat restricted on certain single shock models due to the location of the shock absorber.
✔ With the rear wheel raised off the ground, grasp the wheel at the highest point and attempt to pull it up **(see illustration 8)**. Any play in the swingarm pivot or suspension linkage bearings will be felt as movement. **Note:** *Do not confuse play with actual suspension movement.* Failure to lubricate suspension linkage bearings can lead to bearing failure **(see illustration 9)**.

✔ With the rear wheel raised off the ground, grasp the swingarm ends and attempt to move the swingarm from side to side and forwards and backwards - any play indicates wear of the swingarm pivot bearings **(see illustration 10)**.

# Brakes, Wheels and Tyres

## Brakes

✔ With the wheel raised off the ground, apply the brake then free it off, and check that the wheel is about to revolve freely without brake drag.
✔ On disc brakes, examine the disc itself. Check that it is securely mounted and not cracked.
✔ On disc brakes, view the pad material through the caliper mouth and check that the pads are not worn down beyond the limit **(see illustration 11)**.
✔ On drum brakes, check that when the brake is applied the angle between the operating lever and cable or rod is not too great **(see illustration 12)**. Check also that the operating lever doesn't foul any other components.

✔ On disc brakes, examine the flexible hoses from top to bottom. Have an assistant hold the brake on so that the fluid in the hose is under pressure, and check that there is no sign of fluid leakage, bulges or cracking. If there are any metal brake pipes or unions, check that these are free from corrosion and damage. Where a brake-linked anti-dive system is fitted, check the hoses to the anti-dive in a similar manner.
✔ Check that the rear brake torque arm is secure and that its fasteners are secured by self-locking nuts or castellated nuts with split-pins or R-pins **(see illustration 13)**.
✔ On models with ABS, check that the self-check warning light in the instrument panel works.
✔ The MOT tester will perform a test of the motorcycle's braking efficiency based on a calculation of rider and motorcycle weight. Although this cannot be carried out at home, you can at least ensure that the braking

systems are properly maintained. For hydraulic disc brakes, check the fluid level, lever/pedal feel (bleed of air if its spongy) and pad material. For drum brakes, check adjustment, cable or rod operation and shoe lining thickness.

## Wheels and tyres

✔ Check the wheel condition. Cast wheels should be free from cracks and if of the built-up design, all fasteners should be secure. Spoked wheels should be checked for broken, corroded, loose or bent spokes.
✔ With the wheel raised off the ground, spin the wheel and visually check that the tyre and wheel run true. Check that the tyre does not foul the suspension or mudguards.

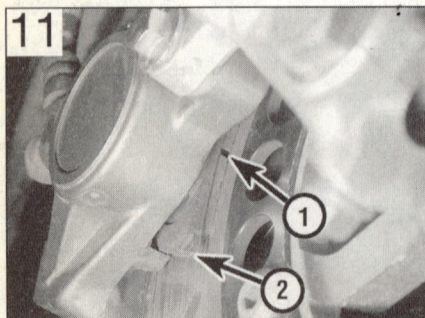

Brake pad wear can usually be viewed without removing the caliper. Most pads have wear indicator grooves (1) and some also have indicator tangs (2)

On drum brakes, check the angle of the operating lever with the brake fully applied. Most drum brakes have a wear indicator pointer and scale.

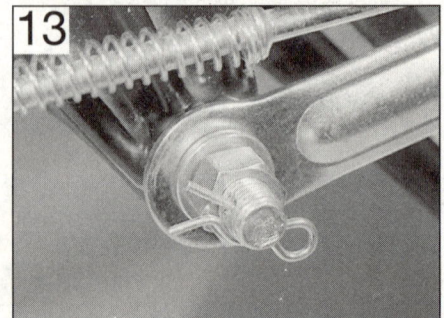

Brake torque arm must be properly secured at both ends

Check for wheel bearing play by trying to move the wheel about the axle (spindle)

Checking the tyre tread depth

Tyre direction of rotation arrow can be found on tyre sidewall

Castellated type wheel axle (spindle) nut must be secured by a split pin or R-pin

Two straightedges are used to check wheel alignment

✔ With the wheel raised off the ground, grasp the wheel and attempt to move it about the axle (spindle) **(see illustration 14)**. Any play felt here indicates wheel bearing failure.

✔ Check the tyre tread depth, tread condition and sidewall condition **(see illustration 15)**.

✔ Check the tyre type. Front and rear tyre types must be compatible and be suitable for road use. Tyres marked NOT FOR ROAD USE, COMPETITION USE ONLY or similar, will fail the MOT.

✔ If the tyre sidewall carries a direction of rotation arrow, this must be pointing in the direction of normal wheel rotation **(see illustration 16)**.

✔ Check that the wheel axle (spindle) nuts (where applicable) are properly secured. A self-locking nut or castellated nut with a split-pin or R-pin can be used **(see illustration 17)**.

✔ Wheel alignment is checked with the motorcycle off the stand and a rider seated. With the front wheel pointing straight ahead, two perfectly straight lengths of metal or wood and placed against the sidewalls of both tyres **(see illustration 18)**. The gap each side of the front tyre must be equidistant on both sides. Incorrect wheel alignment may be due to a cocked rear wheel (often as the result of poor chain adjustment) or in extreme cases, a bent frame.

# General checks and condition

✔ Check the security of all major fasteners, bodypanels, seat, fairings (where fitted) and mudguards.

✔ Check that the rider and pillion footrests, handlebar levers and brake pedal are securely mounted.

✔ Check for corrosion on the frame or any load-bearing components. If severe, this may affect the structure, particularly under stress.

# Sidecars

A motorcycle fitted with a sidecar requires additional checks relating to the stability of the machine and security of attachment and swivel joints, plus specific wheel alignment (toe-in) requirements. Additionally, tyre and lighting requirements differ from conventional motorcycle use. Owners are advised to check MOT test requirements with an official test centre.

# Preparing for storage

## Before you start

If repairs or an overhaul is needed, see that this is carried out now rather than left until you want to ride the bike again.

Give the bike a good wash and scrub all dirt from its underside. Make sure the bike dries completely before preparing for storage.

## Engine

● Remove the spark plug(s) and lubricate the cylinder bores with approximately a teaspoon of motor oil using a spout-type oil can **(see illustration 1)**. Reinstall the spark plug(s). Crank the engine over a couple of times to coat the piston rings and bores with oil. If the bike has a kickstart, use this to turn the engine over. If not, flick the kill switch to the OFF position and crank the engine over on the starter **(see illustration 2)**. If the nature on the ignition system prevents the starter operating with the kill switch in the OFF position,

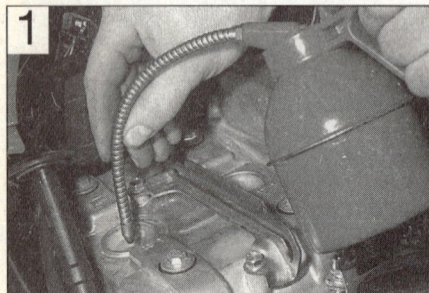
**Squirt a drop of motor oil into each cylinder**

**Flick the kill switch to OFF . . .**

**. . . and ensure that the metal bodies of the plugs (arrows) are earthed against the cylinder head**

remove the spark plugs and fit them back in their caps; ensure that the plugs are earthed (grounded) against the cylinder head when the starter is operated **(see illustration 3)**.

⚠️ *Warning: It is important that the plugs are earthed (grounded) away from the spark plug holes otherwise there is a risk of atomised fuel from the cylinders igniting.*

**HAYNES HINT** *On a single cylinder four-stroke engine, you can seal the combustion chamber completely by positioning the piston at TDC on the compression stroke.*

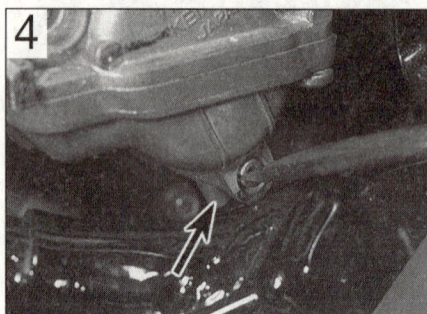
**Connect a hose to the carburettor float chamber drain stub (arrow) and unscrew the drain screw**

● Drain the carburettor(s) otherwise there is a risk of jets becoming blocked by gum deposits from the fuel **(see illustration 4)**.

● If the bike is going into long-term storage, consider adding a fuel stabiliser to the fuel in the tank. If the tank is drained completely, corrosion of its internal surfaces may occur if left unprotected for a long period. The tank can be treated with a rust preventative especially for this purpose. Alternatively, remove the tank and pour half a litre of motor oil into it, install the filler cap and shake the tank to coat its internals with oil before draining off the excess. The same effect can also be achieved by spraying WD40 or a similar water-dispersant around the inside of the tank via its flexible nozzle.

● Make sure the cooling system contains the correct mix of antifreeze. Antifreeze also contains important corrosion inhibitors.

● The air intakes and exhaust can be sealed off by covering or plugging the openings. Ensure that you do not seal in any condensation; run the engine until it is hot, then switch off and allow to cool. Tape a piece of thick plastic over the silencer end(s) **(see illustration 5)**. Note that some advocate pouring a tablespoon of motor oil into the silencer(s) before sealing them off.

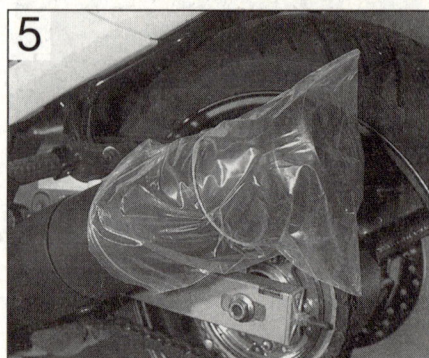
**Exhausts can be sealed off with a plastic bag**

## Battery

● Remove it from the bike - in extreme cases of cold the battery may freeze and crack its case **(see illustration 6)**.

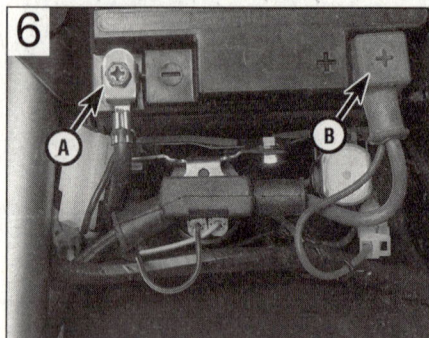
**Disconnect the negative lead (A) first, followed by the positive lead (B)**

● Check the electrolyte level and top up if necessary (conventional refillable batteries). Clean the terminals.

● Store the battery off the motorcycle and away from any sources of fire. Position a wooden block under the battery if it is to sit on the ground.

● Give the battery a trickle charge for a few hours every month **(see illustration 7)**.

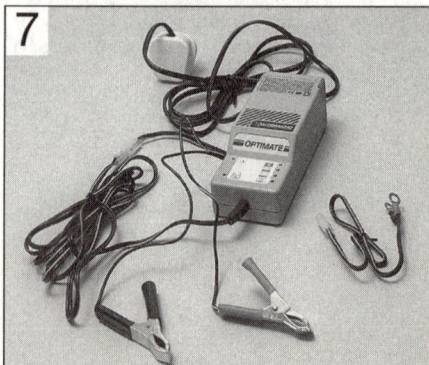
**Use a suitable battery charger - this kit also assess battery condition**

## Tyres

● Place the bike on its centrestand or an auxiliary stand which will support the motorcycle in an upright position. Position wood blocks under the tyres to keep them off the ground and to provide insulation from damp. If the bike is being put into long-term storage, ideally both tyres should be off the ground; not only will this protect the tyres, but will also ensure that no load is placed on the steering head or wheel bearings.

● Deflate each tyre by 5 to 10 psi, no more or the beads may unseat from the rim, making subsequent inflation difficult on tubeless tyres.

## Pivots and controls

● Lubricate all lever, pedal, stand and footrest pivot points. If grease nipples are fitted to the rear suspension components, apply lubricant to the pivots.
● Lubricate all control cables.

## Cycle components

● Apply a wax protectant to all painted and plastic components. Wipe off any excess, but don't polish to a shine. Where fitted, clean the screen with soap and water.
● Coat metal parts with Vaseline (petroleum jelly). When applying this to the fork tubes, do

not compress the forks otherwise the seals will rot from contact with the Vaseline.
● Apply a vinyl cleaner to the seat.

## Storage conditions

● Aim to store the bike in a shed or garage which does not leak and is free from damp.
● Drape an old blanket or bedspread over the bike to protect it from dust and direct contact with sunlight (which will fade paint). This also hides the bike from prying eyes. Beware of tight-fitting plastic covers which may allow condensation to form and settle on the bike.

# Getting back on the road

## Engine and transmission

● Change the oil and replace the oil filter. If this was done prior to storage, check that the oil hasn't emulsified - a thick whitish substance which occurs through condensation.
● Remove the spark plugs. Using a spout-type oil can, squirt a few drops of oil into the cylinder(s). This will provide initial lubrication as the piston rings and bores comes back into contact. Service the spark plugs, or fit new ones, and install them in the engine.
● Check that the clutch isn't stuck on. The plates can stick together if left standing for some time, preventing clutch operation. Engage a gear and try rocking the bike back and forth with the clutch lever held against the handlebar. If this doesn't work on cable-operated clutches, hold the clutch lever back against the handlebar with a strong elastic band or cable tie for a couple of hours (see illustration 8).

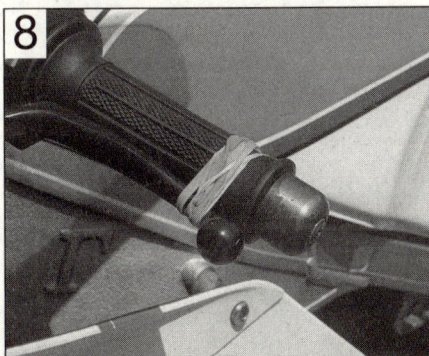

Hold clutch lever back against the handlebar with elastic bands or a cable tie

● If the air intakes or silencer end(s) were blocked off, remove the bung or cover used.
● If the fuel tank was coated with a rust preventative, oil or a stabiliser added to the fuel, drain and flush the tank and dispose of the fuel sensibly. If no action was taken with

the fuel tank prior to storage, it is advised that the old fuel is disposed of since it will go off over a period of time. Refill the fuel tank with fresh fuel.

## Frame and running gear

● Oil all pivot points and cables.
● Check the tyre pressures. They will definitely need inflating if pressures were reduced for storage.
● Lubricate the final drive chain (where applicable).
● Remove any protective coating applied to the fork tubes (stanchions) since this may well destroy the fork seals. If the fork tubes weren't protected and have picked up rust spots, remove them with very fine abrasive paper and refinish with metal polish.
● Check that both brakes operate correctly. Apply each brake hard and check that it's not possible to move the motorcycle forwards, then check that the brake frees off again once released. Brake caliper pistons can stick due to corrosion around the piston head, or on the sliding caliper types, due to corrosion of the slider pins. If the brake doesn't free after repeated operation, take the caliper off for examination. Similarly drum brakes can stick due to a seized operating cam, cable or rod linkage.
● If the motorcycle has been in long-term storage, renew the brake fluid and clutch fluid (where applicable).
● Depending on where the bike has been stored, the wiring, cables and hoses may have been nibbled by rodents. Make a visual check and investigate disturbed wiring loom tape.

## Battery

● If the battery has been previously removal and given top up charges it can simply be reconnected. Remember to connect the positive cable first and the negative cable last.
● On conventional refillable batteries, if the battery has not received any attention,

remove it from the motorcycle and check its electrolyte level. Top up if necessary then charge the battery. If the battery fails to hold a charge and a visual checks show heavy white sulphation of the plates, the battery is probably defective and must be renewed. This is particularly likely if the battery is old. Confirm battery condition with a specific gravity check.
● On sealed (MF) batteries, if the battery has not received any attention, remove it from the motorcycle and charge it according to the information on the battery case - if the battery fails to hold a charge it must be renewed.

## Starting procedure

● If a kickstart is fitted, turn the engine over a couple of times with the ignition OFF to distribute oil around the engine. If no kickstart is fitted, flick the engine kill switch OFF and the ignition ON and crank the engine over a couple of times to work oil around the upper cylinder components. If the nature of the ignition system is such that the starter won't work with the kill switch OFF, remove the spark plugs, fit them back into their caps and earth (ground) their bodies on the cylinder head. Reinstall the spark plugs afterwards.
● Switch the kill switch to RUN, operate the choke and start the engine. If the engine won't start don't continue cranking the engine - not only will this flatten the battery, but the starter motor will overheat. Switch the ignition off and try again later. If the engine refuses to start, go through the fault finding procedures in this manual. **Note:** *If the bike has been in storage for a long time, old fuel or a carburettor blockage may be the problem. Gum deposits in carburettors can block jets - if a carburettor cleaner doesn't prove successful the carburettors must be dismantled for cleaning.*

● Once the engine has started, check that the lights, turn signals and horn work properly.
● Treat the bike gently for the first ride and check all fluid levels on completion. Settle the bike back into the maintenance schedule.

This Section provides an easy reference-guide to the more common faults that are likely to afflict your machine. Obviously, the opportunities are almost limitless for faults to occur as a result of obscure failures, and to try and cover all eventualities would require a book. Indeed, a number have been written on the subject.

Successful troubleshooting is not a mysterious 'black art' but the application of a bit of knowledge combined with a systematic and logical approach to the problem. Approach any troubleshooting by first accurately identifying the symptom and then checking through the list of possible causes, starting with the simplest or most obvious and progressing in stages to the most complex.

Take nothing for granted, but above all apply liberal quantities of common sense.

The main symptom of a fault is given in the text as a major heading below which are listed the various systems or areas which may contain the fault. Details of each possible cause for a fault and the remedial action to be taken are given, in brief, in the paragraphs below each heading. Further information should be sought in the relevant Chapter.

## 1 Engine doesn't start or is difficult to start

- [ ] Starter motor does not rotate
- [ ] Starter motor rotates but engine does not turn over
- [ ] Starter works but engine won't turn over (seized)
- [ ] No fuel flow
- [ ] Engine flooded
- [ ] No spark or weak spark
- [ ] Compression low
- [ ] Stalls after starting
- [ ] Rough idle

## 2 Poor running at low speed

- [ ] Spark weak
- [ ] Fuel/air mixture incorrect
- [ ] Compression low
- [ ] Poor acceleration

## 3 Poor running or no power at high speed

- [ ] Firing incorrect
- [ ] Fuel/air mixture incorrect
- [ ] Compression low
- [ ] Knocking or pinging
- [ ] Miscellaneous causes

## 4 Overheating

- [ ] Cooling system not operating properly
- [ ] Firing incorrect
- [ ] Fuel/air mixture incorrect
- [ ] Compression too high
- [ ] Engine load excessive
- [ ] Lubrication inadequate
- [ ] Miscellaneous causes

## 5 Clutch problems

- [ ] Clutch slipping
- [ ] Clutch not disengaging completely

## 6 Gearshifting problems

- [ ] Doesn't go into gear or lever doesn't return
- [ ] Jumps out of gear
- [ ] Overshifts

## 7 Abnormal engine noise

- [ ] Knocking or pinging
- [ ] Piston slap or rattling
- [ ] Valve noise
- [ ] Other noise

## 8 Abnormal driveline noise

- [ ] Clutch noise
- [ ] Transmission noise
- [ ] Chain or final drive noise

## 9 Abnormal frame and suspension noise

- [ ] Front end noise
- [ ] Shock absorber noise
- [ ] Disc brake noise

## 10 Oil pressure warning light comes on

- [ ] Engine lubrication system
- [ ] Electrical system

## 11 Excessive exhaust smoke

- [ ] White smoke
- [ ] Black smoke
- [ ] Brown smoke

## 12 Poor handling or stability

- [ ] Handlebar hard to turn
- [ ] Handlebar shakes or vibrates excessively
- [ ] Handlebar pulls to one side
- [ ] Poor shock absorbing qualities

## 13 Braking problems

- [ ] Front brakes are spongy, don't hold
- [ ] Brake lever or pedal pulsates
- [ ] Brakes drag

## 14 Electrical problems

- [ ] Battery dead or weak
- [ ] Battery overcharged

# 1 Engine doesn't start or is difficult to start

### Starter motor does not rotate

- ☐ Engine kill switch Off.
- ☐ Fuse blown. Check fuse block (Chapter 9).
- ☐ Battery voltage low. Check and recharge battery (Chapter 9).
- ☐ Starter motor defective. Make sure the wiring to the starter is secure. Make sure the starter solenoid (relay) clicks when the start button is pushed. If the solenoid clicks, then the fault is in the wiring or motor.
- ☐ Starter solenoid (relay) faulty. Check it according to the procedure in Chapter 9.
- ☐ Starter button not contacting. The contacts could be wet, corroded or dirty. Disassemble and clean the switch (Chapter 9).
- ☐ Wiring open or shorted. Check all wiring connections and harnesses to make sure that they are dry, tight and not corroded. Also check for broken or frayed wires that can cause a short to ground/earth (see *Wiring diagrams*, Chapter 9).
- ☐ Ignition switch defective. Check the switch according to the procedure in Chapter 9. Replace the switch with a new one if it is defective.
- ☐ Engine kill switch defective. Check for wet, dirty or corroded contacts. Clean or replace the switch as necessary (Chapter 9).
- ☐ Faulty starter lockout switch. Check the wiring to the switch and the switch itself according to the procedures in Chapter 9.

### Starter motor rotates but engine does not turn over

- ☐ Starter motor clutch defective. Inspect and repair or replace (Chapter 2).
- ☐ Damaged idler or starter gears. Inspect and replace the damaged parts (Chapter 2).

### Starter works but engine won't turn over (seized)

- ☐ Seized engine caused by one or more internally damaged components. Failure due to wear, abuse or lack of lubrication. Damage can include seized valves, rocker arms, camshafts, pistons, crankshaft, connecting rod bearings, or transmission gears or bearings. Refer to Chapter 2 for engine disassembly.

### No fuel flow

- ☐ No fuel in tank.
- ☐ Fuel tap vacuum hose broken or disconnected.
- ☐ Tank cap air vent obstructed. Usually caused by dirt or water. Remove it and clean the cap vent hole.
- ☐ Fuel filter clogged. Inspect, and if necessary replace the filter (Chapter 4).
- ☐ Fuel line clogged. Pull the fuel line loose and carefully blow through it.
- ☐ Inlet needle valves clogged. For both the valves to be clogged, either a very bad batch of fuel with an unusual additive has been used, or some other foreign object has entered the tank. Many times after a machine has been stored for many months without running, the fuel turns to a varnish-like liquid and forms deposits on the inlet needle valves and jets. The carburetors should be removed and overhauled if draining the float bowls does not alleviate the problem.

### Engine flooded

- ☐ Float level too high. Check and adjust as described in Chapter 4.
- ☐ Inlet needle valve worn or stuck open. A piece of dirt, rust or other debris can cause the inlet needle to seat improperly, causing excess fuel to be admitted to the float bowl. In this case, the float chamber should be cleaned and the needle and seat inspected. If the needle and seat are worn, then the leaking will persist and the parts should be replaced with new ones (Chapter 4).
- ☐ Starting technique incorrect. Under normal circumstances (i.e., if all the carburetor functions are sound) the machine should start with little or no throttle. When the engine is cold, the choke should be operated and the engine started without opening the throttle. When the engine is at operating temperature, only a very slight amount of throttle should be necessary. If the engine is flooded, turn the fuel tap off and hold the throttle open while cranking the engine. This will allow additional air to reach the cylinders. Remember to turn the fuel back on after the engine starts.

### No spark or weak spark

- ☐ Ignition switch Off.
- ☐ Engine kill switch turned to the Off position.
- ☐ Battery voltage low. Check and recharge battery as necessary (Chapter 9).
- ☐ Spark plug dirty, defective or worn out. Locate reason for fouled plug(s) using spark plug condition chart and follow the plug maintenance procedures in Chapter 1.
- ☐ Spark plug cap or secondary (HT) wiring faulty. Check condition. Replace either or both components if cracks or deterioration are evident (Chapter 5).
- ☐ Spark plug cap not making good contact. Make sure that the plug cap fits snugly over the plug end.
- ☐ IC igniter defective. Check the unit, referring to Chapter 5 for details.
- ☐ Pickup coil defective. Check the unit, referring to Chapter 5 for details.
- ☐ Ignition coil(s) defective. Check the coils, referring to Chapter 5.
- ☐ Ignition or kill switch shorted. This is usually caused by water, corrosion, damage or excessive wear. The switches can be disassembled and cleaned with electrical contact cleaner. If cleaning does not help, replace the switches (Chapter 9).
- ☐ Wiring shorted or broken between:
  - a)   Ignition switch and engine kill switch
  - b)   IC igniter and engine kill switch
  - c)   IC igniter and ignition coil
  - d)   Ignition coil and plug
  - e)   IC igniter and pickup coil
- ☐ Make sure that all wiring connections are clean, dry and tight. Look for chafed and broken wires (Chapters 5 and 9).

### Compression low

- ☐ Spark plug loose. Remove the plug and inspect the threads. Reinstall and tighten to the specified torque (Chapter 1).
- ☐ Cylinder head not sufficiently tightened down. If the cylinder head is suspected of being loose, then there's a chance that the gasket or head is damaged if the problem has persisted for any length of time. The head bolts should be tightened to the proper torque in the correct sequence (Chapter 2).
- ☐ Improper valve clearance. This means that the valve is not closing completely and compression pressure is leaking past the valve. Check and adjust the valve clearances (Chapter 1).
- ☐ Cylinder and/or piston worn. Excessive wear will cause compression pressure to leak past the rings. This is usually accompanied by worn rings as well. A top end overhaul is necessary (Chapter 2).
- ☐ Piston rings worn, weak, broken, or sticking. Broken or sticking piston rings usually indicate a lubrication or carburetion problem that causes excess carbon deposits to form on the pistons and rings. Top end overhaul is necessary (Chapter 2).
- ☐ Piston ring-to-groove clearance excessive. This is caused by excessive wear of the piston ring lands. Piston replacement is necessary (Chapter 2).
- ☐ Cylinder head gasket damaged. If the head is allowed to become loose, or if excessive carbon build-up on the piston crown and combustion chamber causes extremely high compression, the head gasket may leak. Retorquing the head is not always sufficient to restore the seal, so gasket replacement is necessary (Chapter 2).

# 1 Engine doesn't start or is difficult to start (continued)

☐ Cylinder head warped. This is caused by overheating or improperly tightened head bolts. Machine shop resurfacing or head replacement is necessary (Chapter 2).

☐ Valve spring broken or weak. Caused by component failure or wear; the spring(s) must be replaced (Chapter 2).

☐ Valve not seating properly. This is caused by a bent valve (from over-revving or improper valve adjustment), burned valve or seat (improper carburetion) or an accumulation of carbon deposits on the seat (from carburetion, lubrication problems). The valves must be cleaned and/or replaced and the seats serviced if possible (Chapter 2).

### Stalls after starting

☐ Improper choke action. Make sure the choke rod is getting a full stroke and staying in the 'out' position. Adjustment of the cable slack is covered in Chapter 1.

☐ Ignition malfunction. See Chapter 5.

☐ Carburetor malfunction. See Chapter 4.

☐ Fuel contaminated. The fuel can be contaminated with either dirt or water, or can change chemically if the machine is allowed to sit for several months or more. Drain the tank and float bowls (Chapter 4).

☐ Intake air leak. Check for loose carburetor-to-intake manifold connections, loose or missing vacuum gauge access port cap or hose, or loose carburetor top (Chapter 4).

☐ Idle speed incorrect. Turn idle speed adjuster screw until the engine idles at the specified rpm (Chapters 1 and 4).

### Rough idle

☐ Ignition malfunction. See Chapter 5.

☐ Idle speed incorrect. See Chapter 1.

☐ Carburetors not synchronized. Adjust carburetors with vacuum gauge set or manometer as outlined in Chapter 1.

☐ Carburetor malfunction. See Chapter 4.

☐ Fuel contaminated. The fuel can be contaminated with either dirt or water, or can change chemically if the machine is allowed to sit for several months or more. Drain the tank and float bowls. If the problem is severe, a carburetor overhaul may be necessary (Chapters 1 and 4).

☐ Intake air leak.

☐ Air cleaner clogged. Service or replace air filter element (Chapter 1).

# 2 Poor running at low speed

### Spark weak

☐ Battery voltage low. Check and recharge battery (Chapter 9).

☐ Spark plug fouled, defective or worn out. Refer to Chapter 1 for spark plug maintenance.

☐ Spark plug cap or high tension wiring defective. Refer to Chapters 1 and 5 for details of the ignition system.

☐ Spark plug cap not making contact.

☐ Incorrect spark plug. Wrong type, heat range or cap configuration. Check and install correct plugs listed in Chapter 1. A cold plug or one with a recessed firing electrode will not operate at low speeds without fouling.

☐ IC igniter defective. See Chapter 5.

☐ Pickup coil defective. See Chapter 5.

☐ Ignition coil(s) defective. See Chapter 5.

### Fuel/air mixture incorrect

☐ Pilot screw(s) out of adjustment (Chapters 1 and 4).

☐ Pilot jet or air passage clogged. Remove and overhaul the carburetors (Chapter 4).

☐ Air bleed holes clogged. Remove carburetor and blow out all passages (Chapter 4).

☐ Air cleaner clogged, poorly sealed or missing.

☐ Air cleaner-to-carburetor boot poorly sealed. Look for cracks, holes or loose clamps and replace or repair defective parts.

☐ Fuel level too high or too low. Adjust the floats (Chapter 4).

☐ Fuel tank air vent obstructed. Make sure that the air vent passage in the filler cap is open (except California models).

☐ Carburetor intake manifolds loose. Check for cracks, breaks, tears or loose clamps or bolts. Repair or replace the rubber boots.

### Compression low

☐ Spark plug loose. Remove the plug and inspect the threads. Reinstall and tighten to the specified torque (Chapter 1).

☐ Cylinder head not sufficiently tightened down. If the cylinder head is suspected of being loose, then there's a chance that the gasket and head are damaged if the problem has persisted for any length of time. The head bolts should be tightened to the proper torque in the correct sequence (Chapter 2).

☐ Improper valve clearance. This means that the valve is not closing completely and compression pressure is leaking past the valve. Check and adjust the valve clearances (Chapter 1).

☐ Cylinder and/or piston worn. Excessive wear will cause compression pressure to leak past the rings. This is usually accompanied by worn rings as well. A top end overhaul is necessary (Chapter 2).

☐ Piston rings worn, weak, broken, or sticking. Broken or sticking piston rings usually indicate a lubrication or carburetion problem that causes excess carbon deposits to form on the pistons and rings. Top end overhaul is necessary (Chapter 2).

☐ Piston ring-to-groove clearance excessive. This is caused by excessive wear of the piston ring lands. Piston replacement is necessary (Chapter 2).

☐ Cylinder head gasket damaged. If the head is allowed to become loose, or if excessive carbon build-up on the piston crown and combustion chamber causes extremely high compression, the head gasket may leak. Retorquing the head is not always sufficient to restore the seal, so gasket replacement is necessary (Chapter 2).

☐ Cylinder head warped. This is caused by overheating or improperly tightened head bolts. Machine shop resurfacing or head replacement is necessary (Chapter 2).

☐ Valve spring broken or weak. Caused by component failure or wear; the spring(s) must be replaced (Chapter 2).

☐ Valve not seating properly. This is caused by a bent valve (from over-revving or improper valve adjustment), burned valve or seat (improper carburetion) or an accumulation of carbon deposits on the seat (from carburetion, lubrication problems). The valves must be cleaned and/or replaced and the seats serviced if possible (Chapter 2).

## Poor acceleration

- [ ] Carburetors leaking or dirty. Overhaul the carburetors (Chapter 4).
- [ ] Timing not advancing. The pickup coil unit or the IC igniter may be defective. If so, they must be replaced with new ones, as they cannot be repaired.
- [ ] Carburetors not synchronized. Adjust them with a vacuum gauge set or manometer (Chapter 1).

- [ ] Engine oil viscosity too high. Using a heavier oil than that recommended in Chapter 1 can damage the oil pump or lubrication system and cause drag on the engine.
- [ ] Brakes dragging. Usually caused by debris which has entered the brake piston sealing boot, or from a warped disc or bent axle. Repair as necessary (Chapter 7).

# 3 Poor running or no power at high speed

## Firing incorrect

- [ ] Air filter restricted. Clean or replace filter (Chapter 1).
- [ ] Spark plug fouled, defective or worn out. See Chapter 1 for spark plug maintenance.
- [ ] Spark plug cap or secondary (HT) wiring defective. See Chapters 1 and 5 for details of the ignition system.
- [ ] Spark plug cap not in good contact. See Chapter 5.
- [ ] Incorrect spark plug. Wrong type, heat range or cap configuration. Check and install correct plugs listed in Chapter 1. A cold plug or one with a recessed firing electrode will not operate at low speeds without fouling.
- [ ] IC igniter defective. See Chapter 5.
- [ ] Ignition coil(s) defective. See Chapter 5.

## Fuel/air mixture incorrect

- [ ] Main jet clogged. Dirt, water and other contaminants can clog the main jets. Clean the fuel tap filter, the float bowl area, and the jets and carburetor orifices (Chapter 4).
- [ ] Main jet wrong size. The standard jetting is for sea level atmospheric pressure and oxygen content.
- [ ] Throttle shaft-to-carburetor body clearance excessive. Refer to Chapter 4 for inspection and part replacement procedures.
- [ ] Air bleed holes clogged. Remove and overhaul carburetors (Chapter 4).
- [ ] Air cleaner clogged, poorly sealed or missing.
- [ ] Air cleaner-to-carburetor boot poorly sealed. Look for cracks, holes or loose clamps, and replace or repair defective parts.
- [ ] Fuel level too high or too low. Adjust the float(s) (Chapter 4).
- [ ] Fuel tank air vent obstructed. Make sure the air vent passage in the filler cap is open.
- [ ] Carburetor intake manifolds loose. Check for cracks, breaks, tears or loose clamps or bolts. Repair or replace the rubber boots (Chapter 2).
- [ ] Fuel filter clogged. Clean, and if necessary, replace the filter (Chapter 1).
- [ ] Fuel line clogged. Pull the fuel line loose and carefully blow through it.

## Compression low

- [ ] Spark plug loose. Remove the plug and inspect the threads. Reinstall and tighten to the specified torque (Chapter 1).
- [ ] Cylinder head not sufficiently tightened down. If the cylinder head is suspected of being loose, then there's a chance that the gasket and head are damaged if the problem has persisted for any length of time. The head bolts should be tightened to the proper torque in the correct sequence (Chapter 2).

- [ ] Improper valve clearance. This means that the valve is not closing completely and compression pressure is leaking past the valve. Check and adjust the valve clearances (Chapter 1).
- [ ] Cylinder and/or piston worn. Excessive wear will cause compression pressure to leak past the rings. This is usually accompanied by worn rings as well. A top end overhaul is necessary (Chapter 2).
- [ ] Piston rings worn, weak, broken, or sticking. Broken or sticking piston rings usually indicate a lubrication or carburetion problem that causes excess carbon deposits or seizures to form on the pistons and rings. Top end overhaul is necessary (Chapter 2).
- [ ] Piston ring-to-groove clearance excessive. This is caused by excessive wear of the piston ring lands. Piston replacement is necessary (Chapter 2).
- [ ] Cylinder head gasket damaged. If the head is allowed to become loose, or if excessive carbon build-up on the piston crown and combustion chamber causes extremely high compression, the head gasket may leak. Retorquing the head is not always sufficient to restore the seal, so gasket replacement is necessary (Chapter 2).
- [ ] Cylinder head warped. This is caused by overheating or improperly tightened head bolts. Machine shop resurfacing or head replacement is necessary (Chapter 2).
- [ ] Valve spring broken or weak. Caused by component failure or wear; the spring(s) must be replaced (Chapter 2).
- [ ] Valve not seating properly. This is caused by a bent valve (from over-revving or improper valve adjustment), burned valve or seat (improper carburetion) or an accumulation of carbon deposits on the seat (from carburetion, lubrication problems). The valves must be cleaned and/or replaced and the seats serviced if possible (Chapter 2).

## Knocking or pinging

- [ ] Carbon build-up in combustion chamber. Use of a fuel additive that will dissolve the adhesive bonding the carbon particles to the crown and chamber is the easiest way to remove the build-up. Otherwise, the cylinder head will have to be removed and decarbonized (Chapter 2).
- [ ] Incorrect or poor quality fuel. Old or improper grades of gasoline (petrol) can cause detonation. This causes the piston to rattle, thus the knocking or pinging sound. Drain old fuel and always use the recommended fuel grade.
- [ ] Spark plug heat range incorrect. Uncontrolled detonation indicates the plug heat range is too hot. The plug in effect becomes a glow plug, raising cylinder temperatures. Install the proper heat range plug (Chapter 1).
- [ ] Improper air/fuel mixture. This will cause the cylinder to run hot, which leads to detonation. Clogged jets or an air leak can cause this imbalance. See Chapter 4.

# 3 Poor running or no power at high speed (continued)

## Miscellaneous causes

☐ Throttle valve doesn't open fully. Adjust the cable slack (Chapter 1).
☐ Clutch slipping. Caused by damaged, loose or worn clutch components. Refer to Chapter 2 for adjustment and overhaul procedures.
☐ Timing not advancing.

☐ Engine oil viscosity too high. Using a heavier oil than the one recommended in Chapter 1 can damage the oil pump or lubrication system and cause drag on the engine.
☐ Brakes dragging. Usually caused by debris which has entered the brake piston sealing boot, or from a warped disc or bent axle. Repair as necessary.

# 4 Overheating

## Cooling system not operating properly

☐ Coolant level low. Check coolant level as described in 'Daily (pre-ride) checks'. If coolant level is low, the engine will overheat.
☐ Leak in cooling system. Check cooling system hoses and radiator for leaks and other damage. Repair or replace parts as necessary (Chapter 3).
☐ Thermostat sticking open or closed. Check and replace as described in Chapter 3.
☐ Faulty radiator cap. Remove the cap and have it pressure checked at a service station.
☐ Coolant passages clogged. Have the entire system drained and flushed, then refill with new coolant.
☐ Water pump defective. Remove the pump and check the components.
☐ Clogged radiator fins. Clean them by blowing compressed air through the fins from the back side.

## Firing incorrect

☐ Spark plug fouled, defective or worn out. See Chapter 1 for spark plug maintenance.
☐ Incorrect spark plug.
☐ Faulty ignition coil(s) (Chapter 5).

## Fuel/air mixture incorrect

☐ Main jet clogged. Dirt, water and other contaminants can clog the main jets. Clean the fuel tap filter, the float bowl area and the jets and carburetor orifices (Chapter 4).
☐ Main jet wrong size. The standard jetting is for sea level atmospheric pressure and oxygen content.
☐ Air cleaner poorly sealed or missing.
☐ Air cleaner-to-carburetor boot poorly sealed. Look for cracks, holes or loose clamps and replace or repair.
☐ Fuel level too low. Adjust the float(s) (Chapter 4).
☐ Fuel tank air vent obstructed. Make sure that the air vent passage in the filler cap is open (except California models).
☐ Carburetor intake manifolds loose. Check for cracks, breaks, tears or loose clamps or bolts. Repair or replace the rubber boots (Chapter 4).

## Compression too high

☐ Carbon build-up in combustion chamber. Use of a fuel additive that will dissolve the adhesive bonding the carbon particles to the piston crown and chamber is the easiest way to remove the build-up. Otherwise, the cylinder head will have to be removed and decarbonized (Chapter 2).
☐ Improperly machined head surface or installation of incorrect gasket during engine assembly. Check Specifications (Chapter 2).

## Engine load excessive

☐ Clutch slipping. Caused by damaged, loose or worn clutch components. Refer to Chapter 2 for overhaul procedures.
☐ Engine oil level too high. The addition of too much oil will cause pressurization of the crankcase and inefficient engine operation. Check Specifications and drain to proper level (Chapter 1).
☐ Engine oil viscosity too high. Using a heavier oil than the one recommended in Chapter 1 can damage the oil pump or lubrication system as well as cause drag on the engine.
☐ Brakes dragging. Usually caused by debris which has entered the brake piston sealing boot, or from a warped disc or bent axle. Repair as necessary.

## Lubrication inadequate

☐ Engine oil level too low. Friction caused by intermittent lack of lubrication or from oil that is 'overworked' can cause overheating. The oil provides a definite cooling function in the engine. Check the oil level (see 'Daily (pre-ride) checks').
☐ Poor quality engine oil or incorrect viscosity or type. Oil is rated not only according to viscosity but also according to type. Some oils are not rated high enough for use in this engine. Check the Specifications section and change to the correct oil (Chapter 1).

## Miscellaneous causes

☐ Modification to exhaust system. Most aftermarket exhaust systems cause the engine to run leaner, which makes it run hotter. When installing an accessory exhaust system, always rejet the carburetors.

# 5 Clutch problems

## Clutch slipping

- ☐ Friction plates worn or warped. Overhaul the clutch assembly (Chapter 2).
- ☐ Metal plates worn or warped (Chapter 2).
- ☐ Clutch springs broken or weak. Old or heat-damaged (from slipping clutch) springs should be replaced with new ones (Chapter 2).
- ☐ Clutch release mechanism defective. Check the mechanism and replace any defective parts (Chapter 2).
- ☐ Clutch hub or housing unevenly worn. This causes improper engagement of the discs. Replace the damaged or worn parts (Chapter 2).

## Clutch not disengaging completely

- ☐ Air in clutch hydraulic system. Bleed the system (Chapter 2).
- ☐ Clutch master or release cylinder worn. Inspect and, if necessary, overhaul the cylinders (Chapter 2).
- ☐ Clutch plates warped or damaged. This will cause clutch drag, which in turn causes the machine to creep. Overhaul the clutch assembly (Chapter 2).

- ☐ Clutch spring tension uneven. Usually caused by a sagged or broken spring. Check and replace the springs (Chapter 2).
- ☐ Engine oil deteriorated. Old, thin, worn out oil will not provide proper lubrication for the discs, causing the clutch to drag. Replace the oil and filter (Chapter 1).
- ☐ Engine oil viscosity too high. Using a heavier oil than recommended in Chapter 1 can cause the plates to stick together, putting a drag on the engine. Change to the correct weight oil (Chapter 1).
- ☐ Clutch housing seized on shaft. Lack of lubrication, severe wear or damage can cause the housing to seize on the shaft. Overhaul of the clutch, and perhaps transmission, may be necessary to repair damage (Chapter 2).
- ☐ Clutch release mechanism defective. Worn or damaged release mechanism parts can stick and fail to apply force to the pressure plate. Overhaul the release mechanism (Chapter 2).
- ☐ Loose clutch hub nut. Causes housing and hub misalignment putting a drag on the engine. Engagement adjustment continually varies. Overhaul the clutch assembly (Chapter 2).

# 6 Gearshifting problems

## Doesn't go into gear or lever doesn't return

- ☐ Clutch not disengaging. See Section 27.
- ☐ Shift fork(s) bent or seized. Often caused by dropping the machine or from lack of lubrication. Overhaul the transmission (Chapter 2).
- ☐ Gear(s) stuck on shaft. Most often caused by a lack of lubrication or excessive wear in transmission bearings and bushings. Overhaul the transmission (Chapter 2).
- ☐ Shift drum binding. Caused by lubrication failure or excessive wear. Replace the drum and bearings (Chapter 2).
- ☐ Shift lever return spring weak or broken (Chapter 2).
- ☐ Shift lever broken. Splines stripped out of lever or shaft, caused by allowing the lever to get loose or from dropping the machine. Replace necessary parts (Chapter 2).
- ☐ Shift mechanism pawl broken or worn. Full engagement and rotary movement of shift drum results. Replace shaft assembly (Chapter 2).

- ☐ Pawl spring broken. Allows pawl to 'float', causing sporadic shift operation. Replace spring (Chapter 2).

## Jumps out of gear

- ☐ Shift fork(s) worn. Overhaul the transmission (Chapter 2).
- ☐ Gear groove(s) worn. Overhaul the transmission (Chapter 2).
- ☐ Gear dogs or dog slots worn or damaged. The gears should be inspected and replaced. No attempt should be made to service the worn parts.

## Overshifts

- ☐ Pawl spring weak or broken (Chapter 2).
- ☐ Shift drum stopper lever not functioning (Chapter 2).
- ☐ Overshift limiter broken or distorted (Chapter 2).

# 7 Abnormal engine noise

## Knocking or pinging

- ☐ Carbon build-up in combustion chamber. Use of a fuel additive that will dissolve the adhesive bonding the carbon particles to the piston crown and chamber is the easiest way to remove the build-up. Otherwise, the cylinder head will have to be removed and decarbonized (Chapter 2).
- ☐ Incorrect or poor quality fuel. Old or improper fuel can cause detonation. This causes the piston to rattle, thus the knocking or pinging sound. Drain the old fuel and always use the recommended grade (Chapter 4).
- ☐ Spark plug heat range incorrect. Uncontrolled detonation indicates that the plug heat range is too hot. The plug in effect becomes a glow plug, raising cylinder temperatures. Install the proper heat range plug (Chapter 1).
- ☐ Improper air/fuel mixture. This will cause the cylinder to run hot and lead to detonation. Clogged jets or an air leak can cause this imbalance. See Chapter 4.

## Piston slap or rattling

- ☐ Cylinder-to-piston clearance excessive. Caused by improper assembly. Inspect and overhaul top end parts (Chapter 2).
- ☐ Connecting rod bent. Caused by over-revving, trying to start a badly flooded engine or from ingesting a foreign object into the combustion chamber. Replace the damaged parts (Chapter 2).
- ☐ Piston pin or piston pin bore worn or seized from wear or lack of lubrication. Replace damaged parts (Chapter 2).
- ☐ Piston ring(s) worn, broken or sticking. Overhaul the top end (Chapter 2).
- ☐ Piston seizure damage. Usually from lack of lubrication or overheating. Replace the pistons and bore the cylinders, as necessary (Chapter 2).
- ☐ Connecting rod bearing and/or piston pin-end clearance excessive. Caused by excessive wear or lack of lubrication. Replace worn parts.

# 7 Abnormal engine noise (continued)

### Valve noise

☐ Incorrect valve clearances. Adjust the clearances by referring to Chapter 1.

☐ Valve spring broken or weak. Check and replace weak valve springs (Chapter 2).

☐ Camshaft or cylinder head worn or damaged. Lack of lubrication at high rpm is usually the cause of damage. Insufficient oil or failure to change the oil at the recommended intervals are the chief causes. Since there are no replaceable bearings in the head, the head itself will have to be replaced if there is excessive wear or damage (Chapter 2).

### Other noise

☐ Cylinder head gasket leaking. This will cause compression leakage into the cooling system (which may show up as air bubbles in the coolant in the radiator). Also, coolant may get into the oil (which will turn the oil gray and foamy). In either case, have the cooling system checked by a dealer service department.

☐ Exhaust pipe leaking at cylinder head connection. Caused by improper fit of pipe(s) or loose exhaust flange. All exhaust fasteners should be tightened evenly and carefully. Failure to do this will lead to a leak.

☐ Crankshaft runout excessive. Caused by a bent crankshaft (from over-revving) or damage from an upper cylinder component failure. Can also be attributed to dropping the machine on either of the crankshaft ends.

☐ Engine mounting fasteners loose. Tighten all engine mounting fasteners to the specified torque (Chapter 2).

☐ Crankshaft bearings worn (Chapter 2).

☐ Camshaft chain tensioner defective. Replace according to the procedure in Chapter 2.

☐ Camshaft chain, sprockets or guides worn (Chapter 2).

☐ Poorly tensioned alternator drive belt (early models) or drive chain (later models).

# 8 Abnormal driveline noise

### Clutch noise

☐ Clutch housing/friction plate clearance excessive (Chapter 2).

☐ Loose or damaged clutch pressure plate and/or bolts (Chapter 2).

### Transmission noise

☐ Bearings worn. Also includes the possibility that the shafts are worn. Overhaul the transmission (Chapter 2).

☐ Gears worn or chipped (Chapter 2).

☐ Metal chips jammed in gear teeth. Probably pieces from a broken clutch, gear or shift mechanism that were picked up by the gears. This will cause early bearing failure (Chapter 2).

☐ Engine oil level too low. Causes a howl from transmission. Also affects engine power and clutch operation (see 'Daily (pre-ride) checks').

### Chain or final drive noise

☐ Chain not adjusted properly (Chapter 1).

☐ Sprocket (engine sprocket or rear sprocket) loose. Tighten fasteners (Chapter 6).

☐ Sprocket(s) worn. Replace sprocket(s) (Chapter 6).

☐ Rear sprocket warped. Replace sprockets and chain as a set (Chapter 6).

☐ Wheel coupling worn. Replace coupling (Chapter 6).

# 9 Abnormal frame and suspension noise

### Front end noise

☐ Low fluid level or improper viscosity oil in forks. This can sound like 'spurting' and is usually accompanied by irregular fork action (Chapter 6).

☐ Spring weak or broken. Makes a clicking or scraping sound. Fork oil, when drained, will have a lot of metal particles in it (Chapter 6).

☐ Steering head bearings loose or damaged. Clicks when braking. Check and adjust or replace as necessary (Chapters 1 and 6).

☐ Fork clamps loose. Make sure all fork clamp pinch bolts are tight (Chapter 6).

☐ Fork tube bent. Good possibility if machine has been dropped. Replace tube with a new one (Chapter 6).

☐ Front axle or axle clamp bolt loose. Tighten them to the specified torque (Chapter 7).

### Shock absorber noise

☐ Fluid level incorrect. Indicates a leak caused by defective seal. Shock will be covered with oil. Replace shock (Chapter 6).

☐ Defective shock absorber with internal damage. This is in the body of the shock and cannot be remedied. The shock must be replaced with a new one (Chapter 6).

☐ Bent or damaged shock body. Replace the shock with a new one (Chapter 6).

### Disc brake noise

☐ Squeal caused by pad shim not installed or positioned correctly (Chapter 7).

☐ Squeal caused by dust on brake pads. Usually found in combination with glazed pads. Clean using brake cleaning solvent (Chapter 7).

☐ Contamination of brake pads. Oil, brake fluid or dirt causing brake to chatter or squeal. Clean or replace pads (Chapter 7).

☐ Pads glazed. Caused by excessive heat from prolonged use or from contamination. Do not use sandpaper, emery cloth, carborundum cloth or any other abrasive to roughen the pad surfaces as abrasives will stay in the pad material and damage the disc. A very fine flat file can be used, but pad replacement is suggested as a cure (Chapter 7).

☐ Disc warped. Can cause a chattering, clicking or intermittent squeal. Usually accompanied by a pulsating lever and uneven braking. Replace the disc (Chapter 7).

☐ Loose or worn wheel bearings. Check and replace as needed (Chapter 7).

# 10 Oil pressure warning light comes on

## Engine lubrication system

- [ ] Engine oil pump defective (Chapter 2).
- [ ] Engine oil level low. Inspect for leak or other problem causing low oil level and add recommended lubricant (see 'Daily (pre-ride) checks').
- [ ] Engine oil viscosity too low. Very old, thin oil or an improper weight of oil used in engine. Change to correct lubricant (Chapter 1).
- [ ] Camshaft or journals worn. Excessive wear causing drop in oil pressure. Replace cam and/or head. Abnormal wear could be caused by oil starvation at high rpm from low oil level or improper oil weight or type ('Daily (pre-ride) checks' and Chapter 1).
- [ ] Crankshaft and/or bearings worn. Same problems as paragraph 4. Check and replace crankshaft and/or bearings (Chapter 2).

## Electrical system

- [ ] Oil pressure switch defective. Check the switch according to the procedure in Chapter 9. Replace it if it is defective.
- [ ] Oil pressure warning light circuit defective. Check for pinched, shorted, disconnected or damaged wiring (Chapter 9).

# 11 Excessive exhaust smoke

## White smoke

- [ ] Piston oil ring worn. The ring may be broken or damaged, causing oil from the crankcase to be pulled past the piston into the combustion chamber. Replace the rings with new ones (Chapter 2).
- [ ] Cylinders worn, cracked, or scored. Caused by overheating or oil starvation. The cylinders will have to be rebored and new pistons installed.
- [ ] Valve oil seal damaged or worn. Replace oil seals with new ones (Chapter 2).
- [ ] Valve guide worn. Perform a complete valve job (Chapter 2).
- [ ] Engine oil level too high, which causes oil to be forced past the rings. Drain oil to the proper level (see 'Daily (pre-ride) checks').
- [ ] Head gasket broken between oil return and cylinder. Causes oil to be pulled into combustion chamber. Replace the head gasket and check the head for warpage (Chapter 2).
- [ ] Abnormal crankcase pressurization, which forces oil past the rings. Clogged breather or hoses usually the cause (Chapter 4).

## Black smoke

- [ ] Air cleaner clogged. Clean or replace the element (Chapter 1).
- [ ] Main jet too large or loose. Compare the jet size to the Specifications (Chapter 4).
- [ ] Choke stuck, causing fuel to be pulled through choke circuit (Chapter 4).
- [ ] Fuel level too high. Check and adjust the float height as necessary (Chapter 4).
- [ ] Inlet needle held off needle seat. Clean float bowl and fuel line and replace needle and seat if necessary (Chapter 4).

## Brown smoke

- [ ] Main jet too small or clogged. Lean condition caused by wrong size main jet or by a restricted orifice. Clean float bowl and jets and compare jet size to Specifications (Chapter 4).
- [ ] Fuel flow insufficient. Fuel inlet needle valve stuck closed due to chemical reaction with old fuel. Float height incorrect. Restricted fuel line. Clean line and float bowl and adjust floats if necessary (Chapter 4).
- [ ] Carburetor intake manifolds loose (Chapter 4).
- [ ] Air cleaner poorly sealed or not installed (Chapter 1).

# 12 Poor handling or stability

## Handlebar hard to turn

- [ ] Steering stem locknut too tight (Chapter 6).
- [ ] Bearings damaged. Roughness can be felt as the bars are turned from side-to-side. Replace bearings and races (Chapter 6).
- [ ] Races dented or worn. Denting results from wear in only one position (e. g., straight ahead), from striking an immovable object or hole or from dropping the machine. Replace races and bearings (Chapter 6).
- [ ] Steering stem lubrication inadequate. Causes are grease getting hard from age or being washed out by high pressure car washes. Disassemble steering head and repack bearings (Chapter 6).
- [ ] Steering stem bent. Caused by hitting a curb or hole or from dropping the machine. Replace damaged part. Do not try to straighten stem (Chapter 6).
- [ ] Front tire air pressure too low ('Daily (pre-ride) checks').

## Handlebar shakes or vibrates excessively

- [ ] Tires worn or out of balance (Chapter 7).
- [ ] Swingarm bearings worn. Replace worn bearings by referring to Chapter 6.
- [ ] Rim(s) warped or damaged. Inspect wheels for runout (Chapter 7).
- [ ] Wheel bearings worn. Worn front or rear wheel bearings can cause poor tracking. Worn front bearings will cause wobble (Chapter 7).
- [ ] Handlebar clamp bolts loose (Chapter 6).
- [ ] Steering stem or fork clamps loose. Tighten them to the specified torque (Chapter 6).
- [ ] Engine mounting bolts loose. Will cause excessive vibration with increased engine rpm (Chapter 2).

## Handlebar pulls to one side

- [ ] Frame bent. Definitely suspect this if the machine has been dropped. May or may not be accompanied by cracking near the bend. Replace the frame (Chapter 6).
- [ ] Wheel out of alignment. Caused by improper location of axle spacers or from bent steering stem or frame (Chapter 6).
- [ ] Swingarm bent or twisted. Caused by age (metal fatigue) or impact damage. Replace the arm (Chapter 6).
- [ ] Steering stem bent. Caused by impact damage or from dropping the motorcycle. Replace the steering stem (Chapter 6).
- [ ] Fork leg bent. Disassemble the forks and replace the damaged parts (Chapter 6).
- [ ] Fork oil level uneven.

## Poor shock absorbing qualities

- [ ] Too hard:
  - a) Fork oil level excessive (Chapter 6).
  - b) Fork oil viscosity too high. Use a lighter oil (see the Specifications in Chapter 6).
  - c) Fork tube bent. Causes a harsh, sticking feeling (Chapter 6).
  - d) Shock shaft or body bent or damaged (Chapter 6).
  - e) Fork internal damage (Chapter 6).
  - f) Shock internal damage.
  - g) Tire pressure too high ('Daily (pre-ride) checks').
- [ ] Too soft:
  - a) Fork or shock oil insufficient and/or leaking (Chapter 6).
  - b) Fork oil viscosity too light (Chapter 6).
  - c) Fork springs weak or broken (Chapter 6).

## 13 Braking problems

### Front brakes are spongy, don't hold

- ☐ Air in brake line. Caused by inattention to master cylinder fluid level or by leakage. Locate problem and bleed brakes (Chapter 7).
- ☐ Pad or disc worn (Chapters 1 and 7).
- ☐ Brake fluid leak. See paragraph 1.
- ☐ Contaminated pads. Caused by contamination with oil, grease, brake fluid, etc. Clean or replace pads. Clean disc thoroughly with brake cleaner (Chapter 7).
- ☐ Brake fluid deteriorated. Fluid is old or contaminated. Drain system, replenish with new fluid and bleed the system (Chapter 7).
- ☐ Master cylinder internal parts worn or damaged causing fluid to bypass (Chapter 7).
- ☐ Master cylinder bore scratched from ingestion of foreign material or broken spring. Repair or replace master cylinder (Chapter 7).
- ☐ Disc warped. Replace disc (Chapter 7).

### Brake lever or pedal pulsates

- ☐ Disc warped. Replace disc (Chapter 7).
- ☐ Axle bent. Replace axle (Chapter 6).

- ☐ Brake caliper bolts loose (Chapter 7).
- ☐ Brake caliper shafts damaged or sticking, causing caliper to bind. Lube the shafts and/or replace them if they are corroded or bent (Chapter 7).
- ☐ Wheel warped or otherwise damaged (Chapter 7).
- ☐ Wheel bearings damaged or worn (Chapter 7).

### Brakes drag

- ☐ Master cylinder piston seized. Caused by wear or damage to piston or cylinder bore (Chapter 7).
- ☐ Lever balky or stuck. Check pivot and lubricate (Chapter 7).
- ☐ Brake caliper binds. Caused by inadequate lubrication or damage to caliper shafts (Chapter 7).
- ☐ Brake caliper piston seized in bore. Caused by wear or ingestion of dirt past deteriorated seal (Chapter 7).
- ☐ Brake pad damaged. Pad material separating from backing plate. Usually caused by faulty manufacturing process or from contact with chemicals. Replace pads (Chapter 7).
- ☐ Pads improperly installed (Chapter 7).

## 14 Electrical problems

### Battery dead or weak

- ☐ Battery faulty. Caused by sulfated plates which are shorted through sedimentation or by low electrolyte level. Also, broken battery terminal making only occasional contact (Chapter 9).
- ☐ Battery cables making poor contact (Chapter 9).
- ☐ Load excessive. Caused by addition of high wattage lights or other electrical accessories.
- ☐ Ignition switch defective. Switch either grounds (earths) internally or fails to shut off system. Replace the switch (Chapter 9).
- ☐ Regulator/rectifier defective (Chapter 9).
- ☐ Stator coil open or shorted (Chapter 9).
- ☐ Wiring faulty. Wiring grounded (earthed) or connections loose in ignition, charging or lighting circuits (Chapter 9).

### Battery overcharged

- ☐ Regulator/rectifier defective. Overcharging is noticed when battery gets excessively warm or 'boils' over (Chapter 9).
- ☐ Battery defective. Replace battery with a new one (Chapter 9).
- ☐ Battery amperage too low, wrong type or size. Install manufacturer's specified amp-hour battery to handle charging load (Chapter 9).

# Fault Finding Equipment

### Checking engine compression

● Low compression will result in exhaust smoke, heavy oil consumption, poor starting and poor performance. A compression test will provide useful information about an engine's condition and if performed regularly, can give warning of trouble before any other symptoms become apparent.
● A compression gauge will be required, along with an adapter to suit the spark plug hole thread size. Note that the screw-in type gauge/adapter set up is preferable to the rubber cone type.

● Before carrying out the test, first check the valve clearances as described in Chapter 1.
1 Run the engine until it reaches normal operating temperature, then stop it and remove the spark plug(s), taking care not to scald your hands on the hot components.
2 Install the gauge adapter and compression gauge in No. 1 cylinder spark plug hole **(see illustration 1)**.
3 On kickstart-equipped motorcycles, make sure the ignition switch is OFF, then open the throttle fully and kick the engine over a couple of times until the gauge reading stabilises.
4 On motorcycles with electric start only, the procedure will differ depending on the nature of the ignition system. Flick the engine kill switch (engine stop switch) to OFF and turn

Screw the compression gauge adapter into the spark plug hole, then screw the gauge into the adapter

the ignition switch ON; open the throttle fully and crank the engine over on the starter motor for a couple of revolutions until the gauge reading stabilises. If the starter will not operate with the kill switch OFF, turn the ignition switch OFF and refer to the next paragraph.

**5** Install the spark plugs back into their suppressor caps and arrange the plug electrodes so that their metal bodies are earthed (grounded) against the cylinder head; this is essential to prevent damage to the ignition system as the engine is spun over **(see illustration 2)**. Position the plugs well away from the plug holes otherwise there is a risk of atomised fuel escaping from the combustion chambers and igniting. As a safety precaution, cover the top of the valve cover with rag. Now turn the ignition switch ON and kill switch ON, open the throttle fully and crank the engine over on the starter motor for a couple of revolutions until the gauge reading stabilises.

All spark plugs must be earthed (grounded) against the cylinder head

**6** After one or two revolutions the pressure should build up to a maximum figure and then stabilise. Take a note of this reading and on multi-cylinder engines repeat the test on the remaining cylinders.

**7** The correct pressures are given in Chapter 2 Specifications. If the results fall within the specified range and on multi-cylinder engines all are relatively equal, the engine is in good condition. If there is a marked difference between the readings, or if the readings are lower than specified, inspection of the top-end components will be required.

**8** Low compression pressure may be due to worn cylinder bores, pistons or rings, failure of the cylinder head gasket, worn valve seals, or poor valve seating.

**9** To distinguish between cylinder/piston wear and valve leakage, pour a small quantity of oil into the bore to temporarily seal the piston rings, then repeat the compression tests **(see illustration 3)**. If the readings show a noticeable increase in pressure this confirms that the cylinder bore, piston, or rings are worn. If, however, no change is indicated, the cylinder head gasket or valves should be examined.

Bores can be temporarily sealed with a squirt of motor oil

**10** High compression pressure indicates excessive carbon build-up in the combustion chamber and on the piston crown. If this is the case the cylinder head should be removed and the deposits removed. Note that excessive carbon build-up is less likely with the used on modern fuels.

## Checking battery open-circuit voltage

⚠️ *Warning: The gases produced by the battery are explosive - never smoke or create any sparks in the vicinity of the battery. Never allow the electrolyte to contact your skin or clothing - if it does, wash it off and seek immediate medical attention.*

● Before any electrical fault is investigated the battery should be checked.
● You'll need a dc voltmeter or multimeter to check battery voltage. Check that the leads are inserted in the correct terminals on the meter, red lead to positive (+ve), black lead to negative (-ve). Incorrect connections can damage the meter.
● A sound fully-charged 12 volt battery should produce between 12.3 and 12.6 volts across its terminals (12.8 volts for a maintenance-free battery). On machines with a 6 volt battery, voltage should be between 6.1 and 6.3 volts.

**1** Set a multimeter to the 0 to 20 volts dc range and connect its probes across the

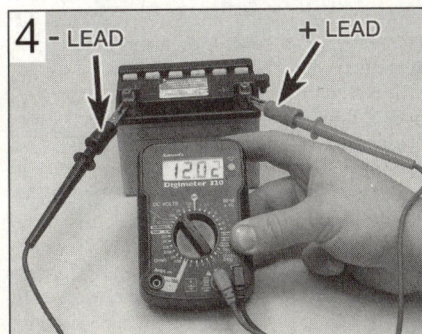

Measuring open-circuit battery voltage

battery terminals. Connect the meter's positive (+ve) probe, usually red, to the battery positive (+ve) terminal, followed by the meter's negative (-ve) probe, usually black, to the battery negative terminal (-ve) **(see illustration 4)**.

**2** If battery voltage is low (below 10 volts on a 12 volt battery or below 4 volts on a six volt battery), charge the battery and test the voltage again. If the battery repeatedly goes flat, investigate the motorcycle's charging system.

## Checking battery specific gravity (SG)

⚠️ *Warning: The gases produced by the battery are explosive - never smoke or create any sparks in the vicinity of the battery. Never allow the electrolyte to contact your skin or clothing - if it does, wash it off and seek immediate medical attention.*

● The specific gravity check gives an indication of a battery's state of charge.
● A hydrometer is used for measuring specific gravity. Make sure you purchase one which has a small enough hose to insert in the aperture of a motorcycle battery.
● Specific gravity is simply a measure of the electrolyte's density compared with that of water. Water has an SG of 1.000 and fully-charged battery electrolyte is about 26% heavier, at 1.260.
● Specific gravity checks are not possible on maintenance-free batteries. Testing the open-circuit voltage is the only means of determining their state of charge.

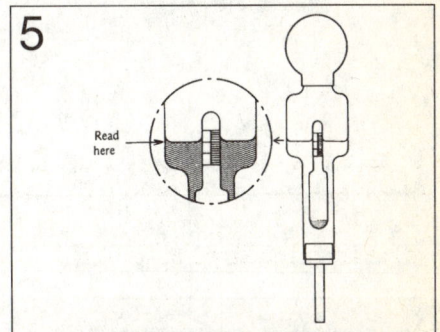

Float-type hydrometer for measuring battery specific gravity

**1** To measure SG, remove the battery from the motorcycle and remove the first cell cap. Draw some electrolyte into the hydrometer and note the reading **(see illustration 5)**. Return the electrolyte to the cell and install the cap.

**2** The reading should be in the region of 1.260 to 1.280. If SG is below 1.200 the battery needs charging. Note that SG will vary with temperature; it should be measured at 20°C (68°F). Add 0.007 to the reading for every 10°C above 20°C, and subtract 0.007 from the reading for every 10°C below 20°C. Add 0.004 to the reading for every 10°F above 68°F, and subtract 0.004 from the reading for every 10°F below 68°F.

3 When the check is complete, rinse the hydrometer thoroughly with clean water.

## Checking for continuity

● The term continuity describes the uninterrupted flow of electricity through an electrical circuit. A continuity check will determine whether an **open-circuit** situation exists.

● Continuity can be checked with an ohmmeter, multimeter, continuity tester or battery and bulb test circuit **(see illustrations 6, 7 and 8)**.

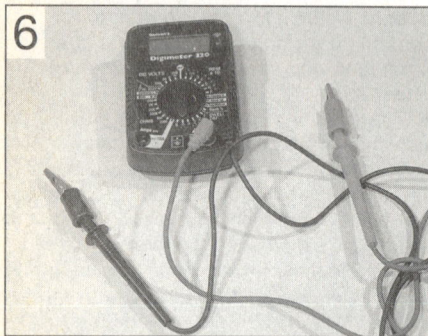
**Digital multimeter can be used for all electrical tests**

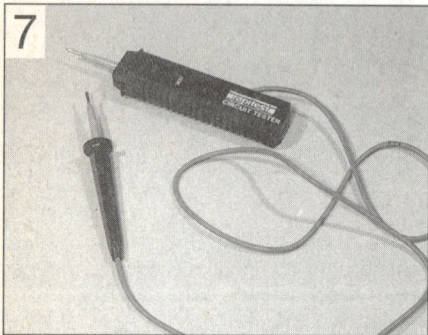
**Battery-powered continuity tester**

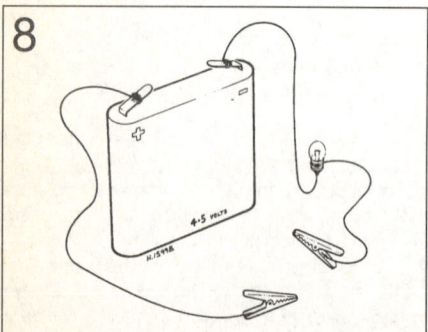
**Battery and bulb test circuit**

● All of these instruments are self-powered by a battery, therefore the checks are made with the ignition OFF.

● As a safety precaution, always disconnect the battery negative (-ve) lead before making checks, particularly if ignition switch checks are being made.

● If using a meter, select the appropriate ohms scale and check that the meter reads infinity (∞). Touch the meter probes together and check that meter reads zero; where necessary adjust the meter so that it reads zero.

● After using a meter, always switch it OFF to conserve its battery.

### Switch checks

1 If a switch is at fault, trace its wiring up to the wiring connectors. Separate the wire connectors and inspect them for security and condition. A build-up of dirt or corrosion here will most likely be the cause of the problem - clean up and apply a water dispersant such as WD40.

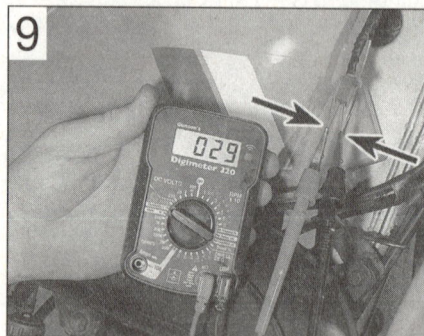
**Continuity check of front brake light switch using a meter - note split pins used to access connector terminals**

2 If using a test meter, set the meter to the ohms x 10 scale and connect its probes across the wires from the switch **(see illustration 9)**. Simple ON/OFF type switches, such as brake light switches, only have two wires whereas combination switches, like the ignition switch, have many internal links. Study the wiring diagram to ensure that you are connecting across the correct pair of wires. Continuity (low or no measurable resistance - 0 ohms) should be indicated with the switch ON and no continuity (high resistance) with it OFF.

3 Note that the polarity of the test probes doesn't matter for continuity checks, although care should be taken to follow specific test procedures if a diode or solid-state component is being checked.

4 A continuity tester or battery and bulb circuit can be used in the same way. Connect its probes as described above **(see illustration 10)**. The light should come on to indicate continuity in the ON switch position, but should extinguish in the OFF position.

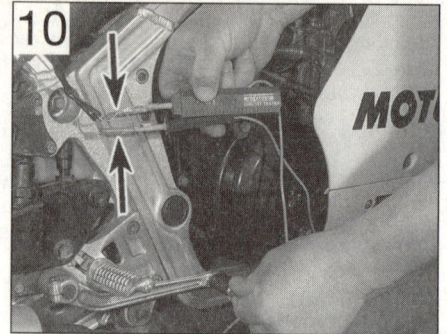
**Continuity check of rear brake light switch using a continuity tester**

### Wiring checks

● Many electrical faults are caused by damaged wiring, often due to incorrect routing or chaffing on frame components.

● Loose, wet or corroded wire connectors can also be the cause of electrical problems, especially in exposed locations.

1 A continuity check can be made on a single length of wire by disconnecting it at each end and connecting a meter or continuity tester across both ends of the wire **(see illustration 11)**.

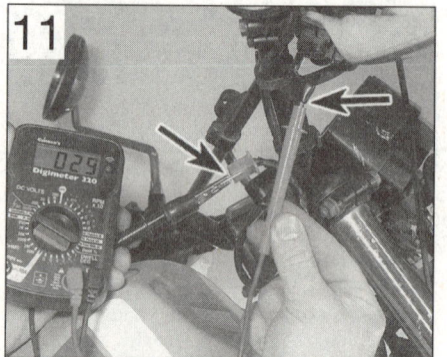
**Continuity check of front brake light switch sub-harness**

2 Continuity (low or no resistance - 0 ohms) should be indicated if the wire is good. If no continuity (high resistance) is shown, suspect a broken wire.

## Checking for voltage

● A voltage check can determine whether current is reaching a component.

● Voltage can be checked with a dc voltmeter, multimeter set on the dc volts scale, test light or buzzer **(see illustrations 12 and 13)**. A meter has the advantage of being able to measure actual voltage.

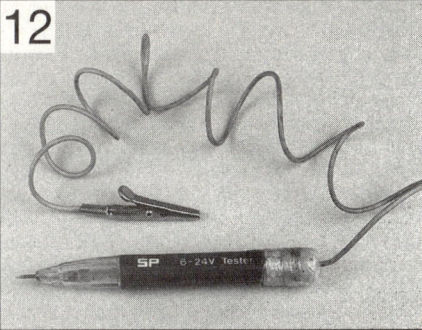

A simple test light can be used for voltage checks

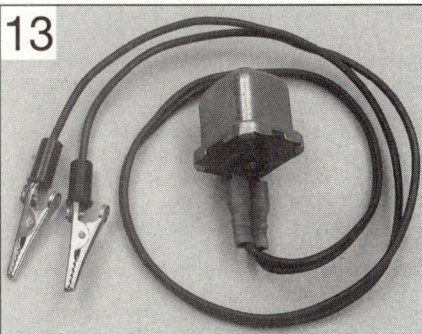

A buzzer is useful for voltage checks

● When using a meter, check that its leads are inserted in the correct terminals on the meter, red to positive (+ve), black to negative (-ve). Incorrect connections can damage the meter.
● A voltmeter (or multimeter set to the dc volts scale) should always be connected in parallel (across the load). Connecting it in series will destroy the meter.
● Voltage checks are made with the ignition ON.

1 First identify the relevant wiring circuit by referring to the wiring diagram at the end of this manual. If other electrical components share the same power supply (ie are fed from the same fuse), take note whether they are working correctly - this is useful information in deciding where to start checking the circuit.

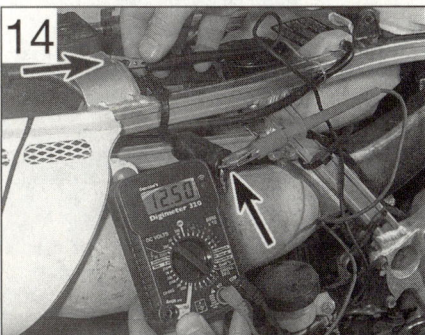

Checking for voltage at the rear brake light power supply wire using a meter . . .

2 If using a meter, check first that the meter leads are plugged into the correct terminals on the meter (see above). Set the meter to the dc volts function, at a range suitable for the battery voltage. Connect the meter red probe (+ve) to the power supply wire and the black probe to a good metal earth (ground) on the motorcycle's frame or directly to the battery negative (-ve) terminal (see illustration 14). Battery voltage should be shown on the meter with the ignition switched ON.
3 If using a test light or buzzer, connect its positive (+ve) probe to the power supply terminal and its negative (-ve) probe to a good earth (ground) on the motorcycle's frame or directly to the battery negative (-ve) terminal (see illustration 15). With the ignition ON, the test light should illuminate or the buzzer sound.

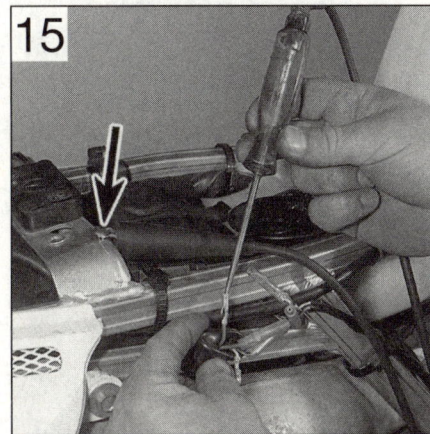

. . . or a test light - note the earth connection to the frame (arrow)

4 If no voltage is indicated, work back towards the fuse continuing to check for voltage. When you reach a point where there is voltage, you know the problem lies between that point and your last check point.

## Checking the earth (ground)

● Earth connections are made either directly to the engine or frame (such as sensors, neutral switch etc. which only have a positive feed) or by a separate wire into the earth circuit of the wiring harness. Alternatively a short earth wire is sometimes run directly from the component to the motorcycle's frame.
● Corrosion is often the cause of a poor earth connection.
● If total failure is experienced, check the security of the main earth lead from the negative (-ve) terminal of the battery and also the main earth (ground) point on the wiring harness. If corroded, dismantle the connection and clean all surfaces back to bare metal.

1 To check the earth on a component, use an insulated jumper wire to temporarily bypass its earth connection (see illustration 16). Connect one end of the jumper wire between the earth terminal or metal body of the component and the other end to the motorcycle's frame.

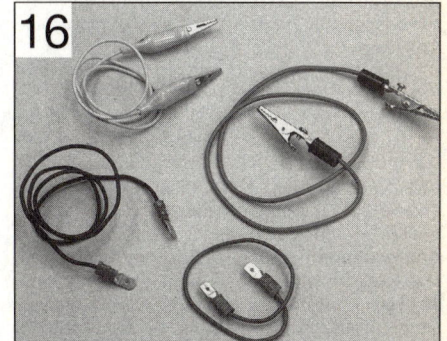

A selection of jumper wires for making earth (ground) checks

2 If the circuit works with the jumper wire installed, the original earth circuit is faulty. Check the wiring for open-circuits or poor connections. Clean up direct earth connections, removing all traces of corrosion and remake the joint. Apply petroleum jelly to the joint to prevent future corrosion.

## Tracing a short-circuit

● A short-circuit occurs where current shorts to earth (ground) bypassing the circuit components. This usually results in a blown fuse.

● A short-circuit is most likely to occur where the insulation has worn through due to wiring chafing on a component, allowing a direct path to earth (ground) on the frame.

1 Remove any bodypanels necessary to access the circuit wiring.

2 Check that all electrical switches in the circuit are OFF, then remove the circuit fuse and connect a test light, buzzer or voltmeter (set to the dc scale) across the fuse terminals. No voltage should be shown.

3 Move the wiring from side to side whilst observing the test light or meter. When the test light comes on, buzzer sounds or meter shows voltage, you have found the cause of the short. It will usually shown up as damaged or burned insulation.

4 Note that the same test can be performed on each component in the circuit, even the switch.

# A

**ABS (Anti-lock braking system)** A system, usually electronically controlled, that senses incipient wheel lockup during braking and relieves hydraulic pressure at wheel which is about to skid.

**Aftermarket** Components suitable for the motorcycle, but not produced by the motorcycle manufacturer.

**Allen key** A hexagonal wrench which fits into a recessed hexagonal hole.

**Alternating current (ac)** Current produced by an alternator. Requires converting to direct current by a rectifier for charging purposes.

**Alternator** Converts mechanical energy from the engine into electrical energy to charge the battery and power the electrical system.

**Ampere (amp)** A unit of measurement for the flow of electrical current. Current = Volts ÷ Ohms.

**Ampere-hour (Ah)** Measure of battery capacity.

**Angle-tightening** A torque expressed in degrees. Often follows a conventional tightening torque for cylinder head or main bearing fasteners **(see illustration)**.

Angle-tightening cylinder head bolts

**Antifreeze** A substance (usually ethylene glycol) mixed with water, and added to the cooling system, to prevent freezing of the coolant in winter. Antifreeze also contains chemicals to inhibit corrosion and the formation of rust and other deposits that would tend to clog the radiator and coolant passages and reduce cooling efficiency.

**Anti-dive** System attached to the fork lower leg (slider) to prevent fork dive when braking hard.

**Anti-seize compound** A coating that reduces the risk of seizing on fasteners that are subjected to high temperatures, such as exhaust clamp bolts and nuts.

**API** American Petroleum Institute. A quality standard for 4-stroke motor oils.

**Asbestos** A natural fibrous mineral with great heat resistance, commonly used in the composition of brake friction materials. Asbestos is a health hazard and the dust created by brake systems should never be inhaled or ingested.

**ATF** Automatic Transmission Fluid. Often used in front forks.

**ATU** Automatic Timing Unit. Mechanical device for advancing the ignition timing on early engines.

**ATV** All Terrain Vehicle. Often called a Quad.

**Axial play** Side-to-side movement.

**Axle** A shaft on which a wheel revolves. Also known as a spindle.

# B

**Backlash** The amount of movement between meshed components when one component is held still. Usually applies to gear teeth.

**Ball bearing** A bearing consisting of a hardened inner and outer race with hardened steel balls between the two races.

**Bearings** Used between two working surfaces to prevent wear of the components and a build-up of heat. Four types of bearing are commonly used on motorcycles: plain shell bearings, ball bearings, tapered roller bearings and needle roller bearings.

**Bevel gears** Used to turn the drive through 90°. Typical applications are shaft final drive and camshaft drive **(see illustration)**.

Bevel gears are used to turn the drive through 90°

**BHP** Brake Horsepower. The British measurement for engine power output. Power output is now usually expressed in kilowatts (kW).

**Bias-belted tyre** Similar construction to radial tyre, but with outer belt running at an angle to the wheel rim.

**Big-end bearing** The bearing in the end of the connecting rod that's attached to the crankshaft.

**Bleeding** The process of removing air from an hydraulic system via a bleed nipple or bleed screw.

**Bottom-end** A description of an engine's crankcase components and all components contained there-in.

**BTDC** Before Top Dead Centre in terms of piston position. Ignition timing is often expressed in terms of degrees or millimetres BTDC.

**Bush** A cylindrical metal or rubber component used between two moving parts.

**Burr** Rough edge left on a component after machining or as a result of excessive wear.

# C

**Cam chain** The chain which takes drive from the crankshaft to the camshaft(s).

**Canister** The main component in an evaporative emission control system (California market only); contains activated charcoal granules to trap vapours from the fuel system rather than allowing them to vent to the atmosphere.

**Castellated** Resembling the parapets along the top of a castle wall. For example, a castellated wheel axle or spindle nut.

**Catalytic converter** A device in the exhaust system of some machines which converts certain pollutants in the exhaust gases into less harmful substances.

**Charging system** Description of the components which charge the battery, ie the alternator, rectifer and regulator.

**Circlip** A ring-shaped clip used to prevent endwise movement of cylindrical parts and shafts. An internal circlip is installed in a groove in a housing; an external circlip fits into a groove on the outside of a cylindrical piece such as a shaft. Also known as a snap-ring.

**Clearance** The amount of space between two parts. For example, between a piston and a cylinder, between a bearing and a journal, etc.

**Coil spring** A spiral of elastic steel found in various sizes throughout a vehicle, for example as a springing medium in the suspension and in the valve train.

**Compression** Reduction in volume, and increase in pressure and temperature, of a gas, caused by squeezing it into a smaller space.

**Compression damping** Controls the speed the suspension compresses when hitting a bump.

**Compression ratio** The relationship between cylinder volume when the piston is at top dead centre and cylinder volume when the piston is at bottom dead centre.

**Continuity** The uninterrupted path in the flow of electricity. Little or no measurable resistance

**Continuity tester** Self-powered bleeper or test light which indicates continuity.

**Cp** Candlepower. Bulb rating common found on US motorcycles.

**Crossply tyre** Tyre plies arranged in a criss-cross pattern. Usually four or six plies used, hence 4PR or 6PR in tyre size codes.

**Cush drive** Rubber damper segments fitted between the rear wheel and final drive sprocket to absorb transmission shocks **(see illustration)**.

Cush drive rubbers dampen out transmission shocks

# D

**Degree disc** Calibrated disc for measuring piston position. Expressed in degrees.

**Dial gauge** Clock-type gauge with adapters for measuring runout and piston position. Expressed in mm or inches.

**Diaphragm** The rubber membrane in a master cylinder or carburettor which seals the upper chamber.

**Diaphragm spring** A single sprung plate often used in clutches.

**Direct current (dc)** Current produced by a dc generator.

**Decarbonisation** The process of removing carbon deposits - typically from the combustion chamber, valves and exhaust port/system.

**Detonation** Destructive and damaging explosion of fuel/air mixture in combustion chamber instead of controlled burning.

**Diode** An electrical valve which only allows current to flow in one direction. Commonly used in rectifiers and starter interlock systems.

**Disc valve (or rotary valve)** A induction system used on some two-stroke engines.

**Double-overhead camshaft (DOHC)** An engine that uses two overhead camshafts, one for the intake valves and one for the exhaust valves.

**Drivebelt** A toothed belt used to transmit drive to the rear wheel on some motorcycles. A drivebelt has also been used to drive the camshafts. Drivebelts are usually made of Kevlar.

**Driveshaft** Any shaft used to transmit motion. Commonly used when referring to the final driveshaft on shaft drive motorcycles.

# E

**Earth return** The return path of an electrical circuit, utilising the motorcycle's frame.

**ECU (Electronic Control Unit)** A computer which controls (for instance) an ignition system, or an anti-lock braking system.

**EGO** Exhaust Gas Oxygen sensor. Sometimes called a Lambda sensor.

**Electrolyte** The fluid in a lead-acid battery.

**EMS (Engine Management System)** A computer controlled system which manages the fuel injection and the ignition systems in an integrated fashion.

**Endfloat** The amount of lengthways movement between two parts. As applied to a crankshaft, the distance that the crankshaft can move side-to-side in the crankcase.

**Endless chain** A chain having no joining link. Common use for cam chains and final drive chains.

**EP (Extreme Pressure)** Oil type used in locations where high loads are applied, such as between gear teeth.

**Evaporative emission control system** Describes a charcoal filled canister which stores fuel vapours from the tank rather than allowing them to vent to the atmosphere. Usually only fitted to California models and referred to as an EVAP system.

**Expansion chamber** Section of two-stroke engine exhaust system so designed to improve engine efficiency and boost power.

# F

**Feeler blade or gauge** A thin strip or blade of hardened steel, ground to an exact thickness, used to check or measure clearances between parts.

**Final drive** Description of the drive from the transmission to the rear wheel. Usually by chain or shaft, but sometimes by belt.

**Firing order** The order in which the engine cylinders fire, or deliver their power strokes, beginning with the number one cylinder.

**Flooding** Term used to describe a high fuel level in the carburettor float chambers, leading to fuel overflow. Also refers to excess fuel in the combustion chamber due to incorrect starting technique.

**Free length** The no-load state of a component when measured. Clutch, valve and fork spring lengths are measured at rest, without any preload.

**Freeplay** The amount of travel before any action takes place. The looseness in a linkage, or an assembly of parts, between the initial application of force and actual movement. For example, the distance the rear brake pedal moves before the rear brake is actuated.

**Fuel injection** The fuel/air mixture is metered electronically and directed into the engine intake ports (indirect injection) or into the cylinders (direct injection). Sensors supply information on engine speed and conditions.

**Fuel/air mixture** The charge of fuel and air going into the engine. See **Stoichiometric ratio**.

**Fuse** An electrical device which protects a circuit against accidental overload. The typical fuse contains a soft piece of metal which is calibrated to melt at a predetermined current flow (expressed as amps) and break the circuit.

# G

**Gap** The distance the spark must travel in jumping from the centre electrode to the side electrode in a spark plug. Also refers to the distance between the ignition rotor and the pickup coil in an electronic ignition system.

**Gasket** Any thin, soft material - usually cork, cardboard, asbestos or soft metal - installed between two metal surfaces to ensure a good seal. For instance, the cylinder head gasket seals the joint between the block and the cylinder head.

**Gauge** An instrument panel display used to monitor engine conditions. A gauge with a movable pointer on a dial or a fixed scale is an analogue gauge. A gauge with a numerical readout is called a digital gauge.

**Gear ratios** The drive ratio of a pair of gears in a gearbox, calculated on their number of teeth.

**Glaze-busting** see **Honing**

**Grinding** Process for renovating the valve face and valve seat contact area in the cylinder head.

**Gudgeon pin** The shaft which connects the connecting rod small-end with the piston. Often called a piston pin or wrist pin.

# H

**Helical gears** Gear teeth are slightly curved and produce less gear noise that straight-cut gears. Often used for primary drives.

**Installing a Helicoil thread insert in a cylinder head**

**Helicoil** A thread insert repair system. Commonly used as a repair for stripped spark plug threads **(see illustration)**.

**Honing** A process used to break down the glaze on a cylinder bore (also called glaze-busting). Can also be carried out to roughen a rebored cylinder to aid ring bedding-in.

**HT High Tension** Description of the electrical circuit from the secondary winding of the ignition coil to the spark plug.

**Hydraulic** A liquid filled system used to transmit pressure from one component to another. Common uses on motorcycles are brakes and clutches.

**Hydrometer** An instrument for measuring the specific gravity of a lead-acid battery.

**Hygroscopic** Water absorbing. In motorcycle applications, braking efficiency will be reduced if DOT 3 or 4 hydraulic fluid absorbs water from the air - care must be taken to keep new brake fluid in tightly sealed containers.

# I

**lbf ft** Pounds-force feet. An imperial unit of torque. Sometimes written as ft-lbs.

**lbf in** Pound-force inch. An imperial unit of torque, applied to components where a very low torque is required. Sometimes written as in-lbs.

**IC** Abbreviation for Integrated Circuit.

**Ignition advance** Means of increasing the timing of the spark at higher engine speeds. Done by mechanical means (ATU) on early engines or electronically by the ignition control unit on later engines.

**Ignition timing** The moment at which the spark plug fires, expressed in the number of crankshaft degrees before the piston reaches the top of its stroke, or in the number of millimetres before the piston reaches the top of its stroke.

**Infinity (∞)** Description of an open-circuit electrical state, where no continuity exists.

**Inverted forks (upside down forks)** The sliders or lower legs are held in the yokes and the fork tubes or stanchions are connected to the wheel axle (spindle). Less unsprung weight and stiffer construction than conventional forks.

# J

**JASO** Quality standard for 2-stroke oils.

**Joule** The unit of electrical energy.

**Journal** The bearing surface of a shaft.

# K

**Kickstart** Mechanical means of turning the engine over for starting purposes. Only usually fitted to mopeds, small capacity motorcycles and off-road motorcycles.

**Kill switch** Handebar-mounted switch for emergency ignition cut-out. Cuts the ignition circuit on all models, and additionally prevent starter motor operation on others.

**km** Symbol for kilometre.

**kph** Abbreviation for kilometres per hour.

# L

**Lambda (λ) sensor** A sensor fitted in the exhaust system to measure the exhaust gas oxygen content (excess air factor).

**Lapping** see **Grinding**.

**LCD** Abbreviation for Liquid Crystal Display.

**LED** Abbreviation for Light Emitting Diode.

**Liner** A steel cylinder liner inserted in a aluminium alloy cylinder block.

**Locknut** A nut used to lock an adjustment nut, or other threaded component, in place.

**Lockstops** The lugs on the lower triple clamp (yoke) which abut those on the frame, preventing handlebar-to-fuel tank contact.

**Lockwasher** A form of washer designed to prevent an attaching nut from working loose.

**LT Low Tension** Description of the electrical circuit from the power supply to the primary winding of the ignition coil.

# M

**Main bearings** The bearings between the crankshaft and crankcase.

**Maintenance-free (MF) battery** A sealed battery which cannot be topped up.

**Manometer** Mercury-filled calibrated tubes used to measure intake tract vacuum. Used to synchronise carburettors on multi-cylinder engines.

**Micrometer** A precision measuring instrument that measures component outside diameters **(see illustration)**.

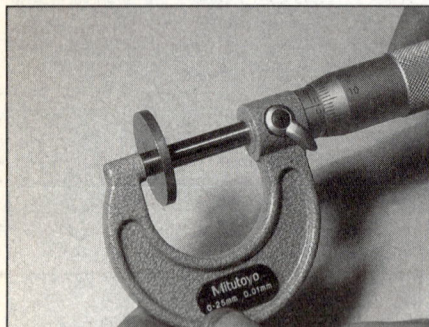

**Tappet shims are measured with a micrometer**

**MON (Motor Octane Number)** A measure of a fuel's resistance to knock.

**Monograde oil** An oil with a single viscosity, eg SAE80W.

**Monoshock** A single suspension unit linking the swingarm or suspension linkage to the frame.

**mph** Abbreviation for miles per hour.

**Multigrade oil** Having a wide viscosity range (eg 10W40). The W stands for Winter, thus the viscosity ranges from SAE10 when cold to SAE40 when hot.

**Multimeter** An electrical test instrument with the capability to measure voltage, current and resistance. Some meters also incorporate a continuity tester and buzzer.

# N

**Needle roller bearing** Inner race of caged needle rollers and hardened outer race. Examples of uncaged needle rollers can be found on some engines. Commonly used in rear suspension applications and in two-stroke engines.

**Nm** Newton metres.

**NOx** Oxides of Nitrogen. A common toxic pollutant emitted by petrol engines at higher temperatures.

# O

**Octane** The measure of a fuel's resistance to knock.

**OE (Original Equipment)** Relates to components fitted to a motorcycle as standard or replacement parts supplied by the motorcycle manufacturer.

**Ohm** The unit of electrical resistance. Ohms = Volts ÷ Current.

**Ohmmeter** An instrument for measuring electrical resistance.

**Oil cooler** System for diverting engine oil outside of the engine to a radiator for cooling purposes.

**Oil injection** A system of two-stroke engine lubrication where oil is pump-fed to the engine in accordance with throttle position.

**Open-circuit** An electrical condition where there is a break in the flow of electricity - no continuity (high resistance).

**O-ring** A type of sealing ring made of a special rubber-like material; in use, the O-ring is compressed into a groove to provide the seal.

**Oversize (OS)** Term used for piston and ring size options fitted to a rebored cylinder.

**Overhead cam (sohc) engine** An engine with single camshaft located on top of the cylinder head.

**Overhead valve (ohv) engine** An engine with the valves located in the cylinder head, but with the camshaft located in the engine block or crankcase.

**Oxygen sensor** A device installed in the exhaust system which senses the oxygen content in the exhaust and converts this information into an electric current. Also called a Lambda sensor.

# P

**Plastigauge** A thin strip of plastic thread, available in different sizes, used for measuring clearances. For example, a strip of Plastigauge is laid across a bearing journal. The parts are assembled and dismantled; the width of the crushed strip indicates the clearance between journal and bearing.

**Polarity** Either negative or positive earth (ground), determined by which battery lead is connected to the frame (earth return). Modern motorcycles are usually negative earth.

**Pre-ignition** A situation where the fuel/air mixture ignites before the spark plug fires. Often due to a hot spot in the combustion chamber caused by carbon build-up. Engine has a tendency to 'run-on'.

**Pre-load (suspension)** The amount a spring is compressed when in the unloaded state. Preload can be applied by gas, spacer or mechanical adjuster.

**Premix** The method of engine lubrication on older two-stroke engines. Engine oil is mixed with the petrol in the fuel tank in a specific ratio. The fuel/oil mix is sometimes referred to as "petroil".

**Primary drive** Description of the drive from the crankshaft to the clutch. Usually by gear or chain.

**PS** Pfedestärke - a German interpretation of BHP.

**PSI** Pounds-force per square inch. Imperial measurement of tyre pressure and cylinder pressure measurement.

**PTFE** Polytetrafluroethylene. A low friction substance.

**Pulse secondary air injection system** A process of promoting the burning of excess fuel present in the exhaust gases by routing fresh air into the exhaust ports.

# Q

**Quartz halogen bulb** Tungsten filament surrounded by a halogen gas. Typically used for the headlight **(see illustration)**.

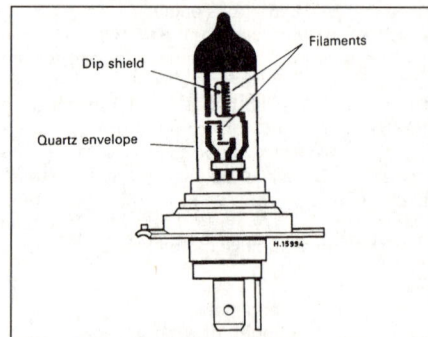

**Quartz halogen headlight bulb construction**

# R

**Rack-and-pinion** A pinion gear on the end of a shaft that mates with a rack (think of a geared wheel opened up and laid flat). Sometimes used in clutch operating systems.

**Radial play** Up and down movement about a shaft.

**Radial ply tyres** Tyre plies run across the tyre (from bead to bead) and around the circumference of the tyre. Less resistant to tread distortion than other tyre types.

**Radiator** A liquid-to-air heat transfer device designed to reduce the temperature of the coolant in a liquid cooled engine.

**Rake** A feature of steering geometry - the angle of the steering head in relation to the vertical **(see illustration)**.

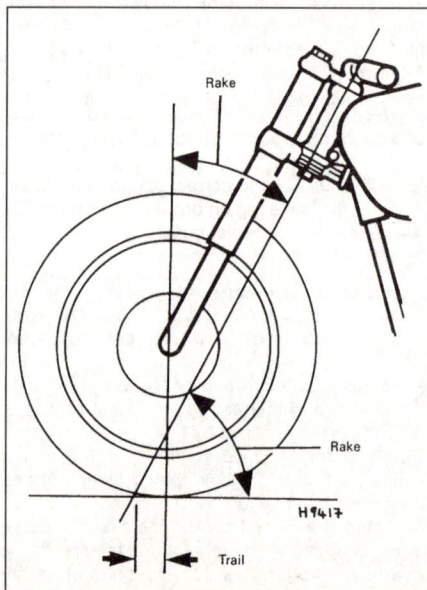

**Steering geometry**

**Rebore** Providing a new working surface to the cylinder bore by boring out the old surface. Necessitates the use of oversize piston and rings.

**Rebound damping** A means of controlling the oscillation of a suspension unit spring after it has been compressed. Resists the spring's natural tendency to bounce back after being compressed.

**Rectifier** Device for converting the ac output of an alternator into dc for battery charging.

**Reed valve** An induction system commonly used on two-stroke engines.

**Regulator** Device for maintaining the charging voltage from the generator or alternator within a specified range.

**Relay** A electrical device used to switch heavy current on and off by using a low current auxiliary circuit.

**Resistance** Measured in ohms. An electrical component's ability to pass electrical current.

**RON (Research Octane Number)** A measure of a fuel's resistance to knock.

**rpm** revolutions per minute.

**Runout** The amount of wobble (in-and-out movement) of a wheel or shaft as it's rotated. The amount a shaft rotates `out-of-true'. The out-of-round condition of a rotating part.

# S

**SAE (Society of Automotive Engineers)** A standard for the viscosity of a fluid.

**Sealant** A liquid or paste used to prevent leakage at a joint. Sometimes used in conjunction with a gasket.

**Service limit** Term for the point where a component is no longer useable and must be renewed.

**Shaft drive** A method of transmitting drive from the transmission to the rear wheel.

**Shell bearings** Plain bearings consisting of two shell halves. Most often used as big-end and main bearings in a four-stroke engine. Often called bearing inserts.

**Shim** Thin spacer, commonly used to adjust the clearance or relative positions between two parts. For example, shims inserted into or under tappets or followers to control valve clearances. Clearance is adjusted by changing the thickness of the shim.

**Short-circuit** An electrical condition where current shorts to earth (ground) bypassing the circuit components.

**Skimming** Process to correct warpage or repair a damaged surface, eg on brake discs or drums.

**Slide-hammer** A special puller that screws into or hooks onto a component such as a shaft or bearing; a heavy sliding handle on the shaft bottoms against the end of the shaft to knock the component free.

**Small-end bearing** The bearing in the upper end of the connecting rod at its joint with the gudgeon pin.

**Spalling** Damage to camshaft lobes or bearing journals shown as pitting of the working surface.

**Specific gravity (SG)** The state of charge of the electrolyte in a lead-acid battery. A measure of the electrolyte's density compared with water.

**Straight-cut gears** Common type gear used on gearbox shafts and for oil pump and water pump drives.

**Stanchion** The inner sliding part of the front forks, held by the yokes. Often called a fork tube.

**Stoichiometric ratio** The optimum chemical air/fuel ratio for a petrol engine, said to be 14.7 parts of air to 1 part of fuel.

**Sulphuric acid** The liquid (electrolyte) used in a lead-acid battery. Poisonous and extremely corrosive.

**Surface grinding (lapping)** Process to correct a warped gasket face, commonly used on cylinder heads.

# T

**Tapered-roller bearing** Tapered inner race of caged needle rollers and separate tapered outer race. Examples of taper roller bearings can be found on steering heads.

**Tappet** A cylindrical component which transmits motion from the cam to the valve stem, either directly or via a pushrod and rocker arm. Also called a cam follower.

**TCS** Traction Control System. An electronically-controlled system which senses wheel spin and reduces engine speed accordingly.

**TDC** Top Dead Centre denotes that the piston is at its highest point in the cylinder.

**Thread-locking compound** Solution applied to fastener threads to prevent slackening. Select type to suit application.

**Thrust washer** A washer positioned between two moving components on a shaft. For example, between gear pinions on gearshaft.

**Timing chain** See **Cam Chain.**

**Timing light** Stroboscopic lamp for carrying out ignition timing checks with the engine running.

**Top-end** A description of an engine's cylinder block, head and valve gear components.

**Torque** Turning or twisting force about a shaft.

**Torque setting** A prescribed tightness specified by the motorcycle manufacturer to ensure that the bolt or nut is secured correctly. Undertightening can result in the bolt or nut coming loose or a surface not being sealed. Overtightening can result in stripped threads, distortion or damage to the component being retained.

**Torx key** A six-point wrench.

**Tracer** A stripe of a second colour applied to a wire insulator to distinguish that wire from another one with the same colour insulator. For example, Br/W is often used to denote a brown insulator with a white tracer.

**Trail** A feature of steering geometry. Distance from the steering head axis to the tyre's central contact point.

**Triple clamps** The cast components which extend from the steering head and support the fork stanchions or tubes. Often called fork yokes.

**Turbocharger** A centrifugal device, driven by exhaust gases, that pressurises the intake air. Normally used to increase the power output from a given engine displacement.

**TWI** Abbreviation for Tyre Wear Indicator. Indicates the location of the tread depth indicator bars on tyres.

# U

**Universal joint or U-joint (UJ)** A double-pivoted connection for transmitting power from a driving to a driven shaft through an angle. Typically found in shaft drive assemblies.

**Unsprung weight** Anything not supported by the bike's suspension (ie the wheel, tyres, brakes, final drive and bottom (moving) part of the suspension).

# V

**Vacuum gauges** Clock-type gauges for measuring intake tract vacuum. Used for carburettor synchronisation on multi-cylinder engines.

**Valve** A device through which the flow of liquid, gas or vacuum may be stopped, started or regulated by a moveable part that opens, shuts or partially obstructs one or more ports or passageways. The intake and exhaust valves in the cylinder head are of the poppet type.

**Valve clearance** The clearance between the valve tip (the end of the valve stem) and the rocker arm or tappet/follower. The valve clearance is measured when the valve is closed. The correct clearance is important - if too small the valve won't close fully and will burn out, whereas if too large noisy operation will result.

**Valve lift** The amount a valve is lifted off its seat by the camshaft lobe.

**Valve timing** The exact setting for the opening and closing of the valves in relation to piston position.

**Vernier caliper** A precision measuring instrument that measures inside and outside dimensions. Not quite as accurate as a micrometer, but more convenient.

**VIN** Vehicle Identification Number. Term for the bike's engine and frame numbers.

**Viscosity** The thickness of a liquid or its resistance to flow.

**Volt** A unit for expressing electrical "pressure" in a circuit. Volts = current x ohms.

# W

**Water pump** A mechanically-driven device for moving coolant around the engine.

**Watt** A unit for expressing electrical power. Watts = volts x current.

**Wear limit** see **Service limit**

**Wet liner** A liquid-cooled engine design where the pistons run in liners which are directly surrounded by coolant **(see illustration).**

**Wet liner arrangement**

**Wheelbase** Distance from the centre of the front wheel to the centre of the rear wheel.

**Wiring harness or loom** Describes the electrical wires running the length of the motorcycle and enclosed in tape or plastic sheathing. Wiring coming off the main harness is usually referred to as a sub harness.

**Woodruff key** A key of semi-circular or square section used to locate a gear to a shaft. Often used to locate the alternator rotor on the crankshaft.

**Wrist pin** Another name for gudgeon or piston pin.

**Note:** References throughtout this index are in the form *"Chapter number"* • *"Page number"*

# Haynes Motorcycle Manuals – The Complete List

| Title | Book No |
|---|---|
| **APRILIA** RS50 (99 - 06) & RS125 (93 - 06) | 4298 |
| Aprilia RSV1000 Mille (98 - 03) | ◆ 4255 |
| Aprilia SR50 | 4755 |
| **BMW** 2-valve Twins (70 - 96) | ◇ 0249 |
| BMW F650 | ◆ 4761 |
| BMW K100 & 75 2-valve Models (83 - 96) | ◆ 1373 |
| BMW R850, 1100 & 1150 4-valve Twins (93 - 04) | ◆ 3466 |
| BMW R1200 (04 - 06) | ◆ 4598 |
| **BSA** Bantam (48 - 71) | 0117 |
| BSA Unit Singles (58 - 72) | 0127 |
| BSA Pre-unit Singles (54 - 61) | 0326 |
| BSA A7 & A10 Twins (47 - 62) | 0121 |
| BSA A50 & A65 Twins (62 - 73) | 0155 |
| Chinese Scooters | 4768 |
| **DUCATI** 600, 620, 750 and 900 2-valve V-Twins (91 - 05) | ◆ 3290 |
| Ducati MK III & Desmo Singles (69 - 76) | ◇ 0445 |
| Ducati 748, 916 & 996 4-valve V-Twins (94 - 01) | ◆ 3756 |
| **GILERA** Runner, DNA, Ice & SKP/Stalker (97 - 07) | 4163 |
| **HARLEY-DAVIDSON** Sportsters (70 - 08) | ◆ 2534 |
| Harley-Davidson Shovelhead and Evolution Big Twins (70 - 99) | ◆ 2536 |
| Harley-Davidson Twin Cam 88 (99 - 03) | ◆ 2478 |
| **HONDA** NB, ND, NP & NS50 Melody (81 - 85) | ◇ 0622 |
| Honda NE/NB50 Vision & SA50 Vision Met-in (85 - 95) | ◇ 1278 |
| Honda MB, MBX, MT & MTX50 (80 - 93) | 0731 |
| Honda C50, C70 & C90 (67 - 03) | 0324 |
| Honda XR80/100R & CRF80/100F (85 - 04) | 2218 |
| Honda XL/XR 80, 100, 125, 185 & 200 2-valve Models (78 - 87) | 0566 |
| Honda H100 & H100S Singles (80 - 92) | ◇ 0734 |
| Honda CB/CD125T & CM125C Twins (77 - 88) | ◇ 0571 |
| Honda CG125 (76 - 07) | ◇ 0433 |
| Honda NS125 (86 - 93) | ◇ 3056 |
| Honda CBR125R (04 - 07) | 4620 |
| Honda MBX/MTX125 & MTX200 (83 - 93) | ◇ 1132 |
| Honda CD/CM185 200T & CM250C 2-valve Twins (77 - 85) | 0572 |
| Honda XL/XR 250 & 500 (78 - 84) | 0567 |
| Honda XR250L, XR250R & XR400R (86 - 03) | 2219 |
| Honda CB250 & CB400N Super Dreams (78 - 84) | ◇ 0540 |
| Honda CR Motocross Bikes (86 - 01) | 2222 |
| Honda CRF250 & CRF450 (02 - 06) | 2630 |
| Honda CBR400RR Fours (88 - 99) | ◇ ◆ 3552 |
| Honda VFR400 (NC30) & RVF400 (NC35) V-Fours (89 - 98) | ◇ ◆ 3496 |
| Honda CB500 (93 - 02) & CBF500 03 - 08 | ◇ 3753 |
| Honda CB400 & CB550 Fours (73 - 77) | 0262 |
| Honda CX/GL500 & 650 V-Twins (78 - 86) | 0442 |
| Honda CBX550 Four (82 - 86) | ◇ 0940 |
| Honda XL600R & XR600R (83 - 08) | ◆ 2183 |
| Honda XL600/650V Transalp & XRV750 Africa Twin (87 to 07) | ◆ 3919 |
| Honda CBR600F1 & 1000F Fours (87 - 96) | ◆ 1730 |
| Honda CBR600F2 & F3 Fours (91 - 98) | ◆ 2070 |
| Honda CBR600F4 (99 - 06) | ◆ 3911 |
| Honda CB600F Hornet & CBF600 (98 - 06) | ◇ ◆ 3915 |
| Honda CBR600RR (03 - 06) | ◆ 4590 |
| Honda CB650 sohc Fours (78 - 84) | 0665 |
| Honda NTV600 Revere, NTV650 and NT650V Deauville (88 - 05) | ◇ ◆ 3243 |
| Honda Shadow VT600 & 750 (USA) (88 - 03) | 2312 |
| Honda CB750 sohc Four (69 - 79) | 0131 |
| Honda V45/65 Sabre & Magna (82 - 88) | 0820 |
| Honda VFR750 & 700 V-Fours (86 - 97) | ◆ 2101 |
| Honda VFR800 V-Fours (97 - 01) | ◆ 3703 |
| Honda VFR800 V-Tec V-Fours (02 - 05) | ◆ 4196 |
| Honda CB750 & CB900 dohc Fours (78 - 84) | 0535 |
| Honda VTR1000 (FireStorm, Super Hawk) & XL1000V (Varadero) (97 - 08) | ◆ 3744 |
| Honda CBR900RR FireBlade (92 - 99) | ◆ 2161 |
| Honda CBR900RR FireBlade (00 - 03) | ◆ 4060 |
| Honda CBR1000RR Fireblade (04 - 07) | ◆ 4604 |
| Honda CBR1100XX Super Blackbird (97 - 07) | ◆ 3901 |
| Honda ST1100 Pan European V-Fours (90 - 02) | ◆ 3384 |
| Honda Shadow VT1100 (USA) (85 - 98) | 2313 |
| Honda GL1000 Gold Wing (75 - 79) | 0309 |

| Title | Book No |
|---|---|
| Honda GL1100 Gold Wing (79 - 81) | 0669 |
| Honda Gold Wing 1200 (USA) (84 - 87) | 2199 |
| Honda Gold Wing 1500 (USA) (88 - 00) | 2225 |
| **KAWASAKI** AE/AR 50 & 80 (81 - 95) | 1007 |
| Kawasaki KC, KE & KH100 (75 - 99) | 1371 |
| Kawasaki KMX125 & 200 (86 - 02) | ◇ 3046 |
| Kawasaki 250, 350 & 400 Triples (72 - 79) | 0134 |
| Kawasaki 400 & 440 Twins (74 - 81) | 0281 |
| Kawasaki 400, 500 & 550 Fours (79 - 91) | 0910 |
| Kawasaki EN450 & 500 Twins (Ltd/Vulcan) (85 - 07) | 2053 |
| Kawasaki EX500 (GPZ500S) & ER500 (ER-5) (87 - 08) | ◆ 2052 |
| Kawasaki ZX600 (ZZ-R600 & Ninja ZX-6) (90 - 06) | ◆ 2146 |
| Kawasaki ZX-6R Ninja Fours (95 - 02) | ◆ 3541 |
| Kawasaki ZX-6R (03 - 06) | ◆ 4742 |
| Kawasaki ZX600 (GPZ600R, GPX600R, Ninja 600R & RX) & ZX750 (GPX750R, Ninja 750R) | ◆ 1780 |
| Kawasaki 650 Four (76 - 78) | 0373 |
| Kawasaki Vulcan 700/750 & 800 (85 - 04) | 2457 |
| Kawasaki 750 Air-cooled Fours (80 - 91) | 0574 |
| Kawasaki ZR550 & 750 Zephyr Fours (90 - 97) | 3382 |
| Kawasaki Z750 & Z1000 (03 - 08) | ◆ 4762 |
| Kawasaki ZX750 (Ninja ZX-7 & ZXR750) Fours (89 - 96) | ◆ 2054 |
| Kawasaki Ninja ZX-7R & ZX-9R (94 - 04) | ◆ 3721 |
| Kawasaki 900 & 1000 Fours (73 - 77) | 0222 |
| Kawasaki ZX900, 1000 & 1100 Liquid-cooled Fours (83 - 97) | ◆ 1681 |
| **KTM** EXC Enduro & SX Motocross (00 - 07) | ◆ 4629 |
| **MOTO GUZZI** 750, 850 & 1000 V-Twins (74 - 78) | 0339 |
| **MZ** ETZ Models (81 - 95) | ◇ 1680 |
| **NORTON** 500, 600, 650 & 750 Twins (57 - 70) | 0187 |
| Norton Commando (68 - 77) | 0125 |
| **PEUGEOT** Speedfight, Trekker & Vivacity Scooters (96 - 08) | ◇ 3920 |
| **PIAGGIO** (Vespa) Scooters (91 - 06) | ◇ 3492 |
| **SUZUKI** GT, ZR & TS50 (77 - 90) | 0799 |
| Suzuki TS50X (84 - 00) | 1599 |
| Suzuki 100, 125, 185 & 250 Air-cooled Trail bikes (79 - 89) | 0797 |
| Suzuki GP100 & 125 Singles (78 - 93) | ◇ 0576 |
| Suzuki GS, GN, GZ & DR125 Singles (82 - 05) | ◇ 0888 |
| Suzuki GSX-R600/750 (06 - 09) | ◆ 4790 |
| Suzuki 250 & 350 Twins (68 - 78) | 0120 |
| Suzuki GT250X7, GT200X5 & SB200 Twins (78 - 83) | ◇ 0469 |
| Suzuki GS/GSX250, 400 & 450 Twins (79 - 85) | 0736 |
| Suzuki GS500 Twin (89 - 06) | ◆ 3238 |
| Suzuki GS550 (77 - 82) & GS750 Fours (76 - 79) | 0363 |
| Suzuki GS/GSX550 4-valve Fours (83 - 88) | 1133 |
| Suzuki SV650 & SV650S (99 - 08) | ◆ 3912 |
| Suzuki GSX-R600 & 750 (96 - 00) | ◆ 3553 |
| Suzuki GSX-R600 (01 - 03), GSX-R750 (00 - 03) & GSX-R1000 (01 - 02) | ◆ 3986 |
| Suzuki GSX-R600/750 (04 - 05) & GSX-R1000 (03 - 06) | ◆ 4382 |
| Suzuki GSF600, 650 & 1200 Bandit Fours (95 - 06) | ◆ 3367 |
| Suzuki Intruder, Marauder, Volusia & Boulevard (85 - 06) | ◆ 2618 |
| Suzuki GS850 Fours (78 - 88) | 0536 |
| Suzuki GS1000 Four (77 - 79) | 0484 |
| Suzuki GSX-R750, GSX-R1100 (85 - 92), GSX600F, GSX750F, GSX1100F (Katana) Fours | ◆ 2055 |
| Suzuki GSX600/750F & GSX750 (98 - 02) | ◆ 3987 |
| Suzuki GSX/GSX1000, 1100 & 1150 4-valve Fours (79 - 88) | 0737 |
| Suzuki TL1000S/R & DL1000 V-Strom (97 - 04) | ◆ 4083 |
| Suzuki GSF650/1250 (05 - 09) | ◆ 4798 |
| Suzuki GSX1300R Hayabusa (99 - 04) | ◆ 4184 |
| Suzuki GSX1400 (02 - 07) | ◆ 4758 |
| **TRIUMPH** Tiger Cub & Terrier (52 - 68) | 0414 |
| Triumph 350 & 500 Unit Twins (58 - 73) | 0137 |
| Triumph Pre-Unit Twins (47 - 62) | 0251 |
| Triumph 650 & 750 2-valve Unit Twins (63 - 83) | 0122 |
| Triumph Trident & BSA Rocket 3 (69 - 75) | 0136 |
| Triumph Bonneville (01 - 07) | ◆ 4364 |
| Triumph Daytona, Speed Triple, Sprint & Tiger (97 - 05) | ◆ 3755 |
| Triumph Triples and Fours (carburettor engines) (91 - 04) | ◆ 2162 |
| **VESPA** P/PX125, 150 & 200 Scooters (78 - 06) | 0707 |
| Vespa Scooters (59 - 78) | 0126 |
| **YAMAHA** DT50 & 80 Trail Bikes (78 - 95) | ◇ 0800 |
| Yamaha T50 & 80 Townmate (83 - 95) | ◇ 1247 |

| Title | Book No |
|---|---|
| **Yamaha** YB100 Singles (73 - 91) | ◇ 0474 |
| Yamaha RS/RXS100 & 125 Singles (74 - 95) | 0331 |
| Yamaha RD & DT125LC (82 - 87) | 0887 |
| Yamaha TZR125 (87 - 93) & DT125R (88 - 07) | ◇ 1655 |
| Yamaha TY50, 80, 125 & 175 (74 - 84) | ◇ 0464 |
| Yamaha XT & SR125 (82 - 03) | ◇ 1021 |
| Yamaha YBR125 | 4797 |
| Yamaha Trail Bikes (81 - 00) | 2350 |
| Yamaha 2-stroke Motocross Bikes 1986 - 2006 | 2662 |
| Yamaha YZ & WR 4-stroke Motocross Bikes (98 - 08) | 2689 |
| Yamaha 250 & 350 Twins (70 - 79) | 0040 |
| Yamaha XS250, 360 & 400 sohc Twins (75 - 84) | 0378 |
| Yamaha RD250 & 350LC Twins (80 - 82) | 0803 |
| Yamaha RD350 YPVS Twins (83 - 95) | 1158 |
| Yamaha RD400 Twin (75 - 79) | 0333 |
| Yamaha XT, TT & SR500 Singles (75 - 83) | 0342 |
| Yamaha XZ550 Vision V-Twins (82 - 85) | 0821 |
| Yamaha FJ, FZ, XJ & YX600 Radian (84 - 92) | 2100 |
| Yamaha XJ600S (Diversion, Seca II) & XJ600N Fours (92 - 03) | ◆ 2145 |
| Yamaha YZF600R Thundercat & FZS600 Fazer (96 - 03) | ◆ 3702 |
| Yamaha FZ-6 Fazer (04 - 07) | ◆ 4751 |
| Yamaha YZF-R6 (99 - 02) | ◆ 3900 |
| Yamaha YZF-R6 (03 - 05) | ◆ 4601 |
| Yamaha 650 Twins (70 - 83) | 0341 |
| Yamaha XJ650 & 750 Fours (80 - 84) | 0738 |
| Yamaha XS750 & 850 Triples (76 - 85) | 0340 |
| Yamaha TDM850, TRX850 & XTZ750 (89 - 99) | ◇ ◆ 3540 |
| Yamaha YZF750R & YZF1000R Thunderace (93 - 00) | ◆ 3720 |
| Yamaha FZR600, 750 & 1000 Fours (87 - 96) | ◆ 2056 |
| Yamaha XV (Virago) V-Twins (81 - 03) | ◆ 0802 |
| Yamaha XVS650 & 1100 Drag Star/V-Star (97 - 05) | ◆ 4195 |
| Yamaha XJ900F Fours (83 - 94) | ◆ 3239 |
| Yamaha XJ900S Diversion (94 - 01) | ◆ 3739 |
| Yamaha YZF-R1 (98 - 03) | ◆ 3754 |
| Yamaha YZF-R1 (04 - 06) | ◆ 4605 |
| Yamaha FZS1000 Fazer (01 - 05) | ◆ 4287 |
| Yamaha FJ1100 & 1200 Fours (84 - 96) | ◆ 2057 |
| Yamaha XJR1200 & 1300 (95 - 06) | ◆ 3981 |
| Yamaha V-Max (85 - 03) | ◆ 4072 |

### ATVs

| Title | Book No |
|---|---|
| Honda ATC70, 90, 110, 185 & 200 (71 - 85) | 0565 |
| Honda Rancher, Recon & TRX250EX ATVs | 2553 |
| Honda TRX300 Shaft Drive ATVs (88 - 00) | 2125 |
| Honda Foreman (95 - 07) | 2465 |
| Honda TRX300EX, TRX400EX & TRX450R/ER ATVs (93 - 06) | 2318 |
| Kawasaki Bayou 220/250/300 & Prairie 300 ATVs (86 - 03) | 2351 |
| Polaris ATVs (85 - 97) | 2302 |
| Polaris ATVs (98 - 06) | 2508 |
| Yamaha YFS200 Blaster ATV (88 - 06) | 2317 |
| Yamaha YFB250 Timberwolf ATVs (92 - 00) | 2217 |
| Yamaha YFM350 & YFM400 (ER and Big Bear) ATVs (87 - 03) | 2126 |
| Yamaha Banshee and Warrior ATVs (87 - 03) | 2314 |
| Yamaha Kodiak and Grizzly ATVs (93 - 05) | 2567 |
| ATV Basics | 10450 |

### TECHBOOK SERIES

| Title | Book No |
|---|---|
| Twist and Go (automatic transmission) Scooters Service and Repair Manual | 4082 |
| Motorcycle Basics TechBook (2nd Edition) | 3515 |
| Motorcycle Electrical TechBook (3rd Edition) | 3471 |
| Motorcycle Fuel Systems TechBook | 3514 |
| Motorcycle Maintenance TechBook | 4071 |
| Motorcycle Modifying | 4272 |
| Motorcycle Workshop Practice TechBook (2nd Edition) | 3470 |

◇ = *not available in the USA*     ◆ = *Superbike*

The manuals on this page are available through good motorcycle dealers and accessory shops.
In case of difficulty, contact: **Haynes Publishing**
(UK) +44 1963 442030     (USA) +1 805 498 6703
(SV) +46 18 124016
(Australia/New Zealand) +61 3 9763 8100

# Notes

# Preserving Our Motoring Heritage

< The Model J Duesenberg Derham Tourster. Only eight of these magnificent cars were ever built – this is the only example to be found outside the United States of America

Almost every car you've ever loved, loathed or desired is gathered under one roof at the Haynes Motor Museum. Over 300 immaculately presented cars and motorbikes represent every aspect of our motoring heritage, from elegant reminders of bygone days, such as the superb Model J Duesenberg to curiosities like the bug-eyed BMW Isetta. There are also many old friends and flames. Perhaps you remember the 1959 Ford Popular that you did your courting in? The magnificent 'Red Collection' is a spectacle of classic sports cars including AC, Alfa Romeo, Austin Healey, Ferrari, Lamborghini, Maserati, MG, Riley, Porsche and Triumph.

## A Perfect Day Out

Each and every vehicle at the Haynes Motor Museum has played its part in the history and culture of Motoring. Today, they make a wonderful spectacle and a great day out for all the family. Bring the kids, bring Mum and Dad, but above all bring your camera to capture those golden memories for ever. You will also find an impressive array of motoring memorabilia, a comfortable 70 seat video cinema and one of the most extensive transport book shops in Britain. The Pit Stop Cafe serves everything from a cup of tea to wholesome, home-made meals or, if you prefer, you can enjoy the large picnic area nestled in the beautiful rural surroundings of Somerset.

> John Haynes O.B.E., Founder and Chairman of the museum at the wheel of a Haynes Light 12.

< The 1936 490cc sohc-engined International Norton – well known for its racing success

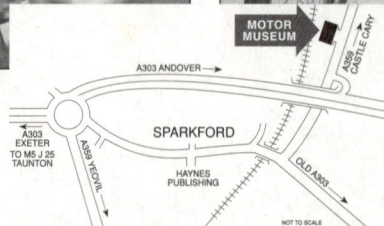

The Museum is situated on the A359 Yeovil to Frome road at Sparkford, just off the A303 in Somerset. It is about 40 miles south of Bristol, and 25 minutes drive from the M5 intersection at Taunton.
Open 9.30am - 5.30pm (10.00am - 4.00pm Winter) 7 days a week, *except Christmas Day, Boxing Day and New Years Day*
Special rates available for schools, coach parties and outings  Charitable Trust No. 292048